Modeling and
of Engineering Systems

Clarence W. de Silva

CRC Press
Taylor & Francis Group
Boca Raton   London   New York

CRC Press is an imprint of the
Taylor & Francis Group, an **informa** business

FIRST INDIAN REPRINT, 2013

CRC Press
Taylor & Francis Group
6000 Broken Sound Parkway NW, Suite 300
Boca Raton, FL 33487-2742

© 2009 by Taylor and Francis Group, LLC
CRC Press is an imprint of Taylor & Francis Group, an Informa business

No claim to original U.S. Government works

Printed and bound in india by Nutech Print Services

International Standard Book Number: 978-1-4200-7686-8

Library of Congress Cataloging-in-Publication Data

De Silva, Clarence W.
    Modeling and control of engineering systems / Clarence W. de Silva.
        p. cm.
    Includes bibliographical references and index.
    ISBN 978-1-4200-7686-8
    1. Automatic control. 2. Control theory. 3. Engineering--Mathematical models. 4. Systems engineering--Mathematical models. I. Title.

TJ213.D465 2010
629.8--dc22   2009025021

Visit the Taylor & Francis Web site at
http://www.taylorandfrancis.com

and the CRC Press Web site at
http://www.crcpress.com

To all my students, present and past

"Education is just the progressive realisation of our ignorance"

–Albert Einstein

Contents

Preface

This is an introductory book in the subject of modeling and control of engineering systems. It serves as both a textbook for undergraduate engineering students and entry-level graduate students, and a reference book for practicing professionals. As a textbook, it is suitable for courses in: modeling of dynamic systems, feedback control systems, control engineering, and design and instrumentation of control systems. There is adequate material in the book for two 14-week courses, one at the junior (third-year undergraduate) or senior (fourth-year undergraduate) level and the other at the first-year graduate level. In view of the analytical techniques, computer and software tools, instrumentation details, design methods, and practical considerations that are presented in the book, and in view of the simplified and snap-shot style presentation of more advanced theory and concepts, the book serves as a useful reference tool for engineers, technicians, project managers, and other practicing professionals in industry and in research laboratories, in the fields of control engineering, mechanical engineering, electrical and computer engineering, manufacturing and production engineering, aeronautical and aerospace engineering, and mechatronics.

A control system is a dynamic system that contains a controller as an integral part. The purpose of the controller is to generate control signals, which will drive the process to be controlled (the plant) in the desired manner—to meet a set of performance specifications. Actuators are needed to perform control actions as well as to directly drive/operate the plant. Sensors and transducers are necessary to measure output signals (process responses) for feedback control; to measure input signals for feedforward control; to measure process variables for system monitoring, diagnosis and supervisory control; and for a variety of other purposes. Design is a necessary part as well, for it is design that enables us to build a control system that meets the performance requirements—starting, perhaps, with basic components such as sensors, actuators, controllers, compensators, and signal modification devices. The book addresses all these issues, starting from the basics and systematically leading to advanced concepts.

Control engineers should be able to model and analyze individual components or an integrated control system, design controllers, identify and select components for a control system, and choose parameter values so that the control system will perform the intended functions of the particular system while meeting a set of specifications. Proper control of an engineering system requires an understanding and a suitable "representation" of the system—a "model" of the system. Any model is an idealization of the actual system. Properties established and results derived are associated with the model rather than the actual system, whereas the excitations are applied to and the output responses are *measured* from the actual system. Modeling is often an essential task in control engineering. For instance, a good understanding of the system to be controlled may be gained through modeling and associated analysis and computer simulation. In fact a controller may be designed and its performance can be studied through modeling and computer simulation even before a physical controller is developed. Such an approach is often more economical and time effective. Furthermore there are control techniques called "model-based control" for which modeling is a requirement.

Important aspects of laboratory experimentation and instrumentation are included in the book. There are numerous worked examples, problems, and exercises, many of which are related to real-life situations and practical applications. Augmenting their traditional role, the problems at the end of each chapter serve as valuable sources of information not found in the main text. In fact, the student is strongly advised to carefully read all the problems in addition to the main text. Complete solutions to the end-of-chapter problems are provided in a *Solutions Manual*, which is available to instructors who adopt the book.

The manuscript for the original book evolved from the notes developed by the author for mandatory undergraduate courses in dynamic system modeling and feedback control, and entry-level graduate courses in control system instrumentation and modern control engineering for students in electrical and computer engineering, mechanical engineering, and chemical engineering at Carnegie Mellon University. During the development of the material for those courses, a deliberate attempt was made as well to cover a major part of the syllabuses for similar courses offered in the Department of Mechanical Engineering at the Massachusetts Institute of Technology. At the University of British Columbia, the original material was further developed, revised, and enhanced for teaching courses in dynamic system modeling, control systems, intelligent control, mechatronics, and control sensors and actuators. The material in the book has acquired an application orientation through the author's industrial experience at places such as IBM Corporation, Westinghouse Electric Corporation, Bruel and Kjaer, and NASA's Lewis and Langley Research Centers.

The material presented in the book provides a firm foundation in modeling and control of engineering systems, for subsequent building up of expertise in the subject—perhaps in an industrial setting or in an academic research laboratory—with further knowledge of control hardware and analytical skills (along with the essential hands-on experience) gained during the process.

Main Features of the Book

There are several shortcomings in existing popular books on modeling and control. For example, some books "pretend" to consider practical applications by first mentioning a real engineering system before posing an analytical or numerical problem. For example, it may describe an automobile (with a graphical sketch and even a photo) and then make a statement such as "let us approximate the automobile by the following transfer function." No effort is made to relate the model to the physical system and to address such issues as why a particular control technique is suitable for controlling the system. Some other books extensively use software tools for modeling and control system analysis without pointing out the fundamentals and the analytical basis behind the methodologies, ways of interpreting and validating the obtained results, and the practical limitations of the tools. While benefiting from the successes of the popular books, the present book makes a substantial effort to overcome their shortcomings. The following are the main features of the book, which will distinguish it from other popular textbooks in the subjects of modeling and control:

- Readability and convenient reference are given priority in the presentation and formatting of the book.
- Key concepts and formulas developed and presented in the book are summarized in windows, tables, and lists, in a user-friendly format, throughout the book, for easy reference and recollection.

- A large number of worked examples are included and are related to real-life situations and the practice of control engineering, throughout the book.

- Numerous problems and exercises, most of which are based on practical situations and applications, and carry additional useful information in modeling and control, are given at the end of each chapter.

- The use of MATLAB® (is a registered trademark of The MathWorks, Inc. For product information, please contact: The MathWorks, Inc., 3 Apple Hill Drive, Natick, MA 01760-2098 USA. Tel: 508 647 7000; Fax: 508-647-7001; E-mail: info@mathworks.com; Web: www.mathworks.com) Simulink®, and LabVIEW®, and associated toolboxes are described and a variety of illustrative examples are given for their use. Many problems in the book are cast for solution using these computer tools. However, the main goal of the book is not simply to train the students in the use of software tools. Instead, a thorough understanding of the core and foundation of the subject as facilitated by the book will enable the student to learn the fundamentals and engineering methodologies behind the software tools; the choice of proper tools to solve a given problem; interpret the results generated by them; assess the validity and correctness of the results; and understand the limitations of the available tools.

- Useful material that cannot be conveniently integrated into the main chapters is given in three separate appendices at the end of the book.

- The subject of modeling is treated using an integrated approach, which is uniformly applicable to mechanical, electrical, fluid, and thermal systems. An inspiration is drawn from the concept of equivalent circuits and Thevenin's theorem in the field of electrical engineering.

- The subject of intelligent control, particularly fuzzy logic control, is introduced. A chapter on control system instrumentation is included, providing practical details for experiments in an undergraduate laboratory.

- An *Instructor's Manual* is available, which provides suggestions for curriculum planning and development, and gives detail solutions to all the end-of-chapter problems in the book.

Clarence W. de Silva
Vancouver, British Columbia, Canada

Acknowledgments

Many individuals have assisted in the preparation of this book, but it is not practical to acknowledge all such assistance here. First, I wish to recognize the contributions, both direct and indirect, of my graduate students, research associates, and technical staff. Particular mention should be made of my PhD student Roland H. Lang, whose research assistance has been very important. I am particularly grateful to Jonathan W. Plant, Senior Editor, CRC Press/Taylor&Francis, for his interest, enthusiasm, and strong support, throughout the project. Other staff at CRC Press and its affiliates, in particular, Jessica Vakili, Arlene Kopeloff, Glenon Butler, Soundar Rajan, and Evelyn Delehanty, deserve special mention. I wish to acknowledge as well the advice and support of various authorities in the field— particularly, Professor Devendra Garg of Duke University, Professor Madan Gupta of the University of Saskatchewan, Professor Mo Jamshidi of the University of Texas (San Antonio), Professors Marcelo Ang, Ben Chen, Tong-Heng Lee, Jim A.N. Poo, and Kok-Kiong Tan of the National University of Singapore, Professor Max Meng of the Chinese University of Hong Kong, Dr. Daniel Repperger of U.S. Air Force Research Laboratory, Professor David N. Wormley of the Pennsylvania State University, and Professor Simon Yang of University of Guelph. Finally, my wife and children deserve much appreciation and apology for the unintentional "neglect" that they may have faced during the latter stages of the preparation of this book.

Author

Dr. Clarence W. de Silva, P.E., Fellow ASME and Fellow IEEE, is a professor of Mechanical Engineering at the University of British Columbia, Vancouver, Canada, and occupies the Tier 1 Canada Research Chair professorship. Prior to that, he has occupied the NSERC-BC Packers Research Chair professorship in Industrial Automation since 1988. He has served as a faculty member at Carnegie Mellon University (1978–1987) and as a Fulbright Visiting Professor at the University of Cambridge (1987/1988).

He has earned PhD degrees from Massachusetts Institute of Technology (1978) and University of Cambridge, England (1998), and an honorary DEng degree from University of Waterloo (2008). De Silva has also occupied the Mobil Endowed Chair Professorship in the Department of Electrical and Computer Engineering at the National University of Singapore and the Honorary Chair Professorship of National Taiwan University of Science and Technology.

Other Fellowships: Fellow Royal Society of Canada; Fellow Canadian Academy of Engineering; Lilly Fellow; NASA-ASEE Fellow; Senior Fulbright Fellow to Cambridge University; Fellow of the Advanced Systems Institute of BC; Killam Fellow; Erskine Fellow.

Awards: Paynter Outstanding Investigator Award and Takahashi Education Award, ASME Dynamic Systems and Control Division; Killam Research Prize; Outstanding Engineering Educator Award, IEEE Canada; Lifetime Achievement Award, World Automation Congress; IEEE Third Millennium Medal; Meritorious Achievement Award, Association of Professional Engineers of BC; Outstanding Contribution Award, IEEE Systems, Man, and Cybernetics Society.

Editorial Duties: Served on 14 journals including *IEEE Transactions on Control System Technology* and *Journal of Dynamic Systems, Measurement and Control, Transactions ASME*; Editor-in-Chief, *International Journal of Control and Intelligent Systems*; Editor-in-Chief, *International Journal of Knowledge-Based Intelligent Engineering Systems*; Senior Technical Editor, *Measurements and Control*; and Regional Editor, North America, *Engineering Applications of Artificial Intelligence— IFAC International Journal of Intelligent Real-Time Automation.*

Publications: 16 technical books, 14 edited books, 32 book chapters, about 180 journal articles, about 200 conference papers.

Research and development Areas: Industrial process monitoring and automation, intelligent multi-robot cooperation, mechatronics, intelligent control, sensors, actuators, and control system instrumentation. Funding of over $5 million, as principal investigator, during the past 15 years.

Further Reading

This book has relied on many publications, directly and indirectly, in its development and evolution. Many of these publications are based on the work of the author and his co-workers. Also, there are some excellent books the reader may refer to for further information and knowledge. Some selected books are listed below.

Brogan, W.L. *Modern Control Theory*. Prentice Hall, Englewood Cliffs, NJ, 1991.

Chen, B.M., Lee, T.H., and Venkataramenan, V. *Hard Disk Drive Servo Systems*. Springer-Verlag, London, UK, 2002.

De Silva, C.W. and Wormley, D.N. *Automated Transit Guideways: Analysis and Design*. D.C. Heath and Co./Simon & Schuster, Lexington, MA, 1983.

De Silva, C.W. *Dynamic Testing and Seismic Quahfication Practice*. Lexington Books, Lexington, MA, 1983.

De Silva, C.W. *Control Sensors and Actuators*. Prentice Hall, Englewood Cliffs, NJ, 1989.

De Silva, C.W. and MacFarlane, A.G.J. *Knowledge-Based Control with Application to Robots*. Springer-Verlag, Berlin, Germany, 1989.

De Silva, C.W. *Intelligent Control—Fuzzy Logic Applications*. CRC Press, Boca Raton, FL, 1995.

De Silva, C.W. (Editor). *Intelligent Machines: Myths and Realities*. Taylor & Francis, CRC Press, Boca Raton, FL, 2000.

De Silva, C.W. *Mechatronics—An Integrated Approach*. Taylor & Francis, CRC Press, Boca Raton, FL, 2005.

De Silva, C.W. (Editor). *Vibration and Shock Handbook*. Taylor & Francis, CRC Press, Boca Raton, FL, 2005.

De Silva, C.W. *Vibration Fundamentals and Practice*, 2nd Edition. CRC Press, Boca Raton, FL, 2007.

De Silva, C.W. *Sensors and Actuators—Control System Instrumentation*. Taylor & Francis, CRC Press, Boca Raton, FL, 2007.

De Silva, C.W. (Editor). *Mechatronic Systems—Devices, Design, Control, Operation, and Monitoring*. Taylor & Francis, CRC Press, Boca Raton, FL, 2007.

Dorf, R.C. and Bishop, R.H. *Modern Control Systems*, 9th Edition. Prentice Hall, Upper Saddle River, NJ, 2001.

Hordeski, M.F. *The Design of Microprocessor, Sensor, and Control Systems*, Reston Publishing Co., Reston, VA, 1985.

Jain, L. and de Silva, C.W. (editors), *Intelligent Adaptive Control: Industrial Applications*. CRC Press, Boca Raton, FL, 1999.

Karray, F. and de Silva, C.W. *Soft Computing Techniques and Their Applications*. Addison Wesley/Pearson, NY/UK, 2004.

Kuo, B.C. *Automatic Control Systems*, 6th Edition. Prentice Hall, Upper Saddle River, NJ, 1991.

Necsulescu, D. *Mechatronics*. Prentice Hall, Upper Saddle River, NJ, 2002.

Ogata, K. *Modern Control Engineering*. Prentice Hall, Upper Saddle Rivver, NJ, 2002.

Rowell, D. and Wormley, D.N. *System Dynamics—An Introduction*. Prentice Hall, Upper Saddle River, NJ, 1997.

Tan, K.K., Lee, T.H., Dou, H., and Huang, S. *Precision Motion Control*, Springer-Verlag, London, UK, 2001.

Van de Vegte, J. *Feedback Control Systems*. Prentice Hall, Englewood Cliffs, NJ, 1986.

Units and Conversions (Approximate)

1 cm	=	1/2.54 in=0.39 in
1 rad	=	57.3°
1 rpm	=	0.105 rad/s
1 g	=	9.8 m/s²=32.2 ft/s²=386 in/s²
1 kg	=	2.205 lb
1 kg·m² (kilogram-meter-square)	=	5.467 oz·in² (ounce-inch-square)=8.85 lb.in.s²
1 N/m	=	5.71×10⁻³ lbf/in
1 N/m/s	=	5.71×10⁻³ lbf/in/s
1 N·m (Newton-meter)	=	141.6 oz·in (ounce-inch)
1 J	=	1 N.m=0.948×10⁻³ Btu=0.278 kWh
1 hp (horse power)	=	746 W (watt)=550 ft·lbf
1 kPa	=	1×10³ Pa= 1×10³ N/m²
	=	0.154 psi= 1×10⁻² bar
1 gal/min	=	3.8 L/min

Metric Prefixes:

giga	G	10⁹
mega	M	10⁶
kilo	k	10³
milli	m	10⁻³
micro	μ	10⁻⁶
nano	n	10⁻⁹
pico	p	10⁻¹²

1

Modeling and Control of Engineering Systems

The purpose of a controller is to make a plant (the system to be controlled) behave in a desired manner, while meeting a set of performance specifications. A control system is one that contains at least a plant and a controller. Design, development, modification, performance evaluation, and control of an engineering system require an understanding and a suitable "representation" of the system. Specifically, a "model" of the system is required. This book will present integrated and unified methodologies for modeling an engineering system that possibly contains multidomain (mechanical, electrical, fluid, thermal, etc.) characteristics. Systematically the book will address the subject of control engineering, while highlighting model-based approaches where relevant. Overall, the book will treat modeling, analysis, simulation, design, instrumentation, and evaluation of control systems. The present introductory chapter sets the stage for this treatment.

1.1 Control Engineering

The purpose of a controller is to make a plant behave in a desired manner—meeting a set of performance specifications. The physical dynamic system (e.g., a mechanical system) whose response (e.g., vibrations or voltage spikes) needs to be controlled is called the *plant* or *process*. The device that generates the signal (or, command) according to some scheme (or, control law) and controls the response of the plant is called the *controller*. The plant and the controller are the two essential components of a *control system*. The system can be quite complex and may be subjected to known or unknown excitations (inputs), as in the case of an aircraft (see Figure 1.1).

Certain *command signals*, or inputs, are applied to the *controller* and the plant is expected to behave in a desirable manner (according to a set of performance specifications) under control. In *feedback control*, the plant has to be monitored and its response needs to be measured using *sensors* and *transducers*, for feeding back into the controller. Then, the controller compares the sensed signal with a desired response as specified externally, and uses the error to generate a proper control signal. Ideally a good control system should be: stable; fast; accurate; insensitive to noise, external disturbances, modeling errors and parameter variations; sufficiently sensitive to control inputs; and be free of undesirable coupling and dynamic interactions. Control engineering concerns development, implementation, operation, and evaluation of control systems. Control engineers should be able to model, analyze, simulate, design, develop, implement, and evaluate controllers and other parts of control systems.

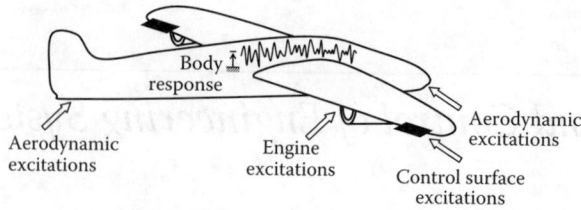

FIGURE 1.1
Aircraft is a complex control system.

1.2 Application Areas

Application areas of control systems are numerous, and typically employ design, sensing, actuation, control, signal conditioning, component interconnection and interfacing, and communication, generally using tools of mechanical, electrical and electronic, computer, and control engineering. For example, control engineering is indispensible in the field of robotics. This is true regardless of the robotic application (e.g., industrial, domestic, security/safety, entertainment, and so on). The humanoid robot shown in Figure 1.2 is a complex and "intelligent" control system. Some other important areas of application of control are indicated below.

Transportation is a broad area where control engineering has numerous applications. In ground transportation in particular, automobiles, trains, and automated transit systems use control for proper operation. They include airbag deployment systems, antilock braking systems (ABS), cruise control systems, active suspension systems, and various devices for monitoring, toll collection, navigation, warning, and control in intelligent vehicular highway systems (IVHS). For example, a modern automated ground transit system such as the one shown in Figure 1.3 will require complex control technologies for a variety of functions such as motion and ride quality control, safety and security, lighting, heating, cooling, and ventilation. In air transportation, modern aircraft designs with advanced materials, structures, and electronics benefit from advanced controllers, flight simulators, flight control systems, navigation systems, landing gear mechanisms, traveler comfort aids, and the like.

Manufacturing and production engineering is another broad field that uses control technologies and systems. Factory robots (for welding, spray painting, assembly, inspection, etc.), automated guided vehicles (AGVs), modern computer-numerical control (CNC) machine tools, machining centers, rapid (and virtual) prototyping systems, and micromachining systems are examples of control applications.

In medical and healthcare applications, robotic technologies for examination, surgery, rehabilitation, drug dispensing, and general patient care are being developed and implemented, which use control technologies. Control applications are needed as well for patient transit devices, various diagnostic probes and scanners, beds, and exercise machines.

In a modern office environment, automated filing systems, multifunctional copying machines (copying, scanning, printing, fax, etc.), food dispensers, multimedia presentation and meeting rooms, and climate control systems incorporate control technologies.

In household applications, home security systems and robots, vacuum cleaners and robots, washers, dryers, dishwashers, garage door openers, and entertainment centers use control devices and technologies.

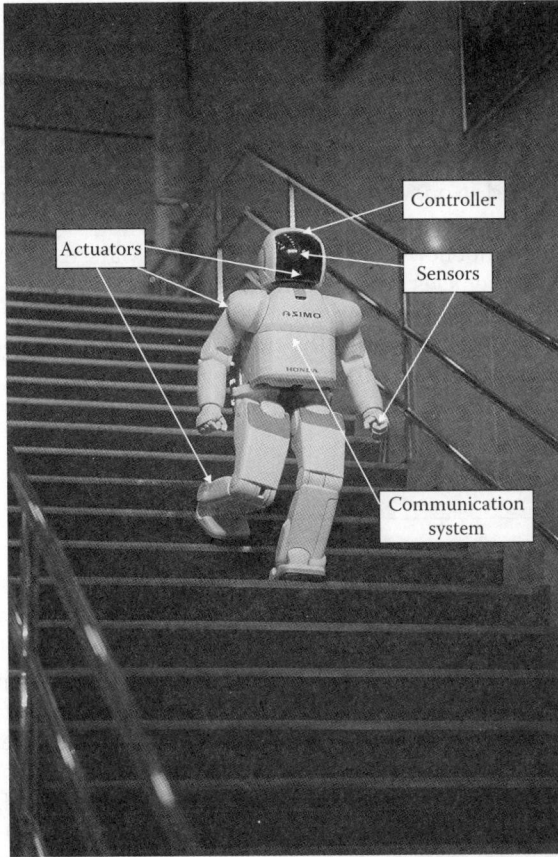

FIGURE 1.2
A humanoid robot is a complex control system. (From American Honda Motor Co. Inc. With permission.)

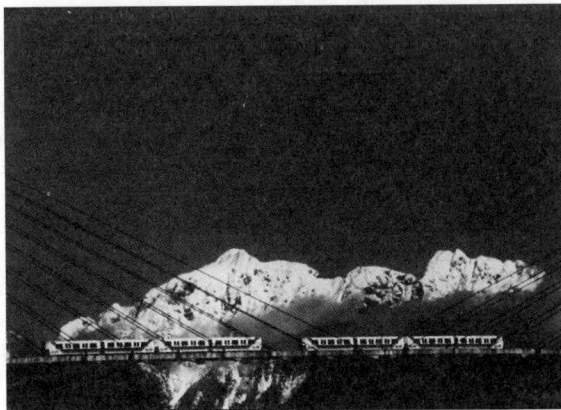

FIGURE 1.3
Skytrain—the high-speed ground transit system in Vancouver, Canada—employs sophisticated control technologies.

In the computer industry, hard disk drives (HDD), disk retrieval, access and ejection devices, and other electromechanical components can considerably benefit from high-precision control. The impact goes further because digital computers are integrated into a vast variety of other devices and applications, which are capable of executing complex control strategies.

In civil engineering applications, cranes, excavators, and other machinery for building, earth removal, mixing and so on, will improve their performance through modeling, analysis, simulation, and automated control.

In space applications, mobile robots such as NASA's Mars exploration Rover, space-station robots, and space vehicles are fundamentally control systems.

It is noted that there is no end to the type of devices and applications that can incorporate control engineering. In view of this, the traditional boundaries between engineering disciplines will become increasingly fuzzy, and the field of control engineering will grow and evolve further through such merging of disciplines.

1.3 Importance of Modeling

In the area of automatic control, models are used in a variety of ways. In particular, an analytical model of the control system is needed for mathematical analysis and computer simulation of the system. A model of the system to be controlled (i.e., plant, process) may be used to develop the performance specifications, based on which a controller is developed for the system. In model-referenced adaptive control, for example, a reference model dictates the desired behavior that is expected under control. This is an implicit way of using a model to represent performance specifications. In model-based control, a dynamic model of the actual process is employed to develop the necessary control schemes. In the early stages of design of a control system, some parts of the desired system do not exist. In this context, a model of the anticipated system, particularly an analytical model or a computer model, can be very useful, economical, and time efficient. In view of the complexity of a design process, particularly when striving for an optimal design, it is useful to incorporate system modeling as a tool for design iteration.

Modeling and design can go hand in hand, in an iterative manner. In the beginning, by knowing some information about the system (e.g., intended functions, performance specifications, past experience and knowledge of related systems) and using the design objectives, it will be possible to develop a model of sufficient (low to moderate) detail and complexity. By analyzing and carrying out computer simulations of the model it will be possible to generate useful information that will guide the design process (e.g., generation of a preliminary design). In this manner design decisions can be made, and the model can be refined using the available (improved) design.

It is unrealistic to attempt to develop a "universal model" that will incorporate all conceivable aspects of the system. The model should be as simple as possible, and may address only a few specific aspects of interest in the particular study or application. For example, in the context of a HDD unit, as shown in Figure 1.4a, if the objective is vibration control, a simplified model as shown in Figure 1.4b and c will be useful.

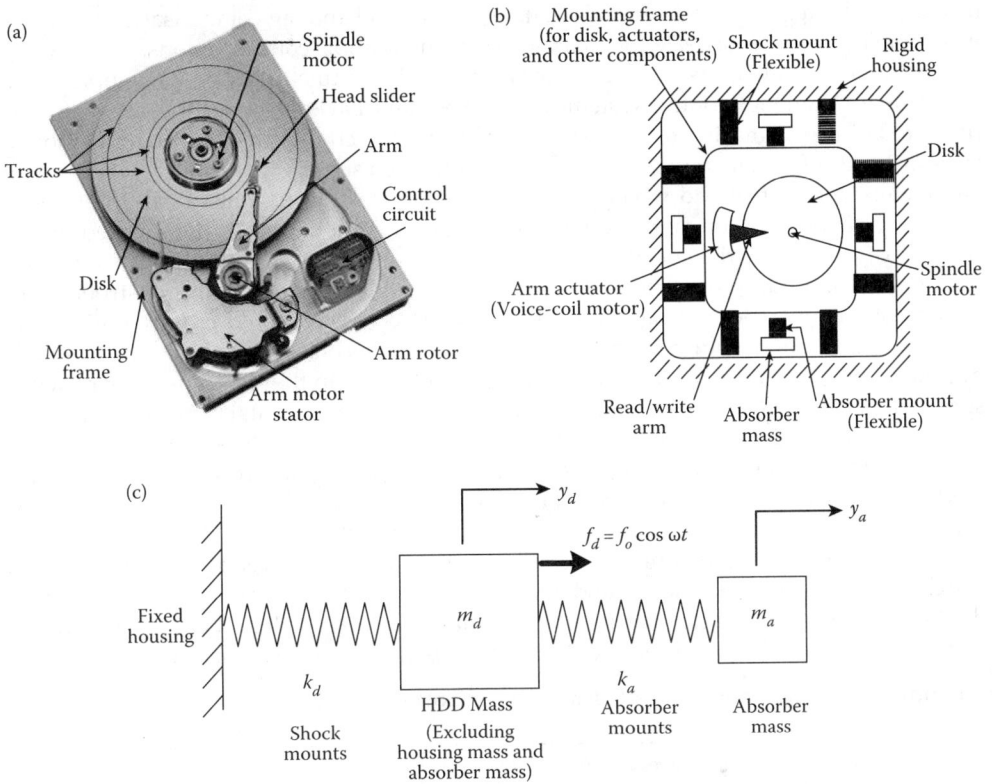

FIGURE 1.4
(a) A hard-disk drive (HDD) unit of a computer. (b) Components for a simplified model. (c) A simplified model for vibration control.

1.4 History of Control Engineering

The demand for servomechanisms in military applications during World War II provided much incentive and many resources for the growth of control technology. Early efforts were devoted to the development of analog controllers—electronic devices or circuits that generate proper drive signals for a plant (process). Parallel advances were necessary in actuators such as motors, solenoids, and valves, which are the driving means of a plant. For feedback control and feedforward control, further developments in sensors and transducers became essential. Innovations and improvements were necessary as well in the devices for signal processing, conditioning, and modification. With added sophistication in control systems, it was soon apparent that analog control techniques had serious limitations. In particular, linear assumptions were used to develop controllers even for highly nonlinear plants. Furthermore, complex and costly circuitry was often needed to generate even simple control signals. Consequently, most analog controllers were limited to on/off and proportional-integral-derivative (PID) actions, where lead and lag compensation networks were employed to compensate for weaknesses in such simple control actions.

The digital computer, first developed for large number-crunching jobs, was employed as a controller in complex control systems in the 1950s and 1960s. Originally, cost constraints restricted its use primarily to aerospace applications, which required the manipulation of large amounts of data (complex models, several hundred signals, and thousands of system parameters) for control and which did not face serious cost restraints. Real-time control requires fast computation, and the required speed of computation is determined by the control bandwidth (or the speed of control) and parameters (e.g., time constants, natural frequencies, and damping constants) of the process (plant) that is being controlled. For instance, prelaunch monitoring and control of a space vehicle would require digital data acquisition at high sampling rates (e.g., 50,000 samples/second). As a result of a favorable decline of computation costs (both hardware and software) in the subsequent years, widespread application of digital computers as control devices (i.e., digital control) has become feasible. Dramatic developments in large-scale integration (LSI) technology and microprocessors in the 1970s resulted in very significant drops of digital processing costs, which made digital control a very attractive alternative to analog control. Today, digital control has become an integral part of numerous systems and applications, including machine tools, robotic manipulators, automobiles, aircraft autopilots, nuclear power plants, traffic control systems, chemical process plants, and general mechatronic systems. Both software-based digital controllers and faster yet less flexible hardware-based digital controllers, which employ digital circuits to perform control functions, are commonly used now.

Landmark developments in the history of control engineering are:

300 BC	Greece (float valves and regulators for liquid level control)
1770	James Watt (steam engine, governor for speed control)
1868	James Maxwell (Cambridge University, theory of governors)
1877	E.J. Routh (stability criterion)
1893	A.M. Lyapunov (Soviet Union, stability theory, basis of state-space formulation)
1927	H.S. Black and H.W. Bode (AT&T Bell Labs, electronic feedback amplifier)
1930	Norbert Wiener (MIT, theory of stochastic processes)
1932	H. Nyquist (AT&T Bell Labs, stability criterion from Nyquist gain/phase plot)
1936	A. Callender, D.R. Hartee, and A. Porter (England, PID control)
1948	Claude Shannon (MIT, mathematical theory of communication)
1948	W.R. Evans (root locus method)
1940s	Theory and applications of servomechanisms, cybernetics, and control (MIT, Bell Labs, etc.)
1959	H.M. Paynter (MIT, bond graph techniques for system modeling)
1960s	Rapid developments in state-space techniques, optimal control, space applications (R. Bellman and R.E. Kalman in USA, L.S. Pontryagin in USSR, NASA)
1965	Theory of fuzzy sets and fuzzy logic (L.A. Zadeh)
1970s	Intelligent control; developments of neural networks; widespread developments of robotics and industrial automation (North America, Japan, Europe)
1980s	Robust control; widespread applications of robotics and flexible automation
1990s	Increased application of smart products; developments in mechatronics, microelectromechanical systems (MEMS)

The current trend and associate challenges of control engineering concern developments, innovations, and applications in such domains as MEMS; nanotechnology; embedded, distributed, and integrated sensors, actuators, and controllers; intelligent multiagent systems; smart and adaptive structures; and intelligent vehicle-highway systems.

1.5 Organization of the Book

The book consists of twelve chapters and three appendices. The chapters are devoted to presenting the fundamentals, analytical concepts, modeling, simulation and design issues, control, instrumentation, computer-based solutions, and applications of control systems. The book uniformly incorporates the underlying fundamentals of modeling and control into the development of useful analytical methods, integrated modeling approaches, and common practical techniques of control, design, and instrumentation in a systematic manner throughout the main chapters. The application of the concepts, approaches, and tools presented in the chapters are demonstrated through numerous illustrative examples and end of chapter problems.

Chapter 1 introduces the field of control engineering with a focus on the use of modeling in control. Application areas of control engineering are indicated. The historical development of the field is presented. This introductory chapter sets the tone for the study, which spans the remaining 11 chapters.

Chapter 2 introduces the subject of modeling of a dynamic system. Analytical models may be developed for mechanical, electrical, fluid, and thermal systems in a rather analogous manner, because some clear analogies exist among these four types of systems. Emphasized in the chapter are model types; the tasks of "understanding" and analytical representation (i.e., analytical modeling) of mechanical, electrical, fluid, and thermal systems; identification of lumped elements (inputs/sources, and equivalent capacitor, inductor, and resistor elements; considerations of the associated variables (e.g., through and across variables; state variables); and the development of state-space models and input–output models.

Chapter 3 studies linearization of a nonlinear system/model in a restricted range of operation, about an operating point. Real systems are nonlinear and they are represented by nonlinear analytical models consisting of nonlinear differential equations. Linear systems (models) are in fact idealized representations, and are represented by linear differential equations. First linearization of analytical models, particularly state-space models and input–output models is treated. Then linearization of experimental models (experimental data) is addressed.

Chapter 4 presents linear graphs—an important graphical tool for developing and representing a model of a dynamic system. State-space models of lumped-parameter dynamic systems; regardless of whether they are mechanical, electrical, fluid, thermal, or multidomain (mixed) systems; can be conveniently developed by using *linear graphs*. The chapter systematically studies the use of linear graphs in the development of analytical models for mechanical, electrical, fluid, and thermal systems.

Chapter 5 treats transfer-function models and the frequency-domain analysis of dynamic systems. A linear, constant-coefficient (time-invariant) time-domain model can be converted into a transfer function, and vice versa, in a simple and straightforward manner. A unified approach is presented for the use of the transfer function approach in the modeling and analysis of multidomain (e.g., mechanical and electrical) systems. Extension of the linear graph approach is presented, with the use of Thevenin and Norton equivalent circuits.

Chapter 6 studies response analysis and simulation. In particular, it provides information regarding how a system responds when excited by an initial condition—free, natural response, or when a specific excitation (i.e., input) is applied—forced response. Such a study may be carried out by the solution of differential equations (analytical) or by computer simulation (numerical). The chapter addresses both approaches. In the latter case, in particular, the use of Simulink® is illustrated.

Chapter 7 introduces common architectures of control systems. A good control system should satisfy a specified set of performance requirements with respect to such attributes as stability, speed, accuracy, and robustness. The chapter presents methods of specifying and analyzing the performance of a control system. Time-domain specifications are emphasized. Steady-state error, error constants, and control system sensitivity are discussed.

Chapter 8 concerns stability of a control system. Both time-domain and frequency-domain techniques of stability analysis are presented. Routh–Hurwitz method, root locus method, Nyquist criterion, and Bode diagram method incorporating gain margin and phase margin are presented for stability analysis of linear time-invariant (LTI) systems.

Chapter 9 deals with design and tuning of controllers and compensators. Designing a control system may involve selection, modification, addition, removal, and relocation of components as well as selection of suitable parameter values for one or more of the components in the control system in order to satisfy a set of *design specifications*. Once a control system is designed and the system parameters are chosen it may be necessary to further tune the parameters in order to achieve the necessary performance levels. Emphasized in the chapter are the frequency-domain and root-locus methods of designing lead and lag compensators. The Ziegler–Nichols method of controller tuning is given.

Chapter 10 studies digital control. In a digital control system, a digital device (e.g., a computer) is used as the controller. The chapter presents relevant issues of data sampling. A convenient way to analyze and design digital control systems is by the z-transform method. The theory behind this method is presented and issues such as stability analysis and controller/compensator design by the z-transform method are described.

Chapter 11 introduces several advanced and popular methodologies of control. What are commonly identified as modern control techniques are time-domain multivariable (multiinput–multioutput [MIMO]) techniques that use the state-space representation for the system. The chapter presents some of the common techniques, particularly in the categories of optimal control and modal control. In this context, linear quadratic regulator (LQR) and pole-placement control are studied. Fuzzy logic control, which has been quite popular in engineering/industrial applications in the context of intelligent control, is also presented.

Chapter 12 introduces the subject of instrumentation, as related to control engineering. It considers "instrumenting" of a control system with sensors, transducers, actuators, and associated hardware. The components have to be properly chosen and interconnected in order to achieve a specified level of performance. Relevant issues are addressed. A representative set of analog and digital sensors are presented. Stepper motor and dc motor are presented as popular actuators in control systems. Procedures of motor selection and control are addressed. The use of the computer software tool LabVIEW® for data acquisition and control, particularly in laboratory experimentation, is illustrated.

Appendix A presents techniques and other useful information on Laplace transform and Fourier transform. Appendix B introduces several popular software tools and environments that are available for simulation and control engineering both at the learning level and at the professional application level. Presented and illustrated in the appendix are Simulink—a graphical environment for modeling, simulation, and analysis of dynamic systems; MATLAB® with its Control Systems Toolbox and Fuzzy Logic Toolbox; and LabVIEW, which is a graphical programming language and a program development environment for data acquisition, processing, display, and instrument control. Appendix C reviews linear algebra—the algebra of sets, vectors, and matrices. It is useful in the study of control systems in general and the state–space approach in particular.

Problems

PROBLEM 1.1

A typical input variable is identified for each of the following examples of dynamic systems. Give at least one output variable for each system.

a. Human body: Neuroelectric pulses
b. Company: Information
c. Power plant: Fuel rate
d. Automobile: Steering wheel movement
e. Robot: Voltage to joint motor

PROBLEM 1.2

According to some observers in the process control industry, early brands of analog control hardware had a product life of about 20 years. New hardware controllers can become obsolete in a couple of years, even before their development costs are recovered. As a control instrumentation engineer responsible for developing an off-the-shelf process controller, what features would you incorporate into the controller in order to correct this problem to a great extent?

PROBLEM 1.3

A soft-drink bottling plant uses an automated bottle-filling system. Describe the operation of such a system, indicating various components in the control system and their functions. Typical components would include a conveyor belt; a motor for the conveyor, with start/stop controls; a measuring cylinder, with an inlet valve, an exit valve, and level sensors; valve actuators; and an alignment sensor for the bottle and the measuring cylinder.

PROBLEM 1.4

One way to classify controllers is to consider their sophistication and physical complexity separately. For instance, we can use an x–y plane with the x-axis denoting the physical complexity and the y-axis denoting the controller sophistication. In this graphical representation simple open-loop on-off controllers (say, opening and closing a valve) would have a very low controller sophistication value and an artificial-intelligence (AI)-based "intelligent" controller would have a high controller sophistication value. Also, a passive device is considered to have less physical complexity than an active device. Hence, a passive spring-operated device (e.g., a relief valve) would occupy a position very close to the origin of the x–y plane and an intelligent machine (e.g., sophisticated robot) would occupy a position diagonally far from the origin. Consider five control devices of your choice. Mark the locations that you expect them to occupy (in relative terms) on this classification plane.

2

Modeling of Dynamic Systems

Design, development, modification, and control of an engineering system require an understanding and a suitable "representation" of the system; specifically, a "model" of the system is required. Any model is an idealization of the actual system. Properties established and results derived are associated with the model rather than the actual system, whereas the excitations are applied to and the output responses are measured from the actual system. This distinction is very important particularly in the context of the present treatment. An engineering system may consist of several different types of component; then, it is termed a *multidomain* (or *mixed*) *system*. Furthermore, it may contain *multifunctional components*; for example, a piezoelectric component which can function as both a sensor and an actuator. It is useful to use analogous procedures for modeling such components. Then the component models can be conveniently and systematically integrated to obtain the overall model. Analytical models may be developed for mechanical, electrical, fluid, and thermal systems in a rather analogous manner, because some clear analogies are present among these four types of systems. In view of the *analogy*, then, a unified approach may be adopted in the analysis, design, and control of engineering systems. Emphasized in this chapter are model types; the tasks of "understanding" and analytical representation (i.e., analytical modeling) of mechanical, electrical, fluid, and thermal systems; identification of lumped elements (inputs/sources, and equivalent capacitor, inductor, and resistor elements); considerations of the associated variables (e.g., through and across variables; state variables); and the development of state-space models and input–output (I/O) models.

2.1 Dynamic Systems

Each interacted component or element of an engineering system will possess an *I/O* (or *cause-effect*, or *causal*) relationship. A *dynamic system* is one whose response variables are functions of time and have nonnegligible "rates" of changes. Also, its present output depends not only on the present input but also on some historical information (e.g., previous input or output). A more formal mathematical definition can be given, but it is adequate to state here that a typical engineering system, which needs to be controlled, is a dynamic system. A model is some form of representation of a practical system. An analytical model (or mathematical model) comprises equations (e.g., differential equations) or an equivalent set of information, which represents the system to some degree of accuracy. Alternatively, a set of curves, digital data (e.g., arrays or tables) stored in a computer, and other numerical data—rather than a set of equations—may be termed a model, strictly a numerical model (or experimental model) from which a representative analytical model can be established or "identified" (the related topic is called "model identification" or "system identification" in the subject of automatic control).

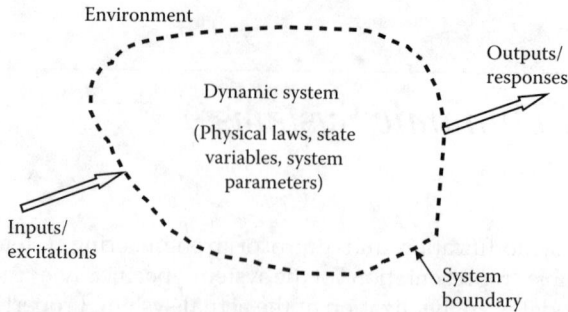

FIGURE 2.1
Nomenclature of a dynamic system.

2.1.1 Terminology

A general representation of a dynamic system is given in Figure 2.1. The system is demarcated by a boundary that may be either real (physical) or imaginary (virtual). What is outside this boundary is the environment of the system. There are inputs that enter the system from the environment and there are outputs that are provided by the system into the environment. Some useful terms are:

System: Collection of interacting components of interest, demarcated by a system boundary.

Dynamic system: A system whose rates of changes of response/state variables cannot be neglected.

Plant or process: The system to be controlled.

Inputs: Excitations (known or unknown) applied to the system.

Outputs: Responses of the system.

State variables: A minimal set of variables that completely identify the "dynamic" state of the system. *Note:* If the state variables at one state in time and the inputs from that state up to a future state in time are known, the future state can be completely determined.

Control system: The system that includes at least the plant and its controller. It may include other subsystems and components (e.g., sensors, signal conditioning and modification).

Dynamic systems are not necessarily engineering, physical, or man-made systems. Some examples of dynamic systems with their representative inputs and outputs are given in Table 2.1. Try to identify several known and deliberately applied inputs; unknown and/or undesirable inputs (e.g., disturbances); desirable outputs; and undesirable outputs for each of these systems.

2.2 Dynamic Models

A dynamic model is a representation of a dynamic system. It is useful in analysis, computer simulation, design, modification, and control of the system. An engineering physical system consists of a mixture of different types (e.g., mechanical, electrical, fluid, thermal) of processes and components; it is typically a multidomain or mixed system. An integrated and unified development of a model is desirable, where all domains are modeled together using similar approaches. Then analogous procedures are used to model

TABLE 2.1

Examples of Dynamic Systems

System	Typical Input	Typical Outputs
Human body	Neuroelectric pulses	Muscle contraction, body movements
Company	Information	Decisions, finished products
Power plant	Fuel rate	Electric power, pollution rate
Automobile	Steering wheel movement	Front wheel turn, direction of heading
Robot	Voltage to joint motor	Joint motions, effector motion

all components, in developing an analytical model. This is the approach that is taken in this book. Analogies which exist in mechanical, electrical, fluid, and thermal systems are exploited for this purpose.

2.2.1 Model Complexity

It is unrealistic to attempt to develop a "universal model" that will incorporate all conceivable aspects of the system. For example, an automobile model that will simultaneously represent ride quality, power, speed, energy consumption, traction characteristics, handling, structural strength, capacity, load characteristics, cost, safety, and so on is not very practical and can be intractably complex. The model should be as simple as possible, and may address only a few specific aspects of interest in the particular study or application. Approximate modeling and model reduction are relevant topics in this context.

2.2.2 Model Types

One way to analyze a system is to impose excitations (inputs) on the system, measure the reactions (outputs) of the system, and fit the I/O data obtained in this manner into a suitable analytical model. This is known as "experimental modeling" or *model identification* or *system identification*. A model determined in this manner is called an experimental model. Another way to analyze a system is by using an analytical model of the system, which originates from the physical (constitutive) equations of the constituent components or processes of the system. Analytical models include *state-space models, linear graphs, bond graphs, transfer function models (in the Laplace domain),* and *frequency domain models.* Since developing a physical model (or prototype) of a system and testing it is often far less economical or practical than analyzing or computer-simulating an analytical model of the system, analytical models are commonly used in practical applications, particularly during the preprototyping stage. Instrumentation (exciters, measuring devices and analyzers) and computer systems for experimental modeling (e.g., modal testing and analyzing systems) are commercially available, and experimental modeling is done, if less often, than analytical modeling.

In general, models may be grouped into the following categories:

1. Physical models (prototypes)
2. Analytical models
3. Computer (numerical) Models (data tables, arrays, curves, programs, files, etc.)
4. Experimental models (use I/O experimental data for model "identification")

Normally, mathematical definitions for a dynamic system are given with reference to an analytical model of the system; form example, a state-space model. In that context the system and its analytical model are somewhat synonymous. In reality, however, an analytical model, or any model for that manner, is an idealization (or approximate representation) of the actual system. Analytical properties that are established and results that are derived would be associated with the model rather than the actual system, whereas the excitations are applied to and the output responses are measured from the actual system. This distinction should be clearly recognized.

Analytical models are quite useful in predicting the dynamic behavior (response) of a system for various types of excitations (inputs). For example, vibration is a dynamic phenomenon and its analysis, practical utilization, and effective control require a good understanding of the vibrating system. Computer-based studies (e.g., computer simulation) may be carried out as well using analytical models in conjunction with suitable values for the system parameters (mass, stiffness, damping, capacitance, inductance, resistance, and so on). A model may be employed for designing an engineering system for proper performance. In the context of *product testing*, for example, analytical models are commonly used to develop test specifications and the input signals that are applied to the exciter in the test procedure. Dynamic effects and interactions in the test object, the excitation system, and their interfaces may be studied in this manner. Product qualification is the procedure that is used to establish the capability of a product to withstand a specified set of operating conditions. In product qualification by testing, the operating conditions are generated and applied to the test object by an exciter (e.g., shaker). In product qualification by analysis, a suitable analytical model of the product replaces the test specimen that is used in product qualification by testing. In the area of automatic control, models are used in a variety of ways. A model of the system to be controlled (plant, process) may be used to develop the performance specifications, based on which a controller is developed for the system. In model-referenced adaptive control, for example, a reference model dictates the desired behavior that is expected under control. This is an implicit way of using a model to represent performance specifications. In process control, a dynamic model of the actual process is employed to develop the necessary control schemes. This is known as *model-based control*.

The main advantages of analytical models (and computer models) over physical models are the following:

1. Modern, high-capacity, high-speed computers can handle complex analytical models at high speed and low cost.
2. Analytical/computer models can be modified quickly, conveniently, and high speed at low cost.
3. There is high flexibility of making structural and parametric changes.
4. Directly applicable in computer simulations.
5. Analytical models can be easily integrated with computer/numerical/experimental models, to generate "hybrid" models.
6. Analytical modeling can be conveniently done well before a prototype is built (in fact this step can be instrumental in deciding whether to prototype).

2.2.3 Types of Analytical Models

The response of an analytical model to an applied excitation may be expressed in either the *time-domain*, where the response value is expressed a function of time (the independent variable is time t), or in the *frequency domain*, where the amplitude and the phase angle of the response is expressed as a function of frequency (the independent variable is frequency ω).

The time-domain response generally involves the solution of a set of differential equations (e.g., state equations). The frequency domain analysis is a special case of Laplace transform analysis (in the Laplace domain) where the independent variable is the Laplace variable s. The corresponding analytical model is a set of transfer functions. A transfer function is the ratio of the Laplace transform of the output variable divided by the Laplace transform of the input variable. In the special case of the frequency domain, $s=j\omega$. We shall see in another chapter that *mobility, admittance, impedance,* and *transmissibility* are convenient transfer-function representations, in the frequency domain. For example, transmissibility is important in vibration isolation, and mechanical impedance is useful in tasks such as cutting, joining, and assembly that employ robots.

There are many types of analytical models. They include the following:

1. Time-domain model: Differential equations with time t as the independent variable.
2. Transfer function model: Laplace transform of the output variable divided by the Laplace transform of the input variable (algebraic equation with the Laplace variable s as the independent variable).
3. Frequency domain model: Frequency transfer function (or frequency response function) which is a special case of the Laplace transfer function, with $s=j\omega$. The independent variable is frequency ω.
4. Nonlinear model: Nonlinear differential equations (principle of superposition does not hold).
5. Linear model: Linear differential equations (principle of superposition holds).
6. Distributed (or continuous)-parameter model: Partial differential equations (Dependent variables are functions of time and space).
7. Lumped-parameter model: Ordinary differential equations (dependent variables are functions of time, not space).
8. Time-varying (or nonstationary or nonautonomous) model: Differential equations with time-varying coefficients (model parameters vary with time).
9. Time-invariant (or stationary or autonomous) model: Differential equations with constant coefficients (model parameters are constant).
10. Random (stochastic) model: Stochastic differential equations (variables and/or parameters are governed by probability distributions).
11. Deterministic model: Nonstochastic differential equations.
12. Continuous-time model: Differential equations (time variable is continuously defined).
13. Discrete-time model: Difference equations (time variable is defined as discrete values at a sequence of time points).
14. Discrete transfer function model: z-transform of the discrete-time output divided by the z-transform of the discrete-time input.

2.2.4 Principle of Superposition

All practical systems can be nonlinear to some degree. If the nonlinearity is negligible, for the situation being considered, the system may be assumed linear. Since linear systems/models are far easier to handle (analyze, simulate, design, control, etc.) than nonlinear systems/models, linearization of a nonlinear model, which may be valid for a limited range or set of conditions of operation, might be required. This subject is studied in detail in Chapter 3.

All linear systems (models) satisfy the principle of superposition. A system is linear if and only if the principle of superposition is satisfied. This principle states that, if y_1 is the system output when the input to the system is u_1, and y_2 is the output when the input is u_2, then $\alpha_1 y_1 + \alpha_2 y_2$ is the output when the input is $\alpha_1 u_1 + \alpha_2 u_2$ where α_1 and α_2 are any real constants. This property is graphically represented in Figure 2.2a.

Another important property that is satisfied by linear systems is the interchangeability in series connection. This is illustrated in Figure 2.2b. Specifically, sequentially connected linear systems (or subsystems or components or elements) may be interchanged without affecting the output of the overall system for a given input. Note that interchangeability in parallel connection is a trivial fact, which is also satisfied.

In this book, we will study/employ the following modeling techniques for response analysis, simulation, and control of an engineering system:

1. State models: They use state variables (e.g., position and velocity of lumped masses, force and displacement in springs, current through an inductor, voltage across a capacitor) to represent the state of the system, in terms of which the system response can be expressed. These are time-domain models, with time t as the independent variable.

2. Linear graphs: They use line graphs where each line represents a basic component of the system, with one end as the point of action and the other end as the point of reference. They are particularly useful in the development of a state model.

3. Transfer function models including frequency domain models.

2.2.5 Lumped Model of a Distributed System

As noted earlier, lumped-parameter models and continuous-parameter models are two broad categories of models for dynamic systems. In a lumped-parameter model, various characteristics of the system are lumped into representative elements located at a discrete set of points in a geometric space. The corresponding analytical models are ordinary differential equations. In most physical systems, the properties are continuously distributed in various components or regions; they have distributed-parameter (or continuous)

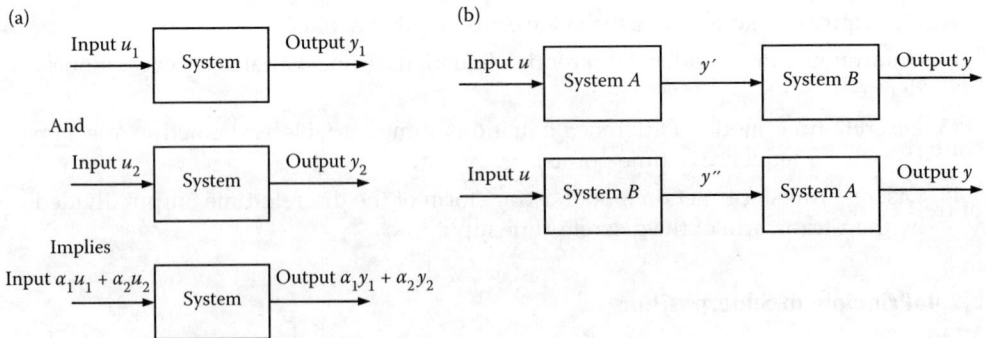

FIGURE 2.2
Properties of a linear system. (a) Principle of superposition. (b) Interchangeability in series connection.

components. To represent system parameters that are continuously distributed in space, we need spatial coordinates.

These dynamic systems have time (t) and space coordinates (e.g., x, y, z) as the independent variables. The corresponding analytical models are partial differential equations. For analytical convenience, we may attempt to approximate such distributed-parameter models into lumped-parameter ones. The accuracy of the model can be improved by increasing the number of discrete elements in such a model; for example, by using finite element techniques. In view of their convenience, lumped-parameter models are more commonly employed than continuous-parameter models. Continuous-parameter elements may be included into otherwise lumped-parameter models in order to improve the model accuracy. Let us address some pertinent issues by considering the case of a heavy spring.

2.2.5.1 Heavy Spring

A coil spring has a mass, an elastic (spring) effect, and an energy-dissipation characteristic, each of which is distributed over the entire coil. These characteristics are distributed phenomena, in general. The distributed mass of the spring has the capacity to store *kinetic energy* by acquiring velocity. Stored kinetic energy can be recovered as work done through a process of deceleration. Furthermore, in view of the distributed flexibility of the coil, each small element in the coil has the capacity to store *elastic potential energy* through reversible (elastic) deflection. If the coil was moving in the vertical direction, there would be changes in *gravitational potential energy*, but we can disregard this in dynamic response studies if the deflections are measured from the static equilibrium position of the system. The coil will undoubtedly get warmer, make creaking noises, and over time will wear out at the joints, clear evidence of its capacity to dissipate energy. A further indication of damping is provided by the fact that when the coil is pressed and released, it will eventually come to rest; the work done by pressing the coil is completely dissipated. Even though these effects are distributed in the actual system, a discrete or lumped-parameter model is usually sufficient to predict the system response to a forcing function. Further approximations are possible under certain circumstances. For instance, if the maximum kinetic energy is small in comparison with the maximum elastic potential energy in general (particularly true for light stiff coils, and at low frequencies of oscillation), and if in addition the rate of energy dissipation is relatively small (determined with respect to the time span of interest), the coil can be modeled by a discrete (lumped) stiffness (spring) element only. These are modeling decisions.

In an analytical model, the individual distributed characteristics of inertia, flexibility, and dissipation of a heavy spring can be approximated by a separate mass element, a spring element, and a damper element, which are interconnected in some parallel–series configuration, thereby producing a lumped-parameter model. Since a heavy spring has its mass continuously distributed throughout its body, it has an infinite number of degrees of freedom. A single coordinate cannot represent its motion. But, for many practical purposes, a lumped-parameter approximation with just one lumped mass to represent the inertial characteristics of the spring would be sufficient. Such an approximation may be obtained by using one of several approaches. One is the energy approach. Another approach is equivalence of natural frequency. Let us consider the energy approach first. Here we represent a distributed-parameter spring by a lumped-parameter "model" such that the original spring and the model have the same net kinetic energy and same potential energy. This energy equivalence is used in deriving a lumped mass parameter for the model. Even though damping (energy dissipation) is neglected in the present analysis, it is not difficult to incorporate that as well in the model.

2.2.5.2 Kinetic Energy Equivalence

Consider the uniform, heavy spring shown in Figure 2.3, with one end fixed and the other end moving at velocity v. Note that m_s=mass of spring; k=stiffness of spring; and l=length of spring.

In view of the linear distribution of the speed along the spring, with zero speed at the fixed end and v at the free end (Figure 2.3b), the local speed of an infinitesimal element δx of the spring is given by $(x/l)/v$. Element mass=$(m_s/l)/\delta x$. Hence, the element kinetic energy $KE = (1/2)(m_s/l)\delta x[(x/l)v]^2$. In the limit time, we have $\delta x \to dx$. Accordingly, by performing the necessary integration, we get

$$\text{Total } KE = \int_0^l \frac{1}{2}\frac{m_s}{l}\,dx\left(\frac{x}{l}v\right)^2 = \frac{1}{2}\frac{m_s v^2}{l^3}\int_0^l x^2 dx = \frac{1}{2}\frac{m_s v^2}{3} \tag{2.1}$$

Hence,

Equivalent lumped mass concentrated at the free end=$1/3\times$spring mass

Note: This derivation assumes that one end of the spring is fixed and, furthermore, the conditions are uniform along the spring.

An example of utilizing this result is shown in Figure 2.4. Here a system with a heavy spring and a lumped mass is approximated by a light spring (having the same stiffness) and a lumped mass.

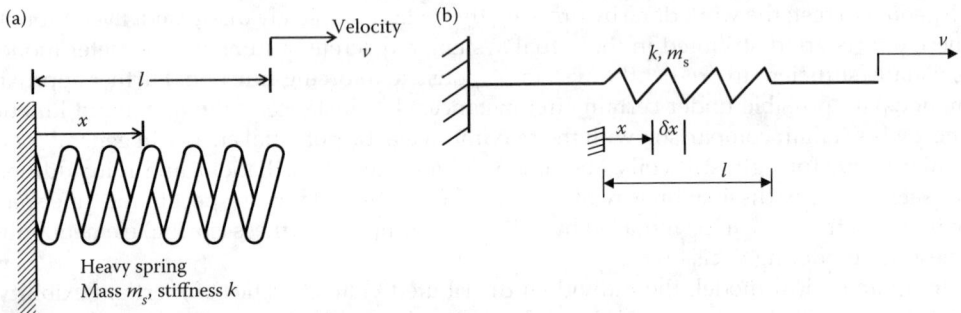

FIGURE 2.3
(a) A uniform heavy spring. (b) Analytical representation.

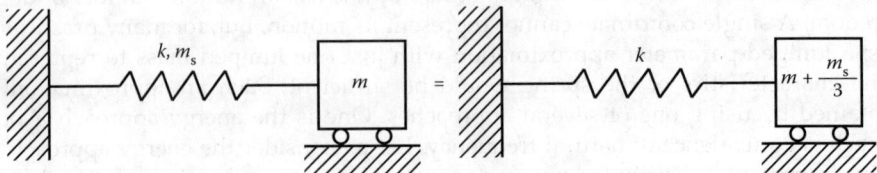

FIGURE 2.4
Lumped-parameter approximation for an oscillator with heavy spring.

2.2.5.3 Natural Frequency Equivalence

Now consider the approach of natural frequency equivalence. Here we derive an equivalent lumped-parameter model by equating the fundamental (lowest) natural frequency of the distributed-parameter system to the natural frequency of the lumped-parameter model (in the one-degree-of-freedom case). The method can be easily extended to multi-degree-of-freedom lumped-parameter models as well. We will illustrate our approach by using an example.

A heavy spring of mass m_s and stiffness k_s with one end fixed and the other end attached to a sliding mass m, is shown in Figure 2.5a. If the mass m is sufficiently larger than m_s, then at relatively high frequencies the mass will virtually stand still. Under these conditions we have the configuration shown in Figure 2.5b where the two ends of the spring are fixed. Also, approximate the distributed mass by an equivalent mass m_e at the mid point of the spring: each spring segment has double the stiffness of the original spring. Hence the overall stiffness is $4k_s$. The natural frequency of the lumped-model is

$$\omega_e = \sqrt{\frac{4k_s}{m_e}} \tag{2.2}$$

It is known from a complete analysis of a heavy spring (which beyond the present scope) that the natural frequency for the fixed-fixed configuration is

$$\omega_s = \pi n \sqrt{\frac{k_s}{m_e}} \tag{2.3}$$

where n is the mode number. Then, for the fundamental (first) mode (i.e., $n=1$), the natural frequency equivalence gives

$$\sqrt{\frac{4k_s}{m_e}} = \pi \sqrt{\frac{k_s}{m_e}}$$

or,

$$m_e = \frac{4}{\pi^2} m_s \approx 0.4 m_s \tag{2.4}$$

Note that since the effect of inertia decreases with increasing frequency, it is not necessary to consider the case of high frequencies.

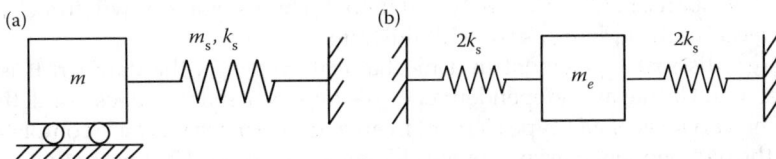

FIGURE 2.5
(a) A lumped mass with a distributed-parameter system. (b) A lumped-parameter model of the system.

The natural frequency equivalence may be generalized as an *eigenvalue* equivalence (*pole* equivalence) for any dynamic system. In this general approach, the eigenvalues of the lumped-parameter model are equated to the corresponding eignevalues of the distributed-parameter system, and the model parameters are determined accordingly.

2.3 Lumped Elements and Analogies

A system may possess various physical characteristics incorporating multiple domains; for example, mechanical, electrical, thermal, and fluid components and processes. The procedure of model development will be facilitated if we understand the similarities of these various domains and in the characteristics of different types of components. This issue is addressed in this section.

The basic system elements in an engineering system can be divided into two groups: energy-storage elements and energy-dissipation elements. The dynamic "state" of a system is determined by its independent energy-storage elements and the associated state variables. Depending on the element we can use either an across variable or a through variable as its state variable.

2.3.1 Across Variables and Through Variables

An across variable is measured across an element, as the difference in the values at the two ends. Velocity, voltage, pressure, and temperature are across variables. A through variable represents a property that appears to flow through an element, unaltered. Force, current, fluid flow rate, and heat transfer rate are through variables. If the across variable of an element is the appropriate state variable for that element, it is termed an *A*-type element. Alternatively, if the through variable of an element is the appropriate state variable for that element, it is termed a *T*-type element.

Analogies exist among mechanical, electrical, hydraulic, and thermal systems/processes. Next we state the physical equations (i.e., constitutive equations) of the basic elements in these four domains, identify appropriate state variables, and recognize analogies that exist across these domains.

2.3.2 Mechanical Elements

For mechanical elements, it will be seen that the velocity (across variable) of each independent mass and the force (through variable) of each independent spring are the appropriate *state variables* (response variables). Hence, mass is an *A*-type element and spring is a *T*-type element. These are energy-storage elements. The corresponding constitutive equations form the "state-space shell" for an analytical model. These equations will directly lead to a *state-space model* of the system, as we will illustrate.

The energy dissipating element in a mechanical system is the damper. It is called a *D*-type element. Unlike an independent energy-storage element, it does not define a state variable. The variables of a *D*-type element in an engineering system are completely determined by the independent energy-storage elements (*A*-type and *T*-type) in the system. The input elements (or source elements) of a mechanical system are the force source, where its force is the independent variable, which is not affected by the changes in the system (while the associated velocity variable—the dependent variable—will be affected); and the velocity source, where its velocity is the independent variable, which is not affected by the changes in the system (while the associated force variable—the dependent variable—will be affected). These are "ideal" sources since in practice the source variable will be affected to some extent by the dynamics of the system, and is not completely "independent."

2.3.2.1 Mass (Inertia) Element

Consider the mass element shown in Figure 2.6a. The constitutive equation (the physical law) of the element is given by Newton's second law:

$$m\frac{dv}{dt} = f \tag{2.5}$$

Here v denotes the velocity of mass m, measured relative to an inertial (fixed on earth) reference, and f is the force applied "through" the mass. Since power$=fv=$rate of change of energy, by substituting Equation 2.5, the energy of the element may be expressed as

$$E = \int fv\, dt = \int m\frac{dv}{dt}v\, dt = \int mv\, dv$$

or

$$\text{Energy } E = \frac{1}{2}mv^2 \tag{2.6}$$

This is the well-known *kinetic energy*. Now by integrating Equation 2.5, we have

$$v(t) = v(0^-) + \frac{1}{m}\int_{0^-}^{t} f\, dt \tag{2.7}$$

By setting $t = 0^+$ in Equation 2.7, we see that as long as force f is finite,

$$v(0^+) = v(0^-) \tag{2.8}$$

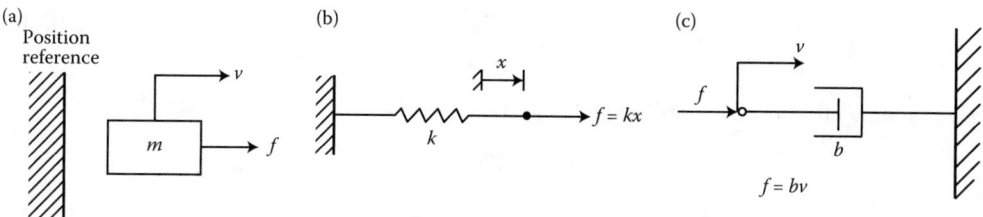

FIGURE 2.6
Basic mechanical elements. (a) Mass (inertia). (b) Spring (stiffness element). (c) Damper (dissipating element).

Note that 0^- denotes the instant just before $t=0$ and 0^+ denotes the instant just after $t=0$. In view of these observations, we may state the following:

1. An inertia is an energy-storage element (kinetic energy).
2. Velocity can represent the state of an inertia element. This is justified by two reasons: First, from Equation 2.7, the velocity at any time t can be completely determined with the knowledge of the initial velocity and the applied force during the time interval 0 to t. Second, from Equation 2.6, the energy of an inertia element can be represented by the variable v alone.
3. Velocity across an inertia element cannot change instantaneously unless an infinite force is applied to it.
4. A finite force cannot cause an infinite acceleration (or step change in velocity) in an inertia element. Conversely, a finite instantaneous (step) change in velocity will need an infinite force. Hence, v is a natural output (or response) variable for an inertia element, which can represent its dynamic state (i.e., state variable), and f is a natural input variable for an inertia element.
5. Since its state variable, velocity, is an across variable, an inertia is an A-type element.

2.3.2.2 Spring (Stiffness) Element

Consider the spring element (linear) shown in Figure 2.6b. The constitutive equation (physical law) for a spring is given by Hooke's law:

$$\frac{df}{dt} = kv \tag{2.9}$$

Here, k is the stiffness of the spring.

Note: We have differentiated the familiar force-deflection Hooke's law, in order to be consistent with the response/state variable (velocity) that is used for its counterpart, the inertia element.

Now following the same steps as for the inertia element, the energy of a spring element may be expressed as

$$E = \int fv\,dt = \int f\frac{1}{k}\frac{df}{dt}dt = \int \frac{1}{k}f\,df$$

or

$$\text{Energy } E = \frac{1}{2}\frac{f^2}{k} \tag{2.10}$$

This is the well-known (elastic) *potential energy*.

Also,

$$f(t) = f(0^-) + k\int_{0^-}^{t} v\,dt \tag{2.11}$$

We see that, as long as the applied velocity is finite,

$$f(0^+) = f(0^-) \tag{2.12}$$

In summary, we have

1. A spring (stiffness element) is an energy-storage element (elastic potential energy).
2. Force can represent the state of a stiffness (spring) element. This is justified by two reasons: First, from Equation 2.11, the force of a spring at any general time t may be completely determined with the knowledge of the initial force and the applied velocity from time 0 to t. Second, from Equation 2.10, the energy of a spring element can be represented in terms of the variable f alone.
3. Force through a stiffness element cannot change instantaneously unless an infinite velocity is applied to it.
4. Force f is a natural output (response) variable, which can represent its dynamic state (i.e., state variable), and v is a natural input variable for a stiffness element.
5. Since its state variable, force, is a through variable, a spring is a T-type element.

2.3.2.3 Damping (Dissipation) Element

Consider the mechanical damper (linear viscous damper or dashpot) shown in Figure 2.6c. It is a D-type element (energy dissipating element). The constitutive equation (physical law) is:

$$f = bv \tag{2.13}$$

where b is the damping constant. Equation 2.13 is an algebraic equation. Hence either f or v can serve as the natural output variable for a damper, and either one can determine its state. However, since the state variables v and f are established by an independent inertial element and an independent spring element, respectively, a damper will not introduce a new state variable.

In summary:

1. Mechanical damper is an energy dissipating element (D-type element).
2. Either force f or velocity v may represent its state.
3. No new state variable is defined by this element.

2.3.3 Electrical Elements

In electrical systems, capacitor is the A-type element, with voltage (across variable) as its state variable; and inductor is the T-type element, with current (through variable) as its state variable. These are energy-storage elements and their constitutive equations are differential equations. The resistor is the energy dissipater (D-type element) and as usual, with an algebraic constitutive equation; it does not define a new state variable. These three elements are discussed below. The input elements (or source elements) of an electrical system are the voltage source, where its voltage is the independent variable, which is not affected by the changes in the system (while the associated current variable—the dependent variable—will be affected); and the current source, where its current is the independent variable, which is not affected by the changes in the system (while the associated voltage variable—the dependent variable—will be affected). These are "ideal" sources since in practice the source variable will be affected to some extent by the dynamics of the system, and is not completely "independent."

2.3.3.1 Capacitor Element

Consider the capacitor element shown in Figure 2.7a. Its constitutive equation (the physical law) is given by the differential equation:

$$C\frac{dv}{dt} = i \tag{2.14}$$

Here v denotes the voltage "across" the capacitor with capacitance C, and i is the current "through" the capacitor. Since power is given by the product iv, by substituting Equation 2.14, the energy in a capacitor may be expressed as

$$E = \int iv\, dt = \int C\frac{dv}{dt}v\, dt = \int Cv\, dv$$

or

$$\text{Energy } E = \frac{1}{2}C\,v^2 \tag{2.15}$$

This is the familiar *electrostatic energy* of a capacitor.
Also,

$$v(t) = v(0^-) + \frac{1}{C}\int_{0^-}^{t} i\, dt \tag{2.16}$$

Hence, for a capacitor with a finite current, we have

$$v(0^+) = v(0^-) \tag{2.17}$$

(a)   (b)   (c)

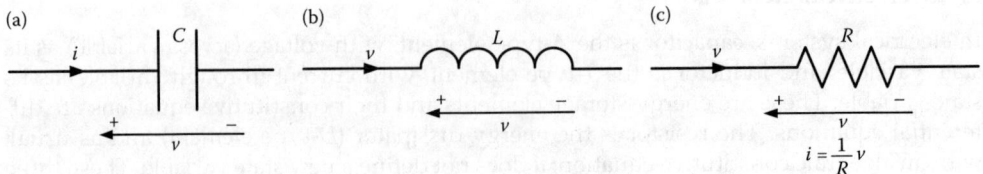

FIGURE 2.7
Basic electrical elements. (a) Capacitor. (b) Inductor. (c) Resistor (dissipating element).

We summarize:

1. A capacitor is an energy-storage element (electrostatic energy).
2. Voltage is an appropriate (natural) response variable (or state variable) for a capacitor element. This is justified by two reasons: First, from Equation 2.16, the voltage at any time t can be completely determined with the knowledge of the initial voltage and the applied current during the time interval 0 to t. Second, from Equation 2.15, the energy of a capacitor element can be represented by the variable v alone.
3. Voltage across a capacitor cannot change instantaneously unless an infinite current is applied.
4. Voltage is a natural output variable and current is a natural input variable for a capacitor.
5. Since its state variable, voltage, is an across variable, a capacitor is an A-type element.

2.3.3.2 Inductor Element

Consider the inductor element shown in Figure 2.7b. Its constitutive equation (the physical law) is given by the differential equation:

$$L\frac{di}{dt} = v \qquad (2.18)$$

Here, L is the inductance of the inductor. As before, it can be easily shown that energy in an inductor is given by

$$E = \frac{1}{2}Li^2 \qquad (2.19)$$

This is the well-known *electromagnetic energy* of an inductor.
Also, by integrating Equation 2.18 we get

$$i(t) = i(0^-) + \frac{1}{L}\int_{0^-}^{t} v \, dt \qquad (2.20)$$

Hence, for an inductor, for a finite voltage we have

$$i(0^+) = i(0^-) \qquad (2.21)$$

To summarize:

1. An inductor is an energy-storage element (electromagnetic energy).
2. Current is an appropriate response variable (or state variable) for an inductor. This is justified by two reasons: First, from Equation 2.20, the current at any time t can be completely determined with the knowledge of the initial current and the applied current during the time interval 0 to t. Second, from Equation 2.19, the energy of an inductor element can be represented by the variable i alone.
3. Current through an inductor cannot change instantaneously unless an infinite voltage is applied.

4. Current is a natural output variable and voltage is a natural input variable for an inductor.

5. Since its state variable, current, is a through variable, an inductor is a *T*-type element.

2.3.3.3 Resistor (Dissipation) Element

Consider the resistor element shown in Figure 2.7c. It is a *D*-type element (energy dissipating element). The constitutive equation (physical law) is the well-known Ohm's law:

$$v = Ri \tag{2.22}$$

where *R* is the resistance of the resistor. Equation 2.22 is an algebraic equation. Hence either *v* or *i* can serve as the natural output variable for a resistor, and either one can determine its state. However, since the state variables *v* and *i* are established by an independent capacitor element and an independent inductor element, respectively, a damper will not introduce a new state variable.

In summary:

1. Electrical resistor is an energy dissipating element (*D*-type element).

2. Either current *i* or voltage *v* may represent its state.

3. No new state variable is defined by this element.

2.3.4 Fluid Elements

In a fluid component, pressure (*P*) is the across variable and the volume flow rate (*Q*) is the through variable. The three basic fluid elements are shown in Figure 2.8 and discussed below. Note the following:

1. The elements are usually distributed, but lumped-parameter approximations are used here.

2. The elements are usually nonlinear (particularly, the fluid resistor), but linear models are used here.

The input elements (or source elements) of a fluid system are the pressure source, where its pressure is the independent variable, which is not affected by the changes in the system (while the associated flow rate variable—the dependent variable—will be affected); and

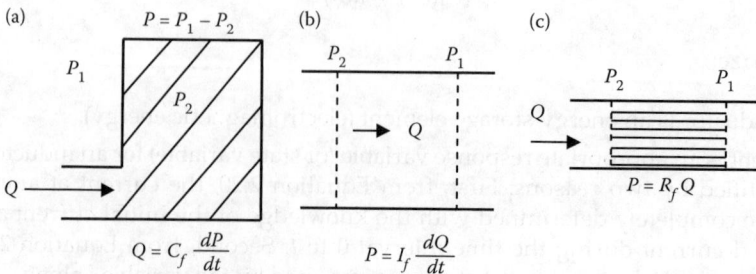

FIGURE 2.8
Basic fluid elements. (a) Capacitor. (b) Inertor. (c) Resistor.

the flow source, where its flow rate is the independent variable, which is not affected by the changes in the system (while the associated pressure variable—the dependent variable—will be affected).

2.3.4.1 Fluid Capacitor or Accumulator (A-Type Element)

Consider a rigid container with a single inlet through which fluid is pumped in at the volume flow rate Q, as shown in Figure 2.8a. The pressure inside the container with respect to the outside is P. We can write the linear constitutive equation

$$Q = C_f \frac{dP}{dt} \tag{2.23}$$

where C_f=fluid capacitance (capacity). Several special cases of fluid capacitance will be discussed later.

A fluid capacitor stores potential energy, given by $1/2\, C_f P^2$. Hence, this element is like a fluid spring. The appropriate state variable is the pressure difference (across variable) P. Contrast here that the mechanical spring is a T-type element.

2.3.4.2 Fluid Inertor (T-Type Element)

Consider a conduit carrying an accelerating flow of fluid, as shown in Figure 2.8b. The associated linear constitutive equation may be written as

$$P = I_f \frac{dQ}{dt} \tag{2.24}$$

where I_f=fluid inertance (inertia).

A fluid inertor stores kinetic energy, given by $1/2\, I_f Q^2$. Hence, this element is a fluid inertia. The appropriate state variable is the volume flow rate (through variable) Q. Contrast here that the mechanical inertia is an A-type element. Energy exchange between a fluid capacitor and a fluid inertor leads to oscillations (e.g., water hammer) in fluid systems, analogous to oscillations in mechanical and electrical systems.

2.3.4.3 Fluid Resistor (D-Type Element)

Consider the flow of fluid through a narrow element such as a thin pipe, orifice, or valve. The associated flow will result in energy dissipation due to fluid friction. The linear constitutive equation is (see Figure 2.8c):

$$P = R_f Q \tag{2.25}$$

2.3.4.4 Derivation of Constitutive Equations

We now indicate the derivation of the constitutive equations for fluid elements.
1. Fluid capacitor
    The capacitance in a fluid element may originate from:
    (a) Bulk modulus effects of liquids
    (b) Compressibility effects of gases

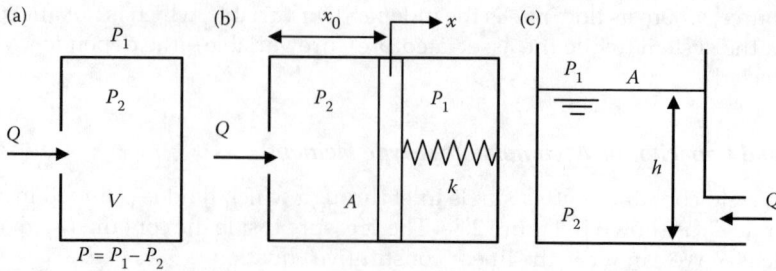

FIGURE 2.9
Three types of fluid capacitance. (a) Bulk modulus or compressibility. (b) Flexibility of container. (c) Gravity head of fluid column.

    (c)   Flexibility of the fluid container itself
    (d)   Gravity head of a fluid column

Derivation of the associated constitutive equations is outlined below.

(a) Bulk modulus effect of liquids
   Consider a rigid container. A liquid is pumped in at the volume flow rate Q. An increase of the pressure in the container will compress the liquid volume, thereby letting in more liquid (see Figure 2.9a). From calculus we have

$$\frac{dP}{dt} = \frac{\partial P}{\partial V}\frac{dV}{dt}$$

where V is the control volume of liquid. Now, the volume flow rate (into the container) is given by

$$Q = -\frac{dV}{dt}.$$

The bulk modulus of liquid is defined by:

$$\beta = -V\frac{\partial P}{\partial V} \tag{2.26}$$

Hence,

$$\frac{dP}{dt} = \frac{\beta}{V}Q \quad \text{or} \quad Q = \frac{V}{\beta}\frac{dP}{dt} \tag{2.27}$$

and the associated capacitance is

$$C_{\text{bulk}} = \frac{V}{\beta} \tag{2.28}$$

(b) Compression of gases

Consider a perfect (ideal) gas, which is governed by the gas law

$$PV = mRT \tag{2.29}$$

where P=pressure (units are pascals: 1 Pa=1 N/m²); V=volume (units are m³); T=absolute temperature (units are degrees Kelvin, °K); m=mass (units are kg); R=specific gas constant (units: kJ/kg/°K where 1 J=1 joule=1 N.m; 1 kJ=1000 J).

Isothermal case: Consider a slow flow of gas into a rigid container (see Figure 2.9a) so that the heat transfer is allowed to maintain the temperature constant (isothermal). Differentiate Equation 2.29 keeping T constant (i.e., RHS is constant):

$$P\frac{dV}{dt} + V\frac{dP}{dt} = 0$$

Noting that $Q = -(dV/dt)$ and substituting into this the above equation and Equation 2.29 we get

$$Q = \frac{V}{P}\frac{dP}{dt} = \frac{mRT}{P^2}\frac{dP}{dt} \tag{2.30}$$

Hence, the corresponding capacitance is given by:

$$C_{comp} = \frac{V}{P} = \frac{mRT}{P^2} \tag{2.31}$$

Adiabatic case: Consider a fast flow of gas (see Figure 2.9a) into a rigid container so that there is no time for heat transfer (adiabatic $\Rightarrow$ zero heat transfer). The associated gas law is known to be

$$PV^k = C \quad \text{with} \quad k = Cp/C_V \tag{2.32}$$

where C_p=specific heat when the pressure is maintained constant; C_v=specific heat when the volume is maintained constant; C=constant; k=ratio of specific heats.

Differentiate Equation 2.32:

$$PkV^{k-1}\frac{dV}{dt} + V^k\frac{dP}{dt} = 0$$

Divide by V^k:

$$\frac{Pk}{V}\frac{dV}{dt} + \frac{dP}{dt} = 0$$

Now use $Q = -(dV/dt)$ as usual, and also substitute Equation 2.29:

$$Q = \frac{V}{kP}\frac{dP}{dt} = \frac{mRT}{kP^2}\frac{dP}{dt} \tag{2.33}$$

The corresponding capacitance is

$$C_{comp} = \frac{V}{kP} = \frac{mRT}{kP^2} \tag{2.34}$$

(c) Effect of flexible container

Without loss of generality, consider a cylinder of cross-sectional area A with a spring-loaded wall (stiffness k) as shown in Figure 2.9b. As a fluid (assumed incompressible) is pumped into the cylinder, the flexible wall will move through x.

$$\text{Conservation of flow:} \ Q = \frac{d(A(x_0 + x))}{dt} = A\frac{dx}{dt} \tag{i}$$

$$\text{Equilibrium of spring:} \ A(P_2 - P_1) = kx \quad \text{or} \quad x = \frac{A}{k}P \tag{ii}$$

Substitute (ii) in (i). We get

$$Q = \frac{A^2}{k}\frac{dP}{dt} \tag{2.35}$$

The corresponding capacitance is

$$C_{elastic} = \frac{A^2}{k} \tag{2.36}$$

Note: For an elastic container and a fluid having bulk modulus, the combined capacitance will be additive:

$$C_{eq} = C_{bulk} + C_{elastic}$$

A similar result holds for a compressible gas and an elastic container.

(d) Gravity head of a fluid column

Consider a liquid column (tank) having area of across section A, height h, and mass density ρ, as shown in Figure 2.9c. A liquid is pumped into the tank at the volume rate Q. As a result, the liquid level rises.

Relative pressure at the foot of the column $P = P_2 - P_1 = \rho gh$

$$\text{Flow rate} \ Q = \frac{d(Ah)}{dt} = A\frac{dh}{dt}$$

Direct substitution gives

$$Q = \frac{A}{\rho g}\frac{dP}{dt} \tag{2.37}$$

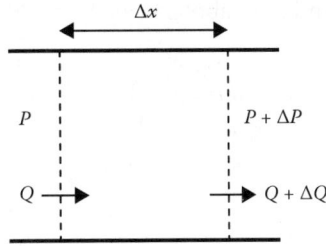

FIGURE 2.10
A fluid flow element.

The corresponding capacitance is

$$C_{grav} = \frac{A}{\rho g} \tag{2.38}$$

2. Fluid inertor

First assume a fluid flow in a conduit, with a uniform velocity distribution across it. Along a small element of length Δx of fluid, as shown in Figure 2.10, the pressure will change from P to $P+\Delta P$, and the volume flow rate will change from Q to $Q+\Delta Q$.

Mass of the fluid element $= \rho A \Delta x$
Net force in the direction of flow $= -\Delta P A$
Velocity of flow $= Q/A$
where $\rho =$ mass density of the fluid; $A =$ area of cross section.
Assuming A to be constant, we have
Fluid acceleration $= 1/A\ (dQ/dt)$.

Hence, Newton's second law gives

$$-\Delta P A = (\rho A \Delta x) \frac{1}{A} \frac{dQ}{dt}$$

or

$$-\Delta P = \frac{\rho \Delta x}{A} \frac{dQ}{dt} \tag{2.39}$$

Hence,

$$\text{Fluid inertance } I_f = \frac{\rho \Delta x}{A} \tag{2.40a}$$

where a nonuniform cross-section, $A = A(x)$, is assumed. Then, for a length L, we have

$$I_f = \int_0^L \frac{\rho}{A(x)} dx \tag{2.40b}$$

For a circular cross-section and a parabolic velocity profile, we have

$$I_f = \frac{2\rho\Delta x}{A} \tag{2.40c}$$

or, in general:

$$I_f = \alpha\frac{\rho\Delta x}{A} \tag{2.40}$$

where α is a suitable correction factor.

3. Fluid resistor

For the ideal case of viscous, laminar flow we have (Figure 2.8c):

$$P = R_f Q \tag{2.41}$$

with

$R_f = 128\ \mu L/\pi d^4$ for a circular pipe of diameter d.

$R_f = 12\ \mu L/wb^3$ for a pipe of rectangular cross section (width w and height b) with $b \ll w$.

where L=length of pipe segment; μ = absolute viscosity of fluid (dynamic viscosity).

Note: Fluid stress = $\mu(du/dy)$, where du/dy is the velocity gradient across the pipe.

$v = \mu/\rho$=kinematic viscosity.

Reynold's number $R_e = uL/v = \rho uL/\mu$

u=fluid velocity along the pipe.

For turbulent flow, the resistance equation will be nonlinear, as given by:

$$P = K_R Q^n \tag{2.42}$$

2.3.5 Thermal Elements

Thermal systems have temperature (T) as the across variable, as it is always measured with respect to some reference (or as a temperature difference across an element), and heat transfer (flow) rate (Q) as the through variable. Heat source and temperature source are the two types of source elements (inputs). The former type of source is more common. The latter type of source may correspond to a large reservoir whose temperature is hardly affected by heat transfer into or out of it. There is only one type of energy (thermal energy) in a thermal system. Hence there is only one type (A-type) energy-storage element (a thermal capacitor) with the associated state variable, temperature. There is no T-type element in a thermal system (i.e., there are no thermal inductors). As a direct result of the absence of two different types of energy-storage elements (unlike the case of mechanical, electrical, and fluid system) a pure thermal system cannot exhibit natural oscillations. It can exhibit "forced" oscillations, however, when excited by an oscillatory input source.

2.3.5.1 Constitutive Equations

The constitutive equations in a thermal system are the physical equations for thermal capacitors (A-type elements) and thermal resistors (D-type elements). There are no

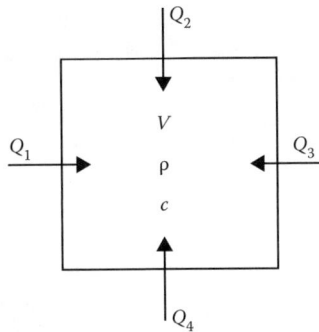

FIGURE 2.11
A control volume of a thermal system.

T-type elements. There are three types of thermal resistance: conduction, convection, and radiation.

1. Thermal capacitor

Consider a control volume of an object, with various heat transfer processes Q_i taking place at the boundary of the object (see Figure 2.11). The level of thermal energy in the object$=\rho V_c T$, where $T=$temperature of the object (assumed uniform); $V=$volume of the object; $\rho=$mass density of the object; $c=$specific heat of the object.

Since the net heat inflow is equal to the rate of change (increase) of thermal energy, the associated constitutive relation is

$$\sum Q_i = \rho V c \frac{dT}{dt} \qquad (2.43)$$

where $\rho V c$ is assumed constant. We write this as

$$Q = C_h \frac{dT}{dl} \qquad (2.44)$$

where $C_h = \rho V c = mc =$thermal capacitance.

Here $m = \rho V$ is the mass of the element.

Note: Thermal capacitance means the "capacity" to store thermal energy in a body.

2. Thermal resistor

A thermal resistor provides resistance to heat transfer in a body or a medium. There are three general types of thermal resistance:

    (a)   Conduction

    (b)   Convection

    (c)   Radiation

We will now give constitutive relations for each of these three types of thermal resistors.

(a) Conduction

The heat transfer in a medium takes place by conduction when the molecules of the medium itself do not move to transfer the heat. Heat transfer takes place from a point of

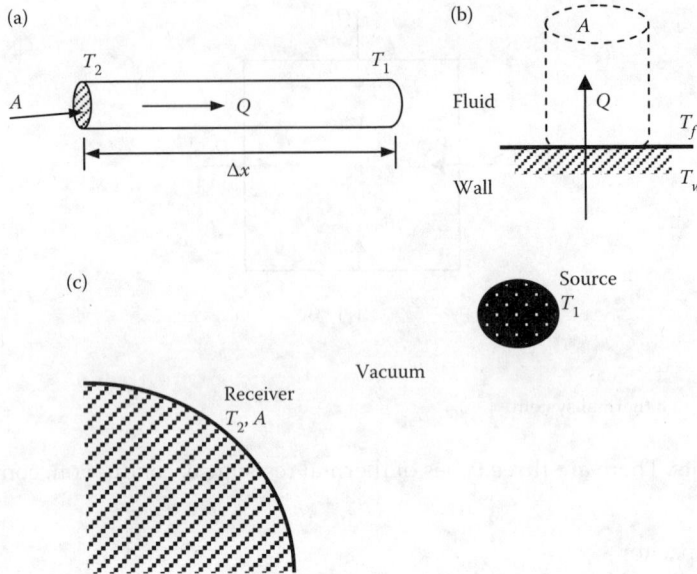

(a)

(b)

(c)

FIGURE 2.12

Three types of thermal resistance. (a) An element of 1-D heat conduction. (b) A control volume for heat transfer by convection. (c) Heat transfer by radiation.

higher temperature to one of lower temperature. Specifically, heat conduction rate is proportional to the negative temperature gradient, and is given by the *Fourier equation*:

$$Q = -kA\frac{\partial T}{\partial x} \tag{2.45}$$

where x=direction of heat transfer; A=area of cross section of the element along which heat transfer takes place; k=thermal conductivity.

Equation 2.45 (Fourier) is a "local" equation. If we consider a finite object of length Δx and cross section A, with temperatures T_2 and T_1 at the two ends, as shown in Figure 2.12a, the one-dimensional heat transfer rate Q can be written according Equation 2.45 as

$$Q = kA\frac{(T_2 - T_1)}{\Delta x} \tag{2.46a}$$

or

$$Q = \frac{1}{R_k}(T_2 - T_1) \tag{2.46b}$$

where

$$R_k = \frac{\Delta x}{kA} = \text{conductive thermal resistance} \tag{2.47}$$

(b) Convection

In convection, the heat transfer takes place by the physical movement of the heat-carrying molecules in the medium. An example is the case of fluid flowing against a wall, as shown in Figure 2.12b. The constitutive equation is

$$Q = h_c A (T_w - T_f) \tag{2.48a}$$

where T_w=wall temperature; T_f=fluid temperature at the wall interface; A=area of cross-section of the fluid control volume across which heat transfer Q takes place; h_c=convection heat transfer coefficient.

In practice h_c may depend on the temperature itself, and hence Equation 2.48a is nonlinear in general. But, by approximating to a linear constitutive equation, we can write

$$Q = \frac{1}{R_c}(T_w - T_f) \tag{2.48b}$$

where

$$R_c = \frac{1}{h_c A} = \text{convective thermal resistance} \tag{2.49}$$

In *natural convention*, the particles in the heat transfer medium move naturally. In *forced convection*, they are moved by an actuator such as a fan or pump.

(c) Radiation

In radiation, the heat transfer takes place from a higher temperature object (source) to a lower temperature object (receiver) through energy radiation, without needing a physical medium between the two objects (unlike in conduction and convection), as shown in Figure 2.12c. The associated constitutive equation is the *Stefan–Boltzman law:*

$$Q = \sigma c_e c_r A (T_1^4 - T_2^4) \tag{2.50a}$$

where A=effective (normal) area of the receiver; c_e=effective emmissivity of the source; c_r=shape factor of the receiver; σ−Stefan–Boltzman constant ($=5.7 \times 10^{-8}$ W/m^2/°K^4).

This corresponds to a nonlinear thermal resistor.

Heat transfer rate is measured in watts (W), the area in square meters (m^2), and the temperature in degrees Kelvin (°K). The relation (Equation 2.50a) is nonlinear, which may be linearized as

$$Q = \frac{1}{R_r}(T_1 - T_2) \tag{2.50b}$$

where R_r=radiation thermal resistance.

Since the slope $\partial Q / \partial T$ at an operating point may be given by $4\sigma c_e c_r A \bar{T}^3$, where $\bar{T}$ is the representative temperature (which is variable) at the operating point, we have

$$R_r = \frac{1}{4\sigma c_e c_r A \bar{T}^3} \tag{2.51a}$$

Alternatively, since $T_1^4 - T_2^4 = (T_1^2 + T_2^2)(T_1 + T_2)(T_1 - T_2)$, we may use the approximate expression:

$$R_r = \frac{1}{\sigma c_e c_r A(\overline{T}_1^2 + \overline{T}_2^2)(\overline{T}_1 + \overline{T}_2)} \tag{2.51b}$$

where the overbar denotes a representative (operating point) temperature.

2.3.5.2 Three Dimensional Conduction

Conduction heat transfer in a continuous 3-D medium is represented by a distributed-parameter model. In this case the Fourier equation (Equation 2.45) is applicable in each of the three orthogonal directions (x, y, z). To obtain a model, the thermal capacitance equation (Equation 2.43) has to be applied as well.

Consider the small 3-D model element of sides dx, dy, and dz, in a conduction medium, as shown in Figure 2.13. First consider heat transfer into the bottom $(dx \times dy)$ surface in the z direction, which according to Equation 2.45 is:

$$-k\, dx\, dy\, \frac{\partial T}{\partial z}.$$

Since the temperature gradient at the top $(dx \times dy)$ surface is $(\partial T / \partial z) + (\partial^2 T / \partial z^2)dx$ (from calculus), the heat transfer out of this surface is $k\, dx\, dy[(\partial T / \partial z) + (\partial^2 T / \partial z^2)dz]$. Hence, the net heat transfer into the element in the z direction is $k\, dx\, dy(\partial^2 T / \partial z^2)dz$ or $k\, dx\, dy\, dz(\partial^2 T / \partial z^2)$. Similarly, the net heat transfer in the x and y directions are $k\, dx\, dy\, dz(\partial^2 T / \partial x^2)$ and $k\, dx\, dy\, dz(\partial^2 T / \partial y^2)$, respectively.

The thermal energy of the element is $\rho\, dx\, dy\, dz\, cT$ where $\rho\, dx\, dy\, dz$ is the mass of the element and c is the specific heat (at constant pressure). Hence, the capacitance equation (Equation 2.43) gives

$$k\, dx\, dy\, dz\left(\frac{\partial^2 T}{\partial x^2} + \frac{\partial^2 T}{\partial y^2} + \frac{\partial^2 T}{\partial z^2}\right) = \rho\, dx\, dy\, dz\, c\frac{\partial T}{\partial t}$$

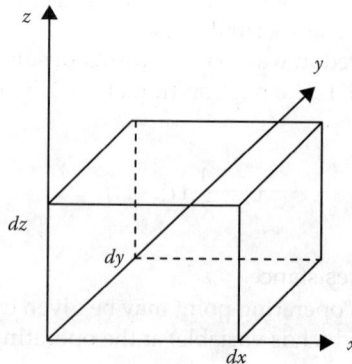

FIGURE 2.13
A 3-D heat conduction element.

or

$$\frac{\partial^2 T}{\partial x^2} + \frac{\partial^2 T}{\partial y^2} + \frac{\partial^2 T}{\partial z^2} = \frac{1}{\alpha}\frac{\partial T}{\partial t} \tag{2.52}$$

where,

$$\alpha = \frac{k}{\rho c} = \text{thermal diffusivity.}$$

Equation 2.52 is called the *Laplace equation*. Note that partial derivatives are used because T is a function of many variables; and derivatives with respect to x, y, z, and t would be needed. Hence, in general, distributed-parameter models have spatial variables (x, y, z) as well as the temporal variable (t) as independent variables, and are represented by partial differential equations.

2.3.5.3 Biot Number

This is a nondimensional parameter giving the ratio: [conductive resistance]/[convective resistance]. Hence from Equations 2.47 and 2.49 we have

$$\text{Biot number} = \frac{R_k}{R_c} = \frac{\Delta x h_c A}{kA} = \frac{h_c \Delta x}{k} \tag{2.53}$$

This parameter may be used as the basis for approximating the distributed-parameter model (Equation 2.52) by a lumped-parameter one. Specifically, divide the conduction medium into slabs of thickness Δx. If the corresponding Biot number ≤ 0.1, a lumped-parameter model may be used for each slab.

Table 2.2 summarizes the linear constitutive relationships, which describe the behavior of translatory-mechanical, electrical, thermal, and fluid elements. The analogy used in Table 2.2 between mechanical and electrical elements is known as the force-current analogy. This follows from the fact that both force and current are *through variables*, which are analogous to fluid flow through a pipe, and furthermore, both velocity and voltage are *across variables*, which vary across the flow direction, as in the case of fluid pressure along a pipe. This analogy appears more logical than a force-voltage analogy, as is clear from Table 2.3. The correspondence between the parameter pairs given in Table 2.3 follows from the relations in Table 2.2. A rotational (rotatory) mechanical element possesses constitutive relations between torque and angular velocity, which can be treated as a generalized force and a generalized velocity, respectively (compare this with a rectilinear or translatory mechanical element as listed in Table 2.2).

2.3.6 Natural Oscillations

Mechanical systems can produce natural (free) oscillatory responses (or, free vibrations) because they can possess two types of energy (kinetic and potential energies). When one type of stored energy is converted into the other type repeatedly, back and forth, the resulting response is oscillatory. Of course, some of the energy will dissipate (through the

TABLE 2.2

Some Linear Constitutive Relations

	Constitutive Relation for		
System	**Energy-Storage Elements**		**Energy Dissipating Elements**
Type	**A-Type (across) Element**	**T-Type (through) Element**	**D-Type (dissipative) Element**
Translatory-mechanical v = velocity f = force	Mass $m\dfrac{dv}{dt}=f$ (Newton's second law) m=mass	Spring $\dfrac{df}{dt}=kv$ (Hooke's law) k=stiffness	Viscous damper $f = bv$ b=damping constant
Electrical v=voltage i=current	Capacitor $C\dfrac{dv}{dt}=i$ C=capacitance	Inductor $L\dfrac{di}{dt}=v$ L=inductance	Resistor $Ri = v$ R=resistance
Thermal T=temperature difference Q=heat transfer rate	Thermal capacitor $C_t\dfrac{dT}{dt}=Q$ C_t=thermal capacitance	None	Thermal resistor $R_t\,Q = T$ R_t=thermal resistance
Fluid P=pressure difference Q=volume flow rate	Fluid capacitor $C_f\dfrac{dP}{dt}=Q$ C_f= fluid capacitance	Fluid inertor $I_f\dfrac{dQ}{dt}=P$ I_f= inertance	Fluid resistor $R_f\,Q = P$ R_f= fluid resistance

TABLE 2.3

Force-current Analogy

System Type	**Mechanical**	**Electrical**
System-response variables:		
Through-variables	Force f	Current i
Across-variables	Velocity v	Voltage v
System parameters	m	C
	k	$1/L$
	b	$1/R$

dissipative mechanism of a D-type element or damper) and the free natural oscillations will decay as a result. Similarly, electrical circuits and fluid systems can exhibit free, natural oscillatory responses due to the presence of two types of energy-storage mechanism, where energy can "flow" back and forth repeatedly between the two types of elements. But, thermal systems have only one type of energy-storage element (A-type) with only one type of energy (thermal energy). Hence, purely thermal systems cannot naturally produce oscillatory responses, unless forced by external means, or integrated with other types of systems that can produce natural oscillations (e.g., fluid systems).

2.4 Analytical Model Development

We have been able to make the following observations concerning analytical dynamic models:

- A dynamic model is a representation of a dynamic system.
- It is useful in analysis, simulation, design, modification, and control of a system.
- In view of the multidomain nature of practical engineering systems, integrated and unified development of models is desirable. Then all domains are modeled together using similar approaches.
- It is desirable to use analogous procedures to model all components in a system.
- Capability to incorporate multifunctional devices (e.g., piezoelectric elements which work as both sensors and actuators) into the modeling framework is desirable.
- Analogies exist in mechanical, electrical, fluid, and thermal systems.

A systematic procedure for the development of a lumped-parameter analytical model of a dynamic system involves formulating three types of equations:

1. Constitutive equations (physical laws for the lumped elements).
2. Continuity equations (or node equations or equilibrium equations) for the through variables.
3. Compatibility equations (or loop equations or path equations) for the across variables.

Among these, the constitutive equations have been studied in the previous section.

A continuity equation is the equation written for the through variables at a junction (i.e., node) connecting several lumped elements in the system. It dictates the fact that there cannot be any accumulation (storage) or disappearance (dissipation) or generation (source) of the through variables at a junction (i.e., what comes in must go out), because node is not an element but a junction that connects elements. Summation of forces (force balance or equilibrium), currents (Kirchhoff's current law), fluid flow rates (flow continuity equation), or heat transfer rates at a junction to zero provides a continuity equation. Note that source elements, which generate inputs to the system, should be included as well in writing these equations.

A compatibility equation is the equation written for the across variables around a closed path (i.e., loop) connecting several lumped elements in the system. It dictates the fact that at a given instant, the value of the across variable at a point in the system should be unique (i.e., cannot have two or more different values). This guarantees the requirement that a closed path is indeed a closed path; there is no breakage of the loop (i.e., compatible). Summation of velocities, voltages (Kirchhoff's voltage law), pressures, or temperatures to zero around a loop of elements provides a compatibility equation. Again, source elements, which generate inputs to the system, should be included as well in writing these equations.

2.4.1 Steps of Model Development

Development of a suitable analytical model for a large and complex system requires a systematic approach. Tools are available to aid this process. The process of modeling can be made simple by following a systematic sequence of steps. The main steps are:

1. Identify the system of interest by defining its *purpose* and the system *boundary*.
2. Identify or specify the *variables* of interest. These include inputs (forcing functions or excitations) and outputs (response).
3. Approximate (or model) various segments (components or processes or phenomena) in the system by *ideal elements*, which are suitably interconnected.
4. Draw a *free-body diagram* for the system where the individual elements are isolated/separated, as appropriate.
5. (a) Write *constitutive equations* (physical laws) for the elements.
   (b) Write *continuity* (or conservation) equations for through variables (equilibrium of forces at joints; current balance at nodes, fluid flow balance, etc.) at junctions (nodes) of the system.
   (c) Write *compatibility* equations for across (potential or path) variables around closed paths linking elements. These are loop equations for velocities (geometric connectivity), voltage (potential balance), pressure drop, etc.
   (d) Eliminate *auxiliary* variables, which are redundant and not needed to define the model.
6. Express the system *boundary conditions* and response *initial conditions* using system variables.

These steps should be self-explanatory, and should be integral with the particular modeling technique that is used. The associated procedures will be elaborated in the subsequent sections and chapters where many illustrative examples will also be provided.

2.4.2 I/O Models

More than one variable may be needed to represent the response of a dynamic system. Furthermore, there may be more than one input variable in a system. Then we have a *multivariable* system or a multiinput–multioutput (MIMO) system. A time-domain analytical model may be developed as a set of differential equations relating the response variables to the input variables. This is specifically a multivariable *I/O model*. Generally, this set of system equations is coupled, so that more than one response variable appears in each differential equation, and each equation cannot be analyzed, solved, or computer simulated separately.

2.4.3 State-Space Models

A particularly useful time-domain representation for a dynamic system is a state-space model. The state variables are a minimal set of variables which can define the dynamic state of a system. In the state-space representation, the dynamics of an *n*th order system is represented by *n* first-order differential equations, which generally are coupled. This is called a *state-space model* or simply a *state model*. An entire set of state equations can be written as a single vector-matrix state equation.

The choice of state variables is not unique: many choices are possible for a given system. Proper selection of state variables is crucial in developing an analytical model (state model) for a dynamic system. A general approach that may be adopted is to use across variables of the independent *A*-type (or, across-type) energy-storage elements and the through variables of the independent *T*-type (or, through-type) energy-storage element as the state variables. Note that if any two elements are not independent (e.g., if two spring elements are directly connected in series or parallel) then only a single state variable should be used to represent both elements. Separate state variables are not needed to represent *D*-type (dissipative) elements because their response can be represented in terms of the state variables of the energy-storage elements (*A*-type and *T*-type). State-space models and their characteristics are discussed in more detail now.

2.4.3.1 State-Space

The word "state" refers to the dynamic status or condition of a system. A complete description of the state will require all the variables that are associated with the time-evolution of the system response (i.e., both "magnitude" and "direction" of the response trajectory with respect to time). The state is a *vector*, which traces out a trajectory in the *state-space*. The associated analytical treatment requires a definition for the "state-space." In particular, a second-order system requires a two-dimensional (or plane) space, a third-order system requires a three-dimensional space, and so on.

2.4.3.2 Properties of State Models

A state vector x is a column vector, which contains a minimum set of state variables $(x_1, x_2, ..., x_n)$ which completely determine the state of a dynamic system. The number of states variables (n), is the *order* of the system.

Property 1
The state vector $x(t_0)$ at time t_0 and the input (forcing excitation) $u[t_0, t_1]$ over the time interval $[t_0, t_1]$, will uniquely determine the state vector $x(t_1)$ any future time t_1. In other words, a transformation g can be defined such that

$$x(t_1) = g(t_0, t_1, x(t_0), u[t_0, t_1]) \qquad (2.54)$$

Note that by the *causality* property of a dynamic system, future states can be determined if all the inputs from the initial time up to that future time are known. This means that the transformation g is *nonanticipative* (i.e., inputs beyond t_1 are not needed to determine $x(t_1)$). Each forcing function $u[t_0, t_1]$ determines corresponding "trajectory" of the state vector— the *state trajectory*. The n-dimensional vector space formed by all possible state trajectories is the *state-space*.

Property 2
The state $x(t_1)$ and the input $u(t_1)$ at any time t_1 will uniquely determine the system output (or response) vector $y(t_1)$ at that time. This can be expressed as:

$$y(t_1) = h(t_1, x(t_1), u(t_1)) \qquad (2.55)$$

This states that the system response (output) at time t_1 depends on the time, the input, and the state vector.

The transformation h has no *memory*—the response at a previous time cannot be determined through the knowledge of the present state and input. Note also that, in general, the system outputs (y) are not identical to the system states (x) even though the former can be uniquely determined by the later.

A state model consists of a set of n first-order ordinary differential equations (time-domain), which are coupled (i.e., interrelated). In the vector form this is expressed as:

$$\dot{x} = f(x, u, t) \tag{2.56}$$

$$y = h(x, u, t) \tag{2.57}$$

Equation 2.56 represents the n *state equations* (first-order ordinary differential equations) and Equation 2.57 represents the *algebraic output equations*. If f is a nonlinear vector function, then the state model is nonlinear, which is the general case.

Summarizing:

- A state model is a set of n first-order differential equations (coupled) using n state variables (an nth order system).
- State equations define the dynamic state of a system.
- Required minimum set of state variables $x_1, x_2, \ldots x_n \rightarrow$ *state vector*.
- The state vector traces out a *trajectory* in the *state-space*.

2.4.3.3 Linear State Equations

Nonlinear state models are difficult to analyze and simulate. Often linearization is necessary, through various forms of approximations and assumptions. An nth order linear, state model is given by the state equations (differential):

$$\dot{x}_1 = a_{11}x_1 + a_{12}x_2 + a_{13}x_3 + \cdots + a_{1n}x_n + b_{11}u_1 + b_{12}u_2 + \cdots + b_{1r}u_r$$

$$\dot{x}_2 = a_{21}x_1 + a_{22}x_2 + a_{23}x_3 + \cdots + a_{2n}x_n + b_{21}u_1 + b_{22}u_2 + \cdots + b_{2r}u_r$$

$$\vdots$$

$$\vdots$$

$$\dot{x}_n = a_{n1}x_1 + a_{n2}x_2 + a_{n3}x_3 + \cdots + a_{nn}x_n + b_{n1}u_1 + b_{n2}u_2 + \cdots + b_{nr}u_r \tag{2.58a}$$

and the output equations (algebraic):

$$y_1 = c_{11}x_1 + c_{12}x_2 + c_{13}x_3 + \cdots + c_{1n}x_n + d_{11}u_1 + d_{12}u_2 + \cdots + d_{1r}u_r$$

$$y_2 = c_{21}x_1 + c_{22}x_2 + c_{23}x_3 + \cdots + c_{2n}x_n + d_{21}u_1 + d_{22}u_2 + \cdots + d_{2r}u_r$$

$$\vdots$$

$$y_m = c_{m1}x_1 + c_{m2}x_2 + c_{m3}x_3 + \cdots + c_{mn}x_n + d_{m1}u_1 + d_{m2}u_2 + \cdots + d_{mr}u_r \tag{2.59a}$$

where $\dot{x} = dx/dt$; $x_1, x_2,..., x_n$ are the n *state variables*; $u_1, u_2,..., u_r$ are the r *input variables*; and $y_1, y_2,..., y_m$ are the m *input variables*. Equation 2.58 simply says that a change in any of the n variables and the r inputs of the system will affect the rate of change of any given state variable. In general, in addition to the states variables, the output variables are needed as well to represent the output variables, as indicated in Equation 2.59. More often, however, the input variables are not present in this set of output equations (i.e., the coefficients d_{ij} are all zero).

This state model may be rewritten in the vector-matrix form as

$$\dot{x} = Ax + Bu \tag{2.58b}$$

$$y = Cx + Du \tag{2.59b}$$

The bold-type upper-case letter indicates that the variable is a *matrix*; a bold-type lower-case letter indicates a *vector*, typically a column vector. Specifically,

$$x = \begin{bmatrix} x_1 \\ x_2 \\ \vdots \\ x_n \end{bmatrix}; \dot{x} = \begin{bmatrix} \dot{x}_1 \\ \dot{x}_2 \\ \vdots \\ \dot{x}_n \end{bmatrix}; A = \begin{bmatrix} a_{11} & a_{12} & \cdots & a_{1n} \\ a_{21} & a_{22} & \cdots & a_{2n} \\ \vdots & \vdots & & \vdots \\ a_{n1} & a_{n2} & \cdots & a_{nn} \end{bmatrix}; B = \begin{bmatrix} b_{11} & b_{12} & \cdots & b_{1r} \\ b_{21} & b_{22} & \cdots & b_{2r} \\ \vdots & \vdots & & \vdots \\ b_{n1} & b_{n2} & \cdots & b_{nr} \end{bmatrix};$$

$$C = \begin{bmatrix} c_{11} & c_{12} & \cdots & c_{1n} \\ c_{21} & c_{22} & \cdots & c_{2n} \\ \vdots & \vdots & & \vdots \\ c_{m1} & c_{m2} & \cdots & c_{mn} \end{bmatrix}; D = \begin{bmatrix} d_{11} & d_{12} & \cdots & d_{1r} \\ d_{21} & d_{22} & \cdots & d_{2r} \\ \vdots & \vdots & & \vdots \\ d_{m1} & d_{m2} & \cdots & d_{mr} \end{bmatrix}$$

where $x = [x_1 \ x_2 \ \cdots \ x_n]^T$ = state vector (nth order); $u = [u_1 \ u_2 \ \cdots \ u_r]^T$ = input vector (rth order); $y = [y_1 \ y_2 \ \cdots \ y_m]^T$ = output vector (mth order); A = system matrix ($n \times n$); B = input distribution matrix ($n \times r$); C = output (or measurement) gain matrix ($m \times n$); D = feedforward input gain matrix ($m \times r$).

Note that $[\ ]^T$ denotes the transpose of a matrix or vector. The system matrix A tells us how the system responds naturally without any external input, and B tells us the input u affects (i.e., how it is amplified and distributed when reaching) the system.

Example 2.1

The concepts of state, output, and order of a system, and the importance of the system's initial state, can be shown using a simple example. Consider the rectilinear motion of a particle of mass m subject to an input force $u(t)$. By Newton's second law, its position x can be expressed as the second-order differential equation:

$$m\frac{d^2x}{dt^2} = u(t) \quad \text{or} \quad m\ddot{x} - u = 0 \tag{i}$$

We consider the following three cases and develop state models and I/O models for each case:

Case 1: Position x is the output.
Case 2: Velocity $\dot{x} = v$ is the output.
Case 3: Both position x and velocity $\dot{x} = v$ are outputs.

I/O Models

Case 1: Here output $y=x$. From Equation (i) $m(d^2y/dt^2) = u(t)$ is indeed the I/O model (a second-order model).
Case 2: Here output $y=\dot{x}=v$. From Equation (i) $m(dy/dt)= u(t)$ is the I/O model (a first order model).
Case 3: Here the two outputs are $y_1=x$, $y_2 = \dot{x} = v$. From Equation (i) the I/O model is

$$m\frac{d^2y_1}{dt^2} = u(t)$$

$$m\frac{dy_2}{dt} = u(t)$$

This is also a second-order model.

State-Space Models

Case 1: Define two state variables $x_1 = x$, $x_2 = \frac{dx}{dt}$
   Then we have

State equations:
$$\dot{x}_1 = x_2$$
$$\dot{x}_2 = \frac{1}{m}u(t)$$

Output equation: $y=x_1$

Case 1b: Another state model
   Define the two state variables according to: $x_1 = -6x$ and $x_2 = -1/2(\dot{x}_1)$
   Then, one state equation is given by one of the definitions itself, and the other state equation is obtained by substituting the two definitions into (i). We have

State equations:
$$\dot{x}_1 = -2x_2$$
$$\dot{x}_2 = \frac{3}{m}u(t)$$

Output equation: $y = -\frac{1}{6}x_1$

Case 2: Define one state variable $x_1 = \frac{dx}{dt}$

Corresponding state equation: $\dot{x}_1 = \frac{1}{m}u(t)$

Output equation: $y=x_1$

Note: Position cannot be determined from this model

Case 3: Here we can use the same state equations as in Case 1 or Case 1b. Let us use Case 1 as an illustration.

State equations:
$$\dot{x}_1 = x_2$$
$$\dot{x}_2 = \frac{1}{m}u(t)$$

$$\text{Output equations:} \quad \begin{bmatrix} y_1 \\ y_2 \end{bmatrix} = \begin{bmatrix} 1 & 0 \\ 0 & 1 \end{bmatrix} \begin{bmatrix} x_1 \\ x_2 \end{bmatrix}$$

Case 3b: Alternatively, if the state variables used in Case 1b are used we have the following state model.

$$\text{State equations:} \quad \begin{aligned} \dot{x}_1 &= -2x_2 \\ \dot{x}_2 &= \frac{3}{m} u(t) \end{aligned}$$

$$\text{Output equations:} \quad \begin{bmatrix} y_1 \\ y_2 \end{bmatrix} = \begin{bmatrix} -1/6 & 0 \\ 0 & 1/3 \end{bmatrix} \begin{bmatrix} x_1 \\ x_2 \end{bmatrix}$$

In this example, it should be noted that the three variables x, $\dot{x}$, and $\ddot{x}$ do not form a state vector because this is not a minimal set. Specifically, $\ddot{x}$ is redundant as it is completely known from u.

Another important aspect can be observed when deriving the system response by directly integrating the system equation (i). When the output is velocity, just one initial condition $\dot{x}(0)$ is adequate, whereas if the output is position, two initial conditions $x(0)$ and $\dot{x}(0)$ are needed, to determine the complete response. In the latter case, just one initial state does not uniquely generate a state trajectory corresponding to a given forcing input. This intuitively clear fact, nevertheless, constitutes an important property of the state of a system: the number of initial conditions needed=order of the system.

Finally, it is also important to understand the nonuniqueness of the choice of state variables, as clear from Case 1 and Case 1b (or Case 3 and Case 3b).

Summarizing:

- State vector is a least (minimal) set of variables that completely determines the dynamic state of system ➔ a state variable cannot be expressed as a linear combination of the remaining state variables
- State vector is not unique; many choices are possible for a given system.
- Output (response) variables can be completely determined from any such choice of state variables.
- State variables may or may not have a physical interpretation.

2.4.4 Time-Invariant Systems

If in Equations 2.56 and 2.57, there is no explicit dependence on time in the functions f and h, the dynamic system is said to be *time-invariant*, or *stationary*, or *autonomous*. In this case, the system behavior is not a function of the time origin for a given initial state and input function. In particular, a linear system is time-invariant if the matrices A, B, C, and D (in Equations 2.58 and 2.59) are constant.

Example 2.2

A torsional dynamic model of a pipeline segment is shown in Figure 2.14a. The free-body diagram in Figure 2.14b shows the internal torques acting at sectioned inertia junctions, for free motion. We will a state model for this system by using the generalized velocities (angular velocities Ω_i) of the independent inertia elements and the generalized forces (torques T_i) of the independent elastic (torsional spring) elements as the state variables. A minimum set of states, which is required for a complete representation, determines the system order.

(a)   (b)

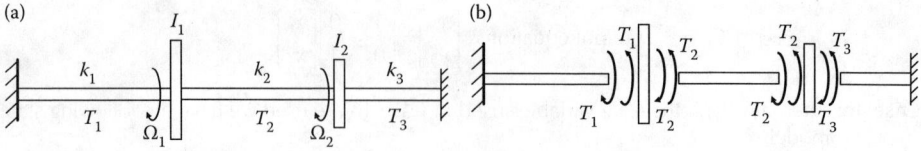

FIGURE 2.14
(a) Dynamic model of a pipeline segment. (b) Free body diagram.

In this system there are two inertia elements and three spring elements—a total of five energy-storage elements. However, the three springs are not independent. The motion of any two springs completely determines the motion of the third. This indicates that the system is a fourth-order system. The state vector is chosen as

$$x = [\Omega_1 \quad \Omega_2 \quad T_1 \quad T_2]^T$$

To develop the state-space model, we first formulate the state-space shell by writing the *constitutive equations* as follows:
Newton's second law gives

$$I_1\dot{\Omega}_1 = -T_1 + T_2$$

$$I_2\dot{\Omega}_2 = -T_2 - T_3 \tag{i}$$

Hooke's law gives

$$\dot{T}_1 = k_1\Omega$$

$$\dot{T}_2 = k_2(\Omega_2 - \Omega_1) \tag{ii}$$

The first, third, and fourth equations above are already in the final state-equation form. Only the second equation has to be further manipulated to eliminate T_3, which is not a state variable. To accomplish this we write the *compatibility equation*—the displacement compatibility relation at inertia I_2:

$$\frac{T_1}{k_1} + \frac{T_2}{k_2} = \frac{T_3}{k_3} \tag{iii}$$

Note: In this example, *continuity equations* (torque balance) are already satisfied, as shown in Figure 2.14b. Hence they need not be explicitly written.
Next, torque T_3 in (i) is substituted in terms of T_1 and T_2, using (iii). This gives the remaining state equation (modified second equation of (i)):

$$I_2\dot{\Omega}_2 = -\left(\frac{k_3}{k_1}\right)T_1 - \left(1 + \frac{k_3}{k_2}\right)T_2$$

The system matrix of the resulting state model is

$$A = \begin{bmatrix} 0 & 0 & -\dfrac{1}{I_1} & \dfrac{1}{I_1} \\ 0 & 0 & -\dfrac{1}{I_2}\left(\dfrac{k_3}{k_1}\right) & -\dfrac{1}{I_2}\left(1 + \dfrac{k_3}{k_2}\right) \\ k_1 & 0 & 0 & 0 \\ -k_2 & k_2 & 0 & 0 \end{bmatrix} \tag{iv}$$

The output (displacement) vector is

$$y = \left[\frac{T_1}{k_1}, \frac{T_1}{k_1} + \frac{T_2}{k_2} \right]^T$$ (v)

which corresponds to the following output-gain matrix:

$$C = \begin{bmatrix} 0 & 0 & \dfrac{1}{k_1} & 0 \\[2mm] 0 & 0 & \dfrac{1}{k_1} & \dfrac{1}{k_2} \end{bmatrix}$$ (vi)

Note: There is no **B** matrix in this example because it concerns free (unforced) motion and there is no input.

2.4.5 Systematic Steps for State Model Development

At this stage it useful to summarize our systematic approach for formulating a state model. *Note:* Inputs (**u**) and outputs (**y**) are given

Step 1: State variable (*x*) selection.

Across variables for independent *A*-type energy-storage elements.

Through variables for independent *T*-type energy-storage elements.

Step 2: Write constitutive equations for all the elements (both energy-storage and dissipative elements).

Step 3: Write compatibility equations and continuity equations.

Step 4: Eliminate redundant variables.

Step 5: Express the outputs in terms of state variables.

Example 2.3

The rigid output shaft of a diesel engine prime mover is running at known angular velocity $\Omega(t)$. It is connected through a friction clutch to a flexible shaft, which in turn drives a hydraulic pump (see Figure 2.15a). A linear model for this system is shown schematically in Figure 2.15b. The clutch is represented by a viscous rotatory damper of damping constant B_1 (units: torque/angular velocity). The stiffness of the flexible shaft is K (units: torque/rotation). The pump is represented by a wheel of moment of inertia J (units: torque/angular acceleration) and viscous damping constant B_2.

a. Write down the two state equations relating the state variables T and ω to the input Ω. Where T=torque in flexible shaft; ω=pump speed.

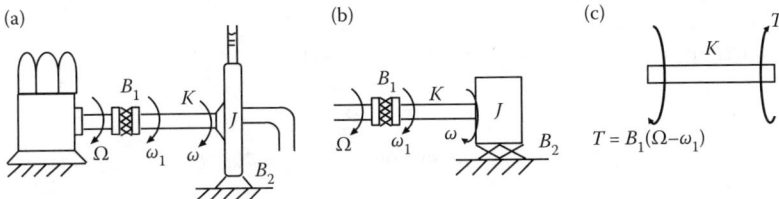

FIGURE 2.15
(a) Diesel engine. (b) Linear model. (c) Free body diagram of the shaft. (d) Free-body diagram of the wheel.

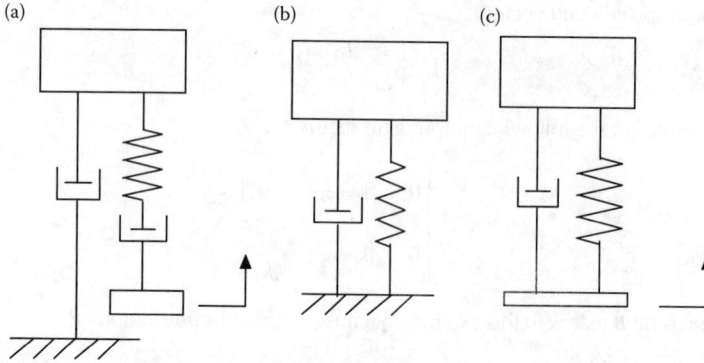

FIGURE 2.16
Three translatory mechanical systems.

Hints:

1. Free body diagram for the shaft is given in Figure 2.15c, where ω_1 is the angular speed at the left end of the shaft.
2. Write down the torque "balance" and "constitutive" relations for the shaft, and eliminate ω_1.
3. Draw the free body diagram for the wheel J and use D'Alembert's principle.
4. Comment on why the compatibility equations and continuity equations are not explicitly used in the development of the state equations.
5. Express the state equations in the vector-matrix form.
6. To complete the state-space model, determine the output equation for: (i) Output=ω; (ii) Output=T; (iii) Output=ω_1.
7. Which one of the translatory systems given in Figure 2.16 is the system in Figure 2.15b analogous to?

Solution

a.

$$\text{Constitutive relation for } K: \frac{dT}{dt} = K(\omega_1 - \omega) \qquad (i)$$

$$\text{Constitutive relation for } B_1: T = B_1\,(\Omega - \omega_1) \qquad (ii)$$

Substitute (ii) into (i):

$$\frac{dT}{dt} = -\frac{K}{B_1}T - K\omega + K\Omega \qquad (iii)$$

This is one state equation.
Constitutive equation for J (D'Alembert's principle, See Figure 2.15d):

$$J\dot{\omega} = T - T_2 \qquad (iv)$$

$$\text{Constitutive relation for } B_2: T_2 = B_2\omega \qquad (v)$$

$$\text{Substitute (v) in (iv): } \frac{d\omega}{dt} = -\frac{B_2}{J}\omega + \frac{1}{J}T \qquad (vi)$$

This is the second state equation.

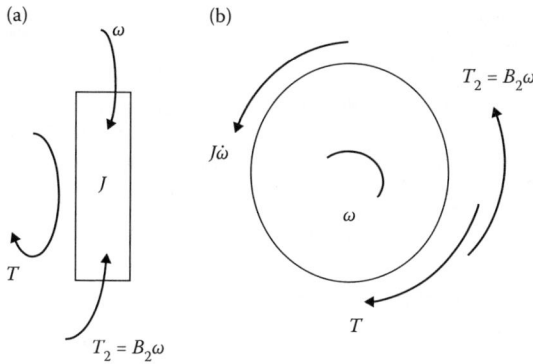

FIGURE 2.17
(a) Model details for writing. (a) Loop equations. (b) Node equations.

Note: For this example it is not necessary to write the continuity and compatibility equations because they are implicitly satisfied by the particular choice of variables as given in Figure 2.15.
Important comments

(1) Generally, some of the continuity equations (node equations) and compatibility equations (loop equations) are automatically satisfied by the choice of the variables. Then, we don't have to write the corresponding equations.
(2) In mechanical systems, typically, compatibility equations are automatically satisfied. This is the case in the present example. In particular, from Figure 2.17a we have:
Loop 1 equation: $\omega + (-\omega) = 0$
Loop 2 equation: $\omega + (\omega - \omega) + (\Omega - \omega_1) + (-\Omega) = 0$
(3) Node equations may be written in further detail by introducing other auxiliary variables into the free-body diagram; and furthermore, the constitutive equations for the damping elements may be written separately. Specifically, from Figure 2.17b, we can write:

Node 1: $T_{d1} - T = 0$
Node 2: $T - T_{J1} = 0$
Node 3: $T_{J2} - T_{d2} = 0$
Damper B_1: $T_{d1} = B_1 (\Omega - \omega_1)$
Damper B_2: $I_{d2} = B_2 \omega$

b. Vector-matrix form of the state equations (iii) and (iv) is:

$$\begin{bmatrix} \dfrac{dT}{dt} \\ \dfrac{d\omega}{dt} \end{bmatrix} = \begin{bmatrix} \dfrac{-K}{B_1} & -K \\ \dfrac{1}{J} & -\dfrac{B_2}{J} \end{bmatrix} \begin{bmatrix} T \\ \omega \end{bmatrix} + \begin{bmatrix} K \\ 0 \end{bmatrix} \Omega$$

with the state vector $x = [T \ \omega]^T$ and the input $u = [\Omega]$.

c.

(i) $C = [0 \ 1]; D = [0]$
(ii) $C = [1 \ 0]; D = [0]$
(iii) Here we use the continuity equation to express the output as

$$\omega_1 = \Omega - \frac{T}{B_1}$$

Then, the corresponding matrices are

$$C = [-1/B_1 \quad 0]; \quad D = [1]$$

In this case, we notice a direct "feedforward" of the input Ω into the output ω_1 through the clutch B_1. Furthermore, as will be clear from the material in Chapter 5, now the system transfer function will have its numerator order equal to the denominator order ($=2$). This is a characteristic of systems with direct feedforward of inputs into the outputs.

d. The translatory system in Figure 2.16a is analogous to the given rotatory system.

2.4.6 I/O Models from State-Space Models

Suppose that Equation 2.57 is substituted into Equation 2.56 to eliminate x and $\dot{x}$, and a set of differential equations for y are obtained (with u and its derivatives present). The result is the I/O model, in the time-domain. If these I/O differential equations are nonlinear, then the system (or strictly, the I/O model) is nonlinear.

From the linear state model given by Equations 2.58 and 2.59, the I/O model is obtained as follows: First differentiate Equation 2.59 and eliminate $\dot{x}$ by substituting Equation 2.58. Then use the resulting equation and Equation 2.59 to eliminate x.

Example 2.4

Consider two water tanks joined by a horizontal pipe with an on–off valve. With the valve closed, the water levels in the two tanks were initially maintained unequal. When the valve was suddenly opened, some oscillations were observed in the water levels of the tanks. Suppose that the system is modeled as two gravity-type capacitors linked by a fluid resistor. Would this model exhibit oscillations in the water levels when subjected to an initial-condition excitation? Clearly explain your answer.

A centrifugal pump is used to pump water from a well into an overhead tank. This fluid system is schematically shown in Figure 2.18a. The pump is considered as a pressure source $P_s(t)$ and the water level h in the overhead tank is the system output. The ambient pressure is denoted by P_a. The following system parameters are given:

L_v, d_v=length and internal diameter of the vertical segment of pipe.
L_h, d_h=length and internal diameter of the horizontal segment of pipe.
A_t=area of cross section of overhead tank (uniform).
ρ=mass density of water.
μ=dynamic viscosity of water.
g=acceleration due to gravity.

Suppose that this fluid system is approximated by the lumped-parameter model shown in Figure 2.18b.

a. Give expressions for the equivalent linear fluid resistance of the overall pipe (i.e., combined vertical and horizontal segments) R_{eq}, the equivalent fluid inertance within the overall pipe I_{eq}, and the gravitational fluid capacitance of the overhead tank C_{grv}, in terms of the system parameters defined above.
b. Treating $x = [P_{3a} \quad Q]^T$ as the state vector, where P_{3a}=pressure head of the overhead tank; Q=volume flow rate through the pipe develop a complete state-space model for the system. Specifically, obtain the matrices A, B, C, and D.
c. Obtain the I/O differential equation of the system.

Solution

Since the inertia effects are neglected in the model, and only two capacitors are used as the energy-storage elements, there exists only one type of energy in this system. Hence this model

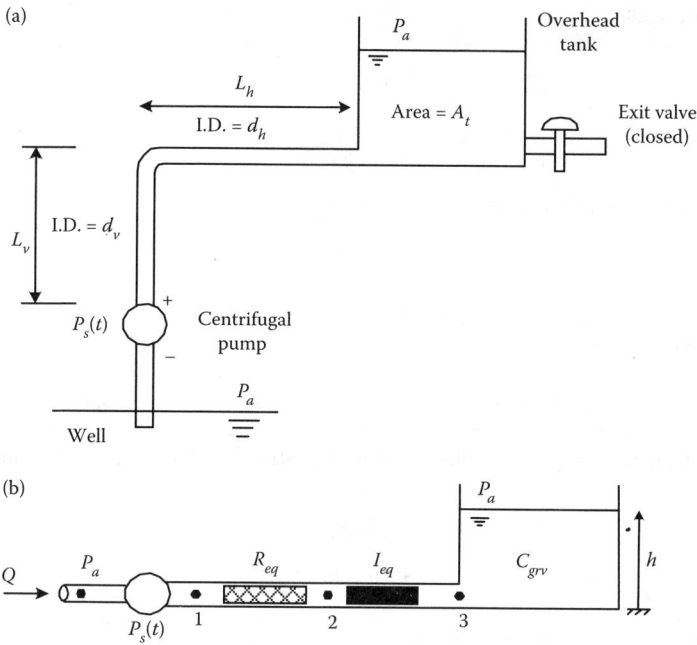

FIGURE 2.18
(a) A system for pumping water from a well into an overhead tank (Note: I.D. means "internal diameter").
(b) A lumped-parameter model of the fluid system.

cannot provide an oscillatory response to an initial condition excitation (i.e., natural oscillations are not possible). But, the actual physical system has fluid inertia, and hence the system can exhibit an oscillatory response.

a. Assuming a parabolic velocity profile, the fluid inertance in a pipe of uniform cross-section A and length L, is given by

$$I = \frac{2\rho L}{A}$$

Since the same volume flow rate Q is present in both segments of piping (continuity) we have, for series connection,

$$I_{eq} = \frac{2\rho L_v}{(\pi/4)d_v^2} + \frac{2\rho L_h}{(\pi/4)d_h^2} = \frac{8\rho}{\pi}\left[\frac{L_v}{d_v^2} + \frac{L_h}{d_h^2}\right]$$

The linear fluid resistance in a circular pipe is

$$R = \frac{128\mu L}{\pi d^4} \quad \text{where } d \text{ is the internal diameter.}$$

Again, since the same Q exists in both segments of the series-connected pipe,

$$R_{eq} = \frac{128\mu}{\pi}\left[\frac{L_v}{d_v^4} + \frac{L_h}{d_h^4}\right]$$

Also

$$C_{grv} = \frac{A_t}{\rho g}$$

b. State-space shell:

$$C_{grv} \frac{dP_{3a}}{dt} = Q$$

$$l_{eq} \frac{dQ}{dt} = P_{23}$$

Remaining constitutive equation:

$$P_{12} = R_{eq} Q$$

Note: Constitutive (node) equations are automatically satisfied.
Compatibility (loop) equations:

$$P_{1a} = P_{12} + P_{23} + P_{3a} \quad \text{with} \quad P_{1a} = P_s(t) \quad \text{and} \quad P_{3a} = \rho g h$$

Now eliminate the auxiliary variable P_{23} in the state-space shell, using the remaining equations.
We get

$$P_{23} = P_{1a} - P_{12} - P_{3a}$$

$$= P_s(t) - R_{eq} Q - P_{3a}$$

Hence, the state-space model is given by
State equations:

$$\frac{dP_{3a}}{dt} = \frac{1}{C_{grv}} Q \tag{i}$$

$$\frac{dQ}{dt} = \frac{1}{l_{eq}} \left[P_s(t) - P_{3a} - R_{eq} Q \right] \tag{ii}$$

Output equation:

$$h = \frac{1}{\rho g} P_{3a} \tag{iii}$$

Corresponding matrices are:

$$A = \begin{bmatrix} 0 & 1/C_{grv} \\ -1/l_{eq} & -R_{eq}/l_{eq} \end{bmatrix}; \quad B = \begin{bmatrix} 0 \\ 1/l_{eq} \end{bmatrix}; \quad C = \begin{bmatrix} \dfrac{1}{\rho g} & 0 \end{bmatrix}; \quad D = 0$$

c. Substitute Equation (i) into (ii):

$$l_{eq} C_{grv} \frac{d^2 P_{3a}}{dt^2} = P_s(t) - P_{3a} - R_{eq} C_{grv} \frac{dP_{3a}}{dt}$$

Now substitute Equation (iii) for P_{3a}:

$$l_{eq} C_{grv} \frac{d^2 h}{dt^2} + R_{eq} C_{grv} \frac{dh}{dt} + h = \frac{1}{\rho g} P_s(t)$$

This is the I/O model.

Problems

PROBLEM 2.1

The use of solar energy is a sustainable way to generate electric power for houses. A schematic arrangement is shown in Figure P2.1a. Radiation from the sun is received at a solar panel, which consists of photovoltaic cells to convert solar energy to electric energy in the form of direct current (dc). Using an inverter, the dc power is converted into alternating current (ac) power of appropriate frequency (60 or 50 Hz) for household use. This supply is connected through a two-way meter to the supply line of the house and to the main electricity grid (Figure 2.1b). In this manner, any excess power from the solar panels can be sold to the grid and when the supply from the solar panel is not adequate (e.g., cloudy days, nights) electricity can be purchased from the grid.

The ac power is used for various household purposes such as operation of appliances, heating, and cooling.

a. Explain why this is a multidomain (i.e., mixed) system.
b. Identify several key components of the system (*Note*: some are shown in Figure P2.1a). Discuss various processes within the components that may be categorized into the mechanical, electrical, fluid, and thermal domains. Indicate applicable modeling issues for the overall system.
c. Sketch the energy flow of the system, indicating relevant stages of energy conversion.

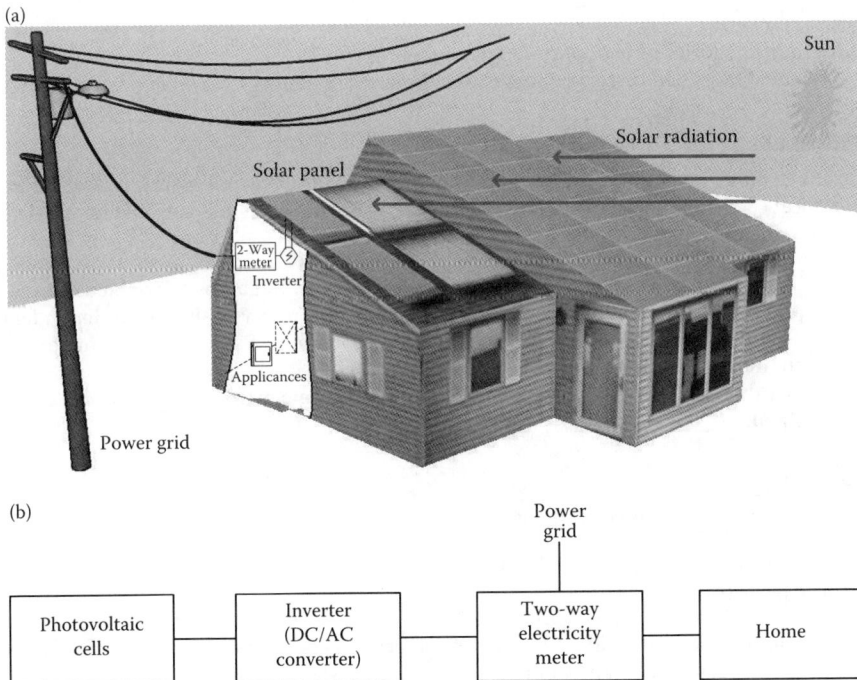

FIGURE P2.1
(a) A solar-powered house. (b) Schematic diagram of the ac power supply.

PROBLEM 2.2

What is a "dynamic" system, a special case of any system?

A typical input variable is identified or each of the following examples of dynamic systems. Give at least one output variable for each system.

a. Human body: Neuroelectric pulses
b. Company: Information
c. Power plant: Fuel rate
d. Automobile: Steering wheel movement
e. Robot: Voltage to joint motor
f. Highway bridge: Vehicle force

PROBLEM 2.3

Real systems are nonlinear. Under what conditions a linear model is sufficient in studying a real system?

Consider the following system equations:

a. $\ddot{y} + (2\sin\omega t + 3)\dot{y} + 5y = u(t)$
b. $3\ddot{y} - 2y = u(t)$
c. $3\ddot{y} + 2\dot{y}^3 + y = u(t)$
d. $5\ddot{y} + 2\dot{y} + 3y = 5u(t)$

(i)   Which ones of these are linear?
(ii) Which ones are nonlinear?
(iii) Which ones are time-variant?

PROBLEM 2.4

Give four categories of uses of dynamic modeling.

List advantages and disadvantages of experimental modeling over analytical modeling.

PROBLEM 2.5

Briefly explain/justify why voltage and not current is the natural state variable for an electrical capacitor; and current and not voltage is the natural state variable for an electrical inductor.

a. List several advantages of using as state variables, the across variables of independent *A*-type energy-storage elements and through variables of independent *T*-type energy-storage elements, in the development of a state-space model for an engineering system.
b. List three things to which the order of an electromechanical dynamic system is equal.

PROBLEM 2.6

What are the basic lumped elements of

i. A mechanical system
ii. An electrical system?

Indicate whether a distributed-parameter method is needed or a lumped-parameter model is adequate in the study of following dynamic systems:

a. Vehicle suspension system (motion)
b. Elevated vehicle guideway (transverse motion)

c. Oscillator circuit (electrical signals)
d. Environment (weather) system (temperature)
e. Aircraft (motion and stresses)
f. Large transmission cable (capacitance and inductance).

PROBLEM 2.7

Write down the order of the systems shown in Figure P2.7.

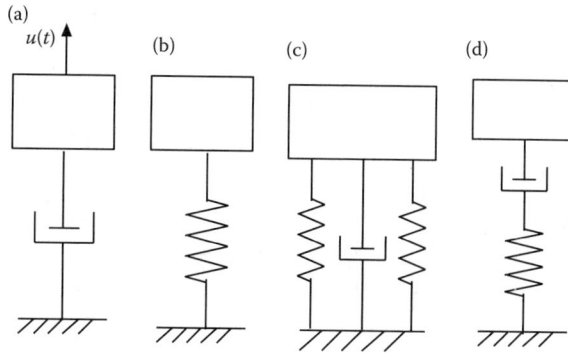

FIGURE P2.7
Models of four mechanical systems.

PROBLEM 2.8

a. Give logical steps of the analytical modeling process for a general physical system.
b. Once a dynamic model is derived, what other information would be needed for analyzing its time response (or for computer simulation)?
c. A system is divided into two subsystems, and models are developed for these subsystems. What other information would be needed to obtain a model for the overall system?

PROBLEM 2.9

Various possibilities of model development for a physical system are shown in Figure P2.9. Give advantages and disadvantages of the SM approach of developing an approximate model in comparison to a combined DM+MR approach.

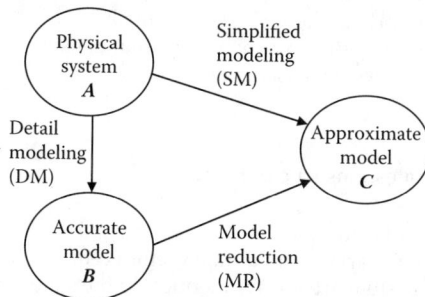

FIGURE P2.9
Approaches of model development.

PROBLEM 2.10

Describe two approaches of determining the parameters of a lumped-parameter model that is (approximately) equivalent to a distributed-parameter (i.e., continuous) dynamic system.

One end of a heavy spring of mass m_s and stiffness k_s is attached to a lumped mass m. The other end is attached to a support that is free to move, as shown in Figure P2.10.

Using the method of natural frequency equivalence, determine an equivalent lumped-parameter model for the spring where the equivalent lumped mass is located at the free end (support end) of the system. The natural frequencies of a heavy spring with one end fixed and the other end free is given by

$$\omega_n = \frac{\pi}{2}(2n-1)\sqrt{\frac{k_s}{m_s}}$$

where n is the mode number.

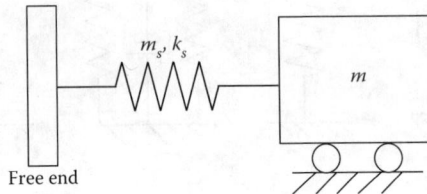

FIGURE P2.10
A mechanical system with a heavy spring and attached mass.

PROBLEM 2.11

a. Why are analogies important in modeling of dynamic systems?
b. In the force-current analogy, what mechanical element corresponds to an electrical capacitor?
c. In the velocity-pressure analogy, is the fluid inertia element analogous to the mechanical inertia element?

PROBLEM 2.12

a. What are through variables in mechanical, electrical, fluid, and thermal system?
b. What are across variables in mechanical, electrical, fluid, and thermal systems?
c. Can the velocity of a mass changes instantaneously?
d. Can the voltage across a capacitor change instantaneously?
e. Can the force in a spring change instantaneously?
f. Can the current in an inductor change instantaneously?
g. Can purely thermal systems oscillate?

PROBLEM 2.13

Answer the following questions true or false:

a. A state-space model is unique.
b. The number of state variables in a state vector is equal to the order of the system.
c. The outputs of a system are always identical to the state variables.
d. Outputs can be expressed in terms of state variables.
e. State model is a time-domain model.

PROBLEM 2.14

Consider a system given by the state equations

$$\dot{x}_1 = x_1 + 2x_2$$
$$\dot{x}_2 = -x_1 + 2u$$

in which x_1 and x_2 are the state variables and u is the input variable. Suppose that the output y is given by

$$y = 2x_1 - x_2.$$

a. Write this state-space model in the vector-matrix form:

$$\dot{x} = Ax + Bu$$

$$y = Cx$$

and identify the elements of the matrices A, B, and C.
b. What is the order of the system?

PROBLEM 2.15

Consider the mass-spring system shown in Figure P2.15. The mass m is supported by a spring of stiffness k and is excited by a dynamic force $f(t)$.

a. Taking $f(t)$ as the input, and position and speed of the mass as the two outputs, obtain a state-space model for the system.
b. What is the order of the system?
c. Repeat the problem, this time taking the compression force in the spring as the only output.
d. How many initial conditions are needed to determine the complete response of the system?

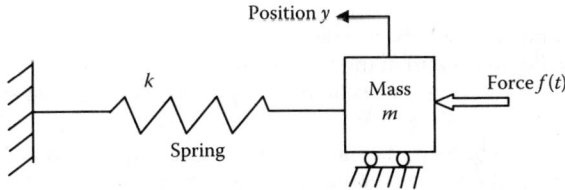

FIGURE P2.15
A mechanical system.

PROBLEM 2.16

a. Briefly explain why a purely thermal system typically does not have an oscillatory response whereas a fluid system can.
b. Figure P2.16 shows a pressure-regulated system that can provide a high-speed jet of liquid. The system consists of a pump, a spring-loaded accumulator, and a fairly long section of piping which ends with a nozzle. The pump is considered as a flow source of value Q_s. The following parameters are important:

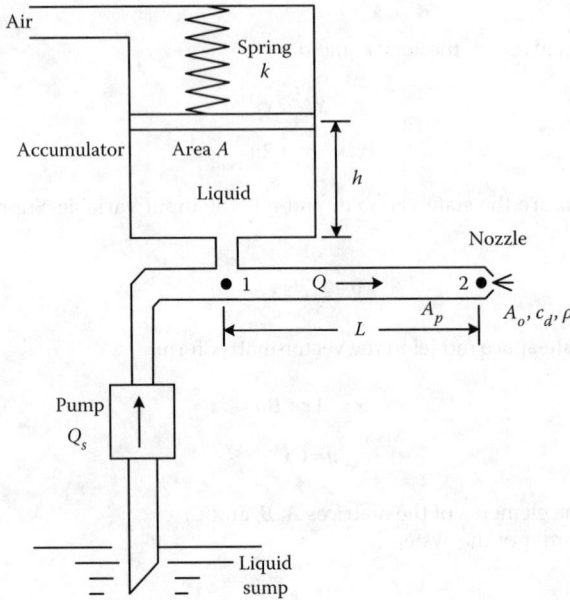

FIGURE P2.16
Pressure regulated liquid jet system.

A =area of cross section (uniform) of the accumulator cylinder
k =spring stiffness of the accumulator piston
L =length of the section of piping from the accumulator to the nozzle
A_p =area of cross section (uniform, circular) of the piping
A_o =discharge area of the nozzle
C_d =discharge coefficient of the nozzle
Q =mass density of the liquid.
Assume that the liquid is incompressible. The following variables are important:
$P_{1r} = P_1 - P_r$=pressure at the inlet of the accumulator with respect to the ambient reference r
Q =volume flow rate through the nozzle
h =height of the liquid column in the accumulator.
Note that the piston (wall) of the accumulator can move against the spring, thereby varying h.

   i. Considering the effects of the movement of the spring loaded wall and also the gravity head of the liquid, obtain an expression for the equivalent fluid capacitance C_a of the accumulator in terms of k, A, ρ, and g. Are the two capacitances which contribute to C_a (i.e., wall stretching and gravity) connected in parallel or in series?

Note: Neglect the effect of bulk modulus of the liquid.

   ii. Considering the capacitance C_a, the inertance I of the fluid volume in the piping (length L and cross section area A_p), and the resistance of the nozzle only, develop a nonlinear state-space model for the system. The state vector $x=[P_{1r} \quad Q]^T$, and the input $u=[Q_s]$.

For flow in the (circular) pipe with a parabolic velocity profile, the inertance $I = \dfrac{2\rho L}{A_p}$

and for the discharge through the nozzle $Q = A_o c_d \sqrt{\dfrac{2P_{2r}}{\rho}}$ in which

P_{2r}=pressure inside the nozzle with respect to the outside reference (r).
c_d=discharge coefficient.

PROBLEM 2.17

A model for the automatic gage control (AGC) system of a steel rolling mill is shown in Figure P2.17. The rolls are pressed using a single acting hydraulic actuator with a valve displacement of u. The rolls are displaced through y, thereby pressing the steel that is being rolled. The rolling force F is completely known from the steel parameters for a given y.

  i. Identify the inputs and the controlled variable in this control system.
 ii. In terms of the variables and system parameters indicated in Figure P2.17, write dynamic equations for the system, including valve nonlinearities.
iii. What is the order of the system? Identify the response variables.
 iv. What variables would you measure (and feed back through suitable controllers) in order to improve the performance of the control system?

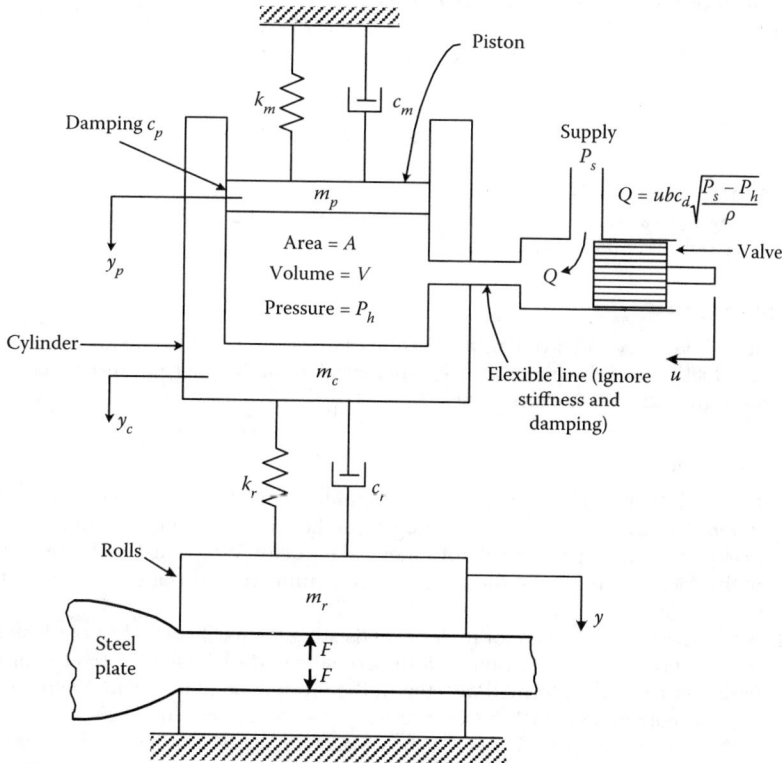

FIGURE P2.17
Automatic gage control (AGC) system of a steel rolling mill.

PROBLEM 2.18

A simplified model of a hotwater heating systemis shown in Figure P2.18.

Q_s=rate of heat supplied by the furnace to the water heater (1000 kW)
T_a =ambient temperature (°C)

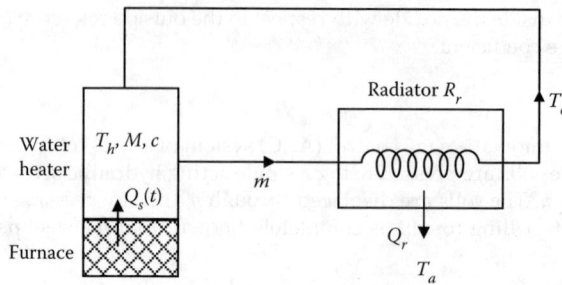

FIGURE P2.18
A household heating system.

T_h = temperature of water in the water heater—assumed uniform (°C)
T_o = temperature of the water leaving the radiator (°C)
Q_r = rate of heat transfer from the radiator to the ambience (kW)
M = mass of water in the water heater (500 kg)
$\dot{m}$ = mass rate of water flow through the radiator (25 kg/min)
c = specific heat of water (4200 J/kg/°C).
The radiator satisfies the equation
$T_h - T_a = R_r Q_r$
where R_r = thermal resistance of the radiator (2×10^{-3} °C/kW)

a. What are the inputs to the system?
b. Using T_h as a state variable, develop a state-space model for the system.
c. Give output equations for Q_r and T_o.

PROBLEM 2.19

Consider a hollow cylinder of length l, inside diameter d_i, and the outside diameter d_o. If the conductivity of the material is k, what the conductive thermal resistance of the cylinder in the radial direction?

PROBLEM 2.20

1. In the electro-thermal analogy of thermal systems, where voltage is analogous to temperature and current is analogous to heat transfer rate, explain why there exists a thermal capacitor but not a thermal inductor. What is a direct consequence of this fact with regard to the natural (free or unforced) response of a purely thermal system?

2. A package of semiconductor material consisting primarily of wafers of crystalline silicon substrate with minute amounts of silicon dioxide is heat treated at high temperature as an intermediate step in the production of transistor elements. An approximate model of the heating process is shown in Figure P2.20.

   The package is placed inside a heating chamber whose walls are uniformly heated by a distributed heating element. The associated heat transfer rate into the wall is Q_i. The interior of the chamber contains a gas of mass m_c and specific heat c_c, and is maintained at a uniform temperature T_c. The temperature of silicon is T_s and that of the wall is T_w. The outside environment is maintained at temperature T_o. The specific heats of the silicon package and the wall are denoted by c_s and c_w, respectively, and the corresponding masses are denoted by m_s and m_w as shown. The convective heat transfer coefficient at the interface of silicon and gas inside the chamber is h_s, and the effective surface area is A_s. Similarly, h_i and h_o denote the convective heat transfer coefficients at the inside and outside surfaces of the chamber wall, and the corresponding surface areas are A_i and A_o, respectively.

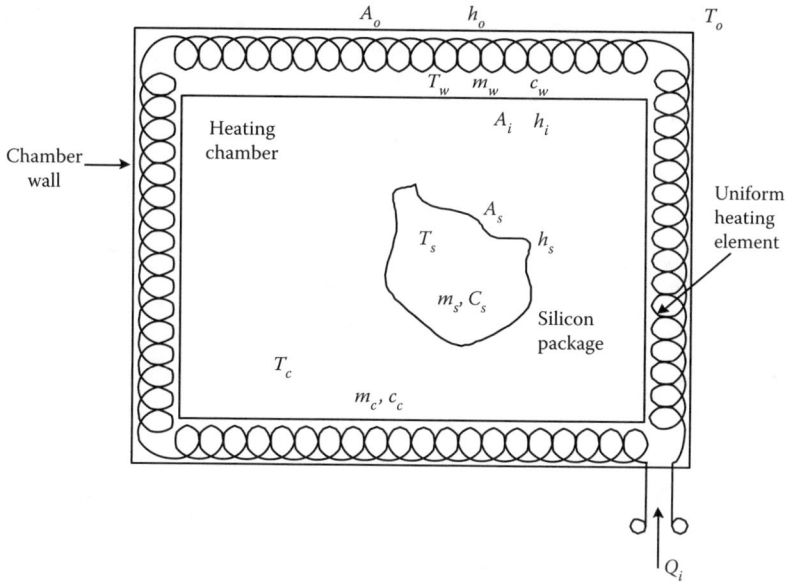

FIGURE P2.20
A model of the heat treatment of a package of silicon.

(a) Using T_s, T_c, and T_w as state variables, write three state equations for the process.
(b) Express these equations in terms of the parameters $C_{hs} = m_s c_s$, $C_{hc} = m_c c_c$, $C_{hw} = m_w c_w$, $R_s = 1/h_s A_s$, $R_i = 1/h_i A_i$, and $R_o = 1/h_o A_o$. Explain the electrical analogy and physical significance of these parameters.
(c) What are the inputs to the process? If T_s is the output of importance, obtain the matrices A, B, C, and D of the state-space model.
(d) Comment on the accuracy of the model in the context of the actual physical process of producing semiconductor elements.

3

Model Linearization

Real systems are nonlinear and they are represented by nonlinear analytical models consisting of nonlinear differential equations. Linear systems (models) are in fact idealized representations, and are represented by linear differential equations. Clearly, it is far more convenient to analyze, simulate, design, and control linear systems. For this reason, nonlinear systems are often approximated by linear models.

It is not possible to represent a highly nonlinear system by a single linear model in its entire range of operation. For small "changes" in the system response, however, a linear model may be developed, which is valid in the neighborhood of an operating point of the system, under small response changes. In this chapter we will study linearization of a nonlinear system/model in a restricted range of operation, about an operating point. First linearization of both analytical models, particularly state-space models and input–output models will be treated. Then linearization of experimental models (experimental data) is addressed.

3.1 Model Linearization

Real systems are nonlinear and are represented by nonlinear analytical models. A device or system is considered linear if it can be modeled by linear differential equations, with time t as the independent variable. Common analytical techniques (e.g., response analysis, frequency domain analysis, eigenvalue problem analysis, simulation, control) use linear models. Furthermore it is far more convenient to solve, simulate, design, and control linear models. In particular, as has been observed in Chapter 2, the *principle of superposition* holds for linear systems, thereby making the analytical procedures far simpler.

Nonlinear devices and systems are often analyzed using linear techniques by considering small excursions about an operating point. This linearization is accomplished by introducing incremental variables for inputs and outputs. If one increment can cover the entire operating range of a device with sufficient accuracy, it is an indication that the device is linear. If the input–output relations are nonlinear algebraic equations, it represents a *static nonlinearity*. Such a situation can be handled simply by using nonlinear calibration curves, which linearize the device without introducing nonlinearity errors. If, on the other hand, the input–output relations are nonlinear differential equations, analysis usually becomes quite complex. This situation represents a *dynamic nonlinearity*. Common manifestations of nonlinearities in devices and systems are: saturation; dead zone; hysteresis; the jump phenomenon; limit cycle response; and frequency creation.

A nonlinear analytical model may contain several nonlinear terms. The approach taken here is to linearize each nonlinear term by using the first order Taylor series approximation, which involves only the first derivative of the nonlinear terms (i.e., the slope of its graphical representation). Note that a nonlinear term may be a function of more than one independent variable. In that case the first derivatives with respect to all its independent variables are needed in the linearization process (i.e., slopes along all orthogonal directions of the coordinate axes which represent the independent variables).

3.1.1 Linearization about an Operating Point

Linearization is carried out about some operating point. This is typically the normal operating condition of the system. By necessity, the normal operating condition (steady-state or the equilibrium state). In a steady-state, by definition, the rates of changes of the system variables are zero. Hence, the steady-state (equilibrium state) is determined by setting the time-derivative terms in the system equations to zero and then solving the resulting algebraic equations. This may lead to more than one solution, since the steady-state (algebraic) equations themselves are nonlinear. The steady-state (equilibrium) solutions can be:

1. Stable (given a slight shift, the system response will eventually return to the original steady-state).
2. Unstable (given a slight shift, the system response will continue to move away from the original steady-state), or
3. Neutral (given a slight shift, the system response will remain in the shifted condition).

Consider a nonlinear function $f(x)$ of the independent variables x. Its Taylor series approximation about an operating point $(\ )_o$, up to the first derivative, is given by

$$f(x) \approx f(x_o) + \frac{df(x_o)}{dx}\delta x \quad \text{with} \quad x = x_o + \delta x \tag{3.1a}$$

Note that δx represents a small change from the operating point.

Now denote operating condition by $(^-)$ and a small increment about that condition by $(^\wedge)$. We have

$$f(\bar{x} + \hat{x}) \approx f(\bar{x}) + \frac{df(\bar{x})}{dx}\hat{x} \tag{3.1b}$$

A graphical illustration of this approach to linearization is given in Figure 3.1.

From Equation 3.1 it is seen that the increment of the function, due to the increment in its independent variable, is

$$\delta f = f(x) - f(x_o) \approx \frac{df(x_o)}{dx}\delta x \tag{3.2a}$$

or

$$\hat{f} = f(\bar{x} + \hat{x}) - f(\bar{x}) \approx \frac{df(\bar{x})}{dx}\hat{x} \tag{3.2b}$$

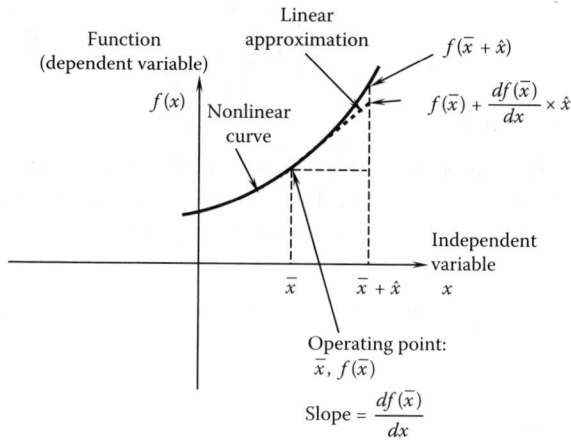

FIGURE 3.1
Linearization about an operating point.

The error resulting from this approximation is

$$\text{Error } e = f(\bar{x}+\hat{x}) - \left[f(\bar{x}) + \frac{df(\bar{x})}{dx}\hat{x} \right] \tag{3.3}$$

This error can be decreased by:

1. Making the nonlinear function more linear
2. Making the change $\hat{x}$ from the operating point smaller

Note: If the term is already linear, as for example, is the case of

$$f = ax$$

where a is the coefficient (constant) of the term, the corresponding linearized incremental term is

$$\delta f = a\,\delta x \tag{3.4a}$$

or

$$\hat{f} = a\hat{x} \tag{3.4b}$$

This is obvious because, for a linear term, the first derivative (slope) is simply its coefficient. Furthermore, the following incremental results hold for the time derivatives of the variables:

$$\delta \dot{x} = \frac{d\hat{x}}{dt} = \dot{\hat{x}} \tag{3.5}$$

$$\delta \ddot{x} = \frac{d^2 \hat{x}}{dt^2} = \ddot{\hat{x}} \qquad\qquad\qquad (3.6)$$

3.1.2 Function of Two Variables

The process of linearization, as presented above, can be easily extended to functions of more than one independent variable. For illustration consider a nonlinear function $f(x,y)$ of two independent variables x and y. The first order Taylor series approximation is

$$f(x,y) \approx f(x_o,y_o) + \frac{\partial f(x_o,y_o)}{\partial x}\delta x + \frac{\partial f(x_o,y_o)}{\partial y}\delta y \quad \text{with} \quad x = x_o + \delta x, \quad y = y_o + \delta y \qquad (3.7a)$$

or

$$f(\bar{x}+\hat{x}, \bar{y}+\hat{y}) \approx f(\bar{x},\bar{y}) + \frac{\partial f(\bar{x},\bar{y})}{\partial x}\hat{x} + \frac{\partial f(\bar{x},\bar{y})}{\partial y}\hat{y} \qquad\qquad (3.7b)$$

where ($^-$) denotes the operating condition and ($^\wedge$) denotes a small increment about that condition, as usual. In this case, for the process of linearization, we need two local slopes $\partial f(\bar{x},\bar{y})/\partial x$ and $\partial f(\bar{x},\bar{y})/\partial y$ along the two orthogonal directions of the independent variables x and y.

It should be clear now that linearization of a nonlinear system is carried out by replacing each term in the system equation by its increment, about an operating point. We summarize below the steps of local linearization about an operating point:

1. Select the operating point (or, reference condition). This is typically a steady-state, which can be determined by setting the time-derivative terms in the system equations to zero and solving the resulting nonlinear algebraic equations.

2. Determine the slopes (first order derivatives) of each nonlinear term (function) in the systems equation at the operating point, with respect to (along) each independent variable.

3. Consider each term in the system equation. If a term is nonlinear, replace it by its slope (at the operating point) times the corresponding incremental variable. If a term is linear, replace it by its coefficient (which is indeed the constant slope of the linear term) times the corresponding incremental variable.

3.2 Nonlinear State-Space Models

Consider a general nonlinear, time-variant, $n$thorder system represented by n first order differential equations, which generally are coupled, as given by

$$\frac{dq_1}{dt} = f_1(q_1, q_2, \ldots, q_n, r_1, r_2, \ldots, r_m, t)$$

$$\frac{dq_2}{dt} = f_2(q_1, q_2, \ldots, q_n, r_1, r_2, \ldots, r_m, t)$$

$$\vdots$$

(3.8a)

$$\frac{dq_n}{dt} = f_n(q_1, q_2, \ldots, q_n, r_1, r_2, \ldots, r_m, t)$$

The state vector is

$$q = [q_1, q_2, \ldots, q_n]^T$$ (3.9a)

and the input vector is

$$r = [r_1, r_2, \ldots, r_m]^T$$ (3.10a)

Equation 3.6a may be written in the vector notation:

$$\dot{q} = f(q, r, t)$$ (3.8b)

3.2.1 Linearization

An equilibrium state of the dynamic system given by Equation 3.8 corresponds to the condition when the rates of changes of the state variables are all zero:

$$\dot{q} = 0$$ (3.11)

This is true because in equilibrium (i.e., at an operating point) the system response remains steady and hence its rate of change is zero. Consequently, the equilibrium states $\bar{q}$ are obtained by solving the set of n nonlinear algebraic equations

$$f(q, r, t) = 0$$ (3.12)

for a particular steady input $\bar{r}$. Usually a system operates in the neighborhood of one of its equilibrium states. This state is the *operating point* of the system. The steady-state of a dynamic system is also an *equilibrium state*.

To study the stability of various equilibrium states of a nonlinear dynamic system, it is first necessary to linearize the system model about these equilibrium states. Linear models are also useful in analyzing nonlinear systems when it is known that the variations of the system response about the system operating point are small in comparison to the maximum allowable variation (*dynamic range*). As noted before, Equation 3.8 can be linearized for small variations δq and δr about an equilibrium point $(\bar{q}, \bar{r})$ by employing up to only the first derivative term (i.e., O(1) term) in the Taylor series expansion of the nonlinear function f. The higher-order terms are negligible for small δq and δr. As explained before, for the scalar case, this method yields

$$\delta \dot{q} = \frac{\partial f}{\partial q}(\bar{q}, \bar{r}, t)\delta q + \frac{\partial f}{\partial r}(\bar{q}, \bar{r}, t)\delta r$$ (3.8)

Denote the state vector and the input vector of the linearized system by

$$\delta q = x = [x_1, x_2, \ldots, x_n]^T \tag{3.9b}$$

$$\delta r = u = [u_1, u_2, \ldots, u_m]^T \tag{3.10b}$$

This results in the linear model

$$\dot{x} = Ax + Bu \tag{3.13}$$

The linear system matrix $A(t)$ and the input distribution (gain) matrix $B(t)$ are given by

$$A(t) = \frac{\partial f}{\partial q}(\bar{q}, \bar{r}, t) \tag{3.14}$$

$$B(t) = \frac{\partial f}{\partial r}(\bar{q}, \bar{r}, t) \tag{3.15}$$

If the dynamic system is a constant parameter (i.e., *stationary* or *time-invariant*) system, or if it can be assumed as such for the time period of interest, then A and B become constant matrices.

3.2.2 Reduction of System Nonlinearities

Under steady conditions, system nonlinearities can be removed through calibration. Under dynamic conditions, however, the task becomes far more difficult. The following are some of the precautions and procedures that can be taken to remove or reduce nonlinearities in dynamic systems:

1. Avoid operating the device over a wide range of signal levels.
2. Avoid operation over a wide frequency band.
3. Use devices that do not generate large mechanical motions.
4. Minimize Coulomb friction and stiction (e.g., through lubrication).
5. Avoid loose joints, gear coupling, etc., which can cause backlash.
6. Use linearizing elements such as resistors and amplifiers.
7. Use linearizing feedback.

Next, we will illustrate model linearization and operating point analysis using several examples, which involve state-space models and input–output models.

Example 3.1

The robotic spray painting system of an automobile assembly plant employs an induction motor and pump combination to supply paint, at an overall peak rate of 15 gal/min, to a cluster of spray-paint heads in several painting booths. The painting booths are an integral part of the production line in the plant. The pumping and filtering stations are in the ground level of the building and the

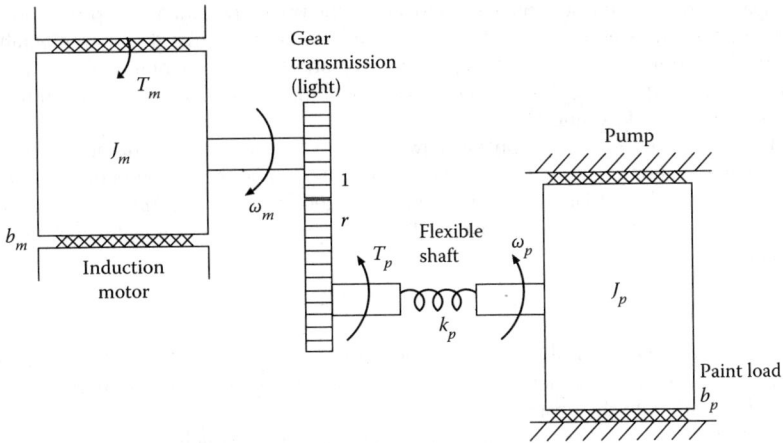

FIGURE 3.2
A model for a paint pumping system in an automobile assembly plant.

painting booths are in an upper level. Not all booths or painting heads operate at a given time. The pressure in the paint supply line is maintained at a desired level (approximately 275 psi or 1.8 MPa) by controlling the speed of the pump, which is achieved through a combination of voltage control and frequency control of the induction motor. An approximate model for the paint pumping system is shown in Figure 3.2.

The induction motor is linked to the pump through a gear transmission of efficiency η and speed ratio 1:r, and a flexible shaft of torsional stiffness k_p. The moments inertia of the motor rotor and the pump impeller are denoted by J_m and J_p, respectively. The gear inertia is neglected (or lumped with J_m). The mechanical dissipation in the motor and its bearings is modeled as a linear viscous damper of damping constant b_m. The load on the pump (the paint load plus any mechanical dissipation) is also modeled by a viscous damper, and the equivalent damping constant is b_p. The magnetic torque T_m generated by the induction motor is given by

$$T_m = \frac{T_0 q \omega_0 (\omega_0 - \omega_m)}{(q \omega_0^2 - \omega_m^2)} \tag{3.16}$$

in which ω_m is the motor speed. The parameter T_0 depends directly (quadratically) on the phase voltage (ac) supplied to the motor. The second parameter ω_0 is directly proportional to the line frequency of the ac supply. The third parameter q is positive and greater than unity, and this parameter is assumed constant in the control system.

a. Comment about the accuracy of the model shown in Figure 3.2.
b. Taking the motor speed ω_m, the pump-shaft torque T_p, and the pump speed ω_p as the state variables, systematically derive the three state equations for this (nonlinear) model. Clearly explain all steps involved in the derivation. What are the inputs to the system?
c. What do the motor parameters ω_0 and T_0 represent, with regard to motor behavior? Obtain the partial derivatives $\partial T_m/\partial \omega_m$, $\partial T_m/\partial T_0$ and $\partial T_m/\partial \omega_0$ and verify that the first of these three expressions is negative and the other two are positive. *Note*: under normal operating conditions $0 < \omega_m < \omega_0$.
d. Consider the steady-state operating point where the motor speed is steady at $\bar{\omega}_m$. Obtain expressions for ω_p, T_p, and T_0 at this operating point, in terms of $\bar{\omega}_m$ and $\bar{\omega}_0$.
e. Suppose that $(\partial T_m/\partial \omega_m)=-b$, $(\partial T_m/\partial T_0)=\beta_1$, and $(\partial T_m/\partial \omega_0)=\beta_2$ at the operating point given in (d). *Note*: Voltage control is achieved by varying T_0 and frequency control by varying ω_0.

Linearize the state model obtained in (b) about the operating pint and express it in terms of the incremental variables $\hat{\omega}_m$, $\hat{T}_p$, $\hat{\omega}_p$, $\hat{T}_0$, and $\hat{\omega}_0$. Suppose that the (incremental) output variables are the incremental pump speed $\hat{\omega}_p$ and the incremental angle of twist of the pump shaft. Express the linear state-space model in the usual form and obtain the associated matrices A, B, C and D.

f. For the case of frequency control only (i.e., $\hat{T}_0 = 0$) obtain the input–output model (differential equation) relating the incremental output $\hat{\omega}_p$ and the incremental input $\hat{\omega}_0$. Using this equation show that if $\hat{\omega}_0$ is suddenly changed by a step of $\Delta\hat{\omega}_0$ then $d^3\hat{\omega}_p/dt^3$ will immediately change by a step of $\beta_2 r k_p / J_m J_p \, \Delta\hat{\omega}_0$, but the lower derivatives of $\hat{\omega}_p$ will not change instantaneously.

Solution

a. Backlash and inertia of the gear transmission have been neglected in the model shown. This is not accurate in general. Also, the gear efficiency η, which is assumed constant here, usually varies with the gear speed.

Usually there is some flexibility in the shaft (coupling), which connects the gear to the drive motor.

Energy dissipation (in the pump load and in various bearings) has been lumped into a single linear viscous damping element. In practice, this energy dissipation is nonlinear and distributed.

b. Motor speed $\omega_m = d\theta_m/dt$

Load (pump) speed $\omega_p = d\theta_p/dt$

where θ_m = motor rotation, and θ_p = pump rotation. Let T_g = reaction torque on the motor from the gear. By definition, gear efficiency is given by

$$\eta = \frac{T_p r \omega_m}{T_g \omega_m} = \frac{\text{Output Power}}{\text{Input Power}}$$

Since r is the gear ratio, $r\omega_m$ is the output speed of the gear. Also, power = torque × speed. We have,

$$T_g = \frac{r}{\eta} T_p \tag{i}$$

The following three constitutive equations can be written:

Newton's second law (torque = inertia × angular acceleration) for the motor:

$$T_m - T_g - b_m \omega_m = J_m \dot{\omega}_m \tag{ii}$$

Newton's second law for the pump:

$$T_p - b_p \omega_p = J_p \dot{\omega}_p \tag{iii}$$

Hooke's law (torque = torsional stiffness × angle of twist) for the flexible shaft:

$$T_p = k_p (r\theta_m - \theta_p) \tag{iv}$$

Equations (ii) through (iv) provide the three state equations. Specifically,

Substitute Equation (i) into Equation (ii): $J_m \dot{\omega}_m = T_m - b_m \omega_m - \dfrac{r}{\eta} T_p \tag{v}$

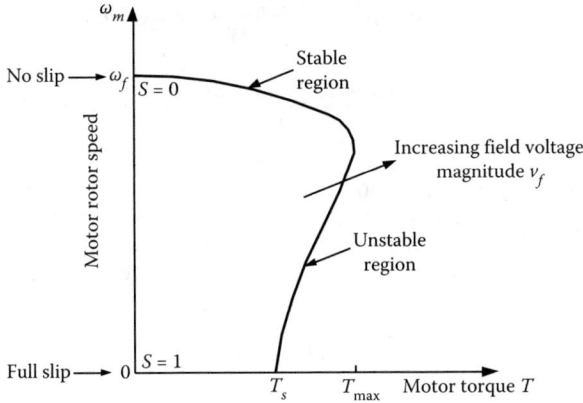

FIGURE 3.3
Torque-speed characteristic curve of an induction motor.

$$\text{Differentiate Equation (iv): } \dot{T}_p = k_p(r\omega_m - \omega_p) \tag{vi}$$

$$\text{Equation (iii): } J_p\dot{\omega}_p = T_p - b_p\omega_p \tag{vii}$$

Equations (v) through (vii) are the three state equations. This is a nonlinear model with the state vector $[\omega_m \; T_p \; \omega_p]^T$. The input is T_m. Strictly there are two inputs, in view of the torque-speed characteristic curve of the motor as given by Equation 3.16 and sketched in Figure 3.3, where the fractional slip S of the induction motor is given by

$$S = \frac{(\omega_0 - \omega_m)}{\omega_0} \tag{3.17}$$

The two inputs are: ω_0, the speed of the rotating magnetic field, which is proportional to the line frequency; and T_0 which depends quadratically on the phase voltage.

c. From Equation 3.16:
   When $\omega_m = 0$ we have $T_m = T_0$. Hence

$$T_0 = \text{starting torque of the motor.}$$

   When $T_m = 0$, we have $\omega_m = \omega_0$. Hence,

$$\omega_0 = \text{no-load speed.}$$

This is the synchronous speed—under no-load conditions, there is no slip in the induction motor (i.e., actual speed of the motor is equal to the speed ω_0 of the rotating magnetic field).
   Differentiate Equation 3.16 with respect to T_0, ω_0, and ω_m. We have

$$\frac{\partial T_m}{\partial T_0} = \frac{q\omega_0(\omega_0 - \omega_m)}{(q\omega_0^2 - \omega_m^2)} = \beta_1 \text{ (say)} \tag{3.18a}$$

$$\frac{\partial T_m}{\partial \omega_0} = \frac{T_0 q[(q\omega_0^2 - \omega_m^2)(2\omega_0 - \omega_m) - \omega_0(\omega_0 - \omega_m)2q\omega_0]}{(q\omega_0^2 - \omega_m^2)^2}$$

$$= \frac{T_0 q\omega_m[(\omega_0 - \omega_m)^2 + (q-1)\omega_0^2]}{(q\omega_0^2 - \omega_m^2)^2} = \beta_2 \text{ (say)} \tag{3.18b}$$

$$\frac{\partial T_m}{\partial \omega_m} = \frac{T_0 q \omega_0 [(q\omega_0^2 - \omega_m^2)(-1) - (\omega_0 - \omega_m)(-2\omega_m)]}{(q\omega_0^2 - \omega_m^2)^2}$$

$$= -\frac{T_0 q \omega_0 [(q-1)\omega_0^2 + (\omega_0 - \omega_m)^2]}{(q\omega_0^2 - \omega_m^2)^2} = -b_e \text{ (say)}$$

(3.19)

We have $\beta_1 > 0$; $\beta_2 > 0$; and $b_e > 0$.

Note: b_e = electrical damping constant of the motor.

d. At a steady-state operating point, the rates of changes of the state variables will be zero. Hence set $\dot{\omega}_m = 0 = \dot{T}_p = \dot{\omega}_p$ in Equations (v) through (vii). We get

$$0 = \bar{T}_m - b_m \bar{\omega}_m - \frac{r}{\eta} \bar{T}_p$$

$$0 = k_p(r\bar{\omega}_m - \bar{\omega}_p)$$

$$0 = \bar{T}_p - b_p \bar{\omega}_p$$

Hence,

$$\bar{\omega}_p = r\bar{\omega}_m \tag{viii}$$

$$\bar{T}_p = b_p r \bar{\omega}_m \tag{ix}$$

$$\bar{T}_m = b_m \bar{\omega}_m + r^2 b_p \bar{\omega}_m / \eta = \frac{\bar{T}_0 q \bar{\omega}_0 (\bar{\omega}_0 - \bar{\omega}_m)}{(q\bar{\omega}_0^2 - \bar{\omega}_m^2)} \quad \text{(from Equation 3.16)}$$

or

$$\bar{T}_0 = \frac{\bar{\omega}_m (b_m + r^2 b_p / \eta)(q\bar{\omega}_0^2 - \bar{\omega}_m^2)}{q\bar{\omega}_0 (\bar{\omega}_0 - \bar{\omega}_m)} \tag{x}$$

e. Take the increments of the state Equations (v) through (vii). We get

$$J_m \dot{\hat{\omega}}_m = -b_m \hat{\omega}_m - \frac{r}{\eta} \hat{T}_p - b_e \hat{\omega}_m + \beta_1 \hat{T}_0 + \beta_2 \hat{\omega}_0 \tag{xi}$$

$$\dot{\hat{T}}_p = k_p(r\hat{\omega}_m - \hat{\omega}_p) \tag{xii}$$

$$J_p \dot{\hat{\omega}}_p = \hat{T}_p - b_p \hat{\omega}_p \tag{xiii}$$

Note:

$$\hat{T}_m = \frac{\partial T_m}{\partial \omega_m} \hat{\omega}_m + \left[\frac{\partial T_m}{\partial T_0}\right] \hat{T}_0 + \left[\frac{\partial T_m}{\partial \omega_0}\right] \hat{\omega}_0 = -b_e \hat{\omega}_m + \beta_1 \hat{T}_0 + \beta_2 \hat{\omega}_0 \tag{xiv}$$

where in each partial derivative, the remaining independent variables are kept constant (by definition).

Equations (xi) through (xiii) subject to (ix) are the three linearized state equations.

Define the linear:

$$\text{State vector } \mathbf{x} = \begin{bmatrix} \hat{\omega}_m & \hat{T}_p & \hat{\omega}_p \end{bmatrix}^T$$

$$\text{Input vector } \mathbf{u} = \begin{bmatrix} \hat{T}_0 & \hat{\omega}_0 \end{bmatrix}^T$$

$$\text{Output vector } \mathbf{y} = \begin{bmatrix} \hat{\omega}_p & \hat{T}_s/k_p \end{bmatrix}^T$$

We have

$$\mathbf{A} = \begin{bmatrix} -(b_e + b_m)/J_m & -r/(\eta J_m) & 0 \\ k_p r & 0 & -k_p \\ 0 & 1/J_p & -b_p/J_p \end{bmatrix}; \mathbf{B} = \begin{bmatrix} \beta_1/J_m & \beta_2/J_m \\ 0 & 0 \\ 0 & 0 \end{bmatrix}; \mathbf{C} = \begin{bmatrix} 0 & 0 & 1 \\ 0 & 1/k_p & 0 \end{bmatrix}; \mathbf{D} = 0$$

Note: b_e = electrical damping constant of the motor; b_m = mechanical damping constant of the motor.

f. For frequency control, $\hat{T}_0 = 0$.

Substitute Equation (xii) into Equation (xi) in order to eliminate $\hat{\omega}_m$. Then substitute Equation (xiii) into the result in order to eliminate $\hat{T}_p$. On simplification we get the input–output model (differential equation):

$$J_m J_p \frac{d^3\hat{\omega}_p}{dt^3} + [J_m b_p + J_p(b_m + b_e)]\frac{d^2\hat{\omega}_p}{dt^2} + \left[k_p\left(J_m + \frac{r^2 J_p}{\eta} \right) + b_p(b_m + b_e) \right]\frac{d\hat{\omega}_p}{dt} +$$

$$k_p\left(\frac{r^2 b_p}{\eta} + b_m + b_e \right)\hat{\omega}_p = \beta_2 r k_p \hat{\omega}_0$$ (xv)

This is a third order differential equation, as expected, since the system is third order. Also, as we have seen, the state-space model is also third order.

Observation from Equation (xv):

When $\hat{\omega}_0$ is changed by "finite" step $\Delta\dot{\omega}_0$, the RHS of Equation (xix) will be finite. Hence the LHS, and particularly highest derivative $(d^3\hat{\omega}_p/dt^3)$ also must change by a finite value.

Further verification: If as a result, $d^2\hat{\omega}_p/dt^2$ or lower derivatives also change by a finite step, then $d^3\hat{\omega}_p/dt^3$ should change by an infinite value (*Note*: derivative of a step = impulse.)

This contradicts the fact that RHS of Equation (xix) is finite. Hence $d^2\hat{\omega}_p/dt^2$, $d\hat{\omega}_p/dt$, and $\hat{\omega}_p$ will not change instantaneously. Only $d^3\hat{\omega}_p/dt^3$ will change instantaneously by a finite value due to finite step change of $\hat{\omega}_0$.

From Equation (xix): Resulting change of $d^3\hat{\omega}_p/dt^3$ is $\beta_2 r k_p/J_m J_p \, \Delta\hat{\omega}_0$

Furthermore, the following somewhat general observations can be made from this example:

1. Mechanical damping constant b_m comes from bearing friction and other mechanical sources of the motor.
2. Electrical damping constant be comes from the electromagnetic interactions in the motor.
3. The two must occur together (e.g., in model analysis, simulation, design, and control). For example, whether the response is underdamped or overdamped depends on the sum $b_m + b_e$ and not the individual components←electro-mechanical coupling.

Example 3.2

An automated wood cutting system contains a cutting unit, which consists of a dc motor and a cutting blade, linked by a flexible shaft and a coupling. The purpose of the flexible shaft is to position the blade unit at any desirable configuration, away from the motor itself. The coupling unit helps with the shaft alignment (compensates for possible misalignment). A simplified, lumped-parameter, dynamic model of the cutting device is shown in Figure 3.4 and the following parameters and variables are used: J_m=axial moment of inertia of the motor rotor; b_m=equivalent viscous damping constant of the motor bearings; k=torsional stiffness of the flexible shaft; J_c=axial moment of inertia of the cutter blade; b_c=equivalent viscous damping constant of the cutter bearings; T_m=magnetic torque of the motor; ω_m=motor speed; T_k=torque transmitted through the flexible shaft; ω_c=cutter speed; T_L=load torque on the cutter from the workpiece (wood).

In comparison with the flexible shaft, the coupling unit is assumed rigid, and is also assumed light. The cutting load is given by

$$T_L = c \, |\omega_c| \omega_c \qquad (3.20)$$

The parameter c, which depends on factors such as the depth of cut and the material properties of the workpiece, is assumed constant in the present analysis.

a. Using T_m as the input, T_L as the output, and $[\omega_m \ T_k \ \omega_c]^T$ as the state vector, develop a complete (nonlinear) state model for the system shown in Figure 3.4. What is the order of the system?

b. Using the state model derived in (a), obtain a single input–output differential equation for the system, with T_m as the input and ω_c as the output.

c. Consider the steady operating conditions where $T_m = \bar{T}_m$, $\omega_m = \bar{\omega}_m$, $T_k = \bar{T}_k$, $\omega_c = \bar{\omega}_c$, $T_L = \bar{T}_L$ are all constants. Express the operating point values $\bar{\omega}_m$, $\bar{T}_k$, $\bar{\omega}_c$, and $\bar{T}_L$ in terms of $\bar{T}_m$ and the model parameters only. You must consider both cases $\bar{T}_m > 0$ and $\bar{T}_m < 0$.

d. Now consider an incremental change $\hat{T}_m$ in the motor torque and the corresponding changes $\hat{\omega}_m$, $\hat{T}_k$, $\hat{\omega}_c$, and $\hat{T}_L$ in the system variables. Determine a linear state model ($\mathbf{A}$, $\mathbf{B}$, $\mathbf{C}$, $\mathbf{D}$) for the incremental dynamics of the system in this case, using $\mathbf{x} = [\hat{\omega}_m \ \hat{T}_k \ \hat{\omega}_c]^T$ as the state vector, $\mathbf{u} = [\hat{T}_m]$ as the input, and $\mathbf{y} = [\hat{T}_L]$ as the output.

e. In the incremental model (see a), if the twist angle of the flexible shaft (i.e., $\theta_m - \theta_c$) is used as the output what will be a suitable state model? What is the system order then?

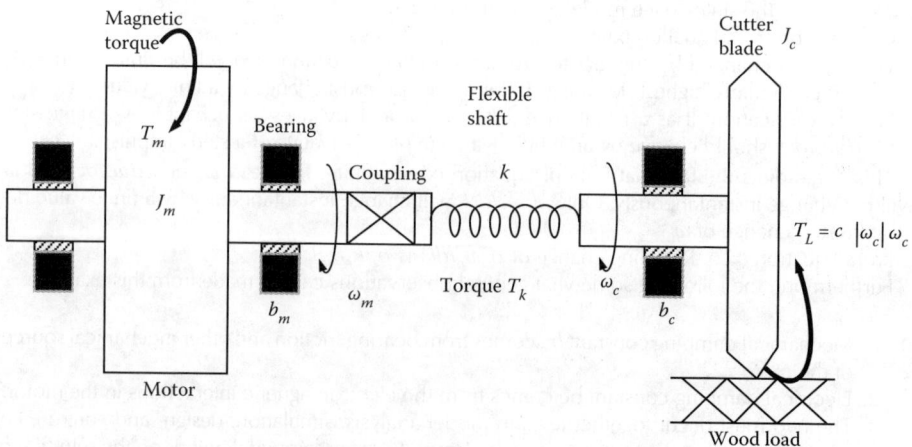

FIGURE 3.4
A wood cutting machine.

f. In the incremental model, if the angular position θ_c of the cutter blade is used as the output variable, explain how the state model obtained in (a) should be modified. What is the system order in this case?

Hint for (b):

$$\frac{d}{dt}\left(|\omega_c|\,\omega_c\right)=2|\omega_c|\,\dot{\omega}_c \tag{3.21}$$

$$\frac{d^2}{dt^2}\left(|\omega_c|\omega_c\right)=2|\omega_c|\ddot{\omega}_c+2\omega_c^2\,\mathrm{sgn}(\omega_c) \tag{3.22}$$

Note: These results may be derived as follows: since $|\omega_c|=\omega_c\,\mathrm{sgn}\,\omega_c$ we have

$$\frac{d}{dt}\left(|\omega_c|\omega_c\right)=\frac{d}{dt}(\omega_c^2\,\mathrm{sgn}\,\omega_c)=2\omega_c\dot{\omega}_c\,\mathrm{sgn}\,\omega_c=2|\omega_c|\dot{\omega}_c$$

and

$$\frac{d^2}{dt^2}\left(|\omega_c|\omega_c\right)=2|\omega_c|\ddot{\omega}_c+2\dot{\omega}_c^2\,\mathrm{sgn}(\omega_c)$$

Note: Since $\mathrm{sgn}(\omega_c)=+1$ for $\omega_c>0;=-1$ for $\omega_c<0$; it is a constant and its time derivative is zero (except at $\omega_c=0$, which is not important here as it corresponds to the static condition).

Solution

a. The free body diagram of the system is shown in Figure 3.5.
   Constitutive equations of the three elements:

$$J_m\dot{\omega}_m=T_m-b_m\omega_m-T_k \tag{i}$$

$$\dot{T}_k=k(\omega_m-\omega_c) \tag{ii}$$

$$J_c\dot{\omega}_c=T_k-b_c\omega_c-c|\omega_c|\omega_c \tag{iii}$$

These are indeed the state equations, with

State vector $=[\omega_m\ \ T_k\ \ \omega_c]^T$

Input vector $=[T_m]$

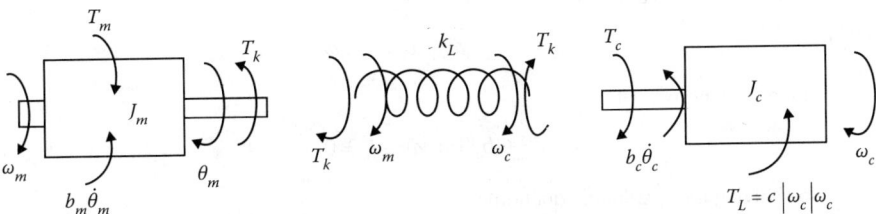

FIGURE 3.5
Free body diagram of the system.

$$\text{Output vector} = [T_l] = c|\omega_c|\omega_c$$

This is a third order system (three state variables; three state equations).

b. Substitute Equation (ii) and its time derivative into Equation (i), to eliminate ω_m and $\dot{\omega}_m$:

$$J_m[\ddot{T}_k/k + \dot{\omega}_c] = T_m - b_m[\dot{T}_k/k + \omega_c] - T_k$$

Now substitute Equation (iii) and its time derivatives in this equation, to eliminate T_k and its time derivatives (using Equations 3.21 and 3.22):

$$\dot{T}_k = J_c\ddot{\omega}_c + b_c\dot{\omega}_c + 2c|\omega_c|\dot{\omega}_c$$

$$\ddot{T}_k = J_c\dddot{\omega}_c + b_c\ddot{\omega}_c + 2c|\omega_c|\ddot{\omega}_c + 2c\dot{\omega}_c^2\,\text{sgn}(\omega_c)$$

We get

$$\frac{J_m}{k}\left[J_c\dddot{\omega}_c + b_c\ddot{\omega}_c + 2c|\omega_c|\ddot{\omega}_c + 2c\dot{\omega}_c^2\,\text{sgn}(\omega_c)\right] + J_m\dot{\omega}_c$$

$$= T_m - \frac{b_m}{k}\left[J_c\ddot{\omega}_c + b_c\dot{\omega}_c + 2c|\omega_c|\dot{\omega}_c\right] - b_m\omega_c - \left[J_c\dot{\omega}_c + b_c\omega_c + c|\omega_c|\omega_c\right]$$

This can be rearranged as the input–output model (differential equation)

$$J_mJ_c\frac{d^3\omega_c}{dt^3} + (J_mb_c + J_cb_m)\frac{d^2\omega_c}{dt^2} + 2cJ_m|\omega_c|\frac{d^2\omega_c}{dt^2} + (J_mk + J_ck + b_mb_c)\frac{d\omega_c}{dt}$$

$$+ 2b_mc|\omega_c|\frac{d\omega_c}{dt} + 2cJ_m\,\text{sgn}(\omega_c)\left(\frac{d\omega_c}{dt}\right)^2 + k(b_m + b_c)\omega_c + kc|\omega_c|\omega_c = kT_m$$

c. At the operating point, rates of changes of the state variables will be zero. Hence, from Equations (i) through (iii) we have

$$0 = \bar{T}_m - b_m\bar{\omega}_m - \bar{T}_k \tag{iv}$$

$$0 = k(\bar{\omega}_m - \bar{\omega}_c) \tag{v}$$

$$0 = \bar{T}_k - b_c\bar{\omega}_c - c|\bar{\omega}_c|\bar{\omega}_c \tag{vi}$$

Case 1: $\bar{T}_m > 0 \Rightarrow \bar{\omega}_c > 0$

Eliminate $\bar{T}_k$ using Equation (iv) and (vi):

$$0 = \bar{T}_m - b_m\bar{\omega}_m - b_c\bar{\omega}_c - c\bar{\omega}_c^2$$

Since $\bar{\omega}_m = \bar{\omega}_c$ we get

$$c\bar{\omega}_c^2 + (b_m + b_c)\bar{\omega}_c - \bar{T}_m = 0$$

or (by solving the quadratic equation)

$$\bar{\omega}_c = \frac{-(b_m + b_c) \pm \sqrt{(b_m + b_c)^2 + 4c\bar{T}_m}}{2c}$$

Take the positive root:

$$\bar{\omega}_c = \frac{\sqrt{(b_m + b_c)^2 + 4c\bar{T}_m}}{2c} - \frac{(b_m + b_c)}{2c} = \bar{\omega}_m$$

From Equation (iv):

$$\bar{T}_k = \bar{T}_m - b_m \left[\frac{\sqrt{(b_m + b_c)^2 + 4c\bar{T}_m}}{2c} - \frac{(b_m + b_c)}{2c} \right]$$

From Equation (vi):

$$\bar{T}_L = c|\bar{\omega}_c|\bar{\omega}_c = \bar{T}_m - (b_m + b_c) \left[\frac{\sqrt{(b_m + b_c)^2 + 4c\bar{T}_m}}{2c} - \frac{(b_m + b_c)}{2c} \right]$$

Case 2: $\bar{T}_m < 0 \Rightarrow \bar{\omega}_c < 0$

In this case

$$0 = \bar{T}_m - b_m \bar{\omega}_c - b_c \bar{\omega}_c + c\bar{\omega}_c^2$$

or

$$c\bar{\omega}_c^2 - (b_m + b_c)\bar{\omega}_c + \bar{T}_m = 0$$

$$\bar{\omega}_c = \frac{(b_m + b_c) \pm \sqrt{(b_m + b_c)^2 - 4c\bar{T}_m}}{2c}$$

Note: $\bar{T}_m < 0$. Use the negative root:

$$\bar{\omega}_c = \frac{(b_m + b_c)}{2c} - \frac{\sqrt{(b_m + b_c)^2 - 4c\bar{T}_m}}{2c}$$

The rest will follow as in Case 1.

d. In linearizing Equations (i) through (iii) we note that the only nonlinear term is $c|\omega_c|\omega_c$ whose slope (derivative) is $d/d\omega_c(|\omega_c|\omega_c) = d/d\omega_c(\omega_c^2 \operatorname{sgn}\omega_c) = 2\omega_c \operatorname{sgn}\omega_c = 2|\omega_c|$

Consequently, by writing the increment of each term in Equations (i) through (iii), we get the linear state model:

$$J_m \dot{\hat{\omega}}_m = \hat{T}_m - b_m \hat{\omega}_m - \hat{T}_k$$

$$\dot{\hat{T}}_k = k(\hat{\omega}_m - \hat{\omega}_c)$$

$$J_c \dot{\hat{\omega}}_c = \hat{T}_k - b_c \hat{\omega}_c - 2c|\bar{\omega}_c|\hat{\omega}_c$$

with the output equation: $\hat{T}_L = 2c|\bar{\omega}_c|\hat{\omega}_c$

$$\text{State vector } \boldsymbol{x} = \begin{bmatrix} \hat{\omega}_m & \hat{T}_k & \hat{\omega}_c \end{bmatrix}^T$$

$$\text{Input vector } \boldsymbol{u} = \begin{bmatrix} \hat{T}_m \end{bmatrix}$$

$$\text{Output vector } \boldsymbol{y} = \begin{bmatrix} \hat{T}_L \end{bmatrix}$$

The corresponding state-model matrices are

$$A = \begin{bmatrix} -b_m/J_m & -1/J_m & 0 \\ k & 0 & -k \\ 0 & 1/J_c & -(b_c + 2c|\bar{\omega}_c|)/J_c \end{bmatrix}; \quad B = \begin{bmatrix} 1/J_m \\ 0 \\ 0 \end{bmatrix}; \quad C = \begin{bmatrix} 0 & 0 & 2c|\bar{\omega}_c| \end{bmatrix}; \quad D = [0]$$

e. The twist angle of the flexible shaft is $y = \theta_m - \theta_c = T_k/k$
   This represents a new output equation: $y = T_k/k$
   Since no new state variables are introduced for this, exactly the same state equations as before are applicable, along with this new output equation.
   System order = 3

f. The new output equation: $y = \theta_c$
   Since θ_c cannot be expressed as an algebraic equation of the three previous state variables, a new state variable θ_c has to be defined. This results in an additional state equation:

$$\frac{d\theta_c}{dt} = \omega_c$$

The system order becomes 4 in view of the extra state variable (and extra state equation).

Discussion

Physically, the new output θ_c is obtained by placing an integrator in front of the old output ω_c. Hence:

New system = old system + integrator at the output

From this series configuration of the old system and an integrator, it should be clear that, even though the system order has increased to 4 due to the new integrator, the basic dynamics of the system is still governed by the original third order system.

In particular, with the unified choice of across variables of independent A-type elements and through variables of independent T-type elements as the state variables, we end up with a unique third order state-space model for this system, which does not allow θ_c as a natural output. To provide the "unnatural" output θ_c there is no other option but to include a new integrator, which increases the system order by 1. The new state-space model corresponds to:

$$A = \begin{bmatrix} [A_{old}] & \begin{matrix} 0 \\ 0 \\ 0 \end{matrix} \\ 0 \quad 0 \quad 1 \quad 0 \end{bmatrix}_{4\times4}; \quad B = \begin{bmatrix} B_{old} \\ 0 \end{bmatrix}_{4\times1}; \quad C = \begin{bmatrix} 0 & 0 & 0 & 1 \end{bmatrix}; \quad D = [0]$$

with the state vector $\boldsymbol{x} = \begin{bmatrix} \hat{\omega}_m & \hat{T}_k & \hat{\omega}_c & \hat{\theta}_c \end{bmatrix}^T$

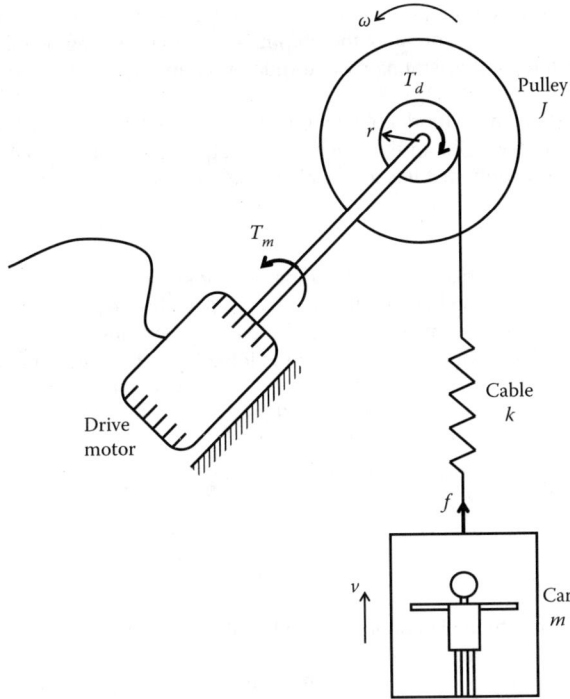

FIGURE 3.6
A simplified model of an elevator.

Example 3.3

A simplified model of an elevator is shown in Figure 3.6. The model parameters are: J=moment of inertia of the cable pulley; r=radius of the pulley; k=stiffness of the cable; m=mass of the car and its occupants.

a. Which system parameters are variable? Explain.
b. Suppose that the damping torque $T_d(\omega)$ at the bearing of the pulley is a nonlinear function of the angular speed ω of the pulley. Suppose that:

$$\text{State vector } \boldsymbol{x}=[\omega \ f \ v]^T$$

with f=tension force in the cable; v=velocity of the car (taken positive upwards).

$$\text{Input vector } \boldsymbol{u}=[T_m]^T$$

with T_m=torque applied by the motor to the pulley (positive in the direction indicated in Figure 3.6).

$$\text{Output vector as } \boldsymbol{y}=[v]$$

Obtain a complete, nonlinear, state-space model for the system.

c. With T_m as the input and v as the output, convert the state-space model into a nonlinear input–output differential equation model. What is the order of the system?
d. Give an equation whose solution provides the steady-state operating speed $\bar{v}$ of the elevator car.

e. Linearize the nonlinear input–output differential-equation model obtained in (c), for small changes $\hat{T}_m$ of the input and $\hat{v}$ of the output, about an operating point.
   Note: $\bar{T}_m$ =steady-state operating point torque of the motor (assumed to be known).
   Hint: Denote dT_d/dt as $b(\omega)$.

f. Linearize the state-space model obtained in (b) and give the model matrices **A**, **B**, **C**, and **D** in the usual notation. Obtain the linear input–output differential equation from this state-space model and verify that it is identical to what was obtained in (e).

Solution

a. The parameter r is a variable due to winding/unwinding of the cable around the pulley. The parameter m is a variable because the car occupancy changes.

b. The state equations are obtained simply by writing the constitutive equations (Newton's second law for the two inertia elements and Hooke's law for the spring element):

$$J\dot{\omega} = T_m - rf - T_d(\omega) \tag{i}$$

$$\dot{f} = k(r\omega - v) \tag{ii}$$

$$m\dot{v} = f - mg \tag{iii}$$

Output $y = v$

c. Eliminate f by substituting Equation (iii) into Equations (i) and (ii):

$$J\dot{\omega} = T_m - rm(\dot{v} - g) - T_d(\omega) \tag{iv}$$

$$m\ddot{v} = k(r\omega - v) \tag{v}$$

Note: To eliminate a quantity we substitute the equation which contains the quantity by itself (not its derivatives or combinations of its nonlinear functions). So, in the above elimination we could also have substituted Equation (i) into Equations (ii) and (iii); but we could not have substituted Equation (ii) into Equations (i) and (iii).
From Equation (v) we have

$$\omega = \frac{1}{r}\left(\frac{m}{k}\ddot{v} + v\right)$$

Differentiate:

$$\dot{\omega} = \frac{1}{r}\left(\frac{m}{k}\dddot{v} + \dot{v}\right)$$

Substitute these into Equation (iv), to eliminate ω and $\dot{\omega}$:

$$\frac{J}{r}\left(\frac{m}{k}\dddot{v} + \dot{v}\right) = T_m - rm(\dot{v} + g) - T_d\left(\frac{1}{r}\left(\frac{m}{k}\ddot{v} + v\right)\right) \tag{vi}$$

This is a third order model (because the highest derivative in Equation (vi) is third order).

d. At steady-state $\dot{v} = 0$. Hence $\ddot{v} = 0$ and $\dddot{v} = 0$ also. Substitute into Equation (vi), to get the steady-state (algebraic) equation:

$$\bar{T}_m - rmg - T_d\left(\frac{\bar{v}}{r}\right) = 0$$

where T $\bar{T}_m$ =steady-state value of the input T_m.

The solution of this nonlinear equation will give the steady-state operating speed $\bar{v}$ of the elevator.

Note: The same result may be obtained from the state Equations (i) through (iii). Specifically, under steady-state conditions:

$$0 = \bar{T}_m - r\bar{f} - T_d(\bar{\omega})$$

$$0 = k(r\bar{\omega} - \bar{v})$$

$$0 = \bar{f} - mg$$

This can be converted into a single equation, by eliminating $\bar{f}$ and $\bar{\omega}$.

e. Linearize Equation (vi) by writing the increment of each term (*Note:* all terms in Equation (vi) are linear; the increment of the constant term is zero):

$$\frac{J}{r}\left(\frac{m}{k}\ddot{\hat{v}} + \dot{\hat{v}}\right) = \hat{T}_m - r m\dot{\hat{v}} - b(\bar{\omega})\hat{\omega} \qquad \text{(vii)}$$

$$\text{where} \quad b(\bar{\omega}) = \frac{dT_m(\omega)}{d\omega}\bigg|_{\omega = \bar{\omega}}$$

Now from Equation (v):

$$m\dot{\hat{v}} = k(r\hat{\omega} - \hat{v}) \qquad \text{(viii)}$$

Substitute Equation (viii) into (vii), to eliminate $\hat{\omega}$. We get

$$\frac{J}{r}\left(\frac{m}{k}\ddot{\hat{v}} + \dot{\hat{v}}\right) = \hat{T}_m - r m\dot{\hat{v}} - b(\bar{\omega})\frac{1}{r}\left(\frac{m}{k}\ddot{\hat{v}} - \hat{v}\right)$$

or

$$\frac{J}{r}\frac{m}{k}\ddot{\hat{v}} + \frac{b(\bar{\omega})m}{rk}\ddot{\hat{v}} + \left(\frac{J}{r} + rm\right)\dot{\hat{v}} + \frac{b(\bar{\omega})}{r}\hat{v} = \hat{T}_m$$

f. Linearize Equations (i) through (iii) by writing the increment of each term in the equations (*Note:* all terms are linear; the increment of the constant term is zero):

$$J\dot{\hat{\omega}} = \hat{T}_m - r\hat{f} - b(\bar{\omega})\hat{\omega} \qquad \text{(ix)}$$

$$\dot{\hat{f}} = k(r\hat{\omega} - \hat{v}) \qquad \text{(x)}$$

$$m\dot{\hat{v}} = \hat{f} \qquad \text{(xi)}$$

$$\text{Output } y = \hat{v}$$

$$\text{Input } u = \hat{T}_m$$

$$\text{State vector } \mathbf{x} = \begin{bmatrix} \hat{\omega} & \hat{f} & \hat{v} \end{bmatrix}^T.$$

The corresponding model matrices are:

$$A = \begin{bmatrix} -\dfrac{b(\bar{\omega})}{J} & -\dfrac{r}{J} & 0 \\ rk & 0 & -k \\ 0 & \dfrac{1}{m} & 0 \end{bmatrix}; \quad B = \begin{bmatrix} 1/J \\ 0 \\ 0 \end{bmatrix}; \quad C = \begin{bmatrix} 0 & 0 & 1 \end{bmatrix}; \quad D = 0$$

Substitute Equation (xi) and its time derivative into Equations (ix) and (x), to eliminate $\hat{f}$ and $\dot{\hat{f}}$:

$$J\dot{\hat{\omega}} = \hat{T}_m - rm\dot{\hat{v}} - b(\bar{\omega})\hat{\omega}$$

$$m\ddot{\hat{v}} = k(r\hat{\omega} - \hat{v})$$

Now eliminate $\hat{\omega}$ by substituting the second equation into the first. We get the same result as before, for the input–output equation.

Example 3.4

A rocket-propelled spacecraft of mass m is fired vertically up (in the Y-direction) from the earth's surface (see Figure 3.7). The vertical distance of the centroid of the spacecraft, measured from the earth's surface, is denoted by y. The upward thrust force of the rocket is $f(t)$. The gravitational pull on the spacecraft is given by $mg\,[R/R+y]^2$, where g is the acceleration due to gravity at the earth's surface and R is the "average" radius of earth (about 6370 km). The magnitude of the aerodynamic drag force resisting the motion of the spacecraft is approximated by $k\dot{y}^2 e^{-y/r}$ where k and r are positive and constant parameters, and $\dot{y} = dy/dt$. Here, the exponential term represents the loss of air density at higher elevations.

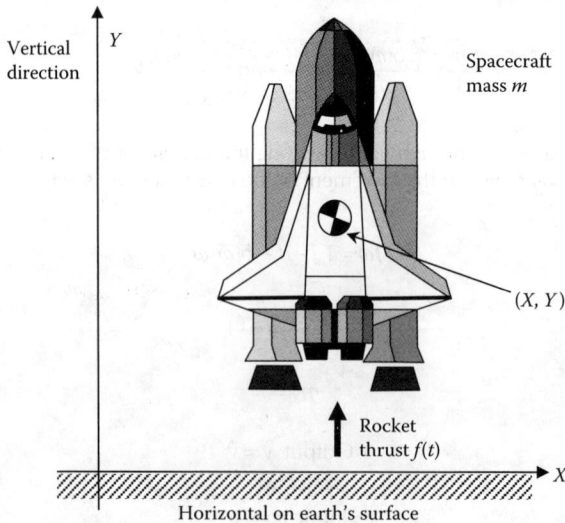

FIGURE 3.7
Coordinate system for the spacecraft problem.

a. Treating f as the input and y as the output, derive the input–output differential equation for the system.
b. The spacecraft accelerates to a height of y_o and then maintains a constant speed v_o, still moving in the same vertical (Y) direction. Determine an expression for the rocket force that is need for this constant-speed motion. Express your answer in terms of y_o, v_o, time t, and the system parameters m, g, R, r, and k. Show that this force decreases as the spacecraft ascends.
c. Linearize the input–output model (a) about the steady operating condition (b), for small variations $\hat{y}$ and $\dot{\hat{y}}$ in the position and speed of the spacecraft, due to a force disturbance $\hat{f}(t)$.
d. Treating y and $\dot{y}$ as state variables and y as the output, derive a complete (nonlinear) state-space model for the vertical dynamics of the spacecraft.
e. Linearize the state-space model in (d) about the steady conditions in (b) for small variations $\hat{y}$ and $\dot{\hat{y}}$ in the position and speed of the spacecraft, due to force disturbance $\hat{f}(t)$.
f. From the linear state model (e) derive the linear input–output model and show that the result is identical to what you obtained in (c).

Solution

a. Newton's second law in the Y-direction (constitutive equation):

$$m\ddot{y} = -mg\left(\frac{R}{R+y}\right)^2 - k|\dot{y}|\dot{y}e^{-y/r} + f(t) \tag{i}$$

b. At constant speed v_o we have,

$$\dot{y} = v_o \tag{ii}$$

$$\ddot{y} = \frac{d}{dt}v_o = 0 \tag{iii}$$

Integrate Equation (ii) and use the initial condition $y=y_o$ at $t=0$. Position under steady conditions:

$$\overline{y} = v_o t + y_o \tag{iv}$$

Substitute Equations (ii) and (iii) into Equation (i). The steady operating condition is given by:

$$0 = -\frac{mg}{\left(1+\dfrac{v_o t + y_o}{R}\right)^2} - k|v_o|v_o e^{-(v_o t + y_o)/r} + f_s(t)$$

where $f_s(t)$ is the force of rocket at constant speed v_o.
  Since v_o is positive, we have

$$f_s(t) = \frac{mg}{\left(1+\dfrac{v_o t + y_o}{R}\right)^2} + k v_o^2 e^{-(v_o t + y_o)/r}$$

It is seen that this expression decreases as t increases, reaching zero in the limit.

c. Derivatives needed for the linearization (O(1) Taylor series terms):

$$\frac{d}{dy}\frac{1}{\left(1+\dfrac{y}{R}\right)^2} = -\frac{2}{R}\frac{1}{\left(1+\dfrac{y}{R}\right)^3}$$

$$\frac{d}{dy}|\dot{y}|\dot{y}e^{-y/r} = -\frac{|\dot{y}|\dot{y}}{r}e^{-y/r}$$

$$\frac{d}{d\dot{y}}|\dot{y}|\dot{y}e^{-y/r} = 2|\dot{y}|e^{-y/r}$$

Note: These derivatives are determined as in Example 3.2.

The input–output Equation (i) is linearized by writing the increment of each term, with the use of the above derivatives (slopes). We get the linearized input–output model

$$m\ddot{\hat{y}} = \frac{2mg}{R}\frac{1}{\left(1+\dfrac{\bar{y}}{R}\right)^3}\hat{y} + \frac{k}{r}|\dot{y}|\dot{y}e^{-\bar{y}/r}\hat{y} - 2k|\dot{y}|e^{-\bar{y}/r}\dot{\hat{y}} + \hat{f}(t) \tag{va}$$

where $\bar{y}$ is as given by Equation (iv).

Since, under steady conditions, $\dot{y} = v_o > 0$ we have the linearized input–output model:

$$m\ddot{\hat{y}} = \frac{2mg}{R}\frac{1}{\left(1+\dfrac{\bar{y}}{R}\right)^3}\hat{y} + \frac{k}{r}v_o^2 e^{-\bar{y}/r}\hat{y} - 2kv_o e^{-\bar{y}/r}\dot{\hat{y}} + \hat{f}(t) \tag{v}$$

Note: Equation (v) represents an unstable system.

d. State vector $\mathbf{x} = \begin{bmatrix} x_1 & x_2 \end{bmatrix}^T = \begin{bmatrix} y & \dot{y} \end{bmatrix}^T$

From Equation (i), the state equations can be written as:

$$\dot{x}_1 = x_2 \tag{vi}$$

$$\dot{x}_2 = -\frac{g}{\left(1+\dfrac{x_1}{R}\right)^2} - \frac{k}{m}|x_2|x_2 e^{-x_1/r} + \frac{1}{m}f(t) \tag{vii}$$

The output equation is

$$y_1 = x_1$$

e. To linearize the state model in (d) we use the derivatives (local lopes) as before:

$$\frac{d}{dx_1}\frac{1}{\left(1+\dfrac{x_1}{R}\right)^2} = -\frac{2}{R}\frac{1}{\left(1+\dfrac{x_1}{R}\right)^3}$$

$$\frac{d}{dx_1}|x_2|x_2 e^{-x_1/r} = -\frac{|x_2|x_2}{r}e^{-x_1/r}$$

$$\frac{d}{dx_2}|x_2|x_2 e^{-x_1/r} = 2|x_2|e^{-x_1/r}$$

.

At the steady operating (constant speed) conditions:

$$\bar{x}_1 = v_o t + y_o \quad \text{and} \quad \bar{x}_2 = v_o > 0$$

Accordingly, the linearized state-space model is obtained by writing the increment of each term in Equations (vi) and (vii):

$$\dot{\hat{x}}_1 = \hat{x}_2 \tag{viii}$$

$$\dot{\hat{x}}_2 = \frac{2g}{R\left(1 + \dfrac{\bar{x}_1}{R}\right)^3}\,\hat{x}_1 + \frac{k}{mr}v_o^2 e^{-(\bar{x}_1)/r}\,\hat{x}_1 - 2\frac{k}{m}v_o e^{-\bar{x}_1/r}\,\hat{x}_2 + \frac{1}{m}\hat{f}(t) \tag{ix}$$

with the output equation

$$\hat{y}_1 = \hat{x}_1 \tag{x}$$

f. Substitute Equation (viii) into (ix). We get

$$\ddot{\hat{x}}_1 = \frac{2mg}{R\left(1 + \dfrac{\bar{x}_1}{R}\right)^3}\,\hat{x}_1 + \frac{k}{mr}v_o^2 e^{-(\bar{x}_1)/r}\,\hat{x}_1 - 2\frac{k}{m}v_o e^{-\bar{x}_1/r}\,\dot{\hat{x}}_1 + \frac{1}{m}\hat{f}(t)$$

This result is identical to Equation (v), since $\hat{x}_1 = \hat{y}$.

3.3 Nonlinear Electrical Elements

The three lumped-parameter passive elements in an electrical system are: capacitor (an *A*-type of element with the *across variable* voltage as the state variable); inductor (a *T*-type element with the *through variable* current as the state variable); and resistor (a *D*-type element representing energy dissipation, and no specific state variable is associated with it). The linear versions of these elements were introduced in Chapter 2. Now let us briefly look into the general, nonlinear versions of these elements.

3.3.1 Capacitor

Electrical charge (q) is a function of the voltage (v) across a capacitor, as given by the non-linear constitutive equation:

$$q = q(v) \tag{3.23a}$$

For the linear case we have

$$q = Cv \tag{3.23b}$$

where C is the *capacitance*. Then the current i, which is given by dq/dt, is expressed by differentiating Equation 3.23b we get

$$i = C\frac{dV}{dt} + V\frac{dC}{dt} \qquad (3.24a)$$

where we have allowed for a time-varying capacitance. If C is assumed constant, we have the familiar linear constitutive equation:

$$i = C\frac{dV}{dt} \qquad (3.24b)$$

3.3.2 Inductor

Magnetic flux linkage (λ) of an inductor is a function of the current (i) through the inductor, as given by the nonlinear constitutive equation:

$$\lambda = \lambda(i) \qquad (3.25a)$$

For the linear case we have

$$\lambda = L\,i \qquad (3.25b)$$

where L is the *inductance*. The voltage induced in an inductor is equal to the rate of change of the flux linkage. Hence, by differentiating Equation 3.25b we get

$$V = L\frac{di}{dt} + i\frac{dL}{dt} \qquad (3.26a)$$

Assuming that the inductance is constant, we have the familiar linear constitutive equation:

$$V = L\frac{di}{dt} \qquad (3.26b)$$

3.3.3 Resistor

In general the voltage across a (nonlinear) resistor is a function of the current through the resistor, as given by

$$v = v(i) \qquad (3.27a)$$

In the linear case, we have the familiar Ohm's law:

$$v = R\,i \qquad (3.27b)$$

where R is the *resistance*, which can be time-varying in general. In most cases, however, we assume R to be a constant.

3.4 Linearization Using Experimental Operating Curves

Linearization of analytical models was studied in the previous sections. In some situations an accurate analytical model may not be readily available for an existing physical system. Yet experiments may be conducted on the system to gather operating curves for the system. These operating curves are useful in deriving a linear model, which can be valuable, for example, in controlling the system. This approach is discussed now, taking an electric motor as the example system.

3.4.1 Torque-Speed Curves of Motors

The speed versus torque curves of motors under steady conditions (i.e., steady-state operating curves) are available from the motor manufacturers. These curves have the characteristic shape that they decrease slowly up to a point and then drop rapidly to zero. An example of an ac induction motor is given in Figure 3.3. The operating curves of dc motors take a similar, not identical characteristic form, as shown in Figure 3.8. The shape of the operating curve depends how the motor windings (rotor and stator) are connected and excited. The torque at zero speed is the "braking torque" or "starting torque" or "stalling torque." The speed at zero torque is the "no-load speed" which, for an ac induction motor, is also the "synchronous speed."

Typically, these experimental curves are obtained as follows. The supply voltage to the motor windings is maintained constant; a known load (torque) is applied to the motor shaft; and once the conditions become steady (i.e., constant speed) the motor speed is measured. The experiment is repeated for increments of torques within an appropriate range. This gives one operating curve, for a specified supply voltage. The experiment is repeated for other supply voltages and a series of curves are obtained.

It should be noted that the motor speed is maintained steady in these experiments as they represent "steady" operating conditions. That means the motor inertia (inertial torque) is not accounted for in these curves, while mechanical damping is. Hence, motor inertia has to be introduced separately when using these curves to determine a "dynamic" model for a motor. Since mechanical damping is included in the measurements, it should not be introduced again. Of course, if the motor is connected to an external load, the damping, inertia, and flexibility of the load all have to be accounted for separately when using experimental operating curves of motors in developing models for motor-integrated dynamic systems.

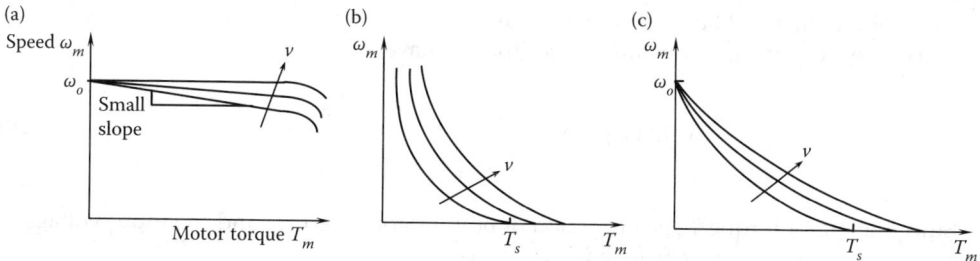

FIGURE 3.8
Torque-speed operating curves of dc motors. (a) Shunt-wound. (b) Series-wound. (c) Compound-wound.

Motor torque

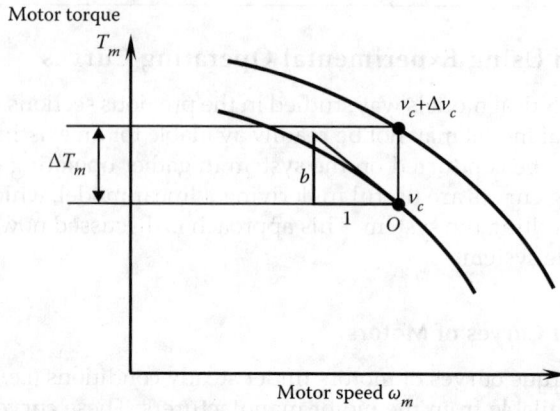

FIGURE 3.9
Two steady-state operating curves of a motor at constant input voltage.

3.4.2 Linear Models for Motor Control

Consider an experimental set of steady-state operating curves for a motor, each obtained at a constant supply/control voltage. In particular consider one curve measured at voltage v_c and the other measured at voltage $v_c + \Delta v_c$, where ΔT_m is the voltage increment from one operating to the other, as shown in Figure 3.9.

Draw a tangent to the first curve at a selected point (operating point O). The slope of the tangent is negative, as shown. Its magnitude b is given by

$$\text{Damping constant } b = -\frac{\partial T_m}{\partial \omega_m}\bigg|_{v_c = \text{constant}} = \text{slope at } O \qquad (3.28)$$

It should be clear that b represents an equivalent rotary damping constant (torque/angular speed) that includes both electro-magnetic and mechanical damping effects in the motor. The included mechanical damping comes primarily from the friction of the motor bearings and aerodynamic effects. Since a specific load is not considered in the operating curve, load damping is not included.

Draw a vertical line through the operating point O to intersect the other operating curve. We get:

ΔT_m = torque intercept between the two curves

Since a vertical line is a constant speed line, we have

$$\text{Voltage gain } k_v = \frac{\partial T_m}{\partial v_c}\bigg|_{\omega_m = \text{constant}} = \frac{\Delta T_m}{\Delta v_c} \qquad (3.29)$$

Since the motor torque T_m is a function of both motor speed ω_m and the input voltage v_c (i.e., $T_m = T_m(\omega_m, v_c)$) we write from basic calculus:

$$\delta T_m = \frac{\partial T_m}{\partial \omega_m}\bigg|_{v_c} \delta \omega_m + \frac{\partial T_m}{\partial v_c}\bigg|_{\omega_m} \delta v_c \qquad (3.30a)$$

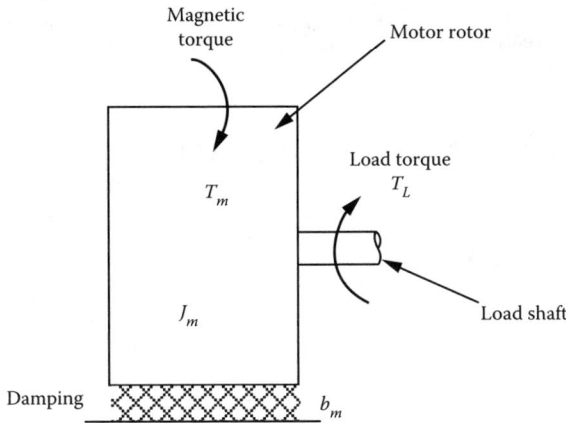

FIGURE 3.10
Mechanical system of the motor.

or

$$\delta T_m = -b\,\delta\omega_m + k_v\,\delta v_c \qquad (3.30)$$

where the motor damping constant b and the voltage gain k_v are given by Equations 3.28 and 3.29, respectively.

Equation 3.30 represents a linearized model of the motor. The torque needed to drive rotor inertia of the motor is not included in this equation (because steady-state curves are used in determining parameters). The inertia term should be explicitly present in the mechanical equation of the motor rotor, as given by Newton's second law (see Figure 3.10), in the linearized (incremental) form:

$$J_m \frac{d\,\delta\omega_m}{dt} = \delta T_m - \delta T_L \qquad (3.31)$$

where J_m is the moment of inertia of the motor rotor and T_L is the load torque (equivalent torque applied on the motor by the load that is driven by the motor).

Note that mechanical damping of the motor, as shown in Figure 3.10, is not included in Equation 3.31 because it (and electro-magnetic damping) is already included in Equation 3.30.

Problems

PROBLEM 3.1

What precautions may be taken in developing and operating a mechanical system, in order to reduce system nonlinearities?

Read about the followings nonlinear phenomena:

   i. Saturation
   ii. Hysteresis

iii. Jump phenomena
iv. Frequency creation
 v. Limit cycle
vi. Deadband

Two types of nonlinearities are shown in Figure P3.1.

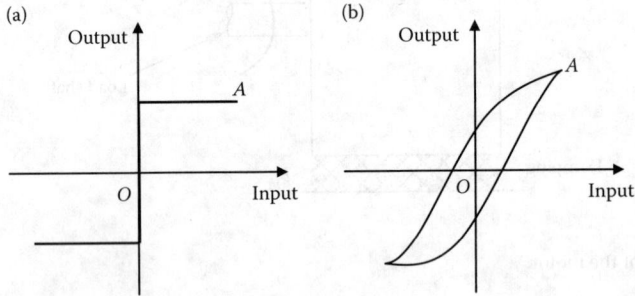

FIGURE P3.1
Two types of nonlinearities. (a) Ideal saturation. (b) Hysteresis.

In each case indicate the difficulties of developing an analytical for operation near:

   i. Point O
  ii. Point A

PROBLEM 3.2
An excitation was applied to a system and its response was observed. Then the excitation was doubled. It was found that the response also doubled. Is the system linear?

PROBLEM 3.3
   a. Determine the derivative $\dfrac{d}{dx} x|x|$.
   b. Linearize the following terms about the operating point $\bar{x} = 2$:

   (i) $3x^3$
   (ii) $|x|$
   (iii) $\dot{x}^2$

PROBLEM 3.4
A nonlinear device obeys the relationship $y = y(u)$ and has an operating curve as shown in Figure P3.4.

   i. Is this device a dynamic system?
A linear model of the form $y = ku$ is to be determined for operation of the device:
   ii. In a small neighborhood of point B
   iii. Over the entire range from A to B.
Suggest a suitable value for k in each case.

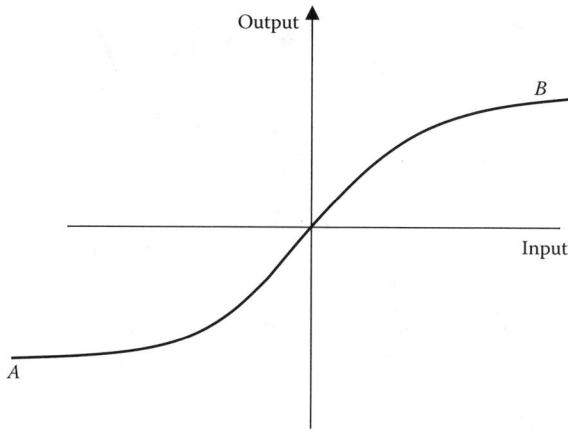

FIGURE P3.4
The characteristic curve of a nonlinear device.

PROBLEM 3.5

A nonlinear damper is connected to a mechanical system as shown in Figure P3.5. The force f, which is exerted by the damper on the system, is $c(v_2 - v_1)^2$ where c is a constant parameter.

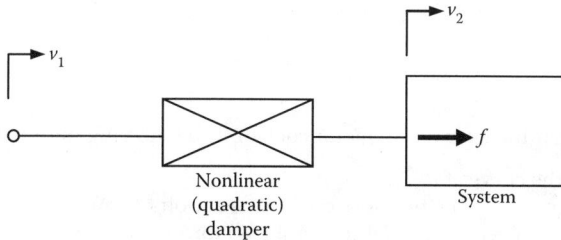

FIGURE P3.5
A nonlinear mechanical system.

- i. Give an analytical expression for f in terms of v_1, v_2, and c, which will be generally valid.
- ii. Give an appropriate linear model.
- iii. If the operating velocities v_1 and v_2 are equal, what will be the linear model about (in the neighborhood of) this operating point?

PROBLEM 3.6

Suppose that a system is in equilibrium under the forces F_i and F_o as shown in Figure P3.6. If the point of application of F_i is given a small "virtual" displacement x in the same direction as the force, suppose that the location of F_o moves through $y = k\,x$ in the opposite direction to F_o.

- i. Determine F_o in terms of F_i (this is a result of the "principle of virtual work").
- ii. What is the relationship between the small changes $\hat{F}_i$ and $\hat{F}_o$, about the operating conditions $\bar{F}_i$ and $\bar{F}_o$, assuming that the system is in equilibrium?

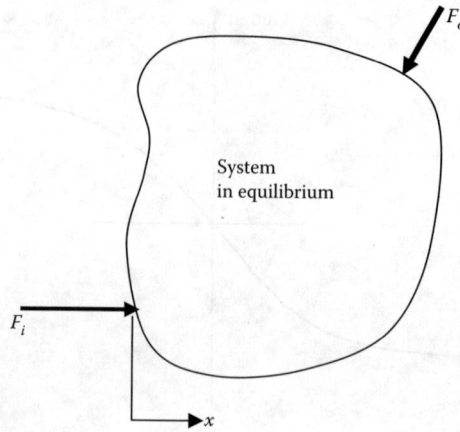

FIGURE P3.6
Virtual displacement of a system in equilibrium.

PROBLEM 3.7

Characteristic curves of an armature-controlled dc motor are as shown in Figure P3.7. These are torque versus speed curves, measured at a constant armature voltage, at steady-state. For the neighborhood of point P, a linear model of the form

$$\hat{\omega} = k_1 \hat{v} + k_2 \hat{T}$$

needs to be determined, for use in motor control. The following information is given:

The slope of the curve at $P = -a$.
The voltage change in the two adjacent curves at point $P = \Delta V$.
Corresponding speed change (at constant load torque through P) $= \Delta \omega$.
Estimate the parameters k_1 and k_2.

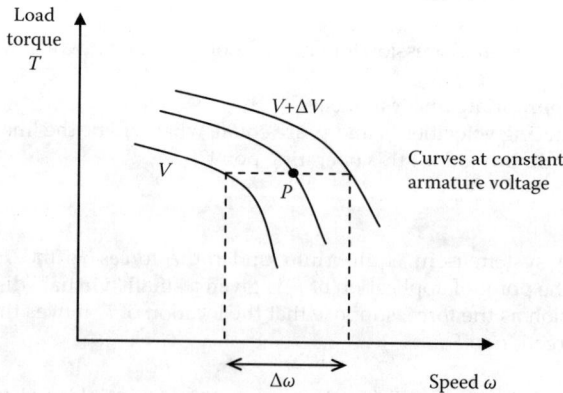

FIGURE P3.7
Characteristic curves of an armature-controlled dc motor.

PROBLEM 3.8

An air circulation fan system of a building is shown in Figure P3.8a, and a simplified model of the system may be developed, as represented in Figure P3.8b.

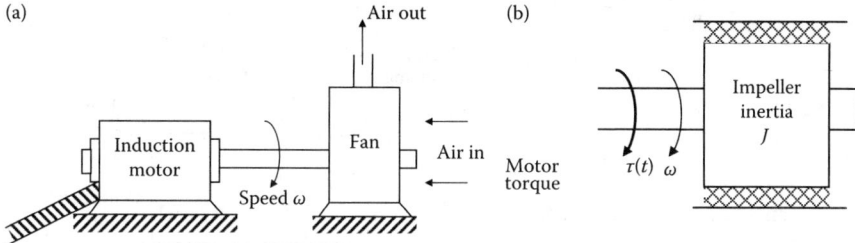

FIGURE P3.8
(a) A motor-fan combination of a building ventilation system. (b) A simplified model of the ventilation fan.

The induction motor is represented as a torque source $\tau(t)$. The speed ω of the fan, which determines the volume flow rate of air, is of interest. The moment of inertia of the fan impeller is J. The energy dissipation in the fan is modeled by a linear viscous damping component (of damping constant b) and a quadratic aerodynamic damping component (of coefficient d).

a. Show that the system equation may be given by

$$J\dot{\omega} + b\omega + d|\omega|\omega = \tau(t)$$

b. Suppose that the motor torque is given by

$$\tau(t) = \bar{\tau} + \hat{\tau}_a \sin \Omega t$$

in which $\bar{\tau}$ is the steady torque and $\hat{\tau}_a$ is a very small amplitude (compared to $\bar{\tau}$) of the torque fluctuations at frequency Ω. Determine the steady-state operating speed $\bar{\omega}$ (which is assumed positive) of the fan.

c. Linearize the model about the steady-state operating conditions and express it in terms of the speed fluctuations $\hat{\omega}$. From this, estimate the amplitude of the speed fluctuations.

PROBLEM 3.9

a. Linearized models of nonlinear systems are commonly used in model-based control of processes. What is the main assumption that is made in using a linearized model to represent a nonlinear system?

b. A three-phase induction motor is used to drive a centrifugal pump for incompressible fluids. To reduce misalignment and associated problems such as vibration, noise, and wear, a flexible coupling is used for connecting the motor shaft to the pump shaft. A schematic representation of the system is shown in Figure P3.9.

Assume that the motor is a "torque source" of torque T_m, which is being applied to the motor of inertia J_m. Also, the following variables and parameters are defined for the system:

J_p =moment of inertia of the pump impeller assembly
Ω_m =angular speed of the motor rotor/shaft

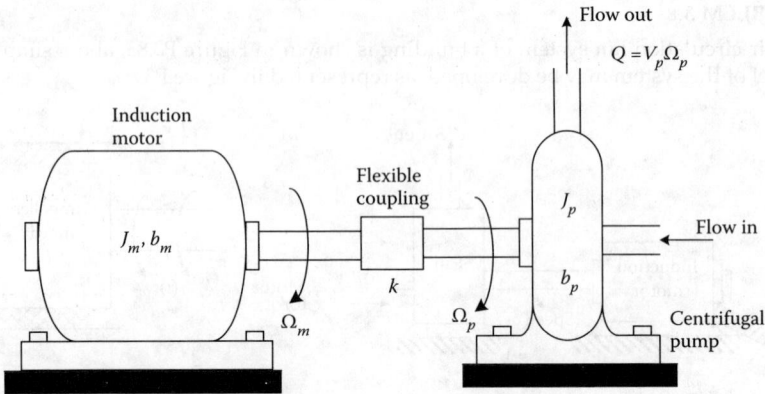

FIGURE P3.9
A centrifugal pump driven by an inductor motor.

Ω_p =angular speed of the pump impeller/shaft
k =torsional stiffness of the flexible coupling
T_f =torque transmitted through the flexible coupling
Q =volume flow rate of the pump
b_m =equivalent viscous damping constant of the motor rotor.

Also, assume that the net torque required at the ump shaft, to pump fluid steadily at a volume flow rate Q, is given by $b_p\Omega_p$, where $Q=V_p\Omega_p$ and V_p=volumetric parameter of the pump (assumed constant).

Using T_m as the input and Q as the output of the system, develop a complete state-space model for the system. Identify the model matrices A, B, C, and D in the usual notation, in this model. What is the order of the system?

c. In (a) suppose that the motor torque is given by

$$T_m = \frac{aSV_f^2}{[1+(S/S_b)^2]}$$

where the fractional slip S of the motor is defined as

$$S = 1 - \frac{\Omega_m}{\Omega_s}$$

Note that a and S_b are constant parameters of the motor. Also,

Ω_s=no-load (i.e., synchronous) speed of the motor
V_f=amplitude of the voltage applied to each phase winding (field) of the motor.

In *voltage control* V_f is used as the input, and in *frequency control* Ω_s is used as the input. For combined voltage control and frequency control, derive a linearized state-space model, using the incremental variables $\hat{V}_f$ and Ω_s, about the operating values $\overline{V}_f$ and Ω_{sy} as the inputs to the system, and the incremental flow $\hat{Q}$ as the output.

PROBLEM 3.10

a. What are *A*-type elements and what are *T*-type elements?
   Classify mechanical inertia, mechanical spring, fluid inertia and fluid capacitor into these two types. Explain a possible conflict that can arise due to this classification.
b. A system that is used to pump an incompressible fluid from a reservoir into an open overhead tank is schematically shown in Figure P3.10. The tank has a uniform across section of area A.

The pump is considered a pressure source of pressure difference $P(t)$. A valve of constant k_v is placed near the pump in the long pipe line, which leads to the overhead tank. The valve equation is $Q=k_v\sqrt{P_1-P_2}$ in which Q is the volume flow rate of the fluid. The resistance to the fluid flow in the pipe may be modeled as $Q=k_p\sqrt{P_2-P_3}$ in which k_p is a pipe flow constant. The effect of the accelerating fluid is represented by the linear equation $I(dQ/dt)=P_3-P_4$ in which I denotes the fluids inertance. Pressures P_1, P_2, P_3, and P_4 are as marked along the pipe length, in Figure P3.10. Also P_0 denotes the ambient pressure.

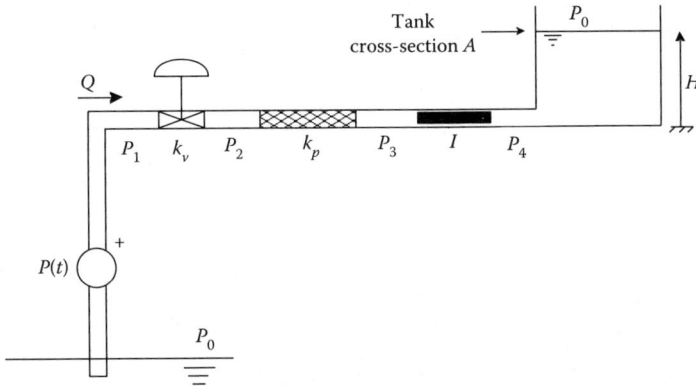

FIGURE P3.10
A pumping system for an overhead tank.

i. Using Q and P_{40} as the state variables, the pump pressure $P(t)$ as the input variable, and the fluid level H in the tank as the output variable, obtain a complete (nonlinear) state-space model for the system. *Note*: $P_{40}=P_4-P_0$. Density of the fluid $=\rho$.
ii. Linearize the state equations about an operating point given by flow rate $\bar{Q}$. Determine the model matrices A, B, C, and D for the linear model.
iii. What is the combined linear resistance of the valve and piping?

PROBLEM 3.11

List several response characteristics of nonlinear systems that are not exhibited by linear systems in general. Also, determine the response y of the nonlinear system

$$\left[\frac{dy}{dt}\right]^{1/3}=u(t)$$

when excited by the input $u(t)=a_1\sin\omega_1 t+a_2\sin\omega_2 t$.
What characteristic of a nonlinear system does this result illustrate?

PROBLEM 3.12

Consider a mechanical component whose response x is governed by the relationship

$$f = f(x, \dot{x})$$

where f denotes the applied (input) force and $\dot{x}$ denotes velocity. The following three special cases are considered:

a. Linear spring: $f = kx$
b. Linear spring with a viscous (linear) damper: $f = kx + b\dot{x}$
c. Linear spring with Coulomb friction: $f = kx + f_c \text{sgn}(\dot{x})$

Suppose that a harmonic excitation of the form $f = f_o \sin \omega t$ is applied in each case. Sketch the force–displacement curves for the three cases at steady-state. Which components exhibit hysteresis? Which components are nonlinear? Discuss your answers.

4

Linear Graphs

Among the graphical tools for developing and representing a model of a dynamic system, linear graphs take an important place. In particular, state-space models of lumped-parameter dynamic systems; regardless of whether they are mechanical, electrical, fluid, thermal, or multidomain (mixed) systems; can be conveniently developed by *linear graphs*. Interconnected line segments (called branches) are used in a linear graph to represent a dynamic model. The term "linear graph" stems from this use of line segments, and does not mean that the system itself has to be linear. Particular advantages of using linear graphs for model development and representation are: they allow visualization of the system structure (even before formulating an analytical model); they help identify similarities (structure, performance, etc.) in different types of systems; they are applicable for multidomain systems (the same approach is used in any domain); and they provide a unified approach to model multifunctional devices (e.g., a piezoelectric device which can function as both a sensors and an actuator). This chapter presents the use of linear graphs in the development of analytical models for mechanical, electrical, fluid, and thermal systems.

4.1 Variables and Sign Convention

Linear graphs systematically use through variables and across variables in providing a unified approach for the modeling of dynamic systems in multiple domains (mechanical, electrical, fluid, thermal). In accomplishing this objective it is important to adhere to standard and uniform conventions so that that there will not be ambiguities in a given linear graph representation. In particular, a standard sign convention must be established. These issues are discussed in this section.

4.1.1 Through Variables and Across Variables

Each branch in the linear graph model has one *through variable* and one *across variable* associated with it. Their product is the power variable. For instance, in a hydraulic or pneumatic system, a pressure "across" an element causes some change of fluid flow "through" the element. The across variable is the pressure, the through variable is the flow. Table 4.1 lists the through and across variable pairs for the four domains considered in the present treatment.

4.1.2 Sign Convention

Consider Figure 4.1 where a general basic element (strictly, a single-port element, as discussed later) of a dynamic system is shown. In the linear graph representation, as shown in Figure 4.1b, the element is shown as a branch (i.e., a line segment). One end of any branch is

TABLE 4.1

Through and Across Variable Pairs in Several Domains

System Type (domain)	Through Variable	Across Variable
Hydraulic/pneumatic	Flow rate	Pressure
Electrical	Current	Voltage
Mechanical	Force/torque	Velocity/angular velocity
Thermal	Heat transfer	Temperature

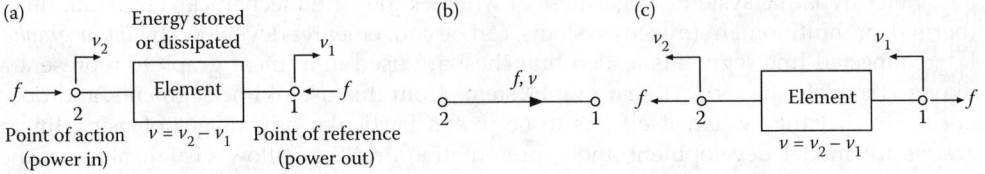

FIGURE 4.1

Sign convention for a linear graph. (a) A basic element and positive directions of its variables. (b) Linear graph branch of the element. (c) An alternative sign convention.

selected as the *point of reference* and the other end automatically becomes the *point of action* (see Figure 4.1a and c). The choice is somewhat arbitrary, and may reflect the physics of the actual system. An *oriented* branch is one to which a direction is assigned, using an arrowhead, as in Figure 4.1b. The arrow head denotes the positive direction of power flow at each end of the element. By convention, the positive direction of power is taken as "into" the element at the point of action, and "out of" the element at the point of reference. According to this convention, the arrowhead of a branch is always pointed toward the point of reference. In this manner the reference point and the action point are easily identified.

The across variable is always given relative to the point of reference. It is also convenient to give the through variable f and the across variable v as an ordered pair $(f\, v)$ on one side of the branch, as in Figure 4.1b. Clearly, the relationship between f and v (the constitutive relation or physical relation, as discussed in Chapters 2 and 3) can be linear or nonlinear. The parameter of the element (e.g., mass, capacitance) is shown on the other side of the branch. It should be noted that the direction of a branch does not represent the positive direction of f or v. For example, when the positive directions of both f and v are changed, as in Figure 4.1a and c, the linear graph remains unchanged, as in Figure 4.1b, because the positive direction of power flow remains the same. In a given problem, the positive direction of any one of the two variables f and v should be preestablished for each branch. Then the corresponding positive direction of the other variable is automatically determined by the convention used to orient linear graphs. It is customary to assign the same positive direction for f (and v) and the power flow at the point of action (i.e., the convention shown in Figure 4.1a is customary, not Figure 4.1c). Then the positive directions of the variables at the point of reference are automatically established.

Note that in a branch (line segment), the through variable (f) is transmitted through the element with no change in value; it is the "through" variable. The absolute value of the across variable, however, changes across the element (from v_2 to v_1, in Figure 4.1a). In fact, it is this change ($v = v_2 - v_1$) across the element that is called the across variable. For example, v_2 and v_1 may represent electric potentials at the two ends of an electric element (e.g., a resistor) and then v represents the voltage across the element. According, the across variable, is measured relative to the point of reference of the particular element.

According to the sign convention shown in Figure 4.1, the work done on the element at the point of action (by an external device) is positive (i.e., power flows in), and work done by the element at the point of reference (on an external load or environment) is positive (i.e., power flows out). The difference in the work done on the element and the work done by the element (i.e., the difference in the work flow at the point of action and the point of reference) is either stored as energy (e.g., kinetic energy of a mass; potential energy of a spring; electrostatic energy of a capacitor; electromagnetic energy of an inductor), which has the capacity to do additional work; or dissipated (e.g., mechanical damper; electrical resistor) through various mechanisms manifested as heat transfer, noise, and other phenomena.

In summary:

1. An element (a single-port element) is represented by a line segment (branch). One end is the point of action and the other end is the point of reference.

2. The through variable f is the same at the point of action and the point of reference of an element; the across variable differs, and it is this difference (value relative to the point of reference) that is called the across variable v.

3. The variable pair (f, v) of the element is shown on one side of the branch. Their relationship (constitutive relation) can be linear or nonlinear. The parameter of the element is shown on the other side of the branch.

4. Power flow p is the product of the through variable and the across variable. By convention, at the point of action, f and p are taken to be positive in the same direction; at the point of reference, f is positive in the opposite direction.

5. The positive direction of power flow p (or energy or work) is into the element at the point of action; and out of the element at the point of reference. This direction is shown by an arrow on the linear graph branch (an oriented branch).

6. The difference in the energy flows at the two ends of the element is either stored (with capacity to do further work) or dissipated, depending on the element type.

Linear graph representation is particularly useful in understanding rates of energy transfer (power) associated with various phenomena, and dynamic interactions in a physical system (mechanical, electrical, fluid, etc.) can be interpreted in terms of power transfer. Power is the product of a through variable (a generalized force or current) and the corresponding across variable (a generalized velocity or voltage). For example, consider a mechanical system. The total work done on the system is, in part, used as stored energy (kinetic and potential) and the remainder is dissipated. Stored energy can be completely recovered when the system is brought back to its original state (i.e., when the cycle is completed). Such a process is *reversible*. On the other hand, dissipation corresponds to irreversible energy transfer that cannot be recovered by returning the system to its initial state. (A fraction of the mechanical energy lost in this manner could be recovered, in principle, by operating a heat engine, but we shall not go into these thermodynamic details which are beyond the present scope). Energy dissipation may appear in many forms including temperature rise (a molecular phenomenon), noise (an acoustic phenomenon), or work used up in wear mechanisms.

4.2 Linear Graph Elements

Many types of basic elements exist, which can be used in the development of a linear graph for a dynamic system. In this section we will discuss two types of basic elements in the categories of single-port elements and two-port elements. Analogous elements in these categories exist across the domains (mechanical, electrical, fluid, and thermal) for the most part.

4.2.1 Single-Port Elements

Single-port (or, *single energy port)* elements are those that can be represented by a single branch (single line segment) of linear graph. These elements possess only one power (or energy) variable; hence the name "single-port." They have two terminals. The general form of these elements is shown in Figure 4.1b.

In modeling mechanical systems we require three passive single-port elements, as shown in Figure 4.2. These lumped-parameter mechanical elements are mass (or inertia), spring, and dashpot/damper. Although translatory mechanical elements are presented in Figure 4.2, corresponding rotary elements are easy to visualize: f denotes an applied torque and v the relative angular velocity in the same direction. Note that the linear graph of an inertia element has a broken line segment. This is because, through an inertia, the force does not physically travel from one end of its linear graph branch to the other end, but rather the force "felt" at the two ends. This will be further discussed in Example 4.1.

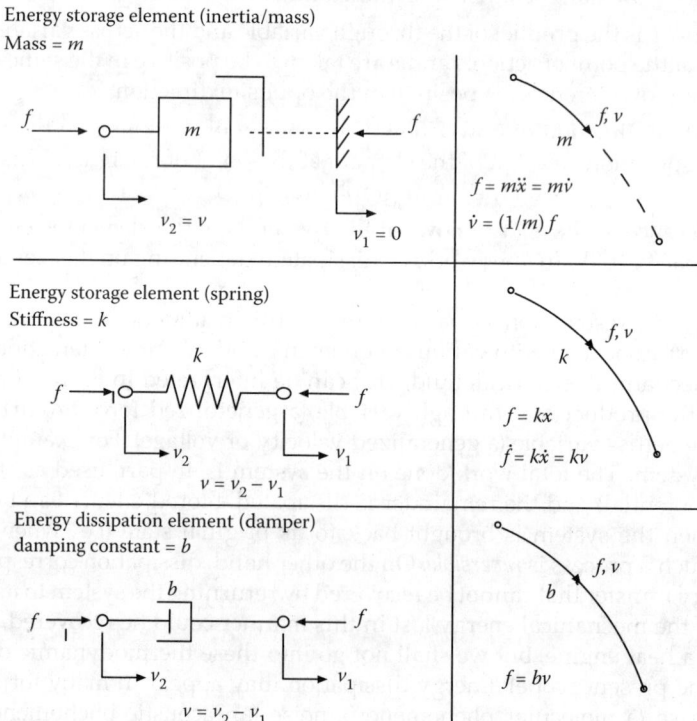

Energy storage element (inertia/mass)
Mass = m

$v_2 = v$ $v_1 = 0$

$f = m\ddot{x} = m\dot{v}$
$\dot{v} = (1/m)f$

Energy storage element (spring)
Stiffness = k

$v = v_2 - v_1$

$f = kx$
$\dot{f} = k\dot{x} = kv$

Energy dissipation element (damper)
damping constant = b

$v = v_2 - v_1$

$f = bv$

FIGURE 4.2
Single-port mechanical system elements and their linear-graph representations.

Energy storage element (capacitor)
Capacitance = C

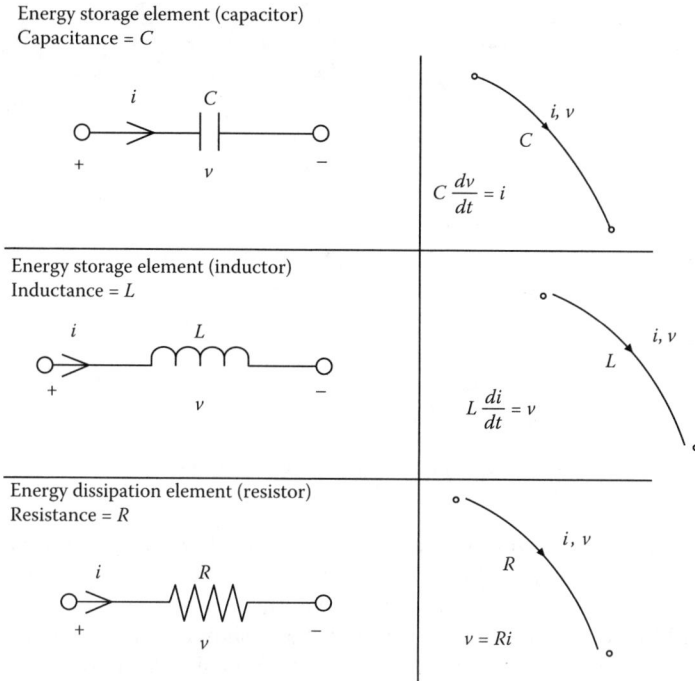

$$C\frac{dv}{dt} = i$$

Energy storage element (inductor)
Inductance = L

$$L\frac{di}{dt} = v$$

Energy dissipation element (resistor)
Resistance = R

$$v = Ri$$

FIGURE 4.3
Single-port electrical system elements and their linear-graph representations.

Analogous single-port electrical elements may be represented in a similar manner. These are shown in Figure 4.3.

4.2.1.1 Source Elements

In linear graph models, system *inputs* are represented by *source elements*. There are two types of sources, as shown in Figure 4.4.

a. *T*-type source (e.g., force source, current source):
The independent variable (i.e., the source output, which is the system input) is the through variable f. The arrow head indicates the positive direction of f.

   Note: For a *T*-type source, the sign convention that the arrow gives the positive direction of f still holds. However, the sign convention that the arrow is from the point of action to the point of reference (or the direction of the drop in the across variable) does not hold.

b. *A*-type source (e.g., velocity source, voltage source):
The independent variable is the across variable v. The arrow head indicates the positive direction of the "drop" in v. *Note*: + and − signs are also indicated, where the drop in v occurs from + to − terminals.

   Note: For an *A*-type source, the sign convention that the arrow is from the point of action to the point of reference (or the direction of the drop in the across variable) holds.

   However, the sign convention that the arrow gives the positive direction of f does not hold.

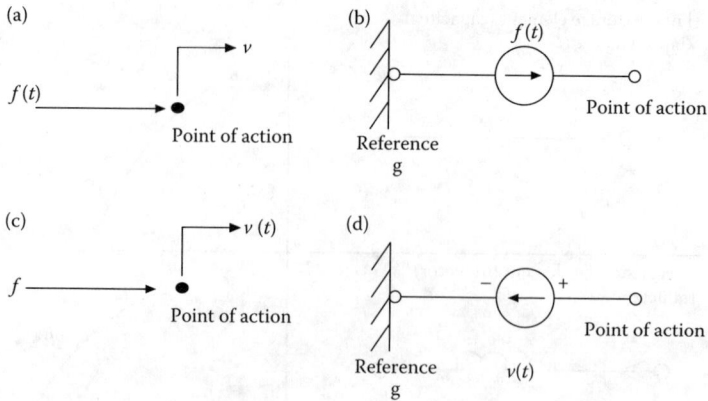

FIGURE 4.4
(a) *T*-type source (through variable input). (b) Linear graph representation of a *T*-type source. (c) *A*-type source. (d) Linear graph representation of an *A*-type source.

An ideal force source (a through-variable source) is able to supply a force input that is not affected by interactions with the rest of the system. The corresponding relative velocity across the force source, however, will vary as determined by the dynamics of the overall system. It should be clear that the direction of $f(t)$ as shown in Figure 4.4a is the applied force. The reaction on the source would be in the opposite direction. An ideal velocity source (across-variable source) supplies a velocity input independent of the system to which it is applied. The corresponding force is, of course, determined by the system dynamics.

4.2.1.2 Effects of Source Elements

We have noted that the source variable (independent variable or input variable) of a source is unaffected by the dynamics of the system to which the source is connected. But the covariable (dependent variable) will change. Another property associated with source elements is identified next.

Source elements can serve as a means of inhibiting interactions between systems. Specifically, it follows from the definition of an ideal source that the dynamic behavior of a system is not affected by connecting a new system in series with an existing *T*-type source (e.g., force source or current source) or in parallel with an existing *A*-type source (e.g., velocity source or voltage source). Conversely, then, the original system is not affected by removing the connected new system, in each case. These two situations are illustrated in Figure 4.5. In general, linking (networking) a subsystem will change the order of the overall system (because new dynamic interactions are introduced) although the two situations in Figure 4.5 are examples where this does not happen. Another way to interpret these situations is to consider the original system and the new system as two uncoupled subsystems driven by the same input source. In this sense, the order of the overall system is the sum of the order of the individual (uncoupled) subsystems.

4.2.2 Two-Port Elements

A two-port element has two energy ports and two separate, yet coupled, branches corresponding to them. These elements can be interpreted as a pair of single-port elements whose net power is zero. A transformer (mechanical, electrical, fluid, etc.) is a two-port

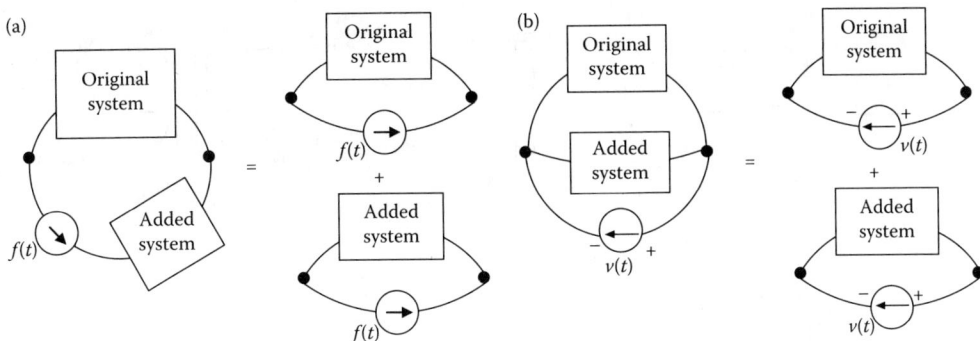

FIGURE 4.5
(a) Two systems connected in series to a *T*-type source. (b) Two systems connected in parallel to an *A*-type source.

element. Also, a mechanical gyrator is a two-port element. Examples of mechanical transformers are a lever and pulley for translatory motions and a meshed pair of gear wheels for rotation. A gyrator is typically an element that displays gyroscopic properties. We shall consider only the linear case; i.e., ideal transformers and gyrators only. The extension to the nonlinear case should be clear.

4.2.2.1 Transformer

In an ideal transformer, the across variables in the two ports (branches) are changed without dissipating or storing energy in the process. Hence the through variables in the two ports will also change. Examples of mechanical, electrical, and fluid transformers are shown in Figure 4.6a through d. The linear graph representation of a transformer is given in Figure 4.6e.

In Figure 4.6e, as for a single-port passive element, the arrows on the two branches (line segments) give the positive direction of power flow (i.e., when the product of the through variable and the across variable for that segment is positive). One of the two ports may be considered the input port and the other the output port. Let

$$v_i \text{ and } f_i = \text{across and through variables at the input port}$$

$$v_o \text{ and } f_o = \text{across and through variables at the output port}$$

The (linear) transformation ratio r of the transformer is given by

$$v_o = r\, v_i \tag{4.1}$$

Due to the conservation of power we have:

$$f_i v_i + f_o v_o = 0 \tag{4.2}$$

By substituting Equation 4.1 into Equation 4.2 gives

$$f_o = -\frac{1}{r} f_i \tag{4.3}$$

Here r is a dimensional parameter. The two constitutive relations for a transformer are given by Equations 4.1 and 4.3.

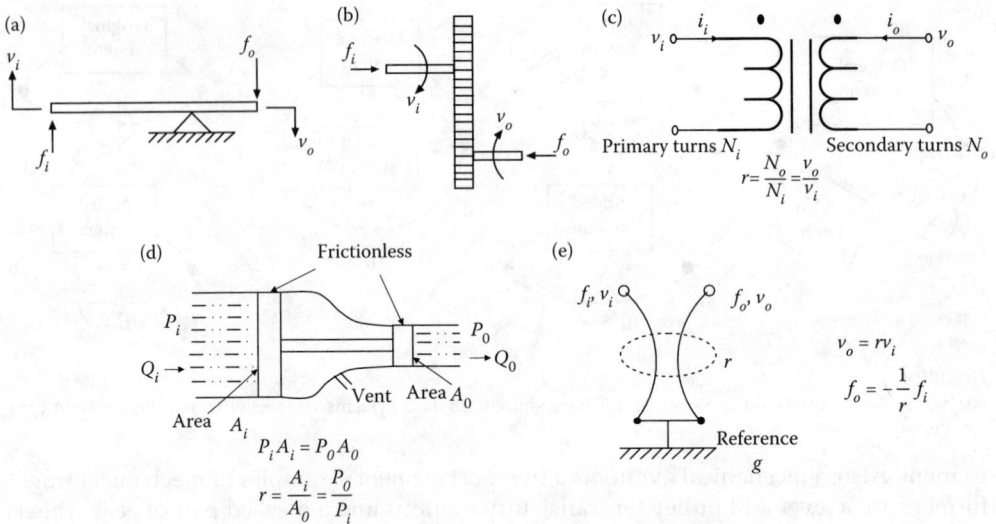

FIGURE 4.6
Transformer. (a) Lever. (b) Meshed gear wheels. (c) Electrical transformer. (d) Fluid transformer. (e) Linear graph representation.

4.2.2.2 Electrical Transformer

As shown in Figure 4.6c, an electrical transformer has a primary coil, which is energized by an ac voltage (v_i), a secondary coil in which an ac voltage (v_o) is induced, and a common core, which helps the linkage of magnetic flux between the two coils. Note that a transformer converts v_i to v_o without making use of an external power source. Hence it is a passive device, just like a capacitor, inductor, or resistor. The turn ratio of the transformer:

$$r = \frac{\text{number of turns in the secondary coil } (N_o)}{\text{number of turns in the primary coil } (N_i)}$$

In Figure 4.6c, the two dots on the top side of the two coils indicate that the two coils are wound in the same direction.

In a pure and ideal transformer, there will be full flux linkage without any dissipation of energy. Then, the flux linkage will be proportional to the number of turns. Hence

$$\lambda_o = r\lambda_i \tag{4.4}$$

where λ denotes the flux linkage in each coil. Differentiation of Equation 4.4, noting that the induced voltage in coil is given by the rate of charge of flux, gives

$$v_o = rv_i \tag{4.1}$$

For an *ideal transformer*, there is no energy dissipation and also the signals will be in phase. Hence, the output power will be equal to the input power; thus,

$$v_o i_o = v_i i_i \tag{4.2b}$$

Hence, the current relation becomes

$$i_o = \frac{1}{r} i_i$$ (4.3b)

4.2.2.3 Gyrator

An ideal gyroscope is an example of a mechanical *gyrator* (Figure 4.7a). It is simply a spinning top that rotates about its own axis at a high angular speed ω (positive in the x direction) and assumed to remain unaffected by other small motions that may be present. If the moment of inertia about this axis of rotation (x in the shown configuration) is J, the corresponding angular momentum is $h = J\omega$, and this vector is also directed in the positive x direction, as shown in Figure 4.7b.

Suppose that the angular momentum vector h is given an incremental rotation $\delta\theta$ about the positive z axis, as shown. The free end of the gyroscope will move in the positive y direction as a result. The resulting change in the angular momentum vector is $\delta h = J\omega\delta\theta$ in the positive y direction, as shown in Figure 4.7b. Hence the rate of change of angular momentum is

$$\frac{\delta h}{\delta t} = \frac{J\omega\delta\theta}{\delta t}$$ (i)

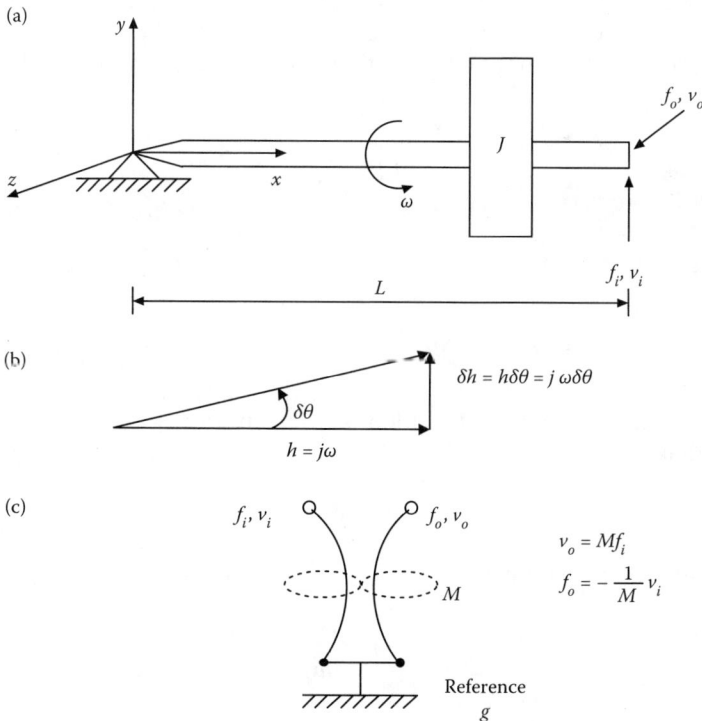

FIGURE 4.7
(a) Gyrator (gyroscope or spinning top)—a two-port element. (b) Derivation of the constitutive equations. (c) Linear-graph representation.

where δt is the time increment of the motion. Hence, in the limit, the rate of change of angular momentum is

$$\frac{dh}{dt} = J\omega\frac{d\theta}{dt} \qquad \text{(ii)}$$

If the velocity given to the free end of the gyroscope, in the positive y direction, to generate this motion is v_i (which will result in a force f_i at that point, in the positive y direction) the corresponding angular velocity about the positive z axis is

$$\frac{d\theta}{dt} = \frac{v_i}{L} \qquad \text{(iii)}$$

in which L is the length of the gyroscope. Substitute Equation (iii) in Equation (ii). The rate of change of angular momentum is

$$\frac{dh}{dt} = \frac{J\omega v_i}{L} \qquad (4.5)$$

about the positive y direction. By Newton's second law, to sustain this rate of change of angular momentum, it will require a torque equal to $J\omega v_i/L$ in the same direction. If the corresponding force at the free end of the gyroscope is denoted by f_o in the positive z-direction, the corresponding torque is $f_o L$ acting about the negative y-direction. It follows that

$$-f_o L = \frac{J\omega v_i}{L} \qquad (4.6)$$

This may be expressed as

$$f_o = -\frac{1}{M}v_i \qquad (4.7)$$

By the conservation of power (Equation 4.2) for an ideal gyroscope, it follows from Equation 4.7 that

$$v_o = Mf_i \qquad (4.8)$$

in which, the gyroscope parameter

$$M = \frac{L^2}{J\omega} \qquad (4.9)$$

Note: M is a "mobility" parameter (velocity/force), as discussed in Chapter 5.

Equations 4.7 and 4.8 are the constitutive equations of a gyrator. The linear graph representation of a gyrator is shown in Figure 4.7c.

4.3 Linear Graph Equations

Three types of equations have to be written to obtain an analytical model from a linear graph:

1. Constitutive equations for all the elements that are not sources (inputs).
2. Compatibility equations (loop equations) for all the independent closed paths.
3. Continuity equations (node equations) for all the independent junctions of two or more branches.

Constitutive equations of elements have been discussed in detail in Chapter 2 and earlier in the present chapter. In the examples in Chapter 2, compatibility equations and continuity equations were not used explicitly because the system variables were chosen to satisfy these two types of equations. In modeling of complex dynamic systems, systematic approaches, which can be computer-automated, will be useful. In that context, approaches are necessary to explicitly write the compatibility equations and continuity equations. The related approaches and issues are discussed next.

4.3.1 Compatibility (Loop) Equations

A loop in a linear graph is a closed path formed by two or more branches. A loop equation (compatibility equation) is obtained by summing all the across variables along the branches of the loop is zero. This is a necessary condition because, at a given point in the linear graph there must be a unique value for the across variable, at a given time. For example, a mass and a spring connected to the same point must have the same velocity at a particular time, and this point must be intact (i.e., does not break or snap); hence, the system is "compatible."

4.3.1.1 Sign Convention

1. Go in the counter-clockwise direction of the loop.
2. In the direction of a branch arrow the across variable drops. This direction is taken to be positive (except in a *T*-source, where the arrow direction indicates an increase in its across variable, which is the negative direction).

The arrow in each branch is important, but we need not (and indeed cannot) always go in the direction of the arrows in the branches that form a loop. If we do go in the direction of the arrow in a branch, the associated across variable is considered positive; when we go opposite to the arrow, the associated across variable is considered negative.

4.3.1.2 Number of "Primary" Loops

Primary loops are a "minimal" set of loops from which any other loop in the linear graph can be determined. A primary loop set is an "independent" set. It will generate all the independent loop equations.

Note: Loops closed by broken-line (inertia) branches should be included as well in counting the primary loops.

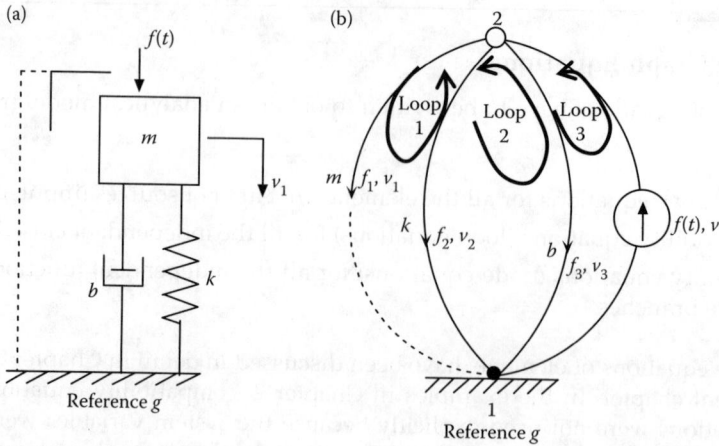

FIGURE 4.8

(a) A mass-spring-damper system. (b) Linear graph having two nodes and three primary loops.

Example 4.1

Figure 4.8 shows a mass-spring-damper system and its linear graph. Each element in the linear graph forms a branch. As noted before, an inertia element is connected to the reference point (ground g) by a dotted line because the mass is not physically connected to ground, but all measurements must be referenced to the ground reference point. This reference point "feels" the inertia force of the mass. To understand this further, suppose that we push a free mass upwards by our hands, imparting it an acceleration. The required force is equal to the inertia force, which is the product of mass and acceleration. An equal force is transmitted to the ground though our feet. Clearly, the mass itself is not directly connected to the ground, yet the force applied to the mass and the force "felt" at the ground are equal. Hence the force "appears" to travel directly through the mass element to the ground. Similarly, in Figure 4.8, the input force from the "force source" also travels to ("felt at") the reference point.

In this example, there are three primary loops. Note that loops closed by broken-line (inertia) branches are included in counting primary loops. The primary loop set can be chosen as ($b-k$, $m-b$, and $m-f$), or as ($b-k$, $m-b$, and $f-k$), or any three closed paths.

One obvious choice of primary loops in this example is what is marked in Figure 4.8: Loop 1 ($m-k$), Loop 2 ($k-b$), Loop 3 ($b-f$). The corresponding loop equations are

$$\text{Loop 1 equation: } v_1 - v_2 = 0$$

$$\text{Loop 2 equation: } v_2 - v_3 = 0$$

$$\text{Loop 3 equation: } v_3 - v = 0$$

Once one has selected a primary set of loops (three loops in this example), any other loop will depend on this primary set. For example, an $m-k$ loop can be obtained by algebraically adding the $m-b$ loop and $b-k$ loop (i.e., subtracting the $b-k$ loop from the $m-b$ loop). Similarly, the $f-m$ loop is obtained by adding the $f-b$ and $b-m$ loops. That is:

$$m-k \text{ loop} = (m-b \text{ loop}) - (b-k \text{ loop}); \text{ or Loop } 1 = (m-b \text{ loop}) - \text{Loop 2}$$

$$f-m \text{ loop} = (f-b \text{ loop}) + (b-m \text{ loop}); \text{ or } f-m \text{ loop} = \text{Loop } 3 + (b-m \text{ loop})$$

We can verify these relations using the three loop equations written before together with the following loop equations:

$$m-b \text{ Loop equation (or } b-m \text{ Loop equation): } v_1 - v_3 = 0$$

$$f-m \text{ Loop equation: } v_1 - v = 0$$

This example illustrates that the primary loop set becomes an "independent" set, which is the minimum number of loops required to obtain all the independent loop equations.

4.3.2 Continuity (Node) Equations

A node is the point where two or more branches meet. A node equation (or, continuity equation) is created by equating to zero the sum of all the through variables at a node. This holds in view of the fact that a node can neither store nor dissipate energy; in effect saying, "what goes in must come out." Hence, a node equation dictates the continuity of the through variables at a node. For this reason one must use proper signs for the variables when writing either node equations or loop equations. The sign convention that is used is: The through variable "into" the node is positive.

The meaning of a node equation in the different domains is:

Mechanical systems: Force balance; equilibrium equation; Newton's third law; etc.

Electrical systems: Current balance; Kirchoff's current law; conservation of charge; etc.

Hydraulic systems: Conservation of matter.

Thermal systems: Conservation of energy.

Example 4.2

Revisit the problem given in Figure 4.8. The system has two nodes. Corresponding node equations are identical, as given below.

$$\text{Node 2 equation: } -f_1 - f_2 - f_3 + f = 0$$

$$\text{Node 1 equation: } f_1 + f_2 + f_3 - f = 0$$

This example illustrates the following important result:
   Required number of node equations = Total number of nodes − 1.

Example 4.3

Consider the *L-C-R* electrical circuit shown in Figure 4.9a. Its linear graph is drawn as in Figure 4.9b. It should be clear that this electrical system is analogous to the mechanical system of Figure 4.8. The system has three primary loops; one primary node; and a voltage source. We may select any three loops as primary loops; for example, (*v-L, L-C, C-R*) or (*v-C, L-C, C-R*) or (*v-L, v-C, v-R*), etc. No matter what set we choose, we will get the same "equivalent" loop equations. In particular, note that the across variables for all four branches of this linear graph are the same.

For example select Loop 1: *L-v*; Loop 2: *C-L*; Loop 3: *R-C* as the primary loops, as shown in Figure 4.9b. The necessary loop equation (three) and the node equations (one) are given below, with our standard sign convention.

$$\text{Loop 1 equation: } -v_1 + v = 0$$

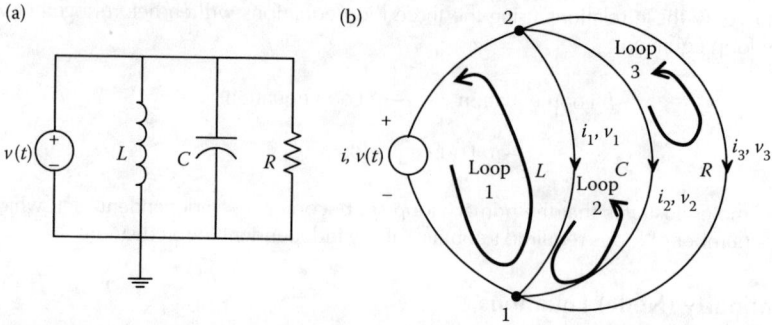

FIGURE 4.9
(a) An *L-C-R* circuit. (b) Its linear graph.

$$\text{Loop 2 equation: } -v_2 + v_1 = 0$$

$$\text{Loop 3 Equation: } -v_3 + v_2 = 0$$

$$\text{Node 2 equation: } i - i_1 - i_2 - i_3 + = 0$$

4.3.3 Series and Parallel Connections

If two elements are connected in series, their through variables are the same but the across variables are not the same (they add algebraically). If two elements are connected in parallel, their across variables are the same but the through variables are not the same (they add algebraically). These facts are given in Table 4.2.

Let us consider two systems with a spring (*k*) and a damper (*b*), and an applied force (*f(t)*). In Figure 4.10a they are connected in parallel, and in Figure 4.10b they are connected in series. Their linear graphs are as shown in the figures. Note that the linear graph in (a) has two primary loops (two elements in parallel with the force source), whereas in (b) it has only one loop, corresponding to all elements in series with the force source. In Table 4.1 we note the differences in their node and loop equations. These observations should be intuitively clear, without even writing the loop and node equations.

4.4 State Models from Linear Graphs

We can obtain a state model of a dynamic system from its linear graph. Each branch in the linear graph is a "model" of an actual system element of the system, with an associated " constitutive relation," As discussed in Chapter 2, for a mechanical system it is justifiable to use the velocities of independent inertia elements and the forces through independent stiffness (spring) elements as state variables. Similarly, for an electrical system, voltages across independent capacitors and currents through independent inductors are appropriate state variables. In general then, in the linear graph approach we use:

State variables: Across variables of independent *A*-type elements and through variables of independent *T*-type elements.

TABLE 4.2

Series-Connected Systems and Parallel-Connected Systems

Series System	Parallel System
Through variables are the same.	Across variables are the same.
Across variables are not the same (they add algebraically).	Through variables are not the same (they add algebraically).

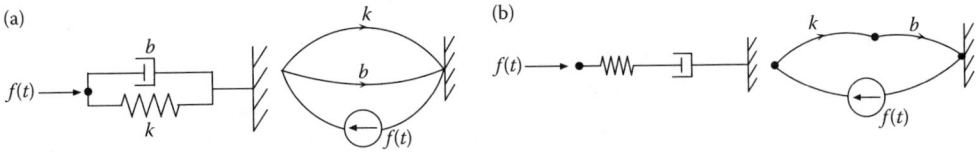

FIGURE 4.10
Spring-damper systems with a force source and their linear graphs. (a) Elements in parallel. (b) Elements in series.

In obtaining an analytical model from the linear graph of a system we write three types of equations:

1. Constitutive equations for all the branches that are not source (input) elements
2. Compatibility equations for the independent loops
3. Continuity equations for the independent nodes

This approach will be further elaborated in this section.

4.4.1 System Order

It is known that A-type elements and T-type elements are energy storage elements. The *system order* is given by the number of independent energy-storage elements in the system. This is also equal to the number state variables; the order of the state-space model; the number of initial conditions required to solve the response of the analytical model; and the order of the input–output differential equation model

The total number of energy storage elements in a system can be greater than the system order because some of these elements might not be independent.

4.4.2 Sign Convention

The important first step of developing a state-space model using linear graphs is indeed to draw a linear graph for the considered system. A sign convention should be established, as discussed before. The sign convention which we use is as follows:

1. Power flows into the action point and out of the reference point of an element (branch). This direction is shown by the branch arrow (which is an oriented branch). *Exception*: In a source element power flows out of the action point.
2. Through variable (f), across variable (v), and power flow (fv) are positive in the same direction at an action point. At reference point, v is positive in the same

direction given by the linear-graph arrow, but f is taken positive in the opposite direction.

3. In writing node equations: Flow into a node is positive.

4. In writing loop equations: Loop direction is counter-clockwise. A potential (A-variable) "drop" is positive (same direction as the branch arrow. *Exception*: In a T-source the arrow is in the direction in which the A-variable increases).

Note: Once the sign convention is established, the actual values of the variables can be positive or negative depending on their actual direction.

4.4.3 Steps of Obtaining a State Model

The following are the systematic steps for obtaining a set of state equations (a state-space model) from a linear graph:

1. Choose as state variables: Across variables for independent A-type elements and through variables for independent T-type elements.

2. Write constitutive equations for independent energy storage elements. This will give the *state-space shell*.

3. Do similarly for the remaining elements (dependent energy storage elements and dissipation—D-type—elements, transformers, etc.).

4. Write compatibility equations for the primary loops.

5. Write continuity equations for the primary nodes (total number of nodes–1).

6. In the state-space shell, retain state and input variables only. Eliminate all other variables using the loop and node equations and extra constitutive equations.

4.4.4 General Observation

Now some general observations are made with regard to a linear graph in terms of its geometric (topological) characteristics (nodes, loops, branches), elements, unknown and known variables, and relevant equations (constitutive, compatibility, and continuity).

First let

$$\text{Number of sources} = s$$

$$\text{Number of branches} = b$$

Since each source branch has one unknown variable (because one variable is the known input to the system—the source output) and all other passive branches have two unknown variables each, we have:

$$\text{Total number of unknown variables} = 2b - s \tag{4.10}$$

Since each branch other than a source branch provides one constitutive equation, we have:

$$\text{Number of constitutive equations} = b - s \tag{4.11}$$

Let

$$\text{Number of primary loops} = \ell$$

Since each primary loop give a compatibility equation, we have:

$$\text{Number of loop (compatibility) equations} = \ell$$

Let

$$\text{Number of nodes} = n$$

Since one of these nodes does not provide an extra node equation, we have:

$$\text{Number of node (continuity) equations} = n - 1 \tag{4.12}$$

Hence,

$$\text{Total number of equations} = (b - s) + \ell + (n - 1) = b + \ell + n - s - 1$$

To uniquely solve the analytical model we must have:

$$\text{Number of unknowns} = \text{Number of equations or } 2b - s = b + \ell + n - s - 1$$

Hence we have the result

$$\ell = b - n + 1 \tag{4.13}$$

This topological result must be satisfied by any linear graph.

4.4.5 Topological Result

As shown before, Equation 4.13 must be satisfied by a linear graph. Now we will prove by induction that this topological result indeed holds for any linear graph.

Consider Figure 4.11. Using the notation: Number of sources$=s$; Number of branches$-b$; Number of nodes$=n$; we proceed with the following steps.

Step 1: Start with Figure 4.11a: For this graph: $\ell = 1, b=2, n=2$. Hence, Equation 4.13 is satisfied.

Step 2: Add new loop to Figure 4.11a by using m nodes and $m+1$ branches, as in Figure 4.11b. For this new graph we have: $\ell=2$; $n=2+m$; $b=2+m+1=m+2$. Hence, Equation 4.13 is still satisfied.

Note: $m=0$ is a special case.

Step 3: Start with a general linear graph having ℓ loops, b ranches, and n nodes that satisfies Equation 4.13. This is the general case of Step 1—Figure 4.11a.

Add a new loop by using m nodes and $m+1$ branches (as in Step 2).

We have

$$\ell \rightarrow \ell+1; \quad n \rightarrow n+m; \quad b \rightarrow b+m+1$$

Equation 4.12 is still satisfied by these new values.

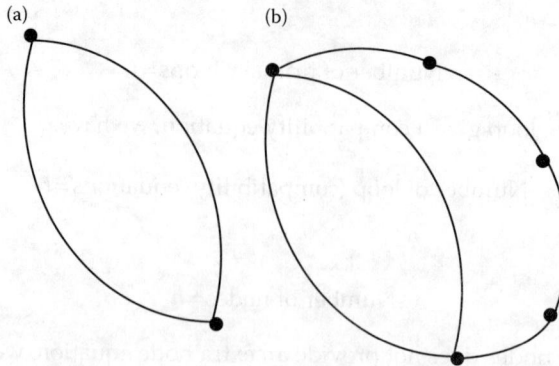

FIGURE 4.11
Proof of topological result. (a) Single loop with two branches. (b) Adding new branches to create a new loop.

Hence, by induction, Equation 4.13 is true in general.

Now we will present five mechanicalsystem examples to illustrate the development of state-space models using linear graphs. Since the approach is unified and uniform across various domains, the same approach is applicable in electrical, fluid, thermal, and multidomain (i.e., mixed) systems, as will be demonstrated later.

Example 4.4

Let us develop a state-space model for the system shown in Figure 4.8, using its linear graph. There are four branches and one source. Thus, $2b - s = 7$; we will need seven equations to solve for unknowns. Note from Figure 4.8 that there are three primary loops. In particular, in this example we have:

> Number of line branches $b = 4$
> Number of nodes $n = 2$
> Number of sources $s = 1$
> Number of primary loops $l = 3$
> Number of unknowns $= v_1, f_1, v_2, f_2, v_3, f_3, v = 7$
> (*Note*: $f(t)$, the input variable, is known)
> Number of constitutive equations (one each for m, k, b) $= b - s = 3$
> Number of node equations $= n - 1 = 1$
> Number of loop equations $= 3$ (because there are three primary loops)

We have:

Total number of equations $=$ constitutive equations $+$ node equations $+$ loop equations $= 3 + 1 + 3 = 7$.

Hence the system is solvable (seven unknowns and seven equations).
The steps of obtaining the state model are given next.

Step 1. Select state variables: Velocity v_1 of mass m and force f_2 of spring k ➔ $x_1 = v_1; x_2 = f_2$
Input variable $=$ applied forcing function (force source) $f(t)$.
Step 2. Constitutive equations for m and k: These generate the state-space shell (model skeleton):

$$\text{From Newton's second law: } \dot{v}_1 = (1/m)f_1 \tag{i}$$

$$\text{Hooke's law for spring: } \dot{f}_2 = kv_2 \tag{ii}$$

Step 3. Remaining constitutive equation (for damper):

$$f_3 = bv_3 \qquad \text{(iii)}$$

Step 4. Node and loop equations:

$$\text{Node equation (for Node 2): } f - f_1 - f_2 - f_3 = 0 \qquad \text{(iv)}$$

$$\text{Loop equation for Loop 1: } v_1 - v_2 = 0 \qquad \text{(v)}$$

$$\text{Loop equation for Loop 2: } v_2 - v_3 = 0 \qquad \text{(vi)}$$

$$\text{Loop equation for loop 3: } v_3 - v = 0 \qquad \text{(vii)}$$

Step 5. Eliminate auxiliary variables:
To obtain state model, retain v_1 and f_2 and eliminate the auxiliary variables f_1 and v_2 in Equations (i) and (ii).
From Equation (v): $v_2 = v_1$
From Equations (iv) and (iii): $f_1 = -f_2 - bv_3 + f \rightarrow f_1 = -f_2 - bv_1 + f$ (from Equations (vi) and (v))
Substituting these into the state-space shell (Equations (i) and (ii)) we get the state model:

$$\dot{v}_1 = -\frac{b}{m}v_1 - \frac{1}{m}f_2 + \frac{1}{m}f \; \rightarrow$$

$$\dot{f}_2 = k\dot{v}$$

with the state vector $x = [x_1 \quad x_2]^T = [v_1 \quad f_2]^T$ and the input vector $u = f(t)$.
The model matrices, in the usual notation, are:

$$A = \begin{bmatrix} -b/m & -1/m \\ k & 0 \end{bmatrix}; \quad B = \begin{bmatrix} 1/m \\ 0 \end{bmatrix}$$

Note that this is a second-order system, as clear from the fact that the state vector x is a second-order vector and, further, from the fact that the system matrix A is a 2×2 matrix. Also, note that in this system, the input vector u has only one element, $f(t)$. Hence it is actually a scalar variable, not a vector.

The velocity (v) of the force source is not a state variable, and we need not use Equation (vii). When v and $f(t)$ are positive, for example, power from the source flows out into Node 2.

Example 4.5

A dynamic absorber is a passive vibration-suppression device, which is mounted on the vibrating area of the dynamic system. By properly tuning (selecting the parameters of) the absorber, it is possible to "absorb" most of the power supplied by an unwanted excitation (e.g., support motion, imbalance in rotating parts) in sustaining the absorber motion such that, in steady operation, the vibratory motions of the main system are inhibited. In practice, there should be some damping present in the absorber to dissipate the energy flowing into the absorber, without generating excessive motions in the absorber mass. In the example shown in Figure 4.12a, the main system and the absorber are modeled as simple oscillators with parameters (m_2, k_2, b_2) and (m_1, k_1, b_1), respectively. The linear graph of this system can be drawn in the usual manner, as shown in Figure 4.12b. The external excitation (system input) is the velocity $u(t)$ of the support. We note the following:

Number of branches $= b = 7$
Number of nodes $= n = 4$
Number of sources $= s = 1$

(a)   (b)

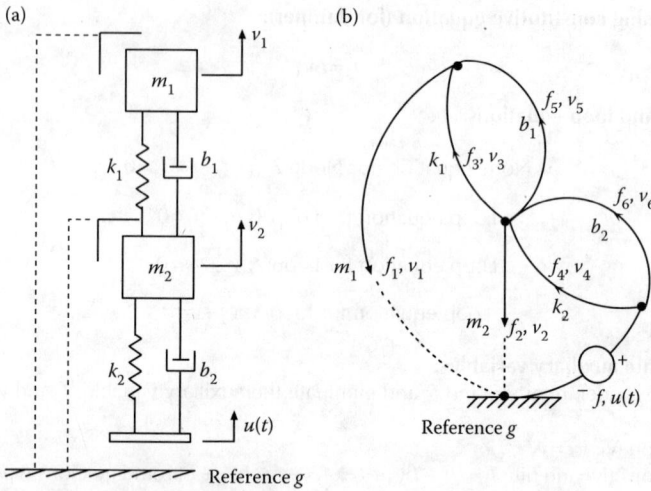

FIGURE 4.12
(a) A mechanical system with a shock absorber. (b) Linear graph of the system.

> Number of independent loops $=l=4$
> Number of unknowns $=2b-s=13$
> Number of constitutive equations $=b-s=6$
> Number of node equations $=n-1=3$
> Number of loop equations $=4$

The four loop equations are provided by the four independent loops of the linear graph.
Check: Number of unknowns $=2b-s=13$

$$\text{Number of equations}=(b-s)+(n-1)+l=6+3+4=13.$$

Hence the analytical model is solvable.

Step 1. Since the system has four independent energy storage elements (m_1, m_2, k_1, k_2) it is a fourth-order system. The state variables are chosen as the across variables of the two masses (velocities v_1 and v_2) and the through variables of the two springs (forces f_1 and f_2). Hence

$$\boldsymbol{x}=[x_1, \quad x_2 \quad x_3 \quad x_4]^T=[v_1, \quad v_2 \quad f_3 \quad f_4]^T$$

The input variable is $u(t)$.

Step 2. The skeleton state equations (model shell) are:

$$\text{Newton's second law for mass } m_1: \ \dot{v}_1=\frac{1}{m_1}f_1$$

$$\text{Newton's second law for mass } m_2: \ \dot{v}_2=\frac{1}{m_2}f_2$$

$$\text{Hooke's law for spring } k_1: \ \dot{f}_3=k_1v_3$$

$$\text{Hooke's law for spring } k_2: \ \dot{f}_4=k_2v_4$$

Step 3. The remaining constitutive equations:

$$\text{For damper } b_1: f_5=b_1v_5$$

$$\text{For damper } b_2: f_6=b_2v_6$$

Step 4. The node equations:

$$-f_1 + f_3 + f_5 = 0$$

$$-f_3 - f_5 - f_2 + f_4 + f_6 = 0$$

$$-f_4 - f_6 + f = 0$$

The loop equations:

$$v_1 - v_2 + v_3 = 0$$

$$v_2 - u + v_4 = 0$$

$$-v_4 + v_6 = 0$$

$$-v_3 + v_5 = 0$$

Step 5. Eliminating the auxiliary variables in the state-space shell.
The following state equations are obtained:

$$\dot{v}_1 = -(b_1/m_1)v_1 + (b_1/m_1)v_2 + (1/m_1)f_3$$
$$\dot{v}_2 = (b_1/m_2)v_1 - [(b_1 + b_2)/m_2]v_2 - (1/m_2)f_3 + (1/m_2)f_4 + (b_2/m_2)u(t)$$
$$\dot{f}_3 = -k_1 v_1 + k_1 v_2$$
$$\dot{f}_4 = -k_2 v_2 + k_2 u(t)$$

This corresponds to:

$$\text{System matrix: } A = \begin{bmatrix} -b_1/m_1 & b_1/m_1 & 1/m_1 & 0 \\ b_1/m_2 & -(b_1+b_2)/m_1 & -1/m_2 & 1/m_2 \\ -k_1 & k_1 & 0 & 0 \\ 0 & k_2 & 0 & 0 \end{bmatrix}$$

$$\text{Input distribution matrix: } B = \begin{bmatrix} 0 \\ b_2/m_2 \\ 0 \\ k_2 \end{bmatrix}$$

Example 4.6

Commercial motion controllers are digitally controlled (microprocessor-controlled) high-torque devices capable of applying a prescribed motion to a system. Such controlled actuators can be considered as velocity sources. Consider an application where a rotatory motion control-ler is used to position an object, which is coupled through a gear box. The system is modeled as in Figure 4.13. We will develop a state-space model for this system using the linear graph approach.

Step 1. Note that the two inertia elements m_1 and m_2 are not independent, and together comprise one storage element. Thus, along with the stiffness element, there are only two independent energy storage elements. Hence the system is second order. Let us choose as state variables, v_1

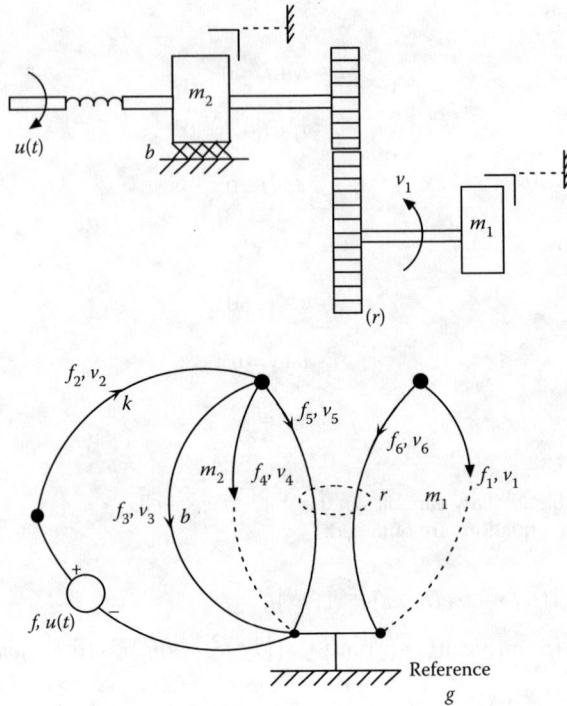

FIGURE 4.13
(a) Rotary-motion system with a gear transmission. (b) Linear graph of the system.

and f_2—the across variable of one of the inertias (because the other inertia will be "dependent") and the through variable of the spring.

$$\text{Let } x_1 = v_1 \quad \text{and} \quad x_2 = f_2$$

$$\text{Hence } [x_1 \ x_2]^T = [v_1 \ f_2]^T$$

Step 2. The constitutive equations for m_1 and k: $\dot{v}_1 = \dfrac{1}{m} f_1$; $\dot{f}_2 = k v_2$
Step 3. The remaining constitutive equations:

$$\text{For damper: } f_3 = b v_3$$

$$\text{For the "dependent" inertia } m_2: \ \dot{v}_4 = \frac{1}{m_2} f_4$$

$$\text{For the transformer (pair of meshed gear wheels): } v_6 = r v_5; \ f_6 = -\frac{1}{r} f_5$$

Step 4. The node equations:

$$-f_6 - f_1 = 0$$

$$f - f_2 = 0$$

$$f_2 - f_3 - f_4 - f_5 = 0$$

The loop equations:

$$v_6 - v_1 = 0$$

$$v_3 - v_4 = 0$$

$$v_4 - v_5 = 0$$

$$-v_2 + u(t) - v_3 = 0$$

Step 5. Eliminate the auxiliary variables.

Using equations from Steps 3 and 4, the auxiliary variable f_1 can be expressed as:

$$f_1 = \frac{1}{r}\left[f_2 - \frac{b}{r}v_1 - \frac{m_2}{r}\dot{v}_1 \right]$$

The auxiliary variable v_2 can be expressed as:

$$v_2 = -\frac{1}{r}v_1 + u(t)$$

By substituting these equations into the state-space shell we obtain the following two state equations:

$$\dot{v}_1 = -\left[\frac{b}{(m_1 r^2 + m_2)} \right]v_1 + \left[\frac{r}{(m_1 r^2 + m_2)} \right]f_2$$

$$\dot{f}_2 = -\frac{k}{r}v_1 + ku(t)$$

Note that the system is second-order; only two state equations are present. The corresponding system matrix and the input-gain matrix (input distribution matrix) are:

$$A = \begin{bmatrix} -b/m & r/m \\ -k/m & 0 \end{bmatrix}; \ B = \begin{bmatrix} 0 \\ k \end{bmatrix}$$

where $m = m_1 r^2 + m_2 =$ *equivalent inertia of* m_1 *and* m_2 *when determined at the location of inertia* m_2.

Example 4.7

a. List several advantages of using linear graphs in developing a state-space model of a dynamic system.

b. Electrodynamic shakers are commonly used in the dynamic testing of products. One possible configuration of a shaker/test-object system is shown in Figure 4.14a. A simple, linear, lumped-parameter model of the mechanical system is shown in Figure 4.14b.

Note that the driving motor is represented by a torque source T_m. Also, the following parameters are indicated:

J_m = equivalent moment of inertia of motor rotor, shaft, coupling, gears, and the shaker platform
r_1 = pitch circle radius of the gear wheel attached to the motor shaft
r_2 = pitch circle radius of the gear wheel rocking the shaker platform
l = lever arm from the center of the rocking gear to the support location of the test object

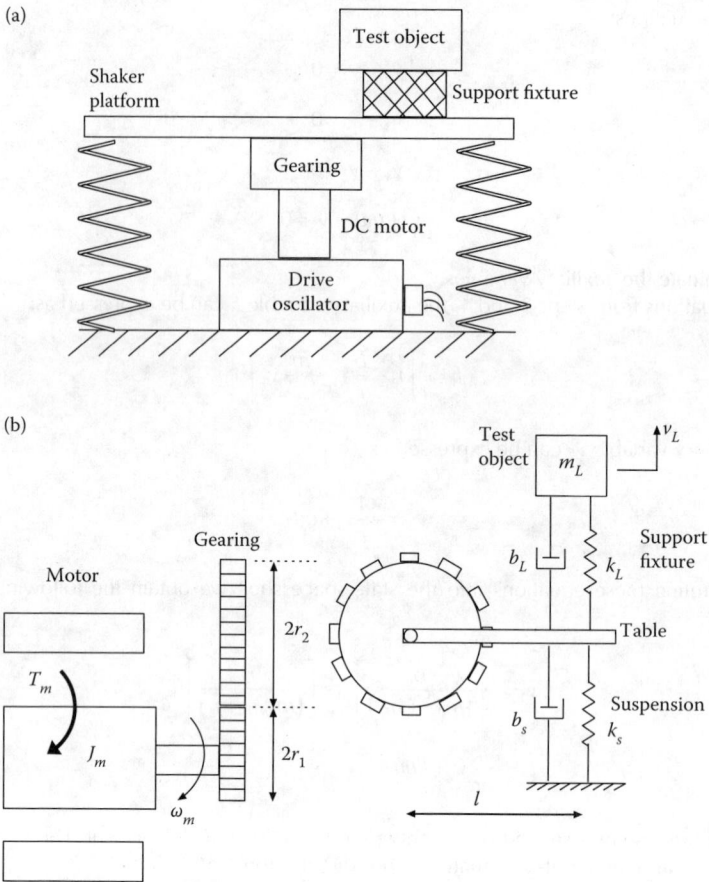

FIGURE 4.14
(a) A dynamic-testing system. (b) A model of the dynamic testing system.

m_L =equivalent mass of the test object and its support fixture
k_L =stiffness of the support fixture
b_L =equivalent viscous damping constant of the support fixture
k_s =stiffness of the suspension system of the shaker table
b_s =equivalent viscous damping constant of the suspension system.

Since the inertia effects are lumped into equivalent elements it may be assumed that the shafts, gearing, platform and the support fixtures are light. The following variables are of interest:

ω_m =angular speed of the drive motor
v_L =vertical speed of motion of the test object
f_L =equivalent dynamic force of the support fixture (force in spring k_L)
f_s =equivalent dynamic force of the suspension system (force in spring k_s).

   i. Obtain an expression for the motion ratio:

$$r = \frac{\text{vertical movement of the shaker table at the test object support location}}{\text{angular movement of the drive motor shaft}}$$

   ii. Draw a linear graph to represent the dynamic model.

iii. Using $x=[\omega_m, f_s, f_L, v_L]^T$ as the state vector, $u=[T_m]$ as the input, and $y=[v_L\ f_L]^T$ as the output vector, obtain a complete state-space model for the system. For this purpose you must use the linear graph drawn in (ii).

Solution

a. Linear graphs:

- Use physical variables as states.
- Provide a generalized and unified approach for mechanical, electrical, fluid, and thermal systems. Hence they can be conveniently used in multidomain (i.e., mixed) systems.
- Provide a unified approach to model multifunctional devices (e.g., a piezoelectric device which can function as both a sensors and an actuator).
- Show the directions of power flow in various parts of the system.
- Provide a graphical representation of the system model.
- Allow visualization of the system structure (even before formulating an analytical model).
- Help identify similarities (structure, performance, etc.) in different types of systems.
- Provide a systematic approach to automatically (using computer) generate state equations.

b. (i) Let θ_m=rotation of the motor (drive gear).
Hence, rotation of the output gear=$(r_1/r_2)\theta_m$
Hence, displacement of the table at the test object support point=$l\ (r_1/r_2)\theta_m$
Hence, $r=l\ (r_1/r_2)$
(ii) The linear graph of the system is drawn as in Figure 4.15.
(iii) Constitutive equations:

State-space shell:

$$J_m \frac{d\omega_m}{dt} = T_2$$

$$\frac{df_s}{dt} = k_s v_5$$

$$\frac{df_L}{dt} = k_L v_L$$

$$m_L \frac{dv_L}{dt} = f_9$$

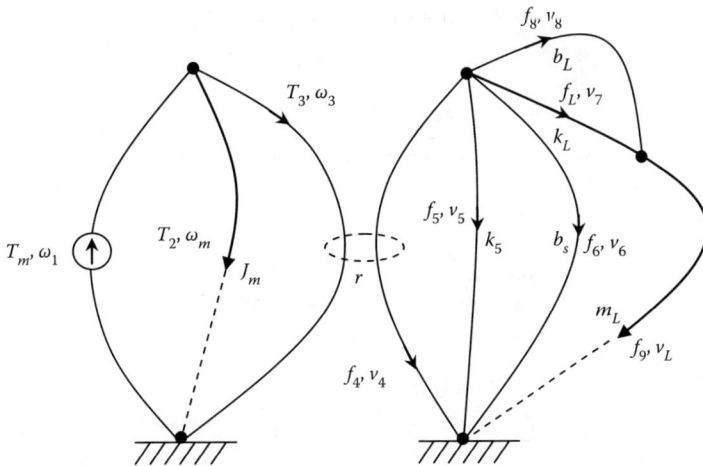

FIGURE 4.15
Linear graph of the shaker system.

Others:

$$v_4 = r\omega \quad \text{where} \quad r = \frac{r_1}{r_2} I$$

$$f_4 = -\frac{1}{r} T_3$$

$$f_6 = b_s v_6$$

$$f_8 = b_L v_8$$

Continuity (Node) equations

$$T_m - T_2 - T_3 = 0$$

$$-f_4 - f_s - f_6 - f_L - f_8 = 0$$

$$f_L + f_8 - f_9 = 0$$

Compatibility (Loop) equations:

$$-\omega_m + \omega_3 = 0$$

$$-v_4 + v_5 = 0$$

$$-v_5 + v_6 = 0$$

$$-v_6 + v_7 + v_L = 0$$

$$-v_7 + v_8 = 0$$

Elimination/substitution results in the following:

$$J_m \frac{d\omega_m}{dt} = T_2 = T_m - T_3 = T_m + rf_4 = T_m + r(-f_s - f_6 - f_L - f_8)$$

$$= T_m - r(f_s + b_s v_6 + f_L + b_L v_8)$$

$$= T_m - r(f_s + f_L + b_s v_6 + b_L v_8)$$

$$v_6 = v_5 = v_4 = r\omega_3 = r\omega_m$$

$$v_8 = v_7 = v_6 - v_L = v_5 - v_L = v_4 - v_L = r\omega_3 - v_L = r\omega_m - v_L$$

Hence,

$$J_m \frac{d\omega_m}{dt} = T_m - r(f_s + f_L) - r^2 b_s \omega_m - r b_L (r\omega_m - v_L) \tag{i}$$

$$\frac{df_s}{dt} = k_s v_5 = k_s v_4 = k_s r\omega_3 = k_s r\omega_m \tag{ii}$$

$$\frac{df_L}{dt} = k_L v_7 = k_L (v_6 - v_L) = k_L (v_4 - v_L) = k_L (r\omega_3 - v_L) = k_L (r\omega_m - v_L) \tag{iii}$$

$$m_L \frac{dv_L}{dt} = f_9 = f_L + f_8 = f_L + b_L v_8 = f_L + b_L (r\omega_m - v_L) \qquad \text{(iv)}$$

In summary, we have the following state equations:

$$J_m \frac{d\omega_m}{dt} = T_m - rf_s - rf_L - r^2(b_s + b_L)\omega_m + rb_L v_L$$

$$\frac{df_s}{dt} = rk_s \omega_m$$

$$\frac{df_L}{dt} = rk_L \omega_m - k_L v_L$$

$$m_L \frac{dv_L}{dt} = f_L + rb_L \omega_m - b_L v_L$$

with v_L and f_L as the outputs.

In the standard notation: $\dot{x} = Ax + Bu$ and $y = Cx + Du$ where

$$x = [\omega_m, f_s, f_L, v_L]^T, \quad u = [T_m], \quad y = [v_L \ \ f_L]^T$$

$$A = \begin{bmatrix} -\dfrac{r^2}{J_m}(b_s + b_L) & -\dfrac{r}{J_m} & -\dfrac{r}{J_m} & \dfrac{rb_L}{J_m} \\ rk_s & 0 & 0 & 0 \\ rk_L & 0 & 0 & -k_L \\ \dfrac{rb_L}{m_L} & 0 & \dfrac{1}{m_L} & -\dfrac{b_L}{m_L} \end{bmatrix}; \quad B = \begin{bmatrix} 1/J_m \\ 0 \\ 0 \\ 0 \end{bmatrix}; \quad C = \begin{bmatrix} 0 & 0 & 0 & 1 \\ 0 & 0 & 1 & 0 \end{bmatrix}; \quad D = 0$$

Example 4.8

A robotic sewing system consists of a conventional sewing head. During operation, a panel of garment is fed by a robotic hand into the sewing head. The sensing and control system of the robotic hand ensures that the seam is accurate and the cloth tension is correct in order to guarantee the quality of the stitch. The sewing head has a frictional feeding mechanism, which pulls the fabric in a cyclic manner away from the robotic hand, using a toothed feeding element. When there is slip between the feeding element and the garment, the feeder functions as a *force source* and the applied force is assumed cyclic with a constant amplitude. When there is no slip, however, the feeder functions as a *velocity source*, which is the case during normal operation. The robot hand has inertia. There is some flexibility at the mounting location of the hand on the robot. The links of the robot are assumed rigid and some of its joints can be locked to reduce the number of degrees of freedom, when desired.

Consider the simplified case of a single-degree-of-freedom robot. The corresponding robotic sewing system is modeled as in Figure 4.16. Here the robot is modeled as a single moment of inertia J, which is linked to the hand with a light rack-and-pinion device with its speed transmission parameter given by:

$$\frac{\text{Rack Tanslatory Movement}}{\text{Pinion Rotatory Movement}} = r$$

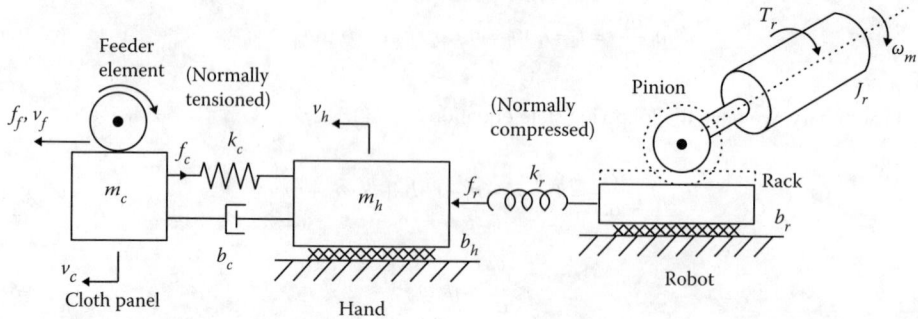

FIGURE 4.16
A robotic sewing system.

The drive torque of the robot is T_r and the associated rotatory speed is ω_r. Under conditions of slip the feeder input to the cloth panel is force f_f, and with no slip the input is the velocity v_f. Various energy dissipation mechanisms are modeled as linear viscous damping of damping constant b (with corresponding subscripts). The flexibility of various system elements is modeled by linear springs with stiffness k. The inertia effects of the cloth panel and the robotic hand are denoted by the lumped masses m_c and m_h, respectively, having velocities v_c and v_h, as shown in Figure 4.16.

Note: The cloth panel is normally in tension with tensile force f_c. In order to push the panel, the robotic wrist is normally in compression with compressive force f_r.

First consider the case of the feeding element with slip:

a. Draw a linear graph for the model shown in Figure 4.16, orient the graph, and mark all the element parameters, through variables and across variables on the graph.
b. Write all the constitutive equations (element physical equations), independent node equations (continuity), and independent loop equations (compatibility). What is the order of the model?
c. Develop a complete state-space model for the system. The outputs are taken as the cloth tension f_c, and the robot speed ω_r, which represent the two variables that have to be measured to control the system. Obtain the system matrices A, B, C, and D.

Now consider the case where there is no slip at the feeder element:

d. What is the order of the system now? How is the linear graph of the model modified for this situation? Accordingly, modify the state-space model obtained earlier to represent the present situation and from that obtain the new model matrices A, B, C and D.
e. Generally comment on the validity of the assumptions made in obtaining the model shown in Figure 4.16 for a robotic sewing system.

Solution

a. Linear graph of the system is drawn as in Figure 4.17. Since in this case the feeder input to the cloth panel is force f_f, a T-source, the arrow of the source element should be retained but the + and – signs (used for an A-source) should be removed.
b. In the present operation f_f is an input. This case corresponds to a fifth-order model, as will be clear from the development given below.
Constitutive equations:

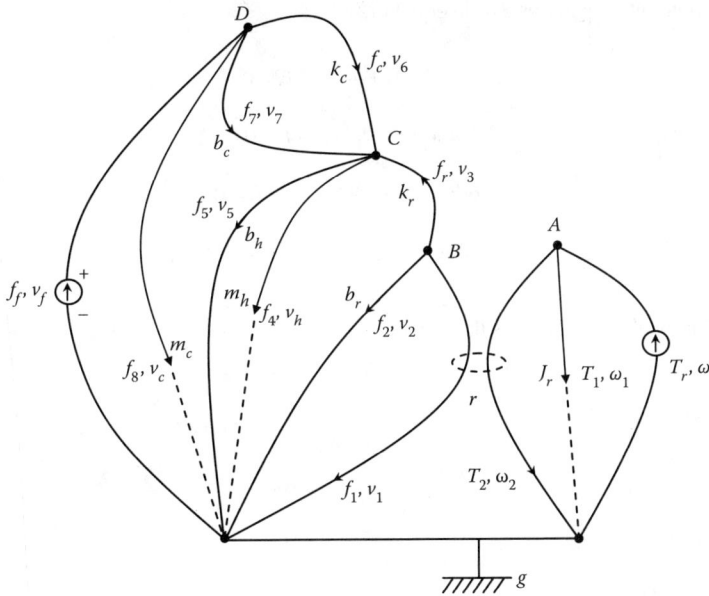

FIGURE 4.17
Linear graph of the robotic sewing system.

$$J_r \frac{d\omega_r}{dt} = T_1$$

$$\frac{df_r}{dt} = k_r v_3$$

$$m_h \frac{dv_h}{dt} = f_4 \quad \left.\right\} \text{State-space shell}$$

$$\frac{df_c}{dt} = k_c v_6$$

$$m_c \frac{dv_c}{dt} = f_8$$

$$v_1 = r\omega_2$$

$$f_1 = -\frac{1}{r}T_2$$

$$f_2 = -b_r v_2$$

$$f_5 = -b_h v_5$$

$$f_7 = -b_c v_7$$

Continuity equations (Node equations):

$$\text{Node } A: T_r - T_1 - T_2 = 0$$

$$\text{Node } B: -f_1 - f_2 - f_r = 0$$

$$\text{Node } C: f_r + f_c + f_7 - f_5 - f_4 = 0$$

$$\text{Node } D: -f_c + f_f - f_8 - f_7 = 0$$

Compatibility equations (Loop equations):

$$-\omega + \omega_1 = 0$$

$$-\omega_1 + \omega_2 = 0$$

$$-v_1 + v_2 = 0$$

$$-v_1 + v_3 + v_h = 0$$

$$-v_h + v_5 = 0$$

$$-v_6 + v_7 = 0$$

$$-v_h - v_7 + v_c = 0$$

$$-v_c + v_f = 0$$

c. Eliminate unwanted variables as follows:

$$T_1 = T_r - T_2 = T_r + rf_1 = T_r + r(-f_2 - f_r)$$

$$= T_r - rb_r v_2 - rf_r = T_r - rb_r v_1 - rf_r$$

$$= T_r - rb_r r\omega_2 - rf_r$$

$$= T_r - r^2 b_r \omega_2 - rf_r$$

$$v_3 = v_1 - v_h = r\omega_2 - v_h = r\omega_r - v_h$$

$$f_4 = f_r + f_c + f_7 - f_5 = f_r + f_c + b_c v_7 - b_h v_5$$

$$= f_r + f_c + b_c(v_c - v_h) - b_h v_h$$

$$v_6 = v_7 = v_c - v_h$$

$$f_8 = f_f - f_c - f_7 = f_f - f_c - b_c v_7 = f_f - f_c - b_c(v_c - v_h)$$

State-space model:

$$J_r \frac{d\omega_r}{dt} = -r^2 b_r \omega_r - r f_r + T_r$$

$$\frac{df_r}{dt} = k_r (r\omega_r - v_h)$$

$$m_h \frac{dv_h}{dt} = f_r - (b_c + b_h)v_h + f_c + b_c v_c$$

$$\frac{df_c}{dt} = k_c(-v_h + v_c)$$

$$m_c \frac{dv_c}{dt} = b_c v_h - f_c - b_c v_c + f_f$$

with $x = \begin{bmatrix} \omega_r & f_r & v_h & f_c & v_c \end{bmatrix}^T$; $u = \begin{bmatrix} T_r & f_f \end{bmatrix}^T$; $y = \begin{bmatrix} f_c & \omega_r \end{bmatrix}^T$

$$A = \begin{bmatrix} -r^2 b_r/J_r & -r/J_r & 0 & 0 & 0 \\ rk_r & 0 & -k_r & 0 & 0 \\ 0 & 1/m_h & -(b_c + b_h)/m_h & 1/m_h & b_c/m_h \\ 0 & 0 & -k_c & 0 & k_c \\ 0 & 0 & b_c/m_c & -1/m_c & -b_c/m_c \end{bmatrix};$$

$$B = \begin{bmatrix} 1/J_r & 0 \\ 0 & 0 \\ 0 & 0 \\ 0 & 0 \\ 0 & 1/m_c \end{bmatrix}; \quad C = \begin{bmatrix} 0 & 0 & 0 & 1 & 0 \\ 1 & 0 & 0 & 0 & 0 \end{bmatrix}; \quad D = 0$$

d. In this case, v_f is an input, which is an *A*-source. The corresponding element in the linear graph given in Figure 4.17 should be modified to account for this. Specifically, the direction of the arrow of this source element should be reversed (because it is an *A*-source) and the + and − signs (used for an *A*-source) should be retained. Furthermore, the inertia element m_c ceases to influence the dynamics of the overall system because, $v_c = v_f$ in this case and is completely specified. This results from the fact that any elements connected in parallel with an *A*-source have no effect on the rest of the system. Accordingly, the branch representing the m_c element should be removed from the linear graph.

Hence, we now have a fourth-order model, with

State vector $x = \begin{bmatrix} \omega_r & f_r & v_h & f_c \end{bmatrix}^T$; Input vector $u = \begin{bmatrix} T_r & v_f \end{bmatrix}^T$

State model:

$$J_r \frac{d\omega_r}{dt} = -r^2 b_r \omega_r - rf_r + T_r$$

$$\frac{df_r}{dt} = k_r(r\omega_r - v_h)$$

$$m_h \frac{dv_h}{dt} = f_r - (b_c + b_h)v_h + f_c + b_c v_f$$

$$\frac{df_c}{dt} = k_c(-v_h + v_f)$$

The corresponding model matrices are:

$$A = \begin{bmatrix} -r^2 b_r/J_r & -r/J_r & 0 & 0 \\ rk_r & 0 & -k_r & 0 \\ 0 & 1/m_h & -(b_c + b_h)/m_h & 1/m_h \\ 0 & 0 & -k_c & 0 \end{bmatrix}; \quad B = \begin{bmatrix} 1/J_r & 0 \\ 0 & 0 \\ 0 & b_c/m_h \\ 0 & k_c \end{bmatrix};$$

$$C = \begin{bmatrix} 0 & 0 & 0 & 1 \\ 1 & 0 & 0 & 0 \end{bmatrix}; \quad D = 0$$

e. In practice, the cloth panel is not a rigid and lumped mass; damping and flexibility effects are nonlinear; and conditions of pure force source and pure velocity source may not be maintained.

4.5 Miscellaneous Examples

Thus far in this chapter we have primarily considered the modeling of lumped-parameter mechanical systems—systems with inertia, flexibility, and mechanical energy dissipation. In view of the analogies that exist between mechanical, electrical, fluid, and thermal components and associated variables, there is an "analytical" similarity between these four types of physical systems. Accordingly, once we have developed procedures for modeling and analysis of one type of systems (say, mechanical systems) the same procedures may be extended (in an "analogous" manner) to the other three types of systems. This fact is exploited in the use of linear graphs in modeling mechanical, electrical, fluid, and thermal systems, in a unified manner, using essentially the same procedures. Furthermore, for this reason, a unified and integrated procedure is provided through linear graphs to model multidomain (mixed systems); for example electro-mechanical or mechatronic systems—systems that use a combination of two or more types of physical components (mechanical, electrical, fluid, and thermal) in a convenient manner. In this section first we will introduce two useful components: amplifier and dc motor, which are useful in electrical, electro-mechanical, and other types of multidomain systems. We will end the section with examples.

4.5.1 Amplifiers

An amplifier is a common component, primarily in an electrical system or electrical subsystem. Purely mechanical, fluid, and thermal amplifiers have been developed and envisaged as well. Two common characteristics of an amplifier are:

1. They accomplish tasks of signal amplification.
2. They are active devices (i.e., they need an external power to operate).

3. They are not affected (ideally) by the load which they drive (i.e., loading effects are small).

4. They have a decoupling effect on systems (i.e., the desirable effect of reducing dynamic interactions between components).

Electrical signals voltage, current, and power are amplified using voltage amplifiers, current amplifiers, and power amplifiers, respectively. Operational amplifiers (opamps) are the basic building block in constructing these amplifiers. Particularly, an opamp, with feedback provides the desirable characteristics of: very high input impedance, low output impedance, and stable operation. For example, due to its impedance characteristics, the output characteristics of a good amplifier are not affected by the device (load) that is connected to its output. In other words electrical loading errors are negligible.

Analogous to electrical amplifiers, a mechanical amplifier can be designed to provide force amplification (a *T*-type amplifier) or a fluid amplifier can be designed to provide pressure amplification (an *A*-type amplifier). In these situations, typically, the device is active and an external power source is needed to operate the amplifier (e.g., to drive a motor-mechanical load combination).

4.5.1.1 Linear Graph Representation

In its linear graph representation, an amplifier is considered as a "dependent source" element or a "modulated source" element. Specifically, the amplifier output depends on (modulated by) the amplifier input, and is not affected by the dynamics of any devices that are connected to the output of the amplifier (i.e., the load of the amplifier). This is the ideal case. In practice some loading error will be present (i.e., the amplifier output will be affected by the load which it drives).

The linear graph representations of an across-variable amplifier (e.g., voltage amplifier, pressure amplifier) and a through-variable amplifier (e.g., current amplifier, force amplifier) are shown in Figure 4.18a and b, respectively. The pertinent constitutive equations in the general and linear cases are given as well in the figures.

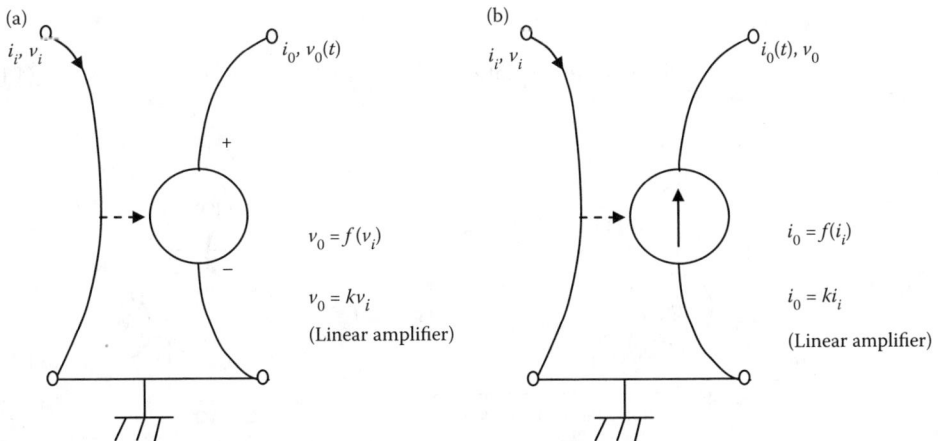

(a)

i_i, v_i $i_0, v_0(t)$

$+$

$-$

$v_0 = f(v_i)$

$v_0 = kv_i$

(Linear amplifier)

(b)

i_i, v_i $i_0(t), v_0$

$i_0 = f(i_i)$

$i_0 = ki_i$

(Linear amplifier)

FIGURE 4.18
Linear graph representation of (a) an across-variable amplifier (*A*-type amplifier) and (b) a through-variable amplifier (*T*-type amplifier).

4.5.2 DC Motor

The dc motor is a commonly used electrical actuator. It converts dc electrical energy into mechanical energy. The principle of operation is based on the fact that when a conductor carrying current is placed in a magnetic field, a force is generated (Lorentz's law). It is this force, which results from the interaction of two magnetic fields, that is presented as the magnetic torque in the rotor of the motor.

A dc motor has a stator and a rotor (armature) with windings which are excited by a field voltage v_f and an armature voltage v_a, respectively. The equivalent circuit of a dc motor is shown in Figure 4.19a, where the field circuit and the armature circuit are shown separately, with the corresponding supply voltages. This is the separately excited case. If the stator filed is provided by a permanent magnet, then the stator circuit that is shown in Figure 4.19a is simply an equivalent circuit, where the stator current i_f can be assumed constant. Similarly, if the rotor is a permanent magnet, what is shown in Figure 4.19a is an equivalent circuit where the armature current i_a can be assumed constant. The magnetic torque of the motor is generated by the interaction of the stator field (proportional to i_f) and the rotor field (proportional to i_a) and is given by

$$T_m = k i_f i_a \tag{4.14}$$

A back-electromotive force (back e.m.f) is generated in the rotor (armature) windings to oppose its rotation when these windings rotate in the magnetic field of the stator (Lenz's law). This voltage is given by

$$v_b = k' i_f \omega_m \tag{4.15}$$

where i_f = field current; i_a = armature current; ω_m = angular speed of the motor.

Note: For perfect transfer of electrical energy to mechanical energy in the rotor we have

$$T_m \omega_m = i_a v_b \tag{4.16}$$

This is an electro-mechanical transformer.

$$\text{Field circuit equation: } v_f = R_f i_f + L_f \frac{d i_f}{dt} \tag{4.17}$$

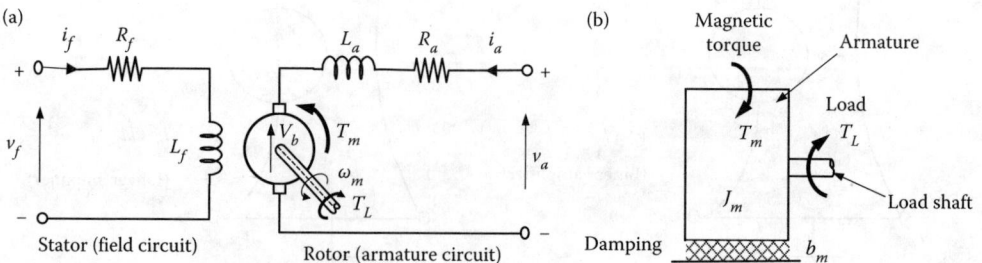

FIGURE 4.19
(a) Equivalent circuit of a dc motor (separately excited). (b) Armature mechanical loading.

where v_f = supply voltage to stator; R_f = resistance of the field windings; L_f = inductance of the field windings.

$$\text{Armature (rotor) circuit equation: } v_a = R_a i_a + L_a \frac{di_a}{dt} + v_b \tag{4.18}$$

where v_a = armature supply voltage; R_a = armature winding resistance; L_a = armature leakage inductance.

Suppose that the motor drives a load whose equivalent torque is T_L. Then from Figure 4.19b.

$$\text{Mechanical (load) equation: } J_m \frac{d\omega_m}{dt} = T_m - T_L - b_m \omega_m \tag{4.19}$$

where J_m = moment of inertia of the rotor; b_m = equivalent (mechanical) damping constant for the rotor; T_L = load torque.

In field control of the motor, the armature supply voltage v_a is kept constant and the field voltage v_f is controlled. In armature control of the motor, the field supply voltage v_f is kept constant and the armature voltage v_a is controlled.

Example 4.9

A classic problem in robotics is the case of robotic hand gripping and turning a doorknob to open a door. The mechanism is schematically shown in Figure 4.20a. Suppose that the actuator of the robotic hand is an armature-controlled dc motor. The associated circuit is shown in Figure 4.20b.

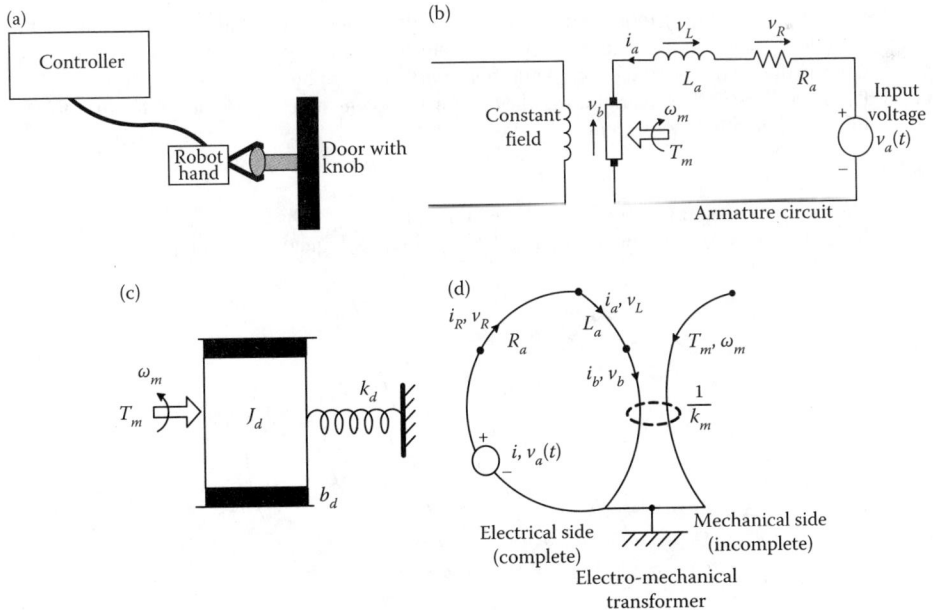

FIGURE 4.20
(a) Robotic hand turning a doorknob. (b) Armature-controlled dc motor of the robotic hand. (c) Mechanical model of the hand-doorknob system. (d) Incomplete linear graph.

The field circuit provides a constant magnetic field to the motor, and is not important in the present problem. The armature (with motor rotor windings) circuit has a back e.m.f. v_b, a leakage inductance L_a, and a resistance R_a. The input signal to the robotic hand is the armature voltage $v_a(t)$ as shown. The rotation of the motor (at an angular speed ω_m) in the two systems of magnetic field generates a torque T_m (which is negative as marked in Figure 4.20b during normal operation). This torque (magnetic torque) is available to turn the doorknob, and is resisted by the inertia force (moment of inertia J_d), the friction (modeled as linear viscous damping of damping constant b_d) and the spring (of stiffness k_d) of the hand-knob-lock combination. A mechanical model is shown in Figure 4.20c. The dc motor may be considered as an ideal electromechanical transducer which is represented by a linear graph transformer. The associated equations are

$$\omega_m = \frac{1}{k_m} v_b \tag{4.20}$$

$$T_m = -k_m i_b \tag{4.21}$$

Note: The negative sign in Equation 4.21 arises due to the specific sign convention. The linear graph may be easily drawn, as shown in Figure 4.20d, for the electrical side of the system.
Answer the following questions:

a. Complete the linear graph by including the mechanical side of the system.
b. Give the number of branches (b), nodes (n), and the independent loops (l) in the completed linear graph. Verify your answer.
c. Take current through the inductor (i_a), speed of rotation of the door knob (ω_d), and the resisting torque of the spring within the door lock (T_k) as the state variables, the armature voltage $v_a(t)$ as the input variable, and ω_d and T_k as the output variables. Write the independent node equations, independent loop equations, and the constitutive equations for the completed linear graph. Clearly show the state-space shell. Also verify that the number of unknown variables is equal to the number of equations obtained in this manner.
d. Eliminate the auxiliary variables and obtain a complete state-space model for the system, using the equations written in (c) above.

Solution

a. The complete linear graph is shown in Figure 4.21.
b. $b=8$, $n=5$, $l=4$ for this linear graph. It satisfies the topological relationship $l=b-n+1$

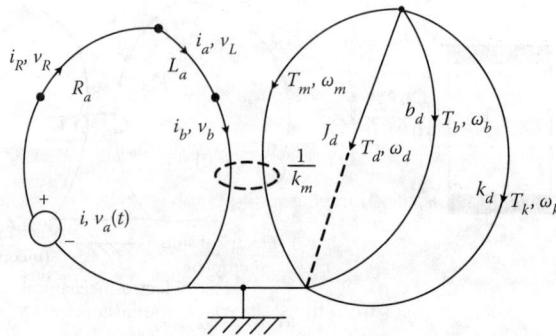

FIGURE 4.21
The complete linear graph of the system.

c. Independent node equations:

$$i - i_R = 0$$

$$i_R - i_a = 0$$

$$i_a - i_b = 0$$

$$-T_m - T_d - T_b - T_k = 0$$

Independent loop equations:

$$v_a(t) - v_R - v_L - v_b = 0$$

$$\omega_m - \omega_d = 0$$

$$\omega_d - \omega_b = 0$$

$$\omega_b - \omega_k = 0$$

Constitutive equations:

$$\left. \begin{array}{l} L_a \dfrac{di_a}{dt} = v_L \\[2mm] J_d \dfrac{d\omega_d}{dt} = T_d \\[2mm] \dfrac{dT_k}{dt} = k_d \omega_k \end{array} \right\} \text{State-space shell}$$

$$\left. \begin{array}{l} v_R = R_a i_R \\[1mm] T_b = b_d \omega_b \end{array} \right\} \text{Auxiliary constitutive equations}$$

$$\left. \begin{array}{l} \omega_m = \dfrac{1}{k_m} v_b \\[2mm] T_m = -k_m i_b \end{array} \right\} \text{Electro-mechanical transformer}$$

Note: There are 15 unknown variables $(i, i_R, i_a, i_b, T_m, T_d, T_b, T_k, v_R, v_L, v_b, \omega_m, \omega_d, \omega_b, \omega_k)$ and 15 equations.

$$\text{Number of unknown variables} = 2b - s = 2 \times 8 - 1 = 15$$

$$\text{Number of independent node equations} = n - 1 = 5 - 1 = 4$$

$$\text{Number of independent loop equations} = l = 4$$

$$\text{Number of constitutive equations} = b - s = 8 - 1 = 7$$

$$\text{Check: } 15 = 4 + 4 + 7$$

d. Eliminate the auxiliary variables from the state-space shell, by substitution:

$$v_L = v_a(t) - v_R - v_b = v_a(t) - R_a i_a - k_m \omega_m$$

$$= v_a(t) - R_a i_a - k_m \omega_d$$

$$T_d = -T_k - T_m - T_b = -T_k + k_m i_b - b_d \omega_b$$

$$= k_m i_a - b_d \omega_d - T_k$$

$$\omega_k = \omega_b = \omega_d$$

Hence, we have the state-space equations:

$$L_a \frac{di_a}{dt} = -R_a i_a - k_m \omega_d + v_a(t)$$

$$J_d \frac{d\omega_d}{dt} = k_m i_a - b_d \omega_d - T_k$$

$$\frac{dT_k}{dt} = k_d \omega_d$$

with $x = [i_a \quad \omega_d \quad T_k]^T$, $u = [v_a(t)]$, and $y = [\omega_d \quad T_k]^T$ we have the state-space model

$$\dot{x} = Ax + Bu$$

$$y = Cx + Du$$

The model matrices are:

$$A = \begin{bmatrix} -R_a/L_a & -k_m/L_a & 0 \\ k_m/J_d & -b_d/J_d & -1/J_d \\ 0 & k_d & 0 \end{bmatrix}; \quad B = \begin{bmatrix} 1/L_a \\ 0 \\ 0 \end{bmatrix}; \quad C = \begin{bmatrix} 0 & 1 & 0 \\ 0 & 0 & 1 \end{bmatrix}; \quad D = 0$$

Note 1: This is a multidomain (electro-mechanical model).
Note 2: Multifunctional devices (e.g., a piezoelectric device that serves as both actuator and sensor) may be modeled similarly, using an electro-mechanical transformer (or, through the use of the "reciprocity principle").

4.5.3 Linear Graphs of Thermal Systems

Thermal systems have temperature (T) as the across variable, as it is always measured with respect to some reference (or as a temperature difference across an element), and heat transfer (flow) rate (Q) as the through variable. Heat source and temperature source are the two types of source elements. The former is more common. The latter may correspond to a large reservoir whose temperature is virtually not affected by heat transfer into or out of it. There is only one type of energy (thermal energy) in a thermal system. Hence there is only one type (A-type) energy storage element with the associated state variable, temperature. As discussed in Chapter 2, there is no T-type element in a thermal system.

4.5.3.1 Model Equations

In developing the model equations for a thermal system, the usual procedure is followed as for any other system. Specifically we write:

1. Constitutive equations (for thermal resistance and capacitance elements)
2. Node equations (the sum of heat transfer rate at a node is zero)
3. Loop equations (the sum of the temperature drop around a closed thermal path is zero)

Finally, we obtain the state-space model by eliminating the auxiliary variables that are not needed.

Example 4.10

A traditional Asian pudding is made by blending roughly equal portions by volume of treacle (a palm honey similar to maple syrup), coconut milk, and eggs, spiced with cloves and cardamoms, and baking in a special oven for about 1 hour. The traditional oven uses charcoal fire in an earthen pit that is well insulated, as the heat source. An aluminum container half filled with water is placed on fire. A smaller aluminum pot containing the dessert mixture is placed inside the water bath and covered fully with an aluminum lid. Both the water and the dessert mixture are well stirred and assumed to have uniform temperatures. A simplified model of the oven is shown in Figure 4.22a.

Assume that the thermal capacitances of the aluminum water container, dessert pot, and the lid are negligible. Also, the following equivalent (linear) parameters and variables are defined:

C_r = thermal capacitance of the water bath
C_d = thermal capacitance of the dessert mixture
R_r = thermal resistance between the water bath and the ambient air
R_d = thermal resistance between the water bath and the dessert mixture
R_c = thermal resistance between the dessert mixture and the ambient air, through the covering lid
T_r = temperature of the water bath
T_d = temperature of the dessert mixture
T_s = ambient temperature
Q = input heat flow rate from the charcoal fire into the water bath.

a. Assuming that T_d is the output of the system, develop a complete state-space model for the system. What are the system inputs?
b. In (a) suppose that the thermal capacitance of the dessert pot is not negligible, and is given by C_p. Also, as shown in Figure 4.22b, thermal resistances R_{p1} and R_{p2} are defined for the two interfaces of the pot. Assuming that the pot temperature is maintained uniform at T_p show how the state-space model of part (a) should be modified to include this improvement. What parameters do R_{p1} and R_{p2} depend on?
c. Draw the linear graphs for the systems in (a) and (b). Indicate in the graph only the system parameters, input variables, and the state variables.

Solution

a. For the water bath:

$$C_w \frac{dT_w}{dt} = Q - \frac{1}{R_w}(T_w - T_a) - \frac{1}{R_d}(T_w - T_d) \qquad (i)$$

For the dessert mixture:

$$C_d \frac{dT_d}{dt} = \frac{1}{R_d}(T_w - T_d) - \frac{1}{R_c}(T_d - T_a) \qquad (ii)$$

Equations (i) and (ii) are the state equations with:

$$\text{State vector } \mathbf{x} = \begin{bmatrix} T_w, & T_d \end{bmatrix}^T$$

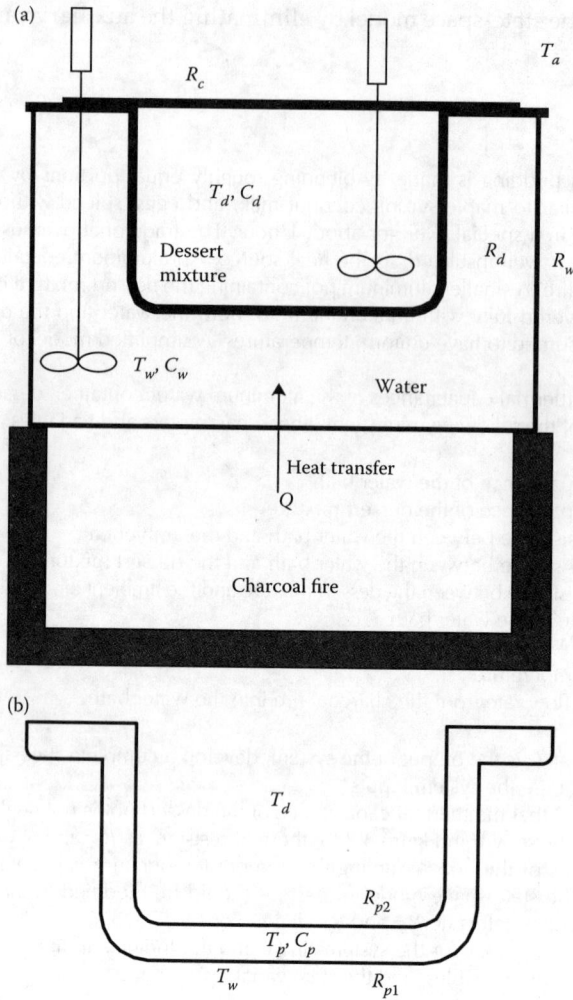

FIGURE 4.22
(a) A simplified model of an Asian dessert oven. (b) An improved model of the dessert pot.

Input vector $\boldsymbol{u} = \begin{bmatrix} Q, & T_a \end{bmatrix}^T$

Output vector $\boldsymbol{y} = \begin{bmatrix} T_d \end{bmatrix}^T$

The corresponding matrices of the state-space model are:

$$\boldsymbol{A} = \begin{bmatrix} -\dfrac{1}{C_w}\left(\dfrac{1}{R_w} + \dfrac{1}{R_d}\right) & \dfrac{1}{C_w R_d} \\[4mm] \dfrac{1}{C_d R_d} & -\dfrac{1}{C_d}\left(\dfrac{1}{R_d} + \dfrac{1}{R_c}\right) \end{bmatrix}; \quad \boldsymbol{B} = \begin{bmatrix} \dfrac{1}{C_w} & \dfrac{1}{C_w R_w} \\[4mm] 0 & \dfrac{1}{C_d R_c} \end{bmatrix}; \quad \boldsymbol{C} = \begin{bmatrix} 0 & 1 \end{bmatrix}; \quad \boldsymbol{D} = \begin{bmatrix} 0 & 0 \end{bmatrix}$$

b. For the dessert pot:

$$C_p \frac{dT_p}{dt} = \frac{1}{R_{p1}}(T_w - T_p) - \frac{1}{R_{p2}}(T_p - T_d)$$ (iii)

Equations (i) and (ii) have to be modified as

$$C_w \frac{dT_w}{dt} = Q - \frac{1}{R_w}(T_w - T_a) - \frac{1}{R_{p1}}(T_w - T_p)$$ (i*)

$$C_d \frac{dT_d}{dt} = \frac{1}{R_{p2}}(T_w - T_d) - \frac{1}{R_c}(T_d - T_a)$$ (ii*)

The system has become third order now, with the state Equations (i*), (ii*), and (iii) and the corresponding state vector:

$$\mathbf{x} = \begin{bmatrix} T_w & T_d & T_p \end{bmatrix}^T$$

But $\mathbf{u}$ and $\mathbf{y}$ remain the same as before. Matrices $\mathbf{A}$, $\mathbf{B}$, and $\mathbf{C}$ have to be modified accordingly. The resistance R_{pi} depends on the heat transfer area A_i and the heat transfer coefficient h_i. Specifically,

$$R_{pi} = \frac{1}{h_i A_i}$$

c. The linear graph for (a) is shown in Figure 4.23a. The linear graph for (b) is shown in Figure 4.23b.

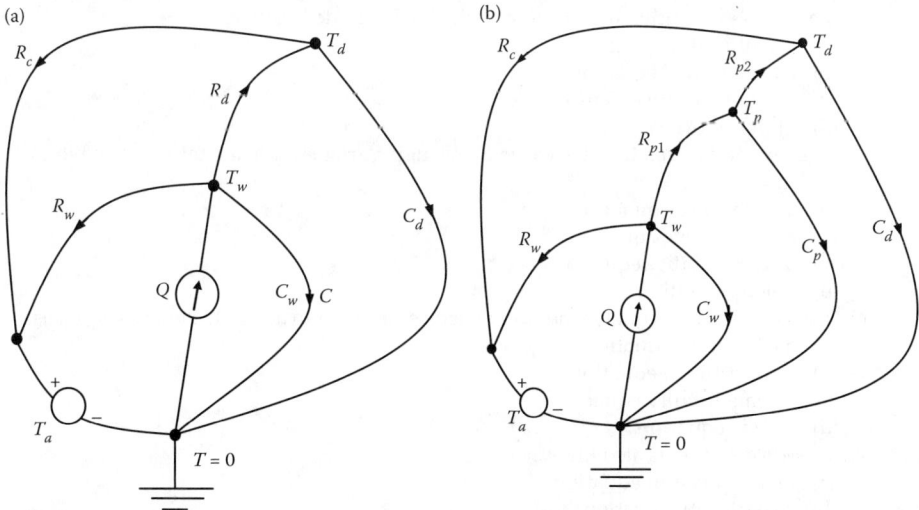

FIGURE 4.23
Linear graph of the (a) simplified model and (b) improved model.

Problems

PROBLEM 4.1

Select the correct answer for each of the following multiple-choice questions:

   i. A through variable is characterized by
      (a) being the same at both ends of the element
      (b) being listed first in the pair representation of a linear graph
      (c) requiring no reference value
      (d) all the above
   ii. An across variable is characterized by
      (a) having different values across the element
      (b) being listed second in the pair representation
      (c) requiring a reference point
      (d) all the above
  iii. Which of the following could be a through variable?
      (a) pressure
      (b) voltage
      (c) force
      (d) all the above
   iv. Which of the following could be an across variable?
      (a) motion (velocity)
      (b) fluid flow
      (c) current
      (d) all the above
    v. If angular velocity is selected as an element's across variable, the accompanying
       through variable is
      (a) force
      (b) flow
      (c) torque
      (d) distance
   vi. The equation written for through variables at a node is called
      (a) a continuity equation
      (b) a constitutive equation
      (c) a compatibility equation
      (d) all the above
  vii. The functional relation between a through variable and its across variable is
       called
      (a) a continuity equation
      (b) a constitutive equation
      (c) a compatibility equation
      (d) a node equation
 viii. The equation that equates the sum of across variables in a loop to zero is known as
      (a) a continuity equation
      (b) a constitutive equation
      (c) a compatibility equation
      (d) a node equation
   ix. A node equation is also known as
      (a) an equilibrium equation
      (b) a continuity equation
      (c) the balance of through variables at the node
      (d) all the above

x. A loop equation is
  (a) a balance of across variables
  (b) a balance of through variables
  (c) a constitutive relationship
  (d) all the above

PROBLEM 4.2

A linear graph has ten branches, two sources, and six nodes.

  i. How many unknown variables are there?
  ii. What is the number of independent loops?
  iii. How many inputs are present in the system?
  iv. How many constitutive equations could be written?
  v. How many independent continuity equations could be written?
  vi. How many independent compatibility equations could be written?
  vii. Do a quick check on your answers.

PROBLEM 4.3

The circuit shown in Figure P4.3 has an inductor L, a capacitor C, a resistor R, and a voltage source $v(t)$. Considering that L is analogous to a spring, and C is analogous to an inertia, follow the standard steps to obtain the state equations. First sketch the linear graph denoting the currents through and the voltages across the elements L, C, and R by (f_1, v_1), (f_2, v_2) and (f_3, v_3), respectively, and then proceed in the usual manner.

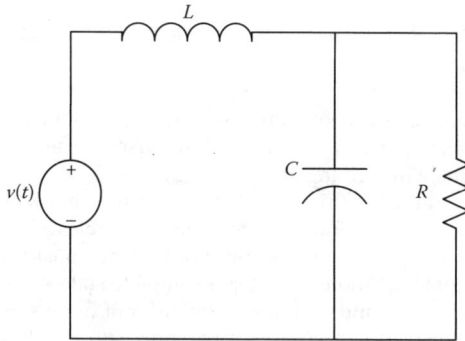

FIGURE P4.3
An electrical circuit.

  i. What is the system matrix and what is the input distribution matrix for your choice of state variables?
  ii. What is the order of the system?
  iii. Briefly explain what happens if the voltage source $v(t)$ is replaced by a current source $i(t)$.

PROBLEM 4.4

Consider an automobile traveling at a constant speed on a rough road, as sketched in Figure P4.4a. The disturbance input due to road irregularities can be considered as a velocity source $u(t)$ at the tires in the vertical direction. An approximate one-dimensional model shown in Figure P4.4b may be used to study the "heave" (up and down) motion

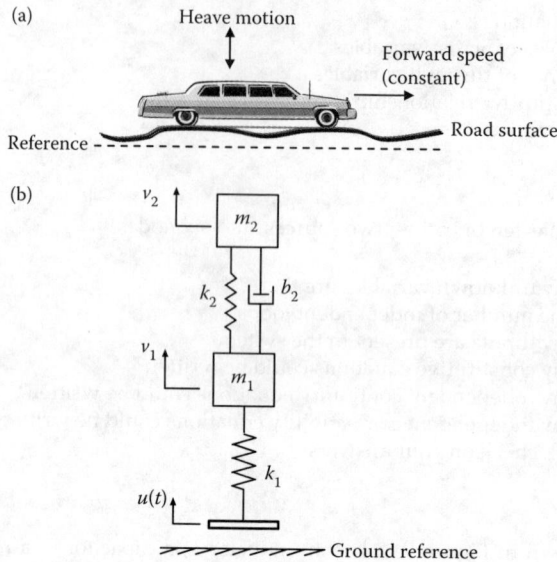

FIGURE P4.4
(a) An automobile traveling at constant speed. (b) A crude model of the automobile for the heave motion analysis.

of the automobile. Note that v_1 and v_2 are the velocities of the lumped masses m_1 and m_2, respectively.

a. Briefly state what physical components of the automobile are represented by the model parameters k_1, m_1, k_2, m_2, and b_2. Also, discuss the validity of the assumptions that are made in arriving at this model.
b. Draw a linear graph for this model, orient it (i.e., mark the directions of the branches), and completely indicate the system variables and parameters.
c. By following the step-by-step procedure of writing constitutive equations, node equations and loop equations, develop a complete state-space model for this system. The outputs are v_1 and v_2. What is the order of the system?
d. If instead of the velocity source $u(t)$, a force source $f(t)$ which is applied at the same location, is considered as the system input, draw a linear graph for this modified model. Obtain the state equations for this model. What is the order of the system now?

Note: In this problem you may assume that the gravitational effects are completely balanced by the initial compression of the springs with reference to which all motions are defined.

PROBLEM 4.5

Suppose that a linear graph has the following characteristics:
   n = number of nodes
   b = number of branches (segments)
   s = number of sources
   l = number of independent loops.

Carefully explaining the underlying reasons, answer the following questions regarding this linear graph:

a. From the topology of the linear graph show that $l=b-n+1$.
b. What is the number of continuity equations required (in terms of n)?
c. What is the number of lumped elements including source elements in the model (expressed in terms of b and s)?
d. What is the number of unknown variables, both state and auxiliary, (expressed in terms of b and s)? Verify that this is equal to the number available equations, and hence the problem is solvable.

PROBLEM 4.6

An approximate model of a motor-compressor combination used in a process control application is shown n Figure P4.6.

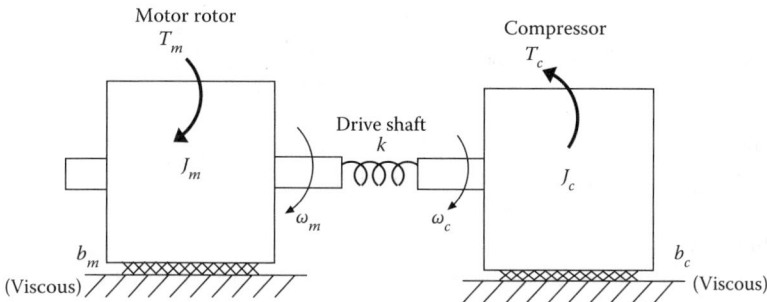

FIGURE P4.6
A model of a motor-compressor unit.

Note that T, J, k, b, and ω denote torque, moment of inertia, torsional stiffness, angular viscous damping constant, and angular speed, respectively, and the subscripts m and c denote the motor rotor and compressor impeller, respectively.

a. Sketch a translatory mechanical model that is analogous to this rotatory mechanical model.
b. Draw a linear graph for the given model, orient it, and indicate all necessary variables and parameters on the graph.
c. By following a systematic procedure and using the linear graph, obtain a complete state-space representation of the given model. The outputs of the system are compressor speed ω_c and the torque T transmitted through the drive shaft.

PROBLEM 4.7

A model for a single joint of a robotic manipulator is shown in Figure P4.7. The usual notation is used. The gear inertia is neglected and the gear reduction ratio is taken as $1{:}r$ (*Note*: $r < 1$).

a. Draw a linear graph for the model, assuming that no external (load) torque is present at the robot arm.
b. Using the linear graph derive a state model for this system. The input is the motor magnetic torque T_m and the output is the angular speed ω_r of the robot arm. What is the order of the system?

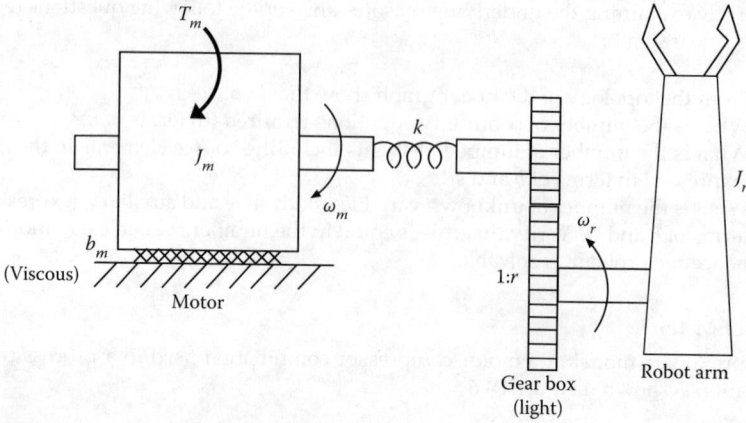

FIGURE P4.7

A model of a single-degree-of-freedom robot.

c. Discuss the validity of various assumptions made in arriving at this simplified model for a commercial robotic manipulator.

PROBLEM 4.8

Consider the rotary feedback control system shown schematically by Figure P4.8a. The load has inertia J, stiffness K and equivalent viscous damping B as shown. The armature circuit for the dc fixed field motor is shown in Figure P4.8b.

The following relations are known:

The back e.m.f. $v_B = K_V \omega$

The motor torque $T_m = K_T i$

a. Identify the system inputs.
b. Write the linear system equations.

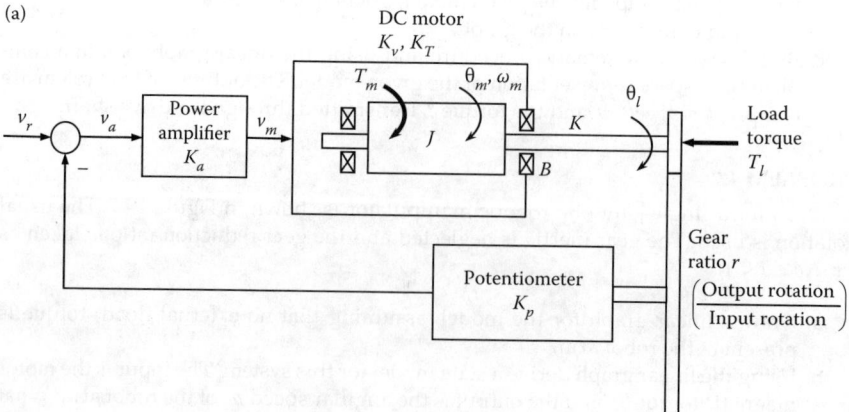

FIGURE P4.8

(a) A rotatory electromechanical system. (b) The armature circuit.

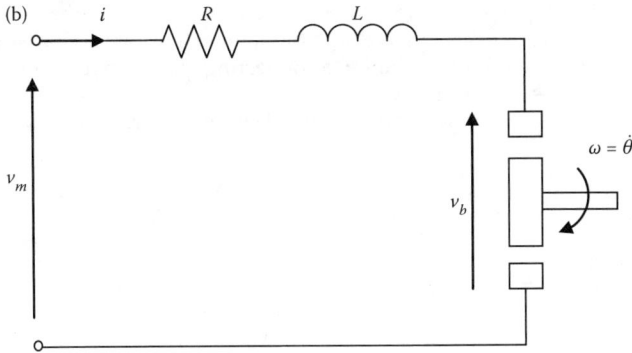

FIGURE P4.8 (continued)

PROBLEM 4.9

a. What is the main physical reason for oscillatory behavior in a purely fluid system? Why do purely fluid systems with large tanks connected by small-diameter pipes rarely exhibit an oscillatory response?

b. Two large tanks connected by a thin horizontal pipe at the bottom level are shown in Figure P4.9a. Tank 1 receives an inflow of liquid at the volume rate Q_i when the inlet valve is open. Tank 2 has an outlet valve, which has a fluid flow resistance of R_o and a flow rate of Q_o when opened. The connecting pipe also has a valve, and when opened, the combined fluid flow resistance of the valve and the thin pipe is R_p. The following parameters and variables are defined:

C_1, C_2=fluid (gravity head) capacitances of Tanks 1 and 2
ρ=mass density of the fluid
g=acceleration due to gravity
P_1, P_2=pressure at the bottom of Tanks 1 and 2
P_0=ambient pressure.

Using $P_{10} = P_1 - P_0$ and $P_{20} = P_2 - P_0$ as the state variables and the liquid levels H_1 and H_2 in the two tanks as the output variables, derive a complete, linear, state-space model for the system.

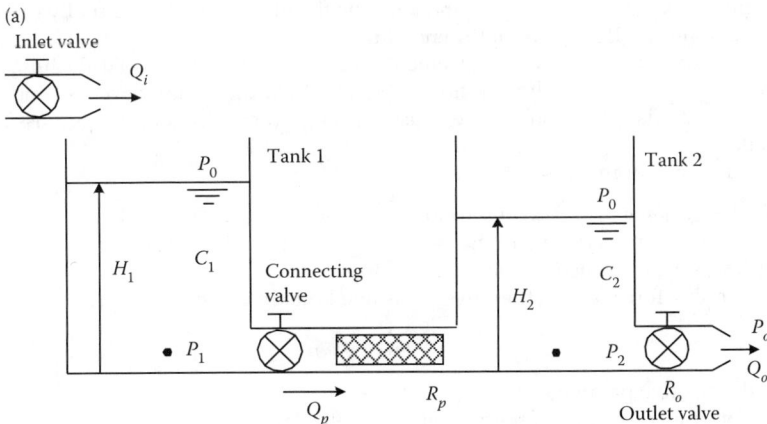

FIGURE P4.9
(a) An interacting two-tank fluid system. (b) A noninteracting two-tank fluid system.

c. Suppose that the two tanks are as in Figure P4.9b. Here Tank 1 has an outlet valve at its bottom whose resistance is R_t and the volume flow rate is Q_t when open. This flow directly enters Tank 2, without a connecting pipe. The remaining characteristics of the tanks are the same as in (b).

Derive a state-space model for the modified system in terms of the same variables as in (b).

(b)

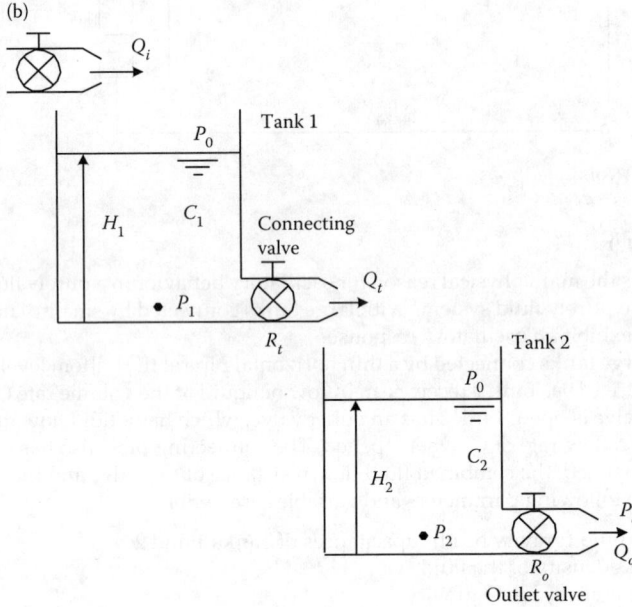

FIGURE P4.9 (continued)

PROBLEM 4.10

Give reasons for the common experience that in the flushing tank of a household toilet, some effort is needed to move the handle for the flushing action but virtually no effort is needed to release the handle at the end of the flush.

A simple model for the valve movement mechanism of a household flushing tank is shown in Figure P4.10. The overflow tube on which the handle lever is hinged, is assumed rigid. Also, the handle rocker is assumed light, and the rocker hinge is assumed frictionless.

The following parameters are indicated in the figure:

$r = l_v / l_h$ = the lever arm ratio of the handle rocker
m = equivalent lumped mass of the valve flapper and the lift rod
k = stiffness of spring action on the valve flapper.
The damping force f_{NLD} on the valve is assumed quadratic and is given by

$$f_{NLD} = a|v_{VLD}|v_{VLD}$$

where the positive parameter:
$a = a_u$ for upward motion of the flapper ($v_{NLD} \geq 0$)
$= a_d$ for downward motion of the flapper ($v_{NLD} < 0$)
with $a_u \gg a_d$

The force applied at the handle is $f(t)$, as shown.

We are interested in studying the dynamic response of the flapper valve. Specially, the valve displacement x and the valve speed v are considered outputs, as shown in Figure P4.10. Note that x is measured from the static equilibrium point of the spring where the weight mg is balanced by the spring force.

a. By defining appropriate through variables and across variables, draw a linear graph for the system shown in Figure P4.10. Clearly indicate the power flow arrows.
b. Using valve speed and the spring force as the state variables, develop a (nonlinear) state-space model for the system, with the aid of the linear graph. Specifically, start with all the constitutive, continuity, and compatibility equations, and eliminate the auxiliary variables systematically, to obtain the state-space model.
c. Linearize the state-space model about an operating point where the valve speed is $\bar{v}$. For the linearized model, obtain the model matrices A, B, C, and D, in the usual notation. The incremental variables $\hat{x}$ and $\hat{v}$ are the outputs in the linear model, and the incremental variable $\hat{f}(t)$ is the input.
d. From the linearized state-space model, derive the input–output model (differential equation) relating $\hat{f}(t)$ and $\hat{x}$.

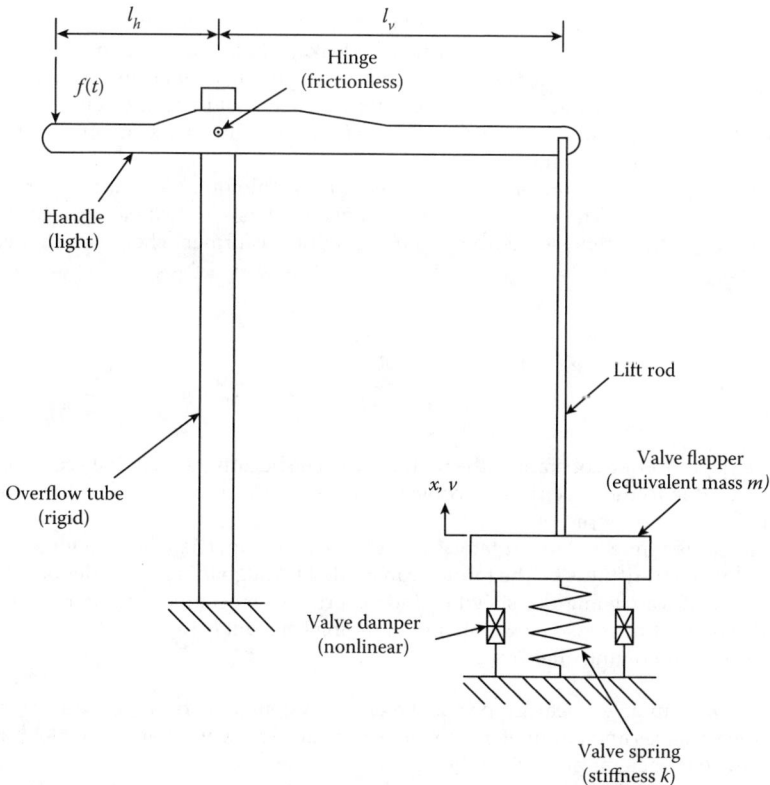

FIGURE P4.10
Simplified model of a toilet-flushing mechanism.

PROBLEM 4.11

A common application of dc motors is in accurate positioning of a mechanical load. A schematic diagram of a possible arrangement is shown in Figure P4.11. The actuator of the system is an armature-controlled dc motor. The moment of inertia of its rotor is J_r and the angular speed is ω_r. The mechanical damping of the motor (including that of its bearings) is neglected in comparison to that of the load.

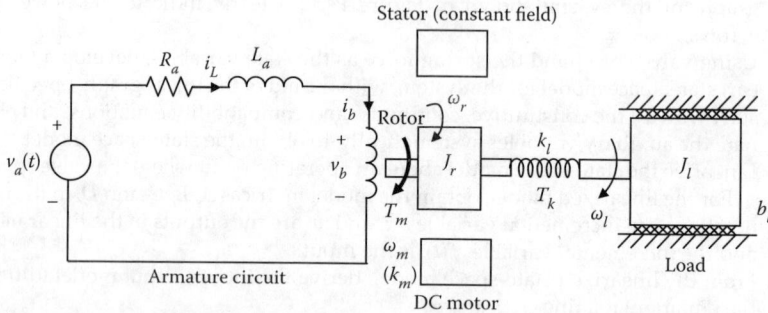

FIGURE P4.11
An electro-mechanical model of a rotatory positioning system.

The armature circuit is also shown in Figure P4.11, which indicates a back e.m.f. v_b (due to the motor rotation in the stator field), a leakage inductance L_a, and a resistance R_a. The current through the leakage inductor is i_L. The input signal is the armature voltage $v_a(t)$ as shown. The interaction of the rotor magnetic field and the stator magnetic field (*Note*: the rotor field rotates at an angular speed ω_m) generates a "magnetic" torque T_m which is exerted on the motor rotor.

The stator provides a constant magnetic field to the motor, and is not important in the present problem. The dc motor may be considered as an ideal electromechanical transducer which is represented by a linear-graph transformer. The associated equations are:

$$\omega_m = \frac{1}{k_m} v_b$$

$$T_m = -k_m i_b$$

where k_m is the torque constant of the motor. *Note*: The negative sign in the second equation arises due to the specific sign convention used for a transformer, in the conventional linear graph representation.

The motor is connected to a rotatory load of moment of inertia J_l using a long flexible shaft of torsional stiffness k_l. The torque transmitted through this shaft is denoted by T_k. The load rotates at an angular speed ω_l and experiences mechanical dissipation, which is modeled by a linear viscous damper of damping constant b_l.

Answer the following questions:

a. Draw a suitable linear graph for the entire system shown in Figure P4.11, mark the variables and parameters (you may introduce new, auxiliary variables but not new parameters), and orient the graph.

b. Give the number of branches (*b*), nodes (*n*), and the independent loops (*l*) in the complete linear graph. What relationship do these three parameters satisfy? How many independent node equations, loop equations, and constitutive equations can be written for the system? Verify the sufficiency of these equations to solve the problem.

c. Take current through the inductor (i_L), speed of rotation of the motor rotor (ω_r), torque transmitted through the load shaft (T_k), and speed of rotation of the load (ω_l) as the four state variables; the armature supply voltage $v_a(t)$ as the input variable; and the shaft torque T_k and the load speed ω_l as the output variables. Write the independent node equations, independent loop equations, and the constitutive equations for the complete linear graph. Clearly show the state-space shell.

d. Eliminate the auxiliary variables and obtain a complete state-space model for the system, using the equations written in (c) above. Express the matrices *A, B, C,* and *D* of the state-space model in terms of the system parameters R_a, L_a, k_m, J_r, k_l, b_l, and J_l only.

PROBLEM 4.12

Consider a multidomain engineering system that you are familiar with (in your projects, research, engineering practice, informed imagination, through literature which you have read, etc.). It should include the mechanical structural domain (i.e., with inertia, flexibility, and damping) and at least one other domain (e.g., electrical, fluid, thermal).

a. Using sketches, describe the system, by giving at least the following information:
   (i) The practical purpose and functions of the system.
   (ii) Typical operation/behavior of the system.
   (iii) System boundary.
   (iv) Inputs and outputs.
   (v) Characteristics of the main components of the system.
b. Sketch a lumped-parameter model of the system, by approximating any significant distributed effects using appropriate lumped elements, and showing how the lumped-parameter elements (including sources) are interconnected. You must justify your choice of elements and approximation decisions. Also, you must retain significant nonlinearities in the original system.
c. Develop an analytical model of the system by writing the necessary constitutive equations, continuity equations, and compatibility equations. The model should be at least fifth-order but not greater than tenth-order.
   Note: Draw a linear graph of the system (particularly if you plan to use the linear graph approach to obtain the analytical model).
d. Approximate the nonlinear elements by suitable linear elements.
e. Identify suitable state variables for the linear system and develop a complete state-space model (i.e., matrices *A, B, C,* and *D*) for the system.

5

Transfer-Function and Frequency-Domain Models

Transfer-function models (strictly, Laplace transfer-functions) are based on the Laplace transform, and are versatile means of representing linear systems with constant (time-invariant) parameters. Strictly, these are dynamic models in the Laplace domain. Frequency-domain models (or frequency transfer-functions) are a special category of Laplace domain models, and they are based on the Fourier transform. However, they are interchangeable—a Laplace domain model can be converted into the corresponding frequency domain model in a trivial manner, and vice versa. Similarly linear, constant-coefficient (time-invariant) time-domain model (e.g., input–output differential equation or a state-space model) can be converted into a transfer-function, and vice versa, in a simple and straightforward manner. A system with just one input (excitation) and one output (response) can be represented uniquely by one transfer-function. When a system has two or more inputs (i.e., an input vector) and/or two or more outputs (i.e., and output vector), its representation needs several transfer-functions (i.e., a transfer-function matrix is needed). The response characteristics at a given location (more correctly, in a given degree of freedom) can be determined using a single frequency-domain transfer-function.

Transfer-function models were widely used in early studies of dynamic systems because they are algebraic functions rather than differential equations. In view of the simpler algebraic operations that are involved in transfer-function approaches, a substantial amount of information regarding the dynamic behavior of a system can be obtained with minimal computational effort. This is the primary reason for the popularity enjoyed by the transfer-function methods prior to the advent of the digital computer. One might think that the abundance of high-speed, low-cost, digital computers would lead to the dominance of time-domain methods, over frequency-domain transfer-function methods. But there is evidence to the contrary in many areas, particularly in dynamic systems and control, due to the analytical simplicity and the intuitive appeal of transfer-function techniques. Only a minimal knowledge of the theory of Laplace transform and Fourier transform is needed to use transfer-function methods in system modeling, analysis, design, and control. Techniques of transfer-function models, both in the Laplace domain and the frequency (Fourier) domain are treated in the present chapter.

5.1 Laplace and Fourier Transforms

The logarithm is a transform which coverts the multiplication operation into an addition and the division operation into a subtraction, thereby making the analysis simpler. In a similar manner the Laplace transform converts differentiation into a multiplication by the Laplace variable s; and integration into a division by s, thereby providing significant analytical convenience. Fourier transform may be considered as a special case of the Laplace transform. The corresponding results can be obtained simply by setting $s = j\omega$, where ω is the frequency variable. Further details are found in Appendix A.

5.1.1 Laplace Transform

The Laplace transform involves the mathematical transformation from the time-domain to the Laplace domain (also termed s-domain or complex frequency domain) according to:

$$Y(s) = \int_0^\infty y(t)\exp(-st)\,dt \quad \text{or} \quad Y(s) = \mathcal{L}y(t) \tag{5.1}$$

where the Laplace operator $= \mathcal{L}$; Laplace variable $s = \sigma + j\omega$ and $j = \sqrt{-1}$.

Note: The real value σ is chosen sufficiently large so that the transform integral (Equation 5.1) is finite even when $\int y(t)dt$ is not finite.

The inverse Laplace transform is:

$$y(t) = \frac{1}{2\pi j} \int_{\sigma - j\infty}^{\sigma + j\infty} Y(s)\exp(st)\,ds \quad \text{or} \quad y(t) = \mathcal{L}^{-1}Y(s) \tag{5.2}$$

which is obtained simply through mathematical manipulation (multiplying both side by appropriate exponential and integration) of the forward transform (Equation 5.1).

5.1.2 Laplace Transform of a Derivative

Using Equation 5.1, the Laplace transform of the time derivative $\dot{y} = (dy/dt)$ may be determined as:

$$\mathcal{L}\dot{y} = \int_0^\infty e^{-st}\frac{dy}{dt}\,dt = sY(s) - y(0) \tag{5.3}$$

Note: Integration by parts: $\int u\,dv = uv - \int v\,du$ is used in obtaining the result (Equation 5.3). Also $y(0)$ is the initial condition (IC) of $y(t)$ at $t = 0$.

By repeatedly applying Equation 5.2 we can get the Laplace transform of the higher derivatives; specifically,

$$\mathcal{L}\ddot{y}(t) = s\mathcal{L}[\dot{y}(t)] - \dot{y}(0) = s[sY(s) - y(0)] - \dot{y}(0)$$

gives the result

$$\mathcal{L}\ddot{y}(t) = s^2\mathcal{L}[y(t)] - sy(0) - \dot{y}(0) \tag{5.4}$$

Similarly we obtain

$$\mathcal{L}\dddot{y} = s^3 Y(s) - s^2 y(0) - s\dot{y}(0) - \ddot{y}(0) \tag{5.5}$$

The general result is:

$$\mathcal{L}\frac{d^n y(t)}{dt^n} = s^n Y(s) - s^{n-1} y(0) - s^{n-2} \dot{y}(0) - \cdots - \frac{d^{n-1} y}{dt^{n-1}}(0) \tag{5.6}$$

Note: With zero ICs, we have:

$$\mathcal{L}\frac{d^n y(t)}{dt^n} = s^n Y(s) \tag{5.7}$$

or, the time derivative corresponds to multiplication by s in the Laplace domain. As a result, differential equations (time-domain models) become algebraic equations (transfer-functions) resulting in easier mathematics. From the result (Equation 5.7) it is clear that the Laplace variable s can be interpreted as the *derivative operator* in the context of a dynamic system.

Note: ICs can be added separately to any model (e.g., Laplace model) after using Equation 5.7. Hence in model transformation, first the ICs are assumed zero.

5.1.3 Laplace Transform of an Integral

The Laplace transform of the time integral $\int_0^t y(\tau)d\tau$ is obtained by the direct application of Equation 5.1 as:

$$\mathcal{L}\int_0^t y(\tau)d\tau = \int_0^\infty e^{-st}\int_0^t y(\tau)d\tau dt = \int_0^\infty \left(-\frac{1}{s}\right)\frac{d}{dt}(e^{-st})\int_0^t y(\tau)d\tau dt$$

Integrate by parts: $\int u dv = uv - \int v du$ gives

$$\mathcal{L}\int_0^t y(\tau)d\tau = \left(-\frac{1}{s}\right)e^{-st}\int_0^t y(\tau)d\tau \Big|_0^\infty - \int_0^\infty \left(-\frac{1}{s}\right)e^{-st}y(t)dt = 0 - 0 + \int_0^\infty \left(\frac{1}{s}\right)e^{-st}y(t)dt$$

The final result is

$$\mathcal{L}\int_0^t y(\tau)d\tau = \frac{1}{s}Y(s) \tag{5.8}$$

It follows that time integration becomes multiplication by $1/s$ in the Laplace domain. In particular $1/s$ can be interpreted as the *integration operator*, in the context of a dynamic system.

5.1.4 Fourier Transform

The Fourier transform involves the mathematical transformation from the time-domain to the frequency domain according to:

$$Y(j\omega) = \int_{-\infty}^\infty y(t)\exp(-j\omega t)\,dt \quad \text{or} \quad Y(j\omega) = \mathcal{F}\,y(t) \tag{5.9}$$

where cyclic frequency variable$=f$; and angular frequency variable$=\omega=2\pi f$

The inverse Fourier transform is:

$$y(t)=\frac{1}{2\pi}\int_{-\infty}^{\infty}Y(j\omega)\exp(j\omega t)\,d\omega \quad \text{or} \quad y(t)=\mathcal{F}^{-1}\,Y(j\omega) \tag{5.10}$$

which is obtained simply through mathematical manipulation (multiplying both side by appropriate exponential and integration) of the forward transform (Equation 5.9).

By examining the transforms (Equations 5.1 and 5.9) it is clear that the conversion from the Laplace domain into the Fourier (frequency) domain may be done simply by setting $s=j\omega$. Strictly, the one-sided Fourier transform is used here (where the lower limit of integration in Equation 5.9 is set to $t=0$) because it then that Equation 5.1 becomes equal to Equation 5.1 with $s=j\omega$.

We summarize these results and observations below.

Laplace transform:

Time-domain → Laplace (complex frequency) domain

Time derivative → Laplace variable s

Differential equations → Algebraic equations (easier math)

Time integration → $1/s$

Fourier transform:

Time-domain → Frequency domain

Conversion from Laplace to Fourier (one-sided): Set $s=j\omega$

In using techniques of Laplace transform, the general approach is to first convert the time-domain problem into a s-domain problem (conveniently, by using Laplace transform tables); perform the necessary analysis (algebra rather than calculus) in the s-domain; and convert the results back into the time-domain (again, conveniently using Laplace transform tables). Further discussion, techniques and Laplace tables are found in Appendix A.

5.2 Transfer-Function

The transfer-function is a dynamic model represented in the Laplace domain. Specifically, the transfer-function $G(s)$ of a linear, time-invariant, single-input–single-output (SISO) system is given by the ratio of the Laplace-transformed output to the Laplace-transformed input, assuming zero initial conditions (zero ICs). This is a unique function, which represents the system (model); it does not depend on the input, the output, or the initial conditions. A physically realizable linear, constant-parameter system possesses a unique transfer-function even if the Laplace transforms of a particular input to the system and the corresponding output do not exist. For example, suppose that the Laplace transform of a particular input $u(t)$ is infinite. Then the Laplace transform of the corresponding output $y(t)$ will also be infinite. But the transfer-function itself will be finite.

Consider the nth-order linear, constant-parameter system given by:

$$a_n \frac{d^n y}{dt^n} + a_{n-1} \frac{d^{n-1} y}{dt^{n-1}} + \cdots + a_0 y = b_0 u + b_1 \frac{du}{dt} + \cdots + b_m \frac{d^m u}{dt^m} \tag{5.11}$$

Note: For the time being we will assume $m < n$, or at worst $m \le n$ when the corresponding systems are said to be *physically realizable*. For systems that possess dynamic delay (i.e., systems whose response does not tend to feel the excitation either instantly or ahead of time, or systems whose excitation or its derivatives are not directly fed forward to the output, we will have $m < n$. These are the systems that concern us most in real applications.

Use the result (Equation 5.7) in Equation 5.11, assuming zero ICs. We obtain the transfer-function:

$$\frac{Y(s)}{U(s)} = G(s) = \frac{b_0 + b_1 s + \cdots + b_m s^m}{a_0 + a_1 s + \cdots + a_n s^n} \tag{5.12}$$

It should be clear from Equations 5.11 and 5.12 that the transfer-function corresponding to a system differential equation can be written simply by inspection, without requiring any knowledge of Laplace-transform theory. Conversely, once the transfer-function is given, the corresponding time-domain (differential) equation should be immediately obvious.

Note: The dominator polynomial of a transfer-function is called the *characteristic polynomial*, and the corresponding equation is called the *characteristic equation*: $a_0 + a_1 s + \cdots + a_n s^n = 0$. These topics will be discussed later (in Chapter 6, in particular).

Transfer-functions are simple algebraic expressions. Differential equations are transformed into simple algebraic relations through the Laplace transform. This is a major advantage of the transfer-function approach. Once the analysis is performed using transfer-functions, the inverse Laplace transform can convert the results into the corresponding time-domain results. This can be accomplished simply by using Laplace transform tables.

5.2.1 Transfer-Function Matrix

Consider the state variable representation of a linear, time-invariant system:

$$\dot{x} = Ax + Bu \tag{5.13}$$

$$y = Cx + Du \tag{5.14}$$

where $x(t)$ is an nth-order state vector; u is an rth-order input (excitation); and y is the mth-order output (response) vector. This is a multiinput–multioutput (MIMO) system. The corresponding transfer-function model relates the output vector y to the input vector u. We will need $m \times n$ transfer-functions, or a transfer-function matrix, to represent this MIMO system. To obtain an expression for this matrix, we first apply Laplace transform to Equations 5.13 and 5.14 with zero ICs for x. We get

$$sX(s) = AX(s) + BU(s) \tag{5.13a}$$

$$Y(s) = CX(s) + DU(s) \tag{5.14a}$$

From Equation 5.13a we have

$$X(s) = (sI - A)^{-1}BU(s) \tag{5.13b}$$

in which I is the nth-order identity matrix (a matrix with 1 as its diagonal elements and 0 for all other elements). By substituting Equation 5.13b into Equation 5.14a we get the transfer-function relation:

$$Y(s) = \left[C\left((sI - A)^{-1}B \right) + D \right] U(s) \tag{5.15a}$$

or

$$Y(s) = G(s)U(s) \tag{5.15}$$

The transfer-function matrix $G(s)$ is an $m \times n$ matrix given by:

$$G(s) = C\left((sI - A)^{-1}B \right) + D \tag{5.16a}$$

In practical systems with dynamic delay, the excitation $u(t)$ is not naturally fed forward to the response y; and as a result its is not instantaneously felt in the response y. Then we have $D = 0$ and Equation 5.16a becomes

$$G(s) = C(sI - A)^{-1}B \tag{5.16}$$

Several examples are presented to illustrate some approaches of obtaining transfer-function models when the time-domain (differential-equation) models are given.

Example 5.1

Consider the simple oscillator (mass-spring-damper) shown in Figure 5.1. Its dynamic equation is obtained in a straightforward manner as:

$$m\ddot{y} + b\dot{y} + ky = ku(t) \tag{5.17}$$

where the response (output) y of the mass is measured from its static equilibrium position (so that the gravitational force is balanced by the initial force in the spring). In Figure 5.1, the input $u(t)$ comes from the force applied to the mass:

$$f(t) = ku(t) \tag{5.18}$$

Alternatively $u(t)$ may be considered the displacement of the support structure (base), without an applied force f.

Take the Laplace transform of the system Equation 5.17 with zero ICs:

$$(ms^2 + bs + k)Y(s) = kU(s)$$

The corresponding transfer-function is:

$$G(s) = \frac{Y(s)}{U(s)} = \frac{k}{(ms^2 + bs + k)} \tag{5.19a}$$

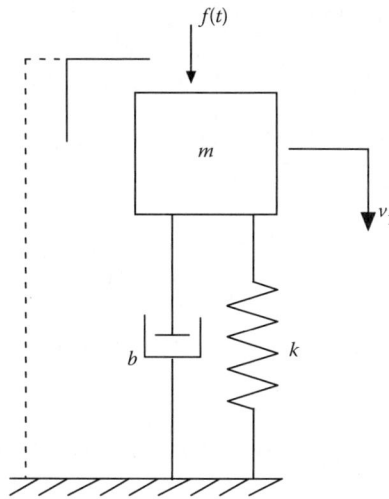

FIGURE 5.1
A damped simple oscillator.

or, by defining $\omega_n^2 = k/m$ and $2\zeta\omega_n = b/m$ where
Undamped natural frequency $= \omega_n$
Damping ratio $= \zeta$
we have the transfer-function:

$$G(s) = \frac{\omega_n^2}{(s^2 + 2\zeta\omega_n s + \omega_n^2)} \tag{5.19}$$

This is the transfer-function corresponding to the displacement output. It follows that the output velocity transfer-function (i.e., the transfer-function if the output is taken to be the velocity) is:

$$\frac{sY(s)}{U(s)} = sG(s) = \frac{s\omega_n^2}{(s^2 + 2\zeta\omega_n s + \omega_n^2)} \tag{5.20}$$

Similarly, the output acceleration transfer-function is:

$$\frac{s^2Y(s)}{U(s)} = s^2G(s) = \frac{s^2\omega_n^2}{(s^2 + 2\zeta\omega_n s + \omega_n^2)} \tag{5.21}$$

In Equation 5.21 we have the numerator order equal to the denominator order: $m=n=2$. This means that the input (applied force) is instantly felt by the acceleration of the mass, which may be verified experimentally by using an accelerometer (sensor). This corresponds to a feedforward of the input, or zero dynamic delay. For example, this is the primary mechanism through which road disturbances are felt inside a vehicle having hard suspensions.
Note: The characteristic equation of the system is

$$s^2 + 2\zeta\omega_n s + \omega_n^2 = 0 \tag{5.22}$$

Example 5.2

Let us consider again the simple oscillator of Example 5.1, given by:

$$\ddot{y} + 2\zeta\omega_n\dot{y} + \omega_n^2 y = \omega_n^2 u(t) \qquad (5.17b)$$

By defining the state variables as:

$$\mathbf{x} = \begin{bmatrix} x_1 & x_2 \end{bmatrix}^T = \begin{bmatrix} y & \dot{y} \end{bmatrix}^T$$

where y=position and $\dot{y}$ =velocity, a state model for this system can be expressed as:

$$\dot{\mathbf{x}} = \begin{bmatrix} 0 & 1 \\ -\omega_n^2 & 2\zeta\omega_n \end{bmatrix} \mathbf{x} + \begin{bmatrix} 0 \\ \omega_n^2 \end{bmatrix} u(t)$$

If we consider both displacement and velocity as outputs, we have:

$$\mathbf{y} = \mathbf{x}$$

Note: The output gain matrix (measurement matrix) $\mathbf{C}=\mathbf{I}$ (the identity matrix) and $\mathbf{D}=0$ in this case. From Equation 5.15a we get:

$$\mathbf{Y}(s) = \begin{bmatrix} s & -1 \\ \omega_n^2 & s+2\zeta\omega_n \end{bmatrix}^{-1} \begin{bmatrix} 0 \\ \omega_n^2 \end{bmatrix} U(s) = \frac{1}{(s^2 + 2\zeta\omega_n s + \omega_n^2)} \begin{bmatrix} s+2\zeta\omega_n & 1 \\ -\omega_n^2 & s \end{bmatrix} \begin{bmatrix} 0 \\ \omega_n^2 \end{bmatrix} U(s)$$

$$= \frac{1}{(s^2 + 2\zeta\omega_n s + \omega_n^2)} \begin{bmatrix} \omega_n^2 \\ s\omega_n^2 \end{bmatrix} U(s)$$

We observe that the transfer-function matrix is:

$$\mathbf{G}(s) = \begin{bmatrix} \omega_n^2/\Delta(s) \\ s\omega_n^2/\Delta(s) \end{bmatrix}$$

in which the *characteristic polynomial* of the system is $\Delta(s) = s^2 + 2\zeta\omega_n s + \omega_n^2$.

The first element in $\mathbf{G}(s)$ is the displacement-output transfer-function, and the second element is the velocity-output transfer-function. These results agree with the expressions obtained in Example 5.1.

Now, let us consider the acceleration $\ddot{y}$ as an output, and denote it by y_3. It is clear from the system Equation 5.17b that:

$$y_3 = \ddot{y} = -2\zeta\omega_n\dot{y} - \omega_n^2 y + \omega_n^2 u(t)$$

or, in terms of the state variables:

$$y_3 = -2\zeta\omega_n x_2 - \omega_n^2 x_1 + \omega_n^2 u(t)$$

Note that this output explicitly contains the input variable. This is a feedforward situation which implies that the matrix $\mathbf{D}$ becomes nonzero when acceleration $\ddot{y}$ is chosen as an output. In this case,

$$Y_3(s) = -2\zeta\omega_n X_2(s) - \omega_n^2 X_1(s) + \omega_n^2 U(s) = -2\zeta\omega_n \frac{s\omega_n^2}{\Delta(s)}U(s) - \omega_n^2 \frac{\omega_n^2}{\Delta(s)}U(s) + \omega_n^2 U(s)$$

which simplifies to

$$Y_3(s) = -2\zeta\omega_n X_2(s) - \omega_n^2 X_1(s) + \omega_n^2 U(s) = \frac{s\omega_n^2}{\Delta(s)}U(s)$$

This confirms the result for the acceleration-output transfer-function obtained in Example 5.1.

Example 5.3

Consider the simplified model of a vehicle shown in Figure 5.2, which can be used to study the heave (vertical up and down) and pitch (front-back rotation) motions due to the road profile and other disturbances. For our purposes, let us assume that the road disturbances exciting the front and back suspensions are independent. The equations of motion for heave (y) and pitch (θ) are written about the static equilibrium configuration of the vehicle model (hence, gravity does not enter into the equations) for small motions:

$$m\ddot{y} = k_1(u_1 - y + l_1\theta) + k_2(u_2 - y + l_2\theta) + b_1(\dot{u}_1 - \dot{y} + l_1\dot{\theta}) + b_2(\dot{u}_2 - \dot{y} + l_2\dot{\theta})$$
$$J\ddot{\theta} = -l_1\left[k_1(u_1 - y + l_1\theta) + b_1(\dot{u}_1 - \dot{y} + l_1\dot{\theta})\right] + l_2\left[k_2(u_2 - y + l_2\theta) + b_2(\dot{u}_2 - \dot{y} + l_2\dot{\theta})\right]$$

Take the Laplace transform of these two equations with zero ICs (i.e., substitute s^2Y for $\ddot{y}$, sY for $\dot{y}$, etc.):

$$\left[ms^2 + (b_1 + b_2)s + (k_1 + k_2)\right]Y(s) + \left[(b_2l_2 - b_1l_1)s + (k_2l_2 - k_1l_1)\right]\theta(s)$$
$$= (b_1s + k_1)U_1(s) + (b_2s + k_2)U_2(s)$$

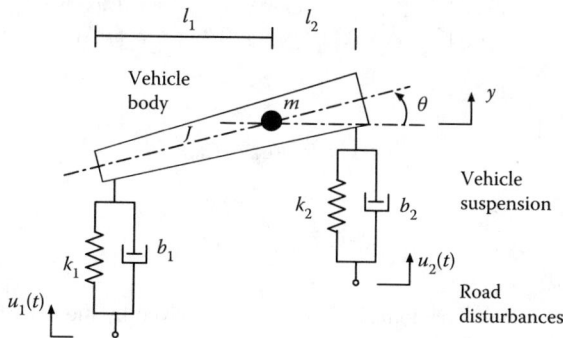

FIGURE 5.2
A model of a vehicle with its suspension system.

$$[(b_2 l_2 - b_1 l_1)s + (k_2 l_2 - k_1 l_1)]Y(s) + [Js^2 + (b_1 l_1^2 + b_2 l_2^2)s + (k_1 l_1^2 + k_2 l_2^2)]\theta(s)$$
$$= -l_1(b_1 s + k_1)U_1(s) + l_2(b_2 s + k_2)U_2(s)$$

Let the coefficients be expressed as:

$$C_1 = m \qquad\qquad C_7 = b_2 s + k_2$$

$$C_2 = b_1 + b_2 \qquad\qquad C_8 = J$$

$$C_3 = k_1 + k_2 \qquad\qquad C_9 = b_1 l_1^2 + b_2 l_2^2$$

$$C_4 = b_2 l_2 - b_1 l_1 \qquad\qquad C_{10} = k_1 l_1^2 + k_2 l_2^2$$

$$C_5 = k_2 l_2 - k_1 l_1 \qquad\qquad C_{11} = -l_1(b_1 s + k_1)$$

$$C_6 = b_1 s + k_1 \qquad\qquad C_{12} = l_2(b_2 s + k_2)$$

Then:

$$[C_1 s^2 + C_2 s + C_3]Y(s) + [C_4 s + C_5]\theta(s) = C_6 U_1(s) + C_7 U_2(s)$$
$$[C_4 s + C_5]Y(s) + [C_8 s^2 + C_9 s + C_{10}]\theta(s) = C_{11} U_1(s) + C_{12} U_2(s)$$

In matrix form:

$$\begin{bmatrix} C_1 s^2 + C_2 s + C_3 & C_4 s + C_5 \\ C_4 s + C_5 & C_8 s^2 + C_9 s + C_{10} \end{bmatrix} \begin{bmatrix} Y(s) \\ \theta(s) \end{bmatrix} = \begin{bmatrix} C_6 \\ C_{11} \end{bmatrix} U_1(s) + \begin{bmatrix} C_7 \\ C_{12} \end{bmatrix} U_2(s)$$

Now, by taking the inverse of the left hand side matrix we get:

$$\begin{bmatrix} Y(s) \\ \theta(s) \end{bmatrix} = \frac{1}{\Delta(s)} \begin{bmatrix} P(s) & Q(s) \\ Q(s) & R(s) \end{bmatrix} \begin{bmatrix} 1 & 1 \\ -l_1 & l_2 \end{bmatrix} \begin{bmatrix} C_6 U_1(s) \\ C_7 U_2(s) \end{bmatrix}$$

in which,

$$P(s) = Js^2 + C_9 s + C_{10}$$
$$Q(s) = -C_4 s - C_5 s$$
$$R(s) = C_1 s^2 + C_2 s + C_3$$

and Δ(s) is the characteristic polynomial of the system as given by the determinant of the transformed system matrix:

$$\Delta(s) = \det \begin{bmatrix} P(s) & -Q(s) \\ -Q(s) & R(s) \end{bmatrix}$$

The transfer-function matrix is given by

$$G(s) = \frac{1}{\Delta(s)} \begin{bmatrix} P(s) & Q(s) \\ Q(s) & R(s) \end{bmatrix} \begin{bmatrix} C_6 & C_7 \\ C_{11} & C_{12} \end{bmatrix}$$

The individual transfer-functions are given by the elements of $G(s)$ as:

$$\frac{Y(s)}{U_1(s)} = \frac{[P(s) - l_1 Q(s)]}{\Delta(s)} C_6 \qquad \frac{\theta(s)}{U_1(s)} = \frac{[Q(s) - l_1 R(s)]}{\Delta(s)} C_6$$

$$\frac{Y(s)}{U_2(s)} = \frac{[P(s) + l_2 Q(s)]}{\Delta(s)} C_7 \qquad \frac{\theta(s)}{U_2(s)} = \frac{[Q(s) + l_2 R(s)]}{\Delta(s)} C_7$$

5.3 Frequency Domain Models

Any transfer-function is defined as the ratio of output to input. If the output and input are expressed in the frequency domain, the frequency transfer-function is given by the ratio of the Fourier transforms of the output to the input. Frequency-domain representations are particularly useful in the analysis, design, control, and testing of electro-mechanical systems. The signal waveforms encountered in such a system can be interpreted and represented as a series of sinusoidal components. Indeed, any waveform can be so represented, and sinusoidal excitation is often used in testing of equipment and components. It is usually easier to obtain frequency-domain models than the associated time-domain models by testing.

5.3.1 Frequency Transfer-Function (Frequency Response Function)

Consider the time-domain system (Equation 5.11) whose transfer-function (in the Laplace-domain) is given by Equation 5.12.

5.3.1.1 Response to a Harmonic Input

Suppose that a harmonic (sinusoidal) input, given in the complex form:

$$u = u_o e^{j\omega t} = u_o (\cos \omega t + j \sin \omega t) \tag{5.23}$$

is applied to the system. After the conditions settle down (i.e., at steady state) the output (response) of the system will also be harmonic, given by:

$$y = y_o e^{j\omega t} = y_o (\cos \omega t + j \sin \omega t) \tag{5.24}$$

By substituting Equations 5.23 and 5.24 in Equation 5.11 and cancelling the common term $e^{j\omega t}$ we get

$$y_o = \left[\frac{b_m (j\omega)^m + b_{m-1} (j\omega)^{m-1} + \cdots + b_0}{a_n (j\omega)^n + a_{n-1} (j\omega)^{n-1} + \cdots + a_0} \right] u_o \tag{5.25a}$$

or, in view of Equation 5.12,

$$y_o = G(j\omega)u_o \tag{5.25b}$$

$$\left(Note: \frac{de^{j\omega t}}{dt} = j\omega e^{j\omega t} \right)$$

Here, the frequency transfer-function (or, frequency response function) is given by

$$G(j\omega) = G(s)|_{s=j\omega} = \frac{b_0 + b_1(j\omega) + \cdots + b_m(j\omega)^m}{a_0 + a_1(j\omega) + \cdots + a_n(j\omega)^n} \tag{5.26}$$

Note: Angular frequency variable (rad/s) $\omega = 2\pi f$ where f=cyclic frequency variable (Hz).

Also, directly from the Laplace-domain result (Equation 5.12) we have the frequency-domain result:

$$G(j\omega) = \frac{Y(j\omega)}{U(j\omega)} \tag{5.26a}$$

where $Y(j\omega) = \mathcal{F}y(t)$ and $U(j\omega) = \mathcal{F}(t)$ with $\mathcal{F}$ denoting the Fourier transform operator.

5.3.1.2 Magnitude (Gain) and Phase

Let us denote the magnitude of $G(j\omega)$ as:

$$|G(j\omega)| = M \tag{5.27a}$$

and the phase angle of $G(j\omega)$ as:

$$\angle G(j\omega) = \phi \tag{5.27b}$$

Then we can write

$$G(j\omega) = M\cos\phi + jM\sin\phi = Me^{j\phi} \tag{5.27c}$$

and from Equations 5.24 and 5.25b:

$$y = u_o Me^{j(\omega t + \phi)} \tag{5.28}$$

Observations:
When a harmonic input of frequency ω is applied to the system:

1. The output is magnified by $M = |G(j\omega)|$
2. The output has a *phase lead* w.r.t. input by $\phi = \angle G(j\omega)$.

Note: For practical systems $\angle G(j\omega)$ is typically a negative phase lead (i.e., output usually lags input).

It follows that $G(j\omega)$ constitutes a complete model for a linear, constant-parameter system, as does $G(s)$.

5.3.2 Bode Diagram (Bode Plot) and Nyquist Diagram

The frequency transfer-function $G(j\omega)$ is in general a complex function of frequency ω (which is a real variable). From the result (Equation 5.27) it should be clear that applying a harmonic (i.e., sinusoidal) excitation and measuring the amplitude gain and the phase change at the output (response) for a series of frequencies, is a convenient method of experimental determination of a system model. This approach of "experimental modeling" is termed *model identification*. Either a *sine-sweep* or a *sine-dwell* excitation may be used with these tests. Specifically, a sinusoidal excitation is applied (i.e., input) to the system and the amplification factor and the phase-lead angle of the resulting response are determined at steady state. The frequency of excitation is varied continuously for a sine sweep, and in steps for a sine dwell. Sweep rate should be sufficiently slow, or dwell times should be sufficiently long, to guarantee achieving steady-state response in these methods. The results are usually presented as either a pair of curves:

$$|G(j\omega)| \text{ versus } \omega$$

$$\angle G(f) \text{ versus } \omega$$

with log axes for magnitude (e.g., decibels) and frequency (e.g., decades). This pair of curves is called the *Bode plot* or *Bode diagram*.

If the same information is plotted on the complex $G(j\omega)$ plane with the real part plotted on the horizontal axis and the imaginary part on the vertical axis, the resulting curve is termed *Nyquist diagram* or *Argand plot* or *polar plot*.

In a Bode diagram the frequency is shown explicitly on one axis, whereas in a Nyquist plot the frequency is a parameter on the curve, and is not explicitly shown unless the curve itself is calibrated. In Bode diagrams, it is customary and convenient to give the magnitude in decibels ($20\log_{10}|G(j\omega)|$) and scale the frequency axis in logarithmic units (typically factors of 10 or decades). Since the argument of a logarithm should necessarily be a dimensionless quantity, $Y(j\omega)$ and $U(j\omega)$ should have the same units, or the ratio of $G(j\omega)$ with respect to some base value such as $G(0)$ should be used.

The arrow on the Nyquist curve indicates the direction of increasing frequency. Only the part corresponding to positive frequencies is actually shown. The frequency response function corresponding to negative frequencies is obtained by replacing ω by $-\omega$ or, equivalently, $j\omega$ by $-j\omega$. The result is clearly the complex conjugate of $G(j\omega)$, and is denoted $G^*(j\omega)$:

$$G^*(j\omega) = |G(s)|_{s=-j\omega} \tag{5.29}$$

Since, in complex conjugation, the magnitude does not change and the phase angle changes sign, it follows that the Nyquist plot for $G^*(j\omega)$ is the mirror image of that for $G(j\omega)$ about the real axis. In other words, the Nyquist plot for the entire frequency range ω $[-\infty, +\infty]$ is symmetric about the real axis.

The shape of these plots for a simple oscillator is shown in Figure 5.3.

(a)   (b)

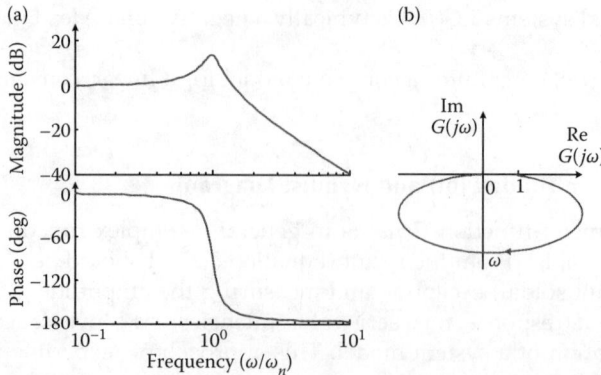

FIGURE 5.3
Frequency domain model of a simple oscillator: (a) Bode plot; (b) Nyquist plot.

5.4 Transfer-Functions of Electro-Mechanical Systems

Impedance is a transfer-function which is useful in both mechanical and electrical systems. Mobility is the inverse of mechanical impedance. Transmissibility is another transfer-function that is useful in mechanical systems. In view of the existing analogies (e.g., force–current analogy) which have been studied in the previous chapters, similar treatments are possible concerning transfer-functions in mechanical and electrical systems. Several relevant topics are addressed next.

5.4.1 Significance of Transfer-Functions in Mechanical Systems

The significance of frequency transfer-function as a dynamic model can be explained by considering the simple oscillator (i.e., a single-degree-of-freedom mass-spring-damper system, as shown in Figure 5.1). Its force–displacement transfer-function, in the frequency domain, can be written as:

$$G(j\omega) = \frac{1}{ms^2 + bs + k} \quad \text{with} \quad s = j\omega \tag{5.30}$$

in which m, b, and k denote mass, damping constant, and stiffness, respectively of the oscillator. When the excitation frequency ω is small in comparison to the system natural frequency $\sqrt{k/m}$, the terms ms^2 and bs can be neglected with respect to k; and the system behaves as a simple spring. When the excitation frequency ω is much larger than the system natural frequency, the terms bs and k can be neglected in comparison to ms^2. In this case the system behaves like a simple mass element. When the excitation frequency ω is very close to the natural frequency (i.e., $s = j\omega \approx j\sqrt{k/m}$), it is seen from Equation 5.30 that the term $ms^2 + k$ in the denominator of the transfer-function (i.e., the characteristic polynomial) becomes almost zero, and can be neglected. Then the transfer-function can be approximated by $G(j\omega) = 1/(bs)$ with $s = j\omega$.

In summary:

1. In the neighborhood of a resonance or natural frequency (i.e., for intermediate values of excitation frequencies), system damping becomes the most important parameter.
2. At low excitation frequencies the system stiffness is the most significant parameter.
3. At high excitation frequencies the mass is the most significant parameter.

Note: In these observations, instead of the physical parameters m, k, and b, we could use natural frequency $\omega_n = \sqrt{k/m}$ and the damping ratio $\zeta = b/(2\sqrt{mk})$ as the system parameters. Then the number of system parameters reduces to two, which is an advantage in parametric and sensitivity studies.

5.4.2 Mechanical Transfer-Functions

Any type of force or motion variable may be used as input and output variables in defining a transfer-function of a mechanical system. We can define several versions of frequency transfer-functions that may be useful in the modeling and analysis of mechanical systems. Some relatively common ones are given in Table 5.1.

In the frequency domain:

$$\text{Acceleration} = (j\omega) \times (\text{Velocity})$$

$$\text{Displacement} = \text{Velocity}/(j\omega)$$

In view of these relations, many of the alternative types of transfer-functions as defined in Table 5.1 are related to mechanical impedance and mobility through a factor of $j\omega$; specifically:

$$\text{Dynamic stiffness} = \text{Force}/\text{displacement} = \text{Impedance} \times j\omega$$

$$\text{Receptance} = \text{Displacement}/\text{force} = \text{Mobility}/(j\omega)$$

$$\text{Dynamic inertia} = \text{Force}/\text{acceleration} = \text{Impedance}/(j\omega)$$

$$\text{Accelerance} = \text{Acceleration}/\text{force} = \text{Mobility} \times j\omega$$

TABLE 5.1

Definitions of useful Mechanical Transfer-Functions

Transfer-Function	Definition (in Laplace or Frequency Domain)
Dynamic stiffness	Force/displacement
Receptance (dynamic flexibility or compliance)	Displacement/force
Mechanical impedance (Z)	Force/velocity
Mobility (M)	Velocity/force
Dynamic inertia	Force/acceleration
Accelerance	Acceleration/force
Force transmissibility (T_f)	Transmitted force/applied force
Motion transmissibility (T_m)	Transmitted velocity/applied velocity

In these definitions the variables force, acceleration and displacement should be interpreted as the corresponding Fourier spectra.

5.4.2.1 Mechanical Impedance and Mobility

In studies of mechanical systems, three types of frequency transfer-functions are particularly useful. They are *mechanical impedance, mobility,* and *transmissibility,* as presented in Table 5.1. In mechanical impedance function, velocity is considered the input variable and the force is the output variable, whereas in the mobility function the converse applies. It is clear that mobility is the inverse of mechanical impedance. Either transfer-function may be used in a given problem, depending on the convenience of analysis, as will be clear from the examples presented in this chapter.

5.4.3 Interconnection Laws

Once the transfer-functions of the system components are known, the interconnection laws may be used to determine the overall transfer-function of the system. Two types of interconnection are useful:

1. Series connection
2. Parallel connection

Determination of the interconnection laws is straightforward in view of the fact that:

1. For series-connected elements: through variable is common and the across variables add.
2. For parallel-connected elements: across variable is common and the through variables add.

5.4.3.1 Interconnection Laws for Mechanical Impedance and Mobility

Since mobility is given by an across variable (velocity) divided by a through variable (force), it is clear (by dividing throughout by the common through variable) that for series-connected elements the mobilities add (or, the inverse of impedance will be additive).

Since mechanical impedance is given by a through variable (force) divided by an across variable (velocity), it is clear (by dividing throughout by the common across variable) that for parallel-connected elements the mechanical impedances add (or, the inverse of mobility will be additive).

These interconnection laws are presented in Table 5.2.

5.4.3.2 Interconnection Laws for Electrical Impedance and Admittance

Since electrical impedance is given by an across variable (voltage) divided by a through variable (current), it is clear (by dividing throughout by the common through variable) that for series-connected elements the electrical impedances add (or, the inverse of admittance will be additive).

TABLE 5.2

Interconnection Laws for Mechanical Impedance (Z) and Mobility (M)

Series Connection	Parallel Connection

Series Connection:

$v = v_1 + v_2$

$$\frac{v}{f} = \frac{v_1}{f} + \frac{v_2}{f}$$

$M = M_1 + M_2$

$$\frac{1}{Z} = \frac{1}{Z_1} + \frac{1}{Z_2}$$

Parallel Connection:

$f = f_1 + f_2$

$$\frac{f}{v} = \frac{f_1}{v} = \frac{f_2}{v}$$

$Z = Z_1 + Z_2$

$$\frac{1}{M} = \frac{1}{M_1} + \frac{1}{M_2}$$

TABLE 5.3

Interconnection Laws for Electrical Impedance (Z)
and Admittance (W)

Series Connections	Parallel Connections
$v = v_1 + v_2$	$i = i_1 + i_2$
$\dfrac{v}{i} = \dfrac{v_1}{i} + \dfrac{v_2}{i}$	$\dfrac{i}{v} = \dfrac{i_1}{v} = \dfrac{i_2}{v}$
$Z = Z_1 + Z_2$	$W = W_1 + W_2$
$\dfrac{1}{W} = \dfrac{1}{W_1} + \dfrac{1}{W_2}$	$\dfrac{1}{Z} = \dfrac{1}{Z_1} + \dfrac{1}{Z_2}$

Since admittance is given by a through variable (current) divided by an across variable (voltage), it is clear (by dividing throughout by the common across variable) that for parallel-connected elements the admittances add (or, the inverse of electrical impedance will be additive). These interconnection laws for electrical are presented in Table 5.3.

5.4.3.3 A-Type Transfer-Functions and T-Type Transfer-Functions

Electrical impedance and mechanical mobility are "A-type transfer-functions" because they are given by: [across variable/through variable]. They follow the same interconnection laws (compare Tables 5.2 and 5.3).

Electrical admittance and mechanical impedance are "T-type transfer-functions" because they are given by: [through variable/across variable]. They follow the same interconnection laws (compare Tables 5.2 and 5.3).

5.4.4 Transfer-Functions of Basic Elements

Since a complex system can be formed through series and parallel interconnections of basic elements, it is possible to systematically generate the transfer-function of a complex system by using the transfer-functions of the basic elements.

In Chapter 2, the linear constitutive relations for the mass, spring and the damper elements were presented as time-domain relations. The corresponding transfer-functions are obtained by replacing the derivative operator d/dt by the Laplace operator s. The frequency transfer-functions are obtained by substituting $j\omega$ or $j2\pi f$ for s. In this manner, the transfer-functions of the basic (linear) mechanical elements: mass, spring, and damper may be obtained, as given in Table 5.4.

Similarly, in Chapter 2, the linear constitutive relations for the electrical capacitor, inductor, and resistor elements were presented as time-domain relations. The corresponding transfer-functions are obtained by replacing the derivative operator d/dt by the Laplace operator s. In this manner, the transfer-functions of the basic (linear) electrical elements may be obtained, as given in Table 5.5.

Three examples are given next to demonstrate the use of impedance and mobility methods in frequency-domain models.

TABLE 5.4

Mechanical Impedance and Mobility of basic Mechanical Elements

Element	Time-domain Model	Impedance	Mobility (Generalized Impedance)
Mass m	$m\dfrac{dv}{dt} = f$	$Z_m = ms$	$M_m = \dfrac{1}{ms}$
Spring k	$\dfrac{df}{dt} = kv$	$Z_k = \dfrac{k}{s}$	$M_k = \dfrac{s}{k}$
Damper b	$f = bv$	$Z_b = b$	$M_b = \dfrac{1}{b}$

TABLE 5.5

Impedance and Admittance of basic Electrical Elements

Element	Time-domain Model	Impedance (Z)	Admittance (W)
Capacitor C	$C\dfrac{dv}{dt} = i$	$Z_C = \dfrac{1}{Cs}$	$W_c = Cs$
Inductor L	$L\dfrac{di}{dt} = v$	$Z_L = Ls$	$W_L = \dfrac{1}{Ls}$
Resistor R	$Ri = v$	$Z_R = R$	$W_R = \dfrac{1}{R}$

Example 5.4: Ground-Based Mechanical Oscillator

Consider the simple oscillator shown in Figure 5.4a. Its mechanical circuit representation is given in Figure 5.4b. The input is the force $f(t)$; accordingly, the source element is a force source (a through-variable source or T-source). The output (response) of the system is the velocity v. In this situation the transfer-function $V(j\omega)/F(j\omega)$ is a mobility function. On the other hand, if the input is the velocity $v(t)$, the source element is a velocity source; and if force f is exerted on the environment, it is the output, and the corresponding transfer-function $F(j\omega)/V(j\omega)$ is an impedance function.

Suppose that using a force source, a known forcing function is applied to this system (with zero ICs) and the velocity response is measured. If we were to move the mass exactly at this predetermined velocity (using a velocity source), the force generated at the source would be identical to the originally applied force. In other words, mobility is the reciprocal (inverse) of impedance, as noted earlier. This reciprocity should be intuitively clear because we are dealing with the same system and same initial conditions. Due to this property, we may use either the impedance representation or the mobility representation, depending on whether the elements are connected in parallel or in series, irrespective of whether the input is a force or a velocity. Once the transfer-function is determined in one form, its reciprocal gives the other form.

In summary:

> From the viewpoint of analysis/modeling of a linear system it is immaterial as to what type of transfer-function is used. In particular, mechanical impedance or mobility may be used without affecting the analytical outcomes. From the physical point of view, however, one transfer-function may not be realizable while another is (*Note*: Physical realizability will be addressed later in this chapter).

In the present example, the three elements are connected in parallel. Hence, as is clear from the impedance circuit shown in Figure 5.4c, the impedance representation (rather than the mobility representation) is more convenient. The overall impedance function of the system is:

$$Z(j\omega) = \frac{F(j\omega)}{V(j\omega)} = Z_m + Z_k + Z_b = \left. ms + \frac{k}{s} + b \right|_{s=j\omega} = \left. \frac{ms^2 + bs + k}{s} \right|_{s=j\omega} \tag{5.31a}$$

The mobility function is the inverse of $Z(j\omega)$:

$$M(j\omega) = \frac{V(j\omega)}{F(j\omega)} = \left. \frac{s}{ms^2 + bs + k} \right|_{s=j\omega} \tag{5.31b}$$

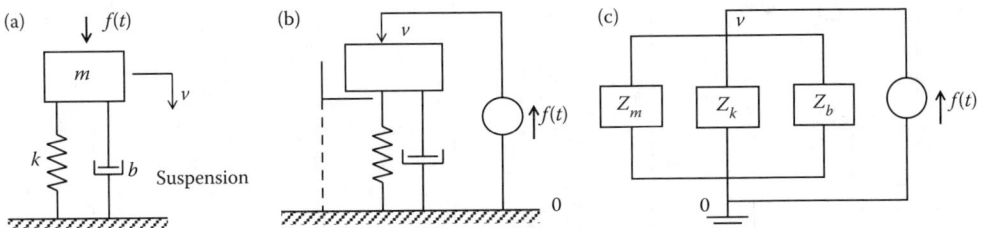

FIGURE 5.4
(a) Ground-based mechanical oscillator. (b) Schematic mechanical circuit. (c) Impedance circuit.

Note that if, physically, the input to the system is the force, the mobility function governs the system behavior. In this case, the characteristic polynomial of the system is s^2+bs+k, which corresponds to a simple oscillator and, accordingly, the (dependent) velocity response of the system would be governed by this characteristic polynomial. If, on the other hand, physically, the input is the velocity, the impedance function governs the system behavior. The characteristic polynomial of the system, in this case, is s, which corresponds to a simple integrator ($1/s$). The (dependent) force response of the system would be governed by an integrator type behavior. To explore this behavior further, suppose the velocity source has a constant value. The inertia force will be zero. The damping force will be constant. The spring force will increase linearly. Hence, the net force will have an integration (linearly increasing) effect. If the velocity source provides a linearly increasing velocity (constant acceleration), the inertia force will be constant, the damping force will increase linearly, and the spring force will increase quadratically. In fact, it will be seen from the later developments in this chapter and elsewhere, the mobility function (as given above), not the impedance function, is the physically realizable transfer-function for the oscillator example in Figure 5.4.

Example 5.5: A Degenerate Case

Consider an intuitively degenerate example of a system as shown in Figure 5.5a. Note that the support motion is not associated with an external force. The mass m has an external force f and velocity v. At this point we shall not specify which of these variables is the input to the system. It should be clear, however, that v cannot be logically considered an input because the application of any arbitrary velocity to the support structure will generate a force at that location and this is not allowed for in the system shown in Figure 5.5. However, since $v_1=v$, it follows from the mechanical circuit representation shown in Figure 5.5b, and its impedance circuit shown in Figure 5.5c, that it is acceptable to indirectly consider v_1 also as the input to the system when v is the input.

When v is the input to the system, the source element in Figure 5.5b becomes a velocity source. The corresponding impedance function is:

$$\frac{F(j\omega)}{V(j\omega)} = Z_m = ms\Big|_{s=j\omega} \tag{5.32a}$$

If, on the other hand, f is the input and v is the output, the mobility function is valid, which given by:

$$\frac{V(j\omega)}{F(j\omega)} = M_m = \frac{1}{ms}\Big|_{s=j\omega} \tag{5.32b}$$

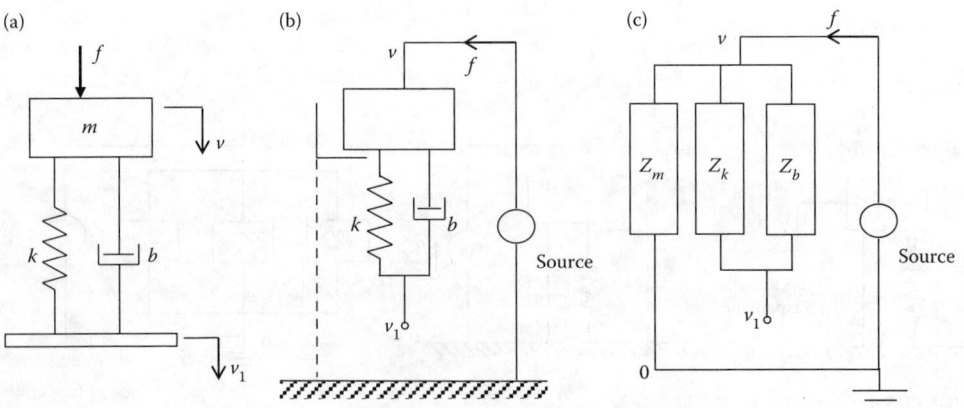

FIGURE 5.5
(a) A mechanical oscillator with support motion. (b) Schematic mechanical circuit. (c) Impedance circuit.

Furthermore, since $v_1 = v$, an alternative impedance function:

$$\frac{F(j\omega)}{V_1(j\omega)} = ms\Big|_{s=j\omega} \qquad (5.33a)$$

and a mobility function:

$$\frac{V_1(j\omega)}{F(j\omega)} = \frac{1}{ms}\Big|_{s=j\omega} \qquad (5.33b)$$

may be defined.

Example 5.6: Oscillator with Support Motion

To show an interesting reciprocity property, consider the system shown in Figure 5.6a. In this example the motion of the mass m is not associated with an external force. The support motion, however, is associated with the force f. A schematic mechanical circuit for the system is shown in Figure 5.6b and the corresponding impedance circuit is shown in Figure 5.6c. They clearly indicate that the spring and the damper are connected in parallel, and the mass is connected in series with this pair. By impedance addition for parallel elements, and mobility addition for series elements, it is seen that the overall mobility function of the system is:

$$\frac{V(j\omega)}{F(j\omega)} = M_m + \frac{1}{(Z_k + Z_b)} = \frac{1}{ms} + \frac{1}{\left(\dfrac{k}{s} + b\right)}\Bigg|_{s=j\omega} = \frac{ms^2 + bs + k}{ms(bs + k)}\Bigg|_{s=j\omega} \qquad (5.34a)$$

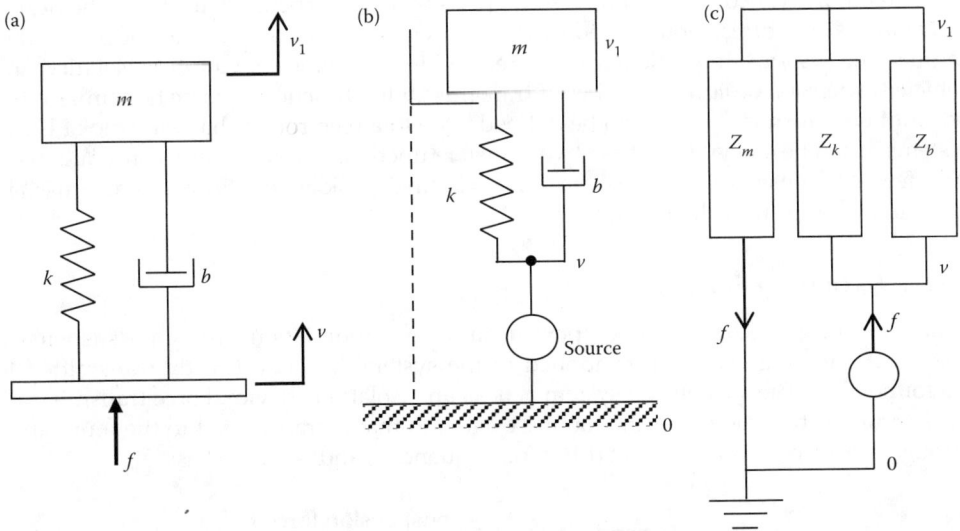

FIGURE 5.6
(a) A mechanical oscillator with support motion. (b) Schematic mechanical circuit. (c) Impedance circuit.

It follows that if force at the support is the input (a force source) and the support velocity is the output, the system characteristic polynomial is $ms(bs+k)$, which is known to be inherently unstable due to the presence of a free integrator, and has a nonoscillatory transient response.

Alternatively, if the support velocity is the input (a velocity source), the corresponding impedance function is the reciprocal of the previous mobility function, and is given by:

$$\frac{F(j\omega)}{V(j\omega)} = \frac{ms(bs+k)}{ms^2+bs+k}\bigg|_{s=j\omega} \tag{5.34b}$$

Furthermore, we have: $\dfrac{V_1(j\omega)}{F(j\omega)} = \dfrac{1}{ms}\bigg|_{s=j\omega}$.

The impedance function $F(j\omega)/V_1(j\omega)$ is not admissible and is physically unrealizable because V_1 cannot be an input for there is no associated force at that location. This is confirmed by the fact that the corresponding transfer-function is a differentiator—a physically nonrealizable device. The mobility function $V_1(j\omega)/F(j\omega)$ corresponds to a simple integrator. Physically, when a force f is applied to the support it transmits to the mass, unchanged, through the parallel spring-damper unit. Accordingly, when f is constant, a constant acceleration is produced at the mass, causing its velocity to increase linearly (an "integration" behavior).

Maxwell's principle of reciprocity is demonstrated by noting that in Examples 5.5 and 5.6 the mobility functions $V_1(f)/F(f)$ are identical. What this means is that the support motion produced by applying a forcing excitation to the mass (system in Figure 5.5a) is equal to the motion of the mass when the same forcing excitation is applied to the support (system in Figure 5.6a), with the same initial conditions.

Note: Maxwell's reciprocity property is valid for linear, constant-parameter systems in general, and is particularly useful in testing of multi-degree-of-freedom mechanical systems; for example, to determine a transfer-function that is difficult to measure, by measuring its symmetrical counterpart in the transfer-function matrix.

5.4.5 Transmissibility Function

Transmissibility functions are transfer-functions that are particularly useful in the design and analysis of fixtures, mounts, and support structures for machinery, vehicles, and other engineering systems. In particular they are used in the studies of vibration isolation and vehicle suspension design. Two types of transmissibility functions—force transmissibility and motion transmissibility—can be defined. Due to a reciprocity characteristic of linear systems, it can be shown that these two transfer-functions are equal and, consequently, it is sufficient to consider only one of them. We will first consider both types of transmissibility functions and show their equivalence.

5.4.5.1 Force Transmissibility

Consider a mechanical system supported on a rigid foundation through a suspension system. If a forcing excitation is applied to the system, it is not directly transmitted to the foundation. The suspension system acts as an "isolation" device. Force transmissibility determines the fraction of the forcing excitation that is transmitted to the foundation through the suspension system at different frequencies, and is defined as:

$$\text{Force transmissibility } T_f = \frac{\text{Suspension force } F_s}{\text{Applied force } F} \tag{5.35}$$

Note: This function is defined in the frequency domain, and accordingly F_s and F should be interpreted as the Fourier spectra of the corresponding forces.

A schematic diagram of a force transmissibility mechanism is shown in Figure 5.7a. The reason for the suspension force f_s not being equal to the applied force f is attributed to the inertia path (broken line in Figure 5.7a) that is present in the mechanical system.

5.4.5.2 Motion Transmissibility

Consider a mechanical system supported through a suspension mechanism on a structure, which may be subjected to undesirable motions (e.g., seismic disturbances, road disturbances, machinery disturbances). Motion transmissibility determines the fraction of the support motion that is transmitted to the system through its suspension at different frequencies. It is defined as:

$$\text{Motion transmissibility } T_m = \frac{\text{System motion } V_m}{\text{Support motion } V} \tag{5.36}$$

Note: The velocities V_m and V are expressed in the frequency domain, as Fourier spectra.

A schematic representation of the motion transmissibility mechanism is shown in Figure 5.7b. Typically, the motion of the system is taken as the velocity of one of its critical masses. Different transmissibility functions are obtained when different mass points (or degrees of freedom) of the system are considered.

Next, two examples are given to show the reciprocity property, which makes the force transmissibility and the motion transmissibility functions identical.

5.4.5.3 Single-Degree-of-Freedom System

Consider the single-degree-of-freedom systems shown in Figure 5.8. In these examples the system is represented by a point mass m, and the suspension system is modeled as a spring of stiffness k and a viscous damper of damping constant b. The model shown in Figure 5.8a is used to study force transmissibility. Its impedance circuit is shown in Figure 5.9a. The

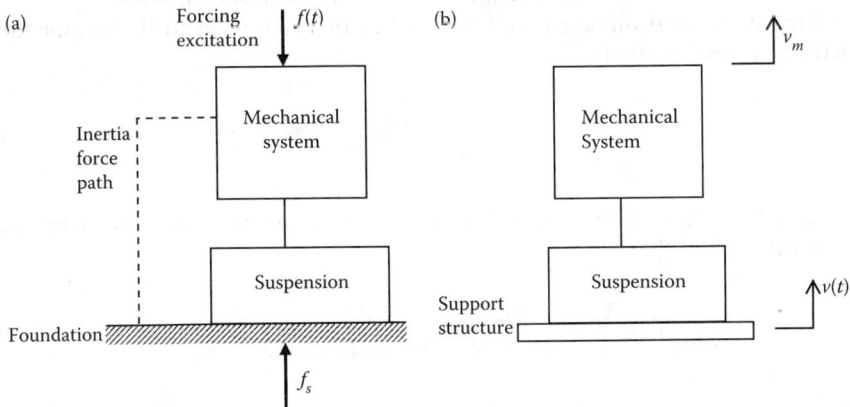

FIGURE 5.7
(a) Force transmissibility mechanism. (b) Motion transmissibility mechanism.

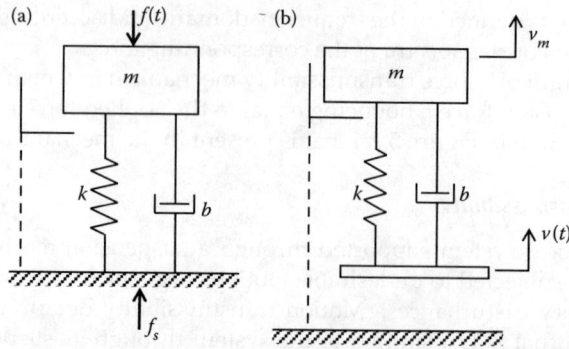

FIGURE 5.8
Single-degree-of-freedom systems. (a) Fixed on ground. (b) With support motion.

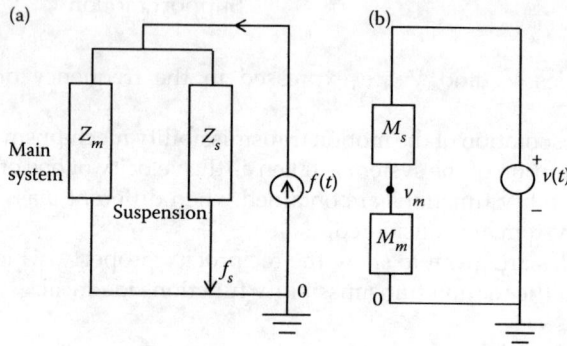

FIGURE 5.9
Impedance circuits of (a) system in Figure 5.8a and (b) system in Figure 5.8b.

model shown in Figure 5.8b is used in determining the motion transmissibility. Its imped-ance (or, mobility) circuit is shown in Figure 5.9b.

Note: mobility elements are suitable for motion transmissibility studies.

Since force is divided among parallel branches in proportion to their impedances it follows from Figure 5.9a that:

$$T_f = \frac{F_s}{F} = \frac{Z_s}{Z_m + Z_b} \tag{5.37a}$$

Since velocity is divided among series elements in proportion to their mobilities, it is clear from Figure 5.9b that:

$$T_m = \frac{V_m}{V} = \frac{M_m}{M_m + M_s} = \frac{1/Z_m}{1/Z_m + 1/Z_s} = \frac{Z_s}{Z_m + Z_b} \tag{5.37b}$$

Consequently, we have:

$$T_f = T_m \tag{5.38}$$

and a distinction between the two types of transmissibility is not necessary. Let us denote them by a common *transmissibility function T*.

Note: It can be concluded that Figure 5.8a and b are complementary systems for transmissibility.

Since, $Z_m = ms$ and $Z_s = k/s + b$, it follows that

$$T = \left[\frac{bs + k}{ms^2 + bs + k} \right]_{s=j\omega}$$ (5.39a)

It is customary to consider only the magnitude of this complex transmissibility function. This, termed *magnitude transmissibility*, is given by:

$$T = \left[\frac{\omega^2 b^2 + k^2}{\omega^2 b^2 + (k - \omega^2 m^2)^2} \right]^{1/2}$$ (5.39b)

5.4.5.4 Two-Degree-of-Freedom System

Consider the two-degree-of-freedom systems shown in Figure 5.10. The main system is represented by two masses linked through a spring and a damper. Mass m_1 is considered the critical mass. (It is equally acceptable to consider mass m_2 as the critical mass.) To determine the force transmissibility, from Figure 5.11a note that the applied force is divided in the ratio of the impedances among the two parallel branches. The mobility of the main right hand side branch is

$$M = \frac{1}{Z_{s1}} + \frac{1}{Z_{m2} + Z_s}$$ (5.40)

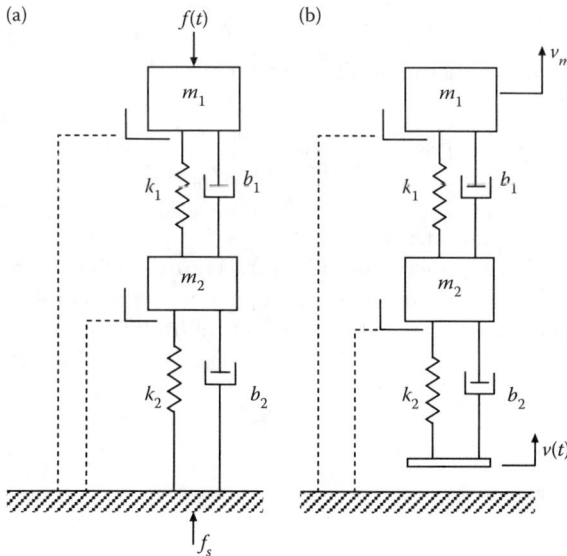

FIGURE 5.10
Systems with two-degree-of-freedom (a) fixed on ground and (b) with support motion.

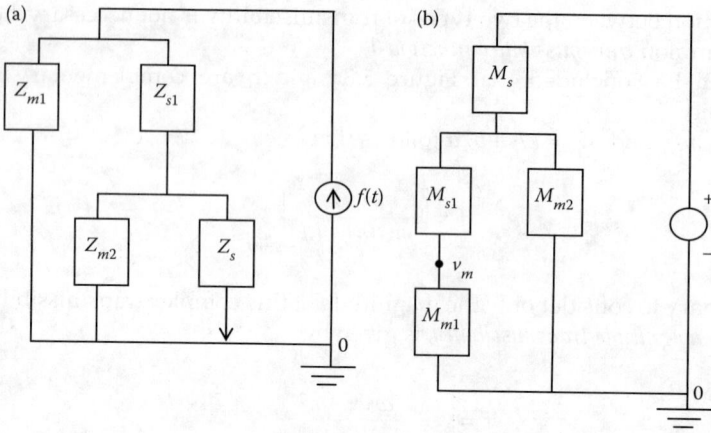

FIGURE 5.11
Impedance circuits of (a) system in Figure 5.10a and (b) system in Figure 5.10b.

and the force through that branch is

$$F' = \left[\frac{\frac{1}{M}}{Z_{m1} + \frac{1}{M}} \right] F = \left[\frac{1}{MZ_{m1} + 1} \right] F$$

The force F_s through Z_s is given by

$$F_s = \left[\frac{Z_s}{Z_{m2} + Z_s} \right] F'$$

Consequently, the force transmissibility is given by:

$$T_f = \frac{F_s}{F} = \left[\frac{1}{MZ_{m1} + 1} \right] \left[\frac{Z_s}{Z_{m2} + Z_s} \right] \tag{5.41}$$

where M is as given in Equation 5.40.

To determine the motion transmissibility, from Figure 5.10b and the associated Figure 5.11b, note that the velocity is distributed in proportion to the mobilities among the series elements. The impedance of the composite series unit in the bottom is

$$Z = \frac{1}{M_{m2}} + \frac{1}{M_{s1} + M_{m1}} \tag{5.42}$$

and the velocity across this unit is

$$V' = \left[\frac{\frac{1}{Z}}{M_s + \frac{1}{Z}} \right] V = \left[\frac{1}{M_s Z + 1} \right] V$$

The velocity V_m of mass m_1 is given by

$$V_m = \left[\frac{M_{m1}}{M_{s1} + M_{m1}} \right] V'$$

As a result, the motion transmissibility can be expressed as

$$T_m = \frac{V_m}{V} = \left[\frac{1}{M_s Z + 1} \right] \left[\frac{M_{m1}}{M_{s1} + M_{m1}} \right] \tag{5.43}$$

where Z is as given by Equation 5.42.

It remains to show that $T_m = T_f$. To this end, let us examine the expression for T_m. Since $Z_s = 1/M_s$, T_m can be written as

$$T_m = \left[\frac{Z_s}{Z + Z_s} \right] \left[\frac{M_{m1}}{M_{s1} + M_{m1}} \right]$$

Note: $Z = \dfrac{1}{M_{s1} + M_{m1}} + Z_{m2}$

Hence,

$$
T_m = \left[\frac{Z_s}{\dfrac{1}{M_{s1} + M_{m1}} + Z_{m2} + Z_s} \right] \left[\frac{M_{m1}}{M_{s1} + M_{m1}} \right] = \left[\frac{M_{m1}}{\dfrac{1}{Z_{m2} + Z_s} + M_{s1} + M_{m1}} \right] \left[\frac{Z_s}{Z_{m2} + Z_s} \right]
$$

$$
= \left[\frac{1}{\dfrac{1}{M_{m1}} \left[\dfrac{1}{Z_{m2} + Z_s} + M_{s1} \right] + 1} \right] \left[\frac{Z_s}{Z_{m2} + Z_s} \right] = \left[\frac{1}{Z_{m1} \left[\dfrac{1}{Z_{m2} + Z_s} + \dfrac{1}{Z_{s1}} \right] + 1} \right] \left[\frac{Z_s}{Z_{m2} + Z_s} \right]
$$

$$\tag{5.43a}$$

This expression is clearly identical to T_f as given in Equation 5.41, in view of Equation 5.40.

Note: It can be concluded that Figure 5.10a and b are complementary systems for transmissibility.

The equivalence of T_f and T_m can be shown in a similar straightforward manner for higher degree-of-freedom systems as well.

5.5 Equivalent Circuits and Linear Graph Reduction

We have observed that transfer-function approaches are more convenient than the differential equation approaches, in dealing with linear systems. This stems primarily from the fact that transfer-function approaches use algebra rather than calculus. Also we have noted that when dealing with circuits (particularly, impedance and mobility circuits), transfer-function approaches are quite natural. Since the circuit approaches are extensively used in electrical systems, and as a result, quite mature procedures are available in that context, it is useful to consider extending such approaches to mechanical systems (and hence, to electro-mechanical systems). In particular, circuit reduction is convenient using Thevenin's equivalence and Norton's equivalence for electrical circuits. Linear graphs, as studied in Chapter 4, can be simplified as well by using transfer-function (frequency domain) approaches and circuit reduction. We will address these issues in this section.

5.5.1 Thevenin's Theorem for Electrical Circuits

Thevenin's theorem provides a powerful approach to reduce a complex circuit segment into a simpler equivalent representation. Two types of equivalent circuits are generated by this theorem:

1. Thevenin equivalent circuit (consists of a voltage source and an impedance Z_e in series).
2. Norton equivalent circuit (consists of a current source and an impedance Z_e in parallel).

The theorem provides means to determine the equivalent source and the equivalent impedance for either of these two equivalent circuits.

Consider a (rather complex) segment of a circuit, consisting of impedances and source elements, as represented in Figure 5.12a. According to the Thevenin's theorem, this circuit segment can be represented by the Thevenin equivalent circuit, as shown in Figure 5.12b or the Norton equivalent circuit, as shown in Figure 5.12c so that for either equivalent circuit, the voltage v and the current i are identical to those at the output port of the considered circuit segment.

Note: The circuit segment of interest (Figure 5.12a) is isolated by "virtually" cutting (separating) a complex circuit into the complex segment of interest and a quite simple (and fully known) segment which is connected to the complex segment. The "virtual" cut is made at the two appropriate terminals linking the two parts of the circuit. The two terminal ends formed by the virtual cutting is the "virtual" output port of the isolated circuit segment. Actually, these terminals are not in open-circuit condition because the cut is "virtual," and a current flows through them.

$V(s) =$ voltage across the cut terminals when entire circuit is complete

$I(s) =$ current through the cut terminals when entire circuit is complete

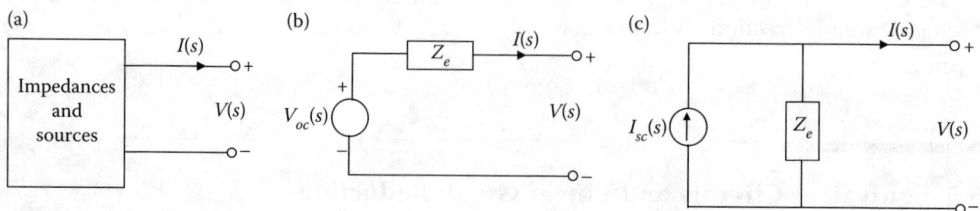

FIGURE 5.12
(a) Circuit segment with impedances and sources. (b) Thevenin equivalent circuit. (c) Norton equivalent circuit.

$V_{oc}(s)$=open-circuit voltage at cut the terminals (i.e., voltage with the terminals open)

$I_{sc}(s)$=short-circuit current at cut terminals (i.e., current when the terminals are shorted)

Z_e=equivalent impedance of the circuit segment with the source killed (i.e., voltage sources shorted and current source opened) = Thevenin resistance

Note 1: Variables are expressed in the Laplace (or frequency domain) using the Laplace variable *s*.

Note 2: For a circuit segment with multiple sources, use superposition (linear system), by taking one source at a time.

Example 5.7: Illustrative Example for Thevenin's Theorem

As usual, we will use electrical impedances, in the Laplace domain. Consider the circuit in Figure 5.13a. We cut it as indicated by the dotted line and determine the Thevenin and Norton equivalent circuits for the left hand side portion.

Determination of the Equivalent Impedance Z_e

First we kill the two sources (i.e., open the current source and short the voltage source so that the source signals become zero). The resulting circuit is shown in Figure 5.13b.

Note the series element and two parallel elements. Since the impedances add in series and inversely in parallel we have

$$Z_e = Z_L + \frac{Z_R Z_C}{Z_R + Z_C} = Ls + \frac{R/(Cs)}{R + 1/(Cs)} = Ls + \frac{R}{RCs + 1} \tag{5.44}$$

Determination of $V_{oc}(s)$ for Thevenin Equivalent Circuit

We find the open-circuit voltage using one source at a time, and then use the principle of super-position to determine the overall open-circuit voltage.

a. With Current Source I(s) Only

The circuit with the current source only (short the voltage source) is shown in Figure 5.13c.

The source current goes through the two parallel elements only, whose equivalent impedance is $(Z_R Z_C / Z_R + Z_C)$. Hence the voltage across it, which is also the open-circuit voltage (since no current and hence no voltage drop along the Inductor), is given by

$$V_{oci} = \frac{Z_R Z_C}{(Z_R + Z_C)} I(s) \tag{5.45a}$$

b. With Voltage Source V(s) Only

The circuit with the voltage source only (i.e., open the current source) is shown in Figure 5.13d.

The voltage drop across R should be equal to that across C, and hence the currents in these two elements must be in the same direction. But, the sum of the currents through these parallel elements must be zero, by the node equation (since the open-circuit current is zero). Hence, each current must be zero and the voltages V_R and V_C must be zero. Furthermore, due to the open-circuit, the voltage V_L across the inductor must be zero. Then from the loop equation, we have $V_{ocv} + V(s) = 0$.

Or,

$$V_{OCV} = -V(s) \tag{5.45b}$$

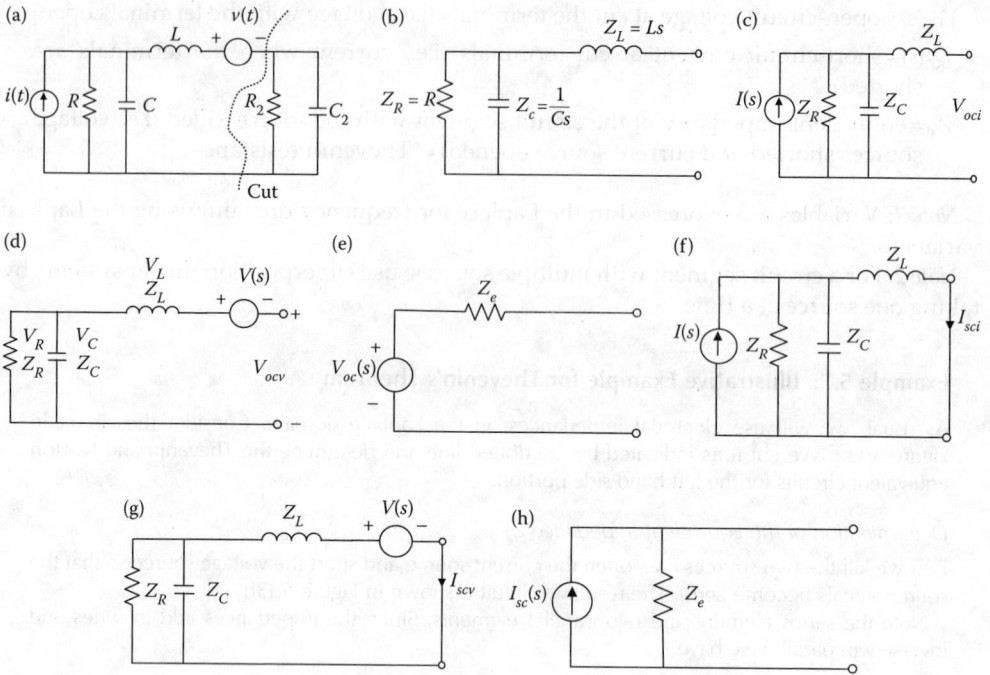

FIGURE 5.13
(a) An electrical impedance circuit. (b) Circuit with the sources killed. (c) Circuit with current source only. (d) Circuit with voltage source only. (e) Thevenin equivalent circuit. (f) Circuit with current source only. (g) Circuit with voltage source only. (h) Norton equivalent circuit.

Note the positive direction of potential drop for the open-circuit voltage, as needed for the Thevenin equivalent voltage source.

By superposition, the overall open-circuit voltage is

$$V_{oc}(s) = V_{oci} + V_{ocv} = \frac{Z_R Z_C}{(Z_R + Z_C)} I(s) - V(s) \tag{5.45}$$

The resulting Thevenin equivalent circuit is shown in Figure 5.13e.

Determination of $I_{sc}(s)$ for Norton Equivalent Circuit

We find the short-circuit current by taking one source at a time, and then using the principle of superposition.

a. With Current Source $I(s)$ Only

The circuit with the current source only (short the voltage source) is shown in Figure 5.13f.

The source current goes through the three parallel elements, and the currents are divided inversely with the respective impedances. Hence the current through the inductor is (note the positive direction as marked, for the Norton equivalent current source)

$$I_{sci} = \frac{1/Z_L}{(1/Z_R + 1/Z_C + 1/Z_L)} I(s) \tag{5.46a}$$

b. With Voltage Source $V(s)$ Only

The circuit with the voltage source only (open the current source) is shown in Figure 5.13g.

Note from the circuit that the short-circuit current is the current that flows through the overall impedance of the circuit (series inductor and a parallel resistor and capacitor combination). According to the polarity of the voltage source, this current is in the opposite direction to the positive direction marked in Figure 5.13g. We have

$$I_{scv}(s) = -\frac{V(s)}{(Z_L + Z_R Z_C/(Z_R + Z_C))} \tag{5.46b}$$

By superposition, the overall short-circuit current is

$$I_{sc}(s) = I_{sci} + I_{scv} = \frac{1/Z_L}{(1/Z_R + 1/Z_C + 1/Z_L)} I(s) - \frac{V(s)}{(Z_L + Z_R Z_C/(Z_R + Z_C))} \tag{5.46}$$

The resulting Norton equivalent circuit is shown in Figure 5.13h.

5.5.2 Mechanical Circuit Analysis Using Linear Graphs

For extending the equivalent-circuit analysis to mechanical systems, we use the force-current analogy (see Chapters 2 and 4), where electrical impedance in analogous to mechanical mobility (A-type transfer-functions) and electrical admittance is analogous to mechanical impedance (T-type transfer-functions). This analogy is summarized in Table 5.6.

Accordingly, the reduction of a linear graph, in the frequency domain, is done by the following two steps:

1. For each branch of the linear graph mark the mobility function (not mechanical impedance).
2. Carry out linear-graph analysis and reduction as if we are dealing with an electrical circuit, in view of the analogy given in Table 5.6.

In particular, we do the following:

1. For parallel branches: mobilities are combined by inverse relation $M = (M_1 M_2)/(M_1 + M_2)$. *Note*: Velocity is common, force is divided inversely to branch mobilities.
2. For series branches: mobilities add ($M = M_1 + M_2$). *Note*: Force is common; velocity is divided in proportion to mobility.
3. Killing a force source means open-circuiting it (so, transmitted force=0).
4. Killing a velocity source means short-circuiting it (so, velocity across=0).

TABLE 5.6

Mechanical and Electrical Transfer-Function Analogy

Mechanical circuit	Electrical circuit analogy
Mobility function	Electrical impedance
Force	Current
Voltage	Velocity

Example 5.8: Ground-Based Mechanical Oscillator (Revisited)

Let us revisit Example 5.4, this time equipped with Thevenin's theorem and linear graphs. The system is shown in Figure 5.14a. Its linear graph is drawn as given in Figure 5.14b.
  The mobility of the suspension unit (spring and damper on which the mass is supported) is

$$M_s = \frac{(s/k)(1/b)}{(s/k)+(1/b)} = \frac{1}{(b+k/s)} = \frac{s}{(bs+k)} \tag{5.47}$$

The mobility of the mass element is

$$M_m = \frac{1}{ms} \tag{5.48}$$

  The linear graph given in Figure 5.14c is identical to that in Figure 5.14b, except that the suspension is shown as a single unit in Figure 5.14c.
  Suppose that the linear graph in Figure 5.14b is cut (virtually) as shown and the part separated to the right is considered.
  Note: It should be obvious that its Norton equivalent circuit is indeed given in Figure 5.14c.
  The Thevenin equivalent circuit of the cut (right hand) segment is as shown in Figure 5.14d. Here, the open-circuit velocity (product of force and mobility of the circuit, with the cut terminals maintained in open-circuit, i.e., zero force) is

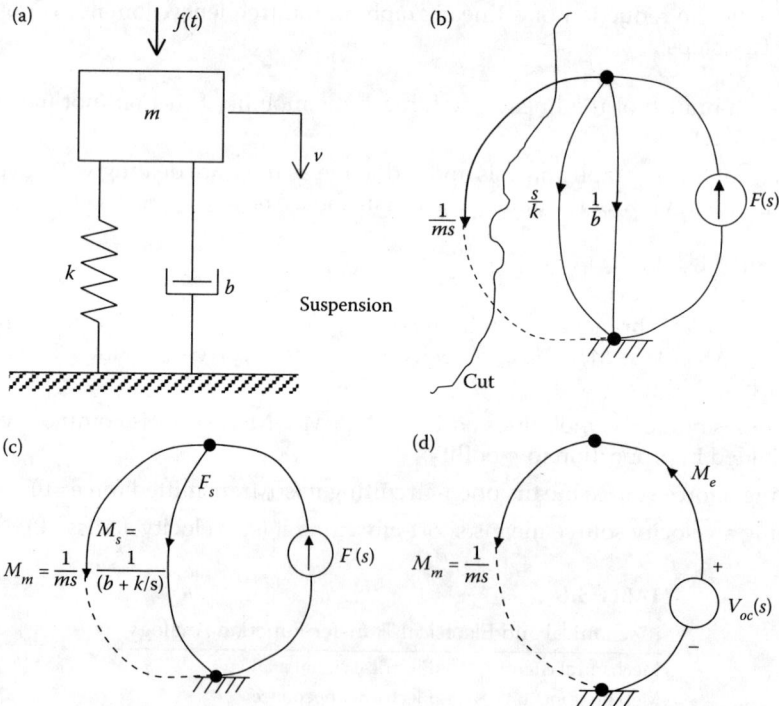

FIGURE 5.14
(a) Ground-based oscillator. (b) Linear graph. (c) Norton equivalent circuit. (d) Thevenin equivalent circuit.

$$V_{oc}(s) = M_s F(s) = \frac{F(s)}{\left(b + \dfrac{k}{s}\right)} \tag{i}$$

The equivalent mobility of the cut (right hand side) circuit, with the source killed (i.e., opened for a force source) is the suspension mobility:

$$M_e = M_s \tag{ii}$$

Since force is divided inversely to mobilities in parallel paths. (Compare: Current is divided inversely to electrical impedances in parallel paths.), we have from Figure 5.14c, force through the suspension:

$$F_s = \frac{M_m}{(M_m + M_s)} F(s) \tag{5.49a}$$

Accordingly, force transmissibility is obtained as

$$T_f = \frac{M_m}{(M_m + M_s)} \tag{5.49}$$

This is identical to what we obtained earlier (Equation 5.37a).

Note: As mentioned before, Norton equivalent circuit is given by the original circuit itself (Figure 5.14c) and will not provide any further useful information.

Now we will check the force through and the velocity across M_m for all circuits.

For Circuits in Figures 5.14b and c:

$$\frac{\text{Velocity across mass element}}{\text{Source force}} = \text{Mobility of circuit}$$

We have

$$\text{Velocity across mass element} = \frac{M_s M_m}{M_s + M_m} F(s) \tag{iii}$$

Note: Suspension and mass are connected in parallel.

$$\text{Force through mass element} = \frac{M_s}{M_s + M_m} F(s) \tag{iv}$$

Note: Equation (iv) is obtained simply by dividing Equation (iii) by the mass mobility or by noting that the source force is divided inversely with the mobilities of the two parallel path.

For Circuit in Figure 5.14d:

Since velocities are divided in proportion to mobilities in a series connection, we have

$$\text{Velocity across mass} = \frac{M_m}{M_e + M_m} V_{oc}(s)$$

Substitute Equation (i) and use Equation (ii):

$$\text{Velocity across mass element} = \frac{M_s M_m}{M_s + M_m} F(s)$$

This is identical to Equation (iii).

$$\text{Force through mass} = \frac{V_{oc}(s)}{M_e + M_m}$$

Substitute Equations (i) and (ii).

$$\text{Force through mass element} = \frac{M_s}{M_s + M_m} F(s)$$

This is identical to Equation (iv), as expected.

Example 5.9: Oscillator with Support Motion (Revisited)

Let us revisit Example 5.6 using Thevenin's theorem and linear graphs. The system is shown in Figure 5.15a. Its linear graph is drawn as given in Figure 5.15b.
As in Example 5.8 we have

$$M_s = \frac{(s/k)(1/b)}{(s/k) + (1/b)} = \frac{1}{(b + k/s)} = \frac{s}{(bs + k)} \tag{5.47}$$

and

$$M_m = \frac{1}{ms} \tag{5.48}$$

The linear graph in Figure 5.15c is identical to that in Figure 5.15b, except that the suspension is shown as a single unit in Figure 5.15c.
Suppose that the circuit in Figure 5.15b is cut as shown and the part separated to the right is considered.
Note: Its Thevenin equivalent circuit is trivial, as shown in Figure 5.15c.
The Norton equivalent circuit is as shown in Figure 5.15d. Here, the short-circuit force is

$$F_{sc}(s) = \frac{V(s)}{M_s} \tag{i}$$

The equivalent mobility of the cut (right-side) circuit, with the source killed (i.e., shorted for a velocity source) is the suspension mobility:

$$M_e = M_s \tag{ii}$$

Since, in a series path, velocity is divided in proportion to mobilities. (Compare: In a series path, voltage is divided in proportion to electrical impedances.), we have from Figure 5.15c, velocity across the mass

$$V_m(s) = \frac{M_m}{M_m + M_s} V(s) \tag{5.50a}$$

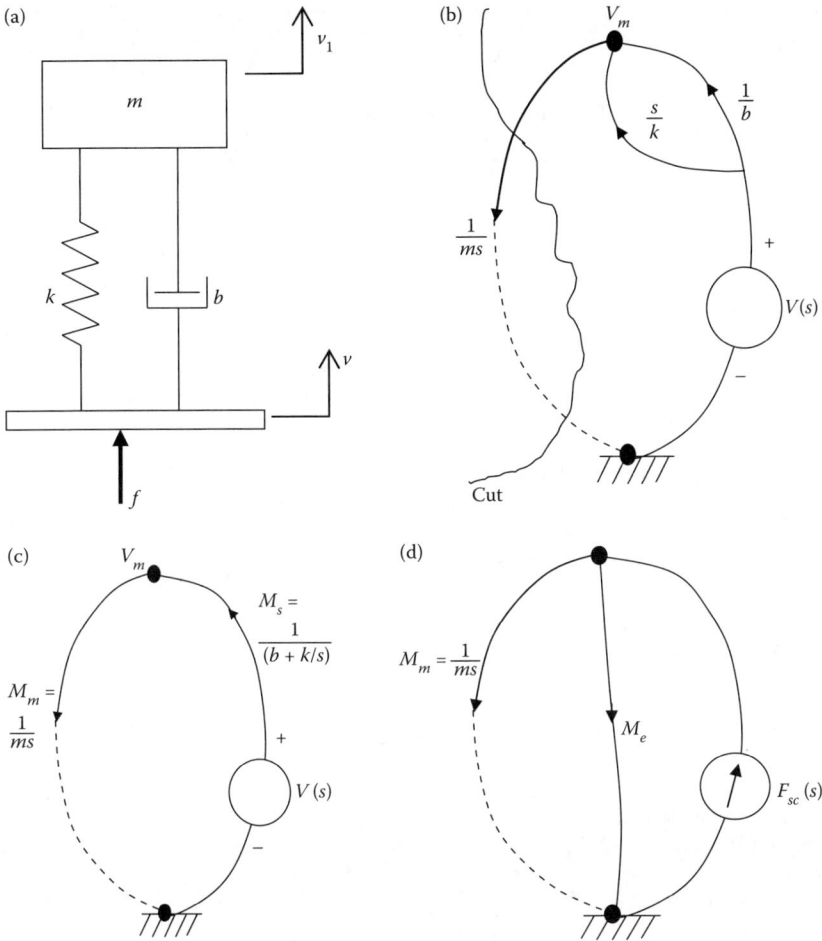

FIGURE 5.15
(a) Oscillator with support motion. (b) Linear graph. (c) Equivalent circuit. (d) Norton equivalent circuit.

Accordingly, motion transmissibility is given by

$$T_m = \frac{M_m}{(M_m + M_s)} \tag{5.50}$$

This is identical to what we obtained earlier (Equation 5.37b).

Note: As mentioned before, Thevenin equivalent circuit is the original circuit itself (Figure 5.15c) and will not provide any further useful information.

Now we will check the force through and the velocity across M_m for all circuits.

For Circuits in Figures 5.15b and c:

$$\text{Velocity across mass element } V_m = \frac{M_m}{M_s + M_m} V(s) \tag{iii}$$

Note: Suspension and mass are connected in series. Hence velocities are divided in proportion to mobilities.

$$\text{Force through mass element} = \frac{V(s)}{M_s + M_m} \qquad \text{(iv)}$$

For Circuit (d):

$$\text{Equivalent mobility of the two parallel elements} = \frac{M_e M_m}{M_e + M_m}$$

$$\text{Velocity across mass} = \frac{M_e M_m}{M_e + M_m} F_{sc}(s)$$

Substitute Equation (i) and use Equation (ii):

$$\text{Velocity across mass element} = \frac{M_m}{M_s + M_m} V(s)$$

This is identical to Equation (iii).

$$\text{Force through mass element} = \frac{M_e}{M_e + M_m} F_{sc}(s)$$

Note: Force is divided inversely to mobilities in a parallel connection. Substitute Equations (i) and (ii):

$$\text{Force through mass element} = \frac{V(s)}{M_s + M_m}$$

This is identical to Equation (iv), as expected.

Example 5.10: Ground-Based Two-Degree-of-Freedom Mechanical Oscillator (Revisited)

Let us revisit the ground-based two-degree-of-freedom oscillator, this time using Thevenin's theorem and linear graphs. The system is shown in Figure 5.16a. Its linear graph is drawn as given in Figure 5.16b. Next Figure 5.16c is drawn by representing each suspension unit by a single branch.

Since we are interested in the force F_s transmitted through the suspension unit M_s, this unit is cut out as in Figure 5.16d in order to determine the Thevenin equivalent circuit of the remaining system, which is shown in Figure 5.16e.

To determine the open-circuit velocity $V_{oc}(s)$ after the cut, note from Figure 5.16d that the force through the second parallel path is (divided inversely with mobilities)

$$F_1 = \frac{M_{m1}}{(M_{s1} + M_{m2} + M_{m1})} F(s)$$

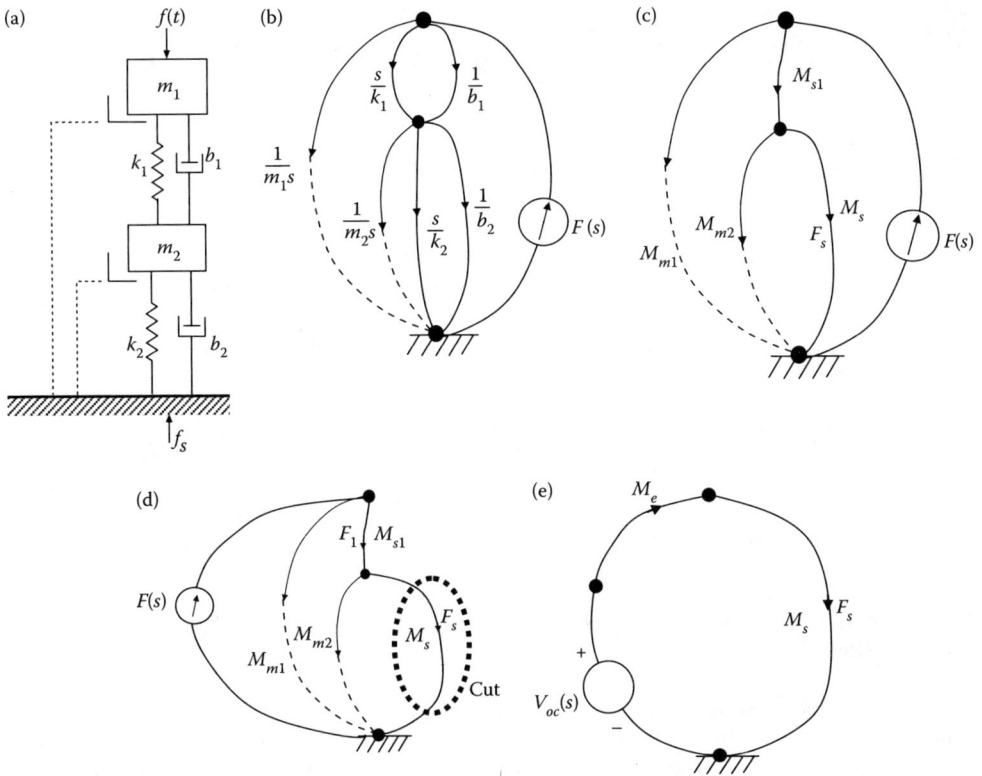

FIGURE 5.16
(a) Ground-based two-degree-of-freedom oscillator. (b) Linear graph. (c) Linear graph showing suspension components. (d) Cutting out the suspension unit. (e) Thevenin equivalent circuit.

Hence, velocity across M_{m2} is (product of force and mobility)

$$V_{oc}(s) = M_{m2}F_1 = \frac{M_{m1}M_{m2}}{(M_{s1} + M_{m2} + M_{m1})}F(s)$$

The Thevenin equivalent mobility of Figure 5.16d is (by open-circuiting the force source and combining the remaining two parallel paths).

$$M_e = \frac{M_{m2}(M_{s1} + M_{m1})}{M_{m2} + (M_{s1} + M_{m1})}$$

From Figure 5.16e, force transmitted through the suspension is

$$F_s = \frac{V_{oc}(s)}{(M_e + M_s)} = \frac{M_{m1}M_{m2}}{(M_{s1} + M_{m2} + M_{m1})}F(s)\frac{1}{\left[\dfrac{M_{m2}(M_{s1} + M_{m1})}{M_{m2} + (M_{s1} + M_{m1})} + M_s\right]}$$

Or

$$F_s = \frac{M_{m1}M_{m2}F(s)}{M_{m2}(M_{s1}+M_{m1})+M_s(M_{s1}+M_{m2}+M_{m1})}$$
(5.51a)

Or

$$\text{Force transmissibility } T_f = \frac{M_{m1}M_{m2}}{M_{m2}(M_{s1}+M_{m1})+M_s(M_{s1}+M_{m2}+M_{m1})}$$
(5.51)

We can show that this result is identical to the earlier result as follows:

$$T_f = \frac{1}{\left[\left(M_{s1}+\dfrac{M_{m2}M_s}{M_{m2}+M_s}\right)\dfrac{1}{M_{m1}}+1\right]\left(\dfrac{M_s}{M_{m2}}+1\right)} = \frac{1}{(M_s+M_{m2})\left[\left(M_{s1}+\dfrac{M_{m2}M_s}{M_{m2}+M_s}\right)+M_{m1}\right]} \cdot M_{m1}M_{m2}$$

$$= \frac{M_{m1}M_{m2}}{M_{s1}(M_s+M_{m2})+M_{m2}M_s+M_{m1}(M_{m2}+M_s)} = \frac{M_{m1}M_{m2}}{M_{m2}(M_{s1}+M_{m1})+M_s(M_{s1}+M_{m2}+M_{m1})}$$

Example 5.11: Two-Degree-of-Freedom Mechanical System with Support Motion (Revisited)

Let us revisit the two-degree-of-freedom mechanical oscillator with support motion, this time armed with Thevenin's theorem and linear graphs. The system is shown in Figure 5.17a. Its linear graph is drawn as given in Figure 5.17b. Next Figure 5.17c is drawn by representing each suspension unit by a single branch.

Since we are interested in the velocity V_m transmitted to the mass element M_{m1}, this unit is cut out as in Figure 5.17c in order to determine the Norton equivalent circuit of the remaining system, which is shown in Figure 5.17d.

To determine the short-circuit force $F_{sc}(s)$ after the cut, note from Figure 5.17c that this will be the force through M_{s1} (after shorting the cut). Since there is a series branch M_s and two parallel branches M_{m2} and M_{s1}, the force provided by the source velocity $V(s)$ is:

$$\frac{V(s)}{\left[\dfrac{M_{s1}M_{m2}}{(M_{s1}+M_{m2})}+M_s\right]}$$

This force is divided inversely according to the mobilities in the two parallel branches M_{m2} and M_{s1}. Hence, the short-circuit force is

$$F_{sc}(s) = \frac{V(s)}{\left[\dfrac{M_{s1}M_{m2}}{(M_{s1}+M_{m2})}+M_s\right]}\frac{M_{m2}}{(M_{s1}+M_{m2})} = \frac{M_{m2}}{M_{s1}M_{m2}+M_s(M_{s1}+M_{m2})}V(s)$$

The Norton equivalent mobility is obtained by short-circuiting the velocity source and combining the remaining two parallel branches M_{m2} and M_s with the series branch M_{s1}. We get

$$M_e = M_{s1}+\frac{M_{m2}M_s}{M_{m2}+M_s} = \frac{M_{s1}(M_{m2}+M_s)+M_{m2}M_s}{M_{m2}+M_s} = \frac{M_{s1}M_{m2}+M_s(M_{s1}+M_{m2})}{M_{m2}+M_s}$$

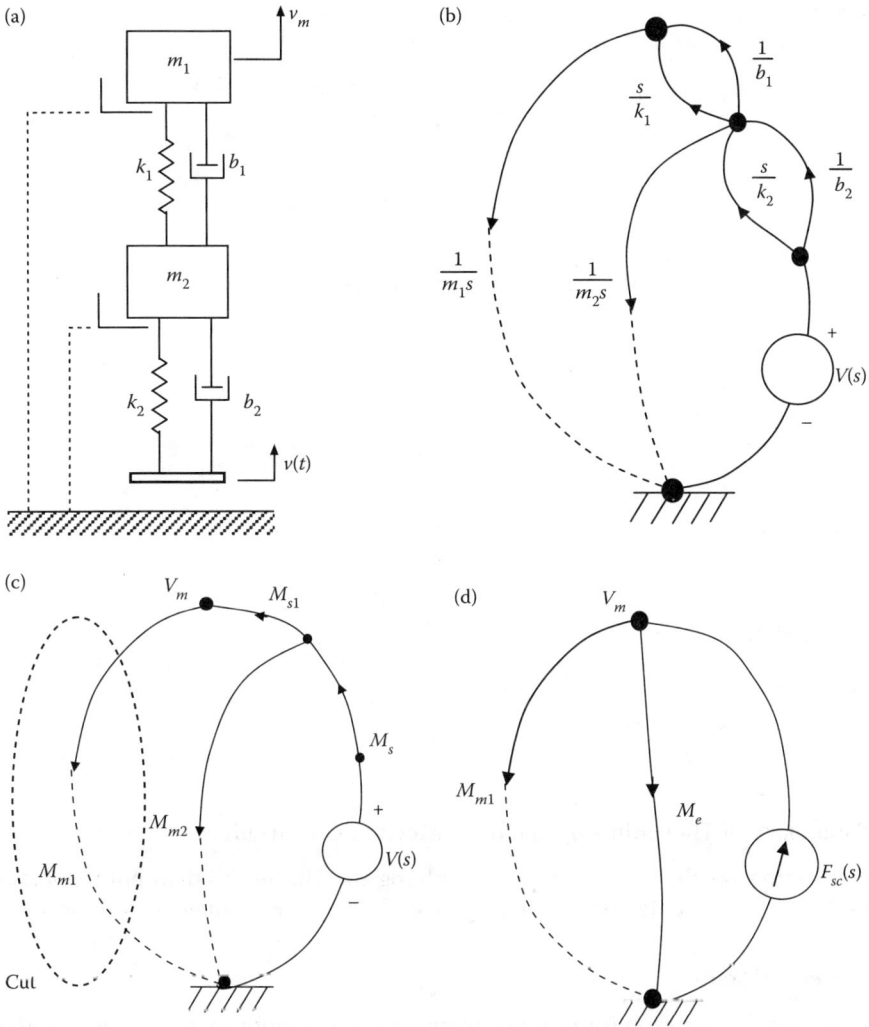

FIGURE 5.17
(a) Two-degree-of-freedom oscillator with support motion. (b) Linear graph. (c) Linear graph showing suspension components. (d) Norton equivalent circuit.

From Figure 5.17d, the velocity V_m is the product of the force and equivalent mobility:

$$V_m = \frac{M_{m1}M_e}{(M_{m1}+M_e)}F_{sc}(s)$$

Note from the previous two results that

$$F_{sc}(s)M_e = \frac{M_{m2}}{(M_{m2}+M_s)}V(s)$$

Hence

$$V_m = \frac{M_{m1}}{\left[M_{m1} + \dfrac{M_{s1}M_{m2} + M_s(M_{s1} + M_{m2})}{M_{m2} + M_s}\right]} \frac{M_{m2}}{(M_{m2} + M_s)} V(s)$$

$$= \frac{M_{m1}M_{m2}}{M_{m1}(M_{m2} + M_s) + M_{s1}M_{m2} + M_s(M_{s1} + M_{m2})} V(s)$$

$$= \frac{M_{m1}M_{m2}}{M_{m2}(M_{m1} + M_{s1}) + M_s(M_{m1} + M_{s1} + M_{m2})} V(s)$$

Or

$$\text{Motion transmissibility } T_m = \frac{M_{m1}M_{m2}}{M_{m2}(M_{m1} + M_{s1}) + M_s(M_{m1} + M_{s1} + M_{m2})} \tag{5.52}$$

We can show that this is identical to the earlier result as follows:

$$T_m = \left[\frac{1}{M_s\left(\dfrac{1}{M_{m2}} + \dfrac{1}{(M_{s1} + M_{m1})}\right) + 1}\right]\left[\frac{M_{m1}}{M_{s1} + M_{m1}}\right] = \frac{M_{m1}}{\dfrac{M_s(M_{s1} + M_{m1})}{M_{m2}} + M_s + M_{s1} + M_{m1}}$$

$$= \frac{M_{m1}M_{m2}}{M_s(M_{s1} + M_{m1}) + M_{m2}(M_s + M_{s1} + M_{m1})} = \frac{M_{m1}M_{m2}}{M_{m2}(M_{m1} + M_{s1}) + M_s(M_{s1} + M_{m1} + M_{m2})}$$

5.5.3 Summary of Thevenin Approach for Mechanical Circuits

We now summarize the general steps in applying the Thevenin's theorem to mechanical circuits that are represented by linear graphs, in the Laplace/frequency domain.

5.5.3.1 General Steps

1. Draw the linear graph for the system and mark the mobility functions for all the branches (except the source elements).
2. Simplify the linear graph by combining branches as appropriate (series branches: add mobilities; parallel branches: inverse rule applies for mobilities) and mark the mobilities of the combined branches.
3. Depending on the problem objective (e.g., determine a particular force, velocity, transfer-function) determine which part of the circuit (linear graph) should be cut (i.e., the variable or function of interest should be associated with the part that is removed from the circuit) so that the equivalent circuit of the remaining part has to be determined.
4. Depending on the problem objective establish whether Thevenin equivalence or Norton equivalence is needed. (Specifically: use Thevenin equivalence if a through variable needs to be determined, because this gives two series elements with a common through variable; use Norton equivalence if an across variable needs to be determined, because this gives two parallel elements with a common across variable.)

5. Determine the equivalent source and mobility of the equivalent circuit.
6. Using the equivalent circuit determine the variable or function of interest.

5.6 Block Diagrams and State-Space Models

The transfer-function model $G(s)$ of a SISO system can be represented by a single block with an input and an output, shown in Figure 5.18a. For a MIMO system the inputs and outputs are vectors u and y. The corresponding information (signal) lines are drawn thicker as in Figure 5.18b to indicate that they represent vectors. One disadvantage of the transfer-function representation is obvious from Figure 5.18: No information regarding how the various elements or components are interconnected within the system can be uniquely determined from the transfer-function. It contains only a unique input–output description. For the same reason a given transfer-function can correspond to different state-space models. We identify the transfer-function of a dynamic model by its inputs and outputs, not by its state variables, which are internal variables. However, the internal structure of a dynamic system can be indicated by a more elaborate graphical representation. One such representation is provided by linear graphs, as we saw in Chapter 4. Another detailed representation can be provided by a block diagram with many blocks representing system elements or components, connected together. Such a detailed block diagram may be used to uniquely indicate the state variables in a particular model.

For example, consider the state-space model Equations 5.13 and 5.14. A block diagram that uniquely possesses this model is shown in Figure 5.19. Note the *feedforward* path corresponding to D. The feedback paths (corresponding to A) do not necessarily represent a feedback control system where "active" feedback paths are generated by an external controller. The internal feedback paths shown in Figure 5.19 are *natural feedback* paths. Strictly speaking thicker signal lines should be used in this diagram since we are dealing with vector variables.

Two or more blocks in cascade can be replaced by a single block having the product of individual transfer-functions. The circle in Figure 5.19 is a *summing junction*. A negative sign at the arrow-head of an incoming signal corresponds to subtraction of that signal. As mentioned earlier, $1/s$ can be interpreted as integration, and s as differentiation.

In generating and simplifying block diagrams, the rules indicated in Table 5.7 are quite useful. All the entries of the table are quite obvious, and may be verified by inspection. We note the following:

1. Circle (summing junction): Two or more signals are added together forming a new signals.

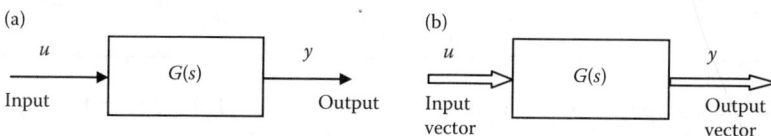

FIGURE 5.18
Block-diagram representation of a transfer-function model. (a) Single-input single-output (SISO) system. (b) Multi-input–multi-output (MIMO) system.

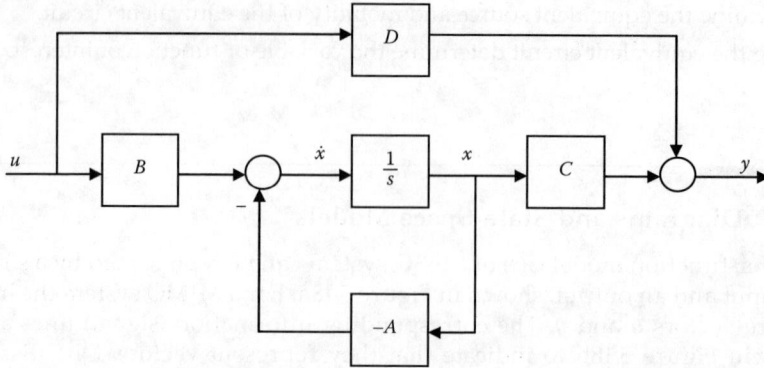

FIGURE 5.19
block-diagram representation of a state-space model.

TABLE 5.7

Basic Relations for Block-diagram Reduction

Description	Equivalent Representation	
Summing junction		$x_3 = x_1 + x_2$
Cascade (series) connection		
Parallel connection		
Shifting signal-pickoff point		
Shifting Signal-application point		
Reduction of feedback loop		

Note: Negative sign at arrow-head of incoming signal ➔ subtract the signal.

2. Two or more blocks in cascade=single block having product of individual transfer-functions.

3. Two or more blocks in parallel=single block having sum of individual transfer-functions.

However, an explanation is appropriate for the last entry of the table, where the equivalent block for a feedback loop is given. The result may be obtained as follows:

The feedback signal at the summing junction $= -Hx_2$

Hence, the signal reaching the block G is $x_1 - Hx_2$

Accordingly, output of the block G is $G(x_1 - Hx_2)$ which is equal to x_2.

We have: $G(x_1 - Hx_2) = x_2$

Straightforward algebra gives:

$$x_2 = \frac{G}{1 + GH} x_1 \tag{5.53}$$

The equivalence of Figure 5.19 and the relations (Equations 5.13 and 5.14) should be obvious. Alternatively, the rules for block diagram reduction (given in Table 5.7) can be used to show that the system transfer-function is given by:

$$\frac{Y(s)}{U(s)} = G(s) = \frac{CB}{(s - A)} + D \tag{5.54}$$

This is the scalar version of the matrix-vector Equation 5.15a.

5.6.1 Simulation Block Diagrams

In a simulation block diagram each block contains either an integrator $(1/s)$ or a constant gain term. The name originates from classical analog computer applications in which hardware modules of summing amplifiers and integrators (along with other units such as potentiometers and resistors) are interconnected to simulate dynamic systems. Recently, the same type of block diagrams has been in wide use for the purpose of computer simulation of dynamic systems; for example, in software tools such as Simulink®.

In summary:

A simulation block diagram consists only of:

1. Integration blocks
2. Constant gain blocks
3. Summing junctions

Also,

- They are useful in computer simulation of dynamic systems.
- They can be obtained from input–output models (see following examples) or state-space models (see previous example or converse of the following examples).
- Can be used to develop state-space models.
- Not unique (see examples given next).

5.6.2 Principle of Superposition

As noted before, for a linear system, the *principle of superposition* applies. In particular if, with zero ICs, x is the response of a system to an input u, then $d^r x/dt^r$ is the response to the input $d^r u/dt^r$. Consequently, by the principle of superposition, $a_1x + a_2 d^r x/dt^r$ is the response to the input $a_1 u + a_2 d^r u/dtr$. This form of the principle of superposition is quite useful in the analytical manipulation of block diagrams.

Using the same example for input–output differential equations, we now illustrate several methods of obtaining state-space models through simulation block diagrams. In a simulation block diagram we have:

1. State variables = Outputs of the integrators
2. State equations = Equations for signals going into the integration blocks
3. Algebraic output equation = Equation for the summing junction that generates y (far right).

Example 5.12: Superposition Method

Consider the time-domain input–output model (differential equation) given by:

$$\dddot{y} + 13\ddot{y} + 56\dot{y} + 80y = \dddot{u} + 6\ddot{u} + 11\dot{u} + 6u \tag{i}$$

The principle of superposition is applied now. Consider the differential equation:

$$\dddot{x} + 13\ddot{x} + 56\dot{x} + 80x = u \tag{ii}$$

This defines the "parent" (or, auxiliary) system. The simulation diagram for Equation (ii) is shown in Figure 5.20. Steps of obtaining this diagram are as follows: start with the highest-order derivative of the response variable (i.e., $\dddot{x}$); successively integrate it until the variable itself (x) is obtained; feed the resulting derivatives of different orders to the summing junction (along with the input variable) to produce the highest-order derivative of the response variable such that the original differential Equation (ii) is satisfied.

By the principle of superposition, it follows from Equations (i) and (ii) that:

$$y = \dddot{x} + 6\ddot{x} + 11\dot{x} + 6x \tag{iii}$$

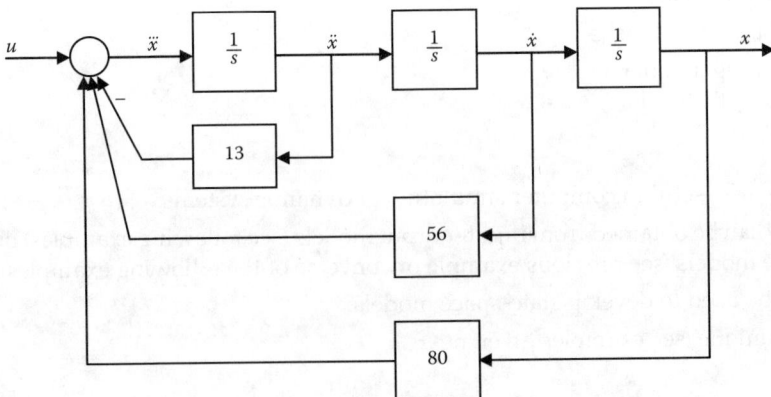

FIGURE 5.20
The simulation diagram of system $\dddot{x} + 13\ddot{x} + 56\dot{x} + 80x = u$.

Hence, the simulation diagram for the original system (Equation (i)) can be derived from Figure 5.20, as shown in Figure 5.21.

In particular, note the resulting feedforward paths. The corresponding state model employs *x* and its derivatives as state variables:

$$\begin{bmatrix} x_1 & x_2 & x_3 \end{bmatrix}^T = \begin{bmatrix} x & \dot{x} & \ddot{x} \end{bmatrix}^T$$

As indicated before, the state variables are the outputs of the integrators (see Figure 5.21). Also, as indicated before, a state equation is written by expressing how the signal that goes into the corresponding integration block (i.e., the first derivative of the corresponding state variable) is formed. Specifically, from Figure 5.21 we have:

$$\dot{x}_1 = x_2$$

$$\dot{x}_2 = x_3 \qquad\qquad\qquad \text{(iv)}$$

$$\dot{x}_3 = -80x_1 - 56x_2 - 13x_3 + u$$

As indicated before, the algebraic output equation is obtained by writing the signal summation equation for the summing junction (far right), which generates *y*. Specifically, from Figure 5.21 we have:

$$y = 6x_1 + 11x_2 + 6x_3 + (-80x_1 - 56x_2 - 13x_3 + u)$$

or,

$$y = -74x_1 - 45x_2 - 7x_3 + u \qquad\qquad\qquad \text{(v)}$$

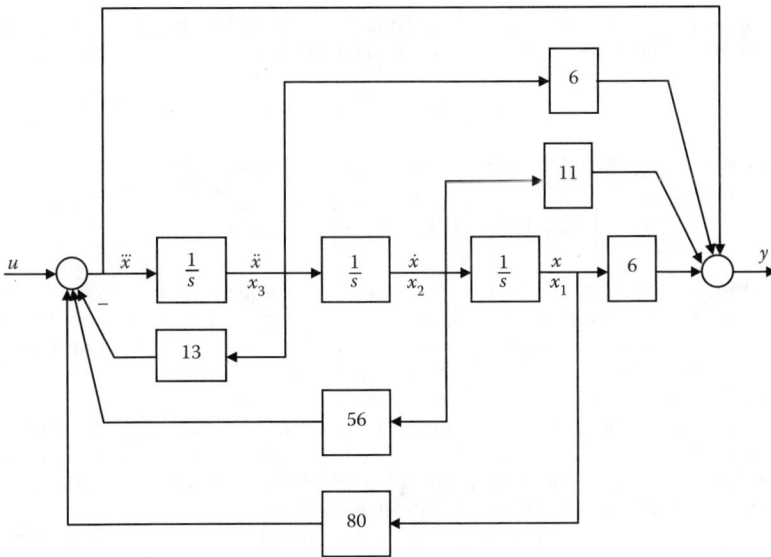

FIGURE 5.21
The simulation diagram of $\dddot{y} + 13\ddot{y} + 56\dot{y} + 80y = \dddot{u} + 6\ddot{u} + 11\dot{u} + 6u$.

The corresponding model matrices are:

$$A = \begin{bmatrix} 0 & 1 & 0 \\ 0 & 0 & 1 \\ -80 & -56 & -13 \end{bmatrix}; \quad B = \begin{bmatrix} 0 \\ 0 \\ 1 \end{bmatrix}; \quad C = \begin{bmatrix} -74 & -45 & -7 \end{bmatrix}; \quad D = 1$$

The system matrix pair (A, B) is said to be in the *companion form* in this state model. *Note:* The system model is third order. Hence the simulation diagram needs three integrators, and the system matrix A is 3×3.

Note further that the "parent" (or, auxiliary) transfer-function (that of Equation (ii)) is given by:

$$\frac{X}{U} = \frac{1}{s^3 + 13s^2 + 56s + 80}$$

From Equation (iii), the output of the original system is given by:

$$Y = s^3 X + 6s^2 X + 11sX + 6X = (s^3 + 6s^2 + 11s + 6)X$$

Hence the transfer-function of the original system is:

$$G(s) = \frac{Y}{U} = \frac{s^3 + 6s^2 + 11s + 6}{s^3 + 13s^2 + 56s + 80}$$

This agrees with the original differential Equation (i). Furthermore, in $G(s)$, since the numerator polynomial is of the same order (third order) as the denominator polynomial (characteristic polynomial), a nonzero feedforward gain matrix D is generated in the state model.

Example 5.13: Grouping Like-Derivatives Method

Consider the same input–output differential Equation (i) as in Example 5.12. By grouping the derivatives of the same order, it can be written in the following form:

$$\dddot{y} = \dddot{u} + (6\ddot{u} - 13\ddot{y}) + (11\dot{u} - 56\dot{y}) + (6u - 80y)$$

By successively integrating this equation three times, we obtain:

$$y = u + \int \left[6u - 13y + \int \left\{ 11u - 56y + \int (6u - 80y) d\tau \right\} d\tau' \right] d\tau'' \tag{i}a$$

Note the three integrations on the right hand side of this equation. Now draw the simulation diagram as follows: Assume that y is available. Form the integrand of the innermost integration in (i) a by feeding forward the necessary u term and feeding back the necessary y term. Perform the innermost integration. The result will form a part of the integrand for the next integration. Complete the integrand through feeding forward the necessary u term and feeding back the necessary y term. Perform this second integration. The result will form a part of the integrand next the next (outermost) integration. Proceed as before to complete the integrand and perform the outermost integration. Feedforward the necessary u term to generate y, which was assumed to be known in the beginning. The result is shown in Figure 5.22.

Note: The "innermost" integration in (i)a forms the "outermost" feedback loop in the block diagram.

As in the previous example, the state variables are defined as the outputs of the integrators. The state equations are written by considering the signals that enter each integration block, to form the first derivative of the corresponding state variable. We get:

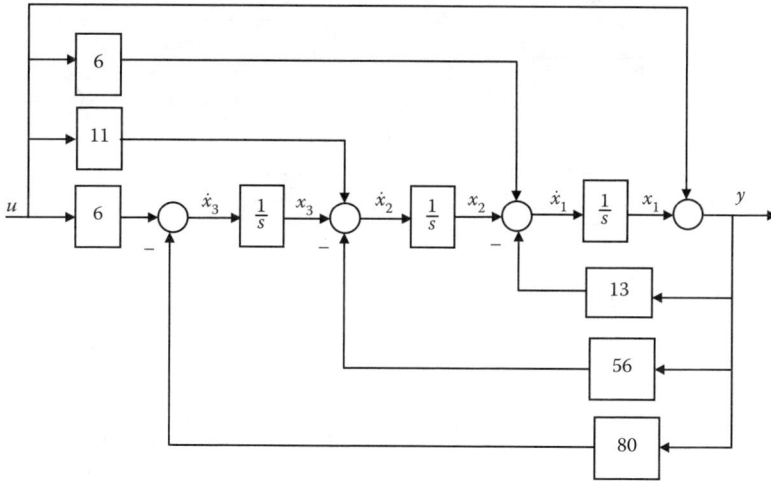

FIGURE 5.22
Simulation diagram obtained by grouping like-derivatives.

$$\dot{x}_1 = 6u + x_2 - 13(x_1 + u) = -13x_1 + x_2 - 7u$$

$$\dot{x}_2 = 11u + x_3 - 56(x_1 + u) = -56x_1 + x_3 - 45u \qquad \text{(iv)a}$$

$$\dot{x}_3 = 6u - 80(x_1 + u) = -80x_1 - 74u$$

The algebraic output equation is written by writing the equation for the summing junction (far right), which generates *y*. We get:

$$y = x_1 + u \qquad \text{(v)a}$$

This corresponds to:

$$A = \begin{bmatrix} -13 & 1 & 0 \\ -56 & 0 & 1 \\ -80 & 0 & 0 \end{bmatrix}; \quad B = \begin{bmatrix} -7 \\ -45 \\ -74 \end{bmatrix}; \quad C = \begin{bmatrix} 1 & 0 & 0 \end{bmatrix}; \quad D = 1$$

This state model is the *dual* of the state model obtained in the previous example.

Example 5.14: Factored-Transfer-Function Method

The method illustrated in this example is appropriate when the system transfer-function is available in the factorized form, with first-order terms expressed in the form:

$$G_1(s) = \frac{(s+b)}{(s+a)}$$

Since the block diagram of the transfer-function $1/(s+a)$ is given by Figure 5.23a, it follows from the superposition method that the block diagram for $(s+b)/(s+a)$ is as in Figure 5.23b. This is one form of the basic block-diagram module, which is used in this method.

An alternative form of block diagram for this basic transfer-function module is obtained by noting the equivalence shown in Figure 5.24a. In other words, when it is needed to supply a

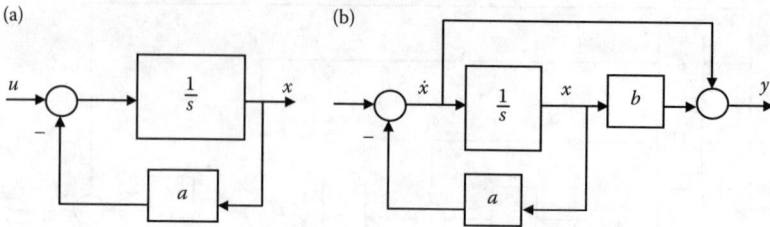

FIGURE 5.23
The simulation diagrams of (a) $1/(s+a)$ and (b) $(s+b)/(s+a)$.

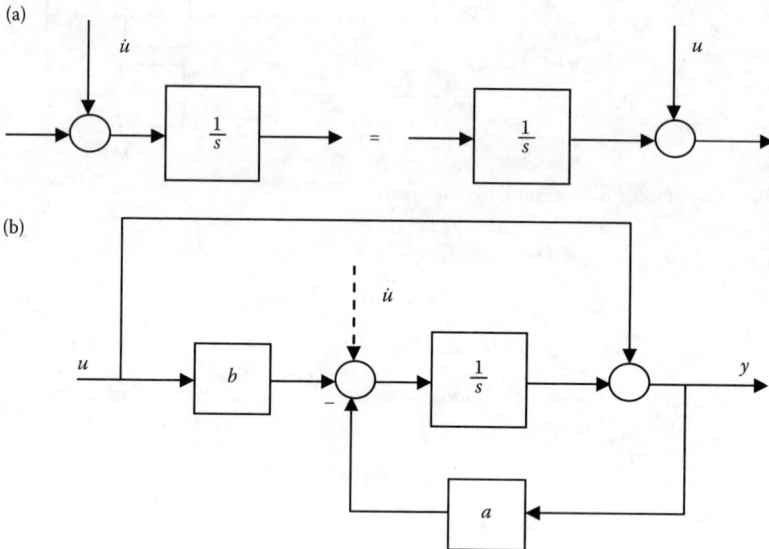

FIGURE 5.24
(a) Two equivalent ways of providing an input derivative $(\dot{u})$. (b) An equivalent simulation diagram for $(s+b)/(s+a)$.

derivative signal $\dot{u}$ at the input to an integrator, instead the signal u itself can be supplied at the output of the integrator. Now, note that the first-order transfer-function unit $(s+b)/(s+a)$ has the terms $\dot{u}+bu$ on the input side. The term bu is generated by cascading a block with simple gain b, as in Figure 5.24b. To provide $\dot{u}$, instead of using the dotted input path in Figure 5.24b, that would require differentiating the input signal, the signal u itself is applied at the output of the integrator. It follows that the block diagram in Figure 5.23b is equivalent to that in Figure 5.24b.

Now, returning to our common example (Equation (i)), the transfer-function is written as:

$$G(s) = \frac{s^3 + 6s^2 + 11s + 6}{s^3 + 13s^2 + 56s + 80}$$

This can be factored into the form:

$$G(s) = \frac{(s+1)}{(s+4)} \times \frac{(s+2)}{(s+4)} \times \frac{(s+3)}{(s+5)} \qquad \text{(i)b}$$

Note that there are two common factors (corresponding to "repeated poles" or "repeated eigen-values") in the characteristic polynomial (denominator). This has no special implications in the present method. The two versions of block diagram for this transfer-function, in the present methods, are shown in Figures 5.25 and 5.26. Here we have used the fact that the product of two transfer-functions corresponds to cascading the corresponding simulation block diagrams. As before, the state variables are chosen as the outputs of the integrators, and the state equations are written for the input terms of the integrator blocks. The output equation comes from the summation block at the far right, which generates the output.

From Figure 5.25, the state equations are obtained as:

$$\dot{x}_3 = u - 4x_3$$

$$\dot{x}_2 = x_3 - 4x_2 + (u - 4x_3) = -4x_2 - 3x_3 + u \qquad\qquad \text{(iv)b}$$

$$\dot{x}_1 = 2x_2 - 5x_1 + (-4x_2 - 3x_3 + u) = -5x_1 - 2x_2 - 3x_3 + u$$

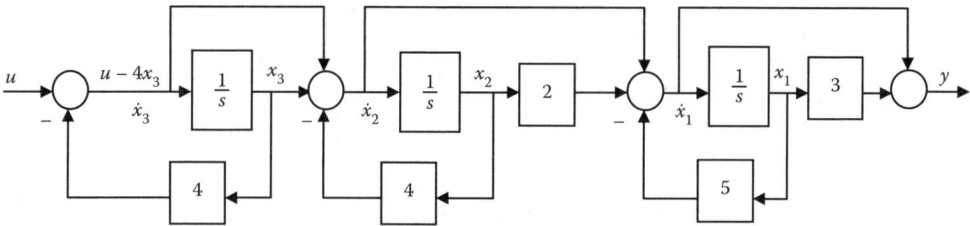

FIGURE 5.25
Simulation block diagram obtained by factorizing the transfer-function.

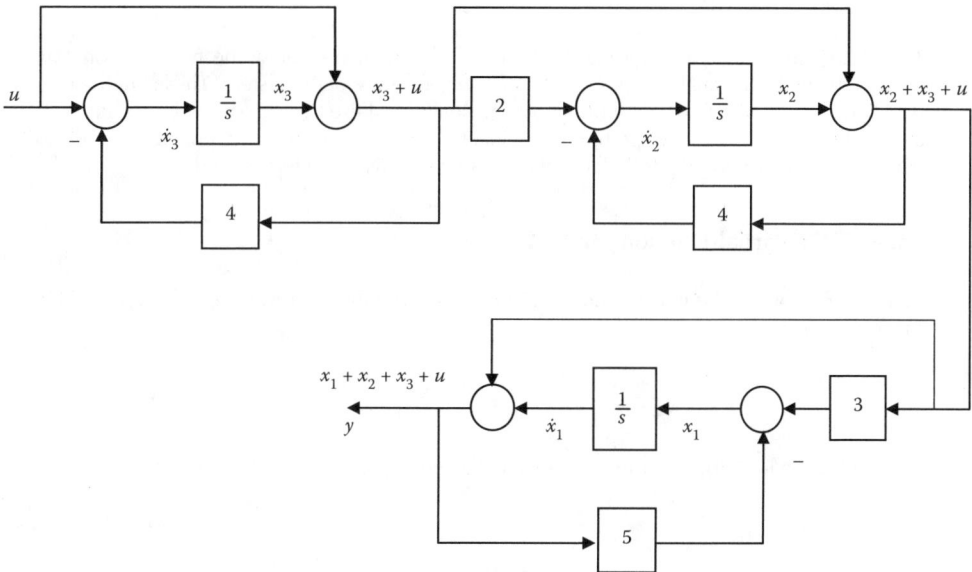

FIGURE 5.26
An alternative simulation diagram obtained by factorizing the transfer-function.

The algebraic output equation is:

$$y = 3x_1 + (-5x_1 - 2x_2 - 3x_3 + u)$$

$$= -2x_1 - 2x_2 - 3x_3 + u$$

(v)b

These corresponds to the state model matrices:

$$A = \begin{bmatrix} -5 & -2 & -3 \\ 0 & -4 & -3 \\ 0 & 0 & -4 \end{bmatrix}; \quad B = \begin{bmatrix} 1 \\ 1 \\ 1 \end{bmatrix}; \quad C = \begin{bmatrix} -2 & -2 & -3 \end{bmatrix}; \quad D = 1$$

The state equations corresponding to Figure 5.26 are:

$$\dot{x}_3 = -4(x_3 + u) + u = -4x_3 - 3u$$

$$\dot{x}_2 = -4(x_2 + x_3 + u) + 2(x_3 + u) = -4x_2 - 2x_3 - 2u$$

(iv)c

$$\dot{x}_1 = -5(x_1 + x_2 + x_3 + u) + 3(x_2 + x_3 + u) = -5x_1 - 2x_2 - 3x_3 - 2u$$

The algebraic output equation is:

$$y = x_1 + x_2 + x_3 + u$$

(v)c

These equations correspond to the state model matrices:

$$A = \begin{bmatrix} -5 & -2 & -2 \\ 0 & -4 & -2 \\ 0 & 0 & -4 \end{bmatrix}; \quad B = \begin{bmatrix} -2 \\ -2 \\ -3 \end{bmatrix}; \quad C = \begin{bmatrix} 1 & 1 & 1 \end{bmatrix}; \quad D = 1$$

Both system matrices are upper-diagonal (i.e., all the elements below the main diagonal are zero), and the main diagonal consists of the poles (eigenvalues) of the system. These are the roots of the characteristic equation. We should note the duality in these two state models (Equations (iv) b and (iv)c). Note also that, if we group the original transfer-function into different factor terms, we get different state models. In particular, the state equations will be interchanged.

Example 5.15: Partial-Fraction Method

The partial fractions of the transfer-function (i) b considered in the previous example are written in the form:

$$G(s) = \frac{s^3 + 6s^2 + 11s + 6}{s^3 + 13s^2 + 56s + 80} = 1 - \frac{a}{(s+4)} - \frac{b}{(s+4)^2} - \frac{c}{(s+5)}$$

By equating the like terms on the two sides of this identity, or by using the fact that:

$$c = -(s+5)G(s)\big|_{s=-5}$$

$$b = -(s+4)^2 G(s)\big|_{s=-4}$$

$$a = -\left[\frac{d}{ds}(s+4)^2 G(s)\right]_{s=-4}$$

we can determine the unknown coefficients; thus

$$a = -17, \quad b = 6, \quad c = 24$$

The simulation block diagram corresponding to the partial-fraction representation of the transfer-function is shown in Figure 5.27. We have used the fact that the sum of two transfer-functions corresponds to combining their block diagrams in parallel. Again the state variables are chosen as the outputs of the integrators. By following the same procedure as before, the corresponding state equations are obtained as:

$$\dot{x}_1 = -5x_1 + u$$

$$\dot{x}_2 = -4x_2 + x_3$$

$$\dot{x}_3 = -4x_3 + u$$

The algebraic output equation is:

$$y = -cx_1 - bx_2 - ax_3 + u$$

This corresponds to the state-model matrices:

$$A = \begin{bmatrix} -5 & 0 & 0 \\ 0 & -4 & 0 \\ 0 & 0 & -4 \end{bmatrix}; \quad B = \begin{bmatrix} 1 \\ 0 \\ 1 \end{bmatrix}; \quad C = \begin{bmatrix} -c & -b & -a \end{bmatrix}; \quad D = 1$$

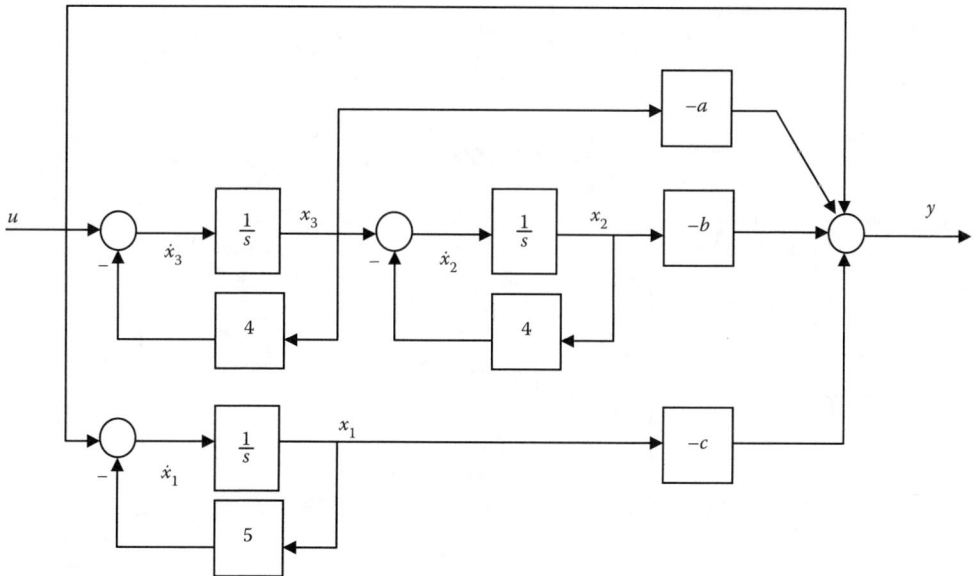

FIGURE 5.27
Simulation block diagram obtained by the partial fraction method.

In this case, the system matrix is said to be in the *Jordan canonical form*. If the eigenvalues are distinct (unequal), the matrix **A**, when expressed in the Jordan form, will be diagonal, and the diagonal elements will be the eigenvalues. When repeated eigenvalues are present, as in the present example, the matrix **A** will consist of diagonal blocks (or *Jordan blocks*) consisting of upper-diagonal submatrices with the repeated eigenvalues lying on the main diagonal, elements of unity at locations immediately above the main diagonal, and zero elements elsewhere. More than one Jordan block can exist for the same repeated eigenvalue. These considerations are beyond the scope of this chapter.

Example 5.16

A manufacturer of rubber parts uses a conventional process of steam-cured molding of latex. The molded rubber parts are first cooled and buffed (polished) and then sent for inspection and packing. A simple version of a rubber buffing machine is shown in Figure 5.28a. It consists of a large hexagonal drum whose inside surfaces are all coated with a layer of bonded emery. The drum is

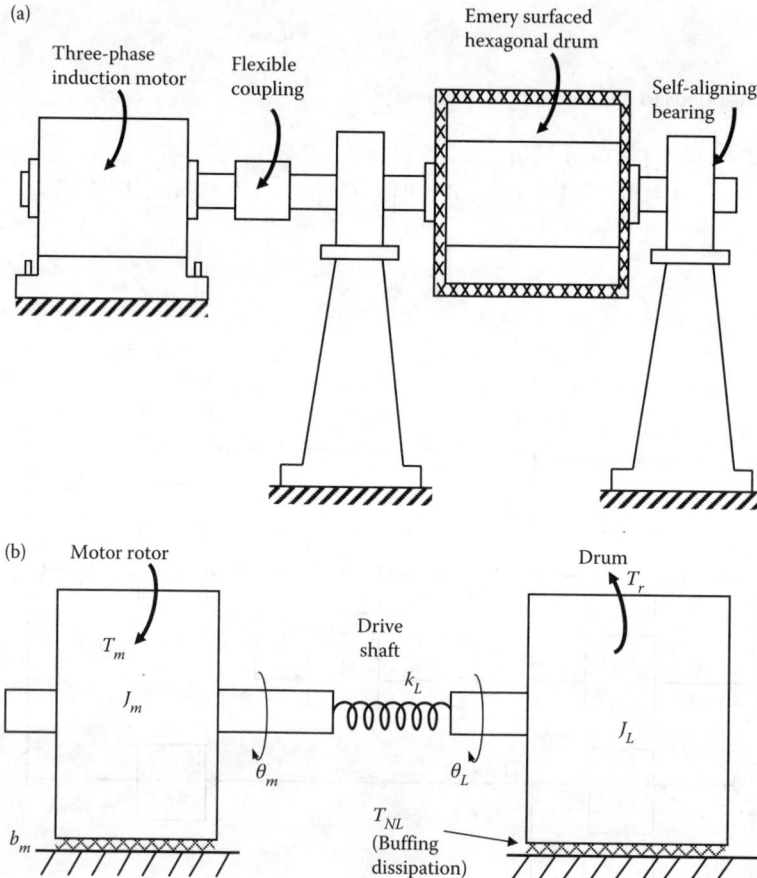

FIGURE 5.28
A rubber buffing machine. (a) Schematic diagram. (b) Dynamic model.

supported horizontally along its axis on two heavy-duty, self-aligning bearings at the two ends, and is rotated using a three-phase induction motor. The drive shaft of the drum is connected to the motor shaft through a flexible coupling, in order to compensate for possible misalignments of the axes. The buffing process consists of filling the drum with rubber parts, steadily rotating the drum for a specified period of time, and finally vacuum-cleaning the drum and its contents. Dynamics of the machine affects the mechanical loading on various parts of the system such as the motor, coupling, bearings, shafts and the support structure.

In order to study the dynamic behavior, particularly at the startup stage and under disturbances during steady-state operation, an engineer develops a simplified model of the buffing machine. This model is shown in Figure 5.28b. The motor is modeled as a torque source T_m, which is applied on the rotor having moment of inertia J_m and resisted by a viscous damping torque of damping constant b_m. The connecting shafts and the coupling unit are represented by a torsional spring of stiffness k_L. The drum and its contents are represented by an equivalent constant moment of inertia J_L. There is a resisting torque on the drum, even at steady operating speed, due to the eccentricity of the contents of the drum. This is represented by a constant torque T_r. Furthermore, energy dissipation due to the buffing action (between the rubber parts and the emery surfaces of the drum) is represented by a nonlinear damping torque T_{NL}, which may be approximated by

$$T_{NL} = c|\dot{\theta}_L|\dot{\theta}_L \quad \text{with} \quad c > 0$$

Note that θ_m and θ_L are the angles of rotation of the motor rotor and the drum, respectively, and these are measured from inertial reference lines which correspond to a relaxed configuration of spring k_L.

a. Comment on the assumptions made in the modeling process of this system and briefly discuss the validity (or accuracy) of the model.

b. Show that the model equations are:

$$J_m\ddot{\theta}_m = T_m - k_L(\theta_m - \theta_L) - b_m\dot{\theta}_m$$

$$J_L\ddot{\theta}_L = k_L(\theta_m - \theta_L) - c|\dot{\theta}_L|\dot{\theta}_L - T_r$$

What are the inputs of this system?

c. Using the speeds $\dot{\theta}_m$ and $\dot{\theta}_L$, and the spring torque T_k as the state variables, and the twist of the spring as the output, obtain a complete state-space model for his nonlinear system.

What is the order of the state model?

d. Suppose that under steady operating conditions, the motor torque is $\overline{T}_m$, which is constant. Determine an expression for the constant speed $\overline{\omega}$ of the drum in terms of $\overline{T}_m$, T_r, and appropriate system parameters under these conditions. Show that, as intuitively clear, we must have $\overline{T}_m > T_r$ for this steady operation to be feasible. Also obtain an expression for the spring twist at steady state, in terms of $\overline{\omega}$, T_r, and the system parameters.

e. Linearize the system equations about the steady operation condition and express the two equations in terms of the following "incremental" variables:

$q_1 =$ variation of θ_m about the steady value
$q_2 =$ variation of θ_L about the steady value
$u =$ disturbance increment of T_m from the steady value $\overline{T}_m$.

f. For the linearized system obtain the input–output differential equation, first considering q_1 as the output and next considering q_2 as the output. Comment about and justify the nature of the homogeneous (characteristic-equation) parts of the two equations. Discuss, by examining the physical nature of the system, why only the derivatives of q_1 and q_2 and not the variables themselves are present in these input–output equations.

Explain why the derivation of the input–output differential equations will become considerably more difficult if a damper is present between the two inertia elements J_m and J_L.

g. Consider the input–output differential equation for q_1. By introducing an auxiliary variable draw a simulation block diagram for this system. (Use integrators, summers, and coefficient blocks only.) Show how this block diagram can be easily modified to represent the following cases:

(i) q_2 is the output.
(ii) $\dot{q}_1$ is the output.
(iii) $\dot{q}_2$ is the output.

What is the order the system (or the number of free integrators needed) in each of the four cases of output considered in this example?

h. Considering the spring twist $(q_1 ñ q_2)$ as the output draw a simulation block diagram for the system. What is the order of the system in this case?

Hint: For this purpose you may use the two linearized second order differential equations obtained in (e).

(i)   Comment on why the "system order" is not the same for the five cases of output considered in (g) and (h).

Solution

a. The assumptions are satisfactory for a preliminary model, particularly because very accurate control is not required in this process. Some sources of error and concern are as follows:

(i)   Since the rubber parts move inside the drum, J_L is not constant and the inertia contribution does not represent a rigid system.
(ii)   Inertia of the shafts and coupling is either neglected or lumped with J_m and J_L.
(iii) Coulomb and other nonlinear types of damping in the motor and the bearings have been approximated by viscous damping.
(iv)   The torque source model (T_m) is only an approximation to a real induction motor.
(v)   The resisting torque of the rubber parts (T_r) is not constant during rotation.
(vi)   Energy dissipation due to relative movements between the rubber parts and the inside surfaces of the drum may take a different form from what is given (a quadratic damping model).

b. For J_m, Newton's second law gives (see Figure 5.29a):

$$J_m \frac{d^2\theta_m}{dt^2} = T_m - b_m \dot{\theta}_m - T_k \tag{i}$$

For spring k_L, Hooke's law gives (see Figure 5.29b):

$$T_k = k_L(\theta_m - \theta_L) \tag{ii}$$

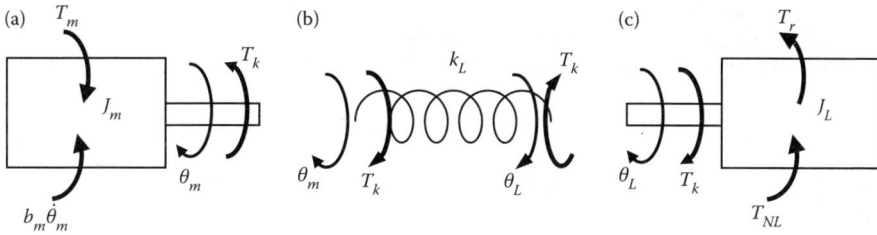

FIGURE 5.29
Figure 5.29 (a) Motor inertia. (b) Drive shaft. (c) Drum inertia.

For J_L, Newtons' second law gives (see Figure 5.29c):

$$J_L \frac{d^2\theta_L}{dt^2} = T_k - T_{NL} - T_r \qquad \text{(iii)}$$

with

$$T_{NL} = c|\dot{\theta}_L|\dot{\theta}_L \qquad \text{(iv)}$$

Substitute Equation (ii) into Equation (i):

$$J_m\ddot{\theta}_m = T_m - b_m\dot{\theta}_m - k_L(\theta_m - \theta_L) \qquad \text{(v)}$$

Substitute Equations (ii) and (iv) into Equation (iii):

$$J_L\ddot{\theta}_L = k_L(\theta_m - \theta_L) - c|\dot{\theta}_L|\dot{\theta}_L - T_r \qquad \text{(vi)}$$

Clearly, T_m and T_r are the inputs to the system (see Equations (v) and (vi)).
c. Let $\dot{\theta}_m = \omega_m$ and $\dot{\theta}_L = \omega_L$

From Equation (i): $\dfrac{d\omega_m}{dt} = -\dfrac{b_m}{J_m}\omega_m - \dfrac{1}{J_m}T_k + \dfrac{1}{J_m}T_m$

From Equations (iii) and (iv): $\dfrac{d\omega_L}{dt} = -\dfrac{c}{J_L}|\omega_L|\omega_L + \dfrac{1}{J_L}T_k - \dfrac{1}{J_L}T_r$

Differentiate Equation (ii): $\dfrac{dT_k}{dt} = k_L\omega_m - k_L\omega_L$

The above three equations are the state equations.

Output $y = $ spring twist $= \theta_m - \theta_L$

Hence, from Equation (ii) we have

$$y = \frac{1}{k_L}T$$

which is the output equation.

The system is third order (three state equations).

d. Under steady conditions:

$$\omega_m = \omega_L = \bar{\omega}, \quad \dot{\omega}_m = 0 = \dot{\omega}_L, \quad T_m = \bar{T}_m$$

and $\theta_m - \theta_L = \Delta\bar{\theta}$; T_r remains a constant.
Then from Equations (v) and (vi):

$$\bar{T}_m - b_m\bar{\omega} - k_L\Delta\bar{\theta} = 0 \tag{vii}$$

and

$$k_L\Delta\bar{\theta} - c\bar{\omega}^2 - T_r = 0 \tag{viii}$$

Without loss of generality, $\bar{\omega}$ is assumed to be positive.
Add the last two Equations (vii) and (viii) to eliminate $k_L\Delta\bar{\theta}$; thus,

$$\bar{T}_m - b_m\bar{\omega} - c\bar{\omega}^2 - T_r = 0$$

or,

$$c\bar{\omega}^2 + b_m\bar{\omega} - (\bar{T}_m - T_r) = 0$$

Hence

$$\bar{\omega} = -\frac{b_m}{2c} \pm \sqrt{\left(\frac{b_m}{2c}\right)^2 + \frac{(\bar{T}_m - T_r)}{c}}$$

The proper solution is the positive one:

$$\bar{\omega} = \sqrt{\left(\frac{b_m}{2c}\right)^2 + \frac{(\bar{T}_m - T_r)}{c}} - \frac{b_m}{2c}$$

and for this to be positive, we must have $\bar{T}_m > T_r$.
Next, from Equation (viii), the steady-state twist of the spring is:

$$\Delta\bar{\theta} = \frac{(c\bar{\omega}^2 - T_r)}{k_L}$$

e. Taylor series expansion up to the first order term gives:

For Equation (v): $J_m\dot{\bar{\omega}} + J_m\ddot{q}_1 = \bar{T}_m + u - b_m\bar{\omega} - b_m\dot{q}_1 - k_L\Delta\bar{\theta} - k_L(q_1 - q_2)$

For Equation (vi): $J_L\dot{\bar{\omega}} + J_L\ddot{q}_2 = k_L\Delta\bar{\theta} + k_L(q_1 - q_2) - c\bar{\omega}^2 - 2c\bar{\omega}\dot{q}_2 - T_r$

The steady-state terms cancel out (also, $\dot{\bar{\omega}} = 0$). Hence, we have the following linearized equations:

$$J_m\ddot{q}_1 = u - b_m\dot{q}_1 - k_L(q_1 - q_2) \tag{ix}$$

$$J_L\ddot{q}_2 = k_L(q_1 - q_2) - 2c\bar{\omega}\dot{q}_2 \qquad \text{(x)}$$

These two equations represent the linear model.

f. From Equation (ix)

$$q_2 = \left[q_1 + \frac{b_m}{k_L}\dot{q}_1 + \frac{J_m}{k_L}\ddot{q}_1 - \frac{u}{k_L} \right] \qquad \text{(xi)}$$

From Equation (x)

$$q_1 = \left[q_2 + \frac{2c\bar{\omega}}{k_L}\dot{q}_2 + \frac{J_L}{k_L}\ddot{q}_2 \right] \qquad \text{(x)}$$

Substitute Equation (xi) into Equation (xii) for q_2:

$$q_2 = \left[q_1 + \frac{b_m}{k_L}\dot{q}_1 + \frac{J_m}{k_L}\ddot{q}_1 - \frac{u}{k_L} \right] + \frac{2c\bar{\omega}}{k_L}\left[\dot{q}_1 + \frac{b_m}{k_L}\ddot{q}_1 + \frac{J_m}{k_L}\dddot{q}_1 - \frac{\dot{u}}{k_L} \right]$$
$$+ \frac{J_L}{k_L}\left[\ddot{q}_1 + \frac{b_m}{k_L}\dddot{q}_1 + \frac{J_m}{k_L}\ddddot{q}_1 - \frac{\ddot{u}}{k_L} \right]$$

which gives

$$\frac{J_m J_L}{k_L^2}\frac{d^4 q_1}{dt^4} + \left(\frac{b_m J_L}{k_L^2} + \frac{2c\bar{\omega}J_m}{k_L^2} \right)\frac{d^3 q_1}{dt^3} + \left(\frac{J_m}{k_L} + \frac{2c\bar{\omega}b_m}{k_L^2} + \frac{J_L}{k_L} \right)\frac{d^2 q_1}{dt^2}$$
$$+ \left(\frac{2c\bar{\omega}}{k_L} + \frac{b_m}{k_L} \right)\frac{dq_1}{dt} = \frac{1}{k_L}u + \frac{2c\bar{\omega}}{k_L^2}\frac{du}{dt} + \frac{J_L}{k_L^2}\frac{d^2 u}{dt^2} \qquad \text{(xiii)}$$

Next, substitute Equation (xii) into Equation (xi) for q_1. We get:

$$\frac{J_m J_L}{k_L^2}\frac{d^4 q_2}{dt^4} + \left(\frac{b_m J_L}{k_L^2} + \frac{2c\bar{\omega}J_m}{k_L^2} \right)\frac{d^3 q_2}{dt^3} + \left(\frac{J_L}{k_L} + \frac{2c\bar{\omega}b_m}{k_L^2} + \frac{J_m}{k_L} \right)\frac{d^2 q_2}{dt^2}$$
$$+ \left(\frac{2c\bar{\omega}}{k_L} + \frac{b_m}{k_L} \right)\frac{dq_2}{dt} = \frac{1}{k_L}u \qquad \text{(xiv)}$$

Observe that the left hand sides (homogenous or characteristic parts) of these two input–output differential equations are identical. The characteristic equation represents the "natural" dynamics of the system and should be common and independent of the input (u). Hence the result is justified. Furthermore, derivatives of u are present only in the q_1 equation. This is justified because motion q_1 is closer than q_2 to the input u. Also, only the derivatives of q_1 and q_2 are present in the two equations. This is a property of a mechanical system that is not anchored (by a spring) to ground. Here the reference value for q_1 or q_2 could be chosen arbitrarily, regardless of the relaxed position of the inter-component spring (k_L) and should not depend on u either. Hence the absolute displacements q_1 and q_2 themselves should not appear in the input–output equations, as clear from Equations (xiii) and (xiv). Such systems are said to possess *rigid body modes*. Even though the differential equations are fourth order, they can be directly integrated once, and the system is actually third order (also see Chapter 3, Example 3.2). The position itself can be defined by an arbitrary reference and should not be used as a state in order to avoid this ambiguity. However, if position (q_1 or q_2 and not the twist $q_1 - q_2$) is chosen as an output, the system has to be treated as fourth order. Compare this to the simple problem of a single mass subjected to an external force, and without any anchoring springs.

If there is a damper between J_m and J_L we cannot write simple expressions for q_2 in terms of q_1, and q_1 in terms of q_2, as in Equations (xi) and (xii). Here, the derivative operator $D = (d/dt)$ has to be introduced for the elimination process, and the solution of one variable by eliminating the other one becomes much more complicated.

   g. Use the auxiliary equation

$$a_4 \frac{d^4 x}{dt^4} + a_3 \frac{d^3 x}{dt^3} + a_2 \frac{d^2 x}{dt^2} + a_1 \frac{dx}{dt} = u$$

where

$$a_4 = \frac{J_m J_L}{k_L}, \; a_3 = \left(\frac{b_m J_L}{k_L} + \frac{2c\bar{\omega}J_m}{k_L} \right), \; a_2 = \left(J_m + \frac{2c\bar{\omega}b_m}{k_L} + J_L \right), \; a_1 = b_m + 2c\bar{\omega}$$

It follows from Equation (xiv) that

$$q_2 = x$$

and from Equation (xiii) that

$$q_1 = x + b_1 \dot{x} + b_2 \ddot{x}$$

where $b_1 = \dfrac{2c\bar{\omega}}{k_L}$, and $b_2 = \dfrac{J_L}{k_L}$.

Hence, we have the block diagram shown in Figure 5.30a for the relationship $u \rightarrow q_1$. Note that four integrators are needed. Hence this is a fourth order system.

  (i) In this case the simulation block diagram is as shown in Figure 5.30b. This also needs four integrators (a fourth order system).

  (ii) In this case the simulation block diagram is as shown in Figure 5.30c. This only needs three integrators (a third order system).

  (iii) By differentiating the expression for q_1, we have $\dot{q}_1 = \dot{x} + b_1 \ddot{x} + b_2 \dddot{x}$. Hence the block diagram in this case is as shown in Figure 5.30d. This needs three integrators (a third order system).

  h. Using Equations (ix) and (x) we get:

$$J_m \ddot{q}_1 = u - b_m \dot{q}_1 - k_L(q_1 - q_2)$$

$$J_L \ddot{q}_1 = k_L(q_1 - q_2) - 2c\bar{\omega}\dot{q}_2$$

Accordingly, we can draw the block diagram shown in Figure 5.30e. There are three integrators in this case. The system is third order.

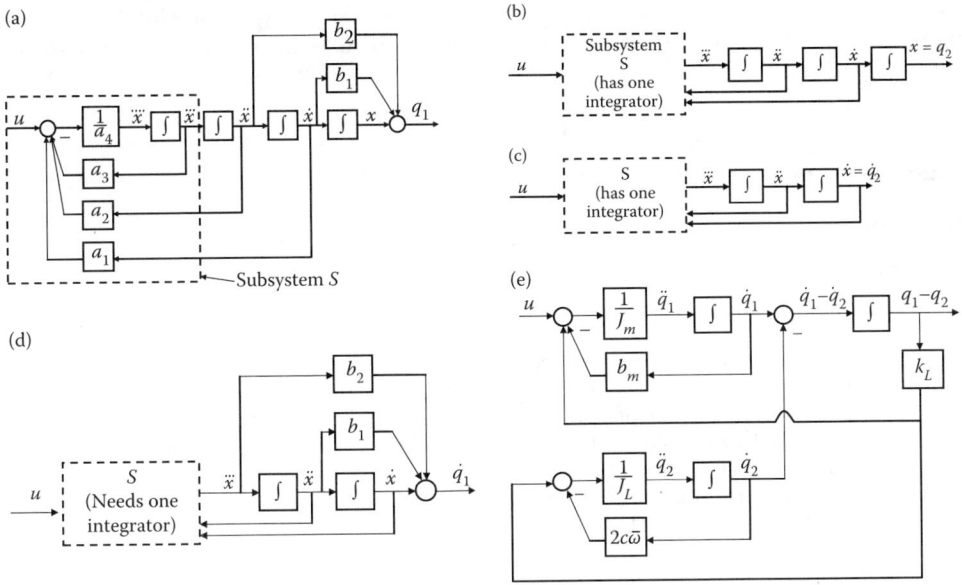

FIGURE 5.30
Simulation block diagram (a) when q_1 is the output; (b) when q_2 is the output; (c) when $\dot{q}_1$ is the output; (d) when $\dot{q}_2$ is the output; and (e) when the spring twist q_1-q_2 is the output.

 i. When q_1 and q_2 are used as outputs, the system order increases to four. But, as discussed in (f), q_1 and q_2 are not realistic state variables for the present problem.

5.6.3 Causality and Physical Realizability

Consider a dynamic system that is represented by the single input–output differential Equation 5.11, with $n>m$. The physical realizability of the system should dictate the causality (cause–effect) of this system that u should be the input and y should be the output. Its transfer-function is given by Equation 5.12. Here, n is the order of the system, $\Delta(s)$ is the characteristic polynomial (of order n), and $N(s)$ is the numerator polynomial (of order m) of the system.

We can prove the above by contradiction. Suppose that $m>n$. Then, if we integrate Equation 5.11 n times, we will have y and its integrals on the left hand side but the right hand side will contain at least one derivative of u. Since the derivative of a step function is an impulse, this implies that a finite change in input will result in an infinite change in the output (response). Such a scenario will require infinite power, and is not physically realizable. It follows that a physically realizable system cannot have a numerator order greater than the denominator order, in its transfer-function. If in fact $m>n$, then, what it means physically is that y should be the system input and u should be the system output. In other words, the causality should be reversed in this case. For a physically realizable system, a simulation block diagram can be established using integrals $(1/s)$ alone, without the need of derivatives (s). Note that pure derivatives are physically not realizable. If $m>n$, the simulation block diagram will need at least one derivative for linking u to y. That will not be physically realizable, again, because it would imply the possibility of producing an infinite response by a finite input. In other words, the simulation block diagram of a physical realizable system will not require feed-forward paths containing pure derivatives.

Problems

PROBLEM 5.1

State whether true (T) or false (F).

a. The output of a system will depend on the input.
b. The output of a system will depend on the system transfer-function.
c. The transfer-function of a system will depend on the input signal.
d. If the Laplace transform of the input signal does not exist (say, infinite), then the transfer-function itself does not exist.
e. If the Laplace transform of the output signal does not exist, then the transfer-function itself does not exist.

PROBLEM 5.2

State whether true (T) or false (F).

a. A transfer-function provides an algebraic expression for a system.
b. The Laplace variable s can be interpreted as time derivative operator d/dt, assuming zero ICs.
c. The variable $1/s$ may be interpreted as the integration of a signal starting at $t = 0$.
d. The numerator of a transfer-function is characteristic polynomial.
e. A SISO, linear, time-invariant system has a unique (one and only one) transfer-function.

PROBLEM 5.3

Consider the system given by the differential equation: $\ddot{y} + 4\dot{y} + 3y = 2u + \dot{u}$

a. What is the order of the system?
b. What is the system transfer-function?
c. Do we need Laplace tables to obtain the transfer-function?
d. What are the poles?
e. What is the characteristic equation?
f. Consider the parent system: $\ddot{x} + 4\dot{x} + 3x = u$
   Express y in terms of x, using the principle of superposition.
g. Using $\begin{bmatrix} x_1 & x_2 \end{bmatrix}^T = \begin{bmatrix} x & \dot{x} \end{bmatrix}^T$ as the state variables, obtain a state-space model for the given original system (not the parent system).
h. Using the superposition approach, draw a simulation block diagram for the system.
i. Express the system differential equation in a form suitable for drawing a simulation diagram by the "grouping like-derivatives" method.
j. From (i) draw the simulation block diagram.
k. Express the transfer-function $(s+2)/(s+3)$ in two forms of simulation block diagrams.
l. Using one of the two forms obtained in (k), draw the simulation block diagram for the original second-order system.

m. What are the partial fractions of the original transfer-function?
n. Using the partial-fraction method, draw a simulation block diagram for the system. What is the corresponding state-space model?
o. Obtain a state-space model for the system using (j).
p. Obtain at least one state model for the system using the block diagram obtained in (l).
q. What can you say about the diagonal elements of the system matrix A in (n) and in (p)?

PROBLEM 5.4

a. List several characteristics of a physically realizable system. How would you recognize the physically realizability of a system by drawing a simulation block diagram, which uses integrators, summing junctions, and gain blocks?
b. Consider the system given by the following input/output differential equation:

$$\dddot{y} + a_2 \ddot{y} + a_1 \dot{y} + a_0 y = b_2 \ddot{u} + b_1 \dot{u} + b_0 u$$

in which u is the input and y is the output.
Is this system physically realizable?
Draw a simulation block diagram for this system using integrators, gains, and summing junctions only.

PROBLEM 5.5

Consider the control system shown in Figure P5.5.
The back e.m.f. $v_B = K_V \omega$
The motor torque $T_m = K_T i$
Draw a simulation block diagram for the system.

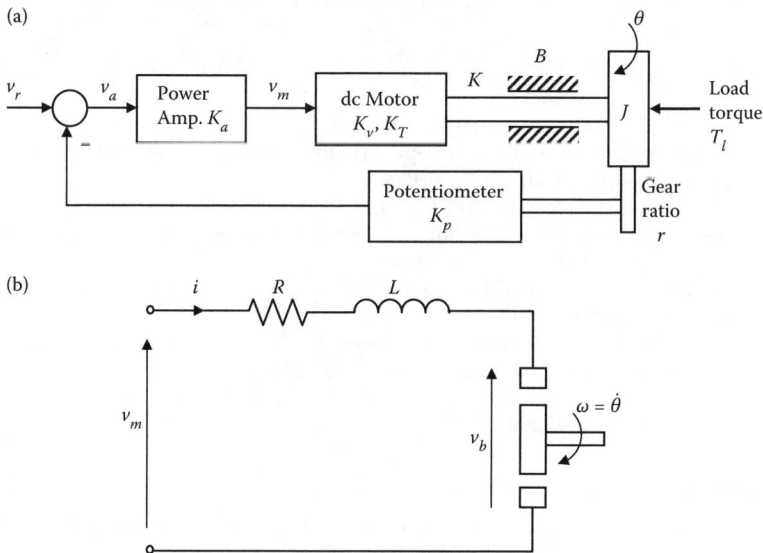

FIGURE P5.5
(a) A rotatory electromechanical system. (b) The armature circuit.

PROBLEM 5.6

It is required to study the dynamics behavior of an automobile during the very brief period of a sudden start from rest. Specifically, the vehicle acceleration a in the direction of primary motion, as shown in Figure P5.6a, is of interest and should be considered as the system output. The equivalent force $f(t)$ of the engine, applied in the direction of primary motion, is considered as the system input. A simple dynamic model that may be used for the study is shown in Figure P5.6b.

Note that k is the equivalent stiffness, primarily due to tire flexibility, and b is the equivalent viscous damping constant, primarily due to dissipations at the tires and other moving parts of the vehicle, taken in the direction of a. Also, m is the mass of the vehicle.

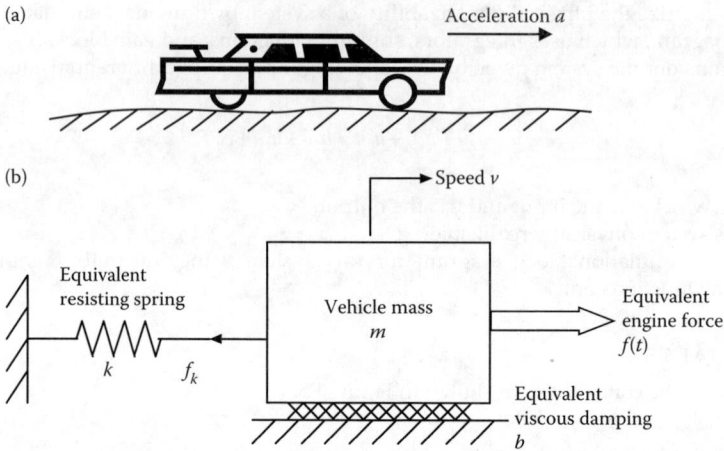

FIGURE P5.6
(a) Vehicle suddenly accelerating from rest. (b) A simplified model of the accelerating vehicle.

a. Discuss advantages and limitations of the proposed model for the particular purpose.

b. Using force f_k of the spring (stiffness k) and velocity v of the vehicle as the state variables, engine force $f(t)$ as the input and the vehicle acceleration a as the output, develop a complete state-space model for the system.
   (Note: You must derive the matrices A, B, C, and D for the model).

c. Draw a simulation block diagram for the model, employing integration and gain blocks, and summation junctions only.

d. Obtain the input–output differential equation of the system. From this, derive the transfer-function (a/f in the Laplace domain).

e. Discuss the characteristics of this model by observing the nature of matrix D, feed-forwardness of the block diagram, input and output orders of the input–output differential equation, and the numerator and denominator orders of the system transfer-function.

PROBLEM 5.7

Consider a dynamic system, which is represented by the transfer-function (output-input):

$$G(s) = \frac{3s^3 + 2s^2 + 2s + 1}{s^3 + 4s^2 + s + 3}$$

System output$=y$; system input$=u$.

a. What is the input–output differential equation of the system? What is the order of the system? Is this system physically realizable?
b. Based on the "superposition method" draw a simulation block diagram for the system, using integrators, constant gain blocks, and summing junctions only. Obtain a state-space model using this simulation block diagram, clearly giving the matrices A, B, C, and D.
c. Based on the "grouping like-derivatives method" draw a simulation block diagram, which should be different from what was drawn in (b), again using integrators, constant gain blocks, and summing junctions only.
Give a state-space model for the system, now using this simulation block diagram. This state-space model should be different from that in (b), which further illustrates that the state-space representation is not unique.

PROBLEM 5.8

The electrical circuit shown in Figure P5.8 has two resistor R_1 and R_2, an inductor L, a capacitor C, and a voltage source $u(t)$. The voltage across the capacitor is considered the output y of the circuit.

a. What is the order of the system and why?
b. Show that the input–output equation of the circuit is given by

$$a_2 \frac{d^2y}{dt^2} + a_1 \frac{dy}{dt} + a_0 y = b_1 \frac{du}{dt} + b_0 u$$

Express the coefficients a_0, a_1, a_2, b_0 and b_1 in terms of the circuit parameters R_1, R_2, L, and C.

c. Starting with the auxiliary differential equation:

$$a_2 \ddot{x} + a_1 \dot{x} + a_0 x = u$$

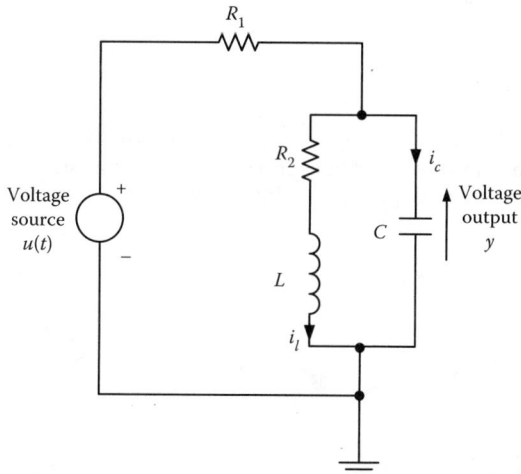

FIGURE P5.8
An *RLC* circuit driven by a voltage source.

and using $x = \begin{bmatrix} x & \dot{x} \end{bmatrix}^T$ as the state vector, obtain a complete state-space model for the system in Figure P5.8. Note that this is the "superposition method" of developing a state model.

d. Clearly explain why, for the system in Figure P5.8, neither the current i_c through the capacitor nor the time derivative of the output $(\dot{y})$ can be chosen as a state variable.

PROBLEM 5.9

Consider an nth order, linear, time-invariant dynamic system with input $u(t)$ and output y. When a step input was applied to this system it was observed that the output jumped instantaneously in the very beginning. Which of the following statements are true for this system?

a. Any simulation block diagram of this system (consisting only of integrators, constant-gain blocks, and summation junctions) will have at least one feedforward path.

b. The D matrix does not vanish (i.e., $D \neq 0$) in its state-space model:

$$\dot{x} = Ax + Bu$$

$$y = Cx + Du$$

c. This is not a physically realizable system.
d. The number of zeros in the system is equal to n.
e. The number of poles in the system is equal to n.

In each case briefly justify your answer.

PROBLEM 5.10

In relation to a dynamic system, briefly explain your interpretation of the terms

a. Causality
b. Physical realizability

Using integrator blocks, summing junctions, and coefficient blocks only, unless it is absolutely necessary to use other types if blocks, draw simulation block diagrams for the following three input–output differential equations:

i. $\quad a_1 \dfrac{dy}{dt} + a_0 y = u$

ii. $\quad a_1 \dfrac{dy}{dt} + a_0 y = u + b_1 \dfrac{du}{dt}$

iii. $\quad a_1 \dfrac{dy}{dt} + a_0 y = u + b_1 \dfrac{du}{dt} + b_2 \dfrac{d^2 u}{dt^2}$

Note that u denotes the input and y denotes the output. Comment about causality and physical realizability of these three systems.

PROBLEM 5.11

The Fourier transform of a position measurement $y(t)$ is $Y(j\omega)$. Answer true (T) or false (F):

  i. The Fourier transform of the corresponding velocity signal is:

  (a) $Y(j\omega)$
  (b) $j\omega Y(j\omega)$
  (c) $Y(j\omega)/(j\omega)$
  (d) $\omega Y(j\omega)$

  ii. The Fourier transform of the acceleration signal is:

  (a) $Y(j\omega)$
  (b) $\omega^2 Y(j\omega)$
  (c) $-\omega^2 Y(j\omega)$
  (d) $Y(j\omega)/(j\omega)$

PROBLEM 5.12

Answer true (T) or false (F):

  i. Mechanical impedances are additive for two elements connected in parallel.
  ii. Mobilities are additive for two elements connected in series.

PROBLEM 5.13

The movable arm with read/write head of a disk drive unit is modeled as a simple oscillator. The unit has an equivalent bending stiffness $k=10$ dyne.cm/rad and damping constant b. An equivalent rotation $u(t)$ radians is imparted at the read/write head. This in turn produces a (bending) moment to the read/write arm, which has an equivalent moment of inertia $J=1\times10^{-3}$ gm.cm^2, and bends the unit at an equivalent angle θ about the centroid.

  a. Write the input output differential equation of motion for the read/write arm unit.
  b. What is the undamped natural frequency of the unit in rad/s?
  c. Determine the value of b for 5% critical damping.
  d. Write the frequency transfer-function of the model.

PROBLEM 5.14

A rotating machine of mass M is placed on a rigid concrete floor. There is an isolation pad made of elastomeric material between the machine and the floor, and is modeled as a viscous damper of damping constant b. In steady operation there is a predominant harmonic force component $f(t)$, which is acting on the machine in the vertical direction at a frequency equal to the speed of rotation (n rev/s) of the machine. To control the vibrations produced by this force, a dynamic absorber of mass m and stiffness k is mounted on the machine. A model of the system is shown in Figure P5.14.

a. Determine the frequency transfer-function of the system, with force $f(t)$ as the input and the vertical velocity v of mass M as the output.

b. What is the mass of the dynamic absorber that should be used in order to virtually eliminate the machine vibration (a tuned absorber)?

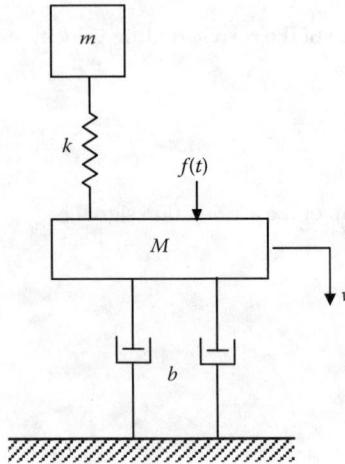

FIGURE P5.14
A mounted machine wit a dynamic absorber.

PROBLEM 5.15

The frequency transfer-function for a simple oscillator is given by

$$G(\omega) = \frac{\omega_n^2}{[\omega_n^2 - \omega^2 + 2j\zeta\omega_n\omega]}$$

a. If a harmonic excitation $u(t) = a\cos\omega_n t$ is applied to this system what is the steady-state response?
b. What is the magnitude of the resonant peak?
c. Using your answers to (a) and (b) suggest a method to measure damping in a mechanical system.
d. At what excitation frequency is the response amplitude maximum under steady-state conditions?
e. Determine an approximate expression for the half-power (3 dB) bandwidth at low damping. Using this result, suggest an alternative method for the damping measurement.

PROBLEM 5.16

a. An approximate frequency transfer-function of a system was determined by Fourier analysis of measured excitation-response data and fitting into an appropriate analytical expression (by curve fitting using the least squares method). This was found to be

$$G(f) = \frac{5}{10 + j2\pi f}$$

What is its magnitude, phase angle, real part, and imaginary part at $f=2$ Hz? If the reference frequency is taken as 1 Hz, what is the transfer-function magnitude at 2 Hz expressed in dB?

b. A dynamic test on a structure using a portable shaker revealed the following: The accelerance between two locations (shaker location and accelerometer location) measured at a frequency ratio of 10 was 35 dB. Determine the corresponding mobility and mechanical impedance at this frequency ratio.

PROBLEM 5.17

Answer true (T) or false (F):

a. Electrical impedances are additive for two elements connected in parallel.
b. Impedance, both mechanical and electrical, is given by the ratio of effort/flow, in the frequency domain.
c. Impedance, both mechanical and electrical, is given by the ratio of across variable/through variable, in the frequency domain.
d. Mechanical impedance is analogous to electrical impedance when determining the equivalent impedance of several interconnected impedances.
e. Mobility is analogous to electrical admittance (current/voltage in the frequency domain) when determining the equivalent value of several interconnected elements.

PROBLEM 5.18

Figure P5.18 shows two systems (a) and (b), which may be used to study force transmissibility and motion transmissibility, respectively. Clearly discuss whether the force transmissibility F_s/F (in the Laplace domain) in System (a) is equal to the motion transmissibility V_m/V (in the Laplace domain) in System (b), by carrying out the following steps:

1. Draw the linear graphs for the two systems and mark the mobility functions for all the branches (except the source elements).
2. Simplify the two linear graphs by combining branches as appropriate (series branches: add mobilities; parallel branches; inverse rule applies for mobilities) and mark the mobilities of the combined branches.
3. Based on the objectives of the problem (i.e., determination of the force transmissibility of System (a) and motion transmissibility of System (b)), for applying Thevenin's theorem, determine which part of the circuit (linear graph) should be cut. (*Note*: The variable of interest in the particular transmissibility function should be associated with the part of the circuit that is cut.)
4. Based on the objectives problem establish whether Thevenin equivalence or Norton equivalence is needed. (Specifically: Use Thevenin equivalence if a through variable needs to be determined, because this gives two series elements with a common through variable; Use Norton equivalence if an across variable needs to be determined, because this gives two parallel elements with a common across variable.)
5. Determine the equivalent sources and mobilities of the equivalent circuits of the two systems.

6. Using the two equivalent circuits determine the transmissibility functions of interest.
7. By analysis, examine whether the two mobility functions obtained in this manner are equivalent.

Note: Neglect the effects of gravity (i.e., assume that the systems are horizontal, supported on frictionless rollers).

Bonus: Extend you results to an *n*-degree-of-freedom system (i.e., one with *n* mass elements), structured as in Figure P5.18a and b.

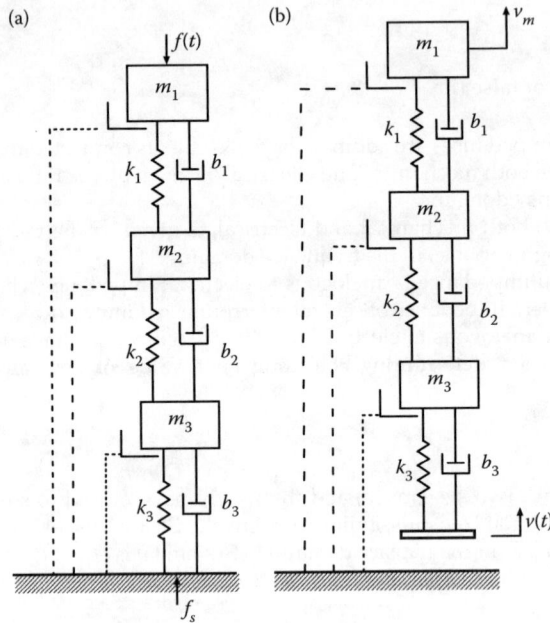

FIGURE P5.18
Figure P5.18 Two mechanical systems (a) for determining force transmissibility and (b) for determining motion transmissibility.

6

Response Analysis and Simulation

An analytical model, which is a set of differential equations, has many uses in the analysis, design, and control of the corresponding system. In particular, it can provide information regarding how the system responds when excited by an initial condition (IC) (i.e., free, natural response) or when a specific excitation (input) is applied (i.e., forced response). Such a study may be carried out by

1. Solution of the differential equations (analytical)
2. Computer simulation (numerical)

In this chapter we will address these two approaches. A response analysis carried out using either approach, is valuable in many applications such as design, control, testing, validation, and qualification. For large-scale and complex systems, a purely analytical study may not be feasible, and one may have to resort to numerical approaches and computer simulation.

6.1 Analytical Solution

The response of a dynamic system may be obtained analytically by solving the associated differential equations, subject to the ICs. This may be done by

1. Direct solution (in the time domain)
2. Solution using Laplace transform

Consider a linear time-invariant model given by the input–output differential equation

$$a_n \frac{d^n y}{dt^n} + a_{n-1} \frac{d^{n-1} y}{dt^{n-1}} + \cdots + a_0 y = u \tag{6.1}$$

At the outset, note that it is not necessary to specifically include derivative terms on the right hand side of Equation 6.1; for example $b_0 u + b_1 (du/dt) + \cdots + b_m (d^m u/dt^m)$, because once we have the solution (say, y_s) for Equation 6.1 we can use the *principle of superposition* to obtain the general solution, which is given by $b_0 y_s + b_1 (dy_s/dt) + \cdots + b_m (d^m y_s/dt^m)$. Hence, we will consider only the case of Equation 6.1.

6.1.1 Homogeneous Solution

The natural characteristics of a dynamic system do not depend on the input to the system. Hence, the natural behavior (or free response) of Equation 6.1 is determined by the homogeneous equation (i.e., with the input=0):

$$a_n \frac{d^n y}{dt^n} + a_{n-1} \frac{d^{n-1} y}{dt^{n-1}} + \cdots + a_0 y = 0 \tag{6.2}$$

Its solution—the homogeneous solution—is denoted by y_h and it depends as the system ICs. For a linear system the natural response is known to take an exponential form given by

$$y_h = c e^{\lambda t} \tag{6.3}$$

where c is an arbitrary constant and, in general, λ can be complex. Substitute Equation 6.3 in Equation 6.2 with the knowledge that

$$\frac{d}{dt} e^{\lambda t} = \lambda e^{\lambda t} \tag{6.4}$$

and cancel the common term $ce^{\lambda t}$, since it cannot be zero at all times. Then we have

$$a_n \lambda^n + a_{n-1} \lambda^{n-1} + \cdots + a_0 = 0 \tag{6.5}$$

This is called the *characteristic equation* of the system. *Note*: the polynomial $a_n \lambda^n + a_{n-1} \lambda^{n-1} + \cdots + a_0$ is called the *characteristic polynomial*. Equation 6.5 has n roots $\lambda_1, \lambda_2, \ldots, \lambda_n$. These are called *poles* or *eigenvalues* of the system. Assuming that they are distinct (i.e., unequal), the overall solution to Equation 6.2 becomes

$$y_h = c_1 e^{\lambda_1 t} + c_2 e^{\lambda_2 t} + \cdots + c_n e^{\lambda_n t} \tag{6.6}$$

The unknown constants $c_1, c_2, \ldots, c_n$ are determined using the necessary n ICs $y(0)$, $\dot{y}(0), \ldots, d^{n-1} y(0)/dt^{n-1}$.

6.1.1.1 Repeated Poles

Suppose that at least two eigenvalues from the solution of Equation 6.5 are equal. Without loss of generality suppose in Equation 6.6 that $\lambda_1 = \lambda_2$. Then the first two terms in Equation 6.6 can be combined into the single unknown $(c_1 + c_2)$. Consequently there are only $n-1$ unknowns in Equation 6.6 but there are n ICs for the system Equation 6.2. It follows that another unknown needs to be introduced for obtaining a complete solution. Since a repeated pole is equivalent to a double integration (i.e., a term $1/(s-\lambda_i)^2$ in the system transfer function), the logical (and correct) solution for Equation 6.5 in the case $\lambda_1 = \lambda_2$ is

$$y_h = (c_1 + c_2 t) e^{\lambda_1 t} + c_3 e^{\lambda_3 t} + \cdots + c_n e^{\lambda_n t} \tag{6.7}$$

This idea can be easily generalized for the case of three or more repeated poles (by adding terms containing t^2, t^3, and so on).

6.1.2 Particular Solution

The homogeneous solution corresponds to the "natural," "free" or "unforced" response of a system and clearly it does not depend on the input function. The effect of the input is incorporated into the *particular solution*, which is defined as one possible function for y that satisfies Equation 6.1. We denote this by y_p. Several important input functions and the corresponding form of y_p which satisfies Equation 6.1 are given in Table 6.1.

The parameters A, B, A_1, A_2, B_1, B_2, and D in Table 6.1 are determined by substituting the pair $u(t)$ and y_p into Equation 6.1 and then equating the like terms. This approach is called the *method of undetermined coefficients*.

The total response of the system Equation 6.1 is given by

$$y = y_h + y_p \tag{6.8}$$

The unknown constants c_1, c_2,..., c_n in this result are determined by substituting into Equation 6.8 the ICs of the system.

Note: It is incorrect to determine c_1, c_2,..., c_n by substituting the ICs into y_h only and then adding y_p to the resulting y_h. This is because the total response is y not y_h. Furthermore, when $u=0$, the homogeneous solution is identical to the free response (which is also the IC response, or the zero-input response). When an input is present, however, the homogeneous solution may not be identical to these other three types of response since they can be influenced by the forcing input as well as the natural dynamics of the system. These ideas are summarized in Box 6.1.

6.1.3 Impulse Response Function

Consider a linear dynamic system. The principle of superposition holds. More specifically, if y_1 is the system response (output) to excitation (input) $u_1(t)$ and y_2 is the response to excitation $u_2(t)$, then $\alpha y_1 + \beta y_2$ is the system response to input $\alpha u_1(t) + \beta u_2(t)$ for any constants α and β and any excitation functions $u_1(t)$ and $u_2(t)$. This is true for both time-variant-parameter linear systems and constant-parameter linear systems.

A unit pulse of width $\Delta\tau$ starting at time $t=\tau$ is shown in Figure 6.1a. Its area is unity. A unit impulse is the limiting case of a unit pulse as $\Delta\tau \to 0$. A unit impulse acting at time $t=\tau$ is denoted by $\delta(t-\tau)$ and is graphically represented as in Figure 6.1b. In mathematical analysis, this is known as the *Dirac delta function*, and is defined by the two conditions:

$$\delta(t-\tau) = 0 \quad \text{for} \quad t \neq \tau \tag{6.9}$$

$$\to \infty \quad \text{at} \quad t = \tau$$

TABLE 6.1

Particular Solutions for useful Input Functions

Input $u(t)$	Particular Solution y_p
c	A
$ct+d$	$At+B$
$\sin ct$	$A_1 \sin ct + A_2 \cos ct$
$\cos ct$	$B_1 \sin ct + B_2 \cos ct$
e^{ct}	De^{ct}

BOX 6.1   SOME CONCEPTS OF SYSTEM RESPONSE

Total response (T) = Homogeneous solution + Particular integral
$$\qquad\qquad\qquad\qquad (H)\qquad\qquad\qquad\qquad (P)$$
$$= \text{Free response} + \text{Forced response}$$
$$(X)\qquad\qquad\qquad (F)$$
$$= \text{IC response} + \text{Zero IC response}$$
$$(X)\qquad\qquad\qquad\qquad (F)$$
$$= \text{Zero-input response} + \text{Zero state response}$$
$$(X)\qquad\qquad\qquad\qquad (F)$$

Note 1: In general, $H \neq X$ and $P \neq F$
Note 2: With no input (no forcing excitation), by definition, $H \equiv X$
Note 3: At steady-state, F becomes equal to P.

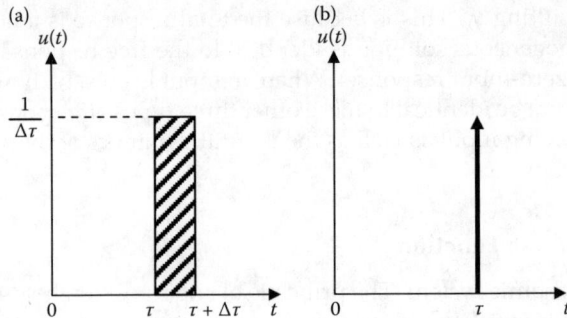

FIGURE 6.1
Illustration of (a) unit pulse; (b) unit impulse.

and

$$\int_{-\infty}^{\infty} \delta(t-\tau)dt = 1 \qquad\qquad (6.10)$$

The Dirac delta function has the following well-known and useful properties:

$$\int_{-\infty}^{\infty} f(t)\delta(t-\tau)dt = f(\tau) \qquad\qquad (6.11)$$

and

$$\int_{-\infty}^{\infty} \frac{d^n f(t)}{dt^n}\delta(t-\tau)dt = \frac{d^n f(t)}{dt^n}\Big|_{t=\tau} \qquad\qquad (6.12)$$

for any well-behaved time function $f(t)$. The system response (output) to a unit impulse excitation (input) acted at time $t=0$, is known as the *impulse response function* and is denoted by $h(t)$.

6.1.3.1 Convolution Integral

The system output (response) to an arbitrary input may be expressed in terms of its impulse response function. This is the essence of the impulse response approach to determining the forced response of a dynamic system. Without loss of generality let us assume that the system input $u(t)$ starts at $t=0$; that is,

$$u(t) = 0 \quad \text{for} \quad t < 0 \tag{6.13}$$

For physically realizable systems (see Chapter 5), the response does not depend on the future values of the input. Consequently,

$$y(t) = 0 \quad \text{for} \quad t < 0 \tag{6.14}$$

and

$$h(t) = 0 \quad \text{for} \quad t < 0 \tag{6.15}$$

where $y(t)$ is the response of the system, to any general excitation $u(t)$.

Furthermore, if the system is a constant-parameter system, then the response does not depend on the time origin used for the input. Mathematically, this is stated as follows: if the response to input $u(t)$ satisfying Equation 6.13 is $y(t)$, which in turn satisfies Equation 6.14, then the response to input $u(t-\tau)$, which satisfies,

$$u(t-\tau) = 0 \quad \text{for} \quad t < \tau \tag{6.16}$$

is $y(t-\tau)$, and it satisfies

$$y(t-\tau) = 0 \quad \text{for} \quad t < \tau \tag{6.17}$$

This situation is illustrated in Figure 6.2. It follows that the delayed-impulse input $\delta(t-\tau)$, having time delay τ, produces the delayed response $h(t-\tau)$.

A given input $u(t)$ can be divided approximately into a series of pulses of width $\Delta\tau$ and magnitude $u(\tau) \cdot \Delta\tau$. In Figure 6.3, as $\Delta\tau \to 0$, the pulse shown by the shaded area becomes an impulse acting at $t=\tau$ having the magnitude $u\tau \cdot d\tau$. This impulse is given by $\delta(t-\tau)u(\tau)d\tau$. In a linear, constant-parameter system, it produces the response $h(t-\tau)u(\tau)d\tau$. By integrating over the entire time duration of the input $u(t)$ (i.e., by using the principle of superposition, since the system is linear) the overall response $y(t)$ is obtained as

$$y(t) = \int_0^\infty h(t-\tau)u(\tau)d\tau \tag{6.18a}$$

Alternatively, by introducing the change of variables $\tau \to t-\tau$ and correspondingly reversing the limits of integration (and changing the sign) we have

$$y(t) = \int_0^\infty h(\tau)u(t-\tau)d\tau \tag{6.18b}$$

Equation 6.18 is known as the *convolution integral*. This is in fact the forced response, under zero ICs. It is also a particular integral (particular solution) of the system.

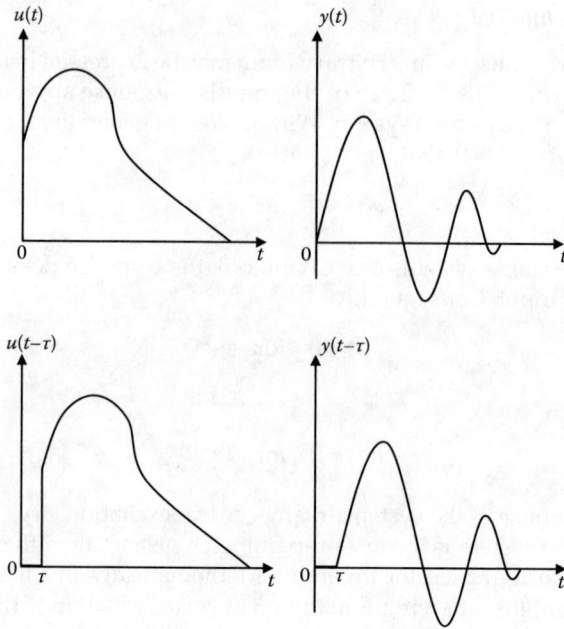

FIGURE 6.2
Response to a delayed input.

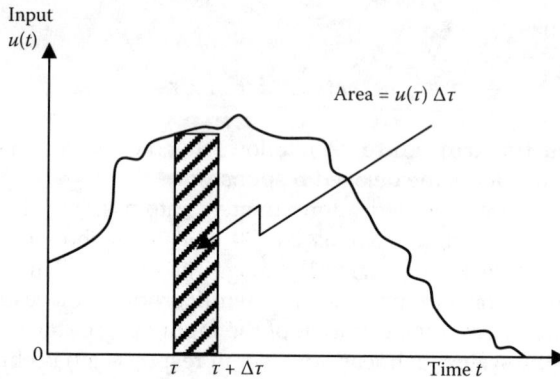

FIGURE 6.3
General input treated as a continuous series of impulses.

Note: The limits of integration in Equation 6.18 can be set in various manner in view of the fact that $u(t)$ and $h(t)$ are zero for $t < 0$ (e.g., the lower limit may be set at τ and the upper limit at t).

6.1.4 Stability

Many definitions are available for stability of a system. For example, a stable system may be defined as one whose natural response (i.e., free, zero-input, IC response) decays to zero. This is in fact the well-known *asymptotic stability*. If the IC response oscillates within finite bounds we say the system is *marginally stable*. For a linear, time-invariant system of the type Equation 6.1, the free response is of the form Equation 6.6, assuming no repeated poles. Hence, if none of the eigenvalues λ_i have positive real parts, the system is considered stable, because in that case, the response (Equation 6.6) does not grow unboundedly. In particular, if the system has a single eigenvalue that is zero, or if the eigenvalues are purely imaginary pairs, the system is marginally stable. If the system has two or more poles that are zero, we will have terms of the form $c_1 + ct$ in Equation 6.6 and hence it will grow polynomially (not exponentially). Then the system will be *unstable*. Even in the presence of repeated poles, however, if the real parts of the eigenvalues are negative, however, the system is stable (because the decay of the exponential terms in the response will be faster than the growth of the polynomial terms—see Equation 6.7 for example).

Note: Since physical systems have real parameters, their eigenvalues must occur as conjugate pairs, if complex.

Since stability is governed by the sign of the real part of the eigenvalues, it can be represented on the eigenvalue plane (or the pole plane, *s*-plane, or root plane). This is illustrated in Figure 6.4.

6.2 First-Order Systems

Consider the first-order dynamic system with time constant τ, input u, and output y, as given by

$$\tau \dot{y} + y = u(t) \tag{6.19}$$

Suppose that the system starts from $y(0) = y_0$ and a step input of magnitude A is applied at that IC. The homogeneous solution is

$$y_h = ce^{-t/\tau}$$

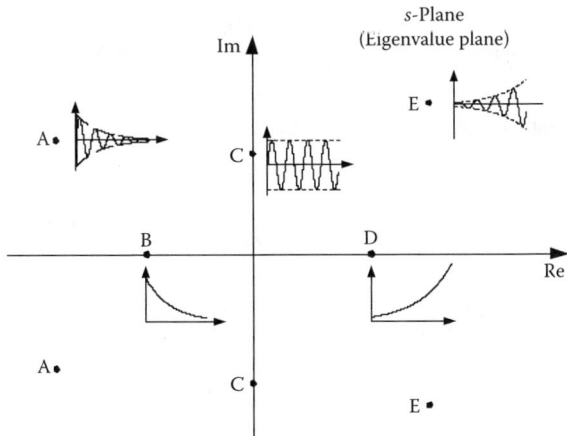

FIGURE 6.4
Dependence of stability on the pole location (A and B are stable pole locations; C is a marginally stable location; D and E are unstable locations).

The particular solution (see Table 6.1) is given by $y_p = A$. Hence, the total response is

$$y = y_h + y_p = ce^{-t/\tau} + A$$

Substitute the IC: $y(0) = y_0$. We get $C + A = y_0$. Hence

$$y_{step} = \underbrace{(y_0 - A)e^{-t/\tau}}_{\substack{\text{Homogeneous} \\ y_h}} + \underbrace{A}_{\substack{\text{Particular} \\ y_p}} = \underbrace{y_0 e^{-t/\tau}}_{\substack{\text{Free response} \\ y_x}} + \underbrace{A(1 - e^{-t/\tau})}_{\substack{\text{Forced response} \\ y_f}} \tag{6.20}$$

The steady-state value is given by $t \to \infty$:

$$y_{ss} = A \tag{6.21}$$

It is seen from Equation 6.20 that the forced response to a unit step input (i.e., $A = 1$) is $(1 - e^{-t/\tau})$. Due to linearity of the system, the forced response to a unit impulse input is $(d/dt)(1 - e^{-t/\tau}) = (1/\tau)e^{-t/\tau}$. Hence, the total response to an impulse input of magnitude P is

$$y_{impulse} = y_0 e^{-t/\tau} + \frac{P}{\tau}e^{-t/\tau} \tag{6.22}$$

This result follows from the fact that

$$\frac{d}{dt}(\text{Step function}) = \text{Impulse function} \tag{6.23}$$

and because, due to linearity, when the input is differentiated, the output is correspondingly differentiated.

Note from Equations 6.20 and 6.22 that if we know the response of a first-order system to a step input, or to an impulse input, the system itself can be determined. This is known as *model identification* or *experimental modeling*. We will illustrate this by an example.

Example 6.1: Model Identification Example

Consider the first-order system (model)

$$\tau \dot{y} + y = ku \tag{i}$$

The system parameters are the time constant τ and the gain parameter k. The IC is $y(0) = y_0$. Using Equation 6.20 we can derive the response of the system to a step input of magnitude A:

$$y_{step} = y_0 e^{-t/\tau} + Ak(1 - e^{-t/\tau}) \tag{ii}$$

Note: Due to linearity, the forced response is magnified by k since the input is magnified by the same factor.

Suppose that the unit step response of a first-order system with zero ICs, was found to be (say, by curve fitting of experimental data)

$$y_{step} = 2.25(1 - e^{-5.2t})$$

Then, it is clear from Equation (ii) that the system parameters are

$$k = 2.25 \quad \text{and} \quad \tau = 1/5.2 = 0.192.$$

These two parameters completely determine the system model.

6.3 Second-Order Systems

A general high-order system can be represented by a suitable combination of first-order and second order models, using the principles of modal analysis. Hence, it is useful to study the response behavior of second-order systems as well. Examples of second-order systems include mass-spring-damper systems and capacitor-inductor-resistor circuits, which we have studied in previous chapters. These are called simple oscillators because they exhibit oscillations in the natural response (free, unforced response) when the level of damping is sufficiently low. We will study both free response and forced response of second-order systems.

6.3.1 Free Response of an Undamped Oscillator

The equation of free (i.e., no excitation force) motion of an undamped simple oscillator is of the general form

$$\ddot{x} + \omega_n^2 x = 0 \tag{6.24}$$

For a mechanical system of mass m and stiffness k, we have the undamped natural frequency (whose meaning will be further discussed later)

$$\omega_n = \sqrt{\frac{k}{m}} \tag{6.25a}$$

For an electrical circuit with capacitance C and inductance L we have

$$\omega_n = \sqrt{\frac{1}{LC}} \tag{6.25b}$$

Note: These results can be immediately established from the electro-mechanical analogy (see Chapter 2) which we use: $m \to C; k \to 1/L; b \to 1/R$.

To determine the time response x of this system, we use the trial solution:

$$x = A \sin(\omega_n t + \phi) \tag{6.26}$$

in which A and ϕ are unknown constants, to be determined by the ICs (for x and $\dot{x}$); say,

$$x(0) = x_o, \quad \dot{x}(0) = v_o \tag{6.27}$$

The parameter A is the amplitude and ϕ is the phase angle of the response, as will be discussed later.

Substitute the trial solution (Equation 6.26) into Equation 6.24. We get

$$(-A\omega_n^2 + A\omega_n^2)\sin(\omega_n t + \phi) = 0$$

This equation is identically satisfied for all t. Hence, the general solution of Equation 6.24 is indeed Equation 6.26, which is periodic and sinusoidal.

This response (Equation 6.26) is sketched in Figure 6.5 (the subscript in ω_n is dropped for convenience). Note that this sinusoidal, oscillatory motion has a *frequency* of oscillation of ω (radians/s). Hence, a system that provides this type of natural motion is called a *simple oscillator*. In other words, the system response exactly repeats itself in time periods of T or at a *cyclic frequency* $f = 1/T$ (cycles/s or Hz).

Note: This fact may be verified by substituting $t = t + T = t + 1/f = t + 2\pi/\omega$ in Equation 6.26, which will give the same x value. The frequency ω is in fact the *angular frequency* given by $\omega = 2\pi f$.

Also, the response has *amplitude A*, which is the peak value of the sinusoidal response. This is verified from Equation 6.26 because the maximum value of a sine function is 1. Now, suppose that we shift the response curve (Equation 6.26) to the right through a time interval of ϕ/ω. Take the resulting curve as the reference signal (whose signal value=0 at $t=0$, and increasing). It should be clear that the response shown in Figure 6.5 leads the reference signal by a time period of ϕ/ω. This may be verified from the fact that the value of the reference signal at time t is the same as that of the signal in Figure 6.5 at time $t - \phi/\omega$. Hence ϕ is termed the *phase angle* of the response, and it is indeed a *phase lead*.

The left hand side portion of Figure 6.5 is the *phasor representation* of a sinusoidal response. In this representation, an arm of length A rotates in the counterclockwise direction at angular speed ω. This arm is the phasor. The arm starts at an angular position ϕ

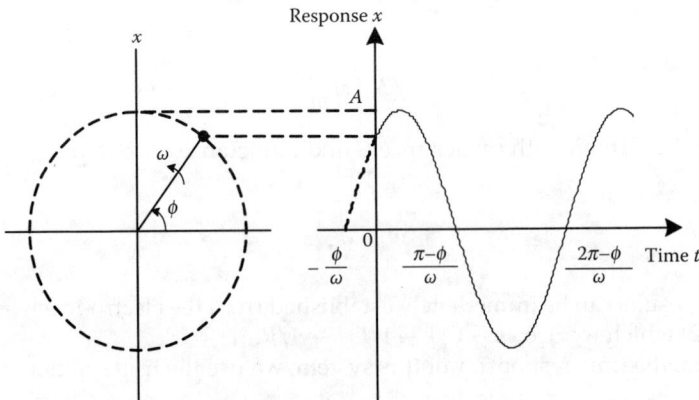

FIGURE 6.5
Free response of an undamped simple oscillator.

from the horizontal axis, at time $t=0$. The projection of the arm onto the vertical (x) axis gives the time response. In this manner, the phasor representation can conveniently indicate the amplitude, frequency, phase angle, and the actual time response (at any time t) of a sinusoidal motion.

6.3.2 Free Response of a Damped Oscillator

Energy dissipation in a mechanical oscillator may be represented by a damping element. For an electrical circuit, a resistor accounts for energy dissipation. In either case, the equation motion of a damped simple oscillator without an input, may be expressed as

$$\ddot{x} + 2\zeta\omega_n\dot{x} + \omega_n^2 x = 0 \tag{6.28}$$

Note that ζ is called the *damping ratio*.
Assume an exponential solution:

$$x = Ce^{\lambda t} \tag{6.29}$$

This is justified by the fact that linear systems have exponential or oscillatory (i.e., complex exponential) free responses (see Equation 6.6). A more convincing justification for this assumption will be provided later.

Substitute, Equation 6.29 into Equation 6.28. We get

$$\left[\lambda^2 + 2\zeta\omega_n\lambda + \omega_n^2\right]Ce^{\lambda t} = 0.$$

Note that $Ce^{\lambda t}$ is not zero for all t; hence, it can be removed from the above equation giving:

$$\lambda^2 + 2\zeta\omega_n\lambda + \omega_n^2 = 0 \tag{6.30}$$

It follows that, when λ satisfies Equation 6.30, then Equation 6.29 will represent a solution of Equation 6.28. As noted before, Equation 6.30 is the *characteristic equation* of the system. This equation depends on the natural dynamics of the system, not the forcing excitation or the ICs. Solution of Equation 6.30 gives the two roots:

$$\lambda = -\zeta\omega_n \pm \sqrt{\zeta^2 - 1}\ \omega_n$$

$$= \lambda_1 \quad \text{and} \quad \lambda_2 \tag{6.31}$$

These are the *eigenvalues* or *poles* of the system. When $\lambda_1 \neq \lambda_2$ (i.e., unequal poles), the general solution of Equation 6.28 is

$$x = C_1e^{\lambda_1} + C_2e^{\lambda_2 t} \tag{6.32}$$

The two unknown constants C_1 and C_2 are related to the integration constants, and can be determined by two ICs which should be known.

If $\lambda_1 = \lambda_2 = \lambda$; we have the case of repeated roots. In this case, as noted before, the general solution (Equation 6.32) does not hold because C_1 and C_2 will no longer be independent constants, which will not require two ICs for their determination. The repetition of the roots suggests that one term of the homogenous solution should have the multiplier t (a result of the associated double integration). Accordingly, the general solution is

$$x = C_1 e^{\lambda t} + C_2 t e^{\lambda t} \tag{6.33}$$

We can identify three ranges of damping, as discussed below, and the nature of the response will depend on the particular range of damping.

6.3.2.1 Case 1: Underdamped Motion ($\zeta < 1$)

In this case it follows from Equation 6.31 that the roots of the characteristic equation are

$$\lambda = -\zeta \omega_n \pm j\sqrt{1-\zeta^2}\,\omega_n = -\zeta \omega_n \pm j\omega_d = \lambda_1 \text{ and } \lambda_2 \tag{6.34}$$

where ω_2 is the *damped natural frequency*, given by

$$\omega_d = \sqrt{1-\zeta^2}\,\omega_n \tag{6.35}$$

Note: λ_1 and λ_2 are complex conjugates, as required. In this case, the response (Equation 6.32) may be expressed as

$$x = e^{-\zeta \omega_n t}[C_1 e^{j\omega_d t} + C_2 e^{-j\omega_d t}] \tag{6.36}$$

The term within the square brackets in Equation 6.36 has to be real, because it represents the time response of a real physical system. It follows that C_1 and C_2 also have to be complex conjugates.

Note: $e^{j\omega_d t} = \cos \omega_d t + j\sin \omega_d t$; $e^{-j\omega_d t} = \cos \omega_d t - j\sin \omega_d t$

So, an alternative form of the general solution would be

$$x = e^{-\zeta \omega_n t}[A_1 \cos \omega_d t + A_2 \sin \omega_d t] \tag{6.37}$$

Here A_1 and A_2 are two unknown real-valued constants. By equating the coefficients in Equations 6.37 and 6.36 it can be shown that

$$A_1 = C_1 + C_2$$

$$A_2 = j(C_1 - C_2) \tag{6.38a}$$

Hence,

$$C_1 = \frac{1}{2}(A_1 - jA_2)$$

$$C_2 = \frac{1}{2}(A_1 + jA_2) \tag{6.38b}$$

ICs:

Let, $x(0) = x_o$, $\dot{x}(0) = v_o$ as before. Then,

$$x_o = A_1 \tag{6.39a}$$

And $v_o = -\zeta\omega_n A_1 + \omega_d A_2$
or,

$$A_2 = \frac{v_o}{\omega_d} + \frac{\zeta\omega_n x_o}{\omega_d} \tag{6.39b}$$

Yet, another form of the solution would be:

$$x = Ae^{-\zeta\omega_n t}\sin(\omega_d t + \phi) \tag{6.40}$$

Here A and ϕ are the two unknown constants with

$$A = \sqrt{A_1^2 + A_2^2} \quad \text{and} \quad \sin\phi = \frac{A_1}{\sqrt{A_1^2 + A_2^2}}. \tag{6.41}$$

$$\text{Also,}\quad \cos\phi = \frac{A_2}{\sqrt{A_1^2 + A_2^2}} \quad \text{and} \quad \tan\phi = \frac{A_1}{A_2} \tag{6.42}$$

Note: The response $x \to 0$ as $t \to \infty$. This means the system is *asymptotically stable*.

6.3.2.2 Case 2: Overdamped Motion ($\zeta > 1$)

In this case, roots λ_1 and λ_2 of the characteristic Equation 6.30 are real and negative. Specifically, we have

$$\lambda_1 = -\zeta\omega_n + \sqrt{\zeta^2 - 1}\,\omega_n < 0 \tag{6.43a}$$

$$\lambda_2 = -\zeta\omega_n - \sqrt{\zeta^2 - 1}\,\omega_n < 0 \tag{6.43b}$$

and the response (Equation 6.32) is nonoscillatory. Also, since both λ_1 and λ_2 are negative (see Equation 6.43), we have $x \to 0$ as $t \to \infty$. This means the system is asymptotically stable.

From the ICs $x(0) = x_o$, $\dot{x}(0) = v_o$ we get:

$$x_o = C_1 + C_2 \tag{i}$$

and

$$v_o = \lambda_1 C_1 + \lambda_2 C_2 \tag{ii}$$

Multiply the first IC (i) by λ_2: $\lambda_2 x_0 = \lambda_2 C_1 + \lambda_2 C_2$ \hfill (iii)

Subtract (ii) from (iii): $v_o - \lambda_2 x_o = C_1(\lambda_1 - \lambda_2)$

We get:

$$C_1 = \frac{v_o - \lambda_2 x_o}{\lambda_1 - \lambda_2} \tag{6.44a}$$

Multiply the first IC (i) by λ_1: $\lambda_1 x_o = \lambda_1 C_1 + \lambda_1 C_2$ (iv)

Subtract (iv) from (ii): $v_o - \lambda_1 x_o = C_2(\lambda_2 - \lambda_1)$
We get:

$$C_2 = \frac{v_o - \lambda_1 x_o}{\lambda_2 - \lambda_1} \tag{6.44b}$$

6.3.2.3 Case 3: Critically Damped Motion ($\zeta = 1$)

Here, we have repeated roots, given by

$$\lambda_1 = \lambda_2 = -\omega_n \tag{6.45}$$

The response, for this case is given by (see Equation 6.33)

$$x = C_1 e^{-\omega_n t} + C_2 t e^{-\omega_n t} \tag{6.46}$$

Since the term $e^{-\omega_n t}$ goes to zero faster than t goes to infinity, we have: $te^{-\omega_n t} \to 0$ as $t \to \infty$. Hence the system is asymptotically stable.

Now use the ICs $x(0) = x_o$, $\dot{x}(0) = v_o$. We have:

$$x_o = C_1$$

$$v_o = -\omega_n C_1 + C_2$$

Hence,

$$C_1 = x_o \tag{6.47a}$$

$$C_2 = v_o + \omega_n x_o \tag{6.47b}$$

Note: When $\zeta = 1$ we have the *critically damped* response because below this value, the response is oscillatory (underdamped) and above this value, the response is nonoscillatory (overdamped). It follows that we may define the damping ratio as

$$\zeta = \text{Damping ratio} = \frac{\text{Damping constant}}{\text{Damping constant for critically damped conditions}}$$

The main results for free (natural) response of a damped oscillator are given in Box 6.2. The response of a damped simple oscillator is shown in Figure 6.6.

BOX 6.2   FREE (NATURAL) RESPONSE OF A DAMPED SIMPLE OSCILLATOR

System equation: $\ddot{x} + 2\zeta\omega_n\dot{x} + \omega_n^2 x = 0$

Undamped natural frequency $\omega_n = \sqrt{\dfrac{k}{m}}$ or $\omega_n = \sqrt{\dfrac{1}{LC}}$

Damping ratio $\zeta = \dfrac{b}{2\sqrt{km}}$ or $\zeta = \dfrac{1}{2R}\sqrt{\dfrac{L}{C}}$

Note: Electro-mechanical analogy $m \to C; k \to 1/L; b \to 1/R$

Characteristic equation: $\lambda^2 + 2\zeta\omega_n\lambda + \omega_n^2 = 0$

Roots (eigenvalues or poles): λ_1 and $\lambda_2 = -\zeta\omega_n \pm \sqrt{\zeta^2 - 1}\,\omega_n$

Response: $x = C_1 e^{\lambda_1 t} + C_2 e^{\lambda_2 t}$ for unequal roots $(\lambda_1 \neq \lambda_2)$

$$x = (C_1 + C_2 t)e^{\lambda t} \text{ for equal roots } (\lambda_1 = \lambda_2 = \lambda)$$

ICs: $x(0) = x_0$ and $\dot{x}(0) = v_0$

Case 1: Underdamped $(\zeta < 1)$

Poles are complex conjugates: $-\zeta\omega_n \pm j\omega_d$

Damped natural frequency $\omega_d = \sqrt{1 - \zeta^2}\,\omega_n$

$$\begin{aligned} x &= e^{-\zeta\omega_n t}[C_1 e^{j\omega_d t} + C_2 e^{-j\omega_d t}] \\ &= e^{-\zeta\omega_n t}[A_1 \cos\omega_d t + A_2 \sin\omega_d t] \\ &= A e^{-\zeta\omega_n t}\sin(\omega_d t + \phi) \end{aligned}$$

$A_1 = C_1 + C_2$ and $A_2 = j(C_1 - C_2)$

$C_1 = \dfrac{1}{2}(A_1 - jA_2)$ and $C_2 = \dfrac{1}{2}(A_1 + jA_2)$

$A = \sqrt{A_1^2 + A_2^2}$ and $\tan\phi = \dfrac{A_1}{A_2}$

ICs give: $A_1 = x_o$ and $A_2 = \dfrac{v_0 + \zeta\omega_n x_o}{\omega_d}$

Logarithmic decrement per radian: $\alpha = \dfrac{1}{2\pi n}\ln r = \dfrac{\zeta}{\sqrt{1 - \zeta^2}}$

where $r = \dfrac{x(t)}{x(t + nT)} = $ decay ratio over n complete cycles. For small ζ: $\zeta \cong \alpha$

Case 2: Overdamped $(\zeta > 1)$

Poles are real and negative: $\lambda_1, \lambda_2 = -\zeta\omega_n \pm \sqrt{\zeta^2 - 1}\,\omega_n$

$$x = C_1 e^{\lambda_1 t} + C_2 e^{\lambda_2 t}$$

$$C_1 = \frac{v_0 - \lambda_2 x_0}{\lambda_1 - \lambda_2} \quad \text{and} \quad C_2 = \frac{v_0 - \lambda_1 x_0}{\lambda_2 - \lambda_1}$$

Case 3: Critically damped $(\zeta = 1)$

Two identical poles: $\lambda_1 = \lambda_2 = \lambda = -\omega_n$

$$x = (C_1 + C_2 t)e^{-\omega_n t} \quad \text{with} \quad C_1 = x_0 \quad \text{and} \quad C_2 = v_0 + \omega_n x_0$$

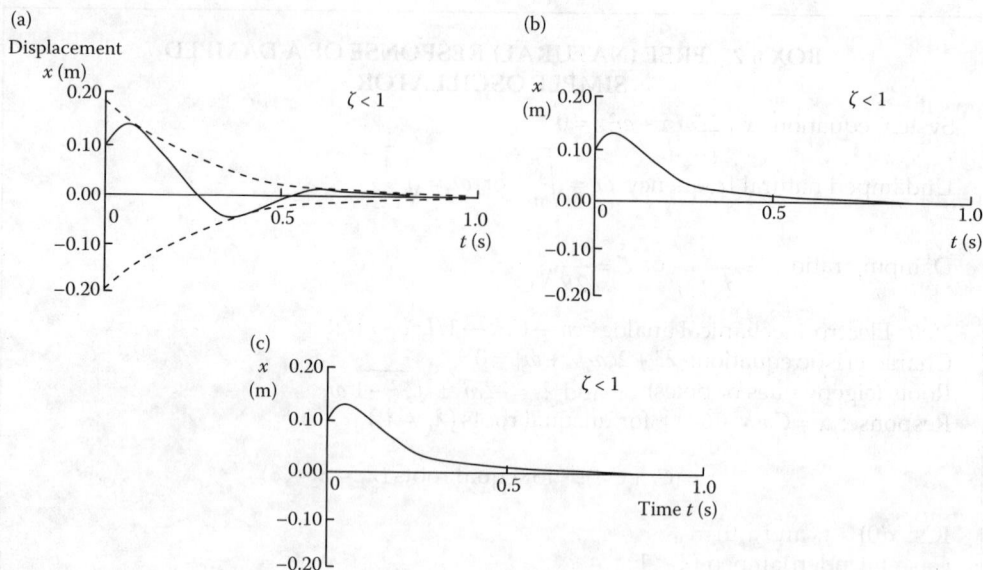

(a)

Displacement

(b)

(c)

FIGURE 6.6
Free response of a damped oscillator. (a) Underdamped. (b) Critically damped. (c) Overdamped.

6.4 Forced Response of a Damped Oscillator

The forced response depends on both the natural characteristics of the system (free response) and the nature of the input. Mathematically, as noted before, the total response is the sum of the homogeneous solution and the particular solution. Consider a damped simple oscillator, with input $u(t)$ scaled such that it has the same units as the response y:

$$\ddot{y} + 2\zeta\omega_n\dot{y} + \omega_n^2 y = \omega_n^2 u(t) \tag{6.48}$$

We will consider the response of this system to three types of inputs:

1. Impulse input
2. Step input
3. Harmonic (sinusoidal) input

6.4.1 Impulse Response

Many important characteristics of a system can be studied by analyzing the system response to a baseline excitation (test excitation) such as an impulse, a step, or a sinusoidal (harmonic) input. Characteristics that may be studied in this manner include: system stability, speed of response, time constants, damping properties, and natural frequencies. Furthermore, models and their parameters can be determined by this method (this subject is known as system identification, model identification or experimental modeling). As well, an insight can be gained into the system response for an arbitrary excitation. Responses to such test inputs can also serve as the basis for system comparison. For example, it is possible to determine the degree of nonlinearity in a system by exciting it with two input intensity levels, separately, and checking whether the proportionality is retained at the output; or when the excitation is harmonic, whether limit cycles are encountered by the response.

The response of the system (Equation 6.48) to a unit impulse input $u(t) = \delta(t)$ may be conveniently determined by the Laplace transform approach (see Section 6.5). However, in the present section we will use a time domain approach, instead. First integrate Equation 6.48, over the almost zero interval from $t = 0^-$ to $t = 0^+$. We get

$$\dot{y}(0^+) = \dot{y}(0^-) - 2\zeta\omega_n\left[y(0^+) - y(0^-)\right] - \omega_n^2\int_{0^-}^{0^+} y\, dt + \omega_n^2\int_{0^-}^{0^+} u(t)dt \qquad (6.49)$$

Suppose that the system starts from rest. Hence, $y(0^-) = 0$ and $\dot{y}(0^-) = 0$. When an impulse is applied over an infinitesimally short time period $[0^-, 0^+]$ the system will not be able to move through a finite distance during that time. Hence, $y(0^+) = 0$, and furthermore, the integral of y on the right hand side of Equation 6.49 also will be zero. Now by definition of a unit impulse, the integral of u on the right hand side of Equation 6.49 will be unity. Hence we have $\dot{y}(0^+) = \omega_n^2$. It follows that as soon as a unit impulse is applied to the system (Equation 6.48) the ICs will become

$$y(0^+) = 0 \text{ and } \dot{y}(0^+) = \omega_n^2 \qquad (6.50)$$

Also, beyond $t = 0^+$ the input is zero ($u(t) = 0$), according to the definition of an impulse. Hence, the impulse response of the system (Equation 6.48) is obtained by its homogeneous solution (as carried out before, for the case of free response), but with the ICs (Equation 6.50). The three cases of damping ratio ($\zeta < 1$, $\zeta > 1$, and $\zeta = 1$) should be considered separately. Then, we can conveniently obtain the following results:

$$y_{impulse}(t) = h(t) = \frac{\omega_n}{\sqrt{1-\zeta^2}}\exp(-\zeta\omega_n t)\sin\omega_d t \quad \text{for} \quad \zeta < 1 \qquad (6.51a)$$

$$y_{impulse}(t) = h(t) = \frac{\omega_n}{2\sqrt{\zeta^2-1}}[\exp\lambda_1 t - \exp\lambda_2 t] \quad \text{for} \quad \zeta > 1 \qquad (6.51b)$$

$$y_{impulse}(t) = h(t) = \omega_n^2 t\exp(-\omega_n t) \quad \text{for} \quad \zeta = 1 \qquad (6.51c)$$

An explanation concerning the dimensions of $h(t)$ is appropriate at this juncture. Note that $y(t)$ has the same dimensions as $u(t)$. Since $h(t)$ is the response to a unit impulse $\delta(t)$, it follows that these two have the same dimensions. The magnitude of $\delta(t)$ is represented by a unit area in the $u(t)$ versus t plane. Consequently, $\delta(t)$ has the dimensions of (1/time) or (frequency). It follows that $h(t)$ also has the dimensions of (1/time) or (frequency).

The impulse response functions given by Equation 6.51 are plotted in Figure 6.7 for some representative values of damping ratio. It should be noted that, for $0 < \zeta < 1$, the angular frequency of damped vibrations is ω_d, which is smaller than the undamped natural frequency ω_n.

6.4.2 The Riddle of Zero ICs

For a second-order system, zero ICs correspond to $y(0)=0$ and $\dot{y}(0)=0$. It is clear from Equations 6.51 that $h(0)=0$, but $\dot{h}(0) \neq 0$, which appears to violate the assumption of zero ICs. This situation is characteristic in a system response to an impulse and its higher derivatives. This may be explained as follows. When an impulse is applied to a system at rest (zero initial state), the highest derivative of the system differential equation momentarily becomes infinity. As a result, the next lower derivative becomes finite (nonzero) at $t = 0^+$. The remaining lower derivatives maintain their original zero values at that instant $t = 0^+$. When an impulse is applied to the mechanical system given by Equation 6.48 for example, the acceleration $\ddot{y}(t)$ becomes infinity and the velocity $\dot{y}(t)$ takes a nonzero (finite) value shortly after its application (i.e., at $t = 0^+$). The displacement $y(t)$, however, would not have sufficient time to change at $t = 0^+$. In this case the impulse input is therefore, equivalent to a velocity IC. This IC is determined by using the integrated form (Equation 6.49) of the system Equation 6.48, as has been done.

6.4.3 Step Response

A unit step excitation is defined by

$$\mathcal{U}(t) = 1 \quad \text{for} \quad t > 0 \tag{6.52}$$

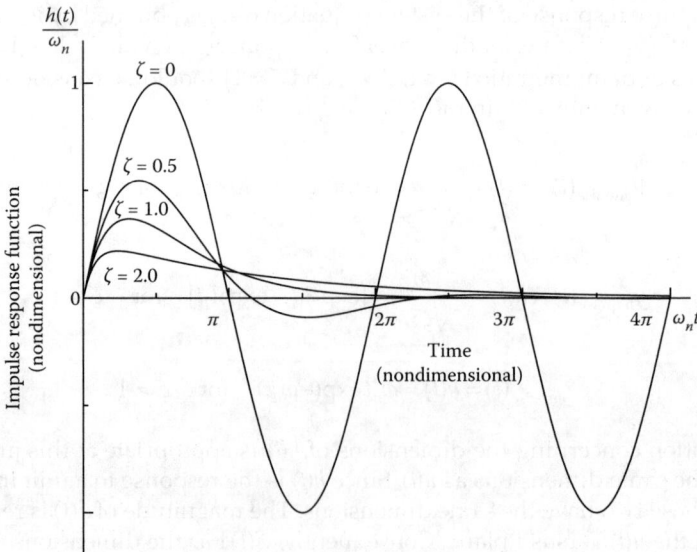

FIGURE 6.7
Impulse response function of a damped oscillator.

$$= 0 \quad \text{for} \quad t \le 0$$

Unit impulse excitation $\delta(t)$ may be interpreted as the time derivative of $\mathcal{U}(t)$:

$$\delta(t) = \frac{d\mathcal{U}(t)}{dt} \tag{6.53}$$

Note: Equation 6.53 re-establishes the fact that for nondimensional $\mathcal{U}(t)$, the dimension of $\delta(t)$ is (time)$^{-1}$. Since a unit step is the integral of a unit impulse, the step response can be obtained directly as the integral of the impulse response:

$$y_{step}(t) = \int_0^t h(\tau) d\tau \tag{6.54}$$

This result also follows from the convolution integral (Equation 6.18b) because, for a delayed unit step, we have

$$\mathcal{U}(t - \tau) = 1 \quad \text{for} \quad \tau < t$$
$$= 0 \quad \text{for} \quad \tau \ge t \tag{6.55}$$

Thus, by integrating Equations 6.51 with zero ICs the following results are obtained for step response:

$$y_{step}(t) = 1 - \frac{1}{\sqrt{1-\zeta^2}} \exp(-\zeta\omega_n t) \sin(\omega_d t + \phi) \quad \text{for} \quad \zeta < 1 \tag{6.56a}$$

$$y_{step} = 1 - \frac{1}{2\sqrt{1-\zeta^2}\,\omega_n} \left[\lambda_1 \exp \lambda_2 t - \lambda_2 \exp \lambda_1 t \right] \quad \text{for} \quad \zeta > 1 \tag{6.56b}$$

$$y_{step} = 1 - (\omega_n t + 1) \exp(-\omega_n t) \quad \text{for} \quad \zeta = 1 \tag{6.56c}$$

with

$$\cos\phi = \zeta \tag{6.57}$$

The step responses given by Equations 6.56 are plotted in Figure 6.8, for several values of damping ratio.

Note: Since a step input does not cause the highest derivative of the system equation to approach infinity at $t=0^+$, the initial conditions which are required to solve the system equation remain unchanged at $t=0^+$, provided that there are no derivative terms on the input side of the system equation. If there is a derivative term in the input side of the system equation, then, a step will be converted into an impulse (due to differentiation), and the response will change accordingly.

The impulse response $h(t)$ assumes a zero initial state. It should be emphasized as well that the response given by the convolution integral (Equation 6.18) is based on the

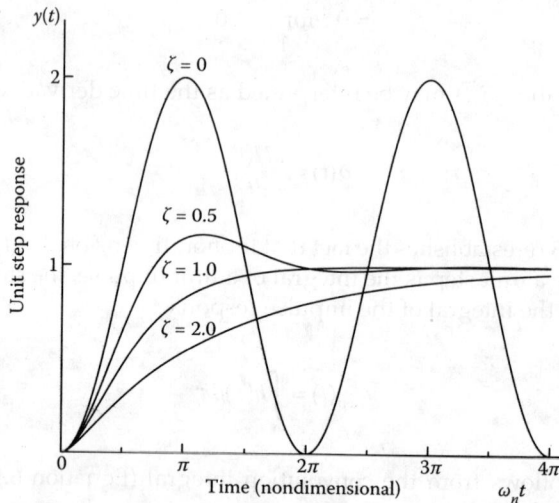

FIGURE 6.8
Unit step response of a damped oscillator.

assumption that the initial state is zero. Hence, it is known as the *zero-state response*. As we have stated before, the zero-state response is not necessarily equal to the "particular solution" in mathematical analysis. Also, as t increases ($t \rightarrow \infty$), this solution approaches the *steady-state response* denoted by y_{ss}, which is typically the particular solution. The impulse response of a system is the inverse Laplace transform of the transfer function. Hence, it can be determined using Laplace transform techniques. Some useful concepts of forced response are summarized in Box 6.3.

6.4.4 Response to Harmonic Excitation

In many engineering problems, the primary excitation typically has a repetitive periodic nature, and in some cases, this periodic input function may even be purely sinusoidal. Examples are excitations due to mass eccentricity and misalignments in rotational components, tooth meshing in gears, and electromagnetic devices excited by ac or periodic electrical signals and frequency generators. In basic terms, the frequency response of a dynamic system is the response to a pure sinusoidal excitation. As the amplitude and the frequency of the excitation are changed, the response also changes. In this manner the response of the system over a range of excitation frequencies can be determined, and this set of input–output data, at steady-state, represents the frequency response. Here we are dealing with the *frequency domain* (rather than the time domain) and frequency (ω) is the independent variable.

Consider the damped oscillator with a harmonic input, as given by

$$\ddot{x} + 2\zeta\omega_n\dot{x} + \omega_n^2 x = a\cos\omega t = u(t) \tag{6.58}$$

The particular solution x_p that satisfies (Equation 6.58) is of the form (see Table 6.1)

$$x_p = a_1\cos\omega t + a_2\sin\omega t \ \{\text{Except for the case: } \zeta = 0 \text{ and } \omega = \omega_n\} \tag{6.59}$$

BOX 6.3   USEFUL CONCEPTS OF FORCED RESPONSE

Convolution integral: Response $y = \int_0^t h(t-\tau)u(\tau)d\tau = \int_0^t h(\tau)u(t-\tau)d\tau$

where $u=$excitation (input) and $h=$impulse response function (response to a unit impulse input).

Damped simple oscillator: $\ddot{y} + 2\zeta\omega_n\dot{y} + \omega_n^2 y = \omega_n^2 u(t)$

Poles (eigenvalues) $\lambda_1, \lambda_2 = -\zeta\omega_n \pm \sqrt{\zeta^2 - 1}\,\omega_n$ for $\zeta \geq 1$

$= -\zeta\omega_n \pm j\omega_d$ for $\zeta < 1$

$\omega_n =$undamped natural frequency, $\omega_d =$damped natural frequency

$\zeta =$damping ratio.

Note: $\omega_d = \sqrt{1-\zeta^2}\,\omega_n$

$$\frac{\text{Impulse Response Function}}{\text{(Zero ICs)}} : h(t) = \frac{\omega_n}{\sqrt{1-\zeta^2}}\exp(-\zeta\omega_n t)\sin\omega_d t \quad \text{for} \quad \zeta < 1$$

$$= \frac{\omega_n}{2\sqrt{\zeta^2-1}}\left[\exp\lambda_1 t - \exp\lambda_2 t\right] \quad \text{for} \quad \zeta > 1$$

$$= \omega_n^2 t\exp(-\omega_n t) \quad \text{for} \quad \zeta = 1$$

$$\frac{\text{Unit Step Response}}{\text{(Zero ICs)}} : \quad y(t)_{step} = 1 - \frac{1}{\sqrt{1-\zeta^2}}\exp(-\zeta\omega_n t)\sin(\omega_d t + \phi) \quad \text{for} \quad \zeta < 1$$

$$= 1 - \frac{1}{2\sqrt{\zeta^2-1}\,\omega_n}\left[\lambda_1\exp\lambda_2 t - \lambda_2\exp\lambda_1 t\right] \quad \text{for} \quad \zeta > 1$$

$$= 1 - (\omega_n t + 1)\exp(-\omega_n t) \quad \text{for} \quad \zeta = 1$$

$$\cos\phi = \zeta$$

Note: Impulse response $= \dfrac{d}{dt}$ (step response).

---

where the constants a_1 and a_2 are determined by substituting Equation 6.59 into the system Equation 6.58 and equating the like coefficient; the *method of undertermined coefficients*. We will consider several important cases.

1. Undamped Oscillator with Excitation Frequency $\neq$ Natural Frequency:
   We have

$$\ddot{x} + \omega_n^2 x = a\cos\omega t \quad \text{with} \quad \omega \neq \omega_n \tag{6.60}$$

$$\text{Homogeneous solution:} \quad x_h = A_1\cos\omega_n t + A_2\sin\omega_n t \tag{6.61}$$

$$\text{Particular solution:} \quad x_p = \frac{a}{(\omega_n^2 - \omega^2)}\cos\omega t \tag{6.62}$$

Note: It can be easily verified that x_p given by Equation 6.62 satisfies the forced system Equation 6.60. Hence it is a particular solution.

Complete solution:

$$\underbrace{x = A_1 \cos \omega_n t + A_2 \sin \omega_n t}_{H \text{ Satisfies the homogeneous equation}} + \underbrace{\frac{a}{\left(\omega_n^2 - \omega^2\right)} \cos \omega t}_{P \text{ Satisfies the forced equation (equation with input)}} \tag{6.63}$$

Now A_1 and A_2 are determined using the ICs:

$$x(0) = x_o \quad \text{and} \quad \dot{x}(0) = v_o \tag{6.64}$$

Specifically, we obtain

$$x_o = A_1 + \frac{a}{\omega_n^2 - \omega^2} \tag{6.65a}$$

$$v_o = A_2 \omega_n \tag{6.65b}$$

Hence, the complete response is

$$x = \underbrace{\left[x_o - \frac{a}{\left(\omega_n^2 - \omega^2\right)} \right] \cos \omega_n t + \frac{v_o}{\omega_n} \sin \omega_n t}_{H \text{ Homogeneous Solution}} + \underbrace{\frac{a}{\omega_n^2 - \omega^2} \cos \omega t}_{P \text{ Particular Solution}} \tag{6.66a}$$

$$= \underbrace{x_o \cos \omega_n t + \frac{v_o}{\omega_n} \sin \omega_n t}_{\substack{X \text{ Free response} \\ \text{(depends only on ICs).} \\ \text{Comes from } x_h; \text{ Sinusodal at } \omega_n.}} + \underbrace{\frac{a}{\left(\omega_n^2 - \omega^2\right)} \underbrace{\left[\cos \omega t - \cos \omega_n t \right]}_{2 \sin \frac{(\omega_n + \omega)t}{2} \sin \frac{(\omega_n - \omega)t}{2}}}_{\substack{F \text{ *Forced response (depends on input).} \\ \text{Comes from both } x_h \text{ and } x_p.}} \tag{6.66b}$$

*Will exhibit a beat phenomenon for small $\omega_n - \omega$; i.e., $(\omega_n + \omega)/2$ wave "modulated" by $(\omega_n - \omega)/2$ wave.

This is a "stable" response in the sense of bounded-input–bounded-output (BIBO) stability, as it is bounded and does not increase steadily.

Note: If there is no forcing excitation, the homogeneous solution H and the free response X will be identical. With a forcing input, the natural free response will be influenced by the input in general, as clear from Equation 6.66b.

2. Undamped Oscillator with $\omega = \omega_n$ (Resonant Condition):

This is the degenerate case given by

$$\ddot{x} + \omega^2 x = a \cos \omega t \tag{6.67}$$

In this case the particular solution x_p that was used before is no longer valid because, then the particular solution would become the same as the homogeneous solution and the former would be completely absorbed into the latter. Instead, in view of the "double-integration" nature of the forced system equation when $\omega = \omega_n$ (see the Laplace transform of Equation 6.67, Section 6.5) we use the following particular solution (P):

$$x_p = \frac{at}{2\omega} \sin \omega t \tag{6.68}$$

This choice of particular solution is justified by the fact that it satisfies the forced system Equation 6.67.

$$\text{Complete solution:} \, x = A_1 \cos \omega t + A_2 \sin \omega t + \frac{at}{2\omega} \sin \omega t \tag{6.69}$$

$$\text{ICs: } x(0) = x_o \quad \text{and} \quad \dot{x}(0) = v_o$$

By substitution of ICs into Equation 6.69 we get:

$$x_o = A_1 \tag{6.70a}$$

$$v_o = \omega A_2 \tag{6.70b}$$

The total response:

$$x = \underbrace{x_o \cos \omega t + \frac{v_o}{\omega} \sin \omega t}_{\substack{X \text{ Free Response (Depends on ICs)} \\ *\text{Sinusoidal with frequency } w.}} \quad \underbrace{+ \frac{at}{2\omega} \sin \omega t}_{\substack{F \text{ Forced Response (Depends on Input)} \\ *\text{Amplitude increases linearly.}}} \tag{6.71}$$

Since the forced response increases steadily, this is an unstable forced response in the BIBO sense. Furthermore, the homogeneous solution H and the free response X are identical, and the particular solution P is identical to the forced response F in this case.

Note: The same system (undamped oscillator) gives a bounded response for some excitations while producing an unstable response (steady linear increase) when the excitation frequency is equal to its natural frequency. Hence, the system is not quite unstable, but is not quite stable either. In fact, the undamped oscillator is said to be marginally stable. When the excitation frequency is equal to the natural frequency it is reasonable for the system to respond in a complementary and steadily increasing manner because this corresponds to the most "receptive" excitation. Specifically, in this case, the excitation complements and reinforces the natural response of the system. In other words, the system is "in resonance" with the excitation. This condition is called a *resonance* and the corresponding frequency is called *resonant frequency*. Later on we will address this aspect for the more general case of a damped oscillator.

Figure 6.9 shows typical forced responses of an undamped oscillator for a large difference in excitation frequency and natural frequency (Case 1); for a small difference in excitation frequency and natural frequency (also Case 1), where a beat-phenomenon is clearly manifested; and for the resonant case where the excitation frequency equals the natural frequency (Case 2).

(a)

Response

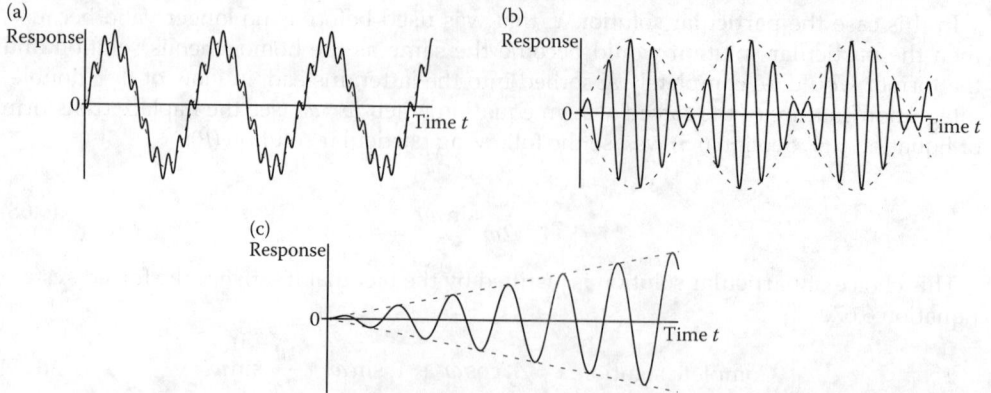

(b)

Response

(c)

Response

FIGURE 6.9

Forced response of a harmonic-excited undamped simple oscillator. (a) For a large frequency difference. (b) For a small frequency difference (beat phenomenon). (c) Response at resonance.

3. Damped Oscillator:

In this case the equation of forced motion is

$$\ddot{x} + 2\zeta\omega_n\dot{x} + \omega_n^2 x = a\cos\omega t \tag{6.72}$$

Particular solution: Since derivatives of both odd order and even order are present in this equation, the particular solution should have terms corresponding to odd and even derivatives of the forcing function (i.e., $\sin \omega t$ and $\cos \omega t$). Hence, the appropriate particular solution will be of the form:

$$x_p = a_1\cos\omega t + a_2\sin\omega t \tag{6.73}$$

We determine the coefficients in Equation 6.73 by the method of undetermined coefficients. Specifically, substitute Equation 6.73 into Equation 6.72. We get:

$$-\omega^2 a_1\cos\omega t - \omega^2 a_2\sin\omega t + 2\zeta\omega_n\left[-\omega a_1\sin\omega t + \omega a_2\cos\omega t\right] + \omega_n^2\left[a_1\cos\omega t + a_2\sin\omega t\right] = a\cos\omega t$$

Equate like coefficients:

$$-\omega^2 a_1 + 2\zeta\omega_n\omega a_2 + \omega_n^2 a_1 = a$$

$$-\omega^2 a_2 - 2\zeta\omega_n\omega a_1 + \omega_n^2 a_2 = 0$$

Hence, we have

$$(\omega_n^2 - \omega^2)\, a_1 + 2\zeta\omega_n\omega a_2 = a \tag{6.74a}$$

$$-2\zeta\omega_n\omega a_1 + (\omega_n^2 - \omega^2)\, a_2 = 0 \tag{6.74b}$$

This can be written in the vector-matrix form:

$$\begin{bmatrix} (\omega_n^2 - \omega^2) & 2\zeta\omega_n\omega \\ -2\zeta\omega_n\omega & (\omega_n^2 - \omega^2) \end{bmatrix} \begin{bmatrix} a_1 \\ a_2 \end{bmatrix} = \begin{bmatrix} a \\ 0 \end{bmatrix}$$

(6.74c)

Its solution is

$$\begin{bmatrix} a_1 \\ a_2 \end{bmatrix} = \frac{1}{D} \begin{bmatrix} (\omega_n^2 - \omega^2) & -2\zeta\omega_n\omega \\ 2\zeta\omega_n\omega & (\omega_n^2 - \omega^2) \end{bmatrix} \begin{bmatrix} a \\ 0 \end{bmatrix}$$

(6.75)

or

$$a_1 = \frac{(\omega_n^2 - \omega^2)}{D} a$$

(6.75a)

$$a_2 = \frac{2\zeta\omega_n\omega}{D} a$$

(6.75b)

with the determinant given by

$$D = (\omega_n^2 - \omega^2)^2 + (2\zeta\omega_n\omega)^2$$

(6.76)

Some useful results on the frequency response of a simple oscillator are summarized in Box 6.4.

6.5 Response Using Laplace Transform

Transfer function concepts are discussed in Chapter 5, and transform techniques are outlined in Appendix A. Once a transfer function model of a system is available, its response can be determined using the Laplace transform approach. The steps are:

1. Using Laplace transform table (Appendix A) determine the Laplace transform ($U(s)$) of the input.
2. Multiply by the transfer function ($G(s)$) to obtain the Laplace transform of the output: $Y(s) = G(s)U(s)$.

Note: The ICs may be introduced in this step by first expressing the system equation in the polynomial form in s and then adding the ICs to each derivative term in the characteristic polynomial.

3. Convert the expression in Step 2 into a convenient form (e.g., by partial fractions).
4. Using Laplace transform table, obtain the inverse Laplace transform of $Y(s)$, which gives the response $y(t)$.

Let us illustrate this approach by determining again the step response of a simple oscillator.

BOX 6.4   HARMONIC RESPONSE OF A SIMPLE OSCILLATOR

Undamped oscillator: $\ddot{x} + \omega_n^2 x = a \cos \omega t$; $x(0) = x_0$, $\dot{x}(0) = v_0$

For $\omega \neq \omega_n$: $x = \underbrace{x_0 \cos \omega_n t + \frac{v_0}{\omega_n} \sin \omega_n t}_{X} + \underbrace{\frac{a}{\omega_n^2 - \omega^2} [\cos \omega t - \cos \omega_n t]}_{F}$

$$\text{For } \omega = \omega_n (\text{resonance}): x = \text{Same } X + \frac{at}{2\omega} \sin \omega t$$

Damped oscillator: $\ddot{x} + 2\zeta \omega_n \dot{x} + \omega_n^2 x = a \cos \omega t$

$$x = H + \underbrace{\frac{a}{\left| \omega_n^2 - \omega^2 + 2j\zeta\omega_n\omega \right|}}_{P} \cos(\omega t - \phi)$$

where $\tan \phi = \dfrac{2\zeta\omega_n\omega}{\omega_n^2 - \omega^2}$; ϕ = phase lag.

Particular solution P is also the steady-state response.
Homogeneous solution $H = A_1 e^{\lambda_1 t} + A_2 e^{\lambda_2 t}$
where, λ_1 and λ_2 are roots of $\lambda^2 + 2\zeta\omega_n\lambda + \omega_n^2 = 0$ (characteristic equation)
A_1 and A_2 are determined from ICs: $x(0) = x_0$, $\dot{x}(0) = v_0$
Resonant frequency: $\omega_r = \sqrt{1 - 2\zeta^2} \, \omega_n$
The magnitude of P will peak at resonance.

Damping ratio: $\zeta = \dfrac{\Delta\omega}{2\omega_n} = \dfrac{\omega_2 - \omega_1}{\omega_2 + \omega_1}$ for low damping

where $\Delta\omega$ = half-power bandwidth = $\omega_2 - \omega_1$

Note: Q-factor = $\dfrac{\omega_n}{\Delta\omega} = \dfrac{1}{2\zeta}$ for low damping

6.5.1 Step Response Using Laplace Transforms

Consider the oscillator system given earlier by Equation 6.48:

$$\ddot{y} + 2\zeta\omega_n\dot{y} + \omega_n^2 y = \omega_n^2 u(t) \tag{6.48}$$

Since $\mathcal{L}U(t) = 1/s$, the unit step response of the dynamic system (Equation 6.48), with zero ICs, can be obtained by taking the inverse Laplace transform of

$$Y_{step}(s) = \frac{1}{s} \frac{\omega_n^2}{(s^2 + 2\zeta\omega_n s + \omega_n^2)} = \frac{1}{s} \frac{\omega_n^2}{\Delta(s)} \tag{6.77a}$$

Here the characteristic polynomial of the system is denoted as

$$\Delta(s) = (s^2 + 2\zeta\omega_n s + \omega_n^2) \tag{6.78a}$$

To facilitate using the Laplace transform table, partial fractions of Equation 6.77a are determined in the form

$$\frac{a_1}{s} + \frac{a_2 + a_3 s}{(s^2 + 2\zeta\omega_n s + \omega_n^2)}$$

in which, the constants a_1, a_2 and a_3 are determined by comparing the numerator polynomial:

$$\omega_n^2 = a_1(s^2 + 2\delta\omega_n s + \omega_n^2) + s(a_2 + a_3 s)$$

We get $a_1 = 1$, $a_2 = -2\zeta\omega_n$, and $a_3 = 1$.
Hence,

$$Y_{\text{Step}}(s) = \frac{1}{s} + \frac{s - 2\zeta\omega_n}{(s^2 + 2\zeta\omega_n s + \omega_n^2)} = \frac{1}{s} + \frac{s - 2\zeta\omega_n}{\Delta(s)} \qquad (6.77b)$$

Next, using Laplace transform tables, the inverse transform of Equation 6.77b is obtained, and verified to be identical to Equation 6.56.

6.5.2 Incorporation of ICs

When the ICs of the system are not zero, they have to be explicitly incorporated into the derivative terms of the system equation, when converting into the Laplace domain. Except for this, the analysis using the Laplace transform approach is identical to that with zero ICs. In fact, the total solution is equal to the sum of the solution with zero ICs and the solution corresponding to the ICs. We will illustrate the approach using two examples.

6.5.2.1 Step Response of a First-Order System

Let us revisit the first-order dynamic system with time constant τ, input u, and output y, as given by

$$\tau\dot{y} + y = u(t) \qquad (6.19)$$

The IC is $y(0) = y_0$. A step input of magnitude A is applied at that IC.
From Laplace tables (see Appendix A), convert each term in Equation 6.19 into the Laplace domain as follows:

$$\tau[sY(s) - y_0] + Y(s) = A/s \qquad (6.79a)$$

Note how the IC is included in the derivative term, as clear from the Laplace tables. On simplification we get

$$Y(s) = \frac{\tau y_0}{(\tau s + 1)} + \frac{A}{s(\tau s + 1)} = \frac{\tau y_0}{(\tau s + 1)} + \frac{A}{s} - \frac{A\tau}{(\tau s + 1)} \qquad (6.79b)$$

Now we use the Laplace tables to determine the inverse Laplace transform each term in Equation 6.79b. We get

$$y_{Step} = y_0 e^{-t/\tau} + Ak(1 - e^{-t/\tau})$$ (6.80)

This is identical to the previous result (Equation 6.20). The response is plotted in Figure 6.10, for different values of the time constant τ. Notice how the response becomes more sluggish (i.e., the response becomes slower) for larger values of the time constant.

6.5.2.2 Step Response of a Second-Order System

As another illustrative example revisit the simple oscillator problem:

$$\ddot{y} + 2\zeta\omega_n\dot{y} + \omega_n^2 y = \omega_n^2 u(t)$$ (6.48)

Only the underdamped case is considered where $0 < \zeta < 1$.

We use the Laplace transform approach to determine the response to a unit step input for the case with the ICs $y(0)$ and $\dot{y}(0)$. First use Laplace tables to convert each term in Equation 6.48 into the Laplace domain, as follows:

$$s^2 Y(s) - sy(0) - \dot{y}(0) + 2\zeta\omega_n[sY(s) - y(0)] + \omega_n^2 Y(s) = \frac{\omega_n^2}{s}$$ (6.81a)

On simplification we have

$$Y(s) = \frac{1}{s}\frac{\omega_n^2}{\Delta(s)} + \frac{sy(0) + \dot{y}(0) + 2\zeta\omega_n y(0)}{\Delta(s)}$$ (6.81b)

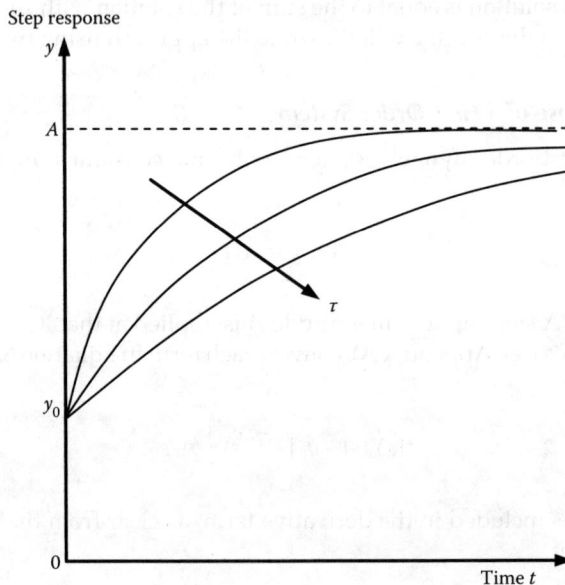

FIGURE 6.10
Step response of a first-order system.

Here the characteristic polynomial is given by

$$\Delta(s) = (s^2 + 2\zeta\omega_n s + \omega_n^2) = (s + \zeta\omega_n)^2 + \omega_d^2 \qquad (6.78\text{b})$$

where the damped natural frequency ω_d is such that

$$\omega_d^2 = (1 - \zeta^2)\omega_n^2 \qquad (6.82)$$

The zero state response (i.e., when the ICs are zero) corresponds to the first term on the right hand side of Equation 6.81b, which can be determined as before by noting that

$$Y_{\text{Forced}}(s) = \frac{1}{s}\frac{\omega_n^2}{\Delta(s)} = \frac{1}{s} - \frac{\zeta\omega_n}{\Delta(s)} - \frac{(s + \zeta\omega_n)}{\Delta(s)} \qquad (6.83)$$

From Laplace tables, the zero state response (i.e., the forced part) is obtained as

$$y_{\text{Forced}}(t) = 1 - \frac{\zeta\omega_n}{\omega_d}\exp(-\zeta\omega_n t)\sin\omega_d t - \exp(-\zeta\omega_n t)\cos\omega_d t \quad \text{for } \zeta < 1 \qquad (6.84\text{a})$$

Now by combining the last two terms on the right hand side we get

$$y_{\text{Forced}}(t) = 1 - \frac{1}{\sqrt{1 - \zeta^2}}\exp(-\zeta\omega_n t)\sin(\omega_d t + \phi) \quad \text{for } \zeta < 1 \qquad (6.84\text{b})$$

$$\text{with } \cos\phi = \zeta$$

This result is identical to what we obtained before.

The response to the ICs is given by the second term on the right hand side of Equation 6.81b. Specifically

$$Y_{\text{IC}}(s) = \frac{sy(0) + \dot{y}(0) + 2\zeta\omega_n y(0)}{\Delta(s)} = \frac{(s + \zeta\omega_n)y(0)}{\Delta(s)} + \frac{\dot{y}(0) + \zeta\omega_n y(0)}{\Delta(s)}$$

The terms in this result are similar to those in Equation 6.83. Term by term conversion into the time domain, using Laplace tables we have

$$y_{\text{IC}}(t) = y(0)\exp(-\zeta\omega_n t)\cos\omega_d t + \left[\frac{\dot{y}(0) + \zeta\omega_n y(0)}{\omega_d}\right]\exp(-\zeta\omega_n t)\sin\omega_d t \quad \text{for } \zeta < 1 \qquad (6.85)$$

The total response is given by the sum of Equations 6.84 and 6.85.

6.6 Determination of ICs for Step Response

When a step input is applied to a system, the initial values of the system variables may change instantaneously. However, not all variables will change in this manner since the value of a state variable cannot change instantaneously. We will illustrate some related considerations using an example.

Example 6.2

The circuit shown in Figure 6.11 consists of an inductor L, a capacitor C, and two resistors R and R_o. The input is the voltage $v_i(t)$ and the output is the voltage v_o across the resistor R_o.

a. Obtain a complete state-space model for the system.
b. Obtain an input–output differential equation for the system.
c. Obtain expressions for undamped natural frequency and the damping ratio of the system.
d. The system starts at steady-state with an input of 5 V (for all $t<0$). Then suddenly, the input is dropped to 1 V (for all $t>0$), which corresponds to a step input as shown in Figure 6.12. For $R=R_o=1\ \Omega$, $L=1$ H, and $C=1$ F, what are the ICs of the system and their derivatives at both $t=0^-$ and $t=0^+$? What are the final (steady-state) values of the state variables and the output variable? Sketch the nature of the system response.

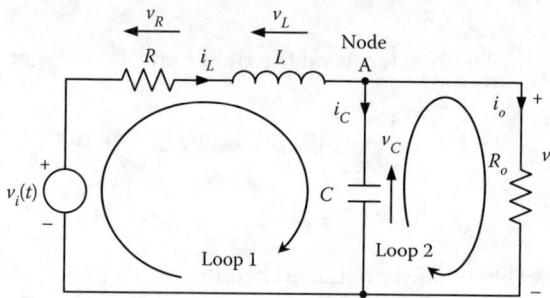

FIGURE 6.11
An electrical circuit.

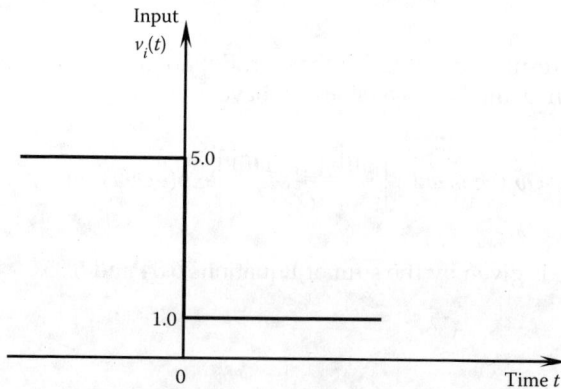

FIGURE 6.12
A step input.

Solution

a. State variables:

Current through independent inductors (i_L); voltage across independent capacitors (v_c).

Constitutive equations:

$$v_L = L\frac{di_L}{dt}; \quad i_C = C\frac{dv_C}{dt}; \quad v_R = Ri_L; \quad v_o = Ri_o$$

First two equations are for independent energy storage elements, and they form the state-space shell.

Continuity equation:

Node A (Kirchhoff's current law): $i_L - i_C - i_o = 0$

Compatibility equations:

Loop 1 (Kirchhoff's voltage law): $v_i - v_R - v_L - v_C = 0$

Loop 2 (Kirchhoff's voltage law): $v_C - v_o = 0$

Eliminate auxiliary variables. We have the state equations:

$$L\frac{di_L}{dt} = v_L = v_i - v_R - v_C = v_i - Ri_L - v_C$$

$$C\frac{dv_C}{dt} = i_C = i_L - i_o = i_L - \frac{v_o}{R_o} = i_L - \frac{v_C}{R_o}$$

State equations:

$$\frac{di_L}{dt} = \frac{1}{L}[-Ri_L - v_C + v_i] \tag{i}$$

$$\frac{dv_C}{dt} = \frac{1}{C}[i_L - \frac{v_C}{R_o}] \tag{ii}$$

Output equation:

$$v_o = v_c$$

Vector-matrix representation

$$\dot{x} = Ax + Bu; \quad y = Cx$$

where

System matrix $A = \begin{bmatrix} -R/L & -1/L \\ 1/C & -1/(R_oC) \end{bmatrix}$; input gain matrix $B = \begin{bmatrix} 1/L \\ 0 \end{bmatrix}$; measurement gain matrix

$C = \begin{bmatrix} 0 & 1 \end{bmatrix}$; state vector $= x = \begin{bmatrix} i_L \\ v_C \end{bmatrix}$; input $= u = [v_i]$; output $= y = [v_o]$

b. From (ii): $i_L = C\frac{dv_C}{dt} + \frac{v_C}{R_o}$

Substitute in (i) for i_L: $L\dfrac{d}{dt}\left(C\dfrac{dv_C}{dt}+\dfrac{v_C}{R_o}\right)=-R\left(C\dfrac{dv_C}{dt}+\dfrac{v_C}{R_o}\right)-v_C+v_i$

This simplifies to the input–output differential equation (since $v_o=v_C$)

$$LC\dfrac{d^2v_o}{dt^2}+\left(\dfrac{L}{R_o}+RC\right)\dfrac{dv_o}{dt}+\left(\dfrac{R}{R_o}+1\right)v_o=v_i \qquad\text{(iii)}$$

c. The input–output differential equation is of the form

$$\dfrac{d^2v_o}{dt^2}+2\zeta\omega_n\dfrac{dv_o}{dt}+\omega_n^2 v_o=\dfrac{1}{LC}v_i$$

Hence:

$$\text{Natural frequency }\omega_n=\sqrt{\dfrac{1}{LC}\left(\dfrac{R}{R_o}+1\right)} \qquad\text{(iv)}$$

$$\text{Damping ratio }\zeta=\dfrac{1}{2\sqrt{LC\left(\dfrac{R}{R_o}+1\right)}}\left(\dfrac{L}{R_o}+RC\right) \qquad\text{(v)}$$

Note: $1/LC$ has units of (frequency)^2. RC and L/R_o have units of "time" (i.e., time constant).

d. ICs:

For $t<0$ (initial steady-state): $\dfrac{di_L}{dt}=0;\quad \dfrac{dv_C}{dt}=0$
   Hence

(i): $\dfrac{di_L(0^-)}{dt}=0=\dfrac{1}{L}[-Ri_L(0^-)-v_C(0^-)+v_i(0^-)]$

(ii): $\dfrac{dv_C(0^-)}{dt}=0=\dfrac{1}{C}\left[i_L(0^-)-\dfrac{v_C(0^-)}{R_o}\right]$

Substitute the given parameter values $R=R_o=1\ \Omega$, $L=1$ H, and $C=1$ F, and the input $v_i(0^-)=5.0$:

$$-i_L(0^-)-v_C(0^-)+5=0$$

$$i_L(0^-)-v_C(0^-)=0$$

We get

$$i_L(0^-)=2.5\text{ A, } v_C(0^-)=2.5\text{ V}$$

State variables cannot undergo step changes (because that violates the corresponding physical laws—constitutive equations). Specifically:

Inductor cannot have a step change in current (needs infinite voltage).
Capacitor cannot have a step change in voltage (needs infinite current).

Hence,

$$i_L(0^+)=i_L(0^-)=2.5\text{ A}$$

$$v_c(0^+)=v_c(0^-)=2.5\text{ V}$$

Note: Since $v_i(0^+) = 1.0$.

(i): $\dfrac{di_L(0^+)}{dt} = -i_L(0^+) - v_C(0^+) + 1.0 = -2.5 - 2.5 + 1.0 = -4.0 \text{ A/s} \neq 0$

(ii): $\dfrac{dv_C(0^+)}{dt} = i_L(0^+) - v_C(0^+) = 2.5 - 2.5 = 0.0 \text{ V/s}$

Final values:
As $t \to \infty$ (at final steady-state)

$$\frac{di_L}{dt} = 0$$

$$\frac{dv_C}{dt} = 0$$

and $v_i = 1.0$
   Substitute:

(i): $\dfrac{di_L(\infty)}{dt} = 0 = -i_L(\infty) - v_C(\infty) + 1.0$

(ii): $\dfrac{dv_C(\infty)}{dt} = 0 = i_L(\infty) - v_C(\infty)$

Solution: $i_L(\infty) = 0.5$ A, $v_c(\infty) = 0.5$ V
For the given parameter values,

(iii): $\dfrac{d^2 v_o}{dt^2} + 2\dfrac{dv_o}{dt} + 2v_o = 1$

Hence, $\omega_n = \sqrt{2}$ and $2\zeta\omega_n = 2$, or, $\zeta = 1/\sqrt{2}$
This is an underdamped system, producing an oscillatory response as a result. The nature of the responses of the two state variables is shown in Figure 6.13. *Note:* Output $v_o = v_c$.

Example 6.3

A system is given by the transfer function $\dfrac{y}{u} = \dfrac{\omega_n^2}{s^2 + 2\zeta\omega_n s + \omega_n^2}$

where $u = $ input; $y = $ output; $s = $ Laplace variable; and ζ, ω_n are system parameters.

   a. Write the input–output differential equation of the system.
   It is well-known that the response of this system to a unit step input with zero ICs: $y(0^-) = 0$ and $\dot{y}(0^-) = 0$ is given by

$$y = 1 - \frac{1}{\sqrt{1-\zeta^2}} e^{-\zeta\omega_n t} \sin(\omega_d t + \phi) \quad \text{for} \quad 0 \leq \zeta < 1$$

where $\omega_d = \sqrt{1-\zeta^2}\,\omega_n$ and $\cos\phi = \zeta$

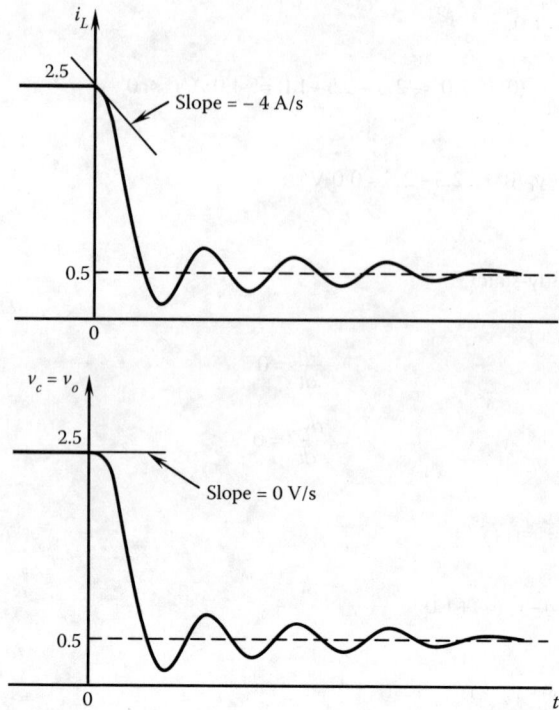

FIGURE 6.13
Responses of the state variables.

b. Determine $y(0^+)$ and $\dot{y}(0^+)$ for this response.
Now consider the system given by the transfer function

$$\frac{y}{u} = \frac{\omega_n^2(\tau s + 1)}{(s^2 + 2\zeta\omega_n s + \omega_n^2)}$$

where τ is an additional system parameter. The remaining parameters are the same as those given for the previous system.

c. Write the input–output differential equation for this modified system.
d. Without using Laplace transform tables, but using the result given for the original system, determine the response of the modified system to a unit step input with zero ICs: $y(0^-) = 0$ and $\dot{y}(0^-) = 0$.

The response must be expressed in terms of the given system parameters (ω_n, ζ, τ).

e. Determine $y(0^+)$ and $\dot{y}(0^+)$ for this response. Comment on your result, if it is different from the values for $y(0^-) = 0$ and $\dot{y}(0^-) = 0$.

Solution

a. To obtain the input–output differential equation, represent the Laplace variable s by the derivative operator d/dt in the given transfer function. We get:

$$\frac{d^2y}{dt^2} + 2\zeta\omega_n \frac{dy}{dt} + \omega_n^2 y = \omega_n^2 u \qquad \text{(i)}$$

b. By direct substitution of $t = 0^+$ in the given response expression we have

$$y(0^+) = 1 - \frac{1}{\sqrt{1-\zeta^2}} \sin\phi = 1 - \frac{1}{\sqrt{1-\zeta^2}} \times \sqrt{1-\zeta^2} = 1 - 1 = 0$$

Now differentiate the given response expression. We have

$$\dot{y} = -\frac{1}{\sqrt{1-\zeta^2}} [e^{-\zeta\omega_n t} \omega_d \cos(\omega_d t + \phi) - \zeta\omega_n e^{-\zeta\omega_n t} \sin(\omega_d t + \phi)]$$

By substituting $t = 0^+$ in this expression we get:

$$\dot{y}(0^+) = -\frac{1}{\sqrt{1-\zeta^2}} [\omega_d \cos\phi - \zeta\omega_n \sin\phi] = -\frac{1}{\sqrt{1-\zeta^2}} \left[\sqrt{1-\zeta^2}\,\omega_n\zeta - \zeta\omega_n\sqrt{1-\zeta^2} \right] = 0$$

c. As before, to obtain the input–output differential equation, represent the Laplace variable s by the derivative operator d/dt in the given transfer function. We get:

$$\frac{d^2y}{dt^2} + 2\zeta\omega_n \frac{dy}{dt} + \omega_n^2 y = \omega_n^2 \left(\tau \frac{du}{dt} + u \right) \qquad \text{(ii)}$$

d. Examine the two differential Equations (i) and (ii). The left hand sides are identical. The right hand sides, which represent the input to the system, are different, but the second corresponds to a linear superposition of the first. Since both systems are linear, from the "principle of super-position" the forced response (i.e., with zero ICs) of system (ii)—call it y—is obtained from the forced response of system (i)—call it y_o—as

$$y = \tau \frac{dy_o}{dt} + y_o \qquad \text{(iii)}$$

It is given that

$$y_o = 1 - \frac{1}{\sqrt{1-\zeta^2}} e^{-\zeta\omega_n t} \sin(\omega_d t + \phi) \qquad \text{(iv)}$$

and from (b)

$$\dot{y}_o = -\frac{1}{\sqrt{1-\zeta^2}} [e^{-\zeta\omega_n t} \omega_d \cos(\omega_d t + \phi) - \zeta\omega_n e^{-\zeta\omega_n t} \sin(\omega_d t + \phi)] \qquad \text{(v)}$$

Substitute Equations (iv) and (v) in Equation (ii). We get

$$y = -\frac{\tau}{\sqrt{1-\zeta^2}}[e^{-\zeta\omega_n t}\omega_d \cos(\omega_d t + \phi) - \zeta\omega_n e^{-\zeta\omega_n t}\sin(\omega_d t + \phi)]$$

$$+1-\frac{1}{\sqrt{1-\zeta^2}}e^{-\zeta\omega_n t}\sin(\omega_d t + \phi)$$

$$=1-\frac{1}{\sqrt{1-\zeta^2}}e^{-\zeta\omega_n t}[\sin(\omega_d t + \phi) + \tau\{\omega_d \cos(\omega_d t + \phi) - \zeta\omega_n \sin(\omega_d t + \phi)\}]$$

$$=1-\frac{1}{\sqrt{1-\zeta^2}}e^{-\zeta\omega_n t}\left[\sin(\omega_d t + \phi) + \tau\omega_n\left\{\sqrt{1-\zeta^2}\cos(\omega_d t + \phi) - \zeta\sin(\omega_d t + \phi)\right\}\right]$$

$$=1-\frac{1}{\sqrt{1-\zeta^2}}e^{-\zeta\omega_n t}[\sin(\omega_d t + \phi) + \tau\omega_n\{\sin\phi\cos(\omega_d t + \phi) - \cos\phi\sin(\omega_d t + \phi)\}]$$

or,

$$y = 1-\frac{1}{\sqrt{1-\zeta^2}}e^{-\zeta\omega_n t}[\sin(\omega_d t + \phi) - \tau\omega_n \sin\omega_d t] \tag{vi}$$

Note: The last step follows from the trigonometric identity

$$\sin(A-B) = \sin A \cos B - \cos A \sin B$$

e. Substitute $t = 0^+$ in Equation (vi). We get

$$y(0^+) = 1-\frac{1}{\sqrt{1-\zeta^2}}[\sin\phi] = 1-\frac{1}{\sqrt{1-\zeta^2}}\left[\sqrt{1-\zeta^2}\right] = 1-1 = 0$$

Next, differentiate Equation (vi). We get

$$\dot{y} = \frac{\zeta\omega_n}{\sqrt{1-\zeta^2}}e^{-\zeta\omega_n t}[\sin(\omega_d t + \phi) - \tau\omega_n \sin\omega_d t]$$

$$-\frac{1}{\sqrt{1-\zeta^2}}e^{-\zeta\omega_n t}[\omega_d \cos(\omega_d t + \phi) - \tau\omega_n\omega_d \cos\omega_d t] \tag{vii}$$

Substitute $t = 0^+$ in Equation (vii). We get

$$\dot{y}(0^+) = \frac{\zeta\omega_n}{\sqrt{1-\zeta^2}}[\sin\phi] - \frac{1}{\sqrt{1-\zeta^2}}[\omega_d \cos\phi - \tau\omega_n\omega_d]$$

$$= \frac{\zeta\omega_n}{\sqrt{1-\zeta^2}}\left[\sqrt{1-\zeta^2}\right] - \frac{1}{\sqrt{1-\zeta^2}}\left[\sqrt{1-\zeta^2}\,\omega_n\zeta - \tau\omega_n^2\sqrt{1-\zeta^2}\right]$$

$$= \zeta\omega_n - [\zeta\omega_n - \tau\omega_n^2]$$

or,

$$\dot{y}(0^+) = \tau\omega_n^2$$

Note: At $t = 0^+$ we have $y(0^+) = 0$ but $\dot{y}(0^+) \neq 0$.
The reason for nonzero rate (or, velocity, if y represents displacement) is as follows. When the input u is a step function, its derivative $\dot{u}$ is an impulse function. In the modified system (ii) it is clear from the right hand side that a linear combination of a step and an impulse are applied to the system (when u is a step function). The impulse input component results in an instantaneous change in $\dot{y}$ (or, instantaneous change in velocity).

6.7 Computer Simulation

Simulation of the response of a dynamic system by using a digital computer is perhaps the most convenient and popular approach to response analysis. An important advantage is that any complex, nonlinear, and time variant system may be analyzed in this manner. The main disadvantage is that the solution is not analytic, and is valid only for a specific excitation, under the particular ICs, over a limited time interval, and so on. Of course, symbolic approaches of obtaining analytical solutions using a digital computer are available as well. We will consider here numerical simulation only.

The key operation of digital simulation is integration over time. This typically involves integration of a differential equation of the form

$$\dot{y} = f(y, u, t) \tag{6.86}$$

where u is the input (excitation) and y is the output (response). Note that the function f is nonlinear and time-variant in general. The most straightforward approach to digital integration of this equation is by using *trapezoidal rule*, which is the Euler's method, as given by

$$y_{n+1} = y_n + f(y_n, u_n, t_n)\Delta t \quad n = 0, 1, \ldots \tag{6.87}$$

Here t_n is the nth time instant, $u_n = u(t_n)$, $y_n = y(t_n)$; and Δt is the integration time step $(\Delta t = t_{n+1} - t_n)$. This approach is generally robust. But depending on the nature of the function f, the integration can be ill behaved. Also, Δt has to be chosen sufficiently small.

For complex nonlinearities in f, a better approach of digital integration is the Runge–Kutta method. In this approach, in each time step, first the following four quantities are computed:

$$g_1 = f(y_n, u_n, t_n) \tag{6.88a}$$

$$g_2 = f\left[\left(y_n + g_1 \frac{\Delta t}{2}\right), u_{n+1/2}, \left(t_n + \frac{\Delta t}{2}\right)\right] \tag{6.88b}$$

$$g_3 = f\left[\left(y_n + g_2\frac{\Delta t}{2}\right), u_{n+1/2}, \left(t_n + \frac{\Delta t}{2}\right)\right]$$ (6.88c)

$$g_4 = f[(y_n + g_3\Delta t), u_{n+1}, t_{n+1}]$$ (6.88d)

Then, the integration step is carried out according to

$$y_{n+1} = y_n + (g_1 + 2g_2 + 2g_3 + g_4)\frac{\Delta t}{6}$$ (6.89)

Note that $u_{n+1/2} = u\left(t_n + \frac{\Delta t}{2}\right)$.

Other sophisticated approaches of digital simulation are available as well. Perhaps the most convenient computer-based approach to simulation of a dynamic model is by using a graphic environment that uses block diagrams. Several such environments are commercially available. One that is widely used is Simulink®, which is an extension to MATLAB®.*

6.7.1 Use of Simulink® in Computer Simulation

Perhaps the most convenient computer-based approach to simulation of a dynamic model is by using a graphic environment that uses block diagrams. Several such environments are commercially available. One that is widely used is Simulink, and is available as an extension to MATLAB®. It provides a graphical environment for modeling, simulating, and analyzing dynamic linear and nonlinear systems. Its use is quite convenient. First a suitable block diagram model of the system is developed on the computer screen, and stored. The Simulink environment provides almost any block that is used in a typical block diagram. These include transfer functions, integrators, gains, summing junctions, inputs (i.e., source blocks) and outputs (i.e., graph blocks or scope blocks). Such a block may be selected and inserted into the workspace as many times as needed, by clicking and dragging using the mouse. These blocks may be converted as required, using directed lines. A block may be opened by clicking on it and the parameter values and text may be inserted or modified as needed. Once the simulation block diagram is generated in this manner, it may be run and the response may be observed through an output block (graph block or scope block). Since Simulink is integrated with MATLAB, data can be easily transferred between programs within various tools and applications.

6.7.1.1 Starting Simulink®

First enter the MATLAB® environment. You will the MATLAB command prompt >>. To start Simulink, enter the command: Simulink. Alternatively, you may click on the "Simulink" button at the top of the MATLAB command window. The Simulink Library Browser window should now appear on the screen. Most of the blocks needed for modeling basic systems can be found in the subfolders of the main Simulink folder.

---

* MATLAB® and Simulink® are properties of The Mathworks, Inc.

6.7.1.2 Basic Elements

There are two types of elements in Simulink®: **blocks** and **lines.** Blocks are used to generate (or input), modify, combine, output, and display signals. Lines are used to transfer signals from one block to another.

Blocks: The subfolders below the Simulink folder show the general classes of blocks available for use. They are

- Continuous: Linear, continuous-time system elements (integrators, transfer functions, state-space models, etc.).
- Discrete: Linear, discrete-time system elements (integrators, transfer functions, state-space models, etc.).
- Functions and tables: User-defined functions and tables for interpolating function values.
- Math: Mathematical operators (sum, gain, dot product, etc.).
- Nonlinear: Nonlinear operators (Coulomb/viscous friction, switches, relays, etc.).
- Signals and systems: Blocks for controlling/monitoring signals and for creating subsystems.
- Sinks: For output or display signals (displays, scopes, graphs, etc.).
- Sources: To generate various types of signals (step, ramp, sinusoidal, etc.).

Blocks may have zero or more input terminals and zero or more output terminals.

Lines: A directed line segment transmits signals in the direction indicated by its arrow. Typically, a line must transmit signals from the output terminal of one block to the input terminal of another block. One exception to this is, a line may be used to tap off the signal from another line. In this manner, the tapped original signal can be sent to other (one or more) destination blocks. However, a line can never inject a signal **into** another line; combining (or, summing) of signals has to be done by using a summing junction. A signal can be either a scalar signal (single signal) or a vector signal (several signals in parallel). The lines used to transmit scalar signals and vector signals are identical; whether it is a scalar or vector is determined by the blocks connected by the line.

6.7.1.3 Building an Application

To build a system for simulation, first bring up a new model window for creating the block diagram. To do this, click on the "New Model" button in the toolbar of the Simulink® Library Browser. Initially the window will be blank. Then, build the system using the following three steps:

1. Gather Blocks

From the Simulink Library Browser, collect the blocks you need in your model. This can be done by simply clicking on a required block and dragging it into your workspace.

2. Modify the Blocks

Simulink allows you to modify the blocks in your model so that they accurately reflect the characteristics of your system. Double-click on the block to be modified. You can modify the parameters of the block in the "Block Parameters" window. Simulink gives a brief explanation of the function of the block in the top portion of this window.

3. Connect the Blocks

The block diagram must accurately reflect the system to be modeled. The selected Simulink blocks have be properly connected by lines, to realize the correct block diagram. Draw the necessary lines for signal paths by dragging the mouse from the starting point of a signal (i.e., output terminal of a block) to the terminating point of the signal (i.e., input terminal of another block). Simulink converts the mouse pointer into a crosshair when it is close to an output terminal, to begin drawing a line, and the pointer will become a double crosshair when it is close enough to be snapped to an input terminal. When drawing a line, the path you follow is not important. The lines will route themselves automatically. The terminals points are what matter. Once the blocks are connected, they can be moved around for neater appearance. A block can be simply clicked dragged to its desired location (the signal lines will remain connected and will re-route themselves).

It may be necessary to branch a signal and transmit it to more than one input terminal. To do this, first placing the mouse cursor at the location where the signal is to be branched (tapped). Then, using either the CTRL key in conjunction with the left mouse button or just the right mouse button, drag the new line to its intended destination.

6.7.1.4 *Running a Simulation*

Once the model is constructed, you are ready to simulate the system. To do this, go to the **Simulation** menu and click on **Start**, or just click on the "Start/Pause Simulation" button in the model window toolbar (this will look like the "Play" button on a VCR). The simulation will be carried out and the necessary signals will be generated.

General tips:

1. You can save your model by selecting **Save** from the file menu and clicking the **OK** button (you should give a name to a file).
2. The results of a simulation can be sent to the MATLAB® window by the use of the **"to workshop"** icon from the Sinks window.
3. Use the **Demux** (i.e., demultiplexing) icon to convert a vector into several scalar lines. The **Mux** icon takes several scalar inputs and multiplexes them into a vector This is useful, for example, when transferring the results from a simulation to the MATLAB workspace).
4. A sign of a **Sum** icon may be changed by double clicking on the icon and changing the sign. The number of inputs to a **Sum** icon may be changed by double clicking on the icon and correctly setting the number of inputs in the window.
5. Be sure to set the integration parameters in the simulation menu. In particular, the default minimum and maximum step sizes must be changed (they should be around 1/100 to 1/10 of the dominant (i.e., slowest) time constant of your system).

Example 6.4

Consider the time domain model given by:

$$\dddot{y} + 13\ddot{y} + 56\dot{y} + 80y = \dddot{u} + 6\ddot{u} + 11\dot{u} + 6u$$

We build the Simulink® model, as given in Figure 6.14a.
The system response to an impulse input is shown in Figure 6.14b.

(a)

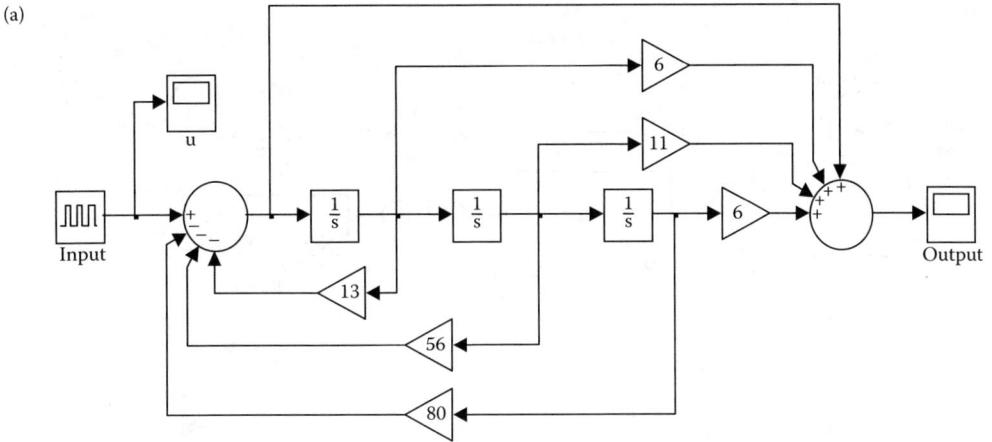

(b)

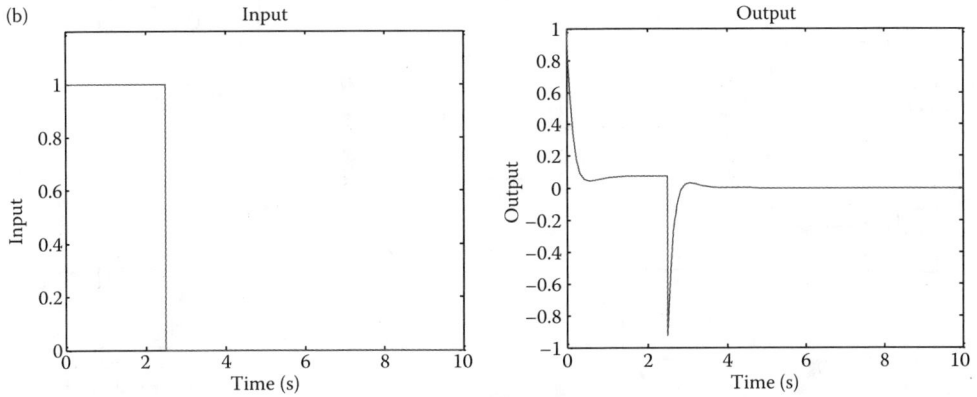

FIGURE 6.14
(a) Simulink® model of the simulation block diagram. (b) System response.

Example 6.5

Consider the model of a robotic sewing system, as studied in Figure 6.15a.

With the state vector $x = [\omega_r \ f_r \ v_h \ f_c \ v_c]^T$; the input vector $u = [T_r \ f_f]^T$; and the output vector $y = [f_c \ \omega_r]^T$, the following state-space model is obtained:

$$\dot{x} = Ax + Bu; \quad y = Cx + Du$$

where

$$A = \begin{bmatrix} -r^2b_r/J_r & -r/J_r & 0 & 0 & 0 \\ rk_r & 0 & -k_r & 0 & 0 \\ 0 & 1/m_h & -(b_c + b_h)/m_h & 1/m_h & b_c/m_h \\ 0 & 0 & -k_c & 0 & k_c \\ 0 & 0 & b_c/m_c & -1/m_c & -b_c/m_c \end{bmatrix}; \quad B = \begin{bmatrix} 1/J_r & 0 \\ 0 & 0 \\ 0 & 0 \\ 0 & 0 \\ 0 & 1/m_c \end{bmatrix}$$

$$C = \begin{bmatrix} 0 & 0 & 0 & 1 & 0 \\ 1 & 0 & 0 & 0 & 0 \end{bmatrix}; \quad D = 0$$

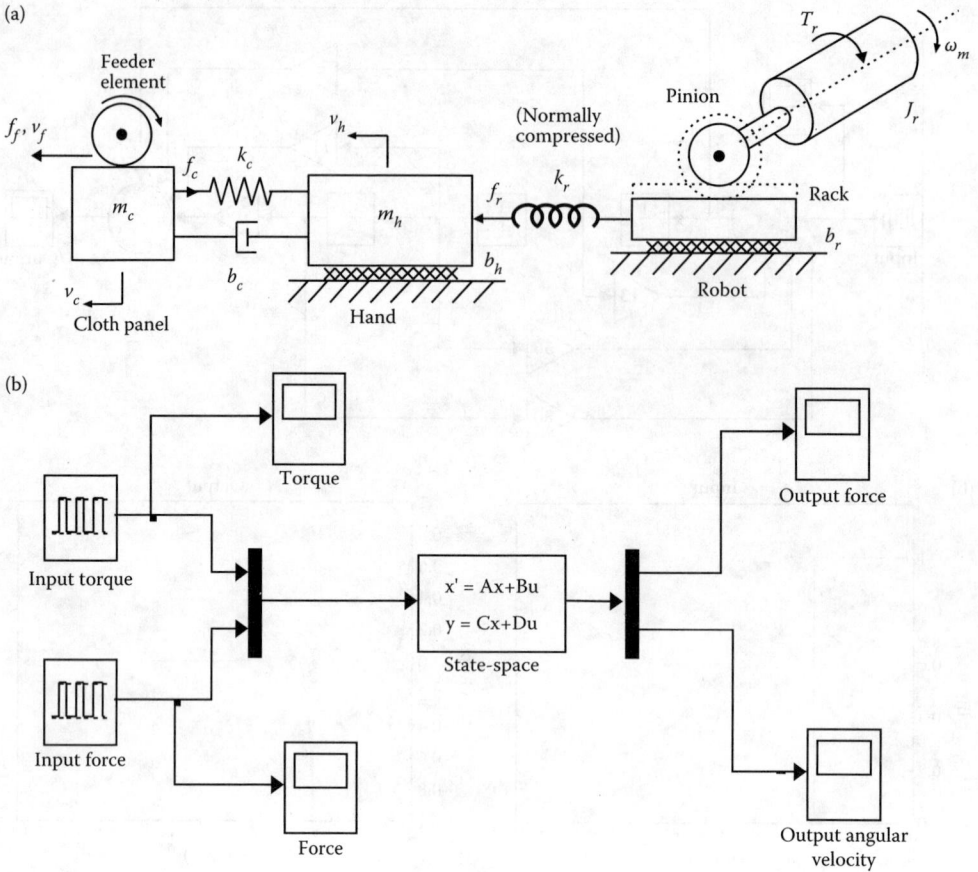

(a)

(b)

FIGURE 6.15
(a) A robotic sewing system. (b) Simulink model of a robotic sewing machine. (c) Simulation results.

To carry out a simulation using Simulink®, we use the following parameter values:

$m_c = 0.6$ kg
$k_c = 100$ N/m
$b_c = 0.3$ N/m/s
$m_h = 1$ kg
$b_h = 1$ N/m/s
$k_r = 200$ N/m
$b_r = 1$ N/m/s
$J_r = 2$ kg.m²
$r = 0.05$ m

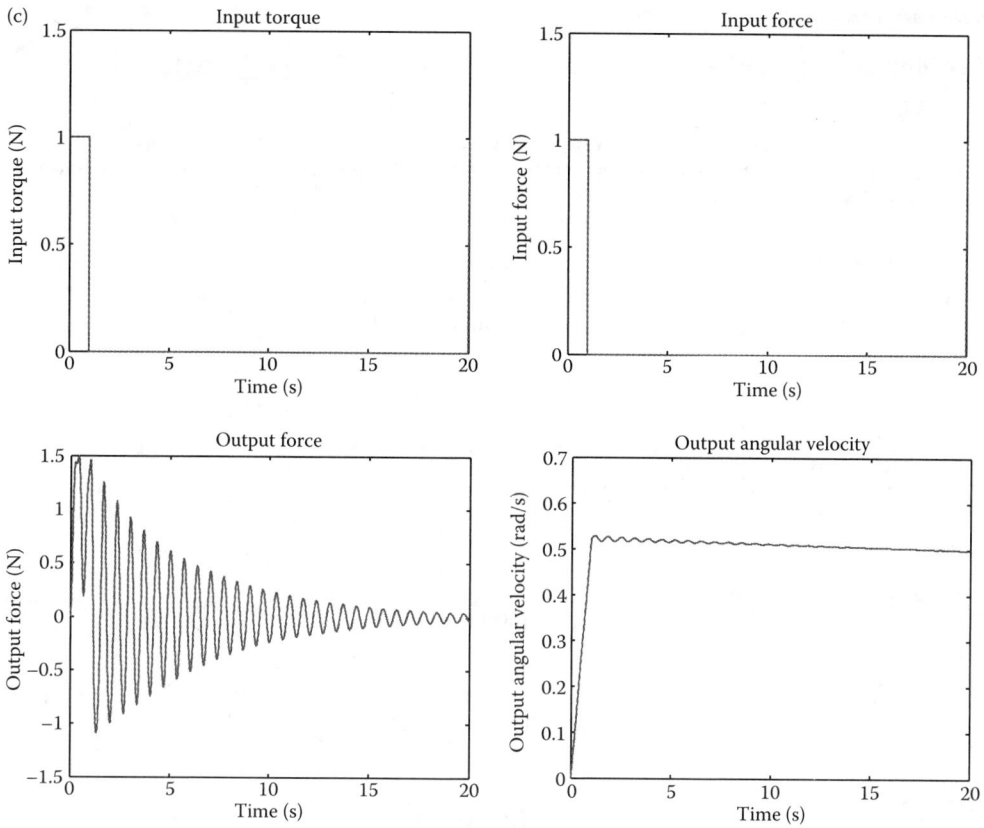

FIGURE 6.15 (continued)

The matrices of the linear model are obtained as:

$$A = \begin{bmatrix} -0.00125 & -0.025 & 0 & 0 & 0 \\ 10 & 0 & -200 & 0 & 0 \\ 0 & 1 & -1.3 & 1 & 0.3 \\ 0 & 0 & -100 & 0 & 100 \\ 0 & 0 & 0.5 & -1.67 & 0.5 \end{bmatrix}; \quad B = \begin{bmatrix} 0.5 & 0 \\ 0 & 0 \\ 0 & 0 \\ 0 & 0 \\ 0 & 1.67 \end{bmatrix};$$

$$C = \begin{bmatrix} 0 & 0 & 0 & 1 & 0 \\ 1 & 0 & 0 & 0 & 0 \end{bmatrix}; \quad D = \begin{bmatrix} 0 & 0 \\ 0 & 0 \end{bmatrix}$$

The Simulink model is built, as shown in Figure 6.15b.
The response of the system to two impulse inputs is shown in Figure 6.15c.

Problems

PROBLEM 6.1

The unit step response of a system, with zero ICs, was found to be $1.5(1-e^{-10t})$. What is the input–output differential equation of the system? What is the transfer function?

PROBLEM 6.2

Discuss why the convolution integrals given below (where u is the input, y is the output, and h is the impulse response function) are all identical.

$$y(t) = \int_0^\infty h(\tau)u(t-\tau)d\tau$$

$$y(t) = \int_{-\infty}^\infty h(t-\tau)u(\tau)d\tau$$

$$y(t) = \int_{-\infty}^\infty h(\tau)u(t-\tau)d\tau$$

$$y(t) = \int_{-\infty}^t h(t-\tau)u(\tau)d\tau$$

$$y(t) = \int_{-\infty}^t h(\tau)u(t-\tau)d\tau$$

$$y(t) = \int_0^t h(t-\tau)u(\tau)d\tau$$

$$y(t) = \int_0^t h(\tau)u(t-\tau)d\tau$$

PROBLEM 6.3

A system at rest is subjected to a unit step input $\mathcal{U}(t)$. Its response is given by

$$y = 2e^{-t}(\cos t - \sin t)\mathcal{U}(t)$$

a. Write the input–output differential equation for the system.
b. What is its transfer function?
c. Determine the damped natural frequency, undamped natural frequency, and the damped ratio.
d. Write the response of the system to a unit impulse and sketch it.

PROBLEM 6.4

Consider the dynamic system given by the transfer function

$$\frac{Y(s)}{U(s)} = \frac{(s+4)}{(s^2 + 3s + 2)}$$

a. Plot the poles and zeros of the systems on the s-plane.
b. Indicate the correct statement among the following:

(i)   The system is stable
(ii)   The system is unstable
(iii) The system stability depends on the input
(iv)   None of the above

c. Obtain the system differential equation.
d. Using the Laplace transfer technique determine the system response $y(t)$ to a unit step input, with zero ICs.

PROBLEM 6.5

A dynamic system is represented by the transfer function

$$\frac{Y(s)}{U(s)} = G(s) = \frac{\omega_n^2}{s^2 + 2\zeta\omega_n s + \omega_n^2}$$

a. Is the system stable?
b. If the system is given an impulse input, at what frequency will it oscillate?
c. If the system is given a unit step input, what is the frequency of the resulting output oscillations? What is its steady-state value?
d. The system is given the sinusoidal input

$$u(t) = a \sin \omega t$$

Determine an expression for the output $y(t)$ at steady-state in terms of a, ω, ω_n, and ζ. At what value of ω will the output $y(t)$ be maximum at steady-state?

PROBLEM 6.6

A system at rest is subjected to a unit step input $\mathcal{U}(t)$. Its response is given by:

$$y = [2e^{-t} \sin t]\mathcal{U}(t)$$

a. Write the input–output differential equation for the system.
b. What is its transfer function?
c. Determine the damped natural frequency, undamped natural frequency, and the damped ratio.
d. Write the response of the system to a unit impulse and find $y(0^+)$.
e. What is the steady-state response for unit step input?

PROBLEM 6.7

a. Define the following terms with reference to the response of a dynamic system:

(i)   Homogeneous solution
(ii)   Particular solution
(iii) Zero-input (or free) response

(iv)   Zero state (or forced) response
(v)   Steady-state response

b. Consider the first-order system

$$\tau \frac{dy}{dt} + y = u(t)$$

in which u is the input, y is the output, and τ is a system constant.

(i)   Suppose that the system is initially at rest with $u=0$ and $y=0$, and suddenly a unit step input is applied. Obtain an expression for the ensuing response of the system. Into which of the above five categories does this response fall? What is the corresponding steady-state response?

(ii)   If the step input in (i) above is of magnitude A what is the corresponding response?

(iii) If the input in (i) above was an impulse of magnitude P what would be the response?

PROBLEM 6.8

An "iron butcher" is a head-cutting machine which is commonly used in the fish processing industry. Millions of dollars worth salmon, is wasted annually due to inaccurate head cutting using these somewhat outdated machines. The main cause of wastage is the "over-feed problem." This occurs when a salmon is inaccurately positioned with respect to the cutter blade so that the cutting location is beyond the collar bone and into the body of a salmon. An effort has been made to correct this situation by sensing the position of the collar bone and automatically positioning the cutter blade accordingly.

A schematic representation of an electromechanical positioning system of a salmon-head cutter is shown in Figure P6.8a. Positioning of the cutter is achieved through a lead screw and nut arrangement, which is driven by a brushless dc motor. The cutter carriage is integral with the nut of the lead screw and the ac motor which derives the cutter blade, and has an overall mass of m (kg). The carriage slides along a lubricated guideway and provides an equivalent viscous damping force of damping constant b (N/m/s). The overall moment of inertia of the motor rotor and the lead screw is J (N/m^2) about the axis of rotation. The motor is driven by a drive system, which provides a voltage v to the stator field windings of the motor. Note that the motor has a permanent magnet rotor. The interaction between the field circuit and, the motor rotor is represented by Figure P6.8b.

The magnetic torque T_m generated by the motor is given by

$$T_m = k_m i_f$$

The force F_L exerted by the lead screw in the y-direction of the cutter carriage is given by

$$F_L = \frac{e}{h} T_{L'}$$

in which

$$h = \frac{\text{Translatory motion of the nut}}{\text{Rotatory motion of lead screw}}$$

and e is the mechanical efficiency of the lead screw-nut unit.

The remaining parameters and variables, as indicated in Figure P6.8, should be self-explanatory.

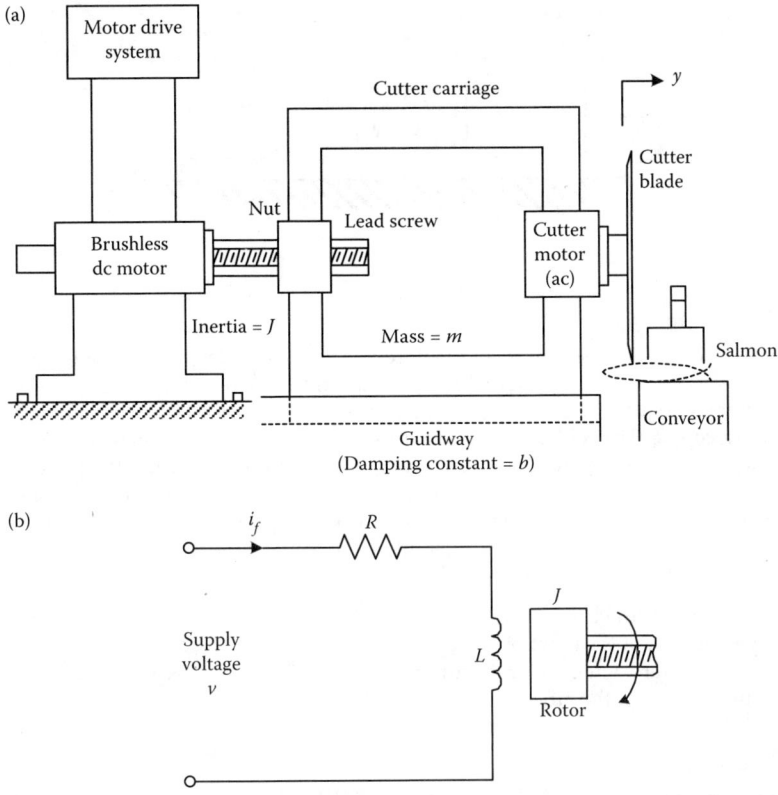

FIGURE P6.8

(a) A positioning system for an automated fish cutting machine. (b) The field circuit of the permanent-magnet rotor dc motor.

a. Write the necessary equations to study the displacement y of the cutter in response to an applied voltage v to the motor. What is the order of the system? Obtain the input–output differential equation for the system and from that determine the characteristic equation. What are the roots (poles or eigenvalues) of the characteristic equation?

b. Using summation junctions, integration blocks, and constant gain blocks only, draw a complete block diagram of the system, with v as the input and y as the output.

c. Obtain a state-space model for the system, using v as the input and y as the output.

d. Assume that L/R ratio is very small and can be neglected. Obtain an expression for the response y of the system to a step input with zero ICs. Show from this expression that the behavior of the system is unstable in the present form (i.e., without feedback control).

PROBLEM 6.9

Consider the two-mass system shown in Figure P6.9.

a. What is the transfer function x_1/f?

b. For a harmonic excitation $f(t)$, at what frequency will m_1 be motionless?

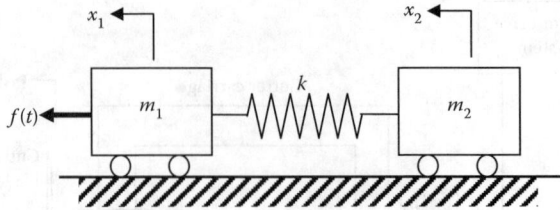

FIGURE P6.9
A two-car train.

PROBLEM 6.10

When two dissimilar metal wires are jointed at the two ends, to form a loop, and one junction is maintained at a different temperature from the other, a voltage is generated between the two junctions. A temperature sensor, which makes use of this property is the thermocouple. The cold junction is maintained at a known temperature (say, by dipping into an ice-water bath). The hot junction is then used to measure the temperature at some location. The temperature of the hot junction (T) does not instantaneously reach that of the sensed location (T_f), in view of the thermal capacitance of the junction. Derive an expression for the thermal time constant of a thermocouple in terms of the following parameters of the hot junction:

> m=mass of the junction
> c=specific heat of the junction
> h=heat transfer coefficient of the junction
> A=surface area of the junction.

PROBLEM 6.11

Consider again Problem 4.9 in Chapter 4 (Figure P4.9).

Defining the time constants $\tau_1 = C_1 R_p$ and $\tau_2 = C_2 R_o$, and the gain parameter $k = R_o/R_p$ express the characteristic equation of the system in terms of these three parameters.

Show that the poles of the system are real and negative but the system is coupled (interacting).

Suppose that the two tanks are as in Figure P4.9b. Here Tank 1 has an outlet valve at its bottom whose resistance is R_t and the volume flow rate is Q_t when open. This flow directly enters Tank 2, without a connecting pipe. The remaining characteristics of the tanks are the same as in (b).

Derive a state-space model for the modified system in terms of the same variables as in (b). With $\tau_1 = C_1 R_t$, $\tau_2 = C_2 R_o$, and $k = R_o/R_t$ obtain the characteristic equation of this system. What are the poles of the system? Show that the modified system is noninteracting.

PROBLEM 6.12

Consider again Problem 4.10 in Chapter 4 (Figure P4.10).

Obtain expressions for the undamped natural frequency and the damping ratio of the linear model, in terms of the parameters a, $\bar{v}$, m, and k. Show that the damping ratio increases with the operating speed.

PROBLEM 6.13

Consider the fluid oscillation problem in Example 2.4 of Chapter 2 (Figure 2.18).

What is the characteristic equation of this system?

Using the following numerical values for the system parameters:

$L_v = 10.0$ m, $\qquad L_h = 4.0$ m, $\qquad d_v = 0.025$ m, $\qquad d_h = 0.02$ m

$\rho = 1000.0$ kg/m³, $\mu = 1.0 \times 10^{-3}$ N.s/m², and tank diameter = 0.5 m

compute the undamped natural frequency ω_n and the damping ratio ζ of the system. Will this system provide an oscillatory natural response? If so what is the corresponding frequency? If not, explain the reasons.

PROBLEM 6.14

The circuit shown in Figure P6.14 consists of an inductor L, a capacitor C, and two resistors R and R_o. The input is the source voltage $v_s(t)$ and the output is the voltage v_o across the resistor R_o.

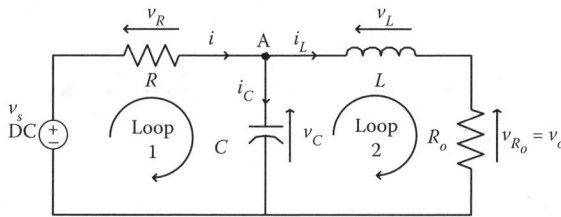

FIGURE P6.14
An electrical circuit with *RLC* elements.

a. Explain why the current i_L through the inductor and the voltage v_C across the capacitor are suitable state variables for this circuit.

b. Using i_L and v_C as the state variables, obtain a complete state-space model for the system. Specifically, express system equations in the vector-matrix form:

$$\dot{x} = Ax + Bu$$

$$y = Cx + Du$$

in the usual notation, where x is the state vector, u is the input vector, and y is the output vector, and determine all the elements of the four matrices A, B, C, and D in terms of the circuit parameters R, R_o, L, and C.

c. The system starts at steady-state with a source voltage of 1 V (for all $t<0$). Then suddenly, the source voltage (i.e., input) is increased to 10 V (for all $t>0$), which corresponds to a step input. For $R=R_o=1$ Ω, $L=1$ H, and $C=1$ F, determine the numerical values of the ICs of the following system variables at both $t=0^-$ and $t=0^+$:

    (i)   Voltage v_L across the inductor
    (ii)   Current i_C through the capacitor
    (iii)   Current i through the resistor R
    (iv)   Current i_L
    (v)   Voltage v_C
    (vi)   Output voltage v_o

Hint: A state variable cannot change its value instantaneously.

PROBLEM 6.15

Consider the linear system with constant coefficients, expressed in the time domain as

$$a_2 \frac{d^2y}{dt^2} + a_1 \frac{dy}{dt} + a_0 y = b_3 \frac{d^3u}{dt^3} + b_2 \frac{d^2u}{dt^2} + b_1 \frac{du}{dt} + b_0 u$$

u = input to the system
y = output of the system
  The coefficients a_i and b_j are constants, and they are the system parameters.
The system may be represented in the block diagram form as in Figure P6.15, with the transfer function $G(s)$.

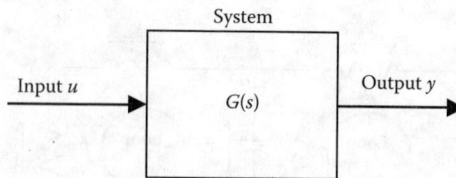

System

Input u → $G(s)$ → Output y

FIGURE P6.15
Block diagram of the system.

a. What is the order of the system? Give reasons.
b. Express the transfer function of the system in terms of the system parameters.
c. What is the characteristic equation of the system? Explain your result.
d. Derive expressions for the poles of the system in terms of the given system parameters.
e. If $a_0 > 0$, $a_1 > 0$, and $a_2 > 0$, discuss the stability of the system.
f. First assume that the coefficients b_3, b_2, and b_1 are zero, and $b_0 = 1$. Then the "forced" (i.e., zero IC) response of this modified system for some input $u(t)$ is denoted by $x(t)$. Now if b_3, b_2, and b_1 are all nonzero, and $b_0 \neq 1$, express in terms of $x(t)$ and the system parameters, the response of this system to the same input $u(t)$ as before (with zero ICs). Clearly indicate the reasons behind your answer.
g. If $b_3 \neq 0$, discuss the "physical realizability" of the system.

  Note: Give all details of your derivations. If you use new parameters or variables or any notation other than what is given in the problem, they have to be defined.

PROBLEM 6.16

a. Answer "true" or "false" for the following:
  The order of a system is equal to

  (i) The number of states in a state-space model of the system.
  (ii) The order of the input–output differential equation of the system.
  (iii) The number of ICs needed to completely determine the time response of the system.
  (iv) The number of independent energy-storage elements in a lumped-parameter model of the system.
  (v) The number of independent energy storage elements and energy dissipation elements in a lumped-parameter model of the system.

b. A fluid pump has an impeller of moment of inertia J and is supported on friction-less bearings. It is driven by a powerful motor at speed ω_m, which may be treated as a velocity source, through a flexible shaft of torsional stiffness K. The fluid load to which the pump impeller is subjected may be approximated by a load torque $c\omega|\omega|$ where ω is the speed of the pump impeller. A schematic diagram of the system is shown in Figure P6.16a and a lumped-parameter model is shown in Figure P6.16b.

Note that the motor speed ω_m is the input to the system. Treat the speed ω of the pump impeller as the output of the system.

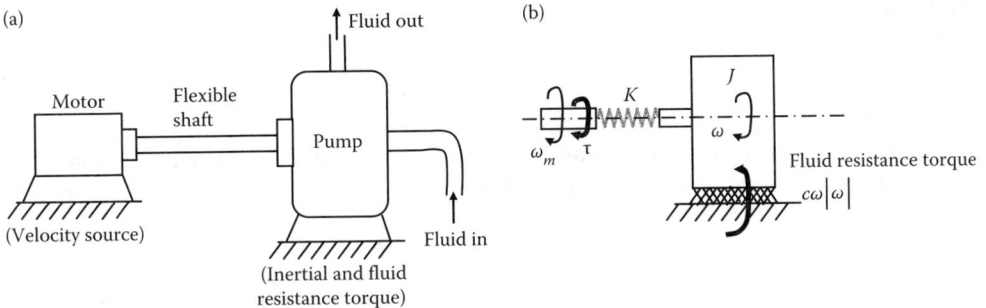

FIGURE P6.16
(a) A pump driven by a powerful motor; (b) A lumped-parameter model.

(i)   Using the torque τ in the drive shaft and the speed ω of the pump as the state variables develop a complete (nonlinear) state-space model of the system.
(ii) What is the order of the system?
(iii) Under steady operating conditions, with constant input ω_m (when the rates of changes of the state variables can be neglected) determine expressions for the operating speed ω_o of the pump and the operating torque τ_o of the drive shaft, in terms of the given quantities (e.g., ω_m, K, J, c).
(iv) Linearize the state-space model about the steady operating conditions in (iii), using the incremental state variables $\hat{\tau}$ and $\hat{\omega}$, and the incremental input variable $\hat{\omega}_m$.
(v)   From the linearized state-space model, obtain a linear input output differential equation (in terms of the incremental input $\hat{\omega}_m$ and incremental output $\hat{\omega}$).
(vi) Obtain expressions for the undamped natural frequency and the damping ratio of the linearized system, in terms of the parameters ω_o, K, J, c.

PROBLEM 6.17

Consider the simple oscillator shown in Figure P6.17, with parameters $m = 4$ kg, $k = 1.6 \times 10^3$ N/m, and the two cases of damping:

1. $b = 80$ N/m/s
2. $b = 320$ N/m/s

Using MATLAB® determine the free response in each case for an IC excitation.

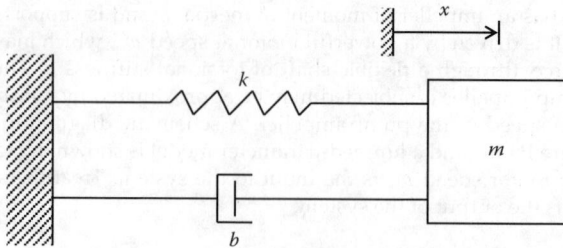

FIGURE P6.17
A damped simple oscillator.

PROBLEM 6.18

Consider the following equation of motion of the single-degree-of-freedom system (damped simple oscillator) shown in Figure P6.17:

$$\ddot{y} + 2\zeta\omega_n\dot{y} + \omega_n^2 y = \omega_n^2 u(t)$$

With an undamped natural frequency of $\omega_n = 10$ rad/s, the step responses may be conveniently determined using Simulink® for the following cases of damping ratio ζ: 0, 0.3, 0.5, 1.0, 2.0.

In particular, the block diagram model for the simulation can be formed as shown in Figure P6.18a, where each case of damping is simulated using the sub-model in Figure P6.18b. Obtain the step for these five cases of damping.

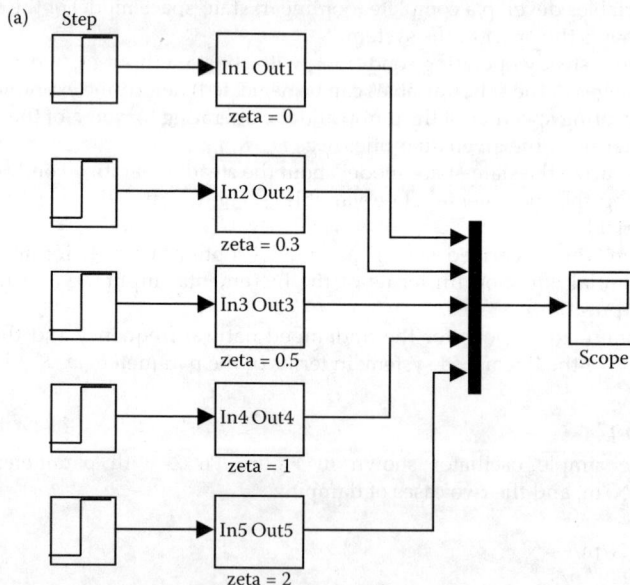

FIGURE P6.18
Use of Simulink to obtain the step response of a simple oscillator. (a) Overall Simulink model. (b) Simulink sub-model for each case of damping.

(b)

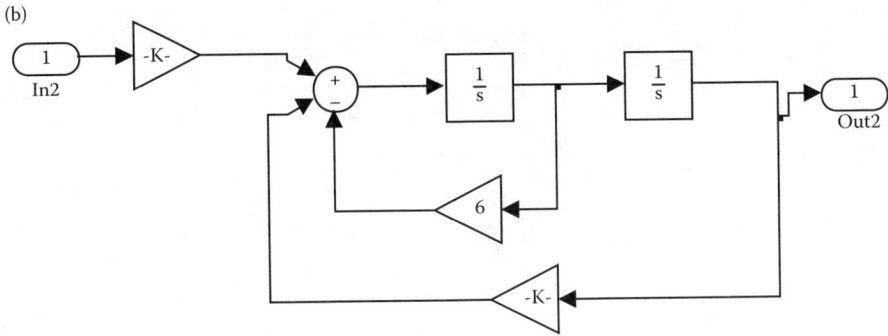

FIGURE P6.18 (continued)

7

Control System Structure and Performance

The purpose of control is to make a plant (i.e., the "dynamic" system to be controlled) behave in a desired manner, according to some *performance specifications.* The overall system that includes at least the plant and the controller is called the control system. The control problem can become challenging due to such reasons as:

- Complex system (many inputs and many outputs, dynamic coupling, nonlinear, etc.)
- Rigorous performance specifications
- Unknown excitations (unknown inputs/disturbances/noise)
- Unknown dynamics (incompletely known plant)

A good control system should satisfy performance requirements concerning such attributes as: *accuracy, stability, speed of response* or *bandwidth, sensitivity,* and *robustness.* This chapter will introduce common architectures of control systems and will present methods of specifying and analyzing the performance of a control system. It is advised to study the concepts of transfer functions and block diagrams as presented in Chapter 5 before proceeding further.

7.1 Control System Structure

A schematic diagram of a control system is shown in Figure 7.1. The physical dynamic system (e.g., a mechanical system) whose response (e.g., motion, voltage, temperature, flow rate) needs to be controlled is called the *plant* or *process.* The device that generates the signal (or, command) according to some scheme (or, control law) and controls the response of the plant, is called the *controller.* The plant and the controller are the two essential components of a control system. Certain *command signals,* or inputs, are applied to the controller and the plant is expected to behave in a desirable manner, under control. In a *feedback control system,* as shown in Figure 7.1, the plant has to be monitored and its response needs to be measured using *sensors* and *transducers,* for feeding back into the controller. Then, the controller compares the sensed signal with a desired response as specified externally, and uses the error to generate a proper control signal.

In the feedback control system in Figure 7.1, the control loop has to be closed, making measurements of the system response and employing that information to generate control signals so as to correct any output errors. Hence, feedback control is also known as *closed-loop control.* In *digital control,* a digital computer serves as the controller. Virtually any control law may be programmed into the control computer.

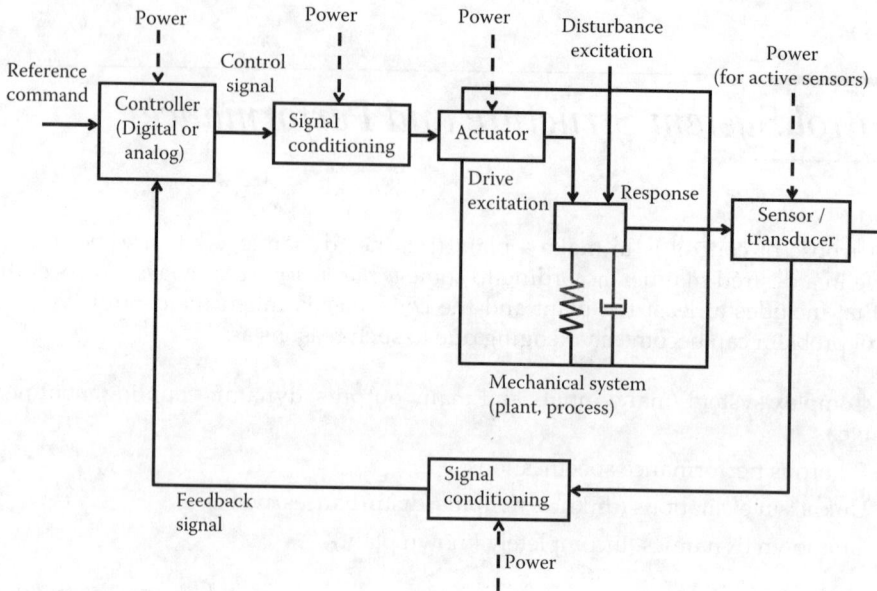

FIGURE 7.1
Schematic diagram of a feedback control system.

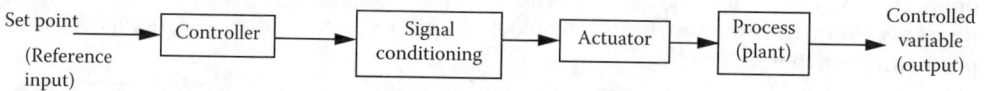

FIGURE 7.2
An open-loop control system.

If the plant is stable and is completely and accurately known, and if the inputs to the plant can be precisely generated (by the controller) and applied, then accurate control might be possible even without feedback control. Under these circumstances a measurement system is not needed (or at least not needed for feedback) and thus we have an *open-loop control* system. In open-loop control, we do not use current information on *system response* to determine the control signals. In other words, there is no feedback. The structure of an open-loop control system is shown in Figure 7.2. Note that a sensor is not explicitly indicated in this open-loop architecture. However, sensors may be employed within an open-loop system to monitor the applied input, the resulting response, and possible disturbance inputs even though feedback control is not used.

Implicit here is the significance of sensors and actuators for a control system. This importance holds regardless of the specific control system architecture that is implemented in a given application. We will now outline several other architectures of control system implementation.

7.1.2 Feedforward Control

Many control systems have inputs that do not participate in feedback control. In other words, these inputs are not compared with feedback (measurement) signals to generate control signals. Some of these inputs might be important variables in the plant (process) itself. Others might be undesirable inputs, such as external disturbances and noise, which are unwanted yet unavoidable. Generally, the performance of a control system can be improved by measuring these (unknown) inputs and somehow using the information to generate control signals.

In *feedforward control*, unknown "inputs" are measured and that information, along with desired inputs, is used to generate control signals that can reduce errors due to these unknown inputs or variations in them. The reason for calling this method feedforward control stems from the fact that the associated measurement and control (and compensation) take place in the forward path of the control system. Note that in feedback control, unknown "outputs" are measured and compared with known (desired) inputs to generate control signals. Both feedback and feedforward schemes may be used in the same control system.

A block diagram of a typical control system that uses feedforward control is shown in Figure 7.3. In this system, in addition to feedback control, a feedforward control scheme is used to reduce the effects of a disturbance input that enters the plant. The disturbance input is measured and fed into the controller. The controller uses this information to modify the control action so as to compensate for the disturbance input, "anticipating" its effect.

As a practical example, consider the natural gas home heating system shown in Figure 7.4a. A simplified block diagram of the system is shown in Figure 7.4b. In conventional feedback control, the room temperature is measured and its deviation from the desired temperature (set point) is used to adjust the natural gas flow into the furnace. On–off control through a thermostat, is used in most such applications. Even if proportional or three-mode (proportional-integral-derivative [PID]) control is employed, it is not easy to steadily maintain the room temperature at the desired value if there are large changes in other (unknown) inputs to the system, such as water flow rate through the furnace, temperature of water entering the furnace, and outdoor temperature. Better results can be obtained by measuring these disturbance inputs and using that information in generating the control action. This is feedforward control. Note that in the absence of feedforward control, any changes in the inputs w_1, w_2, and w_3 in Figure 7.4b would be detected only through their

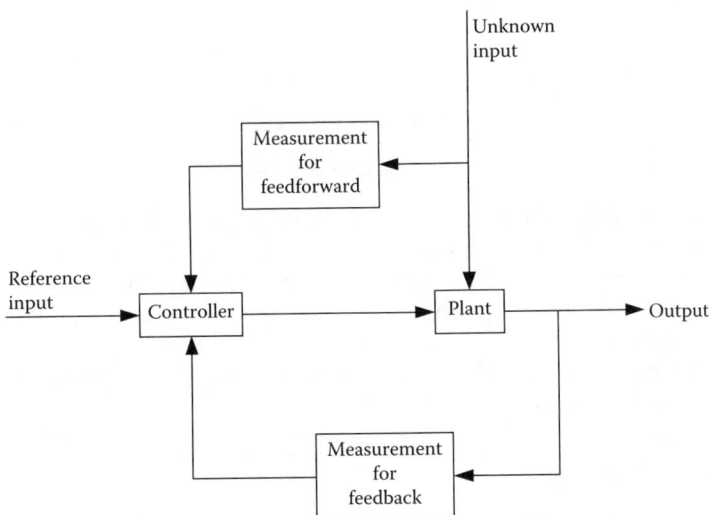

FIGURE 7.3
A system with feedback and feedforward control.

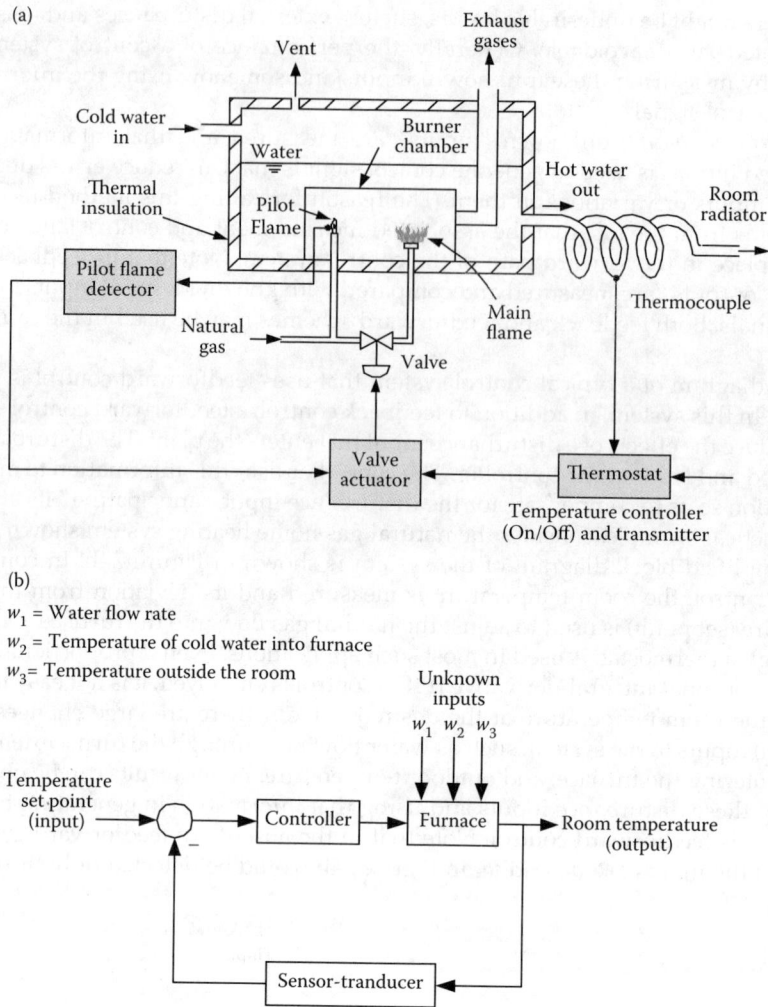

(a)

(b)

w_1 = Water flow rate
w_2 = Temperature of cold water into furnace
w_3 = Temperature outside the room

FIGURE 7.4
(a) A natural gas home heating system. (b) A block diagram representation of the system.

effect on the feedback signal (room temperature). Hence, the subsequent corrective action can considerably lag behind the cause (Note: the cause is the change in w_i). This delay will lead to large errors and possible instability problems. With feedforward control, information on the disturbance input w_i will be available to the controller immediately, and its effect on the system response can be anticipated, thereby speeding up the control action and also improving the response accuracy. Faster action and improved accuracy are two very desirable effects of feedforward control.

7.1.2.1 Computed-Input Control

In some applications, control inputs are computed by using the desired outputs and accurate dynamic models for the plants, and the computed inputs are used for control purposes. This is the *inverse model* (or *inverse dynamics*) approach because the input is computed using the output and a model (inverse model). This is a popular way for controlling robotic manipula-tors, for example. In some literature this method is also known as feedforward control. To avoid confusion, however, it is appropriate to denote this method as *computed-input control*.

Example 7.1

Consider the system shown by the block diagram in Figure 7.5a. Note:

$G_p(s)$=plant transfer function
$G_c(s)$=controller transfer function
$H(s)$=feedback transfer function
$G_f(s)$=feedforward compensation transfer function.

The disturbance input w is measured, compensated using G_f, and fed into the controller, along with the driving input u.

a. In the absence of the disturbance input w, obtain the transfer function relationship between the output y and the driving input u.
b. In the absence of the driving input u, obtain the transfer relationship between y and w.
c. From (a) and (b), write an expression for y in terms of u and w.
d. Show that the effect of disturbance is fully compensated if the feedforward compensator is given by

$$G_f(s) = \frac{1}{G_c(s)}$$

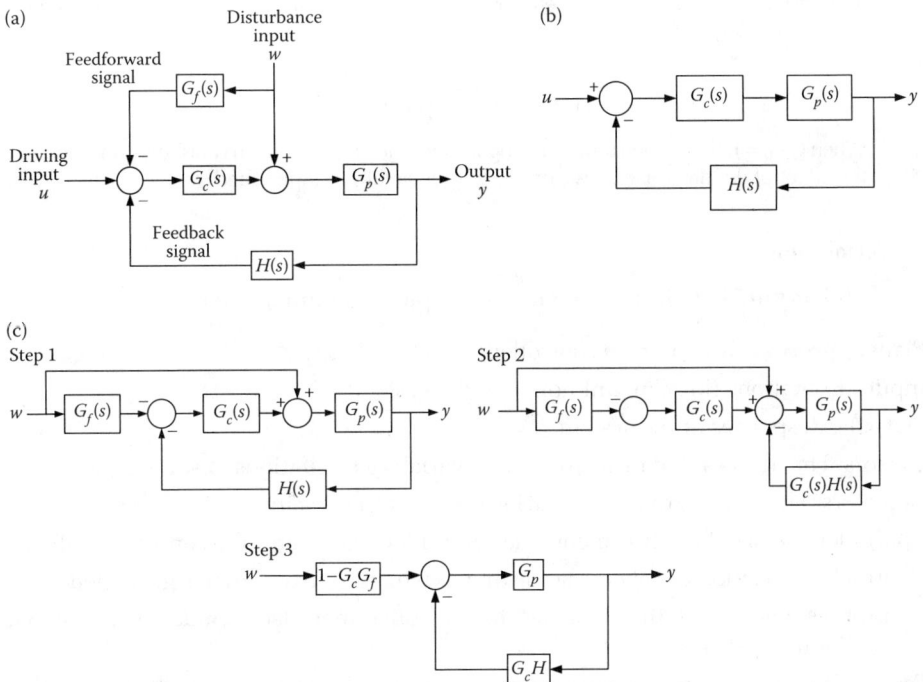

FIGURE 7.5
(a) A block diagram for a system with feedforward control. (b) The block diagram when the disturbance is removed. (c) Steps of block diagram reduction in the absence of the driving input.

Solution

a. When $w=0$, we have the system shown in Figure 7.5b.

Then, from the usual block diagram analysis (see Chapter 5, Table 5.7), we have the transfer function

$$\frac{y}{u} = \frac{G_c G_p}{1 + G_c G_p H} \tag{i}$$

b. When $u=0$, the block diagram may be reduced using the concepts presented in Chapter 5 (see Table 5.7), in four steps as shown in Figure 7.5c. In Step 1, the block diagram is redrawn after removing u. In Step 2, the feedback point is moved forward by a block and as a result the feedback block is multiplied by the transfer function of the forward block (see Table 5.7). In Step 3 we have used the following facts:
   (1) In a serial path the transfer functions multiply.
   (2) In parallel path the transfer functions add.

Also note how the sign of the incoming signal at the first summing junction is taken care of in the reduced diagram. Finally, in Step 4, the feedback loop is represented by a single block (see Table 5.7).
From the last step, we have the transfer function

$$\frac{y}{w} = \frac{\left(1 - G_f G_c\right) G_p}{1 + G_p G_c H} \tag{ii}$$

c. Since the system is linear, the *principle of superposition* applies. Accordingly, the overall transfer function relation when both inputs u and w are present, is given by adding Equations (i) and (ii):

$$y = \frac{G_c G_p}{\left(1 + G_c G_p H\right)} u + \frac{\left(1 - G_f G_c\right) G_p}{1 + G_p G_c H} w \tag{iii}$$

d. When $G_f G_c = 1$, the second term of the right hand side of Equation (iii) vanishes. Consequently, the effect of the disturbance (w), on the output, is fully compensated.

7.1.3 Terminology

Some useful terminology introduced in this chapter is summarized below.

Plant or process: System to be controlled.

Inputs: Excitations (known, unknown) to the system.

Outputs: Responses of the system.

Sensors: The devices that measure system variables (excitations, responses, etc.).

Actuators: The devices that drive various parts of the system.

Controller: Device that implements the control law (generates the control signal).

Control law: Relation or scheme according to which control signal is generated.

Control system: At least the plant and the controller (may also include sensors, signal conditioning, etc.).

Feedback control: Plant response is measured and fed back into the controller. Control signal is determined according to error (between the desired and actual responses).

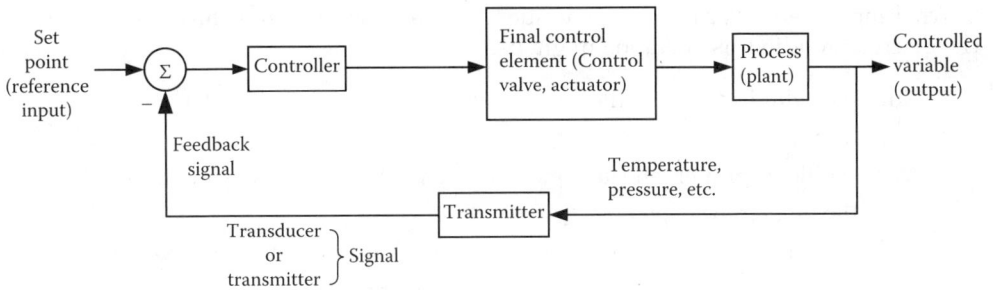

FIGURE 7.6
A feedback system in the process control practice.

Closed-loop control: Same as feedback control. There is a feedback loop (closed-loop).

Open-loop control: Plant response is not used to determine the control action.

Feedforward control: Control signal is determined according to plant "excitation."

The terminology that is particular to the process control practice (e.g., chemical processes) is indicated in the feedback control system shown in Figure 7.6. In particular we note:

Actuator+valve together corresponds to the "actuator" in the conventional control terminology.

Control actuators: Torque motors in servovalves, etc.

Final control element corresponds to an actuator (typically a control valve).

Transmitter transmits the sensed signal to the controller (this can be integrated into sensor/transducer).

7.1.4 Programmable Logic Controllers (PLCs)

A PLC is essentially a digital-computer-like system that can properly sequence a complex task, consisting of many discrete operations and involving several devices, which needs to be carried out in a sequential manner. PLCs are rugged computers typically used in factories and process plants, to connect input devices such as switches to output devices such as valves, at high speed at appropriate times in a task, as governed by a program. Internally, a PLC performs basic computer functions such as logic, sequencing, timing, and counting. It can carry out simpler computations and control tasks such as PID control. Such control operations are called *continuous-state control*, where process variables are continuously monitored and made to stay very close to desired values. There is another important class of controls, known as *discrete-state control*, where the control objective is for the process to follow a required sequence of states (or steps). In each state, however, some form of continuous-state control might be operated, but it is not quite relevant to the discrete-state control task. PLCs are particularly intended for accomplishing discrete-state control tasks.

There are many control systems and industrial tasks that involve the execution of a sequence of steps, depending on the state of some elements in the system and on some external input states. For example, consider an operation of turbine blade manufacture. The discrete steps in this operation might be:

1. Move the cylindrical steel billets into furnace.
2. Heat the billets.
3. When a billet is properly heated, move it to the forging machine and fixture it.
4. Forge the billet into shape.
5. Perform surface finishing operations to get the required aerofoil shape.
6. When the surface finish is satisfactory, machine the blade root.

Note that the entire task involves a sequence of events where each event depends on the completion of the previous event. In addition, it may be necessary for each event to start and end at specified time instants. Such *time sequencing* would be important for coordinating the operation with other activities, and perhaps for proper execution of each operation step. For example, activities of the parts handling robot have to be coordinated with the schedules of the forging machine and milling machine. Furthermore, the billets will have to be heated for a specified time, and machining operation cannot be rushed without compromising product quality, tool failure rate, safety, etc. Note that the task of each step in the discrete sequence might be carried out under continuous-state control. For example, the milling machine would operate using several direct digital control (DDC) loops (say, PID control loops), but discrete-state control is not concerned with this except for the starting point and the end point of each task.

A process operation might consist of a set of two-state (on–off) actions. A PLC can handle the sequencing of these actions in a proper order and at correct times. Examples of such tasks include sequencing the production line operations, starting a complex process plant, and activating the local controllers in a distributed control environment. In the early days of industrial control solenoid-operated electromechanical relays, mechanical timers, and drum controllers were used to sequence such operations. An advantage of using a PLC is that the devices in a plant can be permanently wired, and the plant operation can be modified or restructured by software means (by properly programming the PLC) without requiring hardware modifications and reconnection.

A PLC operates according to some "logic" sequence programmed into it. Connected to a PLC are a set of input devices (e.g., pushbuttons, limit switches, and analog sensors such as RTD temperature sensors, diaphragm-type pressure sensors, piezoelectric accelerometers, and strain-gauge load sensors) and a set of output devices (e.g., actuators such as dc motors, solenoids, and hydraulic rams, warning signal indicators such as lights, alphanumeric LED displays and bells, valves, and continuous control elements such as PID controllers). Each such device is assumed to be a two-state device (taking the logical value 0 or 1). Now, depending on the condition of each input device and according to the programmed-in logic, the PLC will activate the proper state (e.g., on or off) of each output device. Hence, the PLC performs a switching function. Unlike the older generation of sequencing controllers, in the case of a PLC, the logic that determines the state of each output device is processed using software, and not by hardware elements such as hardware relays. Hardware switching takes place at the output port, however, for turning on or off the output devices are controlled by the PLC.

7.1.4.1 PLC Hardware

As noted before, a PLC is a digital computer that is dedicated to perform discrete-state control tasks. A typical PLC consists of a microprocessor, RAM, and ROM memory units, and interface hardware, all interconnected through a suitable bus structure. In addition, there will be a keyboard, a display screen, and other common peripherals. A basic PLC system can be expanded by adding expansion modules (memory, input–output modules, etc.) into the system rack.

A PLC can be programmed using a keyboard or touch-screen. An already developed program could be transferred into the PLC memory from another computer or a peripheral mass-storage medium such as a hard disk. The primary function of a PLC is to switch (energize or de-energize) the output devices connected to it, in a proper sequence, depending on the states of the input devices and according to the logic dictated by the program. A schematic representation of a PLC is shown in Figure 7.7. Note the sensors and actuators in the PLC.

In addition to turning on and off the discrete output components in a correct sequence at proper times, a PLC can perform other useful operations. In particular, it can perform simple arithmetic operations such as addition, subtraction, multiplication, and division on input data. It is also capable of performing counting and timing operations, usually as part of its normal functional requirements. Conversion between binary and binary-coded decimal (BCD) might be required for displaying digits on an LED panel, and for interfacing the PLC with other digital hardware (e.g., digital input devices and digital output devices). For example, a PLC can be programmed to make a temperature measurement and a load measurement, display them on an LED panel, make some computations on these (input) values, and provide a warning signal (output) depending on the result.

The capabilities of a PLC can be determined by such parameters as the number of input devices (e.g., 16) and the number of output devices (e.g., 12) which it can handle, the number of program steps (e.g., 2000), and the speed at which a program can be executed (e.g., 1 M steps/s). Other factors such as the size and the nature of memory and the nature

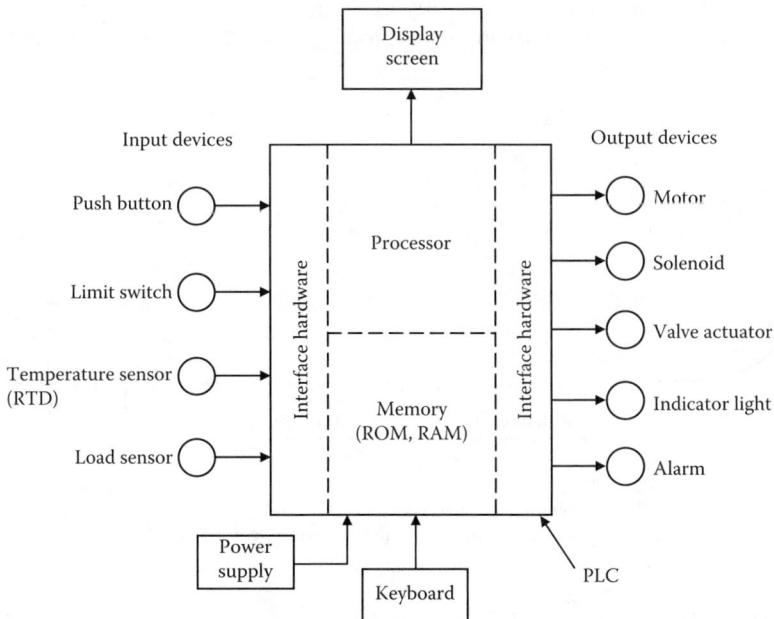

FIGURE 7.7
Schematic representation of a PLC.

of timers and counters in the PLC, signal voltage levels, and choices of outputs, are all important factors.

7.1.5 Distributed Control

For complex processes with a large number of input–output variables (e.g., a chemical plant, a nuclear power plant) and with systems that have various and stringent operating requirements (e.g., the space shuttle), centralized DDC is quite difficult to implement. Some form of distributed control is appropriate in large systems such as manufacturing workcells, factories, and multicomponent process plants. A distributed control system (DCS) will have many users who would need to use the resources simultaneously and, perhaps, would also wish to communicate with each other. Also, the plant will need access to shared and public resources and means of remote monitoring and supervision. Furthermore, different types of devices from a variety of suppliers with different specifications, data types and levels may have to be interconnected. A communication network with switching nodes and multiple routes is needed for this purpose.

In order to achieve connectivity between different types of devices having different origins, it is desirable to use a standardized bus that is supported by all major suppliers of the needed devices. The Foundation Fieldbus or Industrial Ethernet may be adopted for this purpose. Fieldbus is a standardized bus for a plant, which may consist of an interconnected system of devices. It provides connectivity between different types of devices having different origins. Also, it provides access to shared and public resources. Furthermore, it can provide means of remote monitoring and supervision.

A suitable architecture for networking an industrial plant is shown in Figure 7.8. The industrial plant in this case consists of many "process devices" (PD), one or more PLCs and a DCS or a supervisory controller. The PDs will have direct input–output with their own components while possessing connectivity through the plant network. Similarly, a PLC may have direct connectivity with a group of devices as well as networked connectivity with other devices. The DCS will supervise, manage, coordinate, and control the overall plant.

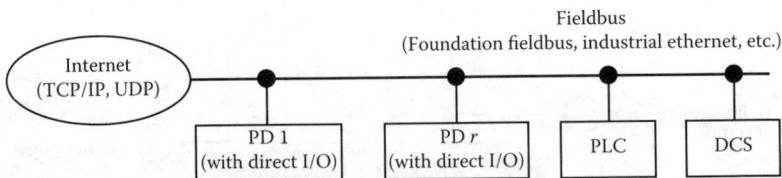

PD = Process device
PLC = Programmable logic controller
DCS = Distributed control system (Supervisory controller)

FIGURE 7.8
A networked industrial plant.

7.1.5.1 A Networked Application

A machine which has been developed by us for head removal of salmon is shown in Figure 7.9a. The system architecture of the machine is sketched in Figure 7.9b. The conveyor, driven by an ac motor, indexes the fish in an intermittent manner. Image of each fish, obtained using a digital charge-coupled device (CCD) camera, is processed to

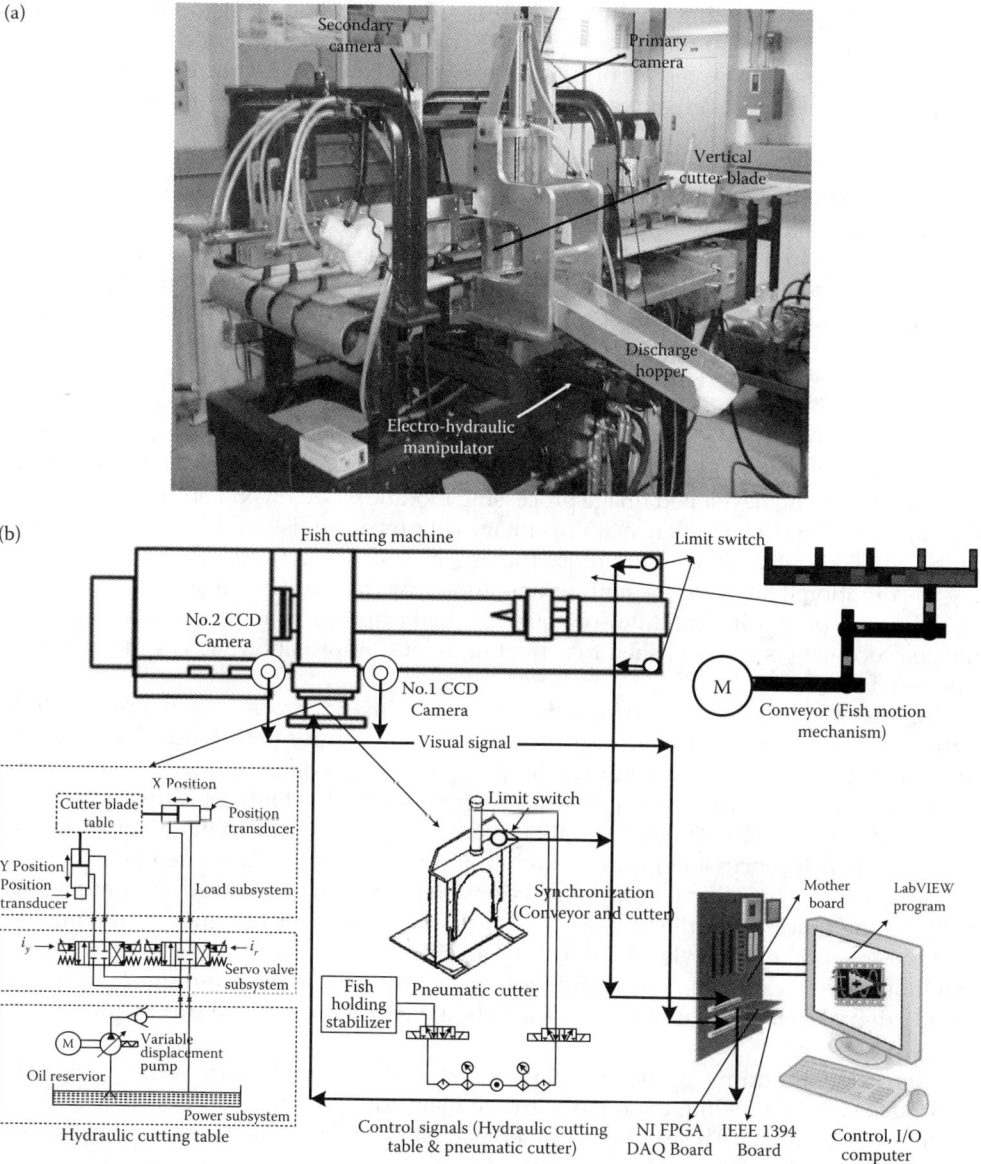

FIGURE 7.9
(a) An intelligent iron butcher. (b) System architecture of the fish processing machine.

determine the geometric features, which in turn establish the proper cutting location. A two-axis hydraulic drive unit positions the cutter accordingly, and the cutting blade is operated using a pneumatic actuator. Position sensing of the hydraulic manipulator is done using linear magnetostrictive displacement transducers, which have a resolution of 0.025 mm when used with a 12-bit analog-to-digital converter. A set of six gage-pressure transducers are installed to measure the fluid pressure in the head and rod sides of each hydraulic cylinder, and also in the supply lines. A high-level imaging system determines the cutting quality, according to which adjustments may be made on-line, to the parameters of the control system so as to improve the process performance. The control system has a hierarchical structure with conventional direct control at the component-level (low level) and an intelligent monitoring and supervisory control system at an upper level.

The primary vision module of the machine is responsible for fast and accurate detection of the gill position of a fish on the basis of an image of the fish as captured by the primary CCD camera. This module is located in the machine host and comprised of a CCD camera, an IEEE 1394 board for image grabbing, a trigger switch for detecting a fish on the conveyor, and an National Instruments (NI) FPGA data acquisition (DAQ) board for analog and digital data communication between the control computer and the electro-hydraulic manipulator. The secondary vision module is responsible for acquisition and processing of visual information pertaining to the quality of the processed fish that leaves the cutter assembly. This module functions as an intelligent sensor in providing high-level information feedback into the control module of the software. The hardware and the software associated with this module are a CCD camera at the exit end for grabbing images of processed fish, and developed image processing module based on NI LabVIEW® for visual data analysis. The CCD camera acquires images of processed fish under the direct control of the host computer, which determines the proper instance to trigger the camera by timing the duration it takes for the cutting operation to complete. The image is then grabbed in the image processing module software for further processing. In this case, however, image processing is accomplished to extract high-level information such as the quality of processed fish.

With the objective of monitoring and control of industrial processes from remote locations, we have developed a universal network architecture, both hardware and software. The developed infrastructure is designed to perform optimally with Fast Ethernet (100Base-T) backbone where each network device only needs a low cost Network Interface Card (NIC). Figure 7.10 shows a simplified hardware architecture, which networks two machines (a fish processing machine and an industrial robot). Each machine is directly connected to its individual control server, which handles networked communication between the process and the web-server, DAQ, sending of control signals to the process, and the execution of low level control laws. The control server of the fish-processing machine contains one or more DAQ boards, which have analog-to-digital conversion (ADC), digital-to-analog conversion (DAC), digital input–output, and frame grabbers for image processing.

Video cameras and microphones are placed at strategic locations to capture live audio and video signals allowing the remote user to view and listen to a process facility, and to communicate with local research personnel. The camera selected in the present application is the Panasonic Model KXDP702 color camera with built-in pan, tilt and 21x zoom (PTZ), which can be controlled through a standard RS-232C communication protocol. Multiple cameras can be connected in daisy-chained manner, to the video-streaming server. For capturing and encoding the audio–video (AV) feed from the camera, the Winnov Videum

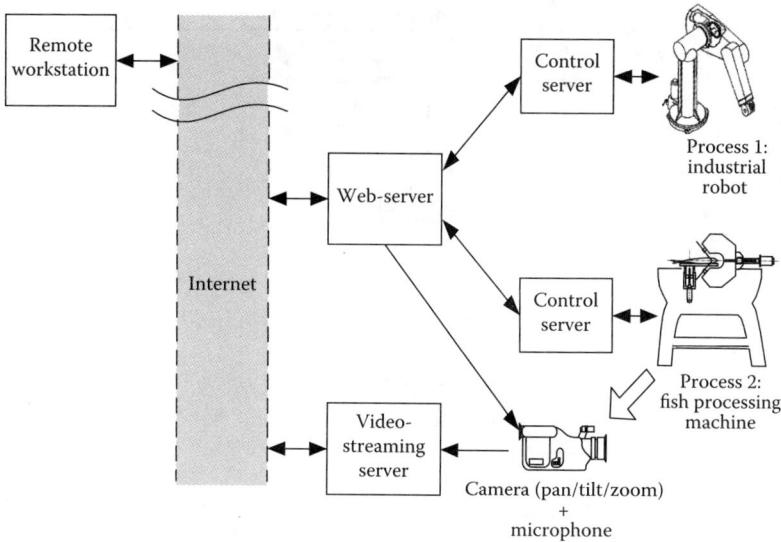

FIGURE 7.10
Network hardware architecture.

1000 PCI board is installed in the video-streaming server. It can capture video signals at a maximum resolution of 640×480 at 30 fps, with a hardware compression that significantly reduces computational overheads of the video-streaming server. Each of the AV capture boards can only support one AV input. Hence multiple boards have to be installed.

7.1.6 Hierarchical Control

A favorite distributed control architecture is provided by hierarchical control. Here, distribution of control is available both geographically and functionally. A hierarchical structure can facilitate efficient control and communication in a complex control system. An example for a three-level hierarchy is shown in Figure 7.11. Management decisions, supervisory control, and coordination between plants in the overall facility are provided by the supervisory control computer, which is at the highest level (level 3) of the hierarchy. The next lower level (intermediate level) generates control settings (or reference inputs) for each control region (subsystem) in the corresponding plant. Set points and reference signals are inputs to the DDCs, which control each control region. The computers in the hierarchical system communicate using a suitable communication network. Information transfer in both directions (up and down) should be possible for best performance and flexibility. In master-slave distributed control, only downloading of information is available.

As another illustration, a three-level hierarchy of an intelligent mechatronic (electromechanical) system (IMS) is shown in Figure 7.12. The bottom level consists of electromechanical components with component-level sensing. Furthermore, actuation and direct feedback control are carried out at this level. The intermediate level uses intelligent preprocessors for abstraction of the information generated by the component-level sensors. The sensors and their intelligent preprocessors together perform tasks of intelligent sensing. State of performance of the system components may be evaluated by this means,

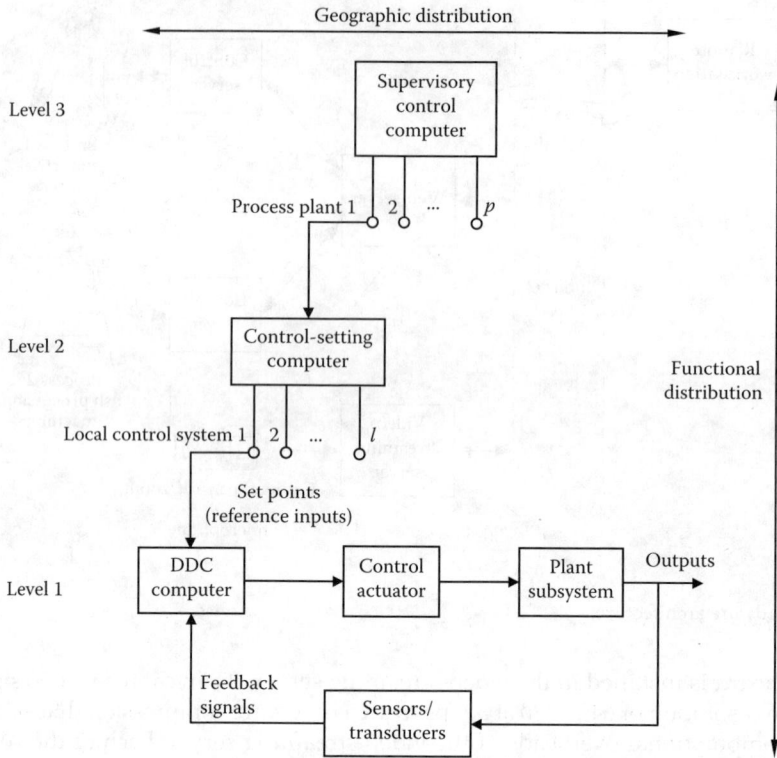

Geographic distribution

Level 3

Supervisory
control
computer

Process plant 1 2 ... *p*

Level 2

Control-setting
computer

Functional
distribution

Local control system 1 2 ... *l*

Set points
(reference inputs)

Level 1

DDC
computer

Control
actuator

Plant
subsystem

Outputs

Feedback
signals

Sensors/
transducers

FIGURE 7.11
A three-layer hierarchical control system.

and component tuning and component-group control may be carried out as a result. The top level of the hierarchy performs task-level activities including planning, scheduling, monitoring of the system performance, and overall supervisory control. Resources such as materials and expertise may be provided at this level and a human-machine interface would be available. Knowledge-based decision-making is carried out at both intermediate and top levels. The resolution of the information that is involved will generally decrease as the hierarchical level increases, while the level of "intelligence" that would be needed in decision-making will increase.

Within the overall system, the communication protocol provides a standard interface between various components such as sensors, actuators, signal conditioners, and controllers, and also with the system environment. The protocol will not only allow highly flexible implementations, but will also enable the system to use distributed intelligence to perform preprocessing and information understanding. The communication protocol should be based on an application-level standard. In essence, it should outline what components can communicate with each other and with the environment, without defining the physical data link and network levels. The communication protocol should allow for different component types and different data abstractions to be interchanged within the same framework. It should also allow for information from geographically removed locations to be communicated to the control and communication system of the IMS.

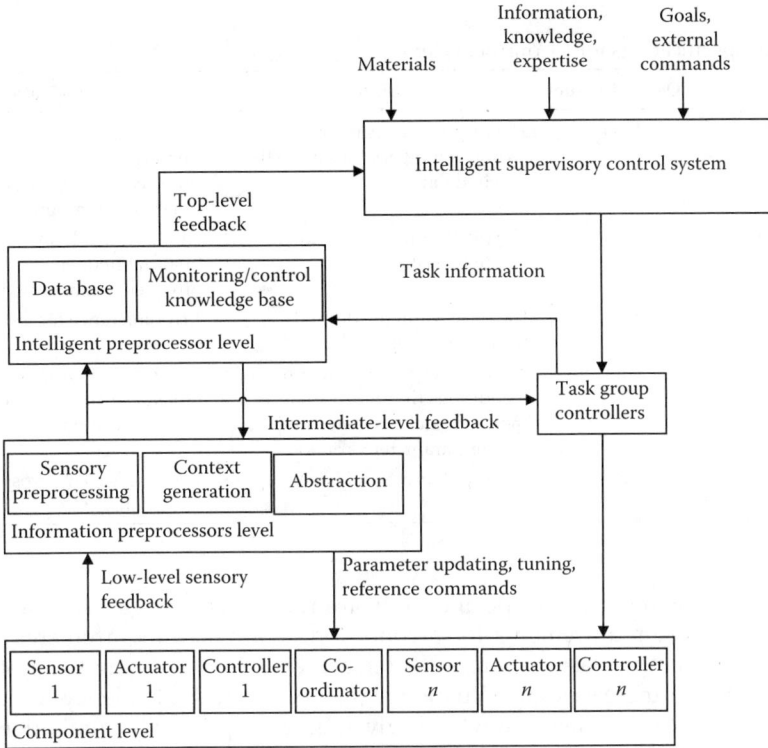

FIGURE 7.12
A hierarchical control/communications structure for an intelligent mechatronic system.

7.2 Control System Performance

A good control system should possess the following performance characteristics:

1. Sufficiently stable response (*stability*). Specifically, the response of the system to an initial-condition (IC) excitation should decay back to the initial steady-state (asymptotic stability). The response to a bounded input should be bounded (bounded-input–bounded-output—BIBO stability).

2. Sufficiently fast response (*speed of response* or *bandwidth*). The system should react quickly to a control input or excitation.

3. Low sensitivity to noise, external disturbances, modeling errors, and parameter variations (*sensitivity* and *robustness*).

4. High sensitivity to control inputs (*input sensitivity*).

5. Low error; for example, tracking error, and steady-state error (*accuracy*).

6. Reduced coupling among system variables (*cross sensitivity* or *dynamic coupling*).

TABLE 7.1

Performance Specifications for a Control System

Attribute	Desired Value	Objective	Specifications
Stability level	High	The response does not grow without limit and decays to the desired value.	Percentage overshoot, settling time, pole (eigenvalue) locations, time constants, phase and gain margins, damping ratios.
Speed of response	Fast	The plant responds quickly to inputs/excitations.	Rise time, peak time, delay time, natural frequencies, resonant frequencies, bandwidth.
Steady-state error	Low	The offset from the desired response is negligible.	Error tolerance for a step input.
Robustness	High	Accurate response under uncertain conditions (input disturbances, noise, model error, etc.) and under parameter variation.	Input disturbance/noise tolerance, measurement error tolerance, model error tolerance.
Dynamic interaction	Low	One input affects only one output.	Cross-sensitivity, cross-transfer functions.

As listed here, some of these specifications are rather general. Table 7.1 summarizes typical performance requirements for a control system. Some requirements might be conflicting. For example, fast response is often achieved by increasing the system gain, and increased gain increases the actuation signal, which has a tendency to destabilize a control system. Note further that what is given here are primarily qualitative descriptions for "good" performance. In designing a control system, however, these descriptions have to be specified in a quantitative manner. The nature of the used quantitative design specifications depends considerably on the particular design technique that is employed.

Some of the design specifications are time-domain parameters and the others are frequency-domain parameters. We will primarily address the time-domain parameters in the present chapter.

7.2.1 Performance Specification in Time-Domain

Speed of response and degree of stability are two commonly used specifications in the conventional time-domain design of a control system. These two types of specifications are conflicting requirements in general. In addition, steady-state error is also commonly specified. Speed of response can be increased by increasing the gain of the control system. This, in turn, can result in reduced steady-state error. Furthermore, steady-date error requirement can often be satisfied by employing integral control. For these reasons, first we will treat speed of response and degree of stability as the requirements for performance specification in time-domain, tacitly assuming that there is no steady-state error. The steady-state error requirement will be treated separately.

Performance specifications in the time-domain are usually given in terms of the response of an oscillatory (under-damped) system to a unit step input, as shown in Figure 7.13. First, assuming that the steady-state error is zero, note that the response will eventually settle at the steady-state value of unity. Then the following performance specifications can be stipulated:

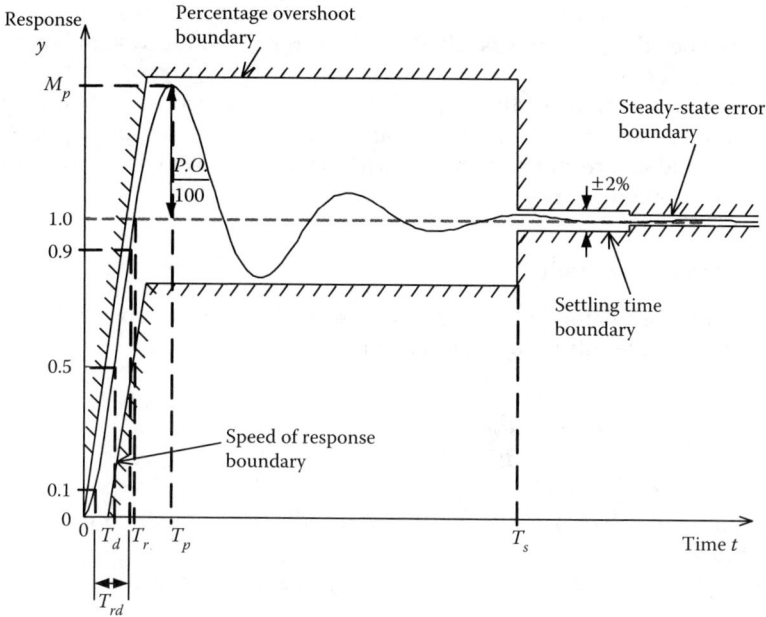

FIGURE 7.13
Conventional performance specifications used in the time-domain design of a control system.

Peak time (T_p): Time at which the response reaches its first peak value.

Rise time (T_r): Time at which the response passes through the steady-state value (normalized to 1.0) for the first time.

Modified rise time (T_{rd}): Time taken for the response to rise from 0.1 to 0.9.

Delay time (T_d): Time taken for the response to reach 0.5 for the first time.

2% settling time (T_s): Time taken for the response to settle within ±2% of the steady-state value (i.e., between 0.98 and 1.02).

Peak magnitude (M_p): Response value at the peak time.

Percentage overshoot (P.O.): This is defined as

$$P.O. = \frac{(\text{Peak magnitude} - \text{Steady-state value})}{\text{Steady-state value}} \times 100\% \qquad (7.1a)$$

In the present case of unity steady-state value, this may be expressed as

$$P.O. = 100(M_p - 1)\% \qquad (7.1b)$$

Note that T_r, T_p, T_{rd}, and T_d are "primarily" measures of the speed of response whereas T_s, M_p, and P.O. are "primarily" measures of the level of stability. Note further that T_r, T_p, M_p, and *P.O.* are not defined for non oscillatory responses. Simple expressions for these time-domain design specifications may be obtained, assuming that the system is approximated by a simple oscillator.

Specifications on the slope of the step-response curve (speed of response), percentage overshoot (stability), settling time (stability), and steady-state error can also be represented as boundaries to the step response curve. This representation of conventional time-domain specifications is shown in Figure 7.8.

7.2.2 Simple Oscillator Model

A damped simple oscillator (mechanical or electrical, as shown in Figure 7.14) may be expressed by the input–output differential equation (see Chapter 6):

$$\frac{d^2y}{dt^2} + 2\zeta\omega_n \frac{dy}{dt} + \omega_n^2 y = \omega_n^2 u \tag{7.2}$$

where u=input (normalized); y=output (normalized); ω_n=undamped natural frequency; ζ=damping ratio.

The corresponding transfer function is given by (see Chapters 5 and 6)

$$\frac{Y(s)}{U(s)} = \frac{\omega_n^2}{\left(s^2 + 2\zeta\omega_n s + \omega_n^2\right)} \tag{7.3}$$

Suppose that a unit step input (i.e., $U(s)=(1/s)$) is applied to the system. As shown in Chapter 6, the resulting response of the oscillator, with zero ICs, is given by

$$y(t) = 1 - \frac{1}{\sqrt{1-\zeta^2}} e^{-\zeta\omega_n t} \sin(\omega_d t + \phi) \text{ for } \zeta < 1 \tag{7.4}$$

where $\omega_d = \sqrt{1-\zeta^2}\,\omega_n$ = damped natural frequency $\tag{7.5}$

$$\cos\phi = \zeta; \sin\phi = \sqrt{1-\zeta^2} \tag{7.6}$$

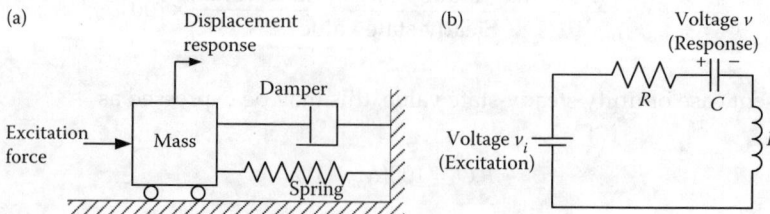

FIGURE 7.14
A damped simple oscillator. (a) Mechanical. (b) Electrical.

Now, using the simple oscillator model let us obtain expressions for some of the design specifications that were defined earlier.

Note: It is clear from Equation 7.4 that the steady-state value of the response (i.e., as $t \to \infty$) is 1. Hence the steady-state error is zero. It follows that the present model does not allow us to address the issue of steady-state error. As indicated before, we will treat steady-state separately. First we will quantify the specifications for stability and the speed of response.

The response given by Equation 7.4 is of the form shown in Figure 7.13. Clearly, the first peak of the response occurs at the end of the first (damped) half cycle. It follows that the *peak time* is given by

$$T_p = \frac{\pi}{\omega_d} \tag{7.7}$$

The same result may be obtained by differentiating Equation 7.4, setting it equal to zero, and solving for the first peak (Exercise: Check the result (Equation 7.7) by this approach).

The *peak magnitude* M_p and the *percentage overshoot* P.O. are obtained by substituting Equation 7.7 into Equation 7.4; thus,

$$M_p = 1 + \exp\left(-\zeta \omega_n T_p\right) \tag{7.8}$$

$$P.O. = 100\exp\left(-\zeta \omega_n T_p\right) \tag{7.9a}$$

Note: In obtaining this result we have used the fact that $\sin\phi = \sqrt{1-\zeta^2}$. Alternatively, by substituting Equations 7.5 and 7.7 into Equation 7.9a we get

$$P.O. = 100\exp\left(-\pi\zeta/\sqrt{1-\zeta^2}\right) \tag{7.9b}$$

The *settling time* is determined by the exponential decay envelope of Equation 7.4. The 2% settling time is given by

$$\exp\left(-\zeta \omega_n T_s\right) = 0.02\sqrt{1-\zeta^2} \tag{7.10}$$

For small damping ratios, T_s is approximately equal to $4/(\zeta\omega_n)$. This should be clear from the fact that $\exp(-4) \approx 0.02$. Note further that the *poles* (*eigenvalues*) of the system, as given by the roots of the *characteristic equation*

$$s^2 + 2\zeta\omega_n s + \omega_n^2 = 0 \tag{7.11}$$

are (see Chapters 5 and 6):

$$p_1, p_2 = -\zeta\omega_n \pm j\omega_d \tag{7.12}$$

It follows that the *time constant* of the system (inverse of the real part of the dominant pole) is

$$\tau = \frac{1}{\zeta\omega_n} \tag{7.13}$$

Hence, an approximate expression for the 2% settling time is

$$T_s = 4\tau \tag{7.14}$$

Rise time is obtained by substituting $y=1$ in Equation 7.14 and solving for t. This gives

$$\sin(\omega_d T_r + \phi) = 0$$

or

$$T_r = \frac{\pi - \phi}{\omega_d} \tag{7.15}$$

in which the *phase angle* ϕ is directly related to the damping ratio, through Equation 7.6.

The expressions for the performance specifications, as obtained using the simple oscillator model, are summarized in Table 7.2. For a higher order system, when applicable, the damping ratio and the natural frequency that are needed to evaluate these expressions may be obtained from the dominant complex pole pair of the system.

In the conventional time-domain design, relative stability specification is usually provided by a limit on P.O. This can be related to damping ratio (ζ) using a design curve. For the simple oscillator approximation, Equation 7.9b is used to calculate ζ when P.O. is specified. This relationship is plotted in Figure 7.15.

TABLE 7.2

Analytical Expressions for Time-Domain Performance Specifications (Simple Oscillator Model)

Performance Specification Parameter	Analytical Expression (Exact for a Simple Oscillator)
Peak time T_p	π/ω_d
Rise time T_r	$(\pi - \phi)/\omega_d$
Time constant τ	$1/(\zeta\omega_n)$
2% Settling time T_s	$-\left(\ln 0.02\sqrt{1-\zeta^2}\right)\tau \approx 4\tau$
Peak magnitude M_p	$1 + \exp\left(-\dfrac{\pi\zeta}{\sqrt{1-\zeta^2}}\right)$
P.O.	$100\exp\left(-\dfrac{\pi\zeta}{\sqrt{1-\zeta^2}}\right)$

P.O.

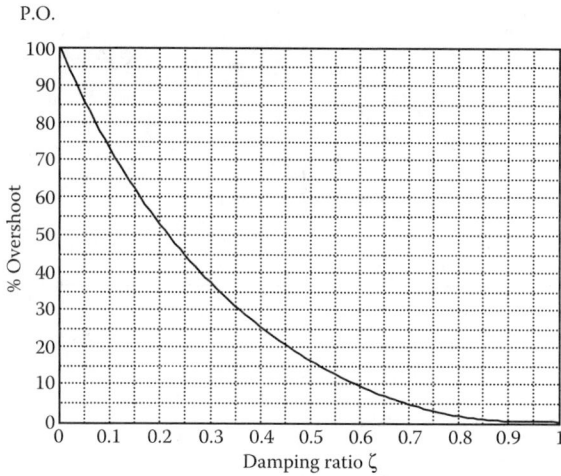

FIGURE 7.15
Damping specification in terms of percentage overshoot (P.O.) using simple oscillator model.

Example 7.2

A control system is required to have a percentage overshoot of less than 10% and a settling time of less than 0.8 seconds. Indicate this design specification as a region on the *s*-plane.

Solution

A 10% overshot means (see Table 7.2): $0.1 = \exp\left(-\dfrac{\pi\zeta}{\sqrt{1-\zeta^2}}\right)$
Hence, $\zeta = 0.60 = \cos\phi$ or: $\phi = 53°$
Next, T_s of 0.8 seconds means (see Table 7.2): $T_s = 4\tau = (4/\zeta\omega_n) = 0.8$ or: $\zeta\omega_n = 5.0$

For the given specifications we require $\phi \leq 53°$ and $\zeta\omega_n \geq 5.0$. The corresponding region on the *s*-plane is given by the shaded area in Figure 7.16.
Note: In this example if we had specified a T_p spec as well, then it would correspond to an ω_d spec. This will result in a horizontal line boundary for the design region in Figure 7.16.

7.3 Control Schemes

By control we mean making a plant (process, machine, etc.), respond to inputs in a desired manner. In a *regulator*-type control system the objective is to maintain the output at a desired (constant) value. In a *servomechanism*-type control system the objective is for the output to follow a desired trajectory (i.e., a specified time response or a path with respect to time). In a control system, in order to meet a specified performance, a suitable control method has to be employed. We will discuss several common control schemes now.

In a *feedback control system*, as shown in Figure 7.17, the control loop has to be closed, by making measurements of the system response and employing that information to generate control signals so as to correct any output errors. Hence, feedback control is also known

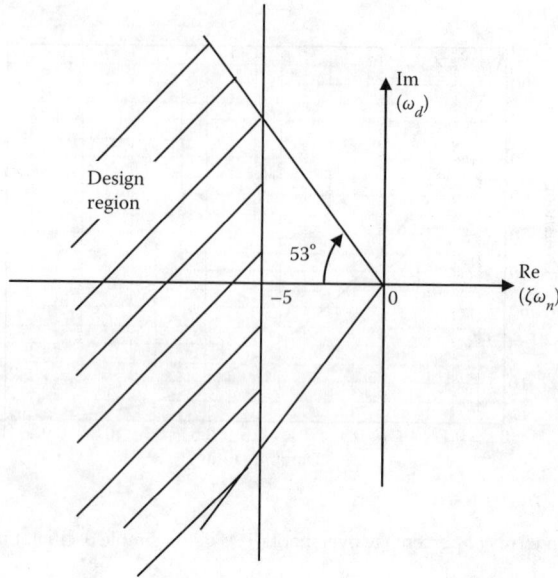

FIGURE 7.16
Design specification on the s-plane.

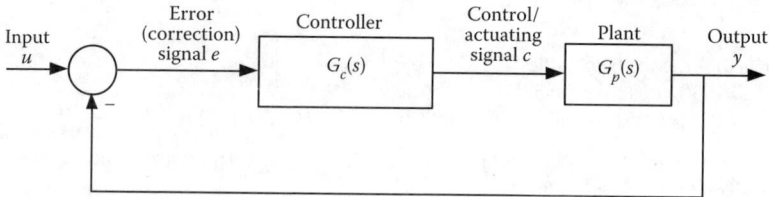

FIGURE 7.17
A feedback control system with unity feedback.

as *closed-loop control*. In Figure 7.17, since the feedback signal is not modified (i.e., gain=1) before subtracting from the reference input, it represents "*unity feedback*."

In selecting a proper controller $G_c(s)$ for an application, the prime factors to be considered are:

1. Precision of the necessary response
2. Difficulty of controlling the plant (process) $G_p(s)$

Although the simplest controller that would produce the desired result should be selected, in some instances the required response cannot be realized even with a sophisticated controller. Selection of the controller may be approached from several angles:

1. Process transfer function (reaction curve or frequency domain) analysis
2. Time domain system (model) analysis

3. Previous experience and human expertise (knowledge-based)

4. Experimental investigation

In the present section we will concentrate only on the first two approaches.

A *control law* is a relationship between the controller output and the plant input. Common control modes are:

1. On–off (bang-bang) control

2. Proportional (P) control

3. Proportional control combined with reset (integral, I) action and/or rate (derivative, D) action (i.e., multimode or multiterm control)

Control laws for commonly used control actions are given in Table 7.3. Some advantages and disadvantages of each control action are also indicated as well in this table. Compare this information with what is given in Table 7.1.

The proportional action provides the necessary speed of response and adequate signal level to drive a plant. Besides, increased proportional action has the tendency to reduce steady-state error. A shortcoming of increased proportional action is the degradation of stability. Derivative action (or rate action) provides stability that is necessary for satisfactory performance of a control system. In the time domain this is explained by the fact that the derivative action tends to oppose sudden changes (large rates) in the system response. Derivative control has its shortcomings, however. For example, if the error signal that drives the controller is constant, the derivative action will be zero and it has no effect on the system response. In particular, derivative control cannot reduce steady-state error in a system. Also, derivative control increases the system bandwidth, which has the desirable effect of increasing the speed of response (and tracking capability) of control system. Derivative action has the drawback of allowing and amplifying high-frequency disturbance inputs and noise components. Hence, derivative action is not practically implemented in its pure analytic form, but rather as a lead circuit, as will be discussed in a later chapter.

The presence of an *offset* (i.e., steady-state error) in the output may be inevitable when proportional control alone is used for a system having finite dc gain. When there is an offset, one way to make the actual steady-state value equal to the desired value would be to change the set point (i.e., input value) in proportion to the desired change. This is known as *manual reset*. Another way to zero out the steady-state error would be to make the dc

TABLE 7.3

Comparison of some Common Control Actions

Control Action	Control Law	Advantages	Disadvantages
On–off	$\frac{c_{max}}{2}\left[\text{sgn}(e)+1\right]$	Simple inexpensive	Continuous chatter Mechanical problems Poor accuracy
Proportional	$k_p e$	Simple Fast response	Offset error (steady-state error) Poor stability
Reset (integral)	$\frac{1}{\tau_i}\int e\, dt$	Eliminates offset Filters out noise	Low bandwidth (Slow response) Reset windup Instability problems
Rate (derivative)	$\tau_d \dfrac{de}{dt}$	High bandwidth (Fast response) Improves stability	Insensitive to dc error. Allows high-frequency noise Amplifies noise Difficult analog implementation

gain infinity. This can be achieved by introducing an integral term (with transfer function $1/s$ as discussed in Chapter 5) in the forward path of the control system (because $1/s \to \infty$ when $s=0$; i.e., at zero frequency because $s=j\omega$ in the frequency domain, as discussed in Chapter 5). This is known as *integral control* or *reset control* or *automatic reset*.

7.3.1 Feedback Control with PID Action

Many control systems employ *three-mode controllers* or *three-term controllers*, which are PID controllers providing the combined action of proportional, integral and derivative modes. The control law for proportional plus PID control is given by

$$c = k_p\left(e + \tau_d \dot{e} + \frac{1}{\tau_i}\int edt\right) \tag{7.16a}$$

or in the transfer function form (see Chapter 5):

$$\frac{c}{e} = k_p\left(1 + \tau_d s + \frac{1}{\tau_i s}\right) \tag{7.16b}$$

in which

$e=$Error signal (controller input)
$c=$Control/actuating signal (controller output or plant input)
$k_p=$Proportional gain
$\tau_d=$Derivative time constant
$\tau_i=$Integral time constant.

Another parameter, which is frequently used in process control, is the *integral rate*. This is defined as

$$r_i = \frac{1}{\tau_i} \tag{7.17}$$

The parameters k_p, τ_d, and τ_i or r_i are the design parameters of the PID controller and are used in controller tuning as well.

Example 7.3

Consider an actuator with transfer function

$$G_p = \frac{1}{s(0.5s+1)} \tag{i}$$

Design:

a. A position feedback controller.
b. A tacho-feedback controller (i.e., position plus velocity feedback controller) that will meet the design specifications $T_p=0.1$ and $P.O.=25\%$.

Solution

a. Position feedback
The block diagram for the position feedback control system is given in Figure 7.18a.

From the standard result for a closed-loop (see Chapter 5, Table 5.7):

$$\text{Closed-loop TF} = \frac{kG_p(s)}{1+kG_p(s)} = \frac{k}{s(0.5s+1)+k} = \frac{2k}{s^2+2s+2k} \tag{ii}$$

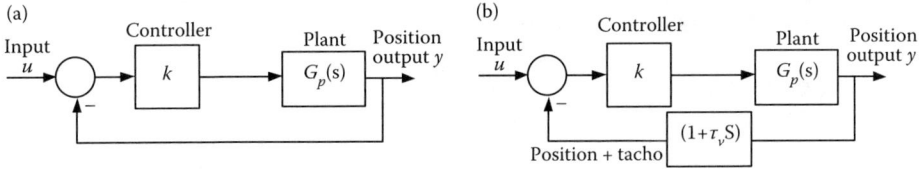

(a)

(b)

FIGURE 7.18
(a) Position feedback control system. (b) Tacho-feedback control system.

TABLE 7.4

Results for Position Control

T_p	ω_d	k	ζ	P.O.
0.1	10π	494.0	0.032	90.5%
1.39	2.264	3.063	0.404	25%

where, k is the gain of the proportional controller in the forward path.
Note: In this case only one parameter (k) is available for specifying two performance requirements. Hence, it is unlikely that both specifications can be met.
To check this further note from the denominator (characteristic polynomial) of Equation (ii) that $\zeta\omega_n=1$ and $\omega_n^2=2k$. Hence,

$$\zeta = \frac{1}{\sqrt{2k}} \tag{iii}$$

and

$$\omega_d = \sqrt{\omega_n^2 - (\zeta\omega_n)^2} = \sqrt{2k-1} \tag{iv}$$

For a given T_p we can compute ω_d using the expression in Table 7.2; k using Equation (iv); ζ using Equation (iii); and finally P.O. using Table 7.2.
Alternatively, for a given P.O., we can determine ζ using Table 7.2; k using Equation (iii); ω_d using Equation (iv); and finally T_p using Table 7.8. These two sets of results are given in Table 7.4.
Note: For $T_p=0.1$ we have P.O.$=90.5\%$; For P.O.$=25\%$ we have $T_p=1.39$.
Hence both requirements cannot be met with the single design parameter k, as expected.
b. Tacho-feedback:
The block diagram for the tacho-feedback system (a) is shown in Figure 7.18b.
Tachometer is a velocity sensor. Customarily, tacho-feedback uses feedback of both position and velocity. Hence, the feedback transfer function is $H=1+\tau_v s$, and from the standard result for a closed-loop system (see Chapter 5, Table 5.7):

$$\text{Closed-loop TF} = \frac{kG_p(s)}{1+kG_p(s)\times(\tau_v s+1)} = \frac{k}{s(0.5s+1)+k(\tau_v s+1)}$$

$$= \frac{2k}{s^2+2(1+k\tau_v)s+2k}$$

where k is the proportional gain and τ_v is the velocity feedback parameter (*tacho gain*). By comparing with the simple oscillator TF, we note

$$\omega_n^2 = 2k \tag{v}$$

$$\zeta\omega_n = 1+k\tau_v \tag{vi}$$

Since two parameters (k and τ_v) are available to meet the two specifications, it is likely that the design goal can be achieved. The computation steps are given below:
As before, for $T_p=0.1$ we have $\omega_d=10\pi$
Also, for $P.O.=25\%$ we have $\zeta=0.404$
Hence, $\omega_n=\omega_d/\sqrt{1-\zeta^2}=10.93\pi$
and $\zeta\omega_n=0.404\times10.93\pi=13.873$
Then we use Equation (v) to compute k.
Substitute in Equation (vi) to compute τ_v.
We get $k=590$ and $\tau_v=0.022$.

7.4 Steady-State Error and Integral Control

It is easy to explain the presence of an *offset* (i.e., *steady-state error*) when proportional control is used for a system having finite dc gain. Consider a unity feedback system under proportional control, with proportional gain k_p. At steady-state (i.e., at zero frequency; $\omega=0$), we have $s=0$ (because $s=j\omega$ in the frequency domain; see Chapter 5). Then the system transfer function $G(s)$ can be represented by its dc gain $G(0)$. The steady-state behavior of this system can be represented by the block diagram in Figure 7.19, with a constant input u (i.e., a steady input).

Note: y_{ss} is the steady-state response of the system and e_{ss} is the steady-state error.

It should be clear that e_{ss} could not be zero because then there would be no actuating (driving) signal for the plant and, consequently, the plant output would be zero, thereby violating the zero error ($u - y_{ss}=0$) condition. Hence, an offset error will always be present in this control system.

To obtain an expression for the offset, note from Figure 7.19 that

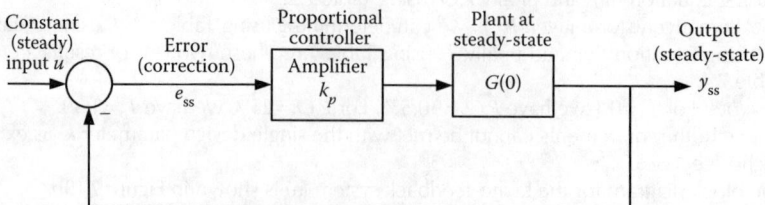

FIGURE 7.19
Presence of offset error in a system under proportional control.

$$e_{ss} = u - y_{ss} \tag{7.18}$$

and

$$y_{ss} = k_p G(0) e_{ss}$$

Hence,

$$e_{ss} = \left[\frac{1}{1 + k_p G(0)} \right] u \tag{7.19}$$

This result further confirms the fact that a steady-state error is always present in the present system, with P control.

Note: An offset could result from an external disturbance such as a change in load as well.

7.4.1 Final Value Theorem (FVT)

The FVT is an important result of Laplace transforms, which is valuable in the analysis of the steady-state behavior of systems. The FVT states:

The steady-state value of a signal $x(t)$ is given by

$$x_{ss} = \lim_{s \to 0} s x(s) \tag{7.20}$$

in which $x(s)$ is the Laplace transform of $x(t)$.

The result (Equation 7.19) can be formally obtained using the FVT, as described below.

Important note: In determining the steady-state value of a response (or, error) the nature of the input in the beginning is not important. What matters is the nature of the input for a sufficiently large time period at the end, inclusive of the settling time (i.e., at steady-state).

Accordingly, in the system shown in Figure 7.19, the constant input can be treated as a "step input" in the steady-state analysis. Hence, in the Laplace domain, the input is u/s (see Chapter 6 and Appendix A). Then, using the steady-state error of the system in Figure 7.19 is given by

$$e_{ss} = \lim_{s \to 0} s \left[\frac{1}{1 + k_p G(s)} \right] \frac{u}{s} = \left[\frac{1}{1 + k_p G(0)} \right] u$$

This is identical to the previous result (Equation 7.19).

7.4.2 Manual Reset

As noted before, one way to make the steady-state value equal to the desired value would be to change the set point (i.e., input value) in proportion to the desired change. This is known as *manual reset*. Strictly, this method does not remove the offset but rather changes the output value. Note further that the percentage overshoot given in Table 7.2 is obtained with respect to the steady-state value, even though this is equal to the steady input value if there is no offset. These concepts can be illustrated by a simple example.

Example 7.4

A feedback control system of a machine tool is shown in Figure 7.20.

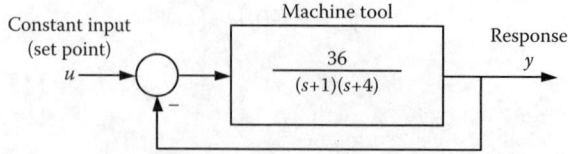

FIGURE 7.20
A feedback control system for a machine tool.

a. Determine the steady-state value y_{ss} of the response and the steady-state error e_{ss}.
b. Determine the percentage overshoot of the system with respect to the set point value.
c. What should be the set point value in order to make the steady-state response value equal to a desired value of y_o?

Solution

a. Forward transfer function (machine tool) is $G(s) = \dfrac{36}{(s+1)(s+4)}$

Note: $G(0)$ is obtained by setting $s=0$. We have $G(0)=9$
In the present system (unity feedback) the closed-loop transfer function is $\dfrac{y}{u} = \dfrac{G(s)}{1+G(s)}$.
For a constant u, the steady-state value of the response is

$$y_{ss} = \frac{G(0)}{1+G(0)}u = \frac{9}{1+9}u = 0.9u \qquad\qquad (i)$$

$$\text{Steady-state error } e_{ss} = \frac{1}{1+G(0)}u = \frac{1}{1+9}u = 0.1u$$

b. Closed-loop transfer function $= \dfrac{G(s)}{1+G(s)} = \dfrac{36}{(s+1)(s+4)+36} = \dfrac{36}{s^2+5s+40} = \dfrac{36}{s^2+2\zeta\omega_n s+\omega_n^2}$

From the characteristic polynomial (denominator of the TF) of the closed-loop system we have: $2\zeta\omega_n = 5$ and $\omega_n^2 = 40$.

$$\text{Hence, } \zeta = \frac{5}{2\sqrt{40}} = 0.395.$$

Then, using the formula in Table 7.2, we have, P.O.$=25.9\%$.
This value is computed with respect to the steady-state value. In order to determine the percentage overshoot w.r.t. the set point value, we proceed as follows:

$$\text{Peak magnitude } M_p = y_{ss}\left[1+\exp\left(-\frac{\pi\zeta}{\sqrt{1-\zeta^2}}\right)\right] = 0.9u\times(1+0.259)$$

P.O. with respect to the set point u is:

$$\frac{M_p-u}{u}\times100 = \frac{0.9u[1+0.259]-u}{u}\times100 = (0.9[1+0.259]-1)\times100 = 13.3\%$$

c. From Equation (i) it is seen that in order to get $y_{ss}=y_o$, we must use a set point of

$$u = \frac{y_o}{0.9} = 1.11 y_o.$$

7.4.3 Automatic Reset (Integral Control)

Another way to zero out the steady-state error in Equation 7.19 is to make $G(0)$ infinity. This can be achieved by introducing an integral term $(1/s)$ in the forward path of the control system (because $1/s \to \infty$ when $s=0$). This is known as *integral control* or *reset control* or *automatic reset*. An alternative way to arrive at the same conclusion is by noting that an integrator can provide a constant output even when the error input is zero, because the initial value of the integrated (accumulated) error signal will be maintained even after the error goes to zero. Integral control is known as reset control because it can reduce the offset to zero, and can counteract external disturbances including load changes. A further advantage of integral control is its low-pass-filter action, which filters out high-frequency noise. Since integral action cannot respond to sudden changes (rates) quickly, it has a destabilizing effect, however. For this reason integral control is practically implemented in conjunction with proportional control, in the form of proportional plus integral (PI) control (a two-term controller), or also including derivative (D) control as PID control (a three-term controller).

7.4.4 Reset Windup

The integral action integrates the error to generate the control action. Suppose that the sign of the error signal remains the same over a sufficiently long period of time. Then, the error signal can integrate into a very large control signal. It can then saturate the device that performs the control action (e.g., a valve actuator can be in the fully open or fully closed position). Then, beyond that point, a change in the control action has no effect on the process. Effectively, the controller has locked in one position while the control signal may keep growing. This "winding up" of the control signal due to error integration, without causing a further control action, is known as *reset windup*.

Due to reset windup, not only will the controller take a longer time to bring the error to zero, but also when the error reaches zero, the integrated error signal will not be zero and the control actuator will remain saturated. As a result, the response will be pushed with the maximum control force, beyond the zero error value (desired value). It will take a further period of time to unsaturate the control actuator and adjust the control action in order to push the response back to the desired value. Large oscillations (and stability problems) can result due to this.

Reset windup problems can be prevented by stopping the reset action before the control actuator becomes saturated. A simple approach is to first determine two limiting values (positive and negative) for the control signal beyond which the control actuator becomes saturated (or very nonlinear). Then, a simple logic implementation (hardware or software) can be used to compare the actual control signal with the limiting values, and deactivate the integral action if the control signal is outside the limits. This is termed *reset inhibit* or *reset windup inhibit*.

7.5 System Type and Error Constants

Characteristics of a system can be determined by applying a known input (*test input*) and studying the resulting response or the error of the response. For a given input, system error will depend on the nature of the system (including its controller). It follows that, error, particularly the steady-state error, to a standard test input, may be used as a parameter for characterizing a control system. This is the basis of the definition of error constants. Before studying that topic we should explain the term "system type."

Consider the general feedback control system shown in Figure 7.21a. The *forward transfer function* is $G(s)$ and the *feedback transfer function* is $H(s)$. The closed-loop transfer function (see Chapter 5, Table 5.7) is $(G/1+GH)$ and the characteristic equation (denominator equation of the transfer function is $1+GH=0$. It should be clear from this result that it is the *loop transfer function GH*, not the individual constituents G and H in the product separately, that primarily determines the dynamic behavior of the closed-loop system (which depends on the characteristic equation). Hence we may assume without loss of generality that the controller elements and any compensators as well as the plant (process) are in the forward path, and their dynamics are jointly represented by $G(s)$; and the feedback transfer function $H(s)$ has the necessary transducers and associated signal conditioning for the sensor that measures the system output y. Consequently it is proper to assume that the dc gain of $H(s)$ to be unity (i.e., $H(0)=1$) because the sensor would be properly calibrated to make correct measurements at least under static conditions.

Customarily, the controller input signal

$$e = u - Hy \tag{7.21}$$

is termed *error signal* (even though the true error is $y-u$), because input u is "compared" with the feedback signal Hy and the difference is used to actuate the corrective measures (i.e., control action) so as to obtain the desired response. Note further that in Equation 7.21 the signals u and Hy should have consistent units. For example in a velocity servo, the input command might be a speed setting on a dial (a physical angle in degrees, for example) and the output would be a velocity (in m/s, for example). But if the velocity is measured using a tachometer, the feedback signal, as typically is the case, would be a voltage. The velocity setting has to be properly converted into a voltage, for example by using a calibrated potentiometer as the input device, so that it could be compared with the feedback voltage signal for (analog) control purposes. Throughout the present discussion we assume consistent units for signals that are compared, and unity dc gain for the feedback transfer function. A further justification for the assumption of unity feedback (or at least unity dc gain in the feedback transfer function $H(s)$) in the following developments is the

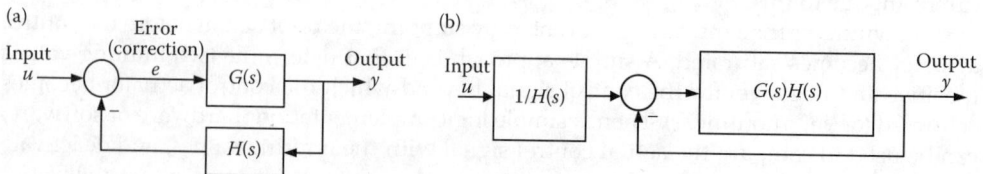

FIGURE 7.21
(a) A general feedback control system; (b) Equivalent representation.

fact that the system shown in Figure 7.21b, which is a unity feedback system, is equivalent to that in Figure 7.21a. This equivalence should be obvious by examining the two systems (or from the block diagram reduction steps presented in Chapter 5, Table 5.7).

7.5.1 Definition of System Type

Assuming that the *feedback transfer function H(s)* has unity dc gain, the system type is defined as the number of free integrators present in the forward transfer function *G(s)*. For example, if there are no free integrators, it is a *Type 0 system*. If there is only one free integrator, it is a *Type 1 system*, and so on.

Obviously, the system type is a system property. Also, the steady-state error to a test input is also a system property. We will see that these two properties are related. Furthermore, system type is a measure of *robustness* with regard to steady-state error, in the presence of variations in system parameters.

7.5.2 Error Constants

When a system is actuated by a "normalized" test signal such as a *unit step, unit ramp,* or *unit parabola*, its steady-state error may be considered a system property. Such system properties may be expressed as error constants. Note, however, that error constants can be defined only if the associated steady-state error is finite. That, in turn, will depend on the number of integrators in *G(s)*, which is known as the system type (or *type number*), as defined before.

To formally define the three types of error constants associated with unit step input, unit ramp input, and unit parabolic input, consider once again the general system shown in Figure 7.21a. The error (correction) signal *e* is given by Equation 7.21. Furthermore, it is clear from Figure 7.21 that

$$y = Ge \tag{7.22}$$

Note: As customary, the same lower case letter is used to denote a time signal and its Laplace transform.

By substituting Equation 7.22 into Equation 7.21 we get

$$e = \left[\frac{1}{1+GH}\right]u \tag{7.23}$$

By applying the FVT to Equation 7.23, the steady-state error may be expressed as

$$e_{ss} = \lim_{s \to 0}\left[\frac{su(s)}{1+GH(s)}\right] = \lim_{s \to 0}\left[\frac{su(s)}{1+G(s)H(0)}\right]$$

Since we assume unity dc gain for the feedback transfer function, we have $H(0)=1$. Hence,

$$e_{ss} = \lim_{s \to 0}\left[\frac{su(s)}{1+G(s)}\right] \tag{7.24}$$

We will use this result to define the three commonly used error constants.

7.5.2.1 Position Error Constant K_p

Consider a *unit step input* defined as

$$u(t) = 1 \quad \text{for} \quad t \geq 0$$
$$= 0 \quad \text{for} \quad t < 0 \tag{7.25a}$$

Its Laplace transform is given by (see Chapter 6 and Appendix A)

$$u(s) = \frac{1}{s} \tag{7.25b}$$

Then from Equation 7.24 we have

$$e_{ss} = \frac{1}{1 + G(0)} \tag{7.26}$$

Now $G(0)$ will be finite only if the system is Type 0. In that case the dc gain of $G(s)$ is defined, and is denoted by K_p. Thus,

$$e_{ss} = \frac{1}{1 + K_p} \tag{7.27}$$

Here, the *position error constant:*

$$K_p = \lim_{s \to 0} G(s) = G(0) \tag{7.28}$$

This is termed position error constant because, for a position control system, a step input can be interpreted as a constant position input. It is seen from Equation 7.26 that for systems of Type 1 or higher, the steady-state error to a step input would be zero, because $G(0) \to \infty$ in those cases.

7.5.2.2 Velocity Error Constant K_v

Consider a *unit ramp input* defined as

$$u(t) = t \quad \text{for} \quad t \geq 0$$
$$= 0 \quad \text{for} \quad t < 0 \tag{7.29a}$$

Its Laplace transform is given by (see Chapter 6 and Appendix A)

$$u(s) = \frac{1}{s^2} \tag{7.29b}$$

Then from Equation 7.24 we get

$$e_{ss} = \lim_{s \to 0} \left[\frac{1}{s + sG(s)} \right]$$

or

$$e_{ss} = \frac{1}{\lim_{s \to 0} sG(s)} \tag{7.30}$$

Now note that:

For a Type 0 system $\lim_{s \to 0} sG(s) = 0$

For a Type 1 system $\lim_{s \to 0} sG(s) = $ constant

For a Type 2 system $\lim_{s \to 0} sG(s) = \infty$

It follows from Equation 7.30 that for a Type 0 system $e_{ss} \to \infty$ and for a Type 2 system $e_{ss} \to \infty$. The steady-state error for a unit ramp is a nonzero constant only for a Type 1 system, and is given by

$$e_{ss} = \frac{1}{K_v} \tag{7.31}$$

Here, the *velocity error constant*:

$$K_v = \lim_{s \to 0} sG(s) \tag{7.32}$$

The constant K_v is termed velocity error constant because for a position control system, a ramp position input is a constant velocity input.

7.5.2.3 Acceleration Error Constant K_a

Consider a *unit parabolic input* defined as a

$$u(t) = \frac{t^2}{2} \quad \text{for} \quad t \geq 0$$

$$= 0 \quad \text{for} \quad t < 0 \tag{7.33a}$$

Its Laplace transform is given by (see Chapter 6 and Appendix A)

$$u(s) = \frac{1}{s^2} \tag{7.33b}$$

Then from Equation 7.24 we have

$$e_{ss} = \frac{1}{\lim_{s \to 0} s^2 G(s)} \tag{7.34}$$

It is now clear that for a Type 0 system or Type 1 system this steady-state error goes to infinity. For a Type 2 system the steady-state error to a unit parabolic input is finite, however, and is given by

$$e_{ss} = \frac{1}{K_a} \tag{7.35}$$

Here, the *acceleration error constant*:

$$K_a = \lim_{s \to 0} s^2 G(s) \tag{7.36}$$

The constant K_a is termed acceleration error constant because, for a position control system, a parabolic position input is a constant acceleration input.

How the steady-state error depends on the system type and the input is summarized in Table 7.5.

Note: For control loops with one or more free integrators (i.e., system Type 1 or higher) the steady-state error to a step input would be zero. This explains why integral control is used to eliminate the offset error in systems under steady inputs, as noted previously.

Example 7.5

Synchro transformer is a feedback sensor that is used in control applications. It consists of a transmitter whose rotor is connected to the input member, and a receiver whose rotor is connected to the output member of the system to be controlled. The field windings of the stator and the rotor field are connected together in the Y-configuration (see Figure 7.22). The transmitter rotor has a single set of windings and is activated by an external ac source. The resulting signal generated in the rotor windings of the receiver is a measure of the position error (difference between the transmitter rotor angle and the receiver rotor angle). Consider a position servo that uses a synchro transformer. When the transmitter rotor turns at a constant speed of 120 rpm, the position error was found to be 2°. Estimate the velocity error constant of the control system. The loop gain of the control system is 50. Determine the required loop gain in order to obtain a steady-state error of 1° at the same input speed.

TABLE 7.5

Dependence on the Steady-State Error on the System Type

Input	$u(t)\ t \geq 0$	$u(s)$	Steady-State Error		
			Type 0 System	Type 1 System	Type 2 System
Unit step	1	$\dfrac{1}{s}$	$\dfrac{1}{1+K_p}$	0	0
Unit ramp	t	$\dfrac{1}{s^2}$	∞	$\dfrac{1}{K_v}$	0
Unit parabola	$\dfrac{t^2}{2}$	$\dfrac{1}{s^3}$	∞	∞	$\dfrac{1}{K_a}$

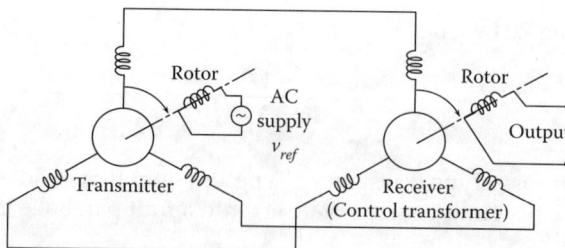

FIGURE 7.22
Synchro transformer.

Solution

$$\text{Input speed} = 120 \text{ rpm} = \frac{120}{60} \times 2\pi \text{ rad/s}$$

$$\text{Steady-state error} = 2° = 2 \times \frac{\pi}{180} \text{rad}$$

Steady-state error for a unit ramp input (i.e., a constant speed of 1 rad/s at the input)

$$e_{ss} = 2 \times \frac{\pi}{180} \bigg/ \frac{120 \times 2\pi}{60} \text{rad/rad/s} = \frac{1}{360} \text{s}$$

From Equation 7.31, the velocity error constant

$$K_v = 360 \text{ s}^{-1}$$

Since K_v is directly proportional to the loop gain of a control system, required loop gain for a specified steady-state error can be determined with the knowledge of K_v. In the present example, in view of Equation 7.31, the K_v required to obtain a steady-state error of 1° at 120 rpm is double the value for a steady-state error of 2°. The corresponding loop gain is 100. Note that the units of loop gain are determined by the particular control system, and is not known for this example even though the units of K_v are known to be s^{-1}.

7.5.3 System Type as a Robustness Property

For a Type 0 system, the steady-state error to a unit step input may be used as a design specification. The gain of the control loop (loop gain) can be chosen on that basis. For a system of Type 1 or higher, this approach is not quite appropriate, however, because the steady-state error would be zero no matter what the loop gain is. An appropriate design specification for a Type 1 system (i.e., a system having a single integrator in the loop) would be the steady-state error under a unit ramp input. Alternatively, the velocity error constant K_v may be used as a design specification. Similarly, for a Type 2 system, the steady-state error under a unit parabolic input, or alternatively the acceleration error constant K_a, may be used as a design specification. Note further that the system type number is a measure of the *robustness* of a system, as clear from Table 7.5. To understand this, we recall the fact that for a Type 0 system, steady-state error to a unit step input depends on plant gain and, hence, error will vary due to variations in the gain parameter. For a system of Type 1 or higher, steady-state error to a step input will remain zero even in the presence of variations in the gain parameter (a robust situation). For a system of Type 2 or higher, steady-state error to a ramp input as well as step input will remain zero in the presence of gain variations. It follows that as the type number increases, the system robustness with respect to the steady-state error to a polynomial input (step, ramp, parabola, etc.), improves.

7.5.4 Performance Specification Using s-Plane

The s-plane is given by a horizontal axis corresponding to the real part of s and a vertical axis corresponding to the imaginary part of s. The poles of a damped oscillator are given by Equation 7.12. Then, the pole location on the s-plane is defined by the real part $-\zeta\omega_n$ and the imaginary part ω_d of the two roots. Note that the magnitude of the real part is the reciprocal of the *time constant* τ (Equation 7.13), and ω_d is the *damped natural frequency*

(Equation 7.5). Now recall the expressions for the performance specifications as given in Table 7.2. The following facts are clear:

- A "constant settling time line" is the same as a "constant time constant line" (i.e., a vertical line on the s-plane).
- A "constant peak time line" is the same as a "constant ω_d line" (i.e., a horizontal line on the s-plane).
- A "constant percentage overshoot line" is the same as a "constant damping ratio line" (i.e., a radial line, cosine of whose angle with reference to the negative real axis is equal to the damping ratio ζ—see Equation 7.6).

These lines are shown in Figure 7.23a, b, and c. Since a satisfactory design is expressed by an inequality constraint on each of the design parameters, we have indicated the acceptable design region in each case.

Next consider an appropriate measure on the s-plane for steady-state error. We recall that for a Type 0 system, the steady-state error to a step input decreases as the loop gain increases. Furthermore, it is also known that the undamped natural frequency ω_n increases with the loop gain. It follows that for a system with variable gain:

A "constant steady-state error line" is a "constant ω_n line" (i.e., a circle on the s-plane, with radius ω_n and centered at the origin of the coordinate system). This line is shown in Figure 7.23d.

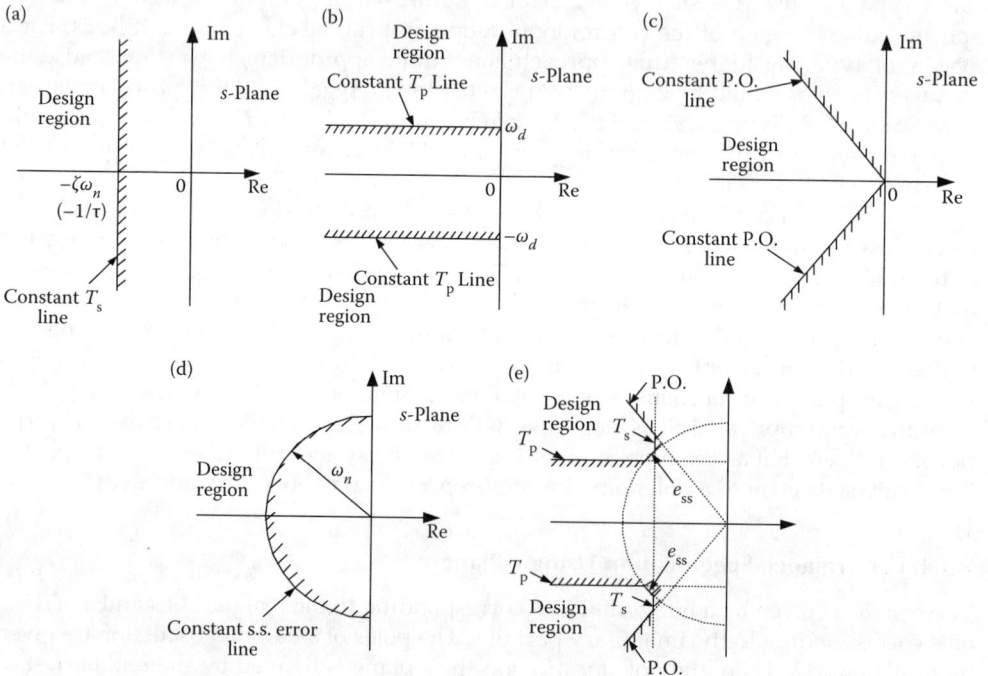

FIGURE 7.23
Performance specification on the s-plane. (a) Settling time. (b) Peak time. (c) Percentage overshoot (P.O.). (d) Steady-state error. (e) Combined specification.

A composite design boundary and a design region (corresponding to a combined design specification) can be obtained by simply overlaying the four regions given in Figure 7.23a through d. This is shown in Figure 7.23e. Note that in Figure 7.23 we have disregarded the right half of the s-plane, at the outset, because it corresponds to an *unstable* system.

Example 7.6

A simplified model of a control system for a paper winding mechanism used in a paper plant is shown in Figure 7.24. The model includes the time constants of the drive motor with load, and the gain of the control amplifier and other circuitry. For a preliminary design it is required to determine the system parameters K, τ_m, and τ_e so as to meet the following design specifications (for the closed-loop system):

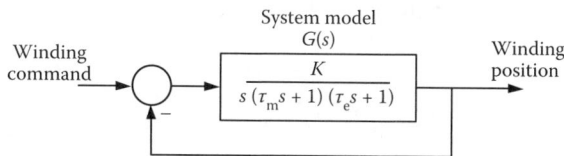

FIGURE 7.24
A simplified control system for a paper rolling mechanism.

    a. 2% Settling time = 4/3 second.
    b. Peak time = $\pi/4$

Determine the minimum value of steady-state error e_{ss} to a unit ramp input and, hence, the maximum value of the velocity error constant K_v that is possible for this control system (closed-loop). Select a value for K so that the closed-loop system behaves like a simple oscillator. Determine the corresponding values of e_{ss} and K_v and system parameters τ_m and τ_e. Estimate the percentage overshoot of this resulting closed-loop system.

Solution

First we note that with the settling time $T_s = 4/3$, the closed-loop system must satisfy (see Table 7.2):

$$\frac{4}{\zeta\omega_n} = \frac{4}{3} \quad \text{or} \quad \zeta\omega_n = 3$$

Also, with a peak time of $T_p = \dfrac{\pi}{4}$ we have

$$\frac{\pi}{\omega_d} = \frac{\pi}{4} \quad \text{or} \quad \omega_d = 4$$

Accordingly, we have

$$\omega_n = 5 \quad \text{and} \quad \zeta = 0.6$$

Since the system is third-order, the closed-loop system should have a transfer function given by

$$\tilde{G}(s) = \frac{\tilde{K}}{(s+p)(s^2 + 2\zeta\omega_n s + \omega_n^2)} \tag{i}$$

with p sufficiently large so that the expressions used for T_s and T_p are sufficiently accurate. Specifically, for large p, its contribution (Ce^{-pt}) to the overall response of the system, will decay very fast. Once this happens, there is no effect from p, and the system will behave like a simple oscillator. With the values computed above, we have

$$\tilde{G}(s) = \frac{\tilde{K}}{(s+p)(s^2+6s+25)} \tag{ii}$$

The corresponding open-loop transfer function

$$G(s) = \frac{\tilde{G}(s)}{1-\tilde{G}(s)} = \frac{\tilde{K}}{(s+p)(s^2+6s+25)-\tilde{K}}$$

Since this transfer function has a free integrator (closed-loop system is Type 1), as given in Figure 7.17, we must have the constant terms in the denominator cancel out; thus

$$\tilde{K} = 25p \tag{iii}$$

Hence,

$$G(s) = \frac{\tilde{K}}{s\left[s^2+(p+6)s+6p+25\right]} \tag{iv}$$

The velocity error constant (Equation 7.32)

$$K_v = \lim_{s \to 0} sG(s) = \frac{\tilde{K}}{6p+25}$$

Now substituting Equation (iii) for p we have

$$K_v = \frac{\tilde{K}}{6\tilde{K}/25+25} \tag{v}$$

This is a monotonically increasing function for positive values of $\tilde{K}$. It follows that the maximum value of K_v is obtained when $\tilde{K} \to \infty$. Hence,

$$K_{v_{max}} = \frac{1}{6/25} = \frac{25}{6}$$

Corresponding steady-state error for a unit ramp input (see Table 7.5):

$$e_{ss_{min}} = \frac{6}{25} = 0.24$$

As noted before, if the closed-loop system is to behave like a simple oscillator, the real pole has to move far to the left of the complex pole pair so that the real pole could be neglected. A rule of thumb is to use a factor of 10 as an adequate distance. Since $\zeta\omega_n=3$, we must then pick

$$p=30$$

From Equation (iii) then $\tilde{K} = 25 \times 30 = 750$. Substituting in Equation (v),

$$K_v = \frac{750}{6 \times 750/25+25} = 3.66$$

Hence,

$$e_{ss} = \frac{1}{3.66} = 0.25$$

Note that this value is quite close to the minimum possible value (0.24), and is quite satisfactory. From Equation (iv)

$$G(s) = \frac{750}{s(s^2 + 36s + 205)} = \frac{750}{s(s + 7.09)(s + 28.91)} = \frac{3.66}{s(0.141s + 1)(0.0346s + 1)}$$

So we have

$$K = 3.66, \quad \tau_m = 0.141, \quad \tau_e = 0.0346$$

Furthermore, since $\zeta = 0.6$ for the closed-loop system, the percentage overshoot is (Table 7.2)

$$P.O. = 100\exp\left(-\pi \times 0.6/\sqrt{1 - 0.36}\right)$$

or

$$P.O. = 9.5\%$$

7.6 Control System Sensitivity

Accuracy of a control system is affected by parameter changes in the control system components and by the influence of external disturbances. It follows that analyzing the sensitivity of a feedback control system to parameter changes and to external disturbances is important.

Consider the block diagram of a typical feedback control system, shown in Figure 7.25. In the usual notation we have:

$G_p(s)$ = transfer function of the plant (or the system to be controlled)
$G_c(s)$ = transfer function of the controller (including compensators)

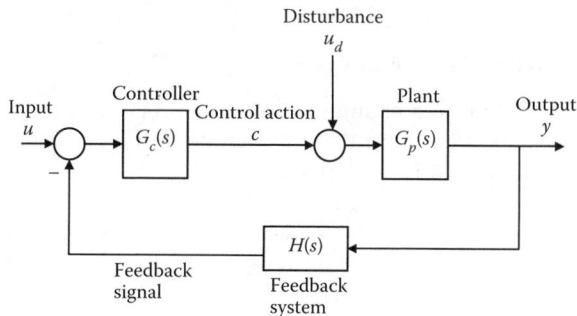

FIGURE 7.25
Block diagram representation of a feedback control system.

$H(s)$=transfer function of the output feedback system (including the measurement system)

u=system input command

u_d=external disturbancy input

y=system output.

Since what we have is a linear system (as necessary in the transfer function representation), the *principle of superposition* applies. In particular, if we know the outputs corresponding to two inputs when applied separately, the output when both inputs are applied simultaneously is given by the sum of the individual outputs.

First set u_d=0:

Then it is straightforward to obtain the input–output relationship:

$$y = \left[\frac{G_c G_p}{1 + G_c G_p H} \right] u \tag{i}$$

Next set u=0:

Then we obtain the input–output relationship:

$$y = \left[\frac{G_p}{1 + G_c G_p H} \right] u_d \tag{ii}$$

By applying the principle of superposition on Equations (i) and (ii), we obtain the overall input–output relationship:

$$y = \left[\frac{G_c G_p}{1 + G_c G_p H} \right] u + \left[\frac{G_p}{1 + G_c G_p H} \right] u_d \tag{7.37}$$

The closed-loop transfer function $\tilde{G}$ is given by y/u, with u_d=0; thus,

$$\tilde{G} = \frac{G_c G_p}{\left[1 + G_c G_p H \right]} \tag{7.38}$$

7.6.1 System Sensitivity to Parameter Change

The sensitivity of the system to a change in some parameter k may be expressed as the ratio of the change in the system output to the change in the parameter; i.e., $\Delta y / \Delta k$. In the nondimensional form, this sensitivity is given by

$$S_k = \frac{k}{y} \frac{\Delta y}{\Delta k} \tag{7.39}$$

Since $y = \tilde{G} u$, with u_d=0, it follows that for a given input u:

$$\frac{\Delta y}{y} = \frac{\Delta \tilde{G}}{\tilde{G}}$$

Consequently, Equation 7.39 may be expressed as

$$S_k = \frac{k}{\tilde{G}} \frac{\Delta \tilde{G}}{\Delta k} \tag{7.40}$$

or, in the limit:

$$S_k = \frac{k}{\tilde{G}} \frac{\partial \tilde{G}}{\partial k} \tag{7.41}$$

Now, by applying Equation 7.41 to Equation 7.38, we are able to determine expressions for the control system sensitivity to changes in various components in the control system. Specifically, by straightforward partial differentiation of Equation 7.38, separately with respect to G_p, G_c, and H, we get

$$S_{Gp} = \frac{1}{\left[1 + G_c G_p H\right]} \tag{7.42}$$

$$S_{Gc} = \frac{1}{\left[1 + G_c G_p H\right]} \tag{7.43}$$

$$S_H = -\frac{G_c G_p H}{\left[1 + G_c G_p H\right]} \tag{7.44}$$

It is clear from these three relations that as the static gain (or, dc gain) of the loop (i.e., $G_c G_p H$, with $s = 0$) is increased, the sensitivity of the control system to changes in the plant and the controller decreases, but the sensitivity to changes in the feedback (measurement) system approaches (negative) unity. Furthermore, it is clear from Equation 7.37 that the effect of the disturbance input can be reduced by increasing the static gain of $G_c H$. By combining these observations, the following design criterion can be stipulated for a feedback control system:

1. Make the measurement system (H) very accurate and stable.
2. Increase the loop gain (i.e., gain of $G_c G_p H$) to reduce the sensitivity of the control system to changes in the plant and controller.
3. Increase the gain of $G_c H$ to reduce the influence of external disturbances.

In practical situations, the plant G_p is usually fixed and cannot be modified. Furthermore, once an accurate measurement system is chosen, H is essentially fixed. Hence, most of the design freedom is available with respect to G_c only. It is virtually impossible to achieve all the design requirements simply by increasing the gain of G_c. The dynamics (i.e., the entire transfer function) of G_c (not just the gain value at $s = 0$) also have to be properly designed in order to obtain the desired performance of a control system.

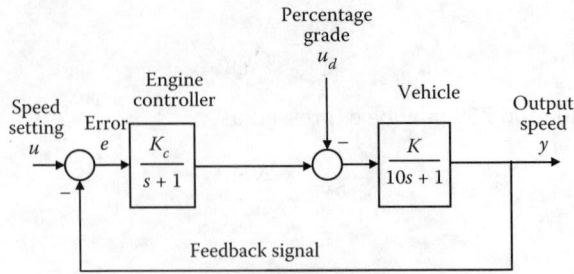

FIGURE 7.26
A cruise control system.

Example 7.7

Consider the cruise control system given by the block diagram in Figure 7.26. The vehicle travels up a constant incline with constant speed setting from the cruise controller.

   a. For a speed setting of $u=u_o$ and a constant road inclination of $u_d=u_{do}$ derive an expressions for the steady-state values y_{ss} of the speed and e_{ss} of the speed error. Express your answers in terms of K, K_c, u_o and u_{do}.

   b. At what minimum percentage grade would the vehicle stall? Use steady-state conditions, and express your answer in terms of the speed setting u_o and controller gain K_c.

   c. Suggest a way to reduce e_{ss}.

   d. If $u_o=4$, $u_{do}=2$ and $K=2$, determine the value of K_c such that $e_{ss}=0.1$.

Solution

   a. For $u_d=0$:

$$y = \dfrac{\dfrac{K_cK}{(s+1)(10s+1)}}{\left[1+\dfrac{K_cK}{(s+1)(10s+1)}\right]}u = \dfrac{K_cK}{\left[(s+1)(10s+1)+K_cK\right]}u$$

   For $u=0$:

$$y = \dfrac{\dfrac{K}{(10s+1)}}{\left[1+\dfrac{K_vK}{(s+1)(10s+1)}\right]}(-u_d) = -\dfrac{K(s+1)}{\left[(s+1)(10s+1)+K_cK\right]}u_d$$

Hence, with both u and u_d present, using the principle of superposition (linear system):

$$y = \dfrac{K_cK}{\left[(s+1)(10s+1)+K_cK\right]}u - \dfrac{K(s+1)}{\left[(s+1)(10s+1)+K_cK\right]}u_d \tag{i}$$

If the inputs are constant at steady-state, the corresponding steady-state output does not depend on the nature of the inputs under starting conditions. Hence, in this problem what matters is the fact that the inputs and the output are constant at steady-state. Hence, without loss of generality, we can assume the inputs to be step functions.

Note: Even if we assume a different starting shape for the inputs, we should get the same answer for the steady-state output, for the same steady-state input values. But the mathematics of getting that answer would be more complex.

Now, using FVT, at steady-state:

$$y_{ss} = \lim_{s \to 0} \left[\frac{K_c K}{[(s+1)(10s+1)+K_c K]} \cdot \frac{u_o}{s} \cdot s - \frac{K(s+1)}{[(s+1)(10s+1)+K_c K]} \cdot \frac{u_{do}}{s} s \right]$$

or,

$$y_{ss} = \frac{K_c K}{(1+K_c K)} u_o - \frac{K}{(1+K_c K)} u_{do} \qquad (ii)$$

Hence, the steady-state error:

$$e_{ss} = u_o - y_{ss} = u_o - \frac{K_c K}{(1+K_c K)} u_o + \frac{K}{(1+K_c K)} u_{do}$$

or,

$$e_{ss} = \frac{1}{(1+K_c K)} u_o + \frac{K}{(1+K_c K)} u_{do} \qquad (iii)$$

b. Stalling condition is $y_{ss}=0$
   Hence, from Equation (ii) we get

$$u_{do} = K_c u_o$$

c. Since K_c is usually fixed (a plant parameter) and cannot be adjusted, we should increase K_c to reduce e_{ss}.

d. Given $u_o=4$, $u_{do}=2$, $K=2$, $e_{ss}=0.1$

   Substitute in Equation (iii): $0.1 = \dfrac{1}{(1+2K_c)} \times 4 + \dfrac{2}{(1+2K_c)} \times 2$

   Hence: $1+2K_c=80$

   or $K_c=39.5$

Problems

PROBLEM 7.1

a. What is an open-loop control system and what is a feedback control system? Give one example of each case.

b. A simple mass-spring-damper system (simple oscillator) is excited by an external force $f(t)$. Its displacement response y (see Figure P7.1a) is given by the differential equation: $m\ddot{y} + b\dot{y} + ky = f(t)$

A block diagram representation of this system is shown in Figure P7.1b. Is this a feedback control system? Explain and justify your answer.

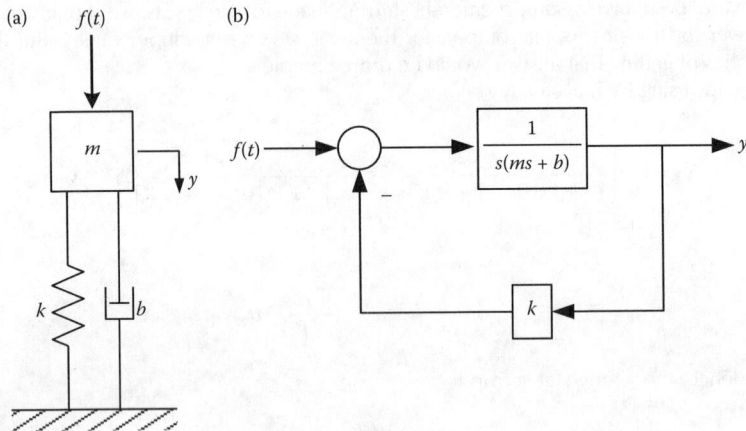

FIGURE P7.1

(a) A mechanical system representing a simple oscillator. (b) A block diagram representation of the simple oscillator.

PROBLEM 7.2

You are asked to design a control system to turn on lights in an art gallery at night, provided that there are people inside the gallery. Explain a suitable control system, identifying the open-loop and feedback functions, if any, and describing the control system components.

PROBLEM 7.3

a. Discuss possible sources of error that can make open-loop control or feedforward control meaningless in some applications.
b. How would you correct the situation?

PROBLEM 7.4

Consider the natural gas home heating system shown Figure 7.4. Describe the functions of various components in the system and classify them into the function groups: controller, actuator, sensor, and signal modification device. Explain the operation of the overall system and suggest possible improvements to obtain more stable and accurate temperature control.

PROBLEM 7.5

In each of the following examples, indicate at least one (unknown) input that should be measured and used for feedforward control to improve the accuracy of the control system.

a. A servo system for positioning a mechanical load. The servo motor is a field-controlled dc motor, with position feedback using a potentiometer and velocity feedback using a tachometer.
b. An electric heating system for a pipeline carrying a liquid. The exit temperature of the liquid is measured using a thermocouple and is used to adjust the power of the heater.

c. A room heating system. Room temperature is measured and compared with the set point. If it is low, a valve of a steam radiator is opened; if it is high, the valve is shut.
d. An assembly robot that grips a delicate part to pick it up without damaging the part.
e. A welding robot that tracks the seam of a part to be welded.

PROBLEM 7.6

Hierarchical control has been applied in many industries, including steel mills, oil refineries, chemical plants, glass works, and automated manufacturing. Most applications have been limited to two or three levels of hierarchy, however. The lower levels usually consist of tight servo loops, with bandwidths on the order of 1 kHz. The upper levels typically control production planning and scheduling events measured in units of days or weeks.

A five-level hierarchy for a flexible manufacturing facility is as follows: The lowest level (level 1) handles servo control of robotic manipulator joints and machine tool degrees of freedom. The second level performs activities such as coordinate transformation in machine tools, which are required in generating control commands for various servo loops. The third level converts task commands into motion trajectories (of manipulator end effector, machine tool bit, etc.) expressed in world coordinates. The fourth level converts complex and general task commands into simple task commands. The top level (level 5) performs supervisory control tasks for various machine tools and material-handling devices, including coordination, scheduling, and definition of basic moves. Suppose that this facility is used as a flexible manufacturing workcell for turbine blade production. Estimate the event duration at the lowest level and the control bandwidth (in hertz) at the highest level for this type of application.

PROBLEM 7.7

The PLC is a sequential control device, which can sequentially and repeatedly activate a series of output devices (e.g., motors, valves, alarms, signal lights) on the basis of the states of a series of input devices (e.g., switches, two-state sensors). Show how a programmable controller and a vision system consisting of a solid-state camera and a simple image processor (say, with an edge-detection algorithm) could be used for sorting fruits on the basis of quality and size for packaging and pricing.

PROBLEM 7.8

It is well known that the block diagram in Figure P7.8a represents a dc motor, for armature control, with the usual notation. Suppose that the load driven by the motor is a pure inertia element (e.g., a wheel or a robot arm) of moment of inertia J_L, which is directly and rigidly attached to the motor rotor.

a. Obtain an expression for the transfer function $\omega_m/v_a = G_m(s)$ for the motor with the inertial load, in terms of the parameters given in Figure P7.8a, and J_L.
b. Now neglect the leakage inductance L_a. Then, show that the transfer function in (a) can be expressed as $G_m(s) = k/(\tau s + 1)$. Give expressions for τ and k in terms of the given system parameters.
c. Suppose that the motor (with the inertial load) is to be controlled using position plus velocity feedback. The block diagram of the corresponding control system is given in Figure P7.8b, where $G_m(s) = k/(\tau s + 1)$. Determine the transfer function of the (closed-loop) control system $G_{CL}(s) = \theta_m/\theta_d$ in terms of the given system parameters (k, k_p, τ, τ_v). Note that θ_m is the angle of rotation of the motor with inertial load, and θ_d is the desired angle of rotation.

(a)

(b)

FIGURE P7.8

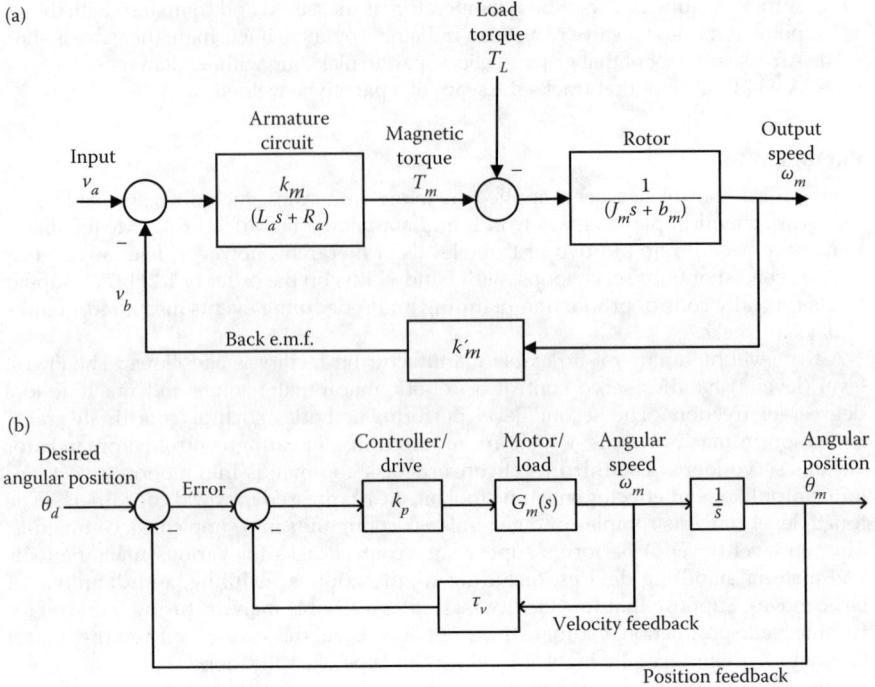

(a) Block diagram of a dc motor for armature control. (b) Motor control with feedback of position and velocity.

PROBLEM 7.9

a. A dc motor has a rotor of moment of inertia J and is supported on bearings which produce a nonlinear damping torque $c\omega^2$ which always acts opposite to the speed of rotation ω of the rotor (see Figure P7.9a). If the magnetic torque T applied on the rotor is the input and the speed ω of the rotor is the output, determine the input–output differential equation of the system.

What is the order of the system?

b. Suppose that the motor in (a) rotates at a positive steady speed of $\bar{\omega}$ under a constant input torque of $\bar{T}$.

Obtain a linear input–output model of the system in (a) to study small variations $\hat{\omega}$ in the output speed for small changes $\hat{T}$ in the input torque.

    (i)   Determine the transfer function $\hat{\omega}/\hat{T}$ of this linear model.

    (ii)   Express the time constant τ of the linear model in terms of the known quantities J, c, and $\bar{\omega}$.

    (iii) Sketch (no derivation needed) the nature of the incremental response $\hat{\omega}$ to a small step change $\hat{T}$ in the input torque.

    (iv) Sketch how the incremental response $\hat{\omega}$ in Equation (iii) changes as τ increases. Using this information discuss what happens to the speed of response of the systems as τ increases.

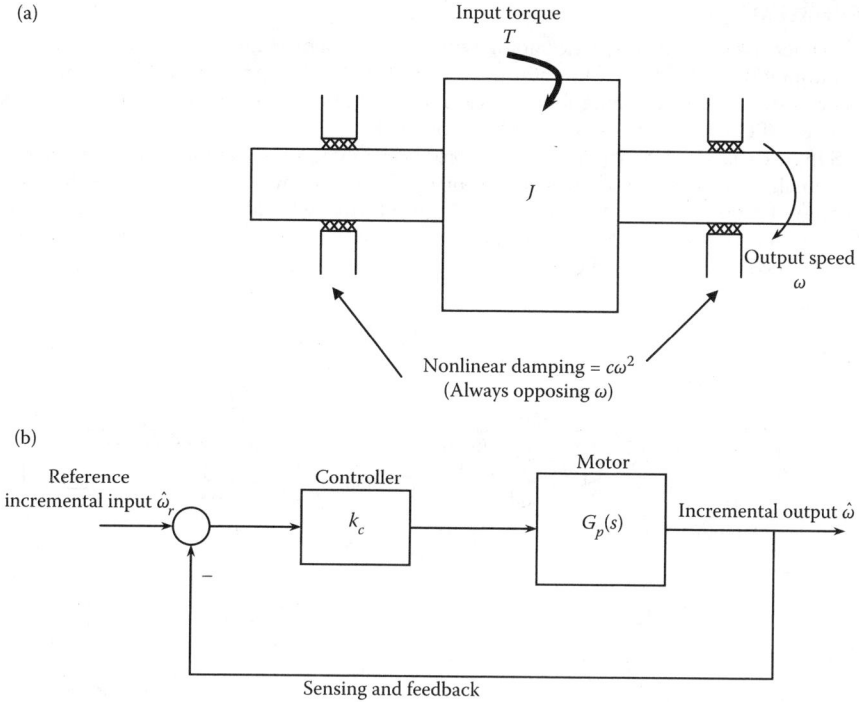

(a)

(b)

FIGURE P7.9
(a) A dc motor supported on nonlinear bearings. (b) A feedback control system for the motor.

c. If in the nonlinear system described in (a), the output is the angle of rotation θ (Note: $\dot{\theta}=\omega$) and the input is T as before, derive a state-space model (nonlinear) for the system. What is the order of this system?

d. For the linearized system in (b) draw a block diagram using a summation junction, an integration ($1/s$) block, and constant-gain blocks only, again with $\hat{T}$ as the input and $\hat{\omega}$ as the output.

Comment on the feedback path present in this block diagram (Specifically, is this a natural/internal feedback path or does this represent a feedback control system?)

Verify that this block diagram corresponds to the transfer function obtained in (b)(i).

e. Denote the transfer function obtained in (b)(i) by $G_p(s)$. Now suppose that a feedback control system is formed as shown in Figure P7.9b.

Here, the incremental speed $\hat{\omega}$ is measured, compared with a derived/reference value $\hat{\omega}_r$ and the error $\hat{\omega}_r-\hat{\omega}$ is multiplied by a control gain K_c to generate the incremental torque $\hat{T}$ to the motor.

(i) What is the transfer function $\hat{\omega}/\hat{\omega}_r$ of the resulting closed-loop system?
(ii) What is the time constant of the closed-loop system?
(iii) What is the effect of the control gain k_c on the speed of response of the control system?

PROBLEM 7.10

Consider a field-controlled dc motor with a permanent-magnet rotor and electronic commutation. In this case the rotor magnetic field may be approximated to a constant. As a result, the motor magnetic torque may be approximately expressed as $T_m = k_m i_f$ where i_f = field current; k_m = motor torque constant.

Suppose that the load driven by the motor is purely inertial, with a moment of inertia J_L, which is connected to the motor rotor (of inertia J_m) by a rigid shaft. A schematic representation of this system is given in Figure P7.10a, where the field (stator) circuit is clearly shown. Note that the field resistance is R_f, the field inductance is L_f, and the input (control) voltage to the field circuit is v_f.

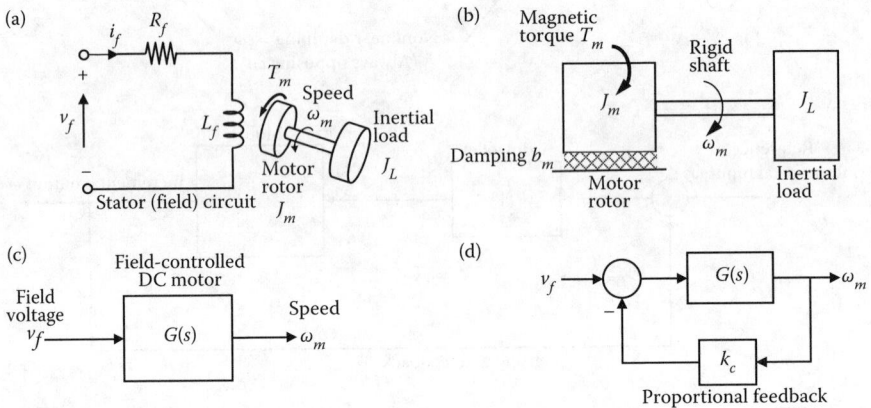

FIGURE P7.10
(a) A field-controlled dc motor with a permanent-magnet rotor. (b) Mechanical system of the motor with an inertial load. (c) Open-loop motor for speed control. (d) Proportional feedback system for speed control.

The mechanical dynamics of the motor system are represented by Figure P7.10b. Here, ω_m is the motor speed and b_m is the mechanical damping constant of the motor. The damping is assumed to be linear and viscous.

a. In addition to the torque equation given above, give the field circuit equation and the mechanical equation (with inertial load) of the motor, in terms of the system parameters given in Figure P7.10a and b. Clearly explain the principles behind these equations.

b. From the equations given in (a) above, obtain an expression for the transfer function $\omega_m/v_f = G(s)$ of the motor with the inertial load. The corresponding open-loop system is shown in Figure P7.10c.

c. Express the mechanical time constant τ_m and the electrical time constant τ_e of the (open-loop) motor system in terms of the system parameters given in Figure P7.10a and b.

d. What are the poles of the open-loop system (with speed as the output)? Is the system stable? Why? Sketch (*Note*: no need to derive) the shape of the output speed of the open-loop system to a step input in field voltage, with zero ICs. Justify the shape of this response.

e. Now suppose that a proportional feedback controller is implemented on the motor system, as shown in Figure P7.10d, by measuring the output speed ω_m and feeding it back with a feedback gain k_c. Express the resulting closed-loop transfer function in terms of τ_m, τ_e, k, and k_c, where $k=k_m/R_f b_m$. What is the characteristic equation of the closed-loop system?

f. For the closed-loop system, obtain expressions for the undamped natural frequency and the damping ratio, in terms of τ_m, τ_e, k, and k_c. Give an expression for the control gain k_c, in terms of the system parameters τ_m, τ_e, and k, such that the closed-loop system has *critical damping*.

g. Assuming that the closed-loop system is *under-damped* (i.e., the response is oscillatory), determine the time constants of this system. Compare them with the time constants of the open-loop system.

PROBLEM 7.11

A dc motor with velocity feedback is given by the block diagram in Figure P7.11 (without the feedforward control path indicated by the broken lines). The input is u, the output is the motor speed ω_m, and the load torque is T_L. The electrical dynamics of the motor are represented by the transfer function $k_e/(\tau_e s+1)$ and the mechanical dynamics of the motor are represented by the transfer function $k_m/(\tau_m s+1)$ where s is the Laplace variable, as usual.

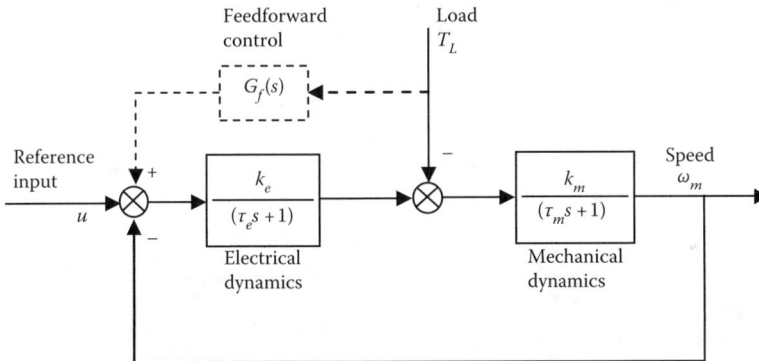

FIGURE P7.11
Control block diagram of a dc motor.

a. Obtain a transfer function equation relating the output ω_m to the two inputs u and T_L in terms of the given parameters and the Laplace variable.

b. Now include the feedfoward controller as shown by the broken line. Obtain an expression for the feedforward control transfer function $G_f(s)$, in terms of the given parameters and the Laplace variable, such that the effects of the load torque would be fully compensated (i.e., not felt in the system response ω_m).

PROBLEM 7.12

Consider a thermal process whose output temperature T is related to the heat input W as follows: $T=0.5W$

The units of T are °C and the units of W are watts. The heat generated by the proportional controller of the process is given by: $W=10(T_0 - T)+1,000$ in which T_0 denotes the temperature set point. Determine the offset for the following three set points:

Case 1: $T_0=500°$

$$\text{Case 2: } T_o = 200°$$

$$\text{Case 3: } T_o = 800°$$

PROBLEM 7.13

A linearized thermal plant is represented by $T = gc$ where, T = temperature (response) of the plant; c = heat input to the plant; and g = transfer function (constant).

A proportional controller is implemented on the plant, to achieve temperature regulation, and it is represented by: $c = ke + \Delta c$ where, $e = T_o - T$ = error signal into the controller; T_o = temperature set point (desired plant temperature); k = controller gain; and Δc = controller output when the error is zero.

Under the given conditions suppose that there is no offset.

   a. Express Δc in terms of T_o and g.
   b. If the plant transfer function changes to g' what is the resulting offset $T_o - T$?
   c. If the set point is changed to T_o' for the original plant (g) what is the resulting offset $T_o' - T$?

PROBLEM 7.14

A block diagram of a speed-control system is shown in Figure P7.14. Here the controlled process (say, an inertia with a damper) has a transfer function given by

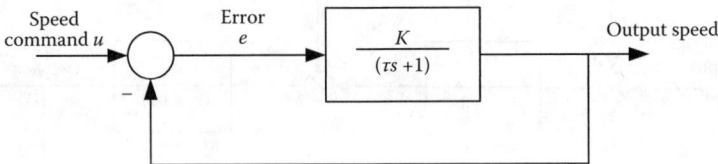

FIGURE P7.14
A speed control system.

$$\frac{1}{\tau s + 1}$$

and the controller is a proportional controller with gain K.

Hence, the combined transfer function process and controller is $K/\tau s + 1$. If the process is given a unit step input, find the final steady-state value of the output. What is the resulting offset? Show that this decreases when K is increased.

PROBLEM 7.15

Fraction (or percentage) of the full scale of controller input that corresponds to the full operating range of the final control element (controller output) is termed proportional band. Hence: $PB = \Delta e / \Delta e \times 100\%$ in which, PB = proportional band (%); ΔE = full scale of controller input; and Δe = range of controller input corresponding to the full range of final control element.

Obtain an equation relating PB to the proportional gain K_p of the controller. Use the additional parameter, Δc = full range of the final control element.

Note: PB is dimensionless but K_p has physical units because Δc and Δe do not have the same units in general.

A feedback control system for a water heater is shown in Figure P7.15. A proportional controller with control law: $W=20e+100$ is used, where W denotes the heat transfer rate into the tank in watts and e denotes the temperature error in °C. The tank characteristic for a given fixed rate of water flow is known to be: $W=20T$, where T is the temperature (°C) of the hot water leaving the tank. The full scale (span) of the controller is 500°C.

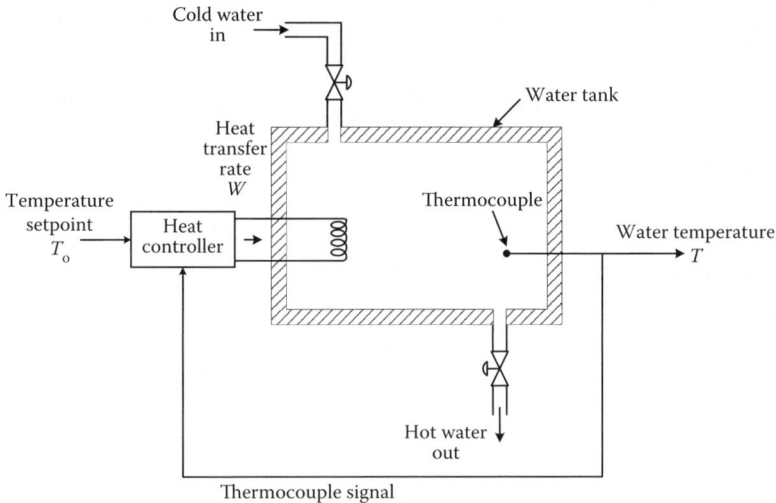

FIGURE P7.15
Feedback control of a water heater.

 a. What is the proportional gain and what is the proportional band of the controller?
 b. What is the set point value for which there is no offset error?
 c. If the set point is 40°C determine the offset.
 d. If the set point is 60°C determine the offset.
 e. If the proportional gain is increased to 80 watts/°C, what is the proportional band and what is the offset in (c)?
 f. Suppose the water flow rate is increased (load increased) so that the process law changes to $W=25T$. Determine the set point corresponding to zero offset noting that this is different from the answer for (b). Also, determine the offset in this case when the set point is 50°C.

PROBLEM 7.16

Consider six control systems whose loop transfer functions (or, forward TFs with unity feedback) are given by:

a. $\dfrac{1}{(s^2+2s+17)(s+5)}$ d. $\dfrac{10(s+2)}{(s^2+2s+101)}$

b. $\dfrac{10(s+2)}{(s^2+2s+17)(s+5)}$ e. $\dfrac{1}{s(s+2)}$

c. $\dfrac{10}{(s^2+2s+101)}$ f. $\dfrac{s}{(s^2+2s+101)}$

Compute the additional gain (multiplication) k needed in each case to meet a steady-state error specification of 5% for a step input.

PROBLEM 7.17

A tachometer is a device that is commonly used to measure speed, both rotatory (angular) and translatory (rectilinear). It consists of a coil which moves in a magnetic field. When the tachometer is connected to the object whose speed is to be sensed, the coil moves with the object and a voltage is induced in the coil. In the ideal case, the generated voltage is proportional to the speed. Accordingly, the output voltage of the tachometer serves as a measure of the speed of the object. High frequency noise that may be present in the tachometer signal can be removed using a low-pass-filter.

Figure P7.17 shows a circuit, which may be used to model the tachometer-filter combination. The angular speed of the object is ω_i and the tachometer gain is k. The leakage inductance in the tachometer is demoted by L and the coil resistance (possibly combined with the input resistance of the filter) is denoted by R. The low-pass-filter has an operational amplifier with a feedback capacitance C_f and a feedback resistor R_f. Since the operational amplifier has a very high gain (typically 10^5–10^9) and the output signal v_o is not large, the voltage at the input node A of the op-amp is approximately zero. It follows that v_o is also the voltage across the capacitor.

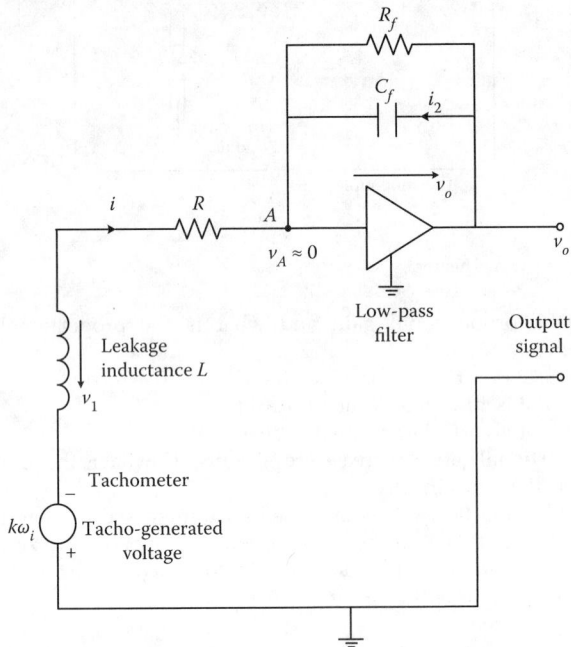

FIGURE P7.17
An approximate model for a tachometer-filter combination.

a. Comment on why the speed of response and the settling time are important in this application. Give two ways of specifying each of these two performance parameters.

b. Using voltage v_o across the capacitor C_f and the current i through the inductor L as the state variables and v_o itself as the output variable, develop a state-space model for the circuit. Obtain the matrices A, B, C, and D for the model.

c. Obtain the input–output differential equation of the model and express the undamped natural frequency ω_n and the damping ratio ζ in terms of L, R, R_f, and C_f. What is the output of the circuit at steady-state? Show that the filter gain k_f is given by R_f/R and the discuss ways of improving the overall amplification of the system.

d. Suppose that the percentage overshoot of the system is maintained at or below 5% and the peak time at or below 1 ms. Also it is known that $L=5.0$ mH and $C_f=10.0$ μF. Determine numerical values for R and R_f that will satisfy the given performance specifications.

PROBLEM 7.18

Consider the proportional plus derivative (PPD) servo system shown in Figure P7.18. The actuator (plant) transfer function is given by

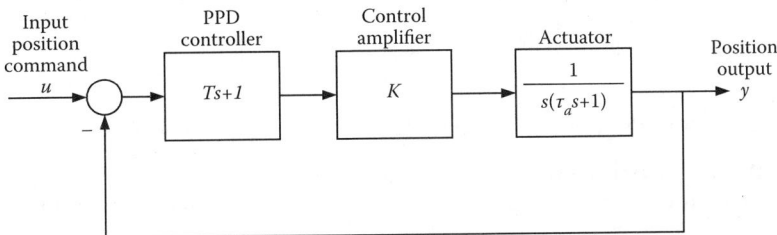

Input position command u — **PPD controller** $Ts+1$ — **Control amplifier** K — **Actuator** $\dfrac{1}{s(\tau_a s+1)}$ — **Position output** y

FIGURE P7.18
An actuator with PPD servo control.

$$G_p(s) = \frac{1}{s(0.5s+1)}.$$

By hand calculation complete the following table:

$\zeta\omega_n T_p$	2.5	2.5	2.5	2.5	2.2	2.1	2.0	2.0	2.0
$\omega_d T_p$	0.9π	0.75π	0.6π	0.55π	0.55π	0.25π	0.55π	0.2π	0.1π
T_p									
P.O.									

Design a PPD controller (i.e., determine the controller parameters T and K) that approximately meets $T_p=0.09$ and P.O.$=10\%$.

PROBLEM 7.19

Compare position feedback servo, tacho-feedback servo, and PPD servo with particular reference to design flexibility, ease of design, and cost.
    Consider an actuator with transfer function

$$G_p = \frac{1}{s(0.5s+1)}$$

Design a position feedback controller and a tacho-feedback controller that will meet the design specifications $T_p=0.09$ and P.O.$=10\%$.

PROBLEM 7.20

Consider the problem of tracking an aircraft using a radar device that has a velocity error constant of 10 s^{-1}. If the airplane flies at a speed of 2,000 km/hr at an altitude of 10 km, estimate the angular position error of the radar antenna that tracks the aircraft.

PROBLEM 7.21

List three advantages and three advantages of open-loop control. *Note*: the disadvantages will correspond to advantages of feedback control.

A process plant is represented by the block diagram shown in Figure P7.21a.

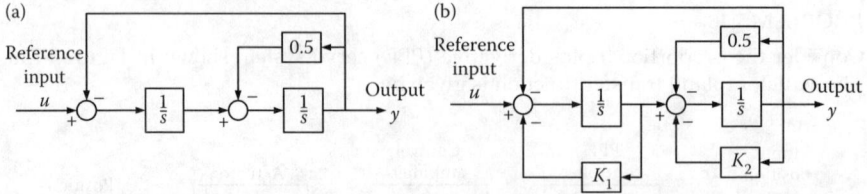

FIGURE P7.21

(a) Simulation block diagram of a process. (b) Plant with a feedback controller.

a. What is the transfer function (y/u) of the plant?
b. Determine the undamped natural frequency and the damping ratio of the plant.

A constant-gain feedback controller with gains K_1 and K_2 is added to the plant as shown in Figure P7.21b.

c. Determine the new transfer function of the overall control system.
d. Determine the values of K_1 and K_2 that will keep the undamped natural frequency at the plant value (of (b)) but will make the system critically damped.
e. Determine the values of K_1 and K_2 such that the overall control system has a natural frequency of $\sqrt{2}$ and a 2% settling time of 4 seconds.

PROBLEM 7.22

Describe the operation of the cruise control loop of an automobile, indicating the *input*, the output, and a *disturbance input* for the control loop. Discuss how the effect of a disturbance input can be reduced using feedforward control.

Synthesis of feedforward compensators is an important problem in control system design. Consider the control system shown in Figure P7.22. Derive the transfer function

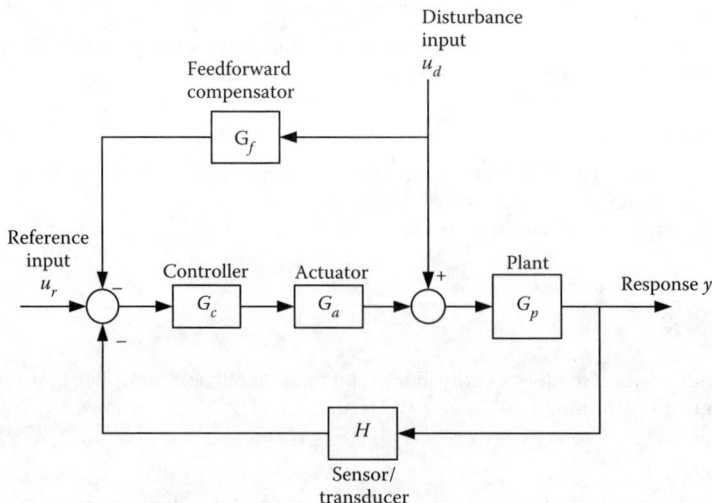

FIGURE P7.22

A feedback control system with feedforward compensation.

relating the disturbance input u_d and the plant output y. If you have the complete freedom to select any transfer function for the feedforward compensator G_f, what would be your choice? If the process bandwidth is known to be very low and if G_f is a pure gain, suggest a suitable value for this gain.

Suppose that a unit step input is applied to the system in Figure P7.22. For what value of step disturbance u_d will the output y be zero at steady-state?

PROBLEM 7.23

The transfer function of a field-controlled dc motor is given by

$$G(s) = \frac{K_m}{s(Js+b)(Ls+R)}$$

for open-loop control, with the usual notation. The following parameter values are given: $J=10$ kg.m², $b=0.1$ N.m/rad/s, $L=1$ Henry, and $R=10$ Ohms.

Calculate the electrical time constant and the mechanical time constant for the motor. Plot the open-loop poles on the s-plane. Obtain an approximate second order transfer function for the motor.

A proportional (position) feedback control system for the motor is shown in Figure P7.23. What is the closed-loop transfer function? Express the undamped natural frequency and damping ratio of the closed-loop system in terms of K, K_m, J, R and b.

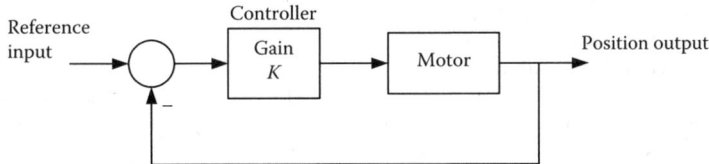

FIGURE P7.23
Block diagram of a position servo.

PROBLEM 7.24

A third-order closed-loop system with unity feedback is known to have the following characteristics:

a. It behaves like a second-order system.
b. Its 2% settling time is 4 seconds.
c. Its peak time is π seconds.
d. Its steady-state value for a unit step input is $=1$.

Determine the corresponding third-order open loop system (i.e., when the feedback is disconnected).

Explain why the steady-state error for a step input is zero for this closed-loop system.

PROBLEM 7.25

A control system with tacho-feedback is represented by the block diagram in Figure P7.25. The following facts are known about the control system:

1. It is a third-order system but behaves almost like a second-order system.
2. Its 2% settling time is 1 second, for a step input.

3. Its peak time is $\pi/3$ seconds, for a step input.
4. Its steady-state error, to a step input, is zero.

    (a)   Completely determine the third-order forward transfer function $G(s)$.
    (b)   Estimate the damping ratio of the closed-loop system.

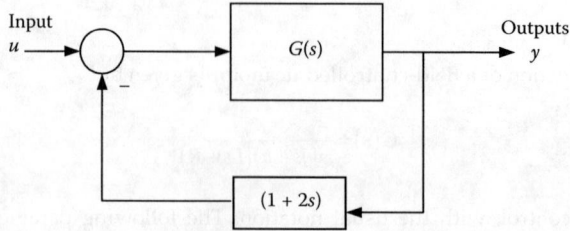

FIGURE P7.25
A control system with tacho-feedback.

PROBLEM 7.26

  a. Define the terms:
    (i)   Open-loop transfer function
    (ii)   Closed-loop transfer function
    (iii) Loop transfer function
    (iv) Forward transfer function
    (v)   Feedback transfer function
  b. Consider the feedback control system given by the block diagram in Figure P7.26. The forward transfer function is denoted by $G(s)$, which represents the plant and the controller.

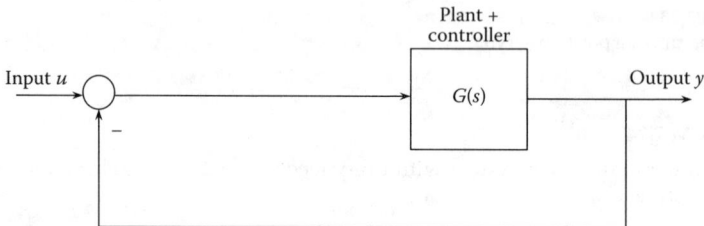

FIGURE P7.26
A feedback control system.

The following information is given to you:

    1. This is a Type 1 system.
    2. The overall closed-loop system is a damped oscillator given by

$$\frac{y}{u} = \frac{K}{s^2 + 2\zeta\omega_n s + \omega_n^2}.$$

    3. For the closed-loop system, the undamped natural frequency is $\sqrt{2}$ rad/s and the 2% settling time is 4 seconds.

Determine (by clearly explaining all your steps):

    i. The transfer function $G(s)$
    ii. Damping ratio ζ
    iii. Damped natural frequency of the closed-loop system
    iv. Steady-state error of the closed-loop system for a unit step input
    v. Velocity error constant (of the closed-loop system)
    vi. Steady-state error of the closed-loop system for a unit ramp input
    vii. Percentage overshoot of the closed-loop system to a step input (*Note*: Just express as an exponential of a numerical quantity. No need to evaluate it)
    viii. Time constant of the closed-loop system
    ix. Peak time of the closed-loop system
    x. Rise time of the closed-loop system

PROBLEMS 7.27

    a. List three parameters each, which can be used to specify the performance of a control system with respect to:
        (i)   Speed of response
        (ii)   Relative stability

        Define these parameters.
    b. Consider the feedback control system given by the block diagram in Figure P7.27. The forward transfer function is denoted by $G(s)$, which represents the plant and part of the controller. The feedback transfer function is given by $H(s)=bs+1$.

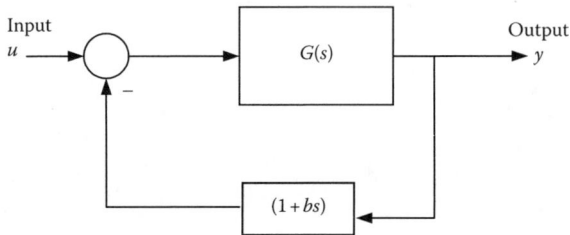

Input u → ⊖ → [$G(s)$] → Output y ; feedback block $(1+bs)$

FIGURE P7.27
A feedback control system.

      (i)   As a hardware component of the system, what does this feedback transfer function represent?
          The following facts are known about the control system:

        (1) It is a third-order system but behaves almost like a second order damped oscillator (with no zeros).
        (2) It is a Type 2 system.
        (3) Its 2% settling time is 4/3 seconds, for a step input.
        (4) Its peak time is $\pi/4$ seconds, for a step input.
      (ii)   Completely determine the third-order forward transfer function $G(s)$ (i.e., the numerical values of all its parameters) and the parameter b in the feedback transfer function.
      (iii) Estimate the damping ratio and the percentage overshoot of the closed-loop system.
      (iv) If a unit ramp input is applied to the system, what is the resulting steady-state error?

8

Stability and Root Locus Method

Stable response is a requirement for any control system. It ensures that the natural response to an initial condition excitation does not grow without bounds (or, more preferably, decays back to the initial condition) and the response to an input excitation (which itself is bounded) does not lead to an unlimited response. Asymptotic stability and bounded-input-bounded-output (BIBO) stability are pertinent in this context. In designing a control system, the required level of stability can be specified in several ways, both in the time domain and the frequency domain. Some ways of performance specification, with regard to stability in the time domain, were introduced in Chapter 7. The present chapter revisits the subject of stability, in time and frequency domains. Routh–Hurwitz method, root locus method, Nyquist criterion, and Bode diagram method incorporating gain margin (GM) and phase margin (PM) are presented for stability analysis of linear time-invariant (LTI) systems.

8.1 Stablility

In this section we will formally study stability of an LTI system, in the time domain. The natural response (homogeneous solution) of a system differential equation is determined by the eigenvalues (poles) of the system. Hence, stability is determined by the system poles.

8.1.1 Natural Response

Consider the LTI system:

$$a_n \frac{d^n y}{dt^n} + a_{n-1} \frac{d^{n-1} y}{dt^{n-1}} + \cdots + a_0 y = b_m \frac{d^m u}{dt^m} + b_{m-1} \frac{d^{m-1} u}{dt^{m-1}} + \cdots + b_0 u \tag{8.1}$$

where u is the input and y is the output. When there is no input, we have the *homogeneous equation*:

$$a_n \frac{d^n y}{dt^n} + a_{n-1} \frac{d^{n-1} y}{dt^{n-1}} + \cdots + a_0 y = 0 \tag{8.2}$$

It is well known (see Chapter 6) that the solution to this equation is of the exponential form

$$y = C e^{\lambda t} \tag{8.3}$$

which can be verified by substituting Equation 8.3 into Equation 8.2. Then, on canceling the common factor $Ce^{\lambda t}$, since $Ce^{\lambda t}$ is not zero for a general t, we get

$$a_n\lambda^n + a_{n-1}\lambda^{n-1} + \cdots + a_0 = 0 \tag{8.4a}$$

If we use s instead of λ in Equation 8.3 we have

$$a_n S^n + a_{n-1}S^{n-1} + \cdots + a_0 = 0 \tag{8.4}$$

Equation 8.4 is called the *characteristic equation* of the system (Equation 8.1) and its roots are called *eigenvalues* or *poles* of the system, as introduced in previous chapters. In general there will be n roots for Equation 8.4, and let us denote them by $\lambda_1, \lambda_2,\ldots, \lambda_n$. The solutions of Equation 8.3 should be formed by combining the contributions of all these roots. Hence, the general solution to Equation 8.2, assuming that there are no repeated roots among $\lambda_1, \lambda_2,\ldots, \lambda_n$, is

$$y_h = C_1 e^{\lambda_1 t} + C_2 e^{\lambda_2 t} + \cdots + C_n e^{\lambda_n t} \tag{8.5}$$

This is the homogeneous solution of Equation 8.1, and it represents the natural response (the response to an initial-condition excitation) of the system. *Note*: This response does not depend on the input.

The constants $C_1, C_2, \ldots, C_n$ are unknowns (integration constants), which have to be determined by using the necessary n initial conditions (say, $y(0), y'(0),\ldots, y^{n-1}(0)$) of the system. If two roots are equal, then the corresponding integration constants C_i are not independent and can be combined into a single constant in Equation 8.5. Then we have the situation of $n-1$ unknowns to be determined from n initial conditions. This is not a valid situation. In order to overcome this, when two roots are identical (repeated) a factor t has to be incorporated into one of the corresponding constants of integration. Without loss of generality, if $\lambda_1 = \lambda_2$ we have

$$y_h = C_1 e^{\lambda_1 t} + C_2 t e^{\lambda_1 t} + \cdots + C_n e^{\lambda_n t} \tag{8.6}$$

Note: Similarly, if there are three repeated roots, terms t and t^2 have to be incorporated into the corresponding constants, and so on.

Next let us see how the natural response of a system depends on the nature of the poles. First, since the system coefficients (parameters) $a_0, a_1,\ldots, a_n$ (in Equations 8.1, 8.2, and 8.4 are all real, any complex roots of Equation 8.4 must occur in complex conjugates (i.e., in pairs of $\lambda_r + j\lambda_i$ and $\lambda_r - j\lambda_i$, where λ_r and λ_i are the real and the imaginary parts, respectively, of a complex root). The corresponding response terms are:

$$e^{(\lambda_r \pm j\lambda_i)t} = e^{\lambda_r t}\left(\cos \lambda_i t \pm j\sin \lambda_i t\right) \tag{8.7}$$

Hence, it is clear that if a root (pole) has an imaginary part, then that pole produces an oscillatory (sinusoidal) natural response. Also, if the real part is negative, it generates an exponential decay and if the real part is positive, it generates an exponential growth (or, unstable response). These observations are summarized in Table 8.1 and further illustrated in Figure 8.1.

Note: Poles are the roots of the characteristic Equation 8.4 and hence they can be marked on the s-plane.

TABLE 8.1

Dependence of Natural Response on System Poles

Pole		Nature of Response	
Real	Negative	Transient	Decaying (stable)
	Positive	(Nonoscillatory)	Growing (unstable)
Imaginary (pair)		Oscillatory with constant amplitude	Steady (marginally stable)
Complex (pair)	Negative real part	Oscillatory with varying amplitude	Decaying (stable)
	Positive real part		Growing (unstable)
Zero value		Constant	Marginally stable
Repeated		Includes a linearly/ploynomially increasing term (unstable if not accompanied by exponential decay)	

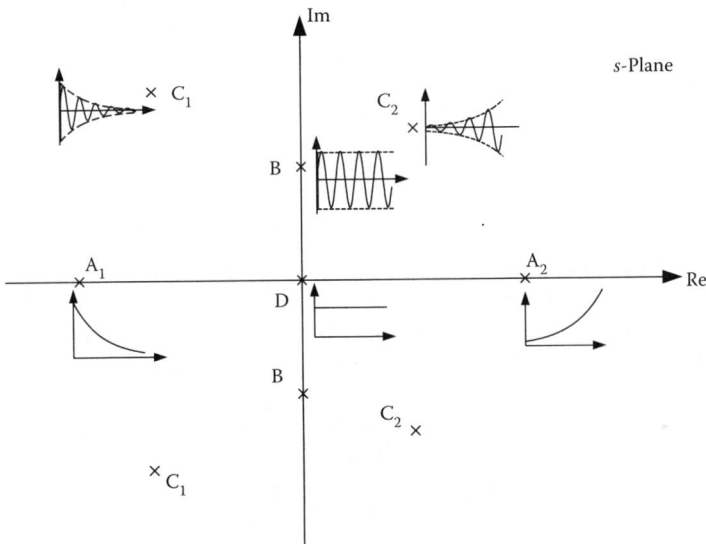

FIGURE 8.1
Pole location on the *s*-plane and the corresponding response.

Note: The case of repeated poles should be given special care. As noted before, the constants of integration will have terms such as t and t^2 (polynomial) in this case. These are growing (unstable) terms unless there are accompanied by decaying exponential terms, which will counteract the polynomial growth (because an exponential decay is stronger than a polynomial growth of any order).

8.2 Routh–Hurwitz Criterion

The Routh test or Routh–Hurwitz stability criterion is a simple way to determine whether a system is stable (i.e., whether none of the poles have positive real parts) by examining the characteristic polynomial, without actually solving the characteristic equation for its roots. If the system is unstable, the Routh test also tells us how many poles are on the right half plane (RHP); i.e., the number of unstable poles. First a Routh array has to be formed in order to perform the test.

8.2.1 Routh Array

The characteristic equation of an nth-order system can be expressed as Equation 8.4. As noted in Chapters 5 and 6, this is also the denominator of the system transfer function, when equated to zero. This has n roots, which are the poles (or eigenvalues) of the system. It is possible to determine the stability of the system without actually finding these n roots, by forming a Routh array, as follows:

	First Column	Second Column	Third Column	...	
s^n	a_n	a_{n-2}	a_{n-4}	...	←First row
s^{n-1}	a_{n-1}	a_{n-3}	a_{n-5}	...	←Second row
s^{n-2}	b_1	b_2	b_3	...	←Third row
s^{n-3}	c_1	c_2	c_3	...	←Fourth row
.	.	.	.	...	
.	.	.	.	...	
s^0	h_1				←Last row

The first two rows are completed first, using the coefficients $a_n, a_{n-1}, ..., a_1, a_0$ of the characteristic polynomial, as shown. Note the use of alternate coefficients in these two rows. Each subsequent row is computed from the elements of the two rows immediately above it, by *cross-multiplying* the elements of those two rows. For example:

$$b_1 = \frac{a_{n-1}a_{n-2} - a_n a_{n-3}}{a_{n-1}}$$

$$b_2 = \frac{a_{n-1}a_{n-4} - a_n a_{n-5}}{a_{n-1}}, \text{ etc.}$$

$$c_1 = \frac{b_1 a_{n-3} - a_{n-1}b_2}{b_1}$$

$$c_2 = \frac{b_1 a_{n-5} - a_{n-1}b_3}{b_1}, \text{ etc.}$$

and so on, until the coefficients of the last row are computed.

The Routh–Hurwitz stability criterion states that for the system to be stable, the following two conditions must be satisfied:

1. All the coefficients $(a_0, a_1, ..., a_n)$ of the characteristic polynomial must be positive (i.e., same sign; because, all signs can be reversed by multiplying by -1).
2. All the elements in the first column of the Routh array must be positive (i.e., same sign).
3. If the system is unstable, the number of unstable poles is given by the number of successive sign changes in the elements of the first column of Routh array.

Example 8.1

Consider a system whose closed-loop transfer function is $G(s)=2/(s^3-s^2+2s+1)$.

Without even completing a Routh array, it is seen that the system is unstable—from Condition 1 of the Routh test (because, a negative coefficient is present in the characteristic polynomial).

Example 8.2

Consider a system having the (closed-loop) transfer function $G(s)=2(s+5)/(3s^3+s^2+4s+2)$.

Its Routh array is formed by examining the characteristic equation:

$$3s^3+s^2+4s+2=0$$

The Routh array is

S^3	3	4
S^2	1	2
S^1	b_1	0
S^0	c_1	0

where

$$b_1 = \frac{1\times 4 - 3\times 2}{1} = -2$$

$$c_1 = \frac{b_1\times 2 - 1\times 0}{b_1} = 2$$

The first column of the array has a negative value, indicating that the system is unstable. Furthermore, since there are "two" sign changes (positive to negative and then back to positive) in the first column, there are two unstable poles in this system.

8.2.2 Auxiliary Equation (Zero-Row Problem)

A Routh array may have a row consisting of zero elements only. This usually indicates a *marginally stable* system (i.e., a pair of purely imaginary poles). The roots of the polynomial equation formed by the row that immediately precedes the row with zero elements, will give the values of these marginally stable poles.

Example 8.3

Consider a plant $G(s)=1/s(s+1)$ and a feedback controller $H(s)=K(s+5)/(s+3)$. Its closed loop characteristic polynomial $(1+GH=0)$ is $1+K(s+5)/s(s+1)(s+3)=0$

$$\text{or: } s(s+1)(s+3)+K(s+5)=0$$

$$\text{or: } s^3+4s^2+(3+K)s+5K=0$$

The Routh array is:

S^3	1	$3+K$
S^2	4	$5K$
S^1	$\dfrac{12-K}{4}$	0
S^0	$5K$	

Note that when $K=12$, the third row (corresponding to s^1) of the Routh array will have all zero elements.

The polynomial equation corresponding to the previous row (s^2) is $4s^2+5K=0$.

With $K=12$, we have the auxiliary equation: $4s^2+5\times 12=0$ or $s^2+15=0$ whose roots are $s=\pm j\sqrt{15}$.

Hence when $K=12$ we have a marginally stable closed-loop system, two of whose poles are $\pm j\sqrt{15}$. The third pole can be determined by comparing coefficients as shown below. This remaining root has to be real (because if it is a complex root, it must occur as a conjugate pair of roots). Call it p. Then, with $K=12$, on combining with the factor corresponding to the auxiliary equation (the complex root pair) the characteristic polynomial will be: $(s-p)(s^2+15)$. This must correspond to the same characteristic equation as given in the problem. Hence:

$$s^3+4s^2+15s+60=(s-p)(s^2+15)=0$$

By comparing coefficients: $60=-15p$ or $p=-4$

Hence the real pole is at -4, which is stable. Since the other two poles are marginally stable, the overall system is also marginally stable.

8.2.3 Zero Coefficient Problem

If the first element in a particular row of a Routh array is zero, a division by zero will be needed when computing the next row. This will create an ambiguity as to the sign of the next element. This problem can be avoided by replacing the zero element by a small positive element ε, and then completing the array in the usual manner.

Example 8.4

Consider a system whose characteristic equation is $s^4+5s^3+5s^2+25+10=0$.

Let us study the stability of the system.

Routh array is:

S^4	1	5	10
S^3	5	25	0
S^2	ε	10	0
S^1	$\dfrac{25\varepsilon-50}{\varepsilon}$	0	
S^0	10		

Note that the first element in the third row (s^2) should be $5\times5-25\times1/5=0$. But we have represented it by ε, which is positive and will tend to zero Then, the first element of the fourth row (s^1) becomes

$25\varepsilon - 50/\varepsilon$. Since ε is very small, 25ε is also very small. Hence, the numerator of this quantity is negative, but the denominator (ε) is positive. It follows that this element is negative (and large). This indicates two sign changes in the first column, and hence the system has two unstable poles.

In applying Routh–Hurwitz stability criterion, we may start with the system differential equation. Consider Example 8.5.

Example 8.5

Consider the differential equation

$$\frac{3d^3y}{dt^3} + \frac{2d^2y}{dt^2} + \frac{dy}{dt} + y = \frac{2du}{dt} + u$$

in which u is the system input and y is the system output. To obtain the transfer function: change d/dt to s. The result of substituting s for d/dt (and s^2 for d^2/dt^2) in the system differential equation is: $(3s^2 + 2s^2 + s + 1)\, y = (2s + 1)u$

The system transfer function (output/input) is: $\dfrac{y}{u} = \dfrac{(2s+1)}{(3s^3 + 2s^2 + s + 1)}$

Routh array for the characteristic polynomial is constructed as:

S^3	3	1
S^2	2	1
S^1	1/2	0
S^0	1	0

This first column has no sign changes. Hence the system is stable.

8.2.4 Relative Stability

Consider a stable system. The pole that is closest to the imaginary axis is the *dominant pole*, because the natural response from remaining poles will decay to zero faster, leaving behind the natural response of this (dominant) pole. It should be clear that the distance of the dominant pole from the imaginary axis is a measure of the "level of stability" or "degree of stability" or "stability margin" or "relative stability" of the system. In other words, if we shift the dominant pole closer to the imaginary axis, the system becomes less stable (i.e., the "relative stability" of the resulting system becomes lower).

The stability margin of a system can be determined by the Routh test. Specifically, consider a stable system. All its poles will be on the left hand plane (LHP). Now, if we shift all the poles to the right by a known amount, the resulting system will be less stable. The stability of the shifted system can be established using the Routh test. If we continue this process of pole shifting in small steps, and repeatedly apply the Routh test, until the resulting system just goes unstable, then the total distance by which the poles have been shifted to the right provides a measure of the stability margin (or, relative stability) of the original system.

Example 8.6

A system has the characteristic equation: $s^3+6s^2+11s+36=0$

a. Using Routh–Hurwitz criterion determine the number of unstable poles in the system.
b. Now move all the poles of the given system to the right of the s-plane by the real value 1. (i.e., add 1 to every pole). Now how many poles are on the RHP?

Note: You should answer this question "without" actually finding the poles (i.e., without solving the characteristic equation).

Solution

a. Characteristic equation: $s^3+6s^2+11s+36=0$
   Routh array:

s^3	1	11
s^2	6	36
s^1	$\dfrac{6\times11-1\times36}{6}=5$	0
s^0	36	

Since the entries of the first column are all positive, there are no unstable poles in the original system.

b. Denote the shifted poles by $\tilde{s}$
   We have $\tilde{s}=s+1$ or $s=\tilde{s}-1$
   Substitute in the original characteristic equation. The characteristic equation of the system with shifted poles is:

$$(\tilde{s}-1)^3 + 6(\tilde{s}-1)^2 +11(\tilde{s}-1)+36 = 0$$

or: $\tilde{s}^3 - 3\tilde{s}^2 + 3\tilde{s} - 1 + 6\tilde{s}^2 - 12\tilde{s} + 6 + 11\tilde{s} - 11 + 36 = 0$

or: $\tilde{s}^3 + 3\tilde{s}^2 + 2\tilde{s} + 30 = 0$

Routh array:

$\tilde{s}^3$	1	2
$\tilde{s}^2$	3	30
$\tilde{s}^1$	$\dfrac{3\times2-1\times30}{3}=-3$	0
$\tilde{s}^0$	30	

There are two sign changes in the first column. Hence, there are two unstable poles.

8.3 Root Locus Method

Root locus is the locus of (i.e., continuous path traced by) the closed-loop poles (i.e., roots of the characteristic equation) of a system, as one parameter of the system (typically the loop gain) is varied. Specifically, the root locus shows how the locations of the poles of a closed-loop system change due to a change in some parameter of the loop transfer function. Hence, it indicates the stability of the closed-loop system as a function of the varied parameter. The method was published by W. R. Evans in 1948 but is used even today as a powerful tool for analysis and design of control systems. Since in the root locus method we use the system transfer function (specifically, the *loop transfer function*) and the associated "algebraic" approach to draw the root locus, we may consider this method as a frequency domain (strictly, Laplace domain) technique. On the other hand, since the closed-loop poles are directly related to the system response (in the time domain) and also to time-domain design specifications such as settling time, peak time, and percentage overshoot, the root locus method can be considered as a time domain technique as well. For these reasons we shall discuss the concepts of root locus without classifying the method into either the time domain or the Laplace domain.

The root locus starts from the open-loop poles (strictly speaking, loop poles). Hence, as the first step, these loop poles must be marked on the complex s-plane.

Consider the feedback control structure, as shown in Figure 8.2. The overall transfer function of this system (i.e., the *closed-loop transfer function*) is

$$\frac{Y(s)}{U(s)} = \frac{G(s)}{1 + G(s)H(s)} \tag{8.8}$$

The stability of the closed-loop system is completely determined by the poles (not zeros) of the closed-loop transfer function (Equation 8.8).

Note: Zeros are the roots of the numerator polynomial equation of a transfer function.

The closed-loop poles are obtained by solving the characteristic equation (the equation of the denominator polynomial):

$$1 + G(s)H(s) = 0 \tag{8.9a}$$

It follows that the closed-loop poles are (and hence, the stability of the closed-loop system is) completely determined by the *loop transfer function G(s)H(s)*. It is clear that the roots of Equation 8.9a depend on both poles and zeros of $G(s)H(s)$. Hence, stability of a closed-loop system depends on the poles and zeros of the loop transfer function.

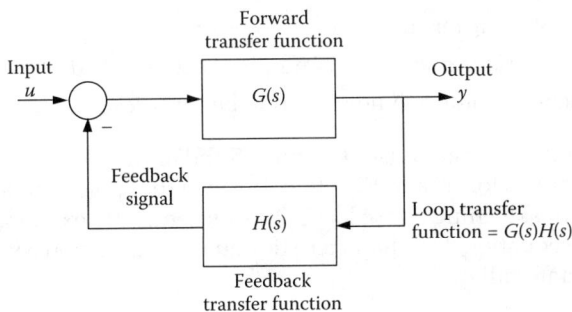

FIGURE 8.2
A feedback control system.

8.3.1 Rules for Plotting Root Locus

In theory, root locus can be plotted by successively varying any one parameter of the system and solving the corresponding closed-loop characteristic equation. This is in fact the method used in most computer programs for root locus plotting. But in many situations an accurate plot is not needed. For preliminary studies it is adequate to roughly sketch the root locus, with the exact numerical values being computed only for several critical parameter values and root locations. In this section we will summarize the rules for sketching a root locus. The principle behind each rule will be explained but for the present purpose it is not necessary to rigorously derive these results.

Since the loop transfer function $G(s)H(s)$ completely determines the closed-loop characteristic equation given by Equation 8.9a, it is the transfer function GH that is analyzed in plotting a root locus. Equation 8.9a can be rewritten in several useful and equivalent forms. First, we have:

$$GH = -1 \qquad\qquad (8.9b)$$

Next, since GH can be expressed as a ratio of two *monic polynomials* (i.e., polynomials whose highest order term coefficient is equal to unity) $N(s)$ and $D(s)$, we can write:

$$K\frac{N(s)}{D(s)} = -1 \quad \text{or} \quad KN(s) + D(s) = 0 \qquad\qquad (8.9c)$$

in which

$N(s)$ = numerator polynomial of the loop transfer function
$D(s)$ = denominator polynomial of the loop transfer function
K = loop gain.

Now since the polynomials can be factorized, we can write:

$$K\frac{(s - z_1)(s - z_2)\cdots(s - z_m)}{(s - p_1)(s - p_2)\cdots(s - p_n)} = -1 \qquad\qquad (8.9d)$$

$$\text{or:} \ (s - p_1)(s - p_2)\cdots(s - p_n) + K(s - z_1)(s - z_2)\cdots(s - z_m) = 0 \qquad (8.9e)$$

in which

z_i = a zero of the loop transfer function
p_i = a pole of the loop transfer function
m = order of the numerator polynomial = number of zeros of GH
n = order of the denominator polynomial = number of poles of GH.

For physically realizable systems (see Chapters 5 and 6) we have $m \leq n$. Equation 8.9c is in the "ratio-of-polynomials form" and Equation 8.9d is in the "pole-zero form." Now we will list the main rules for sketching a root locus, and subsequently explain each rule.

Note: It is just one equation, the characteristic equation (Equation 8.9) of the closed-loop system, which generates all these rules.

8.3.1.1 Complex Numbers

Before stating the rules we need to understand some basic mathematics of complex numbers. A complex number has a real part and an imaginary part. This can be represented

(a)

Im

$\underline{r} = r\cos\theta + jr\sin\theta$ (b)
$= re^{j\theta}$

$\underline{r}$

r

θ

$\rightarrow$ Re

0

Im

$\underline{r}$

$\underline{r} - \underline{a}$

$\underline{a}$

$\rightarrow$ Re

0

FIGURE 8.3

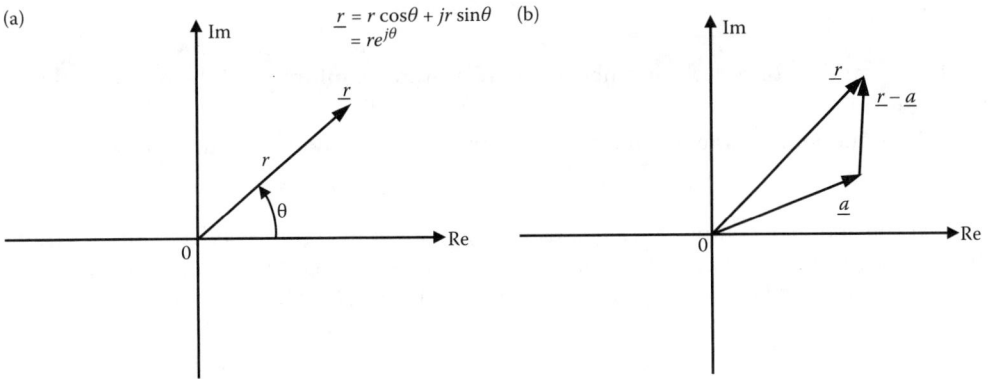

(a) Complex number represented as a two-dimensional vector. (b) The complex number subtraction.

by a vector (or a directed line) on the two-dimensional plane formed by a real axis and an orthogonal imaginary axis, as shown in Figure 8.3a. In particular, a complex number $\underline{r}$ can be expressed as

$$\underline{r} = r\cos\theta + jr\sin\theta = re^{j\theta} \qquad \text{(i)}$$

Note:

1. r denoted by $|\underline{r}|$ is the *magnitude* of the complex number (or vector) $\underline{r}$
2. θ denoted by $\angle\underline{r}$ is the *phase angle* of the complex number (or vector) $\underline{r}$

Now for two complex numbers

$$\underline{r_1} = r_1\cos\theta_1 + jr_1\sin\theta_1 = r_1 e^{j\theta_1} \qquad \text{(ii)}$$

$$\underline{r_2} = r_2\cos\theta_2 + jr_2\sin\theta_2 = r_2 e^{j\theta_2} \qquad \text{(iii)}$$

the product is

$$\underline{r_1}\,\underline{r_2} = (r_1\cos\theta_1 + jr_1\sin\theta_1)(r_2\cos\theta_2 + jr_2\sin\theta_2)$$
$$= r_1 r_2[(\cos\theta_1\cos\theta_2 - \sin\theta_1\sin\theta_2) + j(\cos\theta_1\sin\theta_2 - \sin\theta_1\cos\theta_2)] = r_1 r_2 e^{j(\theta_1+\theta_1)} \qquad \text{(iv)}$$

and the quotient is

$$\underline{r_1}/\underline{r_2} = (r_1\cos\theta_1 + jr_1\sin\theta_1)/(r_2\cos\theta_2 + jr_2\sin\theta_2)$$

Multiply the numerator and the denominator by $(r_2\cos\theta_2 - jr_2\sin\theta_2)$ and simplify. We get:

$$\frac{r_1}{r_2} = \frac{r_1}{r_2}(\cos\theta_1 + j\sin\theta_1)(\cos\theta_2 - j\sin\theta_2)$$

$$= \frac{r_1}{r_2}[(\cos\theta_1\cos\theta_2 + \sin\theta_1\sin\theta_2) + j(\sin\theta_1\cos\theta_2 - \cos\theta_1\sin\theta_2)] = \frac{r_1}{r_2}e^{j(\theta_1-\theta_1)} \qquad \text{(v)}$$

Hence:

1. In a product of complex numbers, the magnitudes multiply and the phase angles add.
2. In a quotient of complex numbers, the magnitudes divide and the phase angles subtract.

Next consider two complex numbers $\underline{s}$ and $\underline{a}$, as shown in Figure 8.3b. The complex number $\underline{s}-\underline{a}$ is given by the vector line starting from the head of $\underline{a}$ and ending at the head of $\underline{s}$, as shown. This can be confirmed by vector addition using triangle of vectors, since:

$$\underline{a}+\underline{s}-\underline{a}=\underline{s}-\underline{a} \qquad \text{(vi)}$$

8.3.1.2 Root Locus Rules

Rule 0 (symmetry): The root locus is symmetric about the real axis on the s-plane.
Rule 1 (number of branches): Root locus has n branches. They start at the n poles of the loop transfer function *GH*. Out of these, m branches terminate at the zeros of *GH* and the remaining $(n-m)$ branches go to infinity, tangential to $n-m$ lines called *asymptotes*.
Rule 2 (magnitude and phase conditions): The *magnitude condition* is:

$$K\frac{\prod\limits_{i=1}^{m}|s-z_i|}{\prod\limits_{i=1}^{n}|s-p_i|}=1 \qquad \text{(8.10a)}$$

The *phase angle condition* is:

$$\sum_{i=1}^{n}\angle(s-p_i)-\sum_{i=1}^{m}\angle(s-z_i)=\pi+2r\pi \qquad \text{(8.10b)}$$

$$r=0,\pm1,\pm2,\dots$$

Rule 3 (root locus on real axis): Pick any point on the real axis. If (#poles−#zeros) of *GH* to the right of the point is odd, the point lies on the root locus.
Rule 4 (asymptote angles): The $n-m$ asymptotes form angles

$$\theta_r=\frac{\pi+2\pi r}{n-m}$$

$$r=0,\pm1,\pm2,\dots \qquad \text{(8.11)}$$

with respect to the positive real axis of the s-plane.
Rule 5 (break points): *Break-in points* and *breakaway points* of root locus are where two or more branches intersect. They correspond to the points of repeated (multiple) poles of the closed-loop system. These points are determined by differentiating the characteristic

equation (Equation 8.9c) with respect to s, substituting for K using Equation 8.9c again, as: $K=-(D(a)/N(s))$. This gives

$$N(s)\frac{dD}{ds}-D(s)\frac{dN}{ds}=0 \tag{8.12}$$

Note: The root-locus branches at a break point are equally spaced (in angle) around the break point.

Rule 6 (intersection with imaginary axis): If the root locus intersects the imaginary axis, the points of intersection are given by setting $s=j\omega$ and solving the characteristic equation:

$$D(j\omega)+KN(j\omega)=0 \tag{8.13}$$

This gives two equations (one for the real terms and the other for the imaginary terms).

Alternatively, the Routh–Hurwitz criterion and the auxiliary equation for marginal stability, may be used to determine these points and the corresponding gain value (K).

Rule 7 (angles of approach and departure): The *departure angle* α of root locus, from a *GH* pole, is obtained using:

$$\alpha+\angle \text{ at other poles}-\angle \text{ at zeros}=\pi+2r\pi \tag{8.14a}$$

The *approach angle* α to a *GH* zero is obtained using

$$\angle \text{ at poles}-\alpha-\angle \text{ at other zeros}=\pi+2r\pi \tag{8.14b}$$

Note: Angles mentioned in Rule 7 are measured by drawing a line from the approach/departure point to the other pole or zero of *GH* that is considered and determining the angle of that line measured from the positive real axis (i.e., horizontal line drawn to the right at the other pole or zero).

Rule 8 (intersection of asymptotes with real axis): Asymptotes meet the real axis at the *centroid* about the imaginary axis, of the poles and zeros of *GH*. Each pole is considered to have a weight of $+1$ and each zero a weight of -1.

8.3.1.3 Explanation of the Rules

Rule 0: Since poles of any real system are either real or occur in complex conjugate pairs, the closed-loop poles must fall either on the real axis or as pairs that are symmetric about the real axis on the s-plane (on which the root locus is plotted). Hence the locus of these (closed-loop) poles must be symmetric about the real axis of the s-plane.

Rule 1: From Equation 8.9e it is clear that when we set $K=0$, the closed-loop poles are identical to the poles of *GH*, and when we set $K\rightarrow\infty$, the closed-loop poles are either equal to the zeros of *GH* or are at infinity. The latter is confirmed by the fact that Equation 8.9e may be written as: $(s-z_1)(s-z_2)\cdots(s-z_m)=-(s-p_1)(s-p_2)\cdots(s-p_n)/K$.

Here, as $K\rightarrow\infty$, the right hand side$=0$ for finite s, or else we must have $s\rightarrow\infty$.

Rule 2: This is obtained directly from Equation 8.9d in view of the fact that the magnitude of -1 is 1, and the phase angle of -1 is $\pi+2r\pi$, where r is any integer ($\pm$).

Rule 3: This is a direct result of the phase angle condition (Equation 8.10b). Unless Rule 3 is satisfied, the right hand side of Equation 8.10b would be an even multiple of π (which is incorrect) rather than an odd multiple of π (correct).

Rule 4: This is also a result of the phase angle condition (Equation 8.10b). But this time we use the fact that at infinity (i.e., $s \to \infty$), a root locus and its asymptote are identical, and the fact that when s is at infinity the finite values z_i and p_i can be neglected in Equation 8.10b. Then any of the angles $\angle(s-p_i)$ or $\angle(s-z_j)$ will be equal to the asymptote angle θ_r, and hence: $\sum_{i=1}^{n} \angle(s-p_i) - \sum_{i=1}^{m} \angle(s-z_i) = (n-m)\theta_r$.

Rule 5: To understand this rule, consider the characteristic polynomial of a closed-loop system that has two identical poles. It will have a factor $(s-p)^2$ where p is the double pole of the closed-loop system. If we differentiate the characteristic polynomial with respect to s, there still will remain a common factor $(s-p)$. This means that, at a double pole, both the characteristic polynomial and its derivative will be equal to zero. Similarly, at a triple pole (three identical poles), the characteristic polynomial and its first and second derivatives will vanish, and so on.

Note: The angle condition (Equation 8.10b) may be used to verify that the root locus branches at a break point are equally spaced (in angle) around the break point.

Rule 6: This rule should be clear from common sense because s is purely imaginary on the imaginary axis. Furthermore, purely imaginary poles are *marginally stable* poles. If all the remaining poles are stable, we have a marginally stable system in this case.

Rule 7: This rule is also a direct consequence of the phase angle condition (Equation 8.10b). Specifically, consider a GH pole p_i and a point (s) on the root locus very close to this pole. Then the angle $\angle(s-p_i)$ is in fact the angle of departure of the root locus from the pole p_i. A similar argument can be made for the angle of approach to a GH zero z_i.

Rule 8: To establish this rule using intuitive notions, consider Equation 8.9e, which can be written in the form

$$(s-p_1)(s-p_2)\cdots(s-p_n) = -K(s-z_1)(s-z_2)\cdots(s-z_m) \tag{8.9f}$$

Now define,

$$\bar{s} = \text{point at which asymptotes meet the real axis.}$$

Then, the vector $s-\bar{s}$ defines an asymptote line, where s denotes any general point on the asymptote.

Note: By definition, at infinity, a root locus branch and its asymptote line become identical. Hence, as $s \to \infty$, the asymptote angle is $s-\bar{s}$ where s is a point on the root locus at infinity.

As $s \to \infty$, we notice that the vector lines $(s-p_i)$ and $(s-z_j)$ all appear to be identical to $(s-\bar{s})$, for any finite p_i and z_j, and they all appear to come from the same point $\bar{s}$ on the real axis. Hence, Equation 8.9f becomes

$$(s-\bar{s})^n = -K(s-\bar{s})^m \tag{8.15}$$

as $s \to \infty$ on root locus.

Substitute Equation 8.15 into Equation 8.9f. We get

$$(s-p_1)(s-p_2)\cdots(s-p_n) = (s-\bar{s})^{n-m}(s-z_1)(s-z_2)\cdots(s-z_m) \tag{8.16}$$

as $s \to \infty$ on root locus.

Hence, by equating the coefficients of s^{n-1} terms in Equation 8.16 we have

$$\sum_{i=1}^{n} p_i = (n-m)\bar{s} + \sum_{i=1}^{m} z_i$$

or

$$\bar{s} = \frac{1}{(n-m)} \left[\sum_{i=1}^{n} p_i - \sum_{i=1}^{m} z_i \right] \tag{8.17}$$

Note: Equation 8.17 is indeed Rule 8. Remember here that we need to consider only the real parts of p_i and z_i because the imaginary parts occur as conjugate pairs in complex poles or zeros, and they cancel out in the summation.

8.3.2 Steps of Sketching Root Locus

Now we list the basic steps of the normal procedure that is followed in sketching a root locus.

Step 1: Identify the loop transfer function and the parameter (gain K) to be varied in the root locus.

Step 2: Mark the poles of GH with the symbol ($\times$) and the zeros of GH with the symbol (o) on the s-plane.

Step 3: Using Rule 3 sketch the root locus segments on the real axis.

Step 4: Compute the asymptote angles using Rule 4 and the asymptote origin using Rule 8, and draw the asymptotes.

Step 5: Using Rule 5, determine the break points, if any.

Step 6: Using Rule 7 compute the departure angles and approach angles, if necessary.

Step 7: Using Rule 6, determine the points of intersection with the imaginary axis, if any.

Step 8: Complete the root locus by appropriately joining the points and segments that have been determined in the previous steps.

Example 8.7

A dc servomotor uses a proportional feedback controller along with a low-pass filter to eliminate signal noise. A block diagram of the control system is shown in Figure 8.4. The component transfer functions are as given in the diagram.

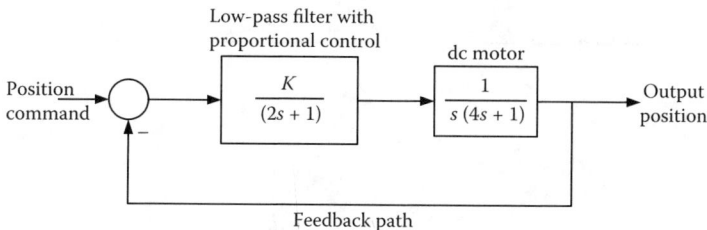

FIGURE 8.4
Block diagram of a dc servomotor.

a. Sketch the root locus for the closed-loop system indicating numerical values of the break points and asymptotes.
b. Determine the value of gain K when the closed-loop system has two equal poles. What is the value of these poles?
c. Determine the range of K for which the closed-loop system is stable.
d. What is the frequency at which the system will oscillate when it is marginally stable?

Solution

a. $GH = \dfrac{K}{s(2s+1)(4s+1)}$

The three GH poles are at: 0, −1/4 and −1/2.

The three branches of the root locus will originate from these locations.

There are no GH zeros. We mark the GH poles on the s-plane, as in Figure 8.5. From Rule 3, the root locus segments on the real axis are between $-\infty$ and −1/2, and between −1/4 and 0, as sketched in Figure 8.5.

Since there are no GH zeros, the three root locus branches will end at asymptotes (infinity).

The asymptote angles are: $\pi \pm 2r\pi/3 = \pm 60°$ and 180° (Rule 4)

Centroid $\bar{s}$ of the poles is given by $3\bar{s} = 0 - 1/4 - 1/2 \Rightarrow \bar{s} = -1/4$

The three asymptotes meet at this centroid (Rule 8). The asymptotes are drawn as in Figure 8.5.

Closed-loop characteristic equation: $s(2s+1)(4s+1)+K=0$

or

$$8s^3 + 6s^2 + s + K = 0 \tag{i}$$

Differentiate to determine break points: $24s^2 + 12s + 1 = 0 \Rightarrow s = (-1/4) \pm (1/4\sqrt{3})$

The correct break point $= (-1/4) + (1/4\sqrt{3})$

Note: The solution $(-1/4) - (1/4\sqrt{3})$ cannot be a break point as it is not on the root locus (from Rule 3, as sketched before).

The complete root locus can be sketched now, as shown in Figure 8.5.

b. As obtained in (a), the repeated roots are given by the break point

$$s = -\frac{1}{4} + \frac{1}{4\sqrt{3}} = -0.106$$

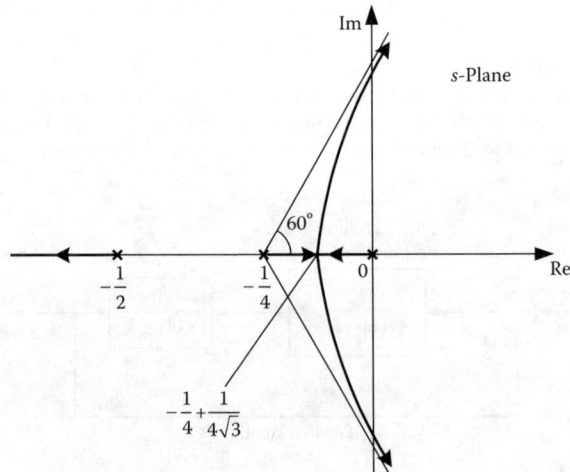

FIGURE 8.5
The root locus of the system in Figure 8.4.

Substitute in Equation (i): $K=-8(-0.106)^3 -6(-0.106)^2 -(-0.106)=0.048$

c. Use Routh–Hurwitz criterion on the closed-loop characteristic equation (i).
Routh array:

s^3	8	1
s^2	6	K
s^1	$\dfrac{6-8K}{6}$	0
s^0	K	

For stability we need $K>0$ and $6-8K>0$.
Hence, the stability region is given by: $0<K<(3/4)$.

d.

Method 1:

From Routh array, the row s^1 becomes null when $K=3/4$. The corresponding auxiliary equation is given by the previous row: $6s^2+K=0 \Rightarrow 6s^2+3/4=0$.
The corresponding imaginary roots are $s=\pm j\,(1/2\sqrt{2})$.
These are the points of intersection of the root locus with the imaginary axis, and they correspond to the frequency of oscillation at marginal stability.

Method 2:

Substitute $s=j\omega$ in the closed-loop characteristic equation (i) for marginal stability. We get

$$8(j\omega)^3+6(j\omega)^2+j\omega+K=0$$

$$\Rightarrow -8j\omega^3-6\omega^2+j\omega+K=0$$

$$\Rightarrow -6\omega^2+K=0 \quad \text{and} \quad -8\omega^3+\omega=0$$

$$\Rightarrow \omega^2=\frac{1}{8} \quad \text{and} \quad K=6\omega^2$$

$$\Rightarrow \omega=\frac{1}{2\sqrt{2}} \quad \text{and} \quad K=\frac{3}{4}.$$

This gives the same result as before.

Example 8.8

Consider the feedback control system shown in Figure 8.6. The following three types of control may be used:

a. Proportional (P) control: $G_c=K$,
b. Proportional+derivative (PD) control: $G_c=K(1+s)$,
c. Proportional+integral (PI) control: $G_c=K(1+1/s)$.

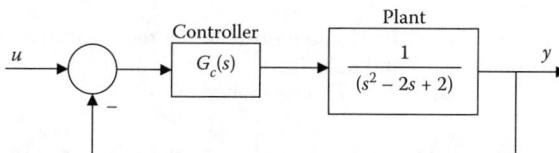

FIGURE 8.6
A feedback control system.

(a)

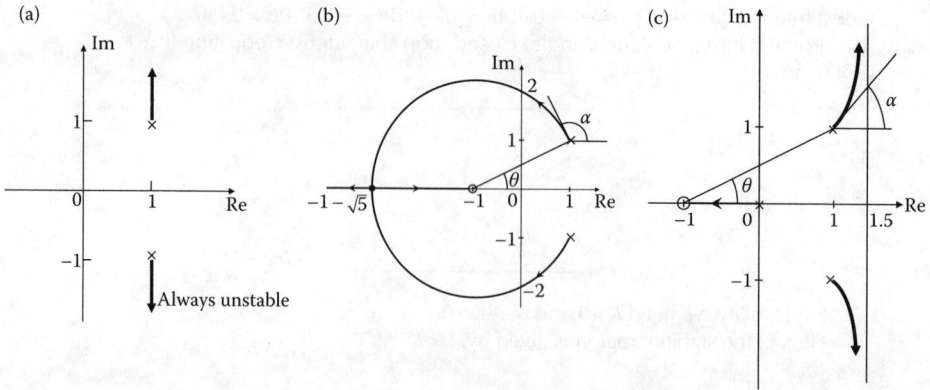

(b)

(c)

FIGURE 8.7
(a) Root locus for the system with P control. (b) Root locus for the system with PD control. (c) Root locus for the system with PI control.

Sketch the root loci for these three cases and compare the behavior of the corresponding controlled systems.

Solution

a. The loop transfer function $GH=K/(s^2-2s+2)$.
   The loop poles are at $s=1\pm j$. The two root locus branches will start from them.
   These are marked on the s-plane in Figure 8.7a.
   There are no loop zeros.
   Hence, according to Rule 3, there are no segments of the root locus on the real axis.
   Since there are no GH zeros, there are two asymptotes where the root locus branches will end (at infinity).
   The asymptote angles are: $\pm 90°$ (Rule 4).
   The pole centroid $\bar{s}=(1\times 2/2)=1$.
   The asymptotes intersect at this centriod (Rule 8), and can be sketched as in Figure 8.7a.
   Note: Even though obvious, the departure angle α at the pole $1+j$ can be determined by:

$$\alpha+90°=180°\Rightarrow\alpha=90° \hspace{2cm} \text{(Rule 7)}$$

   The complete root locus for this case (P control) is sketched in Figure 8.7a.
   It is seen that the system is always unstable.

b. The loop transfer function $GH=K(1+s)/(s^2-2s+2)$.
   There are two loop poles, at: $s=1\pm j$ from which the two branches of root locus originate.
   There is a loop zero at: -1 where one of the root locus branches terminates.
   These are marked on the s-plane in Figure 8.7b.
   According to Rule 3, the root locus lies on the real axis between $-\infty$ and -1, as sketched in Figure 8.7b.
   Since there are two loop poles and one loop zero, the root locus has one asymptote with asymptote angle $180°$ (Rule 1 and Rule 4).
   The departure angle α from pole $1+j$ is determined by Rule 7:

$$\alpha+90°-\theta=180° \text{ (where } \tan\theta=\frac{1}{2} \text{ or } \theta=26.6°)\Rightarrow\alpha=116.6°.$$

Note: The departure angle from pole $1 - j$ may be determined simply by the symmetry of the root locus (Rule 0) or by Rule 7 as:

$$\alpha + (-90°) - (-\theta) = 180° \text{ (where } \theta = 26.6° \text{ as before)} \Rightarrow \alpha = 243.4° \text{ or } -116.6°$$

Break points (Rule 5):
Here, $N(s) = (1 + s)$ and $D(s) = s^2 - 2s + 2$.
Hence, the break point is given by (Equation 8.12):

$$(1 + s)(2s - 2) - (s^2 - 2s + 2) = 0 \Rightarrow s^2 + 2s - 4 = 0 \Rightarrow s = -1 \pm \sqrt{5}$$

The correct break point must be on the root locus (i.e., < -1). Hence we pick

$$s = -1 - \sqrt{5}.$$

The complete root locus for the present case of PD control is sketched in Figure 8.7b.
It is seen that the system is unstable for low values of gain K, starting from 0, but becomes stable beyond a certain gain value. This is due to the inclusion of derivative control (or, a loop zero on the LHP), which has a stabilizing effect. The gain value and the frequency of marginal stability can be determined as usual. In the present case this is a relatively simple exercise. The closed-loop characteristic equation (Equation 8.9c) is:

$$\frac{K(1 + s)}{(s^2 - 2s + 2)} = -1 \Rightarrow (s^2 - 2s + 2) + K(1 + s) = 0$$

$$\Rightarrow s^2 + (K - 2)s + K + 2 = 0.$$

Hence, for stability, we must have $K > 2$. The gain for marginal stability is $K = 2$. The corresponding characteristic equation is $s^2 + K + 2 = 0$, and the resulting marginally stable closed-loop poles are $s = \pm j2$

c. The loop transfer function $GH = K(1 + s)/s(s^2 - 2s + 2)$.
   Now there are three loop poles, at: $s = 0$ and $1 \pm j$ from which the three branches of root locus originate.
   There is a loop zero at -1 where one of the root locus branches terminates.
   These are marked on the s-plane in Figure 8.7c.
   According to Rule 3, the root locus lies on the real axis from 0 to -1, as sketched in Figure 8.7b.
   Since there are three loop poles and one loop zero, the root locus has two asymptotes with asymptote angles $= \pm 90°$ (Rule 1 and Rule 4).
   The pole centroid $\bar{s} = (1 \times 2 - 1 \times (-1))/3 - 1 = 1.5$.
   The asymptotes intersect at this point (Rule 8), as sketched in Figure 8.7c.
   The departure angle α from the pole $1 + j$ is determined by:

$$\alpha + 90° + 45° - \theta = 180° \quad \text{where} \quad \tan\theta = \frac{1}{2} \quad \text{or} \quad \theta = 26.6° \Rightarrow \alpha = 71.6°$$

Note: As usual, the departure angle from the other (conjugate) pole $1 - j$ is determined by symmetry as: $\alpha = -71.6°$.
The complete root locus for the case with PI control is sketched in Figure 8.7c.
It is seen that the system is always unstable. In fact, the system is more unstable than with P control alone, and becomes worse as the gain K is increased. This is due to the presence of the integral (I) action in the controller, which has a destabilizing effect.

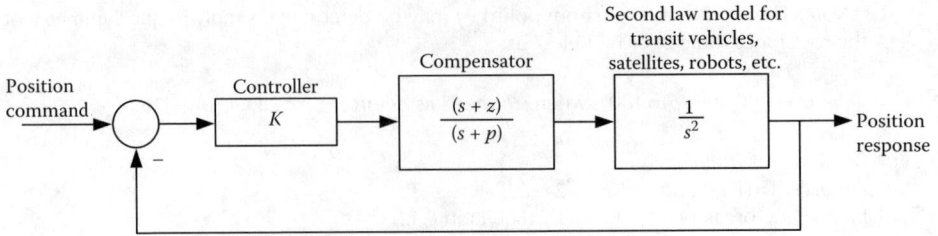

FIGURE 8.8
Feedback control of unconstrained motion of a mechanical device.

Example 8.9

Newton's second law suggests that a very approximate model for an unconstrained rigid body is a double integrator. Aircraft, satellites, guideway vehicles, and robotic manipulators, in their direction of motion, can be crudely approximated by this model. Consider position feedback control of such a system. A compensator is used in the forward path of the control system as shown in Figure 8.8. The controller gain K includes gain in other components such as the plant, sensors, and the compensator. The parameters p and z, which determine the compensator pole and zero, are assumed positive. We are interested in studying the behavior of the control system for different values of the controller gain. This is easily accomplished by sketching the root locus of the system. Show that:

a. There is only *one* break point if $p<z$
b. There is only *one* break point if $z<p<9z$
c. There are *two* break points if $p=9z$
d. There are *three* break points if $p>9z$

Sketch the root locus for each of the four cases mentioned above and discuss stability of the closed-loop system in these cases.

Solution

The loop transfer function for this example is

$$GH = \frac{K(s+z)}{s^2(s+p)} \tag{i}$$

The parameter that is varied in the root locus is the loop gain K. Before sketching the root locus, let us examine the possible break points, using Rule 5. Since

$$N(s) = s+z \tag{ii}$$

and

$$D(s) = s^2(s+p) \tag{iii}$$

we have the condition for break points:

$$(s+z)(3s^2+2sp) - s^2(s+p) = 0$$

which simplifies to:

$$s[2s^2+(p+3z)s+2pz] = 0 \tag{iv}$$

We observe from Equation (iv) that the point $s=0$ is always a break point. The remaining break points are obtained by solving:

$$2s^2 + (p+3z)s + 2pz = 0 \tag{v}$$

which gives

$$s = -\frac{1}{4}(p+3z) \pm \frac{1}{4}\sqrt{(p+3z)^2 - 16pz} \tag{vi}$$

The number of possible break points will depend on the sign of the *discriminant*—the expression under the square root sign: $\Delta = (p+3z)^2 - 16pz$
This expression can be expanded and factorized as:

$$\Delta = (p-z)(p-9z). \tag{vii}$$

Five cases can be identified for examining the sign of Δ. However, since the case $p=z$ provides a "pole-zero cancellation" in the compensator, it is not considered here. The remaining four cases correspond to those stated in the problem.

First, note from Equation (vi) that the two roots are always negative because the square root term is always less than $(p+3z)$. Now, let us examine the four possible cases.

Case 1: $p<z$

In this case we have $\Delta>0$ and we get two real (and negative) roots from Equation (vi). We can show that these two points do not lie on the root locus and, hence, are not valid break points. First note from Step 3 of root locus sketching that the root locus segment on the real axis extends from $-p$ to $-z$ only. Now in the present case we have

$$z = rp \quad \text{with} \quad r>1 \tag{viii}$$

Then from the coefficient of s^0 in Equation (v) we note that the product of the roots is $pz = rp^2$. If one of these roots is to the right of $-p$ then, the other root will be to the left of $-rp$ (i.e., to the left of $-z$) and consequently both roots fall outside the root locus segment. Therefore, it is adequate to show that one root of Equation (v) falls outside the root locus segment on the real axis, because then the other root also will fall outside the root locus, in the present case. Now substitute Equation (viii) into Equation (vi) with:

$$r = 1+\delta, \ \delta>0 \tag{ix}$$

We have:

$$s = -p - \frac{3\delta p}{4} \pm \frac{p}{4}\sqrt{9\delta^2 + 8\delta} \tag{x}$$

The positive square root in Equation (x) corresponds to a root that is to the right of $-p$. Hence, the other root will be to the left of $-z$ and both roots are not acceptable as break points. In this case the only valid break point is $s=0$. The asymptote angles are $90°$ and $-90°$ (Step 4) and the asymptote origin is

$$\bar{s} = \frac{-p+z}{(3-1)} = (r-1)\frac{p}{2} = \delta\frac{p}{2} > 0 \tag{xi}$$

on the positive real axis. Hence, the asymptotes are on the RHP. The root locus for this case is sketched in Figure 8.9a.

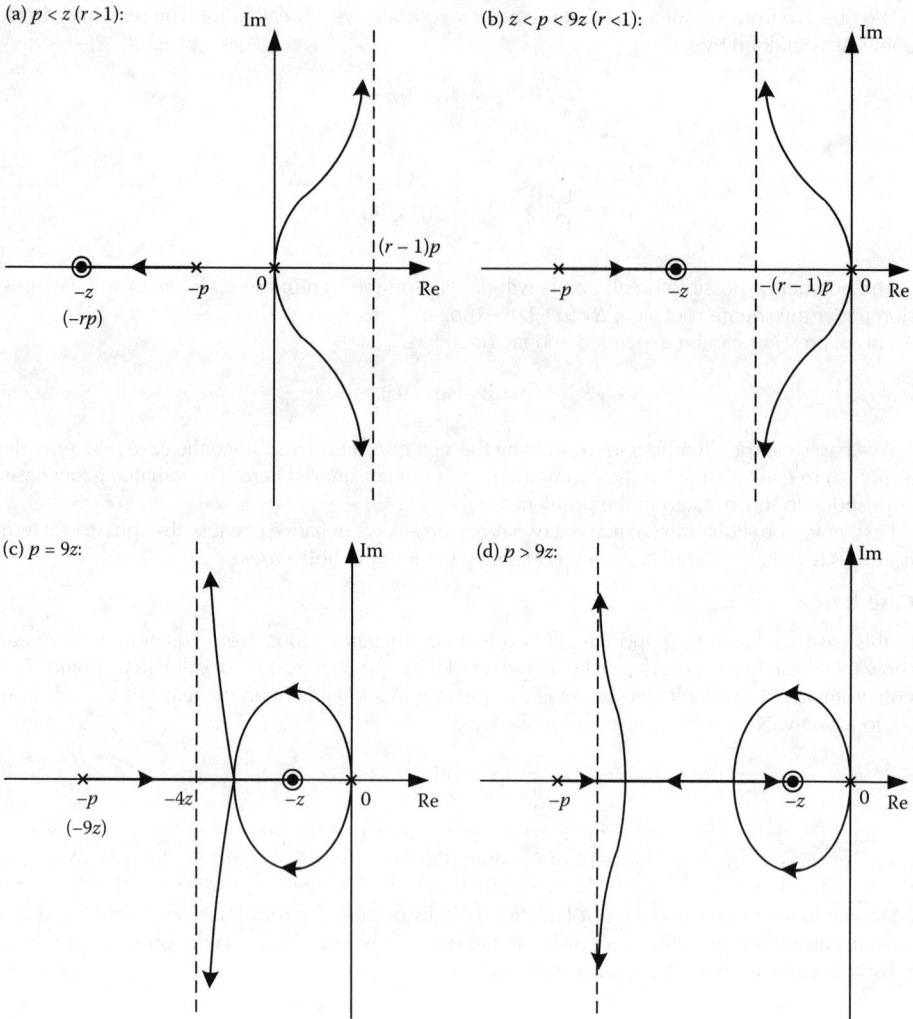

FIGURE 8.9
Root loci for the system in Figure 8.8. (a) $p<z$; (b) $z<p<9z$; (c) $p=9z$; (d) $p>9z$. Note: $z=rp$.

In the present case, the closed-loop system is unstable for all values of gain K because two branches of the root locus are entirely on the RHP. This is to be expected because the plant (double integrator) is marginally stable (actually unstable—see Table 8.1) and the compensator is a *lag compensator* ($p<z$), which has a destabilizing effect on systems.

Note: Further details on lag compensators are found later this chapter and in Chapter 9.

Case 2: $z<p<9z$

In this case, Δ as given by Equation (vii) will be negative. Hence, the roots of Equation (v) will be complex. It follows that the only possible break point in this case is $s=0$, as in the previous case. The asymptote angles are 90° and −90° as before. Also, with $z=rp$ where $r<1$, we have:

$$\bar{s} = (1-r)\frac{p}{2} < 0 \qquad\qquad (\text{xii})$$

This indicates that the asymptotes are on the left half plane.

Note: $\bar{s}$ could lie anywhere from 0 to $-p/2$, and could be on or outside the root locus segment on the real axis. What is shown in Figure 8.9b is the situation when $\bar{s}$ is to the right of $-z$.

The system is stable in the present case, which is to be expected because $z<p$ corresponds to a *lead compensator*.

Note: Further details on lead compensators are found later in this chapter and in Chapter 9.

Case 3: $p=9z$

In this case, $\Delta=0$ and we have two identical roots for Equation (v) at:

$$s = -\frac{1}{4}(p+3z) = -3z$$

Since this point falls within the root locus segment on the real axis (i.e., between $-z$ and $-p$) this is an acceptable break point. Furthermore, if the break point condition gives two identical roots, the characteristic equation has to have three identical roots (at $s=-3z$). Hence, in this case, there are two break points, one at $s=0$ (two identical roots) and the other at $s=-3z$ (three identical roots).

The origin (point of intersection) of the asymptotes is:

$$\bar{s} = \frac{-p+z}{(3-1)} = \frac{-9z+z}{2} = -4z$$

The corresponding root locus is sketched in Figure 8.9c. In this case as well the system is stable for all values of K, which is to be expected because we have a lead compensator.

Note: Further details on lead compensators are found later in this chapter and in Chapter 9.

Case 4: $p>9z$

In this case we note from Equation (vii) that $\Delta>0$. Hence, we get two real roots for Equation (v). We can show using the same arguments as in Case 1 that both these roots are valid break points, both roots falling between $-z$ and $-p$. It follows that there are three break points in this case. The root locus for this case is shown in Figure 8.9d. Once again the system is stable, for we are employing a lead compensator.

This example cautions that one should not rush to sketch a root locus without using complete information. In particular, one should obtain the asymptote origin and the break points before sketching a root locus. Depending on the nature of this information, the root loci can vary significantly even when the system transfer functions are very similar.

Another observation we can make from the present example is that a dominant zero on the LH) will tend to attract the root locus to the LI IP (a stabilizing effect), and a dominant pole will tend to push the root locus away from the LHP (a destabilizing effect).

Example 8.10

Sketch the root locus for a system with the loop transfer function:

$$GH = \frac{K}{s(s+2)(s^2+2s+2)}$$

Discuss the nature of the branches of the root locus at the break point.

Solution

This example may be solved manually in a straightforward manner. First we note the following.

There are four GH poles, at: $0, -2, -1\pm-j$.

There are no GH zeros.

The root locus will be on the real axis between -2 and 0.

There will be four asymptotes, with asymptote angles 45°, 135°, 225°, and 315°.

The point of intersection of the asymptotes is −1 on the real axis, as determined by the centroid condition.

It can be shown that there is a break point at −1, where there will be eight branches of the root locus equally spaced at 45°.

Here we use MATLAB® Control Systems Toolbox (see Appendix B) to plot the root locus, as shown in Figure 8.10. The MATLAB code used for this purpose is:

```
% ---Root locuts plot GH=K/(s(s+2)(s^2+2s+2))---
num= [0 0 0 0 1];
den= [1 4 6 4 0];
rlocus(num,den)
title('Root locus plot GH=K/(s(s+2)(s^2+2s+2)')
xlabel('Real Axis')
ylabel('Imagne Axis')
```

Example 8.11

Consider the feedback control system shown in Figure 8.11. The plant transfer function is

$$G_p(s) = \frac{1}{(s^2 + s - 2)}$$

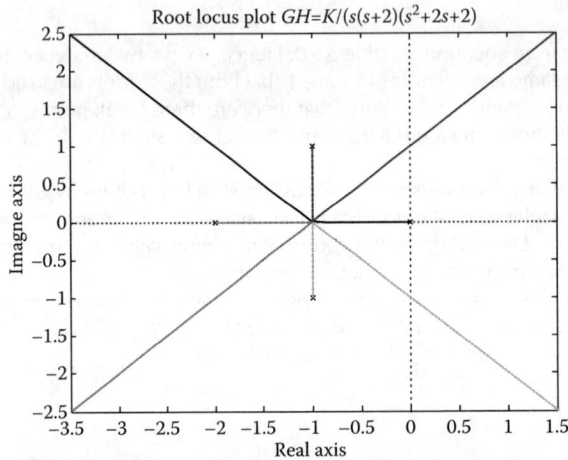

FIGURE 8.10
Root locus plot obtained using MATLAB.

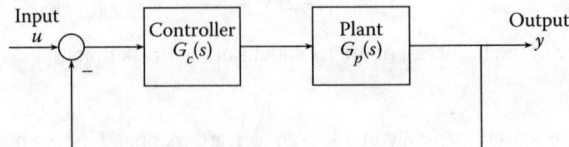

FIGURE 8.11
A feedback control system.

a. Is the plant stable?
   What are its poles?
   What are its zeros?
   The controller transfer function is $G_c(s)=K/s(s^2+2s+1)$.
b. What kind of a controller is this? Explain.
   What are it's poles?
   What are it's zeros?
c. Explaining all your steps, sketch the root locus of the given feedback control system as K changes from 0 to ∞.
   You must indicate the starting and the ending values of all the branches of the root locus.
d. Determine the value of K when the closed-loop system becomes marginally stable. Determine the values of the corresponding marginally stable poles.

Solution

a. The transfer function of the plant is:

$$G_p(s) = \frac{1}{(s^2+s-2)} \tag{i}$$

The characteristic equation of the system is: $s^2+s-2=0$.
Since one coefficient is negative (and the rest positive), it can be determined even before solving the characteristic equation that the system is unstable.
The poles of the plant are given by the solution of: $s^2+s-2=(s+2)(s-1)=0$.
They are: $s=-2$ and $s=1$.
From the numerator of Equation (i) it is clear that the system has no zeros.
b. The controller transfer function is

$$G_c(s) = \frac{K}{s}(s^2+2s+1) \tag{ii}$$

This may be expressed as: $G_c(s)=K(2+s+1/s)$.
Clearly, this is a PID controller.
From Equation (ii) it is noted that the controller has a pole at $s=0$.
The zeros are given by the roots of the numerator equation:

$$s^2+2s+1=(s+1)^2=0$$

It is seen that here are two equal zeros at –1.
c. From the block diagram in Figure 8.11 it is clear that the loop transfer function of the system is $GH=G_cG_p$. From Equations (i) and (ii), we have:

$$GH = \frac{K(s+1)^2}{s(s-1)(s+2)} \tag{iii}$$

We observe the following:
(i)   The loop transfer function GH has three poles at 0, 1, and -2. The three branches of the root locus start from these points.
(ii) The GH has two zeros (identical) at –1. Two branches of the root locus end at this point.
(iii) The root locus will have $3-2=1$ asymptote, whose asymptote angle is $\pi\pm2r\pi/(3-2)=\pm\pi$. Hence, the third branch of the root locus will go to $-\infty$.

(iv) From the condition of odd [#poles−#zeros] to the right, we note that on the real axis, the segment from $s=1$ to 0 and the segment from $s=-2$ to $-\infty$ are on the root locus.

(v)   Break points correspond to repeated roots. They are given by the roots of $N(dD/ds)-D(dN/ds)=0$ where $N=(s+1)^2$ is the numerator polynomial of Equation (iii) and $D=s(s-1)(s+2)$ is the denominator polynomial of Equation (iii). This gives

$$(s+1)^2\times(3s^2+2s-2)-(s^3+s^2-2s)\times2(s+1)=0$$

$$\text{or, } (s+1)[(s+1)(3s^2+2s-2)-2(s^3+s^2-2s)]=0$$

$$\text{or, } s=-1 \quad \text{and} \quad (s+1)(3s^2+2s-2)-2(s^3+s^2-2s)=0.$$

The former root corresponds to the obvious break-in point at the double zero. The latter equation gives the remaining break points (in fact, from the root locus sketch, it should be clear that there is only one other break point, and it has to occur between 0 and 1) correspond to the roots of $s^3+3s^2+4s-2=0$ From MATLAB® (see Appendix B), the three roots are: $1.689\pm j1.558$ and 0.379. The only valid break point is the real root, which is between 0 and 1. This is given by $s=0.379$. The corresponding gain value (K) is obtained by substituting this break point value $(s=0.379)$ into the characteristic equation:

$$1+GH=1+\frac{K(s+1)^2}{s(s-1)(s+2)}=0.$$

The gain value at the break point is $K=0.294$.

(vi) The approach angle (to the double zero) θ_a is obtained using the angle condition $\pi+\pi-2\theta_a=\pi+2r\pi$, or $\theta_a=\pm\pi/2$. This also follows from the symmetry of a root locus about the real axis.

(vii) The break point angle must be $\pm\pi/2$, by the requirement of equal angular spacing of the breaking branches (see Rule 5) or in view of the symmetry of a root locus about the real axis and the continuity of the slope of a root locus branch.

From this information, the root locus may be sketched, as shown in Figure 8.12a.

The root locus plot obtained using the following MATLAB code (see Appendix B) is shown in Figure 8.12b:

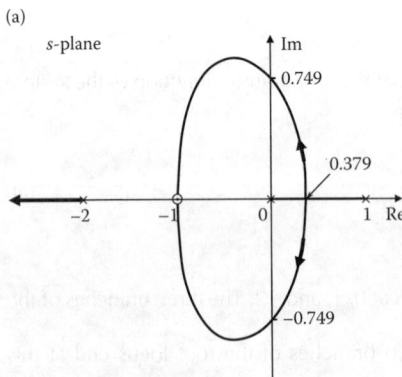

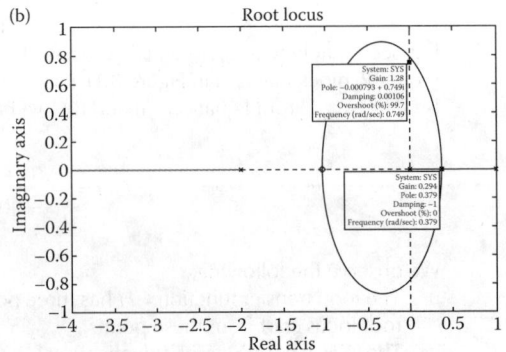

FIGURE 8.12
The root locus of the system. (a) Manual sketch. (b) MATLAB plot.

```
% File name: plotlocus.m
% Aug 15, 2008
% Mech466 solutions.
% ( 1 )
% This is the Matlab m-file used to plot
% root locus of:
%          K(s+1)^2
% GH = ------------
%        s(s-1)(s+2)
clear all
% Specify transfer function directly as
% rational expressions in s (Laplace variable).
  s=tf('s');
% Create the continuous-time transfer function SYS
  SYS= (s+1)^2/(s*(s-1)*(s+2))
% Compute and plot the root locus of the SISO LTI model SYS.
  rlocus(SYS)
% ( 2 )
% Computes the roots of the polynomial whose coefficients
% are the elements of the vector A
  A= [1 3 4 -2];
  roots(A)
```

d. The points of intersection of the root locus with the imaginary axis correspond to marginally stable closed-loop poles. Their locations and the corresponding value of the gain K may be obtained by the Routh array method, as given below.

Closed-loop characteristic equation is given by $1+GH=0$. Now in view of Equation (iii) we have:

$$1+GH=1+\frac{K(s+1)^2}{s(s-1)(s+2)}=0 \;\Rightarrow\; s(s-1)(s+2)+K(s+1)^2=0 \;\Rightarrow$$

$$s^3+s^2-2s+K(s^2+2s+1)=0 \;\Rightarrow\; s^3+(K+1)s^2+2(K-1)s+K=0$$

Routh array:

s^3	1	$2(K-1)$
s^2	$(K+1)$	K
s^1	$\dfrac{2(K^2-1)-K}{K+1}$	0
s^0	K	

The third row becomes null when: $2(K^2-1)-K=0 \;\Rightarrow\; 2K^2-K-2=0$,

whose roots are: $K=(1\pm\sqrt{1+4\times2\times2})/2\times2=1\pm\sqrt{17}/4$.

The only valid root (since K varies from 0 to ∞) is: $K=(1+\sqrt{17})/4=1.281$.

For stability we must have $K>(1+\sqrt{17})/4$.

Note: This is a rare example where increasing the gain results in stability. This is caused by the double zero, which attract the unstable root locus branches to the left half plane (toward -1), there by producing stability.

Under marginally stabile conditions (i.e., $K=(1+\sqrt{17})/4=1.281$), the corresponding marginally stable poles are given by the auxiliary equation:

$$(K+1)s^2+K=0$$

This gives the imaginary poles

$$s = \pm j\sqrt{\frac{K}{K+1}} = \pm j\sqrt{\frac{(1+\sqrt{17})/4}{(1+\sqrt{17})/4+1}} = \pm 0.749j.$$

8.3.4 Variable Parameter in Root Locus

The variable parameter in a root locus does not necessarily have to be the loop gain. Some other parameter (e.g., a time constant) may be put in the required form of a loop transfer function GH, with the parameter factored out, for the purpose of plotting the root locus. As a related point it should be clear that a root locus corresponds to a unique closed-loop characteristic equation, it does not correspond to a unique loop transfer function. Specifically, two or more different G and H combination will result in the same closed-loop characteristic equation and hence the same root locus. We will illustrate this by an example.

Example 8.12

Consider the feedback control system shown in Figure 8.2. You are given two cases of this system with two different loop transfer functions, as follows:

$$\text{System 1: } GH = \frac{K}{s(s+2)(s+4)}$$

$$\text{System 2: } GH = \frac{(s+K)}{s(s^2+6s+7)}.$$

a. For System 1 determine the characteristic equation of the closed-loop system.
b. For System 2 determine the characteristic equation of the closed-loop system.
c. For System 1 determine the root locus (of course, of the closed-loop system) as the parameter K changes from 0 to ∞.
  You must by first determine the:
  (i)   Segments of the root locus on the real axis.
  (ii)   Angles of the asymptotes and the location where the asymptotes intersect the real axis.
  (iii) Break points.
  (iv)   Points at which the root locus intersects with the imaginary axis, and the corresponding value of K (by using the Routh array method).
  (v)   The range of values of K for which the closed-loop system is stable.
d. For System 2 determine the root locus (of course, of the closed-loop system) as the parameter K changes from 0 to ∞.

Solution

a. Characteristic equation: $1+GH(s)=0 \Rightarrow 1+(K/s(s+2)(s+4))=0 \Rightarrow$

$$s(s+2)(s+4)+K=0 \Rightarrow s(s^2+6s+8)+K=0 \Rightarrow s^3+6s^2+8s+K=0$$

b. Characteristic equation: $1+(s+K/s(s^2+6s+7))=0 \Rightarrow s(s^2+2s+7)+s+K=0 \Rightarrow$

$$s^3+6s^2+7s+s+K=0 \Rightarrow s^3+6s^2+8s+K=0$$

Note: The results in (a) and (b) are identical. Hence the two systems have exactly the same closed-loop poles as a function of K, and hence identical root loci, as K is varied from 0 to ∞.

c. $GH = \dfrac{K}{s(s+2)(s+4)}$

#poles $n=3$; #zeros $m=0$

(i) The segment $s=0$ to -2 and $s=-4$ to $-\infty$ are on the root locus, because there are an odd number of poles-zeros to the right of these segments.

(ii) Asymptote angles $= \pi + 2r\pi/n - m$ for $r=0,\pm1,\ldots$

$$= \frac{\pi + 2r\pi}{3} = \pm\frac{\pi}{3} \quad \text{and} \quad \pi$$

The point of intersection of the asymptotes on the real axis is $\bar{s}$ which is given by

$$\bar{s} = \frac{\Sigma p_i - \Sigma z_i}{n-m} = \frac{0-2-4}{3} = -2$$

(iii) Break points are given by the repeated poles of the closed system $\Rightarrow$

$$N\frac{dD}{ds} - D\frac{dN}{ds} = 0$$

where $\dfrac{N(s)}{D(s)} = \dfrac{1}{s(s+2)(s+4)}$

or: $N(s)=1$ and $D(s)=s^3+6s^2+8s$

Hence: $1 \times \dfrac{d}{ds}[s^3+6s^2+8s] = 0 \Rightarrow 3s^2+12s+8=0$

Roots are: $s = \dfrac{-12 \pm \sqrt{12^2 - 4\times3\times8}}{2\times3} = -2 \pm \dfrac{2}{\sqrt{3}}$

The correct break point must be on the root locus. Hence from Equation (i), the break point $= -2 + (2/\sqrt{3}) = -0.845$.

(iv) Closed-loop characteristic equation:

$$s^3 + 6s^2 + 8s + K = 0$$

Routh array:

s^3	1	8
s^2	6	K
s^1	$\dfrac{6\times8 - K\times1}{6}$	0
s^0	K	

The auxiliary equation, which is obtained when the row corresponding to s^1 has all zeros (i.e., when $K=6\times8=48$) is: $6s^2+K=0 \Rightarrow 6s^2+48=0 \Rightarrow s=\pm j2\sqrt{2}$.

These are the purely imaginary closed-loop poles. Hence, the root locus intersects the imaginary axis at $s=\pm j2\sqrt{2}$.

This occurs when $K=48$.

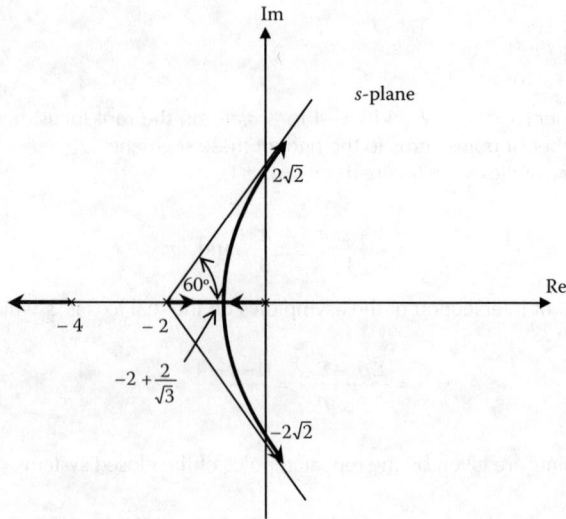

FIGURE 8.13
Root locus of System 1 and System 2.

    (v)   From the Routh array, for stability we must have $48 - K > 0$ and $K > 0$. Hence the stability region is given by: $0 < K < 48$

            With these results, the root locus for System 1 may be sketched as in Figure 8.13.

  d. Since the closed-loop characteristic equations of the two systems are identical, as a function of K (see (a) and (b)), their root loci must be identical, by definition.

8.4 Stability in the Frequency Domain

The concept of transfer function and frequency domain models have been discussed in Chapter 5 and response analysis in the frequency domain has been studied in Chapter 6. Some related concepts of Laplace and Fourier transforms are outlined in Appendix B. Now we will specifically use the concept of *frequency transfer function* (FTF) (or, *frequency response function* [FRF]), where the independent variable is frequency ω radians/s or f cycles/s (or Hertz), to develop some useful techniques in the stability analysis of control systems. In particular, we will show that:

  1. Peak magnitude (and associated Q-factor and half-power bandwidth)
  2. PM
  3. GM

  May be used as measures of relative stability, in the frequency domain. First we will summarize some fundamentals of FTF, which we have covered in Chapters 5 and 6.

8.4.1 Response to a Harmonic Input

Consider an LTI system, given in the time domain (linear ordinary differential equation with constant coefficients) by

$$a_n \frac{d^n y}{dt^n} + a_{n-1} \frac{d^{n-1} y}{dt^{n-1}} + \cdots + a_0 y = b_m \frac{d^m u}{dt^m} + b_{m-1} \frac{d^{m-1} u}{dt^{m-1}} + \cdots + b_0 u. \tag{8.18}$$

The system parameters (coefficients) $a_0, a_1, \ldots, a_n$ and $b_0, b_1, \ldots, b_m$ are constant (time invariant) by definition. The differential equation (Equation 8.18a) is an input–output model where u=system input, and y=system output. The order of the system is n.

The system transfer function (i.e., output-input ratio, in the Laplace domain), is

$$G(s) = \frac{Y(s)}{U(s)} = \frac{b_m s^m + b_{m-1} s^{m-1} + \cdots + b_0}{a_n s^n + a_{n-1} s^{n-1} + \cdots + a_0} \tag{8.18b}$$

Note that when we know Equation 8.18a we can immediately write down Equation 8.18b, and vice versa. This confirms that the two representations (Equations 8.18a and 8.18b) are completely equivalent.

Suppose that a sinusoidal input of amplitude u_o and frequency ω is applied to the system (Equation 8.18a). This input may be represented in the "complex" form

$$u = u_o e^{j\omega t} = u_o(\cos \omega t + j \sin \omega t) \tag{8.19}$$

Actually what we are applying to the system is the real part of the right hand side of Equation 8.19. But in view of the relative ease of manipulating an exponential function in comparison to a sinusoidal function, we use the entire Equation 8.19 and then at the end take the real part of the result. The simplicity of analysis by using the exponential function stems particularly from the fact that $d/dt e^{st} = s e^{st}$ and hence, after differentiation, the original exponential function remains (albeit with a multiplication factor). This is easier than, say, using $d/dt \sin \omega t = \omega \cos \omega t$ and $d/dt \cos \omega t = -\omega \sin \omega t$ where the function type changes on differentiation.

It is reasonable to assume (and, in fact, it can be verified through experiments and practical observations of real systems) that when a harmonic excitation is applied to a system (strictly, to a "linear" system) the response after a while becomes harmonic as well, oscillating at the same frequency (*Note*: If experiments are carried out to observe this property, the system has to be "stable" as well). The amplitude of the resulting response will not be the same as that of the input, in general. Hence, the steady-state harmonic response of Equation 8.18a may be expressed as

$$y = y_o e^{j\omega t} \tag{8.20a}$$

We will show later that not only the amplitude but also the "phase" of the output will be different from that of the input. Now substitute Equations 8.19 and 8.20 into Equation 8.18a and cancel the common term $e^{j\omega t}$ which is not zero for a general value of t. Then we have

$$y_o = \left[\frac{b_m (j\omega)^m + b_{m-1} (j\omega)^{m-1} + \cdots + b_0}{a_n (j\omega)^n + a_{n-1} (j\omega)^{n-1} + \cdots + a_0} \right] u_o \tag{8.21a}$$

In view of Equation 8.18b, we note that what is in the square brackets is indeed the transfer function $G(s)$ with s substituted by $j\omega$. This is the familiar FTF (or, FRF in the terminology of mechanical vibration) as discussed in Chapter 5. Hence,

$$y_o = G(s)\big|_{s=j\omega} u_o \tag{8.21b}$$

$$\text{or } y_o = G(j\omega)u_o \tag{8.21c}$$

The amplitude u_o of the input is clearly a real value. Also, $G(j\omega)$ is a complex number in general, which has a real part and an imaginary; or, a magnitude and a phase angle. Suppose that this magnitude is M and the phase angle is ϕ. Then

$$\text{Magnitude of } G(j\omega) = |G(j\omega)| = M \tag{8.22a}$$

$$\text{Phase angle of } G(j\omega) = \angle G(j\omega) = \phi \tag{8.22b}$$

Furthermore,

$$G(j\omega) = M\cos\phi + jM\sin\phi = Me^{j\phi} \tag{8.23}$$

Substitute Equation 8.23c into Equation 8.21c, and use Equation 8.20. We get

$$y = u_o Me^{j(\omega t + \phi)} \tag{8.20b}$$

The result (Equation 8.20b) states, in view of Equation 8.19 that for an input of $u_o \cos \omega t$ the output will be $u_o M \cos(\omega t + \phi)$ and similarly for an input of $u_o \sin \omega t$ the output will be $u_o M \sin(\omega t + \phi)$. In summary, when a harmonic input is applied to the system having transfer function $G(s)$, we have the following:

1. The output will be magnified by magnitude $|G(j\omega)|$.
2. The output will have a *phase lead* equal to $\angle G(j\omega)$, with respect to the input. In fact $\angle G(j\omega)$ is typically a negative phase lead. Hence the output usually lags the input, as a result of system dynamics.

8.4.2 Complex Numbers

A complex number is represented by a magnitude and a phase angle or a real part and an imaginary part, on the complex plane (where the horizontal axis is the real axis and the vertical axis is the imaginary axis). Clearly, complex numbers are important in the response analysis in the frequency domain. Some related results have been summarized previously, under the root locus method. In particular:

1. In a product of complex numbers, the magnitudes multiply and the phase angles add.
2. In a quotient of complex numbers, the magnitudes divide and the phase angles subtract.

Furthermore, in dealing with the phase angle of a complex number, it is critical to know the proper quadrant on the complex plane where the number is located, as indicated in Figure 8.14.

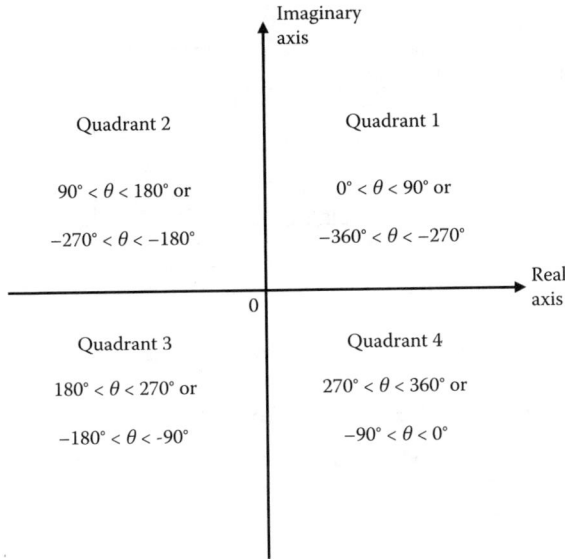

FIGURE 8.14
Phase angles corresponding to the four quadrants.

In particular, note the following:

3. Proper quadrant of a complex number (correct phase angle) is determined by the signs of the real part and the imaginary part of the complex number.
4. An integer multiple of 360° may be added to or subtracted from a complex number without consequence.

8.4.3 Resonant Peak and Resonant Frequency

Resonant peak=peak point of a transfer function magnitude curve.
Resonant frequency=frequency at resonant peak.
Peak magnitude=value of the transfer function magnitude at resonance.
   A second-order system (simple oscillator) can have only one resonant peak while higher order systems can have many resonant peaks and resonant frequencies.

Example 8.13

Consider the mass-spring system shown in Figure 8.15a.
   By Newton's second law, the equation of motion to a forcing input u is given by:

$$m\ddot{y} + ky = u \tag{i}$$

where u denotes the force input applied to the mass m, and y denotes the displacement output (response) of the mass. The spring constant (stiffness) is k. The overall system transfer function of the system (i.e., y/u in the Laplace domain) is

$$\tilde{G}(s) = \frac{1}{ms^2 + k} \tag{ii}$$

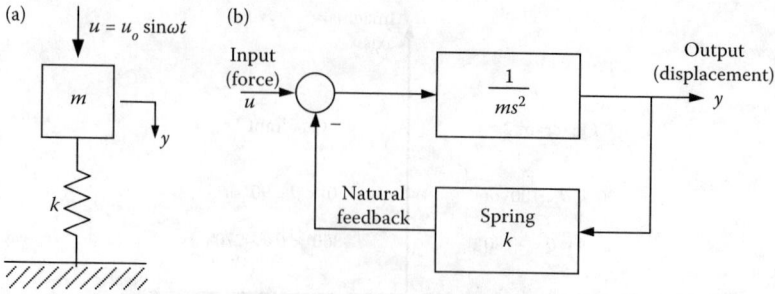

FIGURE 8.15
(a) An undamped simple oscillator. (b) Block diagram representation with a natural feedback.

Suppose that the forcing function is harmonic, as given by:

$$u = u_o \sin \omega t \tag{iii}$$

Note that

$$\tilde{G}(j\omega) = \frac{1}{k - \omega^2 m} \tag{iv}$$

Hence:

$$\text{Magnitude } M = |\tilde{G}(j\omega)| = \frac{1}{|k - \omega^2 m|} \tag{v}$$

and

$$\text{Phase } \phi = \angle\tilde{G}(j\omega) = 0 \quad \text{for} \quad \omega < \sqrt{k/m}$$

$$= -\pi \quad \text{for} \quad \omega > \sqrt{k/m} \tag{vi}$$

The harmonic response of the system (in steady-state) is (see Chapter 6):

$$y = u_o M \sin(\omega t + \phi) \tag{vii}$$

It is clear from Equation (vi) that when $\omega < \sqrt{k/m}$ the response will be in phase with the input, and when $\omega > \sqrt{k/m}$ the response will be 180° out of phase with the input. The phase will switch from 0° to 180° at the specific excitation frequency $\omega = \sqrt{k/m}$.

Furthermore, the response amplitude will increase with the excitation frequency, up to the frequency $\omega = \sqrt{k/m}$. Beyond that, the response amplitude will decrease with the excitation frequency. In particular, at $\omega = \sqrt{k/m}$ the response amplitude will be infinity. This is called a *resonance* and corresponding frequency is called *resonant frequency*. These results are shown in the plots of magnitude $|G(j\omega)|$ and phase lead $\angle G(j\omega)$ versus frequency ω, in Figure 8.16.

Note: An important observation can be made about the system in this example, with reference to Figure 8.15b. The system has a natural feedback as a result of the spring, as shown (which can be established by writing Equation (i) in the form $u - ky = m\ddot{y}$). According to this model then, the forward transfer function is:

$$G(s) = \frac{1}{ms^2} \tag{viii}$$

and the feedback transfer function is:

$$H(s) = k \tag{ix}$$

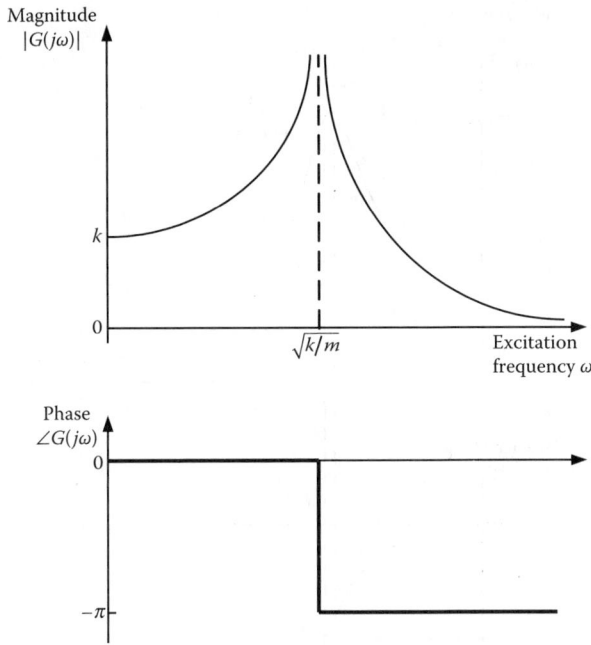

FIGURE 8.16
The magnitude and phase plots of an undamped simple oscillator.

The overall (closed-loop) transfer function is:

$$\tilde{G}(s) = \frac{G}{1+GH} = \frac{1}{ms^2 + k}$$ (x)

which is the same as Equation (ii). The loop transfer function is:

$$GH = \frac{k}{ms^2}$$ (xi)

It is GH that should be used to sketch the root locus of this simple oscillator. In view of the double pole at the origin ($s=0$) in the loop transfer function (xi) we notice that the root locus falls entirely on the imaginary axis for any k. Hence, the system is always marginally stable.

8.4.4.1 Damped Simple Oscillator

If we include linear viscous damping in the oscillator shown in Figure 8.15a, the resulting damped simple oscillator has the transfer function (see Chapters 5 and 6):

$$G(s) = \left[\frac{\omega_n^2}{s^2 + 2\zeta\omega_n s + \omega_n^2} \right]$$ (8.24a)

where $\omega_n = \sqrt{(k/m)}$ = undamped natural frequency; $\zeta = b/2\sqrt{k/m}$ damping ratio.
   The FTF, FRF is given by:

$$G(j\omega) = \left[\frac{\omega_n^2}{s^2 + 2\zeta\omega_n s + \omega_n^2} \right]_{s=j\omega} = \frac{\omega_n^2}{\omega_n^2 - \omega^2 + j2\zeta\omega_n\omega}$$ (8.24b)

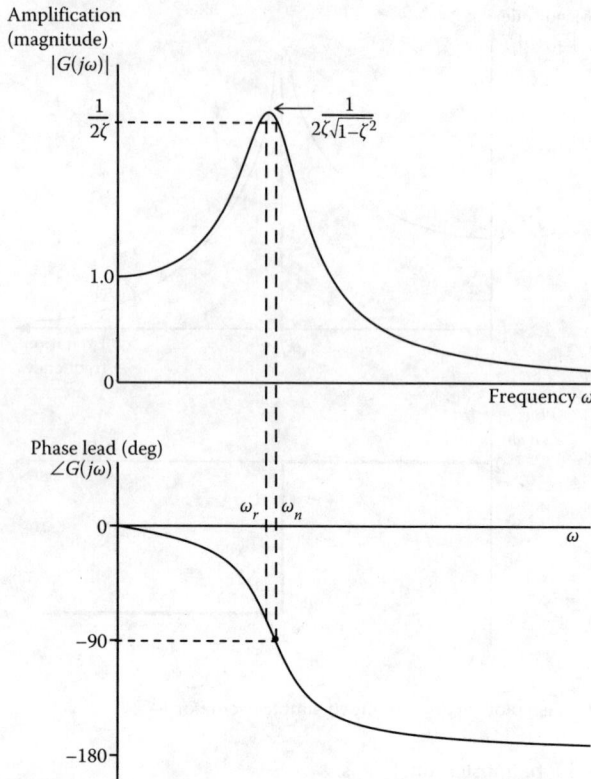

FIGURE 8.17
The magnitude and phase plots of a damped simple oscillator.

Its magnitude and phase are sketched in Figure 8.17.

Denominator of Equation 8.24b: $\Delta = \omega_n^2 - \omega^2 + j2\zeta\omega_n\omega$

And $|\Delta|^2 = \left(\omega_n^2 - \omega^2\right)^2 + \left(2\zeta\omega_n\omega\right)^2 = D$

The resonant peak corresponds to minimum value of D, which is obtained by:

$$\frac{dD}{d\omega} = 2\left(\omega_n^2 - \omega^2\right)(-2\omega) + 2\left(2\zeta\omega_n\right)^2 \omega = 0$$

By solving this we get:

$$\text{Resonant frequency: } \omega_r = \sqrt{1 - 2\zeta^2}\,\omega_n \approx \omega_n \text{ for low damping} \qquad (8.25)$$

Note: In the result (Equation 8.25) we must have $\zeta < 1/\sqrt{2}$ in order to get a valid resonant frequency.

The resonant peak is obtained by substituting the resonant frequency (Equation 8.25) into Equation 8.24b and finding the magnitude of the result. We get

$$\text{Resonant peak: } |G(j\omega)|_{\omega = \omega_r} = \frac{1}{2\zeta\sqrt{1 - \zeta^2}} \qquad (8.26a)$$

For small values of damping (i.e., $\zeta \ll 1.0$) we can approximate Equation 8.26a as the magnitude at the undamped natural frequency:

$$|G(j\omega)|_{\omega=\omega_n} = \frac{1}{2\zeta} \qquad (8.26b)$$

Note: Phase of $G(j\omega)$ at $\omega = \omega_n$ is $-\pi/2$.

8.4.3.2 Peak Magnitude

Some important parameters for the damped oscillator, in the frequency domain are indicated in Figure 8.17. In particular note the resonant frequency and the corresponding peak magnitude, which depends only on the damping ratio ζ. Since damping is a measure of *relative stability*, the peak magnitude also may be used as a measure of relative stability, in the frequency domain.

8.4.4 Half-Power Bandwidth

The half-power bandwidth is defined as the frequency interval (bandwidth) at $(1/\sqrt{2})$ ×resonant peak in the magnitude curve $|G(j\omega)|$.

   Note: Since (Voltage)$^2 \propto$ Power, we have: $(1/2) \times$ Power $\Rightarrow (1/\sqrt{2}) \times$ Voltage. This is the rationale for the terminology.

8.4.4.1 Damped Simple Oscillator

Let us determine the half-power bandwidth corresponding to the damped simple oscillator (Equation 8.24). See Figure 8.18. By definition, we solve for ω in:

$$\frac{1}{\sqrt{2}}\frac{1}{2\zeta} = \left|\frac{\omega_n^2}{\omega_n^2 - \omega^2 + 2j\zeta\omega_n\omega}\right| = \left|\frac{1}{1 - \left(\dfrac{\omega}{\omega_n}\right)^2 + 2j\zeta\left(\dfrac{\omega}{\omega_n}\right)}\right|.$$

First we assume that $\zeta < 1/\sqrt{2}$. Strictly speaking, we should assume that $\zeta < 1/(2\sqrt{2})$, as will be clear from the final result. By squaring the previous equation we have:

$$\frac{1}{2 \times 4\zeta^2} = \frac{1}{\left[1 - r^2\right]^2 + 4\zeta^2 r^2}$$

where r is the normalized (nondimensionalized) excitation frequency given by: $r = \omega/\omega_n$.

Hence: $r^4 - 2r^2 + 1 + 4\zeta^2 r^2 = 8\zeta^2 \rightarrow r^4 - 2(1 - 2\zeta^2)r^2 + \underbrace{(1 - 8\zeta^2)}_{>0} = 0.$ (i)

Now assume that $\zeta^2 < 1/8$, which means $\zeta < 1/(2\sqrt{2})$. Otherwise, we will not get two positive roots for r^2 in Equation (i). Solution of Equation (i) for r^2 will give two roots r_1^2 and r_2^2

Amplification
(magnitude)
$|G(j\omega)|$

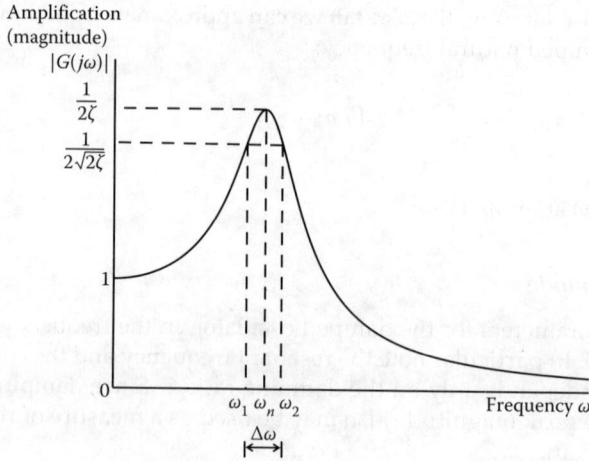

FIGURE 8.18
Peak magnitude, half-power bandwidth, and Q-factor as measures of relative stability in the frequency domain.

for r^2. Next, assume that $r_2^2 > r_1^2$. Now compare $(r^2 - r_1^2)(r^2 - r_2^2) = 0$ with Equation (i). We have

$$\text{Sum of roots: } r_2^2 + r_1^2 = 2(1 - 2\zeta^2) \tag{ii}$$

$$\text{Product of roots: } r_2^2 r_1^2 = (1 - 8\zeta^2) \tag{iii}$$

$$\text{Hence: } (r_2 - r_1)^2 = r_2^2 + r_1^2 - 2r_2 r_1 = 2(1 - 2\zeta^2) - 2\sqrt{1 - 8\zeta^2}$$

$$= 2 - 4\zeta^2 - 2\left[1 - (1/2) \times 8\zeta^2 + O(\zeta^4)\right]$$
$$\text{(by Taylor series expansion)}$$

$$\cong 2 - 4\zeta^2 - 2 + 8\zeta^2 \cong 4\zeta^2$$
$$\text{(because } O(\zeta^4) \to 0 \text{ for small } \zeta\text{)}$$

Consequently, the half-power bandwidth ($\Delta\omega$), as shown in Figure 8.18, is given in the normalized form as:

$$r_2 - r_1 = \frac{\omega_2 - \omega_1}{\omega_n} = \frac{\Delta\omega}{\omega_n} \cong 2\zeta \tag{8.27}$$

Note: Damping ratio may be obtained once the magnitude of the frequency response function $G(j\omega)$ is experimentally determined, using Equation 8.27 as:

$$\zeta \cong \frac{(\omega_2 - \omega_1)}{2\omega_n} = \frac{\Delta\omega}{2\omega_n} \cong \frac{\omega_2 - \omega_1}{\omega_2 + \omega_1} \tag{8.28}$$

The *Q-factor*, which measures the sharpness of resonant peak (see Figure 8.18), is defined by

$$Q\text{-factor} = \frac{\omega_n}{\Delta\omega} = \frac{1}{2\zeta} \tag{8.29}$$

The term originated from the field of electrical tuning circuits where the sharpness of the resonant peak is a desirable thing (*quality factor*). It follows that the Q-factor (inversely) as well as half-power bandwidth (directly) and peak magnitude (inversely) may be used as measures of relative stability (and damping), in the frequency domain.

8.4.5 Marginal Stability

If a dynamic system oscillates steadily in the absence of a steady external excitation, this condition represents a state of *marginal stability*. The "distance" to a state of marginal stability is a measure of the level of stability, and is called a *stability margin*. In the present section we will formally develop the concepts of marginal stability and stability margin using a frequency transfer function model. First we will address marginal stability in qualitative terms and then develop an analytical basis.

8.4.5.1 The (1,0) Condition

Consider a feedback control system represented by the block diagram in Figure 8.19. We have assumed unity feedback, but this can be generalized later. In fact, without loss of generality we can interpret G in Figure 8.19 as the loop transfer function GH, because a system with a general feedback transfer function H can be reduced to a unity feedback system, through block diagram reduction, by placing GH as the forward transfer function (as discussed in Chapter 7 and also in the present chapter under the subject of root locus method).

Suppose that the open-loop transfer function $G(s)$ is such that at a specific frequency of operation ω, we have:

1. Magnitude $|G(j\omega)| = 1$
2. Phase angle $\angle G(j\omega) = -\pi$

Then, if an error signal of frequency ω is injected into the loop (due to noise, disturbance, initial excitation, etc.) its amplitude will not change while passing through $G(s)$, but the phase angle will reduce by π. Hence the output signal y will have the same amplitude as the error signal e, but y will "lag" e by π. Since y is fed back into the loop with a negative feedback (*Note*: -1 corresponds to a further phase lag of π) the overall phase lag in the feedback signal, when reaching the forward path of the loop, will be 2π. Since a phase change of 2π is the same as no phase change, the feedback signal will have the same amplitude as the forward signal (i.e., gain $= 1$) and the same phase angle as the forward signal (i.e., phase $= 0$). This is called the (1,0) *condition*. Under this condition, it is clear that even in the

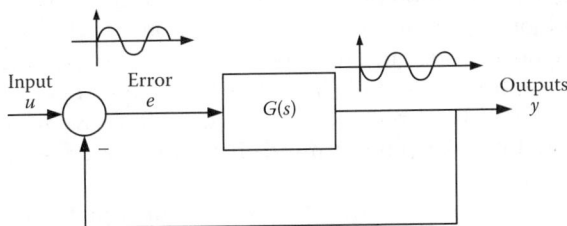

FIGURE 8.19
A feedback control system with unity feedback.

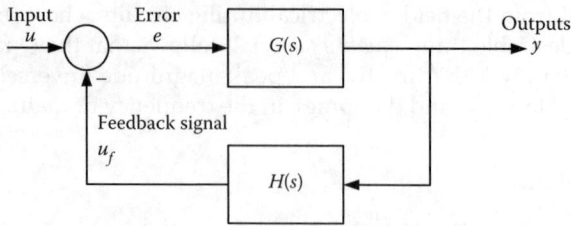

FIGURE 8.20
A control system with nonunity feedback.

absence of an external input u, a harmonic signal of a specific frequency ω can sustain in the loop without growing or decaying. This is a state of self-sustained steady oscillation. If such a condition of steady oscillation is possible, in the absence of a steady external input, the system is said to be marginally stable.

Note: the specific frequency ω at which this condition of steady oscillation would be feasible, is itself a property of the system, and depends on system parameters. In the simple oscillator example, which we discussed earlier, this frequency is $\sqrt{(k/m)}$.

Next, consider a system with nonunity feedback with a feedback transfer function of $H(s)$, as shown in Figure 8.20.

It should be clear that in applying the (1,0) condition for marginal stability, what matters is the overall gain and phase shift in the entire loop. Hence, in Figure 8.20 we need to consider the overall loop transfer function $G(s)H(s)$ and not the individual components. The (1,0) condition for *marginal stability* is, at a specific frequency of operation ω:

1. Magnitude $|G(j\omega)H(j\omega)| = 1$
2. Phase angle $\angle G(j\omega)H(j\omega) = -\pi$ (8.30a)

With respect to the closed-loop system shown in Figure 8.20, the following nomenclature should be remembered:

$G(s)$ = open-loop transfer function (or, forward transfer function)
$H(s)$ = feedback transfer function
$G(s)H(s)$ = loop transfer function

$$\frac{G}{1+GH} = \tilde{G} = \text{closed-loop transfer function.}$$

Note: The characteristic equation of the closed-loop system $\tilde{G}$ is $GH + 1 = 0$. Hence, the stability of a closed-loop system is completely determined by the loop transfer function GH, as already concluded under the root locus method. Specifically, now we need to study the magnitude and the phase of $GH(j\omega)$, in the frequency domain. In the special case of unity feedback ($H=1$) we need to study $G(j\omega)$. For convenience, in these studies we denote GH simply by G, keeping in mind that then G represents the loop transfer function GH.

Bode diagram (Bode plot) and Nyquist diagram (Nyquist plot) are convenient ways of graphically representing transfer functions (see Chapter 5). These plots are valuable in the stability of dynamic systems, in the frequency domain.

8.4.6 PM and GM

Stability margin is an important performance specification, as noted in Chapter 7. In the conventional frequency domain design of control systems, stability requirement is specified using PM or GM. We will revisit the concept of marginal stability to introduce these two stability margins.

We have observed that stability of a closed-loop system is completely determined by the loop transfer function GH. Consider the feedback control system shown in Figure 8.20. The characteristic equation of the closed-loop system is $G(s)H(s) = -1$. The system is marginally stable if one pair of roots of the characteristic equation is purely imaginary (i.e., $\pm j\omega$) while the remaining roots are not unstable. Hence the condition for marginal stability is, there exists a frequency ω such that:

$$G(j\omega)\, H\,(j\omega) = -1 \qquad (8.30b)$$

In fact the two conditions given by Equation 8.30a are exactly equivalent to this single complex equation (Equation 8.30b), because -1 has a magnitude of 1 and a phase angle of $-\pi$. It follows that if the plot of the loop transfer function GH in the complex plane (i.e., the polar plot of imaginary $GH(j\omega)$ vs. real $GH(j\omega)$—the Nyquist plot—see Chapter 5), as ω changes, passes through the point $(-1,0)$, then the control system is *marginally stable*.

8.4.6.1 GM

Suppose that at a particular operating frequency ω, the phase $\phi = -180°$ but the gain (magnitude) M is less than unity (i.e., $M < 1$). Then, if the external u in Figure 8.20 is disconnected, the amplitude of the feedback signal will steadily decay. This, of course, corresponds to a stable system. The smaller the value of M, the more stable the system. Hence, a stability margin known as the *gain margin* g_m can be defined as:

$$g_m = \frac{1}{|G(j\omega)H(j\omega)|} \qquad (8.31a)$$

at the frequency ω where $\angle G(j\omega)H(j\omega) = -180°$.

Note: If the magnitude (i.e., gain) of the transfer function $GH(j\omega)$ is increased by a factor of g_m at this frequency, then the marginal stability conditions (Equation 8.30) will be satisfied. Hence, g_m is the margin by which the gain of a stable system may be increased so that the system becomes just unstable. It follows that the larger the g_m, the better the degree of stability. It is convenient to express g_m in decibels (dB) because the transfer function magnitude (in the frequency domain) is usually expressed in dB (particularly in Bode diagrams—see Chapter 5). Then,

$$g_m = -20 \log_{10} |G(j\omega)\, H(j\omega)| \qquad (8.31b)$$

at the frequency ω where $\angle G(j\omega)H(j\omega) = -180°$.

8.4.6.2 PM

Suppose that there exists some frequency ω at which the magnitude (i.e., gain) of the loop transfer function $GH(j\omega)$ is $M = 1$ (i.e., 0 dB) but the phase ϕ lies between 0 and $-180°$. This frequency ω_c is called the *crossing frequency* or *crossover frequency*, because it corresponds to the point where the gain (magnitude) curve crosses the unity (0 dB) line. Since the phase angle decreases with frequency, there will be a higher frequency at which the phase is –180° but the gain (magnitude) will be less than unity (because the loop transfer function magnitude usually decreases with increasing frequency, at high frequencies). This, as noted under the topic of GM, corresponds to a stable system. The amount by which the phase of the loop transfer function at gain=0 dB, may be decreased (i.e., the phase lag in freased) until it reaches the –180° value, is termed PM (ϕ_m). Specifically, the PM is defined as:

$$\phi_m = 180° + \angle G\,(j\omega)\,H\,(j\omega) \tag{8.32}$$

at the frequency ω where $|G(j\omega)H(j\omega)| = 1$.

The larger the PM, the more stable the system.

In summary, GM tells us the amount (margin) by which the gain may be increased at a phase of –180°, before the system becomes marginally stable; and PM tells us the amount (margin) by which the phase may be "decreased" (i.e., "phase lag" increased) at unity-gain, before the system becomes marginally unstable. A more rigorous development of the concepts of GM and PM requires a knowledge of the *Nyquist stability criterion*, as presented in a separate section.

8.4.7 Bode and Nyquist Plots

As discussed in Chapter 5, Bode diagram (Bode plot) and Nyquist diagram (Nyquist plot) are convenient graphical representations of transfer functions, in the frequency domain. Specifically, the Bode plot of a transfer function $G(s)$ constitutes the following pair of curves:

Magnitude $|G(j\omega)|$ versus frequency ω,
Phase angle $\angle G(j\omega)$ versus frequency ω.

Note that the Bode plot requires two curves—one for gain and one for phase. These two curves can be represented as a single curve by using a so-called *polar plot* with a real axis and an imaginary axis. A polar plot is a way to represent both magnitude and phase of a rotating vector, with one curve. When phase=0°, the vector (which represents the complex number) points to the right; when phase=90°, the vector points up; and when phase=–180°, the vector points to the left, etc.

The solid curve in Figure 8.21a represents the path of the tip of the directed line (two-dimensional vector) representing the frequency transfer function (i.e., complex transfer function in the frequency domain) as the frequency varies. Any one point represents both the *amplitude* (distance to origin) and *phase* (angle measured from the positive real axis), at one given frequency. Thus all the information in the two curves of the Bode plots (Figure 8.21b) is represented by this single curve called *Nyquist plot* (or *polar plot* or *argand plot*).

The marginal stability condition is: (a) gain=1 or 0 dB, and (b) phase lag=180° at some specific operating frequency. A gain of 1 is represented by a vector of unit length. Its tip traces a unit circle with its center at the origin of the coordinate frame (see the broken-line circle in Figure 8.21a). A phase of 180° corresponds to a horizontal vector pointing to the left from the origin (i.e., the negative real axis). The intersection of the Nyquist plot with the unity-gain circle gives the *phase* at the 0 dB point; the length of the vector (distance from origin) of the point where the Nyquist plot intersects the negative side of the real axis gives the gain at the critical 180° phase (lag) point.

Now formal definitions for GM g_m and PM ϕ_m may be given using either the Nyquist diagram or the Bode diagram, as in Figure 8.21. Consider a stable closed-loop system with

(a)

(b)

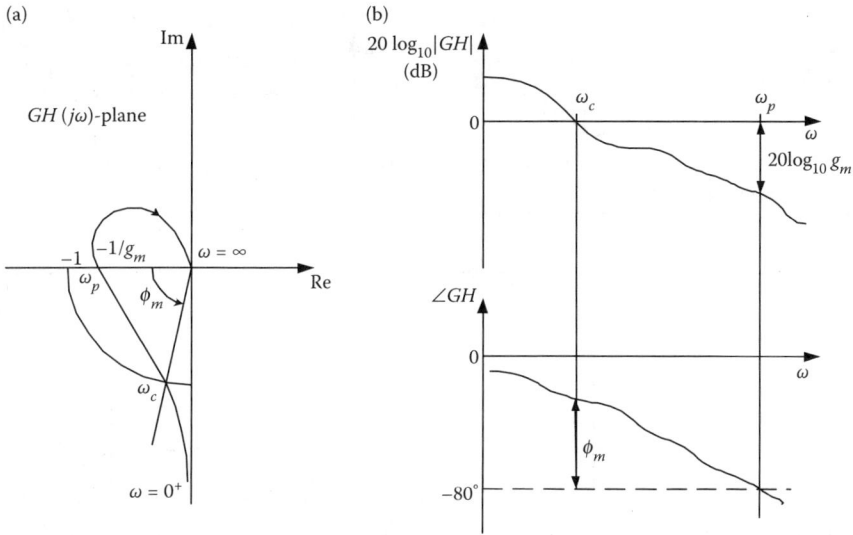

FIGURE 8.21
Definition of gain margin and phase margin. (a) Using Nyquist diagram (b) Using Bode diagram.

the transfer function $\tilde{G} = G/1 + GH$. First we plot the Nyquist diagram for the loop transfer function GH as, for example, shown in Figure 8.21a. The factor by which the Nyquist curve should be expanded (say, by increasing its gain) in order to make the system marginally stable (i.e., to pass through the critical point $(-1,0)$) measures the relative stability (or, *stability margin*) of the closed-loop system. The stability margins may be similarly defined using the Bode plot in Figure 8.21b.

GM: Gain margin g_m is the reciprocal of the magnitude of the loop transfer function $GH(j\omega)$ at the frequency (ω) where the phase angle of the loop transfer function is $-180°$. It follows that the larger the g_m, the larger the separation of the Nyquist curve from the critical point (-1) and the better the closed-loop stability.

PM: Phase margin ϕ_m is the sum of $180°$ and the phase angle (in degrees) of the loop transfer function $GH(j\omega)$ at the frequency (ω) where the magnitude of $GH(j\omega)$ is unity (or 0 dB).

These definitions follow from Equations 8.31 and 8.32.

The relative stability (stability margin) of a control system can be improved by adding a compensator so as to increase and GM. Since, in general, the GM of a system automatically improves when the PM of the system is improved, in design specifications it is adequate to consider only the PM. This subject is addressed in Chapter 9.

Example 8.14

Figures 8.22a and b show Bode and Nyquist plots (of the loop transfer functions GH) for two systems. The one on the left is stable because the phase lag is less than $180°$ at the critical 0 dB (i.e., where gain=1) point, and the gain is less than 1 at the critical phase lag point (i.e., where phase lag=$180°$). Note that the amount by which the gain is less than 0 dB at the phase-crossover $(-180°)$ point is the GM. Similarly, the amount by which the phase lag is less than $180°$ at the gain-crossover point (0 db) is the PM.

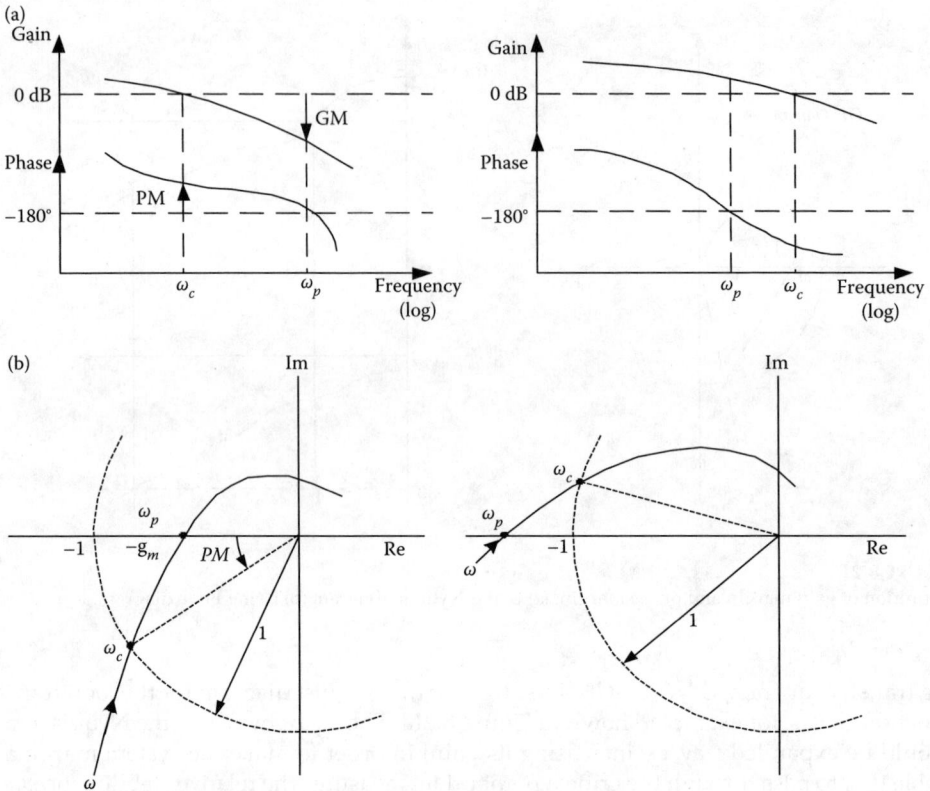

FIGURE 8.22
(a) Bode plots. (b) Nyquist plots, of a stable system (left) and an unstable system (right) (GM=gain margin, PM=phase margin).

8.4.8 PM and Damping Ratio Relation

For a closed-loop system that can be approximated by a simple oscillator, a reasonably accurate relationship for damping ratio (closed-loop) may be established as described now. First note that the loop transfer function for a damped oscillator is:

$$GH(s) = \frac{\omega_n^2}{s(s+2\zeta\omega_n)} \quad \text{with} \quad H=1$$

This is easily verified in view of: $G/(1+GH) = \dfrac{\omega_n^2}{s^2 + 2\zeta\omega_n s + \omega_n^2}$

By definition, crossover frequency is given by: $|GH(j\omega)| = 1 \rightarrow \dfrac{\omega_n^2}{\omega\sqrt{\omega^2 + 4\zeta^2\omega_n^2}} = 1$

It can be shown that the positive solution of this equation is: $\omega_c = a\omega_n$

where $a = \sqrt{\sqrt{4\zeta^2 + 1} - 2\zeta^2} \approx 1$

Now, since $GH(j\omega) = \dfrac{\omega_n^2}{j\omega(j\omega + 2\zeta\omega_n)}$, its phase angle is:

$$\angle GH(j\omega) = 0 - \left(\frac{\pi}{2} + \tan^{-1}\frac{\omega}{2\zeta\omega_n}\right) = -\frac{\pi}{2} - \tan^{-1}\frac{\omega}{2\zeta\omega_n}$$

PM:

$$\phi_m = \pi - \frac{\pi}{2} - \tan^{-1}\frac{\omega_c}{2\zeta\omega_n} = \frac{\pi}{2} - \tan^{-1}\frac{\omega_c}{2\zeta\omega_n} = \tan^{-1}\frac{2\zeta\omega_n}{\omega_c}$$

$$= \tan^{-1}\frac{2\zeta}{a} \approx \tan^{-1}2\zeta \approx 2\zeta \text{ radians (for small } \zeta) = 2\zeta \times \frac{180°}{\pi}$$

We have: $\phi_m = 100\zeta$ degrees; $\zeta = 0.01\,\phi_m$ (8.33)

in which ϕ_m is the PM in degrees, as determined from the loop transfer function. This relationship is acceptable in the damping range $0 \leq \zeta \leq 0.6$, and it provides a slightly conservative estimate for damping ratio in terms of the PM.

8.5 Bode Diagram Using Asymptotes

Any transfer function may be factorized into first order terms of the form $(s+a)$ and the second order oscillatory terms $(s^2 + 2\zeta\omega_n s + \omega_n^2)$, $0 \leq \zeta < 0$ in its denominator and the numerator. Then, with the knowledge of the Bode plot of each of these terms, the Bode plot for the entire transfer function may be constructed (by using additions and subtractions only of the component plots). The rationale for this is (as noted before):

1. In a product of complex numbers, the magnitudes multiply and the phase angles add.

2. In a quotient of complex numbers, the magnitudes divide and the phase angles subtract.

Note further that if a log scale is used for the magnitudes the multiplications and divisions of the magnitudes may be "transformed" into additions and subtractions. This is indeed the normal case for Bode plots, where magnitude is given in decibels. The advantage of the log scale for magnitude is that the Bode diagram for a product of several transfer functions can be obtained by simply adding the Bode plots for the individual transfer functions. In this manner, the Bode plot of a complex system can be conveniently obtained with the knowledge of the Bode plots of its components (plant, controller, actuator, sensor, etc.)

When a log scale is used for both magnitude and frequency, it emphasizes the lower values in a range. The x-axis (frequency axis) of the Bode plot is marked in units of frequency, which may be incremented by factors of 2 (*octaves*) or factors of 10 (*decades*). Typically in a Bode plot, the frequency axis is graduated in *decades*. This is a $\log_{10}$ scale. The amplitude axis is given in decibels, which is also a $\log_{10}$ scale, specifically $20\log_{10}(\ )$.

Note: $20 \log_{10}(\;) = 10 \log_{10}(\;)^2$. Since power and energy are represented by the square of a signal such as voltage, current, velocity, and force, we observe that 10 dB corresponds to a power (or energy) increase by a factor of 10 or a signal increase by a factor of $\sqrt{10}$. Similarly, 20 dB corresponds to a signal increase by a factor of 10 or a power increase by a factor of 100.

In a Bode plot, a linear scale is used to represent the phase angle.

The exercise of sketching a Bode diagram may be further simplified by first sketching the asymptotes of the elementary terms $(s+a)$ and $(s^2+2\zeta\omega_n s +\omega_n^2)$, $0 \leq \zeta < 0$ and then approximating the actual curves, which will approach the asymptotes in the limit. This approach is illustrated now using examples.

Example 8.15

Sketch the Bode plot of the transfer function of an armature controlled dc motor [output speed/input voltage] given by $G(s)=K/(\tau s+1)$

where:

K=Gain parameter (depends on motor constants, armature resistance, and dampers).

τ=Time constant (depends on motor inertia, motor constants, armature resistance, and damping).

Solution

This is a first order system. The frequency transfer function corresponding to the given TF is:

$$G(j\omega)=\frac{K}{(\tau j\omega+1)} \quad \text{or} \quad G(j2\pi f)=\frac{K}{(\tau j2\pi f+1)} \tag{i}$$

Here ω is the *angular frequency* (in rad/s) and f is the *cyclic frequency* (in cycles/s or Hz).

Note: The complex functions $G(j\omega)$ and $G(j2\pi f)$ may be denoted by $G(\omega)$ and $G(f)$, respectively, for notational convenience (even though contrary to strict mathematical meanings).

The numerator term in the TF is a constant. The asymptotes for the numerator and the denominator of the TF are determined now. First we define a critical frequency (discussed later):

$$f_b = \frac{1}{2\pi\tau} \tag{ii}$$

$$\text{When } f \ll f_b: G(f) \approx K \tag{iii}$$

The corresponding magnitude is K (or $\log_{10} K$ dB). This asymptote is a horizontal line as shown in Figure 8.23. The phase angle of this asymptote is zero.

$$\text{When } f \gg f_b : G(f) \approx \frac{K}{\tau j2\pi f}. \tag{iv}$$

The magnitude of this function is $K/(\tau 2\pi f)$. It monotonically decreases with frequency. If decibel scale (i.e., $20\log_{10}(\;)$ dB) is used for the magnitude axis and decade scale (i.e., multiples of 10) for the frequency axis, the slope of this asymptote is -20 dB/decade. The phase angle of this asymptote is $90°$.

The two asymptotes intersect at $f=\phi_b$. This frequency is known as the *break frequency* (or *corner frequency*). Since a significant magnitude attenuation takes place for input signal frequencies greater than f_b and in view of the fast decay of the natural response for large f_b, it is appropriate to consider f_b, given by Equation (ii), as a measure of *bandwidth* for a dc motor.

The asymptotes are drawn and the approximate Bode plots are sketched based on them (to approach them in the limit) as shown in Figure 8.23. For the normalized case of $K=1$ the bode diagram of the given transfer function is shown in Figure 8.24, where angular frequency (ω rad/s) is used instead of cyclic frequency (f Hz).

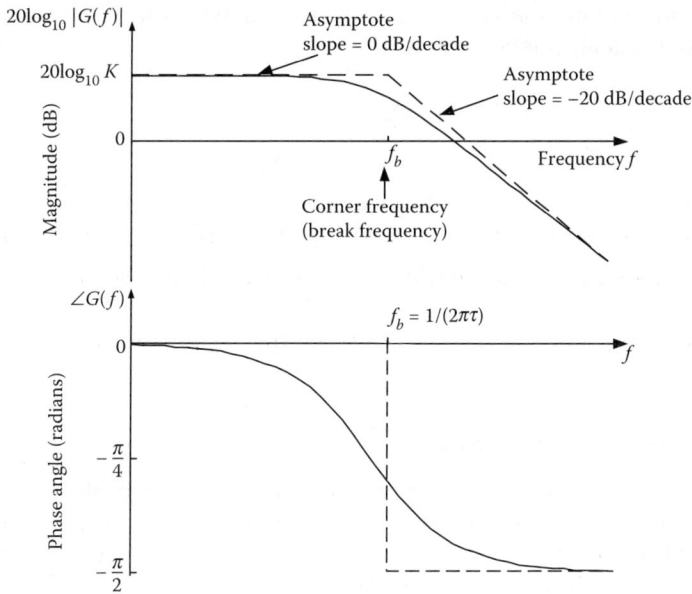

FIGURE 8.23
Bode diagram of a dc motor transfer function.

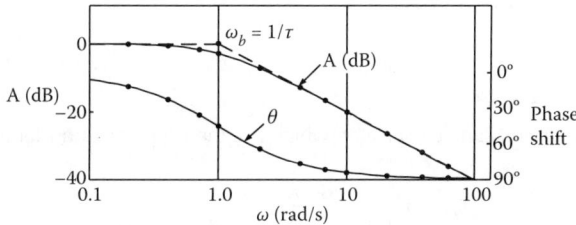

FIGURE 8.24
Bode diagram (plot of amplitude versus frequency and phase vs. frequency) of a first order system.

Suppose that a sinusoidal signal is used as the input test signal to the plant (dc motor). As the input frequency is raised, it is found that the output amplitude decreases and the phase-lag increases. This confirms the shape of the Bode plot.

Note 1: Similarly it can be shown that for a transfer function component of the form $G(s) = K(\tau s + 1)$, the second magnitude (gain) asymptote (beyond the break point of $\omega_b = 1/\tau$) will have a slope of $+20$ dB/decade and the second phase angle asymptote will be a constant at $+90°$.

Note 2: The advantages of using a log scale for frequency are the fact that a wide range of frequencies can be accommodated in a limited plotting area, and that asymptotes to the magnitude curve become straight lines with slopes differing by fixed increments (by ± 20 dB/decade if decibel scale is used for magnitude and decade scale is used for frequency).

8.5.1 Slope-Phase Relationship for Bode Magnitude Curve

H. W. Bode obtained an equation relating to phase angle the slope of the Bode magnitude (gain) curve. This approximate relationship is valid for a system whose loop transfer function does not contain poles or zeros on the right hand plane (i.e., for a *minimum-phase system*). The approximate relationship is

$$\phi = r \times 90° \tag{8.34}$$

in which

ϕ = phase angle (in degrees) at frequency ω.
r = normalized slope of the Bode magnitude curve at ω (dB/decade/20dB/decade).

Note: r is obtained by determining the slope in decibels per decade and dividing the result by 20.

This approximate relation becomes exact when the asymptote curves (for both magnitude and phase) are used, as clear from Example 8.15.

8.5.1.1 Nonminimum-Phase Systems

A transfer function having no poles or zeros on the right hand plane is called a *minimum-phase system*. A system having at least one zero or a pole on the right hand plane is a nonminimum phase system. To under this, consider the factor $(s+a)$ and another factor $(s-a)$ in a transfer function, where a is a positive real quantity. The corresponding frequency transfer function factors are $(j\omega+a)$ and $(j\omega-a)$ The first factor has a phase angle between 0° and 90° and the second factor has a phase angle 90° and 180°, which is larger and provides a nonminimum phase. A nonminimum phase systems can result in added complications to the system behavior. For example, a stable nonminimum phase system can have negative phase and GMs.

Example 8.16

Consider an underdamped simple oscillator, which has the frequency transfer function (FTF):

$$G(j\omega) = \frac{K}{\left(\omega_n^2 - \omega^2 + 2j\zeta\omega_n\omega\right)} \, 0 < \zeta < 1 \tag{i}$$

Note: Underdamped means the damping ratio ζ is less than 1.

The break point for the asymptotes is the undamped natural frequency ω_n.

For $\omega \ll \omega_n$ the frequency transfer function (i) can be approximated by the static gain (i.e., the zero-frequency magnitude):

$$G(j\omega) \approx \frac{K}{\omega_n^2}. \tag{ii}$$

The magnitude of this transfer function is a constant and hence the slope is zero ($r=0$). The phase angle is zero as well, in this region. The corresponding gain and phase asymptote pair (for $\omega=0$ to ω_n) is shown in Figure 8.25.

For $\omega \gg \omega_n$ the frequency transfer function (i) can be approximated by:

$$G(j\omega) \approx -\frac{K}{\omega^2} \tag{iii}$$

In this region the magnitude in decibels is $20\log_{10}(K/K_o) - 40\log_{10}(\omega/\omega_o)$dB

Note: K and ω are nondimensionalized because mathematically it is not correct to obtain the logarithm of a dimensional quantity. An important observation, however, is that when the

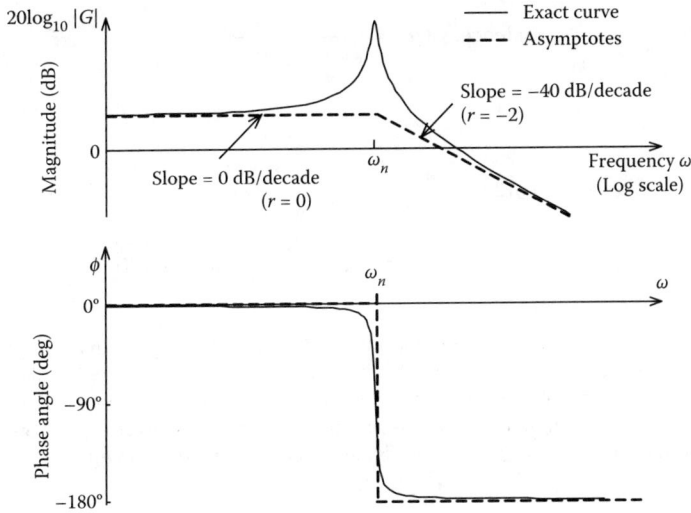

FIGURE 8.25
An example to illustrate the Bode slope-phase relationship.

frequency changes by 1 decade (i.e., when $\omega/\omega_o = 10$), the magnitude of this expression changes by -40 dB.

Hence, the slope of this asymptote is -40 dB/decade. Since Equation (iii) represents a negative real quantity, its phase angle is $-180°$. The corresponding gain and phase asymptote pair (for $\omega = \omega_n$ to ∞) is shown in Figure 8.25.

Note: For this asymptote we have $r = -2$ and it satisfies Bode's slope-phase relationship (Equation 8.34) as expected.

Example 8.17

The open-loop transfer function of a control system with unity feedback (i.e., the loop transfer function), is given by

$$G(s) = \frac{(s+3)}{(s^2 + 4s + 16)}$$

a. Tabulate the magnitude $|G(j\omega)|$ and phase angle $\angle G(j\omega)$ values for about six points of frequency in the range $\omega = 0$ to $\omega = 5$.
b. Plot the Nyquist diagram for G.
c. Plot the Bode diagram for G, and indicate the asymptotes.
d. If the open-loop system (G) is given the sinusoidal input $u = 2 \cos 2t$ what is the output at steady-state?
e. Explain, using the Nyquist plot, why the closed-loop system $\tilde{G}$ given by $\tilde{G} = G/1 + G$ is stable.

Solution

By setting $s = j\omega$ we get the FTF:

$$G(j\omega) = \frac{j\omega + 3}{16 - \omega^2 + 4j\omega} \tag{i}$$

$$\text{Hence:} |G(j\omega)| = \sqrt{\frac{3^2 + \omega^2}{\left(16 - \omega^2\right)^2 + 16\omega^2}} \tag{ii}$$

and

$$\angle G(j\omega) = \tan^{-1}\frac{\omega}{3} - \tan^{-1}\frac{4\omega}{16 - \omega^2} \quad \text{for} \quad \omega < 4$$

$$= \tan^{-1}\frac{\omega}{3} - \pi + \tan^{-1}\frac{4\omega}{\omega^2 - 16} \quad \text{for} \quad \omega > 4 \tag{iii}$$

Note: When $\omega > 4$, the real part of the denominator of the FTF (i) is negative (and the imaginary part is positive). Hence, the denominator term is in quadrant two of the complex plane (see Figure 8.14). Its phase angle $= \pi -$ [phase angle obtained by using positive real part] $= \pi - \tan^{-1}(4\omega/\omega^2 - 16)$. This has to be subtracted (because it corresponds to the denominator of Equation (i)) from the numerator phase angle ($\tan^{-1}(\omega/3)$). This gives the second part in Equation (iii).

a.

Frequency ω	0	1	2	3	4	5	∞
Magnitude	3/16	0.204	0.25	0.305	0.3125	0.266	0
$|G(j\omega)|$ (dB)	(−14.5)	(−13.8)	(−12)	(−10.3)	(−10.1)	(−11.5)	(−∞)
Phase $\angle G(j\omega)(°)$	0	3.5	0	−14.7	−36.8	−55.2	−90

b. The Nyquist curve is now plotted as shown in Figure 8.26.
   Note: For negative frequencies ($\omega = 0^-$ to $-\infty$) the indicated curve ($\omega = 0^+$ to $+\infty$) will be mirror-imaged about the real axis.

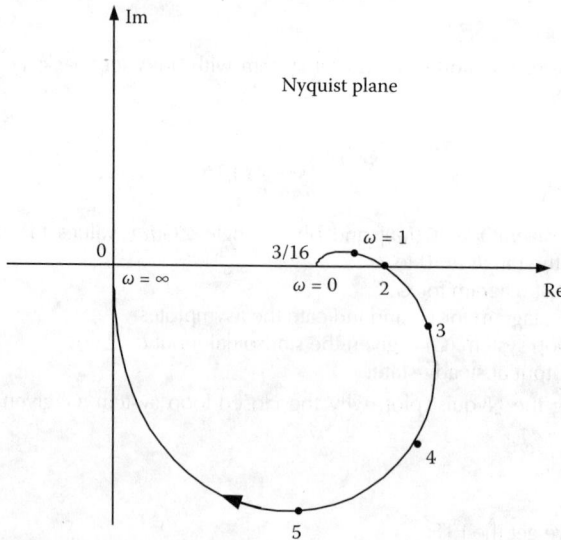

FIGURE 8.26
Nyquist curve of the example.

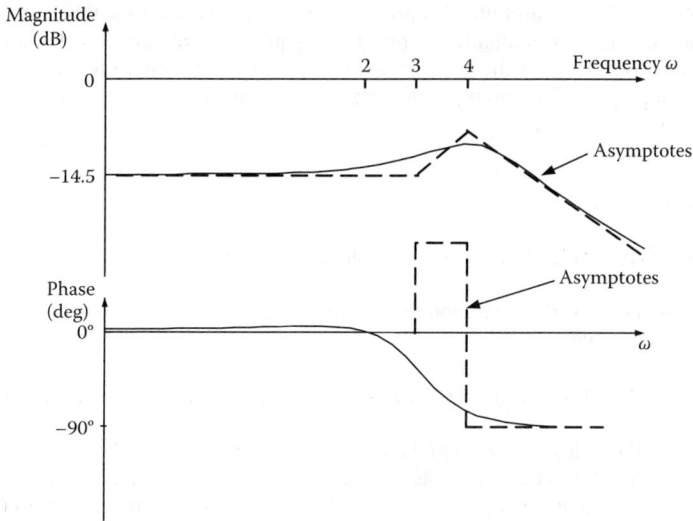

FIGURE 8.27
Bode diagram of the example.

c. The Bode plot is shown in Figure 8.27.

For small frequencies $\omega \ll 3$, the magnitude is approximately constant and equal to the static gain −14.5 dB and the phase angle is approximately zero. At $\omega=3$, the first break point occurs due to the zero in the loop TF. As a result, a slope of +20 dB/decade is added to the magnitude asymptote and a phase "lead" of 90° is added to the phase asymptote, as shown. At $\omega=4$, the second break point occurs, due to the second order (simple oscillator) term in the loop transfer function (see Example 8.16). This adds a slope of −40 dB/decade to the magnitude asymptote (resulting in a net slope of −20 dB/decade) and a phase angle of −180° to the phase asymptote (resulting in a net phase angle of −90°, which is a phase lag of 90°), as shown in Figure 8.27.

Note: An asymptote (particularly a phase asymptote) does not provide good approximation to the actual Bode curve in a narrow frequency range (as clear in this example, for the frequency interval [2,4]).

d. For the open-loop system, $G(j\omega) = \dfrac{j\omega+3}{\left(16-\omega^2+4j\omega\right)}$

$$\text{At } \omega=2: |G(j\omega)|=0.25$$

$$\angle G(j\omega)=0°=0 \text{ rad}$$

Hence, the steady-state response for an input of $u=2\cos2t$ is

$$y=2\times0.25\ \cos(2t+0)$$

$$\text{or: } y=0.5\cos 2t$$

e. Note from the Nyquist plot that, as the frequency increases, the magnitude (of the loop transfer function) remains less than 1 (or, 0 dB) and the phase "lag" never increases beyond 90°, in the entire frequency range (from 0 to ∞). Hence, the system is stable.

Even though the PM and the GM are not numerically defined for this system, both may be taken positive in qualitative terms. The Nyquist plot remains entirely to the right of the critical point −1 on the transfer-function plane. Hence there is no possibility of the closed-loop system becoming even marginally stable (i.e., Nyquist plot will not cross the critical point).

Example 8.18

A system with unity feedback has a loop transfer function whose Bode diagram is as shown in Figure 8.28.

Note: What is shown is the asymptotic magnitude curve with break points at 2 and 6, and the actual phase angle curve.

1. Determine the closed-loop transfer function of the system. What is order of the system? Why?
2. Determine the poles and zeros of the closed-loop system.
3. Determine the PM of the system. Is the system stable? Explain your answer.
4. Describe a practical way for increasing the PM of the given control system to 60°.

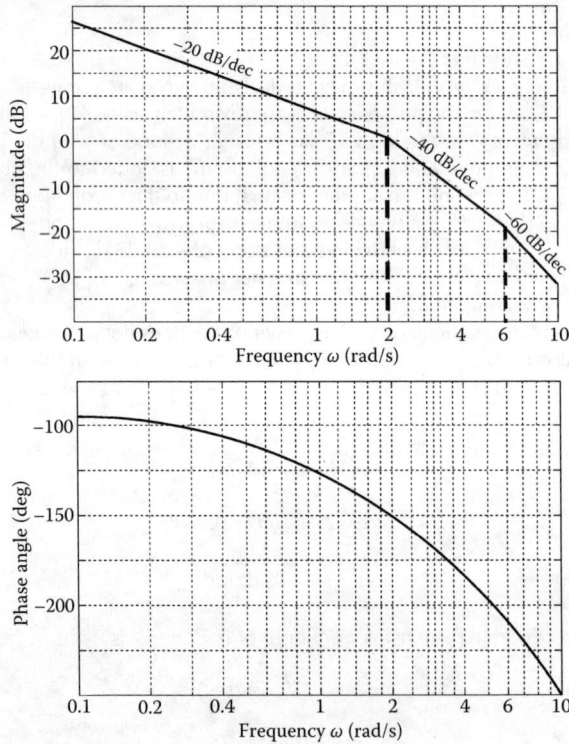

FIGURE 8.28
Bode magnitude asymptotes and phase curve of the loop transfer function.

Solution

1. In the magnitude asymptotic curve:

(a) There is an initial slope of -20 dB/decade. This corresponds to a free integrator ($1/s$).

(b) There is a break at $s=2$, with a change in the slope by -20 dB/decade. This corresponds to a pole term $(s+2)$. *Note*: This term cannot be $(s-2)$ since the Bode phase angle curve keeps monotonically decreasing.

(c) There is a second break at $s=6$, with a further change in the slope by -20 dB/decade. This corresponds to a pole term $(s+6)$. As before, the term cannot be $(s-6)$ since the Bode phase angle curve keeps monotonically decreasing.

Accordingly, the loop transfer function of the system is given by:

$$GH(s) = \frac{K}{s(s+2)(s+6)}$$

Only the gain parameter K needs to be determined. This can be done simply by noting that at the frequency value $\omega=0.2$ rad/s, the magnitude is approximately 20 dB, which is equal to 10. Substitute in

$$GH(j\omega) = \frac{K}{j\omega(j\omega+2)(j\omega+6)}$$

We have

$$\frac{K}{0.2|\,j0.2+2|\,|\,j0.2+6|} = 10$$

By neglecting 0.04 compared to 4, we get, approximately, $K=24$.

$$\text{Hence: } GH(s) = \frac{24}{s(s+2)(s+6)}$$

With unity feedback we have $H=1$.

The Bode and Nyquist plots for the loop transfer function (GH) may be generated using the following MATLAB® code:

```
%  ---Bode plot GH=24/(s(s+2)(s+6))---
num= [0  0  0  24];
den= [1  8  12  0];
SYS=tf(num,den);
bode(SYS)


%  ---Nyquist plot GH=24/(s(s+2)(s+6))---
num= [0  0  0  24];
den= [1  8  12  0];
SYS=tf(num,den);
nyquist(SYS,(0:100))
```

The MATLAB generated plots are shown in Figure 8.29.

Note: Selecting the proper scale is important in obtaining proper plots, particularly the Nyquist diagram. This is a particular disadvantage of computer-generated plots over those that are manually generated (using the first principles). In particular, Figure 8.29b does not show the fact that near zero (positive) frequency the phase angle of GH is about $-90°$ and the Nyquist curve touches the negative imaginary axis on the left hand side, and that at very

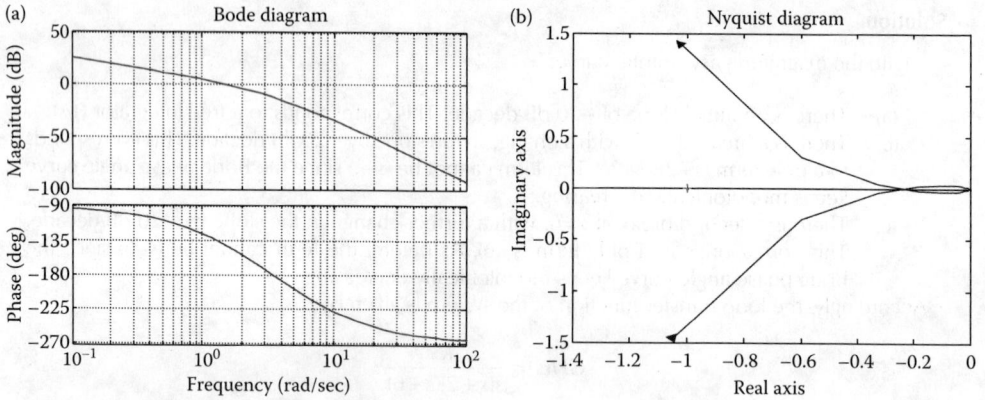

FIGURE 8.29
MATLAB generated (a) Bode diagram, (b) Nyquist diagram.

high (infinite) positive frequencies the phase angle of *GH* is about −270° and the Nyquist curve touches the positive imaginary axis on the left hand side. Both low-frequency and high-frequency segments cannot be properly shown on computer-generated Nyquist plots. Use of a log scale can somewhat improve this situation. It should be cautioned that the log of the real and imaginary parts should not be taken (as the real and the imaginary parts can be negative). Instead, the log of the magnitude (actual magnitude, not the dB value) should be used, while the actual phase angle (linear scale) should be used.

The closed-loop transfer function is:

$$\tilde{G}(s) = \frac{G}{1+G} = \frac{\dfrac{24}{s(s+2)(s+6)}}{1+\dfrac{24}{s(s+2)(s+6)}} = \frac{24}{s(s+2)(s+6)+24}$$

Note: The denominator (characteristic) polynomial is third order, giving three closed-loop poles. Hence the closed-loop system is third order.

2. Note from the numerator of $\tilde{G}(s)$ that the closed-loop system has no zeros. The poles (three) are given by the characteristic equation

$$s(s+2)(s+6)+24=0$$

Its roots are obtained using the following MATLAB® code:

```
%Polynomial roots (s(s+2)(s+6)+24)
P= [1 8 12 24];
roots(P)
```

The three roots are: −6.7488, −0.6256±*j*1.7790
Note that they all on the LHP, indicating a stable system.

3. From the given Bode curve pair it is seen that at the magnitude of 0 db, the phase angle of $GH(j\omega)$ is approximately −150°. Hence,

$$PM=180°-150°=30°$$

Note the positive PM, indicating a stable system.

This can be further confirmed by using Routh–Hurwitz criterion on the closed-loop characteristic equation: $s(s+2)(s+6)+24=0$ or $s^3+8s^2+12s+24=0$.

Routh array:

s^3	1	12
s^2	8	24
s^1	$\dfrac{8\times12-1\times24}{8}=9$	0
s^0	$\dfrac{9\times24-8\times0}{9}=24$	

It is seen that all the coefficients of the characteristic polynomial are positive and that there are no sign changes in the first column of the array. Hence the system is stable.

4. The PM of the system can be increased by including a lead compensator in the feedback path of the control system. The compensator transfer function is of the form

$$\frac{\tau s+1}{\alpha \tau s+1} \quad \text{with} \quad 0<\alpha<1$$

The compensator parameters τ and α can be determined by using established methods of compensator design.

8.5.2 Ambiguous Cases of GM and PM

GM and PM may not be defined for some loop transfer functions. Specifically, if the magnitude curve does not cross the 0 dB line, the PM of the system is not defined. Similarly, if the phase angle curve does not cross the $(-180°)$ line, the GM is not defined. It is possible as well (for rather complex and high order systems) to have multiple crossings of the 0 dB line and/or the $-180°$ line. In some such cases it is still possible to use the concepts of PM and GM to establish stability of the associated closed-loop systems.

Case 1: If the magnitude curve of the loop transfer function (GH) stays below 0 dB throughout the entire frequency range of operation (operating bandwidth) of the control system, then a PM is not defined. Yet, the control system is considered stable (amplitude stabilization) if the phase angle at the lowest magnitude value in the frequency range is between $0°$ and $-180°$. In this case a positive GM can be established using the gain at the frequency where the phase angle is closest to $-180°$ (typically, at the high-frequency end of the operating bandwidth). Also, a positive PM can be established by using the phase angle at the lowest value of the magnitude. On the other hand, if the magnitude curve of GH remains greater than 0 dB throughout the operating bandwidth, the system is considered unstable. Then a negative GM can be defined using the gain value when the phase angle is closest to $-180°$.

Case 2: If the phase angle of GH remains between $0°$ and $-180°$ within the entire operating bandwidth of the system, then a GM is not defined. In this case, the closed-loop control system is considered stable, assuming that the magnitude of GH is less than 0 dB at the high-frequency end of the operating bandwidth, particularly in the region where the is closest to $-180°$. Then, a positive GM is defined using the magnitude value where the phase angle is closest to $-180°$, and a positive PM can be defined using the phase angle at the smallest magnitude value of the operating bandwidth (typically at the high-frequency end).

Case 3: If there are multiple crossings of the 0 dB line, a unique PM is not defined. Similarly, if there are multiple crossings of the (−180°) line, a unique GM is not defined. In these cases, a single PM or a single GM may be defined by taking the worst case (i.e., the smallest of the stability margins) or by considering a limited operating bandwidth that contains only a single crossing.

8.5.3 Destabilizing Effect of Time Delays

Time delays, which are inherently present in control systems, can have a destabilizing effect on the system response. Time delays can result from various causes including transport lags in systems such as chemical processes (e.g., flow changes under transient pressure conditions and temperature changes due to mixing of fluids), measurement delays due to large time constants in sensors, and dynamic delays in mechanical systems with high inertia and damping.

A block diagram representation of a time delay is shown in Figure 8.30. Here a signal $x(t)$ undergoes a pure delay by time τ. Since the Laplace transform of the delayed signal is given by

$$\mathcal{L}x(t-\tau)=\exp(-\tau s)\mathcal{L}x(t) \tag{8.35}$$

it is clear that the transfer function for a pure delay is

$$G(s)_{delay}=\exp(-\tau s) \tag{8.36}$$

In the frequency domain $(s=j\omega)$ the magnitude of this transfer function is unity, and the phase angle is negative and monotonically decreasing with frequency ω:

$$|G(j\omega)|_{delay}=1 \tag{8.37}$$

$$\angle G(j\omega)_{delay}=-\tau\omega \tag{8.38}$$

It follows that due to a pure delay, the system magnitude is unchanged, but the phase angle is decreased. Consequently, the PM and GM of the system are reduced; a destabilizing effect. Note further that the condition gets worse as the frequency increases. It follows that a system operating at high frequencies is more likely to become unstable due to time delays. In control system design, specified PM should allow for the time delays that are present in various components in the control loop.

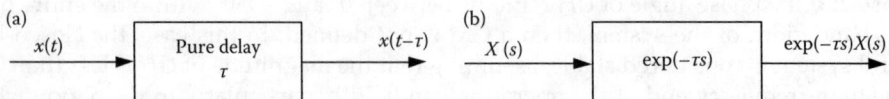

(a)

$x(t)$ → | Pure delay τ | → $x(t-\tau)$

(b)

$X(s)$ → | $\exp(-\tau s)$ | → $\exp(-\tau s)X(s)$

FIGURE 8.30
Representation of a time delay. (a) Time domain representation. (b) Transfer function.

8.6 Nyquist Stability Criterion

Let us revisit the feedback control system shown in Figure 8.20. The closed-loop transfer function is:

$$\tilde{G}(s) = \frac{G}{1+GH} \qquad \bullet \tag{8.39}$$

As noted previously, system stability is determined by the poles (eigenvalues) of $\tilde{G}$. These are the roots of the characteristic equation $1+G(s)H(s)=0$. If all the poles are located on the left hand s-plane (i.e., if the real parts of the roots are all negative), the closed-loop system is stable. If there is at least one pole of $\tilde{G}$ on the right hand s-plane (RHP), the system is unstable. If there is a pole on the imaginary axis of the s-plane (including, of course, the origin), the closed-loop system is considered marginally stable, provided that there are no poles on the right half of the s-plane.

Note: Strictly, if there are two or more identical poles (i.e., repeated poles) on the imaginary axis, the system is unstable.

Nyquist stability criterion follows from *Cauchy's theorem* on complex mapping. Using this criterion, the stability of a closed-loop system can be determined simply by sketching the Nyquist diagram of the corresponding loop transfer function *GH*. As discussed before, to obtain the Nyquist plot we first set $s=j\omega$ in *GH*(s) and plot the resulting function using the imaginary part as the y-coordinate and the real part as the x-coordinate, while varying the frequency parameter ω from $-\infty$ to $+\infty$. This is explained in Figure 8.31.

Consider the area to the right of the imaginary axis on the s-plane, which is enclosed by the closed contour as ω is varied from $-\infty$ to $+\infty$ and the contour is completed in the clockwise (right-handed) sense at infinity. As shown in Figure 8.31, the corresponding mapping of $GH(j\omega)$ on the GH-plane is also a closed contour, which is the Nyquist diagram.

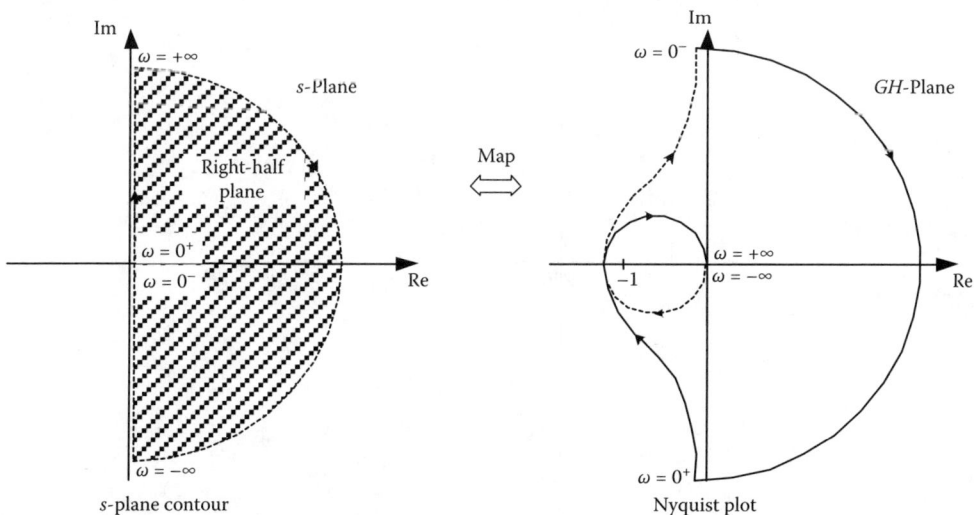

FIGURE 8.31
Generation of a Nyquist diagram.

Note: Changing j to $-j$ in a function of $j\omega$ amounts to the same thing as changing ω to $-\omega$. Furthermore, changing j to $-j$ corresponds to changing the sign of the imaginary part of the complex function (i.e., *complex conjugation*) which forms a mirror image about the real axis. It follows that the Nyquist plot for negative frequencies is the mirror image of that for positive frequencies, about the real axis. Hence, it is only necessary to plot the Nyquist diagram for the positive frequencies (i.e., $\omega=0$ to $+\infty$).

8.6.1 Nyquist Stability Criterion

The Nyquist stability criterion states that

$$\tilde{p} - p = N \tag{8.40}$$

in which

$\tilde{p}$ =number of unstable poles in the closed-loop transfer function $\tilde{G}$
p=number of unstable poles in the loop transfer function GH
N=number of clockwise encirclements of point -1 on the real axis by the Nyquist plot.

For stability of the closed-loop system we need $\tilde{p}=0$. Hence, we should have $N=-p$. For example, in Figure 8.31 we have two clockwise encirclements of the point -1. Hence $N=2$. From Equation 8.40 we have $\tilde{p}>0$ and the closed-loop system will always be unstable (even if GH is stable; i.e., $p=0$).

Note: p and $\tilde{p}$ are nonnegative integers while the integer N can be positive, negative or zero.

The gain of GH can be changed (usually decreased) to shrink the Nyquist plot so that the point (-1) is no longer encircled, however. This situation is shown in Figure 8.32. Then,

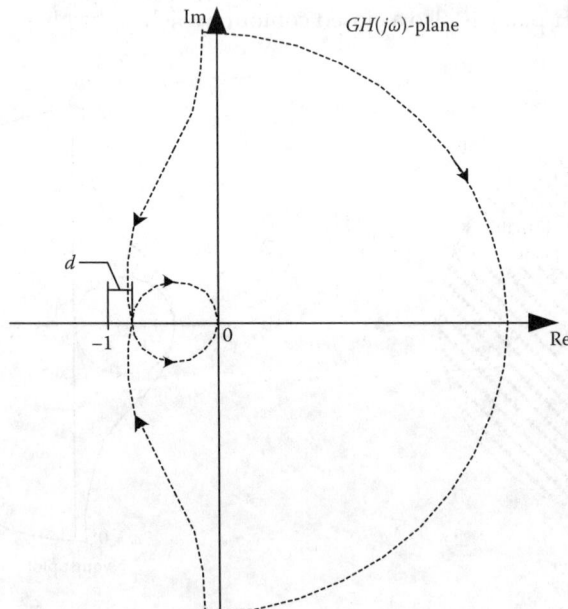

FIGURE 8.32
Gain margin represented on the Nyquist plane.

$N=0$. Hence, it follows from Equation 8.40 that if GH is stable ($p=0$) to begin with, the closed-loop system will remain stable. This indicates how, in many cases, an unstable system can be stabilized by decreasing the gain, and vice versa. The *degree of stability* (or *stability margin*) can be measured by the distance d through which the Nyquist plot must be expanded along the real axis in order to have an encirclement of the point −1 (see Figure 8.32). The concepts of GM and PM are based on this fact.

Note: Typically we are interested in stable closed-loop systems ($\tilde{G}$), not marginally stable ones. Hence we will consider only those situations where the Nyquist curve (of GH) does not pass through −1, when using the Nyquist stability criterion.

8.6.2 Loop Poles on the Imaginary Axis

Caution should be exercised when there are poles of GH on the imaginary axis of the s-plane. In this case, by properly choosing the contour on the s-plane, these marginally stable poles of GH can be excluded from the right half of the s-plane. Consequently, they are not counted in p of Equation 8.40.

A related problem arises due to the fact that at a pole of GH, the magnitude of GH will become infinite (because the denominator—characteristic polynomial—of GH will become zero, by definition of a pole). In generating the Nyquist curve, as ω is varied from $-\infty$ to $+\infty$ (see the left hand side figure of Figure 8.31) the $j\omega$ will move along the imaginary axis of the s-plane. If there are poles (of GH) on the imaginary axis, the will be crossed by this path. At such a pole, the magnitude of $GH(j\omega)$ will become infinite. In avoiding that pole, by looping around it on the right, the question arises as to in what sense (clockwise or counterclockwise) the corresponding infinite ends of the Nyquist curve should be connected. How the correct sense and connectivity are determined can be explained using examples.

Note: Even though the contour we follow in covering the right-hand s-plane is clockwise (see left hand side figure of Figure 8.31) the corresponding contour of the GH-plane (i.e., the Nyquist curve) may not necessary move in the clockwise direction. In fact, some parts of the Nyquist curve may traverse clockwise while some other parts may traverse counterclockwise.

8.6.3 Steps for Applying the Nyquist Criterion

In applying the Nyquist stability criterion, the following systematic steps may be followed.

Step 1: Establish the s-plane contour encompassing the RHP while excluding any poles (of the loop transfer function GH) on the imaginary axis. Count the number of loop poles (unstable) p inside this contour RHP.

Step 2: Plot/sketch the Nyquist curve for the positive frequency range: $\omega=0^+$ to $+\infty$.

Step 3: In view of symmetry of the Nyquist curve about the real axis, plot/sketch the Nyquist for the negative frequency range: $\omega=-\infty$ to 0^-.

Step 4: Close the loop of the Nyquist curve by connecting the open ends of the two segments (for the positive and negative frequencies).

Step 5: Count the number of clockwise encirclements N of the point −1 by the complete Nyquist plot as obtained in Step 4. *Note*: Counterclockwise encirclements are counted as negative.

Step 6: Determine the number of unstable poles $\tilde{p}$ of the closed-loop transfer function using the Nyquist criterion: $\tilde{p} = p + N$. If $\tilde{p} > 0$ the closed-loop system is unstable.

Example 8.19

Sketch the Nyquist curve of the loop transfer function

$$GH(s) = \frac{2(s+1)}{s(s-1)} \tag{i}$$

over the complete frequency range $\omega = -\infty$ to $+\infty$. Using the Nyquist stability criterion determine the stability of the closed-loop system $G/1+GH$.

Solution

We have: $GH(j\omega) = \dfrac{2(j\omega+1)}{j\omega(j\omega-1)}$ \hfill (ii)

Hence: $\left|GH(j\omega)\right| = \dfrac{2\sqrt{\omega^2+1}}{\omega\sqrt{\omega^2+1}} = \dfrac{2}{\omega}$ \hfill (iii)

$$\angle GH(j\omega) = \tan^{-1}\omega - [\pi/2 + (\pi - \tan^{-1}\omega)] = 2\tan^{-1}\omega - 3\pi/2 \tag{iv}$$

Note: The denominator term $j\omega-1$ in Equation (ii) is in the second quadrant and its phase angle is in the range: $90° < \theta < 180°$ (see Figure 8.14). This leads to the proper selection of its phase angle as $(\pi - \tan^{-1}\omega)$ in Equation (iv).

1. Behavior of the Nyquist curve at $\omega = 0^+$:
    From Equation (iii): Magnitude $|GH| \to \infty$
    See Equation (ii):
    The phase angle of $j\omega+1$ (Quadrant 1) becomes a small positive angle (say δ)

    The phase angle of $j\omega-1$ (Quadrant 2) becomes $\pi-\delta$
    $\to$ Phase angle of $(j\omega+1)/(j\omega-1)$ becomes $-\pi+2\delta$
    $\to$ Phase angle of $(j\omega+1)/(j\omega(j\omega-1))$ becomes $-3\pi/2+2\delta$

    It follows that at $\omega = 0^+$ the Nyquist curve is in the second quadrant, touching the positive imaginary axis at ∞.
2. Behavior of the Nyquist curve at $\omega \to +\infty$:
    As $\omega \to +\infty$ we can neglect the constant terms in Equation (ii) in comparison.
    Then, $\tan^{-1}\omega \to \pi/2$. In particular:

    The phase angle of $j\omega+1$ becomes slightly smaller than $\pi/2$
    The phase angle of $j\omega-1$ becomes slightly bigger than $\pi/2$
    $\to$ Phase angle of $(j\omega+1)/(j\omega-1)$ becomes a small negative angle (say $-\delta$)
    $\to$ Phase angle of $(j\omega+1)/(j\omega(j\omega-1))$ becomes $-\pi/2-\delta$
    The magnitude $|GH(j\omega)| \to 0$

This means: As $\omega \to +\infty$ the Nyquist curve will approach 0 (origin) touching the negative imaginary axis from the left (third quadrant).

It is also clear from the above two steps that the Nyquist curve will intersect the negative real axis. At this point, $\angle GH(j\omega) = -\pi$. To determine the corresponding value of the magnitude $|GH(j\omega)|$ note from Equation (iii) that this occurs when $\tan^{-1}\omega = \pi/4$ or $\omega = 1$. Then from Equation (ii) we have the corresponding magnitude as $|GH(j\omega)| = 2$. It follows that the Nyquist curve crosses the real axis at -2.

With this information we are now able to sketch the segment of the Nyquist curve for positive frequencies: $\omega = 0^+$ to $+\infty$, as the solid line shown in Figure 8.33b.

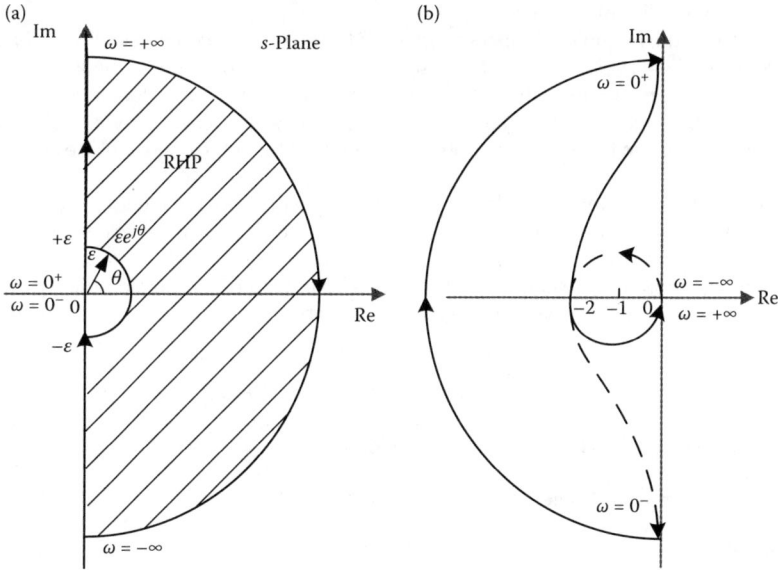

FIGURE 8.33
(a) Contour covering the RH s-plane while avoiding the GH poles on the imaginary axis. (b) Nyquist curve showing one counterclockwise encirclement of –1.

Next, in view of the symmetry of a Nyquist plot about the real axis, we are able to sketch the segment of the Nyquist curve for negative frequencies: $\omega=-\infty$ to 0^-, as the broken line shown in Figure 8.33b.

Now, in order to connect the open ends of the two segments of the Nyquist curve obtained above, we need to determine the behavior of the Nyquist curve near $\omega=0$, which corresponds to a pole of $GH(s)$. First, in order to avoid the pole of GH at $s=0$, note the contour that we follow on the s-plane, as shown in Figure 8.33a. Specifically, we follow the edge of the vector (complex) given by:

$$s=\varepsilon e^{j\theta} \tag{8.41}$$

where ε a very small positive number (=magnitude of the complex number=radius of the vector) while the phase angle θ changes from $-\pi/2$ through 0 to $+\pi/2$.

For this (small) s, we can neglect the s terms in Equation (i) compared to 1. Then, from Equation (i) we have: $GH(s)\approx1/s(-1)$. Since $-1=\varepsilon^{j\pi}$ we get:

$$GH(s)\approx\frac{1}{\varepsilon}e^{-j(\theta+\pi)} \tag{8.42}$$

as θ changes from $-\pi/2$ through 0 to $+\pi/2$.

It is clear from Equation 8.42 that, when the phase of s is θ in this contour (of very mall magnitude ε), the phase of GH is $-(\theta+\pi)$ in a corresponding contour (of very large magnitude $1/\varepsilon$). Some usefully values of these phase angle pairs are tabulated below:

$\angle s=\theta$	$\angle GH=-\theta-\pi$
$-\pi/2+\delta$	$-\pi/2-\delta$
0	$-\pi$
$+\pi/2-\delta$	$-3\pi/2+\delta$

Note: δ is a very small positive angle.

It follows that as the frequency changes from $\omega=0^-$ to $\omega=0^+$ in the counterclockwise sense of the s-plane contour, while avoiding 0, the Nyquist curve will move from $-j\infty$ (in the third quadrant—touching the imaginary axis from the left) to $+j\infty$ (in the second quadrant—touching the imaginary axis from the left), in the clockwise direction, passing the negative real axis (where phase$=-\pi$).

The Nyquist curve can be completed now as in Figure 8.33b for the entire s-plane contour given in Figure 8.33a.

To determine stability of the closed-loop system, we apply the Nyquist criterion. Specifically, from Figure 8.33b we notice that the Nyquist curve has one counterclockwise encirclement of –1 giving:

$N=$number of clockwise encirclements of point –1 on the real axis$=-1$.

Also, from Equation (i) it is seen that $GH(s)$ is unstable (with one unstable pole +1 inside the RHP contour). *Note:* We have avoided the marginally stable pole $s=0$ when choosing our s-plane contour. Hence:

$$p=\text{number of unstable poles in } GH=1$$

Then from the Nyquist stability criterion (Equation 8.40):

$$\text{Number of unstable poles in the closed-loop transfer function}=\tilde{p}=p+N=1-1=0.$$

It follows that the closed-loop system is stable.

Example 8.20

Consider a plant given by the transfer function $G_p=K/(s^2(\tau s+1))$. We can show using the Nyquist criterion that the closed-loop system will always be unstable under proportional feedback control.

If we include PD control, the loop transfer function will be $GH=K(\tau_d s+1)/(s^2(\tau s+1))$. Using the Nyquist criterion we can show that

When $\tau_d<\tau$ the closed-loop system remains unstable (due to inadequate derivative action).
When $\tau_d<\tau$ the closed-loop system becomes stable (adequate derivative action).

These two cases are left as exercises. We will only show here that the plant is unstable under proportional action.

Specifically, consider the loop transfer function

$$GH(s)=\frac{K}{s^2(s+1)} \tag{i}$$

Note: There are two poles of GH at the origin (i.e., on the imaginary axis). The contour shown in Figure 8.34a avoids these poles while encompassing the RHP. There no poles inside this contour. Hence $p=0$

Next we sketch the Nyquist plot for positive frequencies, as follows:

$$\text{On the imaginary axis of the s-plane we have: } s=j\omega \tag{ii}$$

$$\text{Substitute Equation (ii) in Equation (i): } GH(j\omega)=\frac{K}{-\omega^2(j\omega+1)} \tag{iii}$$

The denominator of Equation (iii) is a product of a negative real quantity (phase$=\pi$) and a complex quantity with positive real and imaginary parts (i.e., Quadrant 1; phase$=\tan^{-1}\omega$).

$$\text{Total phase of the denominator}=\pi+\tan^{-1}\omega$$

The numerator of Equation (iii) is a positive real quantity (phase$=0$). Hence,

$$\text{Phase of } GH(j\omega)=0-[\pi+\tan^{-1}\omega]=-\pi-\tan^{-1}\omega \tag{iv}$$

(a)

(b)

$$GH = \frac{K}{s^2(\tau s+1)}$$

$\tau > 0$

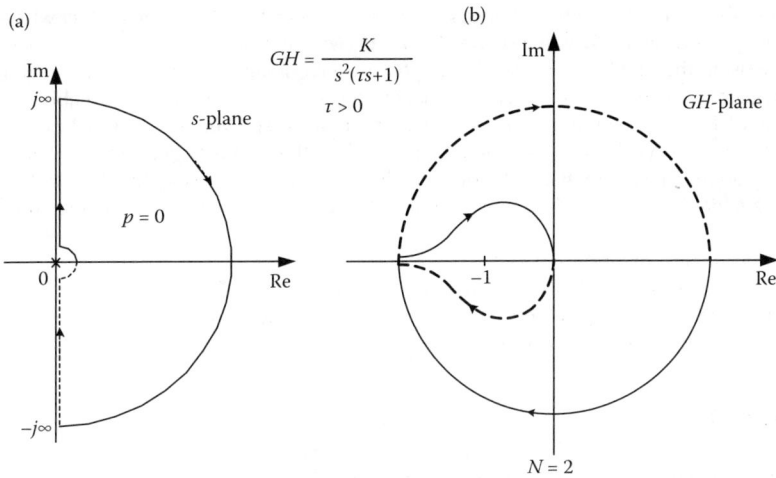

FIGURE 8.34
(a) Contour covering the RH s-plane while avoiding the *GH* poles on the imaginary axis. (b) Nyquist curve for the system with P control, showing two clockwise encirclements of –1.

From Equations (iii) and (iv) we observe the following for $GH(j\omega)$ at the two limits of positive frequency:

At $\omega=0^+$: magnitude$=\infty$; phase$=-\pi-\delta_1$

At $\omega=+\infty$: magnitude$=0$; phase$=-\pi-(\pi/2-\delta_2)=-3\pi/2+\delta_2$

Note: Here δ_1 and δ_2 are very small positive angles.

It follows that the Nyquist curve for the positive frequencies is entirely in Quadrant 2 and: at $\omega=0^+$ it touches the negative real axis at $-\infty$ from above (second quadrant), and at $\omega=+\infty$ it touches the left side of the positive imaginary axis (second quadrant) near the origin.

With this information, the Nyquist curve segment for positive frequencies can be sketched as the solid line in Figure 8.34b.

As well, we can now sketch the Nyquist curve segment for negative frequencies, which is the mirror image about the real axis—in view of the symmetry about the real axis (see the broken line in Figure 8.34b).

We connect the two free ends of these two segments of the Nyquist curve as follows:

These two connecting points are represented by the transfer function *GH* on the tiny circle which avoids the two poles at the origin. This tiny circle may be represented by:

$$s=\varepsilon e^{j\theta} \qquad (v)$$

where ε is a small radius (positive).

$$\text{Substitute Equation (v) in Equation (i): } GH(s)=\frac{K}{\varepsilon^2 e^{j2\theta}} \qquad (vi)$$

Note 1: For small s we have $s+1\approx1$

It follows from Equation (vi) that when the phase of s is θ, in the tiny circular contour (v) near the origin of the s-plane, the phase of *GH* is -2θ in a corresponding contour of very large magnitude K/ε^2. Some usefully values of these phase angle pairs are tabulated below:

$\angle s=\theta$	$\angle GH=-2\theta$
$-\pi/2$	π
$-\pi/4$	$\pi/2$
0	0
$\pi/4$	$-\pi/2$
$\pi/2$	$-\pi$

It follows that as the frequency changes from $\omega=0^-$ to $\omega=0^+$ in the counterclockwise sense of the s-plane contour while avoiding 0, the Nyquist curve will move from $-\infty$ (just below the negative real axis) in the clockwise sense (passing the positive imaginary axis, positive real axis, and then the negative imaginary axis) at infinite magnitude, and will return to $-\infty$ (just above the negative real axis) The corresponding complete Nyquist curve is sketched in Figure 8.33b.

Note: There are two clockwise encirclements of -1 by the Nyquist curve. Hence $N=2$.

From Equation (i) it is seen that $GH(s)$ does not have any unstable inside the RHP contour). *Note:* We have avoided the marginally stable pole pair at $s=0$ when choosing our s-plane contour.

Hence:

$p=$number of unstable poles in $GH=0$.

Then from the Nyquist stability criterion (Equation 8.40):

Number of unstable poles in the closed-loop transfer function$=\tilde{p}=p+N=0+2=2$.

It follows that the closed-loop system is unstable.

Example 8.21

Consider the two examples with loop transfer functions

$$GH(s)=\frac{s(s+1)}{(s+2)(s^2+1)} \tag{i}$$

and

$$GH(s)=\frac{s(s+2)}{(s+1)(s^2+1)}$$

These two problems have two entirely different Nyquist curves. We will consider the first problem here and leave the second one as an exercise.

$$\text{FTF of Equation (i): } GH(j\omega)=\frac{j\omega(j\omega+1)}{(j\omega+2)(-\omega^2+1)} \tag{ii}$$

$$\text{Magnitude: } |GH(j\omega)|=\frac{\omega\sqrt{\omega^2+1}}{\sqrt{\omega^2+4}\,|-\omega^2+1|} \tag{iii}$$

$$\text{Phase: } \angle GH(j\omega)=90°+\tan^{-1}\omega-\tan^{-1}\omega/2 \quad \text{if} \quad \omega^2<1$$

$$=90°+\tan^{-1}\omega-\tan^{-1}\omega/2-180° \quad \text{if} \quad \omega^2>1 \tag{iv}$$

Note: When $\omega=1^-$ we have $\omega^2<1$; $\omega=1^+$ we have $\omega^2>1$. These facts are useful when applying Equation (iv) in the neighborhood of $\omega=1$.

From Equations (ii) through (iv) we can determine the following facts.

At $\omega=0$: magnitude$=0$; phase$=90°$
At $\omega=1^-$: magnitude$=\infty$; phase$=90°+45°-26.6°-0°=108.4°$
At $\omega=1^+$: magnitude$=\infty$; phase$=90°+45°-26.6°-180°=-71.6°$
At $\omega=+\infty$: magnitude$=0$; phase$=90°+90°-90°-180°=-90°$

From this information we can sketch the positive frequency segment of the Nyquist curve, as shown by the solid curve in Figure 8.35b.

By symmetry of a Nyquist curve about the real axis, we can also sketch the negative frequency segment of the Nyquist curve. This is sketched as shown by the broken-line curve in Figure 8.35b.

To exclude the two poles of GH on the imaginary axis we use the contour shown in Figure 8.35a, which covers the RHP of GH. There are no unstable poles of GH in this contour. Hence: $p=0$.

(a) (b) *GH*-plane

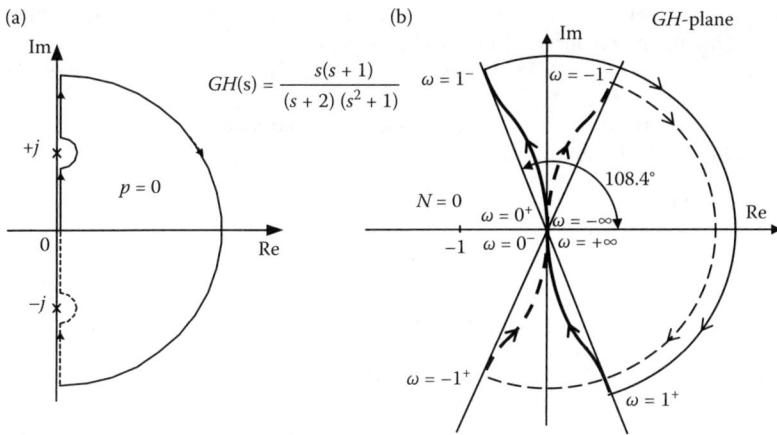

FIGURE 8.35
(a) Contour covering the RH *s*-plane while avoiding the <u>GH poles</u> on the imaginary axis. (b) Nyquist curve for the system (No encirclements of –1).

Behavior of the Nyquist Curve Near $\omega=1$:

To connect the open ends of the Nyquist curve segments at $\omega=1^-$ and $\omega=1^+$ we proceed as follows.

The tiny semicircular contour that excludes the pole $+j$ on the *s*-plane (see Figure 8.35a) is given by:

$$s=j+\varepsilon e^{j\theta} \tag{v}$$

where ε a very small positive number ($=$radius of the semicircle) while the phase angle θ changes from $-\pi/2$ through 0 to $+\pi/2$.

Substitute Equation (v) in Equation (i):

$$GH(s)=\frac{(j+\varepsilon e^{j\theta})(j+\varepsilon e^{j\theta}+1)}{(j+\varepsilon e^{j\theta}+2)((j+\varepsilon e^{j\theta})^2+1)} \simeq \frac{j(j+1)}{(j+2)2j\varepsilon e^{j\theta}} \quad \text{for small } \varepsilon \tag{vi}$$

It is seen from Equation (vi) that near $\omega=1$:

$$\text{Magnitude } |GH(s)|\rightarrow\infty \tag{vii}$$

$$\text{Phase } \angle GH=90°+45°-(26.6°+90°+\theta)=18.4°-\theta \tag{viii}$$

To connect the ends of the Nyquist curve at $\omega=1^-$ to that at $\omega=1^+$, we use Equation (viii) and prepare the following table:

θ	$\angle GH=18.4°-\theta$
$-90°$	108.4°
0	18.4°
$+90°$	$-71.6°$

It is seen that as the frequency changes from $\omega=1^-$ to $\omega=1^+$ in the counterclockwise sense in the *s*-plane contour, while avoiding 1, the Nyquist curve will move in the clockwise sense from Quadrant 2 to Quadrant 4 through Quadrant 1, as shown in Figure 8.35b.

By using symmetry of the Nyquist curve about the real axis, the two open ends of the negative frequency segment (broken line) may be connected now.

To determine stability of the closed-loop system, we notice that the Nyquist curve does not have any encirclement of –1 giving:

N=number of clockwise encirclements of point –1 on the real axis=0

Then from the Nyquist stability criterion (Equation 8.40):

Number of unstable poles in the closed-loop transfer function= $\tilde{p}$ =$p+N$=0–0=0

It follows that the closed-loop system is stable.

8.6.4 Relative Stability Specification

Characteristic equation of a closed-loop system is given by $1+G(s)H(s)=0$. Hence, as noted before, the condition for marginal stability is: $1+G(j\omega)H(j\omega)=0$.

Clearly this latter equation corresponds to the characteristic equation with a purely imaginary pole $j\omega$. Then, assuming that there are no poles on the right hand plane, the closed-loop system will be marginally stable. The margins of stability PM and GM indicate how far the closed-loop system is from reaching the state of marginal stability. As a rule of thumb a GM of at least 6 dB and a PM of at least 30° are known to be adequate for good stability.

The foregoing concept of stability margins uses the state of marginal stability as the reference. For more stringent stability specifications, a specified stable state can be used as the reference for measuring relative stability. Specifically, suppose that instead of plotting the conventional Nyquist curve governed by $G(j\omega)H(j\omega)$, we plot $G(-\sigma+j\omega)H(-\sigma+j\omega)$ on the complex GH plane, where σ is a known real positive parameter. If this modified Nyquist curve passes through the point –1 on the real axis, then what it means is $s=-\sigma+j\omega$ is a pole of the system. This is a stable pole. It follows that, by plotting the Nyquist curve of $G(-\sigma+j\omega)H(-\sigma+j\omega)$ and using the same procedures of Nyquist curve analysis as before, it is possible to study the proximity of the poles of the closed-loop system to the stable pole $s=-\sigma+j\omega$. In particular, if the Nyquist stability criterion is satisfied by this "modified" Nyquist curve of GH, then all the poles of the closed-loop system will be to the left of $-\sigma$ on the s-plane. By specifying σ then we can specify the stability margin.

8.7 Nichols Chart

Root locus, Nyquist, and Bode procedures use the loop transfer function GH (or the open-loop transfer function G, assuming unity feedback) in studying the stability of the closed-loop system $\tilde{G}=G/(1+GH)$. It is useful, then, to have a graphical procedure that will determine the closed-loop transfer function $\tilde{G}$ with the knowledge of the open-loop transfer function. Nichols chart provides such a tool, which is described now.

8.7.1 Graphical Tools for Closed-Loop Frequency Response

Without loss of generality, consider a unity feedback control system with unity feedback (i.e., $H=1$) as shown in Figure 8.36. The closed-loop transfer function is

$$\tilde{G} = \frac{G}{1+G} \tag{8.43}$$

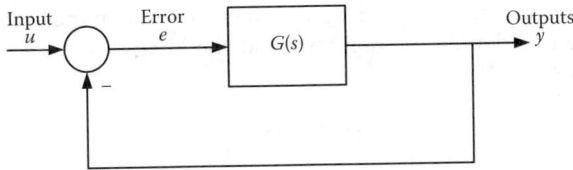

FIGURE 8.36
A closed-loop system with unity feedback.

In analysis procedures involving root locus, Bode diagram, and Nyquist plot, what we use is the loop transfer function G to determine the characteristics of the closed-loop transfer function $\tilde{G}$. In fact, $\tilde{G}$ can be uniquely determined from G, either analytically or computationally, using Equation 8.43. Since it is the loop frequency response function $G(j\omega)$ that is used in the procedures involving Bode and Nyquist plots, it is desirable to have a graphical tool that can generate the closed-loop frequency response function $\tilde{G}(j\omega)$ directly from $G(j\omega)$. The applicable relation is

$$\tilde{G}(j\omega) = \frac{G(j\omega)}{1+G(j\omega)} \tag{8.44}$$

The result is a frequency domain model of the closed-loop system. It carries a useful body of knowledge regarding the closed-loop system, and provides such information as system bandwidth, resonant frequencies, resonant peaks, and damping level, and also the steady-state response of the closed-loop system to a sinusoidal input (i.e., the frequency response), which are all useful in the analysis and design of a control system.

The Nyquist curve of $G(j\omega)$ itself can serve as the graphical tool for determining $\tilde{G}(j\omega)$. Specifically:

- Vector from the coordinate origin (on the $G(j\omega)$-plane) to a point on the Nyquist curve gives $G(j\omega)$ at the corresponding frequency.
- Vector from the point -1 on the real axis to a point on the Nyquist curve gives $1+G(j\omega)$.

Each of these two vectors has a magnitude (length of the line segment) and a phase angle (angle made by the line segment with respect to the positive real axis). The magnitude of $\tilde{G}(j\omega)$ is the ratio of these two lengths, and the phase angle of $\tilde{G}(j\omega)$ is the difference in the phase angles of the two vectors, for the particular frequency (i.e., particular point on the Nyquist curve).

Graphical tools are available where these computations are already performed and marked on a grid, and hence more convenient, particularly when a manual method is needed for a quick analysis or design. In these tools, a grid is provided to plot $G(j\omega)$, and a set of contours is provided on the grid to determine $\tilde{G}(j\omega)$. One such tool is the pair: *M circles* and *N circles*; and the other is the *Nichols chart*. Of course, the two representations are equivalent.

8.7.2 *M* Circles and *N* Circles

An M circle is a curve of constant magnitude of the closed-loop frequency response function (frequency transfer function) plotted on the Nyquist plane. Similarly, an N circle is a curve of constant phase angle of the closed-loop frequency response function (frequency transfer function) plotted on the Nyquist plane. It can be shown that both sets of curves are circles. To illustrate this denote the real part and the imaginary part of the loop transfer function explicitly as

$$G(j\omega) = X + jY \tag{8.45}$$

Note that Y denotes the imaginary axis and X denotes the real axis of the Nyquist plane. Then from Equation 8.44 we have:

$$\tilde{G}(j\omega) = \frac{X + jY}{1 + X + jY} \tag{8.46}$$

A constant magnitude curve for a particular magnitude value M of $\tilde{G}(j\omega)$, as plotted on the $X-Y$ (Nyquist) plane, is obtained from the relation:

$$M = \frac{|X + jY|}{|1 + X + jY|} \tag{8.47}$$

This simplifies to

$$\left(X + \frac{M^2}{M^2 - 1}\right)^2 + Y^2 = \frac{M^2}{(M^2 - 1)^2} \tag{8.48}$$

This a circle with its center at $-M^2/M^2-1$ on the real axis, and radius $|M/M^2-1|$. A family of M circles is shown in Figure 8.37.

The phase angle of the closed-loop frequency response function $\tilde{G}(j\omega)$ is obtained from Equation 8.46 as

$$\angle \tilde{G}(j\omega) = \tan^{-1}\frac{Y}{X} - \tan^{-1}\frac{Y}{1 + X}$$

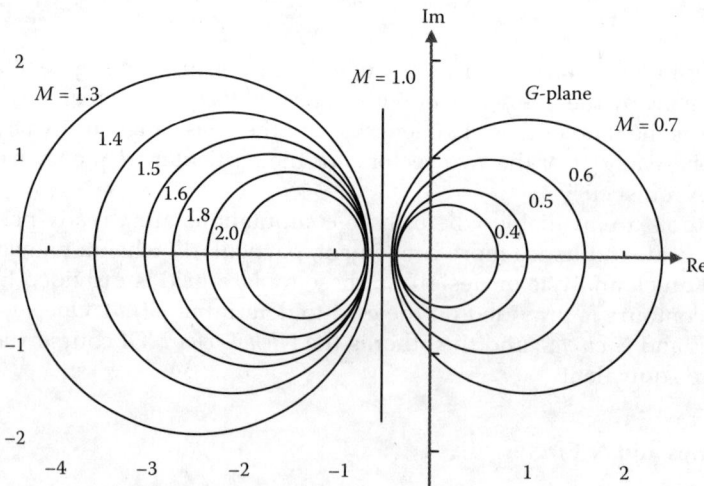

FIGURE 8.37
A family of M circles.

A constant phase angle curve of $\tilde{G}(j\omega)$, corresponding to a specific "tan" value N of the phase angle, is given by

$$N = \tan\left[\tan^{-1}\frac{Y}{X} - \tan^{-1}\frac{Y}{1+X}\right] \tag{8.49}$$

This can be simplified as

$$\left(X+\frac{1}{2}\right)^2 + \left(Y-\frac{1}{2N}\right)^2 = \frac{1}{4} + \left(\frac{1}{2N}\right)^2 \tag{8.50}$$

This too is a circle. A family of N circles is shown in Figure 8.38.

Once the Nyquist plot for a particular loop transfer function $G(j\omega)$ is drawn on the M circle plane of Figure 8.37, the magnitude value of $\tilde{G}(j\omega)$ for each point on the Nyquist curve can be easily obtained at the points of intersection with the M circles, and may be plotted against frequency to generate the magnitude curve of the closed-loop frequency response function. Similarly, the phase angle versus frequency curve for $\tilde{G}(j\omega)$ can be obtained by plotting the Nyquist curve of $G(j\omega)$ on Figure 8.38, and reading off the phase values at the points of intersection with the N circles.

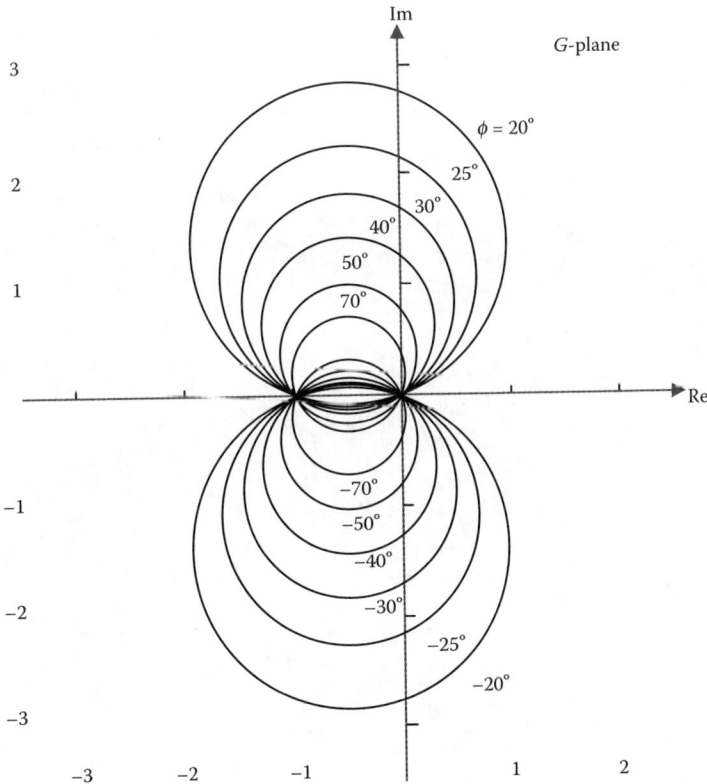

FIGURE 8.38
A family of N circles.

8.7.3 Nichols Chart

The Nichols chart essentially combines both M circles and N circles onto a single chart. There are some differences, however. The two axes of the plane are no longer the real part (X) and the imaginary part (Y) of the loop transfer function $G(j\omega)$. Instead, the vertical axis gives the magnitude of $G(j\omega)$ in decibels and the horizontal axis gives the phase angle of $G(j\omega)$ in degrees. Contours of constant magnitude (in dB) and constant phase angle (in degrees) of the closed loop frequency response function $\tilde{G}(j\omega)$ are given on a Nichols chart, as shown in Figure 8.39, which are obtained according to Equation 8.44.

As in the case of M circles and N circles, the frequency response plot of the loop transfer function can be drawn on a Nichols chart (say using Bode or Nyquist data) and used to determine the magnitude and the phase versus frequency curves (i.e., the frequency response function) of the corresponding closed loop system $\tilde{G}(j\omega)$.

Example 8.22

Consider the loop transfer function $G(s)=1/(s(2s+1)(0.5s+1))$ with unity feedback. The magnitude and phase values of this transfer function $(G(j\omega))$ can be computed and the Nichols chart may be plotted for a series of frequencies, using the following MATLAB® code:

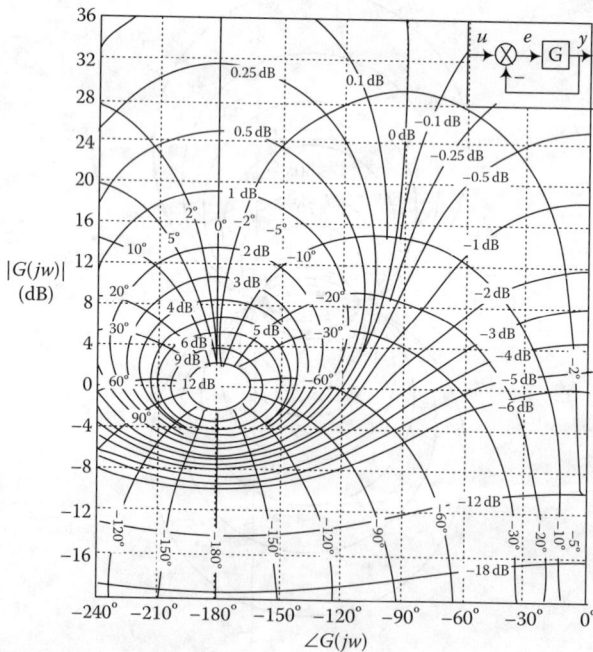

FIGURE 8.39
A Nichols chart.

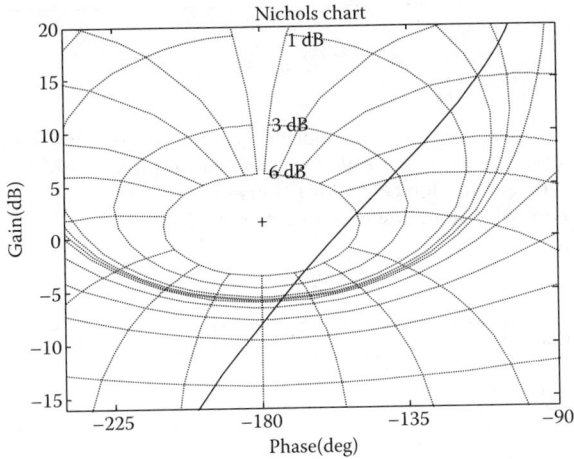

FIGURE 8.40
Application example of Nichols chart.

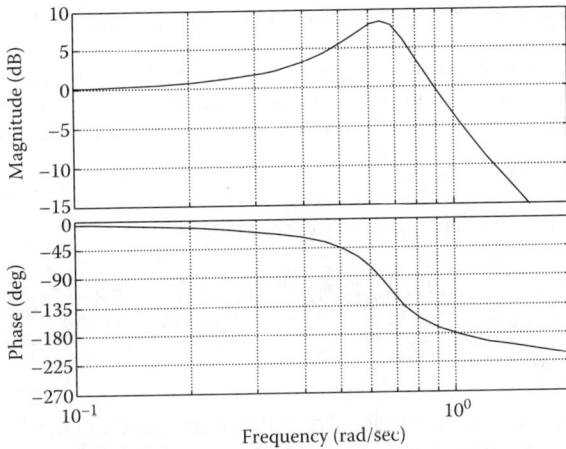

FIGURE 8.41
Closed-loop frequency response determined using Nichols chart.

```
% ---Nichols plot GH=1/(s(0.5s+1)(2s+1))---
num= [0 0 0 1];
den= [1 2.5 1 0];
SYS=tf(num,den);
[mag,phase,w] =nichols(SYS);
nichols(SYS); ngrid
xlabel('Phase(deg)');
ylabel('Gain(dB)');
```

The result is given by the solid curve in Figure 8.40.

Next, from the points of intersection of the magnitude contours, we can obtain the magnitude curve of the closed-loop transfer function ($|\tilde{G}(j\omega)|$) and from the points of intersection with the phase angle contours we can obtain the phase angle curve of the closed-loop transfer function ($\angle \tilde{G}(j\omega)$) as shown in Figure 8.41.

Problems

PROBLEM 8.1

A satellite-tracking system (typically a position control system) having the plant transfer function $G_p(s)=1/s(2s+1)$ is controlled by a control amplifier with *compensator*, having the combined transfer function $G_c(s)=K(s+1)/(\tau s+1)$ and unity feedback ($H=1$). A block diagram of the control system is shown in Figure P8.1.

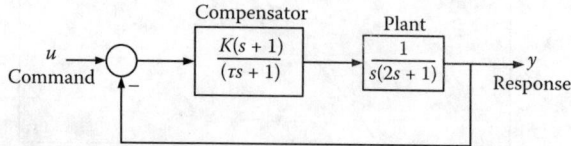

FIGURE P8.1
Block diagram of a satellite-tracking system.

  a. Write the closed-loop characteristic equation (in polynomial form).
  b. Using the Routh–Hurwitz criterion for stability, determine the conditions that should be satisfied by the compensation parameter τ and the controller gain K in order to maintain stability in the closed-loop system.
  c. Sketch this stability region using K as the horizontal axis and τ as the vertical axis.
  d. When $K=5$ and $\tau=3$ find the poles (i.e., eigenvalues or roots) of the closed-loop system. What is the natural frequency of the system for these parameters values?

PROBLEM 8.2

A system is given by the input–output differential equation.

$$\frac{d^3y}{dt^3}+6\frac{d^2y}{dt^2}+11\frac{dy}{dt}+6y=2\frac{du}{dt}+6u$$

where u=input, y=output.

  a. Using Routh–Hurwitz criterion (and without solving the characteristic equation), determine how many poles of the system are on the LHP. Is the system stable?
  b. For a unit step input, determine the steady-state value of the response, by using the differential equation and explaining your rationale. Next, verify your answer using FVT.
    Suppose that all the poles of the given system are moved to the right by 1 (and the system zeros are not changed).
  c. Using Routh–Hurwitz criterion (and without actually solving the characteristic equation) determine the stability of the new system (with moved poles).

PROBLEM 8.3

A system is given by the input–output differential equation:

$$\frac{d^4y}{dt^4}+3\frac{d^3y}{dt^3}+6\frac{d^2y}{dt^2}+12\frac{dy}{dt}+8y=2\frac{du}{dt}+5u$$

where u=input, y=output.

a. By using the Routh–Hurwitz criterion (without solving the characteristic equation, and without using a calculator), determine the stability of the system.
b. Suppose that all the poles of the given system are moved to the right by a distance of 1. By using the Routh–Hurwitz criterion (and without actually solving the characteristic equation) determine how many poles of the system are on the RHP. Is the new system stable?

Note: You must explain all your steps.

PROBLEM 8.4

a. The poles of a system are given in the following five examples:

(i) $-2, -4\pm j5, -3$
(ii) $-2, -4\pm j5, +3$
(iii) $-2, -4\pm j5, \pm j3$
(iv) $-2, +4\pm j5, -3$
(v) $-4\pm j5, 0, 0$

In each case state giving reasons whether the system is stable, unstable, or marginally stable.

b. A system has the characteristic equation:

$$s^3 + 12s^2 + 61s + 150 = 0$$

(i)   Using Routh–Hurwitz criterion determine whether the system is stable, unstable, or marginally stable.
(ii) Now move all the poles of the given system to the right of the s-plane by the real value 3 (i.e., add 3 to every pole of the original system). Using Routh–Hurwitz criterion determine whether this modified system is stable, unstable, or marginally stable. Justify your answer.
(iii) Using the result in (ii) and without directly solving the characteristic equation determine all three poles of the original system.

Note: Give all the details of obtaining your results. You should answer this question without directly solving a third order characteristic equation.

PROBLEM 8.5

Consider the six transfer functions:

a. $\dfrac{1}{(s^2 + 2s + 17)(s + 5)}$

d. $\dfrac{10(s+2)}{(s^2 + 2s + 2)}$

b. $\dfrac{10(s+2)}{(s^2 + 2s + 17)(s + 5)}$

e. $\dfrac{s}{(s^2 + 2s + 2)}$

c. $\dfrac{10}{(s^2 + 2s + 2)}$

f. $\dfrac{1}{s(s+2)}$

Suppose that these transfer functions are the plant transfer functions of five control systems under proportional feedback control. If the loop gain is variable, sketch the root loci of the five systems and discuss their stability.

PROBLEM 8.6

The loop transfer function of a feedback control system is given by

$$GH = \frac{K}{s(s+1)(s+2)}$$

a. Sketch the root locus of the closed-loop system by first determining the:

   (i)   location and the angles of the asymptotes
   (ii) break points
   (iii) points at which the root locus intersects with the imaginary axis, and the corresponding gain value.

b. Fully justifying your answer, state whether the system is stable for $K=10$.
c. Suppose that a zero at -3 is introduced to the control loop so that

$$GH = \frac{K(s+3)}{s(s+1)(s+2)}$$

Sketch the root locus of the new system.

PROBLEM 8.7

Explain why the variable parameter in a root locus does not necessarily have to be the loop gain. Sketch root loci for systems with the following loop transfer functions (*GH*), as the unknown parameter varies from 0 to ∞.

a. $\dfrac{2(s+z)}{(s^2+s+2)}$

b. $\dfrac{2(s+4)}{(\tau s+10)}$

c. $\dfrac{1}{s^2+(2a+3)s+3a+1}$

PROBLEM 8.8

A control system has an unstable plant given by the transfer function

$$G_p(s) = \frac{1}{(s^2-s+1)}$$

By sketching root locus, discuss whether the plant can be stabilized using

a. proportional (P) feedback control
b. proportional plus derivative (PPD) control
c. proportional plus integral (PI) control.

Are these observations intuitively clear?

PROBLEM 8.9

Consider the feedback (closed-loop) control system shown in Figure P8.9.
   You are given the following loop transfer function for the system:

$$GH = \frac{1}{s^3 + 3s^2 + (K+2)s + 3K - 1}$$

where K is a control system parameter which can be varied.
   Determine the root locus of the closed-loop system as the parameter K changes from 0 to ∞. Specifically, you must by first determine the:

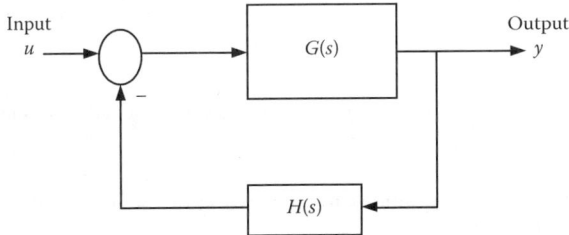

FIGURE P8.9
A feedback control system.

   i. Segments of the root locus on the real axis.
   ii. Angles of the asymptotes and the location where the asymptotes intersect the real axis.
   iii. Break points (as numerical expressions, which need not be evaluated).
   iv. Points at which the root locus intersects with the imaginary axis, if it does.
   v. The range of values of K for which the closed loop system is stable.

Note: You must give details and justify all your steps.

PROBLEM 8.10

An interesting issue of force feedback control in robotic manipulators is discussed in the literature by Eppinger and Seering of the Massachusetts Institute of Technology. Consider the two models representing a robotic manipulator, which interacts with a workpiece, as shown in Figure P8.10a and b. In (a) robot is modeled as a rigid body (without flexibility) connected to ground through a viscous damper, and the workpiece is modeled as a mass-spring-damper system. In this case only the rigid body mode of the robot is modeled. The robot interacts with the workpiece through a compliant device (e.g., remote center compliance—RCC device or a robot hand), which has an effective stiffness and damping. In (b) the robot model has flexibility, and the workpiece is modeled as a clamped rigid body, which cannot move. Note that in this second case a flexible (vibrating) mode as well as a rigid body mode of the robot are modeled. The interaction between the robot and the workpiece is represented the same way as in case (a). In both cases, the employed force feedback control strategy is to sense the force f_c transmitted through the compliant terminal device (the force in spring k_c), compare it with a desired force f_d, and use the error to generate the actuator force f_a. The controller (with driving actuator) is represented by a simple gain k_f. This gain is adjusted in designing or tuning the feedback control system shown in Figure P8.10c.

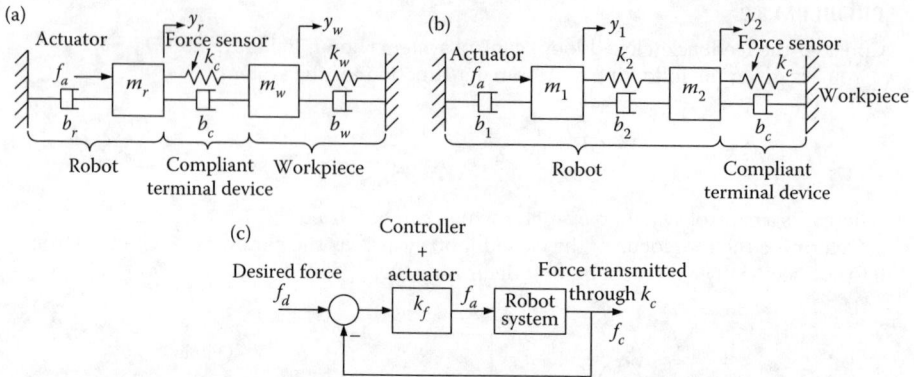

FIGURE P8.10
(a) A model for robot–workpiece interaction. (b) An alternative model. (c) A simple force feedback scheme.

a. Derive the dynamic equations for the two systems and obtain the closed-loop characteristic equations.
b. Give complete block diagrams for the two cases showing all the transfer functions.
c. Using the controller gain k_f as the variable parameter, sketch the root loci for the two cases.
d. Discuss stability of the two feedback control systems. In particular, discuss how stability may be affected by the location of the force sensor. (*Note*: The two models are analytically identical except for the force sensor location.)

PROBLEM 8.11

Sketch the Bode magnitudes plots (asymptotes only when the exact curve needs numerical computation) of the following common system elements:

a. Derivative controller (τs)
b. Integral controller ($1/\tau s$)
c. First order simple lag network ($1/\tau s+1$)
d. PPD controller ($\tau s+1$).

Also, sketch the polar plots (Nyquist plots) for these elements.

PROBLEM 8.12

The open-loop transfer function of a control system is given by

$$G(s) = \frac{s+1}{(s^2 + s + 4)}$$

a. With this transfer function, if the loop is closed through a unity feedback, determine the PM. You should use direct computation rather than a graphical approach.
b. If the input to the open-loop system is $u=3 \cos 2t$ determine the output y under steady conditions.

PROBLEM 8.13

Consider the feedback control system shown in Figure P8.13.

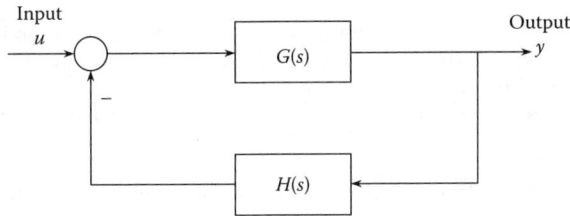

FIGURE P8.13
A feedback control system.

a. Clearly explain why the characteristic equation of the closed-loop system is given by:

$$G(s)\, H(s) = -1 \qquad\qquad (i)$$

Now suppose that the following equation is satisfied by the system:

$$G(j\omega)\, H(j\omega) = -1 \qquad\qquad (ii)$$

for a particular real parameter value ω.

b. Comment, with justification, on the stability of the system.
   What is the GM of the system? Explain.
   What is the PM of the system? Explain.

Suppose that a sinusoidal input: $u = u_0 \sin \omega t$ is applied to the feedback control system, which satisfies Equation (ii).

c. After a sufficiently long time what would be the nature of the response y? Clearly explain the reason for this behavior of the response.

PROBLEM 8.14

A system is shown in Figure P8.14. It was found to have the following properties:

1. The system transfer function $G(s)$ has two zeros and three poles.
2. The product of the three poles is -4.
3. When the system was excited with a sinusoidal input u (as shown in Figure P8.14) at frequency $\omega = 4$, the output y at steady-state was found to be zero (i.e., no response).

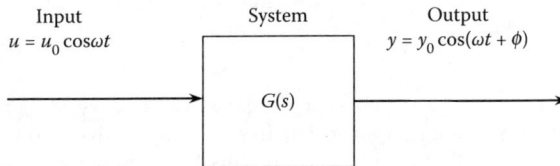

FIGURE P8.14
An open-loop system.

4. When the system was excited with a sinusoidal input u (as shown in Figure P8.14) at frequency $\omega=2$, the output y at steady-state was found to have a phase lag of $180°$ with respect to the input (i.e., the response was in the opposite direction to the input).
5. When the system was excited with a sinusoidal input u (as shown in Figure P8.14) at frequency $\omega=\sqrt{2}$ the output y at steady-state was found to have a phase lag of $90°$ with respect to the input.
6. The dc gain of the system (i.e., the magnitude of the frequency transfer function at zero-frequency) is 8.

Determine the complete transfer function $G(s)$ of the system (i.e., the numerical values of the five parameters in $G(s)$).

PROBLEM 8.15

A system was found to have the following properties:

1. It is a second-order system
2. It has a zero at $s=-z$ where $z>0$.
3. Its dc gain (i.e., the magnitude of the frequency transfer function at zero frequency) is K.
4. When the system is excited at its undamped natural frequency, the magnitude of the frequency transfer function is given by rK, where $r>0$, and the phase angle is $-90°$

In terms of the given parameters z, K, and r, determine the following:

a. Undamped natural frequency of the system.
b. Damping ratio of the system.
c. Complete transfer function of the system.

PROBLEM 8.16

Consider a plant given by the transfer function $G_p=K/s^2(\tau s+1)$. We can show using the Nyquist stability criterion that the closed-loop system will always be unstable under proportional feedback control.

If we include proportional+derivative (PD or PPD) control instead, the loop transfer function will be $GH=(K(\tau_d s+1))/(s^2(\tau s+1))$. Using the Nyquist criterion show that:

a. When $\tau_d>\tau$ the closed-loop system becomes stable (adequate derivative action).
b. When $\tau_d<\tau$ the closed-loop system remains unstable (due to inadequate derivative action).

PROBLEM 8.17

Consider a system with loop transfer function: $GH(s)=(s(s+2))/(s+1)(s^2+1)$.

Sketch the Nyquist curve for this loop transfer function and using the Nyquist stability criterion determine the stability of the closed-loop system.

PROBLEM 8.18

a. Define PM and GM of a system. For what type of linear system, the PM and GM considerations may not be appropriate in assessing relative stability?
b. An approximate relationship for PM in terms of damping ratio ζ is given by:

$$\phi_m=100\zeta \text{ degrees}$$

Give the main steps of deriving this result using a damped oscillator model.

c. A position control system, which uses a dc motor to drive an inertial load, is represented by the block diagram given in Figure P8.18.

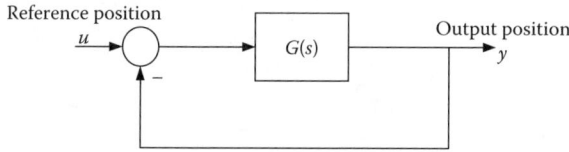

FIGURE P8.18
A position control system.

The forward transfer function is given by $G(s)=2/s(2s+1)$

(i) Sketch the Nyquist diagram of G. On this basis, comment on the stability of the closed-loop system.
(ii) Compute the PM and GM of the closed-loop system.
(iii) Determine the exact damping ratio of the closed-loop system and check whether the result agrees with the approximate relation given in (a).
(iv) A reference position input of $u=3\sin t$ is applied to the system. Determine the position response y at steady-state.

PROBLEM 8.19

Draw the Nichols curve for the open-loop transfer function in the frequency range $\omega=0.1$ to 20 rad/s:

$$G = \frac{2(s+1)}{s^2(0.5s+1)}$$

From that determine and sketch the Bode magnitude and phase curves for the closed-loop transfer function $\hat{G} =G/1+G$.

PROBLEM 8.20

Draw the Nichols curves for the open-loop transfer functions in the frequency range $\omega=0.1-20$ rad/s:

a. $G(s) = \dfrac{s(s+1)}{(s+2)(s^2+1)}$

b. $G(s) = \dfrac{s(s+2)}{(s+1)(s^2+1)}$

From them determine and sketch the Bode magnitude and phase curves for the corresponding closed-loop transfer functions with unity feedback.

9

Controller Design and Tuning

A control system, in general, consists of five types of components:

- *Plant*—the process to be controlled
- *Actuators* for driving the plant and for introducing control actions
- *Measuring devices* consisting of sensors and transducers for feedback control
- *Signal modification units* for conditioning and changing the form of signals in the control system
- *Controllers* for generating control signals to the plant

Designing a control system may involve selection, modification, addition, removal, and relocation of all these components as well as selection of suitable parameter values for one or more of the components in the control system in order to satisfy a set of *design specifications*. Once a control system is designed and the system parameters are chosen it may be necessary to further tune the parameters in order to achieve the necessary performance levels. Tuning and retuning may be needed even before the system is prototyped, but commonly after the system is built (prototyped) and is in operation. We have already encountered several concepts and examples of designing a controller of a control system so as to satisfy a set of performance specifications. These concepts are further examined and enhanced in the present chapter.

9.1 Controller Design and Tuning

The performance of a control system can be improved in many ways. Three common approaches are as follows:

1. Redesign or modify the plant
2. Substitute, add, modify, or relocate sensors, actuators and associated hardware
3. Introduce a new controller (or control scheme) or add a compensator for an existing controller

Plant redesign includes modification of the general structure of the plant as well as adjustment of parameters. Plant redesign is usually a costly way to achieve design improvement, particularly when structural modifications are involved. Furthermore, there are serious limitations to the degree of plant modification that is feasible and the level of performance improvement that can be achieved by this method both from practical and economical points of view. This approach is, in general, unsuitable when quick and short-term solutions are called for.

Adding or improving sensors and actuators normally result in improvements in the flexibility and versatility of a control system. This is a very practical method of design improvement. At least in theory, system controllability and reliability can be improved by adding sensors. For example, state-space design techniques often provide superior controller designs when all state variables in the plant are measurable. This is often not feasible, however, because all responses might not be available for measurement and the increased sensor cost might not justify the expected performance improvement. System controllability can be improved by adding new actuators or improving the existing ones as well. Often, power requirements of the plant alone may call for improved actuators. As in the case of sensors, increased cost and system complexity are two major drawbacks of this approach.

Controller improvement (redesign) is usually a rather economical and convenient method of design implementation. This may be accomplished either by analog means or by using a digital controller. In the former case, appropriately designed analog circuitry (analog controllers and compensators) is added at suitable locations in the control system. The modified control action that is generated in this manner can "compensate" for system weaknesses, thereby improving the overall performance. Proportional, integral, and derivative (PID) control actions and lead and lag compensation methods are commonly employed in industrial control systems. Compensators are analog or digital components that are added to an existing controller in order to "compensate for" weaknesses of the controller.

9.1.1 Design Specifications

Typical step of designing a control system are:

- Establish specifications for system performance (performance specifications or design specifications). These may be provided by the customer or have to be developed by the control engineer.

- Analyze and/or test the plant (or the original system in the case of design modifications). This step may involve system modeling.

- Select suitable components and determine the parameter values to meet the performance specifications. Commonly available (off the shelf) components should be used whenever possible. Many parameter values are available from the product data sheets provided by the manufacturers or vendors.

- Analyze and/or test the overall system to evaluate and verify its performance. Testing should involve the users (e.g., personnel from an industrial facility) of the designed control system when feasible, and should be done under realistic operating conditions.

- Repeat the design iteration if the specifications are not satisfied.

The design process is greatly influenced by the nature of the controller and the performance specifications used. Control systems can be designed by using either time-domain techniques or frequency-domain techniques. Many of the conventional methods of control system design are frequency-domain techniques and are particularly convenient when designing single-input–single-output (SISO) systems. State-space design techniques primarily use time domain concepts. These latter techniques are useful with multiinput, multioutput (MIMO) or multivariable systems. Since poles and zeros of a transfer function determine the time response of the corresponding system, it is difficult to classify some design techniques into time and frequency domains. For example, pole assignment using state-space techniques is considered a time domain approach whereas root locus design may be considered a frequency domain (actually, Laplace domain) approach. Design specifications may concern such attributes of a control system as: stability, bandwidth (speed of response), sensitivity and robustness, input sensitivity, and accuracy. In designing a control system, these attributes or descriptions have to be specified in quantitative terms. The nature of the design specifications used depends considerably on the type of the controller and the particular design technique that is employed. As discussed in several previous chapters, performance specifications can be made in both time domain and frequency domain.

9.1.2 Time-Domain Design Techniques

Common techniques of controller design in the time domain are listed below:

1. Conventional design of proportional (P), derivative (D) and integral (I) controllers (Design specifications: percentage overshoot, rise time, delay time, peak time, settling time, time constants, damping ratio, steady-state error.)
2. Optimal control using state-space approach (Specifications: expressed as a performance function that will be optimized in the design process; e.g., final time, weighted quadratic integral of response variables, inputs, states, error variables, and inputs.)
3. Pole assignment using state-space approach (Specifications: required pole locations.)

9.1.3 Frequency-Domain Design Techniques

Common techniques of controller design in the frequency domain are listed below:

1. Bandwidth design (Design specifications: resonant peak, bandwidth, resonant frequency.)
2. Bode and Nyquist design (Specifications: phase margin, gain margin, steady-state error, gain crossover frequency, slope of the transfer-function magnitude at gain crossover.)
3. Ziegler–Nichols tuning (Specifications: For example, a decay ratio of four in the closed-loop response.)
4. Root locus design (Specifications: pole locations, or any other parameter such as error constant, gain, natural frequency, damping ratio, settling time, peak time, and percentage overshoot that can be expressed in terms of pole locations.)

In the next section, we will describe a conventional method of controller design in the time domain. Common approaches of controller design in the frequency domain and a popular approach of controller tuning are presented in the subsequent sections.

9.2 Conventional Time-Domain Design

Speed of response, degree of stability, and steady-state error are the three specifications that are most commonly used in the conventional time-domain design of a control system. Speed of response can be increased by increasing the control system gain. This, in turn, can result in reduced steady-state error. Furthermore, steady-state error requirement can often be satisfied by employing integral control. For these reasons, one can treat speed of response and degree of stability as the only requirements in conventional time-domain design, and treat the steady-state error requirement separately.

The time-domain design of a proportional controller and a position plus velocity servo has been discussed in Chapter 7. The approach has to be modified for the design of a proportional plus derivative (PD or PPD) controller. This issue is treated next.

9.2.1 Proportional Plus Derivative Controller Design

Actuators with proportional plus derivative (PD or PPD) error control are commonly used as position servos. The two main parameters that can be adjusted in a PPD control element are the control gain K and the derivative time constant τ_d. Values for these two parameters can be chosen to provide specified levels of stability and speed of response, in a control system that employs a PPD servo.

The time-domain design problem is quite straightforward for proportional control and position plus velocity (tachometer) feedback control, as discussed in a previous chapter (see Chapter 7 in particular). This is so because, in both cases, the closed-loop transfer function does not contain finite zeros. With PPD error control, however, a finite zero enters into the closed-loop transfer function, making the design problem more difficult.

A common practice in the classical time-domain design of PPD controllers is to use the same design equations or curves as for position plus velocity feedback. This approach, however, may result in large errors when the finite zero that is introduced by the PPD controller becomes dominant. Hence a modification to the previous design approach is needed, which is discussed now.

Consider the PPD servo system represented by the block diagram in Figure 9.1. The derivative time constant of the PPD controller is denoted by τ_d. The closed-loop transfer function of the system is

$$G(s) = \frac{K(\tau_d s + 1)}{\tau \tau_a s^2 + (\tau + K\tau_d)s + K}$$

(9.1)

which is of the form

$$G(s) = \frac{\omega_n^2(\tau_d s + 1)}{(s^2 + 2\zeta\omega_n s + \omega_n^2)}$$

(9.2)

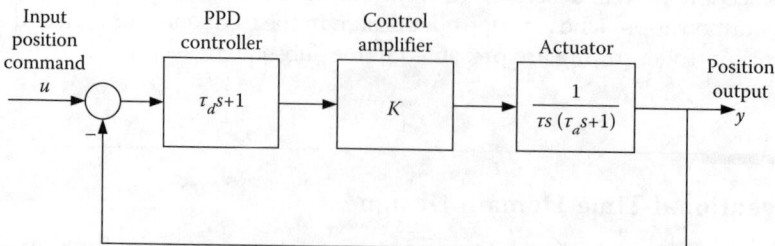

FIGURE 9.1
Block diagram of a PPD position servo system.

The undamped natural frequency ω_n of the closed-loop system is given by

$$\omega_n^2 = \frac{K}{\tau \tau_m} \tag{9.3}$$

and the damping ration ζ is given by

$$2\zeta\omega_n = \frac{\tau + K\tau_d}{\tau \tau_m} \tag{9.4}$$

Equation 9.2 has a finite zero (at $s=-1/\tau_d$). It follows that (by using the principle of super-position, which is valid for linear systems) the step response of the PPD servo is given by:

$$y = y^* + \tau_d \frac{dy^*}{dt} \tag{9.5}$$

in which y^* is the simple oscillator step response (see Chapter 6) as given by:

$$y^* = 1 - \frac{\exp(-\zeta\omega_n t)}{\sqrt{1-\zeta^2}} \sin(\omega_d t + \phi) \tag{9.6}$$

By substituting Equation 9.6 into Equation 9.5 we obtain

$$y = 1 - \frac{\omega_n \tau_d}{\sin \eta} \exp(-\zeta\omega_n t) \sin(\omega_d t + \phi + \eta) \tag{9.7}$$

in which

$$\eta = \tan^{-1}\left[\frac{\omega_d \tau_d}{1 - \zeta\omega_n \tau_d}\right] \tag{9.8}$$

Peak time T_p is determined by the condition: $(dy/dt)=0$. This gives:

$$\tan(\omega_d T_p + \phi + \eta) = \frac{\omega_d}{\zeta\omega_n} = \tan\phi$$

or

$$\eta = \pi - \omega_d T_p \tag{9.9}$$

By substituting Equation 9.9 into Equation 9.7 we get the peak response (at $t=T_p$),

$$M_p = 1 + \frac{\omega_d \tau_d}{\sin \omega_d T_p} \exp(-\zeta\omega_n T_p) \tag{9.10}$$

9.2.2 Design Equations

There are four equations that govern the design of a PPD position servo. They can be obtained directly from the theory outlined above. With straightforward manipulation of the previous equations we get:

$$\frac{1}{\tau_d} = \zeta\omega_n - \frac{\omega_d}{\tan\omega_d T_p} \tag{9.11}$$

$$\frac{K}{\tau\tau_a} = (\zeta\omega_n)^2 + \omega_d^2 \tag{9.12}$$

$$\frac{1}{\tau_a} + \frac{K\tau_d}{\tau\tau_a} = 2\zeta\omega_n \tag{9.13}$$

Percentage overshoot

$$P.O. = \frac{100\omega_d\tau_d}{\sin\omega_d T_p}\exp(-\zeta\omega_n T_p) \tag{9.14}$$

These results can be expressed graphically and used in the design process. Alternatively, the equations can be solved using a nonlinear equation solver, after substituting the specifications and known parameter values.

Example 9.1

Consider PPD control of a dc motor represented in Figure 9.1, with the mechanical time constant $\tau_a = 0.5$ s and the gain parameter $\tau = 1$. Suppose that the design specifications are $T_p = 0.09$ s and $P.O. = 12\%$. The plant transfer function for this example is

$$G_p(s) = \frac{1}{s(0.5s+1)}$$

It can be verified that a PPD controller with parameters $K = 277$ and $\tau_d = 0.078$ will meet the design specifications.

9.3 Compensator Design in the Frequency Domain

Once the components such as actuators, sensors, and transducers are chosen for a control system and the control strategy (e.g., proportional feedback control) is decided upon, the design of the control system can be accomplished by determining the parameter values for the controller (and possibly for other components) that will bring about the desired performance. If the design specifications cannot be met by adjusting the parameters of the existing controller, then new components (called *compensators*) may have to be added to the control system in order to meet the required performance. This process is known as control system compensation.

Lead compensation and lag compensation are the most commonly used methods of compensation in the conventional frequency domain design of control systems. These compensators improve the system performance by modifying the frequency response of the original control system. Both types of compensators can provide improved stability. The way this improvement is brought about by a lag compensator is not quite the same as the way it is achieved by a lead compensator, however. Lead compensation improves the speed of response (or bandwidth) as well, of the control system. Lag compensation improves the low-frequency performance (steady-state accuracy, in particular). Unfortunately, lag compensation has the disadvantage of decreasing the system bandwidth. In general, combined lead-lad compensation, and perhaps several stages of this, may be needed for achieving large improvements in performance. Now we will study the behavior of lead compensation and lag compensation and how these compensators may be designed into a control system.

To get the loop transfer function of the compensated system, the compensator transfer function is multiplied by the loop transfer function GH of the uncompensated system. It follows that the compensator element may be added at any point in the control loop, since the end result of loop transfer function will be the same. Specifically, the compensator may be included either in the forward path or in the feedback path of the loop. In either case the analysis and the underlying design procedures are the same.

The design of a lead compensator or a lag compensator consists of two basic steps. They are:

a. Select the system gain to meet the steady-state accuracy specification. This will result in improved speed of response as well.

b. Choose the zero and the pole of the compensator to meet the phase margin specification (for relative stability).

The compensator design amounts to the selection of appropriate parameters (gains, poles, and zeros) for the compensator elements. Bode diagrams are particularly useful in the frequency-domain design of compensators (see Chapters 6 and 8). A particular advantage is the fact that the Bode plot for the compensated system is obtained by simply adding the Bode plot for the compensator to the Bode plot of the original uncompensated system, where the transfer function magnitudes are expressed in decibels. This follows from the fact that the phase angles of a transfer function product are additive, and the magnitudes are additive as well when a log scale (or decibel scale) is used.

9.3.1 Lead Compensation

Lead compensation is a conventional frequency-domain design approach, which employs the derivative action (lead action) of a compensator circuit (or software) to improve stability by directly increasing the phase margin. Also, lead compensation increases the system bandwidth, thereby improving the speed of response. The speed of response may be further increased and the steady-state accuracy improved as well by increasing the dc gain (zero-frequency gain) of the control loop. Lead compensation primarily modifies the high-frequency region of the system bandwidth, thereby improving transient characteristics of the control system.

The transfer function of a lead compensator is expressed as

$$G_d(s) = \left[\frac{\tau s + 1}{a \tau s + 1}\right] 0 < a < 1 \tag{9.15}$$

Nyquist and Bode diagrams for this compensator are shown in Figure 9.2. The maximum phase (lead) angle is obtained at a point within the frequency range $[1/T, 1/(aT)]$, and is given by:

$$\Delta \phi_m = \sin^{-1}\left[\frac{1-a}{1+a}\right] \tag{9.16}$$

This is obtained by drawing a tangent line to the Nyquist plot (which is a semicircle) as shown in Figure 9.2a. It can be shown that the corresponding frequency is:

$$\omega_c = \frac{1}{\sqrt{a\tau}} \tag{9.17}$$

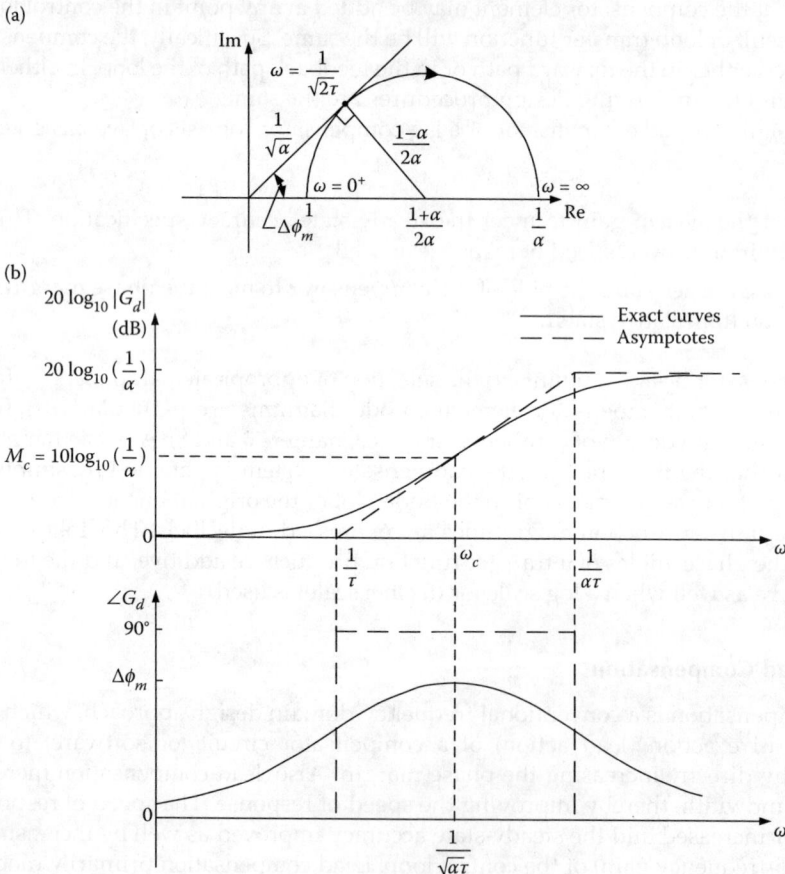

(a)

(b)

FIGURE 9.2
Frequency transfer function of a lead compensator. (a) Nyquist plot. (b) Bode plot.

and the transfer-function magnitude is: $1/\sqrt{a}$. In decibels this magnitude can be expressed as:

$$M_c = 20\log_{10}\left(\frac{1}{\sqrt{a}}\right)\text{dB} \tag{9.18}$$

To verify these results note that in Figure 9.2a, the transfer-function magnitude at the maximum phase angle is given by the length of the tangent to the Nyquist curve:

$$\sqrt{\left(\frac{1+a}{2a}\right)^2 - \left(\frac{1-a}{2a}\right)^2} = \frac{1}{\sqrt{a}}$$

This verifies Equation 9.18. Now if we substitute $s = j/(\sqrt{a}\tau)$ into Equation (9.15) we get:

$$G_d(s)\Big|_{s=\frac{j}{\sqrt{a}\tau}} = \frac{1}{\sqrt{a}}\frac{(j+\sqrt{a})}{(\sqrt{a}j+1)}$$

This has a magnitude of $1/\sqrt{a}$, thus verifying Equation 9.17.

Notice from Figure 9.2b that when a lead compensator is added to a system, its gain is amplified, particularly at high frequencies. This increases the crossing frequency. The phase angle of the uncompensated system at this new crossing frequency is typically more negative than that at the original crossing frequency. This means more phase lead (than would be required if the compensator had not produced a magnitude increase) has to be provided by the lead compensator in order to meet a specified phase margin. Usually, a correction angle $\delta\phi$ of a few degrees should be added to the required phase lead of the compensator (given by the difference: specified phase margin–phase margin of the uncompensated system). The actual correction that is required depends on the rate at which the phase angle of the original system changes in the vicinity of the crossing frequency, but it is not precisely known beforehand.

The final value theorem (see Chapter 7) dictates that the lead compensator given by Equation 9.15 does not affect the steady-state accuracy. The reason is that the compensator has a unity dc gain—the magnitude at zero frequency, and it is the dc gain that determines the steady-state response of a transfer function. Note further that since the crossover frequency increases through lead compensation, the system bandwidth also increases.

Using the results outlined above, an iterative procedure for lead compensator design can be stated. The procedure is considered optimal because each iteration utilizes the maximum phase lead that is offered by the compensator.

9.3.1.1 Design Steps for a Lead Compensator

The main steps of an iterative procedure for designing a lead compensator are now listed. *Note*: Just one iteration is adequate for most purposes. Each step in the design procedure is explained at the end of the listing. The control system configuration considered is shown in Figure 9.3.

Design Specifications: PM_{spec}, e_{ss}

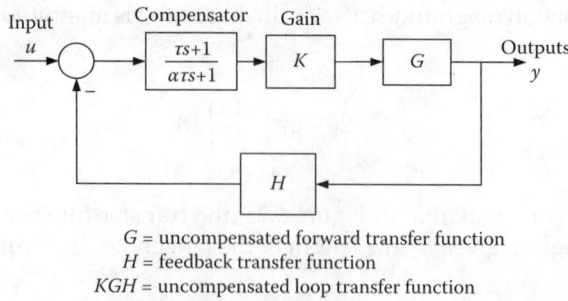

G = uncompensated forward transfer function
H = feedback transfer function
KGH = uncompensated loop transfer function

FIGURE 9.3
Forward (cascade) compensation of a control system.

Step 1: Compute the loop gain K that is needed to meet the steady-state error specification. Obtain the Bode curves for the uncompensated system with this gain included in the loop. Determine the phase margin PM_o

Step 2: Compute the required phase margin improvement:

$$\Delta\phi_m = PM_{spec} - PM_o + \delta\phi \tag{9.19}$$

Step 3: Compute the lead compensator parameter:

$$a = \frac{1 - \sin\Delta\phi_m}{1 + \sin\Delta\phi_m} \tag{9.20}$$

Step 4: Compute the compensator gain at the maximum phase lead:

$$M_c = 10\log_{10}\left(\frac{1}{a}\right) \text{dB} \tag{9.21}$$

Step 5: From the uncompensated Bode gain curve, determine the frequency ω_c at which gain is $(-M_c)$dB

Step 6: Compute the remaining compensator parameter:

$$\tau = \frac{1}{\sqrt{a}\omega_c} \tag{9.22}$$

Step 7: Compute the Bode curves for the designed system (which includes the compensator) and determine the PM of this compensated system

Step 8: If $|PM_{spec} - PM| \leq \delta\phi_o$, stop. If the number of iterations exceeds the limit, stop. Otherwise, increase $\delta\phi$ by $(PM_{spec} - PM)$ and Go To Step 2.

Typically, a phase margin specification PM_{spec} and a steady-state error specification e_{ss} are specified in the compensator design. The gain K that is required to meet the e_{ss} (for a step input) is computed in Step 1. *Note*: If the uncompensated loop has a free integrator (i.e., Type 1 system—see Chapter 7), the steady-state error for a step input will be zero. Then, there is no need to change the system gain, unless the steady-state error specification is based on a rate input such as a ramp or a parabola. The steady-state response to a unit step input $(1/s)$ is given by:

$$y_{ss} = \lim_{s \to 0} \frac{s}{s} \tilde{G}(s) = \frac{KG(0)}{1 + KG(0)H(0)} = \frac{K}{1 + K} \qquad (9.23)$$

This result follows from the *final value theorem*—see Chapter 7. It assumes that the transfer functions $G(s)$ and $H(s)$ do not contain free integrators and have unity dc gains. The steady-state error is given by:

$$e_{ss} = 1 - y_{ss} = 1 - \frac{K}{1 + K} = \frac{1}{1 + K} \qquad (9.24)$$

The system gain that is computed in this manner, is added to the uncompensated system. *Note:* If the uncompensated loop has one or more free integrators, the original gain is unchanged. The phase margin PM_o of the uncompensated system (with gain K included) is determined in Step 1.

In Step 2 the required increase in phase margin is computed. A typical starting value for the correction angle $\delta\phi$ is $5°$. It will be changed in subsequent iterations. One of the compensator parameters (a) is computed in Step 3. Equation 9.20 follows from Equation 9.16. *Note:* If $\Delta\phi_m$ is excessive, a single compensator may not be adequate. For example, if the typical value of 1 decade is used as the separation between the zero and the pole of the lead compensator, then $a = 0.1$ and

$$\Delta\phi_m = \sin^{-1}\left[\frac{1 - 0.1}{1 + 0.1}\right] = 50°$$

A phase margin improvement of better than $50°$ would be very demanding on a single compensator. In general, a single compensator should not be used to obtain a phase increase of more than $70°$.

Equation 9.21, which is used in Step 4 to compute the gain at maximum phase lead of the compensator, follows directly from Equation 9.18. Frequency w_c will be the crossing frequency of the compensated system. Manual determination of w_c can be done simply by noting the frequency at which the gain is $-M_c$ on the uncompensated Bode gain plot. For computer-based determination of this quantity, one has to add M_c to the uncompensated gain value and then obtain the crossing frequency of the resulting modified gain curve.

The remaining compensator parameter τ is computed using Equation 9.22 in Step 6. This equation is identical to Equation 9.17. *Note:* w_c is equal to the geometric mean of the compensator zero and the compensator pole. In Step 7, the compensator transfer function is included in the loop transfer function; the new Bode curves are computed; and the new phase margin PM is determined. The absolute error in the new phase margin is computed in Step 8. If this is less than a prechosen error tolerance $\delta\phi_o$, the design is concluded. Otherwise, the phase margin correction $\delta\phi$ is increased by an amount that is equal to the current error, and the design is repeated. An error tolerance of $1°$ is usually adequate.

If a single lead compensator is unable to provide the necessary gain increase, two or more compensators in cascade should be used. In that case the design could be done with one compensator at a time. The first compensator is designed for its optimum performance.

It is then included in the loop transfer function and the second compensator is designed for this modified loop transfer function, and so on. The computer-aided design procedure that was outlined previously can be easily extended to this case of designing higher order compensators in sequence.

When a lead compensator is added, the resulting crossing frequency ω_c of the compensated system will be larger than that of the uncompensated system. This should be obvious from the gain curve in Figure 9.2b. This means that the bandwidth of the compensated system is larger. This has the favorable effect of increasing the speed of response of the control system. Unfortunately, a lead compensator is also a high-pass filter. This means that the compensated system can allow higher-frequency noise, which will distort the signals in the control system. This is a shortcoming of lead compensation.

9.3.2 Lag Compensation

If a control system has more than adequate bandwidth, a lag compensator can be used to simultaneously improve both steady-state accuracy and stability of the system. Since a lag compensator adds a phase lag to the loop, it actually has a destabilizing effect. But since the crossing frequency (hence, the system bandwidth) is reduced by a lag compensator, the phase lag will be lower in the neighborhood of the new crossing frequency than the phase lag near the old crossing frequency, thereby increasing the phase margin and improving system stability. It follows that even though a lead compensator and a lag compensator both improve system stability, the way they accomplish this is quite different. Also, a lag compensator inherently improves the low-frequency behavior, steady-state accuracy in particular, of the system. Furthermore, since a lag compensator is essentially a low-pass filter, it has the added advantage of filtering out high-frequency noise.

A lag compensator is given by the transfer function:

$$G_g(s) = \left[\frac{\tau s + 1}{\beta \tau s + 1} \right] \beta > 1 \tag{9.25}$$

Its Bode diagram is shown in Figure 9.4. Since this compensator adds a negative slope (in the frequency range $1/(\beta\tau)$ to $1/\tau$) to the original loop, the crossing frequency will decrease. If the phase margin at this new crossing frequency is adequate to meet the *PM* specification, the lag compensator will provide the required stability. Otherwise, a lead compensator should be used to further improve the stability of the system.

The phase lag contribution from a lag compensator is primarily limited to the frequency interval $1/(\beta\tau)$ to $1/\tau$. This range should be shifted far enough to the left of the crossing frequency of the compensated system so that the phase lag of the compensator has a negligible effect on the phase-margin potential of the original system. One way to accomplish this would be to make $1/\tau$ a small fraction (typically 0.1) of the required crossing frequency. Using these considerations, a design procedure for a lag compensator is outlined below.

9.3.2.1 Design Steps for a Lag Compensator

The main steps of designing a lag compensator are given now. The considered control structure is the same as what is shown in Figure 9.3.

Design Specifications: PM_{spec}, e_{ss}

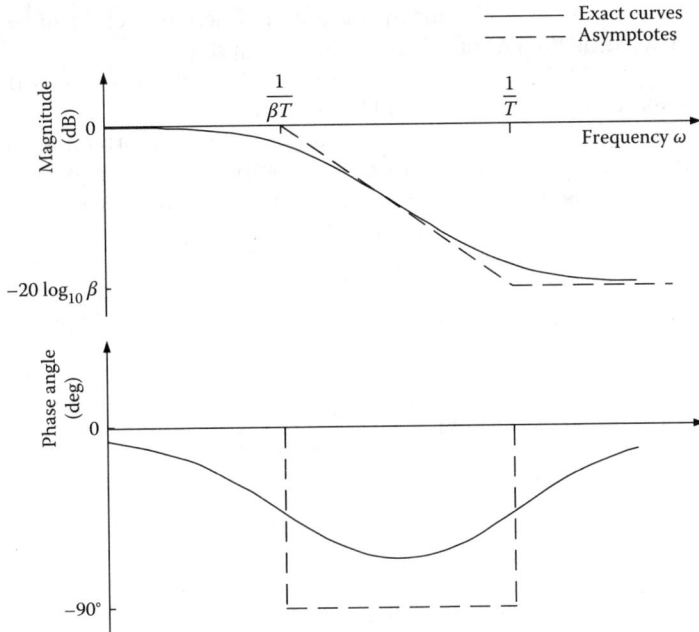

FIGURE 9.4
Bode diagram of a lag compensator.

Step 1: Compute the loop gain K that is needed to meet the steady-state error specification. Obtain the Bode curves for the uncompensated system with this gain included in the loop.

Step 2: Compute the phase angle required at the new crossing frequency:

$$\phi = PM_{spec} - 180° + \delta\phi \tag{9.26}$$

Step 3: From the uncompensated Bode phase curve determine the frequency ω_c where the phase angle is ϕ.

Step 4: Compute the lag compensator parameter:

$$\tau = \frac{10}{\omega_c} \tag{9.27}$$

Step 5: From the uncompensated Bode gain curve determine the magnitude M_c dB at ω_c.

Step 6: Compute the remaining compensator parameter β as follows:

$$a = \frac{M_c}{10} + \log_{10} 101 \tag{9.28}$$

$$\beta = 0.1\sqrt{10^a - 1} \tag{9.29}$$

Step 7: Compute the Bode curves for the designed system (which includes the lag compensator) and determine the *PM* of this compensated system.

Step 8: If $\left|PM_{spec}-PM\right|\leq\delta\phi$, stop. If the number of iterations exceeds the limit, stop. Otherwise increase $\delta\phi$ by $(PM_{spec}-PM)$ and Go To Step 2.

Most of the steps given above are self-explanatory. In Step 2, a correction angle $\delta\phi$ is used to account for the phase lag contributed by the lag compensator at frequency ω_c. In the first iteration, a correction angle of 5° is typically used. The required correction angle is small if the magnitude of the compensator zero $(1/\tau)$ is quite small compared to the crossing frequency (ω_c). This is guaranteed in Step 4, through Equation 9.27. The magnitude of the uncompensated system (M_c) at ω_c (see Step 5) has to be exactly cancelled by the magnitude of the lag compensator at ω_c, in order to force ω_c to be the crossing frequency of the compensated system. The necessary condition is:

$$-M_c = 20\log\frac{\left|\tau j\omega_c+1\right|}{\left|\beta\tau j\omega_c+1\right|} \qquad (9.30)$$

By substituting Equation 9.27 into Equation 9.30 we get:

$$M_c = 20\log_{10}\frac{\left|10\beta j+1\right|}{\left|10j+1\right|} \qquad (9.31)$$

Equations 9.28 and 9.29 as given in Step 6 are obtained directly from Equation 9.31. An approximate relation is obtained by neglecting 1 compared to 10, as:

$$\beta = 10^{M_c/20} \qquad (9.32)$$

Usually, acceptable results are obtained in just one design iteration. In computer-aided design routines a maximum number of design iterations (typically five) should be specified. Then, if the design does not converge, the design computation is stopped when the maximum number of iterations is exceeded, and the best design among performed the several iterations is presented as the final design. *Note*: If the phase margin potential (i.e., the difference: 180°—minimum phase lag angle of the uncompensated system at a frequency above the required bandwidth) is smaller than the phase margin specification, the lag compensator will be unable to meet the phase margin specification.

In some designs it may be necessary to modify both low frequency region and high frequency region of the loop transfer function in order to: simultaneously reduce the steady-state error to a desired level; improve the slow transients; filter out high-frequency noise; increase the speed of response; and improve the input-tracking capability. This may be achieved by using one or more lead-lag compensator stages.

For a good compensator design, the slope of the loop gain curve at the crossover frequency should be approximately equal to (−20 dB/decade). Often, this condition is given as a design specification. It can be shown that if the slope at crossover is substantially smaller (algebraically) than (−20 dB/decade), it is quite difficult to accurately meet a phase-margin specification. To illustrate this, consider an underdamped oscillator. In this example (see Chapter 8) the slope at crossover is (−40 dB/decade) which is considerably smaller than the required (−20 dB/decade), and the phase angle changes rapidly from 0° to −180° in the neighborhood of the natural frequency ω_n. Now, recall the fact that we need to include a correction angle $\delta\phi$ in the design of a lead compensator because the phase lag angle of the uncompensated system at the compensated (new) crossover frequency is different from (usually larger than) that at the uncompensated (old) crossover frequency. Also, a correction angle $\delta\phi$ has to be included in the design of a lag compensator because the compensator adds a small phase lag in the neighborhood of the compensated crossover frequency. Should the phase angle of the uncompensated system changes rapidly (which is the case when the slope of the gain curve is (−40 dB/decade) or smaller), the design would be very sensitive to the phase angle correction $\delta\phi$. Then it would be very difficult to meet the *PM* specification, usually resulting in either a substantially over-designed compensator or a substantially under-designed compensator. For example, for a lightly damped simple oscillator, if the compensated crossover frequency is less than ω_n (which will be the case with a lag compensator), the uncompensated loop itself will provide a phase margin of nearly 180° (an over-designed case). On the other hand, if the compensated crossover frequency is greater than ω_n, the uncompensated loop will have a very small phase margin (approximately zero). Hence, a lead compensator will have to provide the entire requirement of design phase margin, which is not usually possible with a single compensator (an under-designed case).

9.3.3 Design Specifications in Compensator Design

More than a phase margin and a steady-state error might be specified in the design of a compensator. For example, a settling time and a bandwidth of the closed-loop system might also be specified. Since settling time is related to stability, one approach to meet this specification would be to first design the compensator to meet the phase margin specification and subsequently check to see whether the design satisfies the settling time specification. If not, the phase margin specification should be increased and the compensator redesigned on that basis. Similarly, the bandwidth of the designed closed-loop system should also be checked. If it is not satisfied, the system gain should be increased and the compensator redesigned.

As noted in Chapter 8, it is possible to relate a phase margin specification (ϕ_m) to a damping ratio specification, using the simple oscillator model, as:

$$\phi_m - \tan^{-1}\frac{2\zeta}{a} \text{ deg.} \tag{9.33}$$

in which

$$a = \sqrt{\sqrt{4\zeta^2 + 1} - 2\zeta^2} \tag{9.34}$$

Equation 9.33 along with Equation 9.34 provides a relationship for the specifications of *PM* and ζ. In particular for small ζ, if we neglect $O(\zeta^2)$ terms compared to unity, Equation 9.34 can be approximated by $a = 1$. Then Equation 9.33 can be approximated as:

$$\phi_m = 2\zeta \text{ radians} = 2\zeta \times \frac{180}{\pi} \text{ deg.}$$

or, approximately:

$$\phi_m = 100\zeta \text{ degrees} \tag{9.35}$$

Input
u
Compensator
Amplifier
K
Motor
1
$(10s + 1)$
Speed
output
y

Filter
1
$(0.1s + 1)$

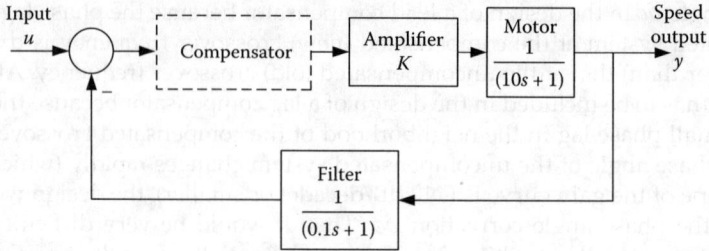

FIGURE 9.5
Compensator design for a velocity servo.

Example 9.2

A speed control system is shown by the block diagram in Figure 9.5. The motor is driven by control circuitry, approximated in this example by an amplifier of gain K. The signal y from the speed sensor is conditioned by a low-pass filter and compared with the speed command u. The resulting error signal is fed into the control amplifier. The controller may be tuned by adjusting the gain K. Since the required performance was not achieved by this adjustment alone, it was decided to add a compensator network into the forward path of the control loop. The design specifications are:

1. Steady-state accuracy of 99.9% for a step input.
2. P.O. of 10%.

Design:

a. A lead compensator.
b. A lag compensator.

to meet these design specifications.

Solution
From the time domain considerations using a damped oscillator model (see Chapter 7) a percentage overshoot of P.O. =10 corresponds to a damping ratio of $\zeta=0.6$. Then in view of Equation 9.35 we have the equivalent phase margin specification:

$$PM_{spec} = 60°$$

Next, note that the dc gain of the filter is $H(0)=1$. Accordingly, we use Equation 9.24 to determine the gain that satisfies the steady-state error specification:

$$\frac{1}{1+K} = \frac{0.1}{100}$$

This gives:

$$K=999$$

With this gain, the transfer function of the uncompensated loop is

$$GH = \frac{999}{(10s + 1)(0.1s + 1)}$$

Bode curve pair for this transfer function is shown in Figure 9.6.

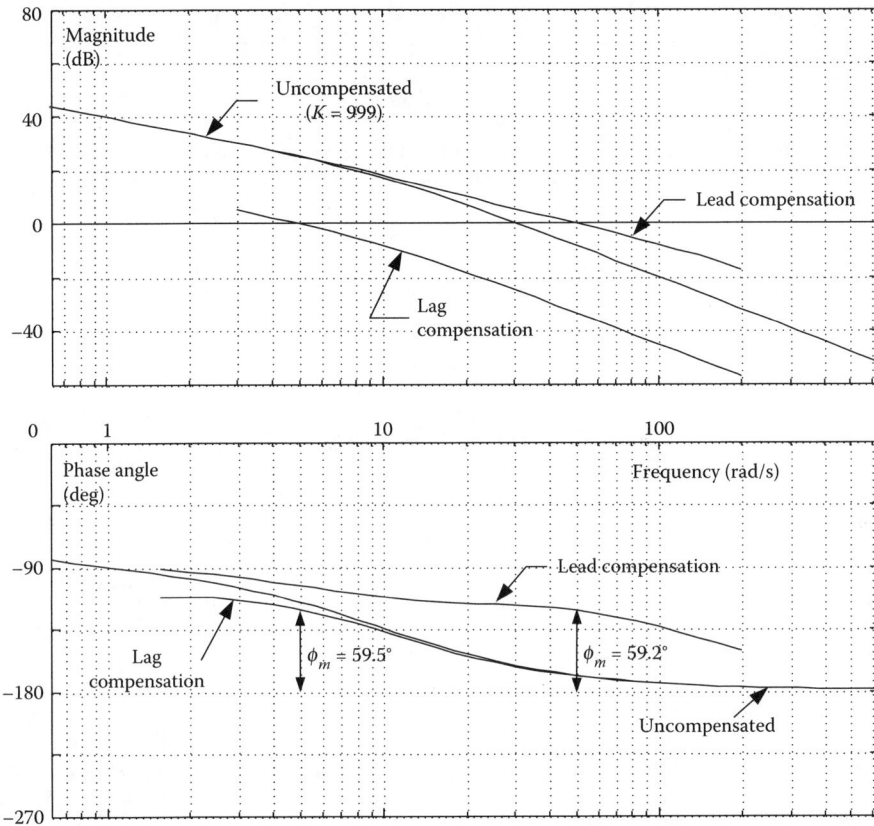

FIGURE 9.6
Bode curves for the compensator design example.

a. Lead Compensation
Here we follow the design steps as given earlier.

Step 1: The Bode plot with $K=999$ gives a phase margin of

$$PM_o = 18° \quad \text{at 31 rad/s (4.9 Hz)}$$

Step 2: Using a correction angle of 6° we have from Equation 9.19

$$\Delta\phi_m = 60 - 18 + 6 = 48°$$

(This correction angle was actually obtained after one design iteration starting with a correction angle of 1°.)

Step 3: From Equation 9.20 we obtain one of the lead compensator parameters:

$$a = \frac{1-\sin 48°}{1+\sin 48°} = 0.15$$

Step 4: From Equation 9.21:

$$M_c = 10\log_{10}\left(\frac{1}{0.15}\right) = 8.2 \text{ dB}$$

Step 5: The frequency corresponding to an uncompensated magnitude of −8.2 dB is obtained from the uncompensated Bode plot. This value (new crossing frequency) is:

$$\omega_c = 51 \text{ rad/s} \ (8.12 \text{ Hz})$$

Step 6: From Equation 9.22, the second parameter is obtained as:

$$\tau = \frac{1}{\sqrt{0.15 \times 51}} = 0.051 \text{ s}.$$

It follows that the transfer function of the lead compensator is:

$$G_d(s) = \left[\frac{0.051s + 1}{0.008s + 1}\right]$$

The Bode curve pair for the compensated loop transfer function is shown in Figure 9.6. It is seen that the phase margin specification has been satisfied. Also note the increased crossover frequency (and hence increased bandwidth).

b. Lag Compensation

By following the steps outlined earlier for lag compensator design, we can obtain a lag compensator that satisfies the design specifications. The pertaining computations are given below.

Step 1: As before we have K=999.
Step 2: Using a correction angle of 5° (this was obtained after one computer iteration) we have from Equation 9.26:

$$\phi = 60 - 180 + 5 = -115°$$

Step 3: From the uncompensated Bode curves, the frequency corresponding to a phase angle of −115° is:

$$\omega_c = 5 \text{ rad/s} \ (0.79 \text{ Hz})$$

Note: This step also verifies that a lag compensator can meet the given design specification (i.e., adequate phase margin is present in the uncompensated system), provided that a low bandwidth (less than 1 Hz) is acceptable.

Step 4: One parameter of the lag compensator is obtained from Equation 9.27:

$$\tau = \frac{10}{5} = 2.0 \text{ s}.$$

Step 5: The magnitude of the uncompensated system at ω_c is (from Bode curve):

$$M_c = 25.2 \text{ dB}$$

Step 6: From Equation 9.28:

$$a = \frac{25.2}{10} + \log_{10} 101 = 4.52$$

From Equation 9.29 the second parameter of the compensator is obtained as:

$$\beta = 0.1\sqrt{10^{4.52} - 1} = 18.3$$

It follows that the transfer function of the lag compensator is

$$G_g(s) = \left[\frac{2.0s + 1}{36.6s + 1} \right]$$

Note from the Bode curves for the lag-compensated system (Figure 9.6) that the phase margin specification has been satisfied. Note further that the crossover frequency has been decreased substantially (from 31 to 5 rad/s) indicating a large reduction in the bandwidth of the control system.

Note: See Appendix B for a MATLAB® treatment of this example.

9.4 Design Using Root Locus

Root locus is the locus of the closed-loop poles as one parameter of the system (typically the loop gain) is varied (see Chapter 8). A set of design specifications can be met by locating the closed-loop poles inside the corresponding design region on the s-plane. Then the design process will involve the selection of parameters such as control gain, compensator poles and compensator zeros, so as to place the closed-loop poles in the proper design region. This can be accomplished by the root locus method.

9.4.1 Design Steps Using Root Locus

Once the design region on the s-plane is chosen, the next step is to check whether the dominant branch (i.e., branch closest to $s=0$) of the root locus (for the closed-loop system) passes through that region. If it does, the corresponding value of the root locus variable (typically the loop gain) is computed using the magnitude condition (see Rule 2 given for sketching a root locus, in Chapter 8). If the steady-state error requirement is already included in the design specification (on the s-plane) as a limit on ω_n, then we do not need to proceed further. Otherwise, the applicable error constant (K_p, K_v, or K_a), as described in Chapter 7, should be computed using the design value of the loop gain, to check whether the steady-state accuracy is adequate. If the design requirements are not met with the existing control loop, a compensator with dynamics (i.e., one having s terms), such as a lead compensator or a lag compensator, should be added to the loop and the compensator parameters should be chosen to satisfy the design specifications. To summarize, the design steps using the root locus method are as follows:

Step 1: Represent the design specifications as a region on the s-plane.

Step 2: Plot the root locus to check whether at least one branch of the root locus passes through the design region while the other branches pass through regions to the left of the design region, for the same parameter values.

Step 3: If it does not satisfy the requirement in Step 2, add a compensator to the control loop and adjust the compensator parameters to achieve the requirement. If it does, then compute the corresponding root locus parameter (typically the loop gain) using the magnitude condition.

Note: It is the dominant poles of the closed-loop system that should fall inside the design region (the remaining poles being to the left of the region, by definition).

The magnitude condition is an important equation in the root locus design. In this regard, a useful result can be obtained by using the closed-loop characteristic equation (see Chapter 8):

$$K\frac{(s-z_1)(s-z_2)\ldots(s-z_m)}{(s-p_1)(s-p_2)\ldots(s-p_n)} = -1 \tag{9.36}$$

Since the coefficient of the second highest power (i.e., coefficient of s^{n-1}) of a monic characteristic polynomial (i.e., a polynomial with the coefficient of s^n equal to 1) is equal to the sum of the roots except for a sign change, we observe from Equation 9.36 the following fact, which can be given as a rule for sketching root locus:

Rule 9: If $m < n-1$, then the (sum of the closed-loop poles)=(sum of the GH poles) and this sum is independent of K.

9.4.2 Lead Compensation

The lead compensator design using the root locus method is illustrated now. Essentially, we follow the three steps for root locus design, as presented earlier.

Lead compensator design in the frequency domain using Bode diagram was discussed previously. The objective of the method was to determine the zero $(-z)$ and the pole $(-p)$ of the lead compensator transfer function

$$G_c = \frac{p}{z}\left[\frac{s+z}{s+p}\right] \quad z < p \tag{9.37}$$

so that the design specifications are satisfied. The steps that are usually followed in the root locus method to design a lead compensator are given below.

Step 1: Select a closed loop pole pair (complex conjugates) that meets the design specifications. This should become the dominant pole pair of the closed-loop system.

Step 2: Locate the compensator zero $(-z)$ vertically below the specified closed-loop pole.

Step 3: Locate the compensator pole $(-p)$, to the left of $-z$ so as to satisfy the angle condition of the root locus (Rule 2 given under root locus method, in Chapter 8).

Step 4: Compute the root locus parameter (usually gain K) at the design pole location, using the magnitude condition (Rule 2 in Chapter 8, under the root locus method).

In Step 1, the design pole pair of the closed-loop system is located in the design region of the s-plane, as discussed earlier. In Step 2, if there is a GH pole at the location where the compensator zero is to be located, we should locate the compensator zero sufficiently left of that location so that the compensator would not drastically alter the dynamic characteristics of the uncompensated system, by producing a closed-loop pole that would dominate over the design poles. Step 3 makes sure that the root locus passes through the design point, and Step 4 provides the root locus parameter value at the design point.

Example 9.3

Sheet steel, which many major industries such as the automobile industry and the household appliance industry depend on, is obtained by either hot rolling or cold rolling of thick plates or slabs of steel castings. The steel is passed through a pair of work rolls, which are driven by heavy duty a motor. The thickness of the rolled steel depends on the roll separation, which is adjusted by a hydraulic actuator (ram). Open-loop adjustment is not satisfactory in this application for reasons such as roll deformation, flexibility of rolled steel, and mill stretch. The thickness of the output steel coil is measured, and the roller separation is corrected accordingly, using feedback control. A simplified model for this control loop is shown in Figure 9.7a. Show that simple proportional control is inadequate for simultaneously meeting the following three specifications:

a. A peak time<0.2 s
b. 2% settling time<0.4 s
c. P.O.<10.

Determine the parameters for a suitable lead compensator that will satisfy these three control specifications. Compute the velocity error constant of the compensated system and the steady-state error to a unit ramp input.

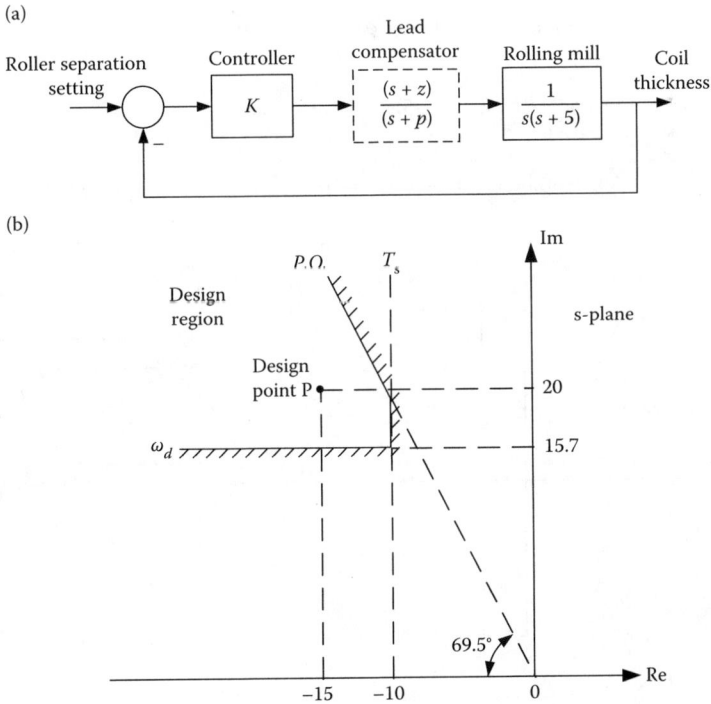

FIGURE 9.7
(a) Block diagram for the control system of a steel rolling mill. (b) Selection of a design point (P) on the s-plane.

Solution

Using the formulas given in Chapter 7 (see Table 7.2) we find that:

$$T_p = 0.2 \text{ s corresponds to } \omega_d = 15.7 \text{ rad/s}$$

$$T_s = 0.4 \text{ s corresponds to } \zeta\omega_n = 10 \text{ rad/s}$$

$$P.O. = 10\% \text{ corresponds to } \zeta = 0.35 \text{ or } \cos^{-1}\zeta = 69.5°$$

These design boundaries are drawn in Figure 9.7b and an acceptable design region is determined as a result. Any point in this region would be satisfactory. Usually, the design point in this region that is closest to the origin of the s-plane is chosen. But since a steady-state error specification is not available, it is not known whether the design gain is satisfactory. As a compromise, the design point P (corresponding to $\omega_d = 20$, $\zeta\omega_n = 15$, $\omega_n = 25$, and $\zeta = 0.6$) is chosen as shown in Figure 9.7b. The root locus of the uncompensated system is sketched in Figure 9.8a. Since it does not pass through the design region as superimposed in Figure 9.8a, it is concluded that proportional control without compensation cannot meet the specifications.

Next, the compensator zero is located vertically below P. This gives: $z = -15$.

We know that the compensator pole lies to the left of this point. Now we are able to sketch the root locus for the compensated system, as in Figure 9.8b. *Note*: We have not yet determined all the parameters that are marked in this figure. We see from the figure that the design pole pair corresponds to the dominant poles of the closed-loop system, the third pole being located to their left (perhaps close to $-p$).

From geometry we can compute the following angles:

$$\text{Angle at the } GH \text{ pole (0): } \phi_1 = 180° - \tan^{-1}\frac{20}{15} = 126.87°$$

$$\text{Angle at the } GH \text{ pole (−5): } \phi_2 = 180° - \tan^{-1}\frac{20}{10} = 116.57°$$

$$\text{Angle at the } GH \text{ pole (−15): } = 90°$$

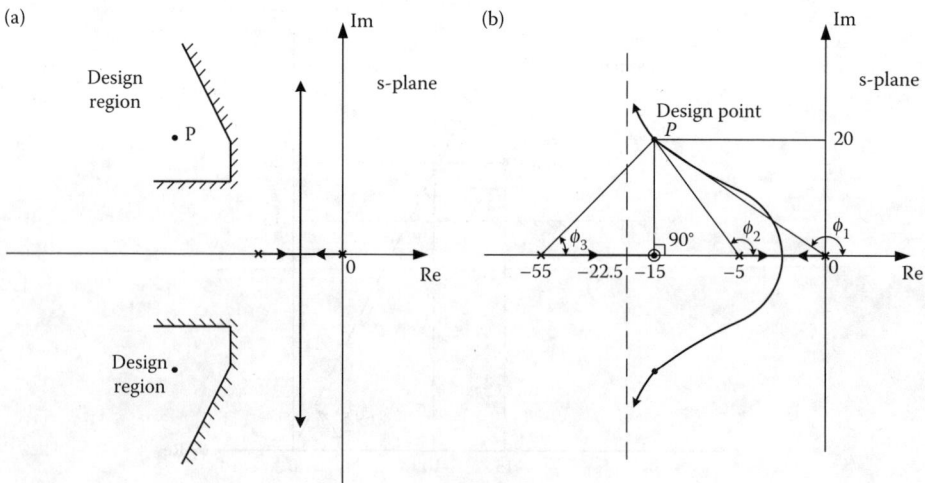

FIGURE 9.8
(a) Design region and the root locus of the uncompensated system. (b) Root locus of the compensated system.

Hence, from the angle condition of plotting a root locus (see Chapter 8) for P to be a point on the root locus we must have the angle at the GH pole $(-p)$, ϕ_3, satisfying the condition:
$\phi_3 + 116.57 + 126.87 - 90 = 180$

Hence: $\phi_3 = 26.56° = \tan^{-1}\dfrac{20}{(p-15)}$

From this we have the compensator pole: $p = 55$

The compensator transfer function is: $G_c(s) = \dfrac{(s+10)}{(s+55)}$

The loop gain K at the design point is obtained using the magnitude condition (Rule 2 of root locus method—see Chapter 8). From geometry of Figure 9.8b we obtain:

$$\text{Distance from } P \text{ to } 0 = \sqrt{20^2 + 15^2} = 25$$

$$\text{Distance from } P \text{ to } -5 = \sqrt{20^2 + 10^2} = 22.36$$

$$\text{Distance from } P \text{ to } -15 = 20$$

$$\text{Distance from } P \text{ to } -55 = \sqrt{20^2 + 40^2} = 44.72$$

Hence the magnitude condition gives: $\dfrac{K \times 20}{25 \times 22.36 \times 44.72} = 1$
Or: $K = 1250$
Now using the result given in Chapter 7, the velocity error constant for the control system is computed as:

$$K_v = \lim_{s \to 0} \frac{s \times 1250(s+15)}{s(s+5)(s+40)} = 93.75$$

Hence, the steady-state error to a unit ramp input is:

$$e_{ss} = \frac{1}{93.75} = 0.011$$

This error is very small, and the design is concluded as satisfactory.

Note: Since the compensator pole is far to the left of the system poles, the real pole of the closed-loop system will not dominate and the compensated system will behave like a second-order system.

Note: See Appendix B for a MATLAB® treatment of this example.

9.4.3 Lag Compensation

We know that lag compensation improves the behavior in low frequency operation (particularly, steady-state accuracy) of a control system. Accordingly, lag compensation is recommended if the uncompensated system has good transient response (i.e., satisfactory moderate-to-high-frequency performance) but has poor steady-state accuracy.

To explain the principle of lag compensation by the root locus method, consider the control system shown in Figure 9.9a. For a controller with gain value $K = K_o$ (i.e., the value of

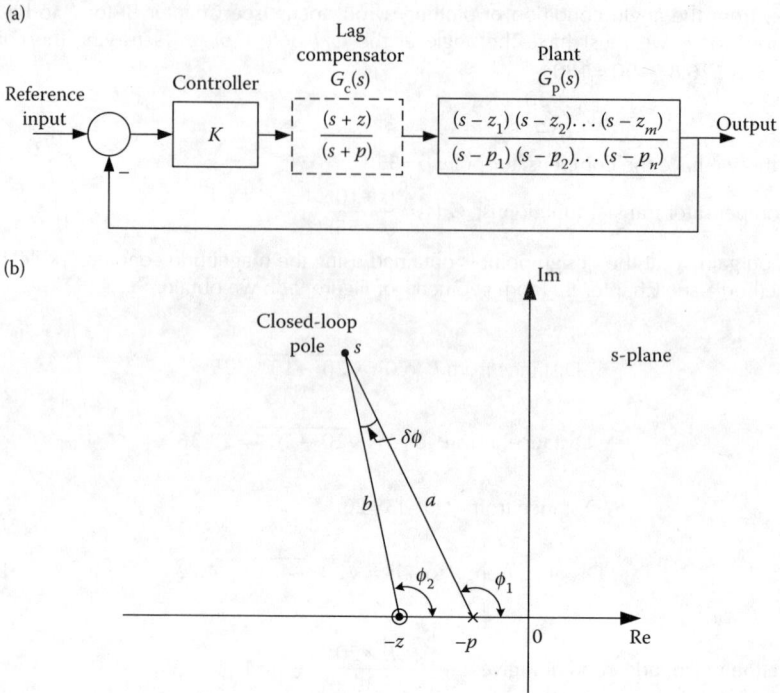

(a)

Lag
compensator
$G_c(s)$

Controller

Reference
input

Plant
$G_p(s)$

K

$\dfrac{(s+z)}{(s+p)}$

$\dfrac{(s-z_1)\,(s-z_2)\ldots(s-z_m)}{(s-p_1)\,(s-p_2)\ldots(s-p_n)}$

Output

(b)

Im

Closed-loop
pole s

s-plane

$\delta\phi$

b a

ϕ_2 ϕ_1

$-z$ $-p$ 0   Re

FIGURE 9.9
(a) Block diagram of a system with a lag compensator. (b) Relationship of pole and zero of a lag compensator to dominant closed-loop pole (S) of uncompensated system.

the root locus parameter), the characteristic equation of the system without lag compensator $G_c(s)$ is:

$$1 + K_o G_p(s) = 0 \tag{9.38}$$

The roots of this characteristic equation give closed-loop poles. Suppose that the dominant pole (one closest to the origin of the s-plane) obtained this way is shown as point S in Figure 9.9b. At point S, the angle condition and the magnitude condition are satisfied. Hence (see Chapter 8):

$$\angle(s-p_1)+\cdots+\angle(s-p_n)-\angle(s-z_1)-\cdots-\angle(s-z_m)=\pi+2r\pi \tag{9.39}$$

$$\frac{|s-p_1|\cdots|s-p_n|}{|s-z_1|\cdots|s-z_m|}=K_o \tag{9.40}$$

Now let us include the lag compensator $G_c(s)$ as shown in Figure 9.9a. Then for the same controller gain K_o (i.e., the root locus parameter value) the closed-loop characteristic equation is:

$$1 + K_o G_c(s) G_p(s) = 0 \tag{9.41}$$

Generally, the roots of this equation are different from the roots of Equation 9.38. In particular, now there may not be a root at point S in Figure 9.9b. But if we select the compensator $G_c(s)$ in an appropriate manner, we can make sure that the roots of Equation 9.38 are "very close" to the roots of Equation 9.41 so that the transient performance of the system is not significantly affected by compensation. In particular, the root locus of the compensated system will be very similar to that of the uncompensated system except in the neighborhood of the compensator pole and zero. To examine the conditions that should be satisfied by the compensator to achieve this, let us write the angle condition and the magnitude condition corresponding to Equation 9.41:

$$\angle(s - p_1) + \cdots + \angle(s - p_n) - \angle(s - z_1) - \cdots - \angle(s - z_m) + \delta\phi = \pi + 2r\pi \tag{9.42}$$

$$\frac{|s - p_1| \cdots |s - p_n|}{|s - z_1| \cdots |s - z_m|} \cdot \frac{a}{b} = K_o \tag{9.43}$$

in which

$$\delta\phi = \angle(s + p) - \angle(s + z) = \phi_1 - \phi_2 \tag{9.44}$$

$$a = |s + p| \tag{9.45}$$

$$b = |s + z| \tag{9.46}$$

It follows that if $\delta\phi$ is very small, Equation 9.44 approximates to Equation 9.39. Similarly, if $a \approx b$, then Equation 9.43 approximates to Equation 9.40.

Now we can conclude that the requirements for the closed-loop poles of the uncompensated system to be "very close" to the closed-loop poles of the compensated system, for the same value of controller gain K_o, are:

$$\text{(i) } \delta\phi \approx 0 \tag{9.47}$$

and

$$\text{(ii) } a \approx b \tag{9.48}$$

It follows that the lag compensator must satisfy these two requirements. But note that the dc gain of the compensated loop is z/p times the dc gain of the uncompensated loop:

$$(\text{dc gain})_{comp} = \frac{z}{p}(\text{dc gain})_{uncomp} \tag{9.49}$$

Since $z > p$ for a lag compensator, this means that the dc gain has increased or, in other words, the error constant has increased. This is the reason for the decreased steady-state error due to lag compensation.

A disadvantage of lag compensation can be easily pointed out. Note that the number of closed-loop poles has increased by one due to compensation. One of these poles (the one that is not close to the uncompensated closed-loop poles) is very close to the compensator pole and zero. Since the compensator pole and zero have to be chosen very close to the origin (and close together) in order to satisfy the conditions (Equations 9.47 and 9.48) it follows that the dominant closed-loop pole now is the one created by the compensator. This is obviously a slow pole producing a slowly decaying transient, even though the magnitude of this transient is usually small. Hence, the settling time will be increased to some extent due to lag compensation.

Usual steps of designing a lag compensator by the root locus method are given below.

Step 1: Determine an appropriate operating point for the uncompensated system so as to satisfy all the performance specifications, except the error constant specification.

Step 2: Select the compensator zero at about $0.1 \times$ real part of the closed-loop operating pole of the uncompensated system.

Step 3: Select the compensator pole to meet the error constant (steady-state error) specification.

Step 4: Check for the margin of error introduced by the compensator.

All these steps should be clear from Example 9.4.

Example 9.4

The motor and the load of a position servo system are represented by the plant transfer function:

$$G_p(s) = \frac{1}{s(s+4)}$$

The controller is represented by a pure gain K along with unity feedback. The system is shown in Figure 9.10a. Sketch the root locus and show that this servo system cannot simultaneously meet the following performance specifications:

   a. $P.O. = 4.3$
   b. $K_v = 10$.

Design a lag compensator to meet these specifications, within a margin of error. Estimate this margin of error. Sketch the root locus of the compensated system.

(a)   (b)

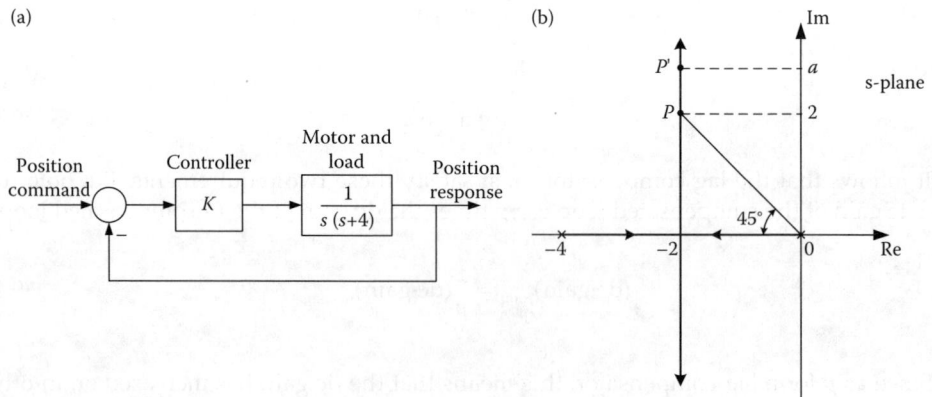

FIGURE 9.10
(a) Block diagram of an uncompensated position servo system. (b) Root locus of the uncompensated system.

Solution

The loop transfer function of the uncompensated system is $s/(s(s+4))$. The root locus is sketched in Figure 9.10b.

A P.O. of 4.3% corresponds to a damping ratio of $\zeta = 0.707$ (this can be verified using the equation given in Chapter 7, Table 7.2). The corresponding closed-loop pole location is shown as point P in Figure 9.10b. The controller gain for these operating conditions is obtained by using the magnitude condition:

$$K = \sqrt{2^2 + 2^2} \times \sqrt{2^2 + 2^2} = 8$$

The corresponding velocity error constant (see Chapter 7) is:

$$K_v = \lim_{s \to 0} s \times \frac{s}{s(s+4)} = \frac{8}{4} = 2$$

This is less than the specified value of 10.

We can meet the K_v specification simply by increasing the controller gain to $K=40$. Then, however, the operating point moves to P' in Figure 9.10b, where the closed-loop poles are $-2 \pm ja$. Now the value of "a" is determined using the magnitude condition: $\sqrt{(a^2+2^2)} \times \sqrt{(a^2+2^2)} = 40 \rightarrow a = 6$.

The corresponding undamped natural frequency $= \sqrt{6^2 + 2^2} = \sqrt{40}$ and the damping ratio $\zeta = 0.316 = 2/\sqrt{40}$. The P.O. with this ζ is 35.1% (from Table 7.2) which is much higher than the specified value. It follows that the specifications cannot be met by adjusting the controller gain alone.

Next, we add a lag compensator to the control loop. Since we want to keep the pole near point P in Figure 9.10b, the pole and the zero of the compensator transfer function:

$$G_c(s) = \frac{(s+z)}{(z+p)} \tag{9.50}$$

should be chosen to be much closer to the origin than the operating point P. This is accomplished by making z equal to 10% of the real part of the operating pole. Accordingly we have: $z = 0.2$

The velocity error constant, with the compensator added, is:

$$K_v = K \frac{\angle 1}{p} \frac{1}{4} = \frac{8 \times 0.2}{p \times 4}$$

We have to make this value equal to 10 (the specification). Hence, we have:

$$\frac{8 \times 0.2}{p \times 4} = 10 \rightarrow p = 0.04$$

The lag compensator transfer function is:

$$G_c(s) = \frac{(s+0.2)}{(s+0.04)}$$

The compensated system is shown in Figure 9.11a and its root locus is sketched in Figure 9.11b. The operating point (with $K=8$) of the system is shown as P' in Figure 9.11b. Note that the damping ratio at this point will be slightly different from the required 0.707, even though the velocity error

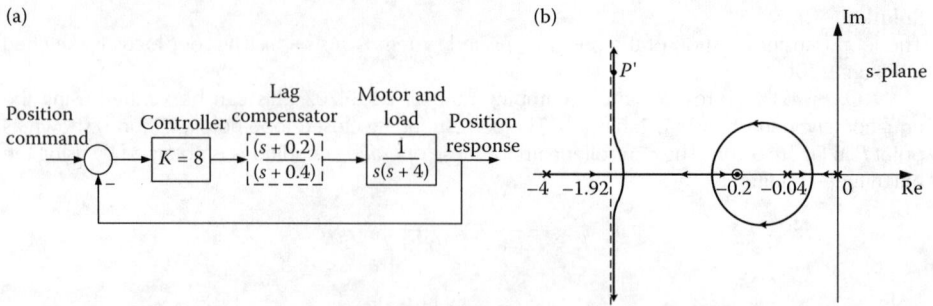

(a)

(b)

FIGURE 9.11
(a) Block diagram of the lag compensated position servo. (b) Root locus of the compensated system (Operating point=P').

constant is exactly met (K_v=10). The error in the damping ratio (or P.O.) can be determined by estimating the phase angle error and magnitude error near the operating point, as introduced by the compensator. Since the operating point (P') is approximately $-2 + j2$, the angle error:

$$\delta\phi \approx \angle(-2 + j2 + 0.04) - \angle(-2 + j2 + 0.2)$$

$$\approx 134.4° - 132.0° = 2.4°$$

The percentage error (since the vector angle of the operating point is approximately 45°) is:

$$\frac{2.4°}{45°} \times 100 = 5.3\%$$

The magnitude error is: $\dfrac{|-2 + j2 + 0.04|}{|-2 + j2 + 0.2|} - 1 = 0.041 = 4.1\%$

It follows that the error will be about 5%.

9.5 Controller Tuning

Ziegler–Nichols tuning is a procedure that is commonly used to set parameters of PID controllers—three-mode controllers or three-term controllers—in industrial control systems. It uses rules of thumb based on practical experience and experimental observation with common types of control systems. We now present this tuning method as it provides quite satisfactory results even though it lacks theoretical rigor.

9.5.1 Ziegler–Nichols Tuning

Adjustment of controller parameters to improve the system performance (response) is known as controller tuning. Selection of parameter values for a controller is an important final step in the control system design. This of course assumes that the design has progressed to the extent that everything about the control system (e.g., control system structure and components, process parameters, controller type) is known except for the parameter values of the controller. Even if the controller parameters are known for the initial design of the controller system, they may have to be further adjusted (or, "fine-tuned") during operation, as further information on the system performance becomes available and as the operating conditions change.

Controller tuning can be accomplished either by analysis of the control system or by testing. Many engineering systems are complex and nonlinear with noisy signals and unknown and time-varying parameter values. Controller tuning by analysis becomes a difficult task for such systems, and consequently controller tuning by testing becomes useful. In their original work, Ziegler and Nichols proposed two empirical methods for tuning three mode (PID) controllers:

a. Reaction curve method

b. Ultimate response method

Both methods of tuning are expected to provide approximately a *quarter decay ratio* (i.e., amplitude decays by a factor of four in each cycle) in the closed-loop system response. On the basis of a simple oscillator model (see Chapter 6) damping ratio may be expressed by the approximate relationship:

$$\zeta = \frac{1}{2\pi r} \ln \frac{A_i}{A_{i+r}} \tag{9.51}$$

in which

A_i = response amplitude in the ith cycle

A_{i+r} = response amplitude in the $(i+r)^{th}$ cycle

We note from Equation 9.51 that the quarter decay ratio corresponds to:

$$\zeta = \frac{1}{2\pi} \ln 4 \approx 0.22 \tag{9.52}$$

or a phase margin of approximately 22° (see Equation 9.35). These are rough estimates, however, because their derivation is based on the simple oscillator model.

9.5.2 Reaction Curve Method

In this method, first the open loop response of the plant alone (without any feedback and control) to a *unit step* input is determined. This response is known as the *reaction curve*. *Note:* If the step input used in the test is not unity, the response curve has to be appropriately scaled (i.e., divided by the magnitude of the step input) in order to obtain the reaction curve.

We assume that the process is *self-regulating*, for open-loop test purposes, implying that it is stable and its (open-loop) step response eventually settles to a steady-state value, even though this assumption is actually not needed in Ziegler–Nichols tuning. The reaction curve of many processes that are self-regulating has the well-known S-shape as represented in Figure 9.12. Note the parameters identified in the figure:

The *lag time L* is also known as *dead time* or *delay time*.

K is the steady-state value of the process variable (process response) for a "unit" step of process demand (process input).

T is termed cycle time.

R is the maximum slope of the process reaction curve.

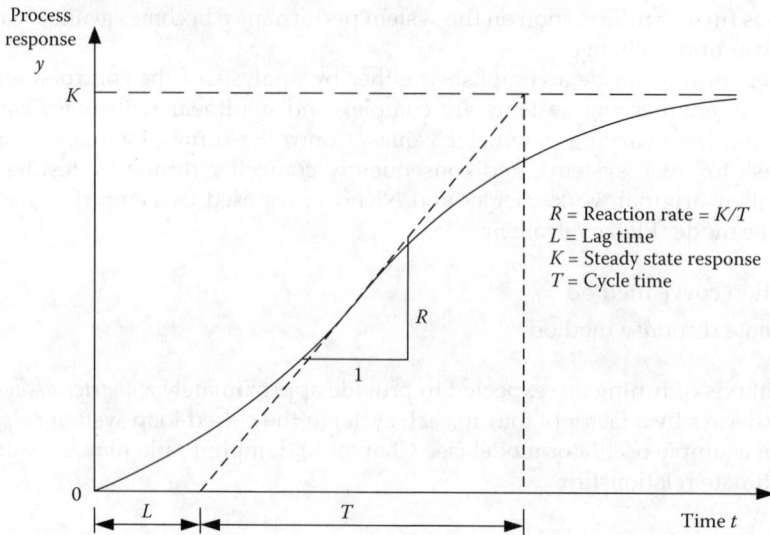

FIGURE 9.12
The reaction curve (response to a unit step input) of an idealized self-regulating process.

These parameters alone completely determine a first-order process with a time delay, as given by the transfer function:

$$G_p(s) = \frac{Ke^{-Ls}}{(Ts+1)} \tag{9.53}$$

This is a self-regulating plant because it has a stable pole at $-1/T$. If instead the pole is at the origin of the s-plane (i.e., it is an integrator), we have a nonself-regulating plant. Higher-order processes have to be approximated by Equation 9.53. Once the parameters L and R are obtained from the experimentally determined reaction curve for the open-loop process, the controller parameter values for proportional (P), proportional plus integral (PI), and PID controllers are determined according to the Ziegler–Nichols method as tabulated in the column named "Reaction curve method" in Table 9.1.

In conducting the step response test, the process should be first maintained steady at normal operating conditions, with the feedback transmitter disconnected and the controller set to "manual." The corresponding process load and other conditions should be kept constant during the test. A step test is conducted by changing the set point by 5% of the full range and recording the response until the steady state is reached. The response curve should be divided by the value of the step in order to get the reaction curve. Since hysteresis effects are usually present, it is a good practice to apply a step change in the reverse direction and determine the corresponding reaction curve and then take the average of the two curves. Accuracy can be improved by conducting several step tests in each direction and taking the average of all measured reaction curves.

The Ziegler–Nichols method is applicable even if the process is nonself-regulating, because the method does not directly depend on the steady-state value K or cycle time T (it depends on the slope K/T). Ziegler–Nichols method is particularly suitable for nonself-regulating processes. In fact, the settings have to be modified when K is of the order of LR, which is the self-regulating case.

9.5.3 Ultimate Response Method

In this method, the closed-loop system with proportional control alone (i.e., integral and derivative control actions turned off) is tested to determine the *ultimate gain* K_u and the corresponding period of oscillation (*ultimate period*) P_u of the process response. Ultimate gain is the controller gain at which the closed-loop system is marginally stable; i.e., when the system response continuously cycles without a noticeable growth or decay. The Ziegler–Nichols controller settings are then determined using K_u and P_u, as given under the "Ultimate response method" in Table 9.1.

In conducting the test, first the process is connected with the controller in the feedback control mode with the *integral rate* r_i (i.e., inverse of the *integral time constant* τ_i) and the *derivative time* τ_d set to zero. Then, the process conditions are maintained at normal operating values. Next, the proportional gain is set to a small value and maintained there until the conditions are steady. Then, a step input change (typically 5% of the full scale) is applied and the process response is noted. It should decay quickly in view of the low value of K_p. The proportional gain is increased in sufficiently large steps and the test repeated to roughly estimate the value of K_p that makes the system marginally stable. Once a rough estimate is found, the test should be repeated in that neighborhood using small changes in proportional gain, in order to obtain a more accurate value for the ultimate gain. With the proportional gain set at this value, the process response (closed-loop) to a step input is recorded and from this data, the ultimate period of oscillations is determined. This step is explained in Figure 9.13.

TABLE 9.1

Ziegler–Nichols Controller Settings

Controller	Parameter	Reaction Curve Method	Ultimate Response Method
P	K_p	$\dfrac{1}{RL}$	$0.5K_u$
PI	K_p	$\dfrac{0.9}{RL}$	$0.45K_u$
	τ_i	$3.3L$	$0.83P_u$
PID	K_p	$\dfrac{1.2}{RL}$	$0.6K_u$
	τ_i	$2L$	$0.5\,P_u$
	τ_d	$0.5L$	$0.125P_u$

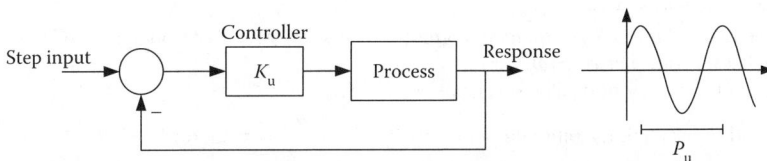

FIGURE 9.13
Test parameters in the ultimate response method of Ziegler–Nichols tuning.

Process

Set point →◯→ Controller → $\dfrac{1}{s(s^2 + s + 4)}$ → Response

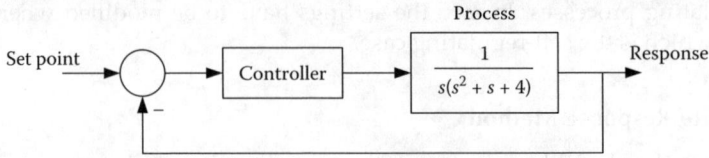

FIGURE 9.14
An example for the ultimate response method of Ziegler–Nichols tuning.

Example 9.5

Consider the feedback control system shown in Figure 9.14. The process transfer function is:

$$G_p(s) = \frac{1}{s(s^2 + s + 4)}$$

With proportional control, determine the proportional gain that will make the system marginally stable. What is the period of oscillations for that condition? Give suitable settings for a:

a. proportional (P) controller
b. PI controller
c. PID controller

Solution

With proportional control (gain= K_p), the closed-loop characteristic equation is:

$$1 + \frac{K_p}{s(s^2 + s + 4)} = 0 \;\rightarrow\; s^3 + s^2 + 4s + K_p = 0$$

To find the condition for marginal stability, we use Routh–Hurwitz method (see Chapter 8). First, form the Routh array:

	Column 1	Column 2
s^3	1	4
s^2	1	K_p
s^1	$4 - K_p$	0
s^0	K_p	

For stability, the terms in the first column should be all positive. Hence we must have $4 - K_p > 0$ and $K_p > 0$. Thus, the stability region is $0 < K_p < 4$. Accordingly, the gain for marginal stability (ultimate gain) is: $K_u = 4$.
The corresponding characteristic equation is: $s^3 + s^2 + 4s + 4 = 0$.
This factorizes into: $(s+1)(s^2+4) = 0$.
Note: $s^2 + 4 = 0$ is the *auxiliary equation*, corresponding to the row s^2 of Routh array with $K_p = K_u = 4$.
The oscillatory root pair is: $\pm j\omega_n = \pm j2$.
Hence, the frequency of oscillations is: $\omega_n = 2$ rad/s.

The period of oscillations (ultimate period) is: $P_u = \dfrac{2\pi}{\omega_n} = \dfrac{2\pi}{2} = \pi$ seconds.

Now, from Table 9.1, we can determine the controller settings.

a. Proportional control:

$$K_p = 0.5 \times 4 = 2$$

b. PI control:

$$K_p = 0.45 \times 4 = 1.8$$

$$\tau_i = 0.83\pi = 2.61\,s$$

c. PID control:

$$K_p = 0.6 \times 4 = 2.4$$

$$\tau_i = 0.5\pi = 1.57\ s$$

$$\tau_d = 0.125\pi = 0.393\ s$$

Note: See Appendix B for a MATLAB treatment of this example.

Problems

PROBLEM 9.1

Sketch an operational amplifier circuit for a PI controller and one for a lag-lead compensator. In each case derive the circuit transfer function.

PROBLEM 9.2

Consider six control systems whose loop transfer functions are given by:

a. $\dfrac{1}{(s^2 + 2s + 17)(s + 5)}$

b. $\dfrac{10(s + 2)}{(s^2 + 2s + 17)(s + 5)}$

c. $\dfrac{10}{(s^2 + 2s + 101)}$

d. $\dfrac{10(s + 2)}{(s^2 + 2s + 101)}$

e. $\dfrac{1}{s(s + 2)}$

f. $\dfrac{s}{(s^2 + 2s + 101)}$

First compute the additional gain (multiplication) k needed in each case to meet a steady-state error specification of 5% for a step input. Plot the Bode curves for the systems with modified gain values. Determine the gain margins and phase margins.

If you were asked to pick one of these systems to design a single lead compensator or a lag compensator so that the compensated system would have a phase margin of exactly 60°, which system would you pick? Discuss your answer by rationalizing why you did not pick the remaining five systems.

PROBLEM 9.3

A control system was found to have poor accuracy at low frequencies, poor speed of response at high frequencies, and a low stability margin in the operating bandwidth. Discuss what type of compensation you would recommend in order to improve the performance of this control system.

PROBLEM 9.4

a. Consider the compensator given by the transfer function

$$G_d(s) = \left[\frac{\tau s + 1}{a\tau s + 1}\right] \quad 0 < a < 1$$

     Is this a lead compensator or lag compensator? Explain your answer by sketching the Bode plot of the transfer function.

b. Derive an expression for the maximum phase angle available from a lead compensator.

c. A feedback control system with two possible locations for a compensator of the form given in (a), is shown in Figure P9.4a.

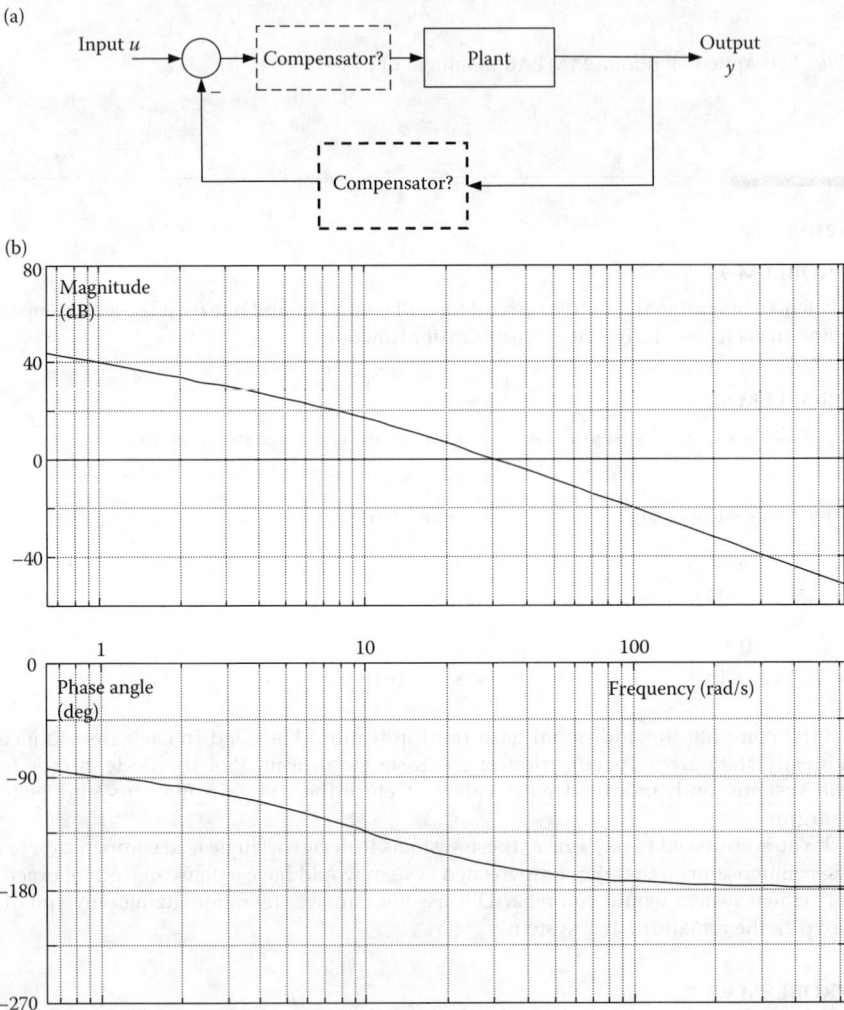

(a)

(b)

FIGURE P9.4
(a) Two possible locations for a compensator in a feedback control system. (b) The Bode diagram of a plant.

Is there any difference in the final effect of the compensator depending on the choice of its location in Figure P9.4? Explain your answer.
  d. In Figure P9.4(a) suppose that the plant transfer function has the Bode diagram shown in Figure P9.4(b).

What is the phase margin and what is the gain margin of the uncompensated feedback system with this plant?

Next suppose that a lead compensator or a lag compensator is added to the system, in the forward path of Figure P9.4(a). In each case, a phase margin of about 40° is required. By sketching how Figure P9.4(b) could be modified using a lead compensator or a lag compensator, explain how this phase margin specification could be achieved.

What are advantages and disadvantages of using a lag compensator to achieve this over a lead compensator?

PROBLEM 9.5

Both frequency domain Bode method and the root locus method can be used for designing lead and lag compensators for control systems. Which method would you prefer in each of the following cases?

  a. Compensator design for a system with large time delays.
  b. Lead compensator design for a stable system with negligible time delay.
  c. Lag compensator design for an unstable system with negligible time delay.
    (i) Even though a lag compensator has a destabilizing effect in general, it can be used to stabilize some unstable systems. Using sketches of Bode diagrams for a typical situation, explain why this is true. What is a major shortcoming of this method of compensation?
    (ii) Specifications on peak time, settling time, and percentage overshoot are commonly used in controller design using the root locus method. Using sketches of design regions on the s-plane, give an example for a situation where all three specifications are necessary and an example where one specification is redundant.

PROBLEM 9.6

A control loop used for controlling roll motion of an aircraft is shown by the block diagram in Figure P9.6. The following specifications must be met by the control system:

  i. A peak time $\leq \pi/2$ seconds
  ii. 2% settling time ≤ 2 seconds
  iii. P.O. $\leq/3\%$.

Show that the settling time specification is redundant.

Using root locus design, determine a set of control parameters (K, z, p) that will satisfy these specifications. What is the acceleration error constant of the system?

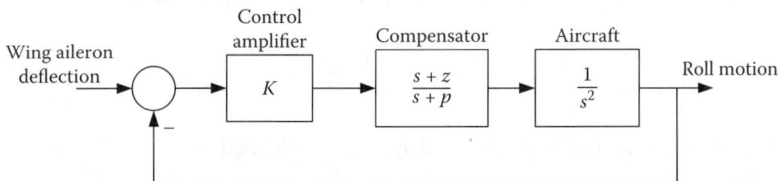

FIGURE P9.6
Roll motion control loop of an aircraft.

PROBLEM 9.7

A speed control system is shown in Figure P9.7. Show that the specifications:

   i. $P.O. \leq 5\%$

  ii. Steady-state error $\leq 2\%$ to a step input

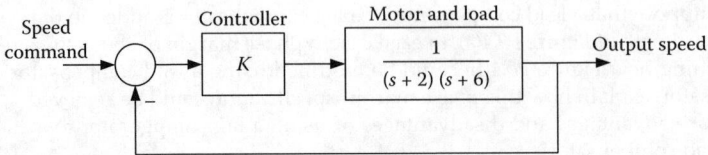

FIGURE P9.7
Block diagram of an uncompensated speed control system.

cannot be simultaneously satisfied by this system. You may use a sketch of the root locus of the system.

    Design a suitable lag compensator to meet both specifications given by the equality conditions (i.e., $P.O.=5\%$ and steady-state error$=2\%$). Estimate the margin of error in the (actual) $P.O.$ of the compensated control system. Sketch the root locus of the compensated system.

PROBLEM 9.8

A microprocessor based loop tuner uses the response signal for a test input to compute in "almost" real-time, controller settings for P, PI, and PID control. Typical tuning specifications include one of the following:

   i. Response with minimum overshoot for a step input.
  ii. Response with 10% overshoot for a step input.
 iii. Response with quarter decay ratio.

    Using schematic diagrams, illustrate the physical connection of such a tuner to a process control loop. Describe the main steps of control loop tuning.

PROBLEM 9.9

What is a self-regulating process? Strictly speaking, Ziegler–Nichols controller settings are applicable to nonself-regulating plants that may be approximated by

$$G_p(s) = \frac{R}{s} e^{-Ls}$$

In the self-regulating case the plant transfer function is approximated by

$$G_p(s) = \frac{K e^{-Ls}}{(Ts+1)}$$

In this latter case an index of self-regulation can be defined as

$$m = \frac{LR}{K}$$

When $m=0$ (i.e., K is large and T is large, but the reaction rate R is finite), we have a nonself-regulating case. Explain how the Ziegler–Nichols controller settings in the reaction curve method should be modified to include the parameter m.

PROBLEM 9.10

The feedback transmitter of the temperature control system of a heating process was disconnected and the set point was changed by 10°C manually with the controller dynamics inactive. The steady-state response was found to be 80°C. The recorded response provided the following values:

$$\text{Lag time } L=0.5 \text{ min}$$

$$\text{Cycle time } T=2.0 \text{ min}$$

Determine suitable settings for the PID controller used in this temperature control system.

PROBLEM 9.11

Explain a situation when the reaction curve method of controller tuning is preferred and a situation when the ultimate response method of controller tuning is preferred.

Consider a process with the transfer function: $G_p(s) = \dfrac{1}{s(s^2+s+25)}$

The control system has unity feedback. Determine suitable parameter settings for a three-mode (PID) controller.

10

Digital Control

In a digital control system, a digital device is used as the controller. The digital controller may be a *hardware* device consisting of permanent logic circuitry or a *software device*—a digital computer. Hardware controllers are inexpensive and fast, but lack flexibility or programmability. A software-based digital controller has programmable memory in addition to a central processor. The control algorithm is "programmed" into the computer memory and is used by the processor in real-time to generate the control signals. The control algorithm in such a controller can be modified simply by reprogramming, without the need for hardware changes. Typically, data are sampled into a digital controller at a fixed sampling period. This chapter will present relevant issues of data sampling. A convenient way to analyze and design digital control systems is by the z-transform method. The theory behind this method will be presented and issues such as stability analysis and controller/compensator design by the z-transform method will be described.

10.1 Digital Control

In a digital control system, a digital device is used as the controller. The digital controller may be a *hardware* device that uses permanent logic circuitry to generate control signals. Such a device is termed a hardware controller. It does not have programmable memory. This type of controller is not flexible in the sense that the control algorithm cannot be modified without replacing hardware and furthermore, implementation of complex control algorithms by this hardware-based method can become difficult and expensive. But the method is typically very fast from the point of view of speed of generating the control signals, and is suitable for simple dedicated controllers. In mass production, hardware controllers are inexpensive. In a *software-based* digital controller a digital computer serves as the controller. A controller of this type has programmable memory devices in addition to a central processor. The control algorithm is stored in the computer memory in machine code (in binary code) and is used by the processor in real-time to generate the control signals, perhaps on the basis of measured outputs from the plant (which is the case in feedback control) and other types of data. This along with associated input/output hardware and driver software forms the digital controller. The control algorithm in such a controller can be modified simply by reprogramming, without the need for hardware changes.

10.1.1 Computer Control Systems

In a computer-based control system, a digital computer serves as the controller. A digital feedback control system is shown in Figure 10.1. The information enters into the control computer in the digital form. Signals generated by the computer are in the digital form. Typically, they have to be converted into the analog form for use in the external purposes

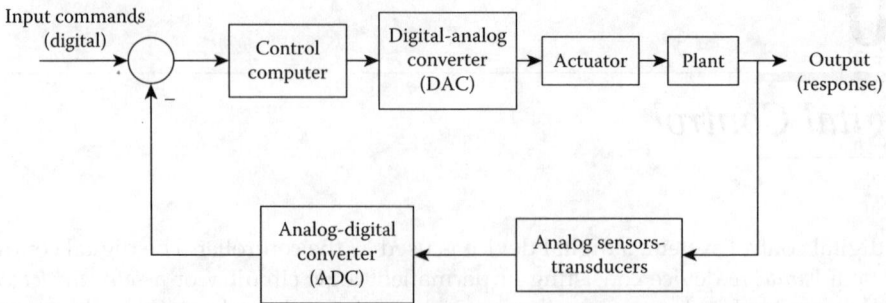

FIGURE 10.1
A digital feedback control system.

such as driving a plant or its actuators. Virtually any control law may be programmed into the control computer. Control computers have to be fast and dedicated machines for real-time operation where processing has to be synchronized with plant operation and actuation requirements. This requires a real-time operating system. Apart from these requirements, control computers are basically no different from general-purpose digital computers. They consist of a processor to perform computations and to oversee data transfer, memory for program and data storage during processing, mass storage devices to store information that is not immediately needed, and input/output devices to read in and send out information.

10.1.2 Components of a Digital Control System

Digital control systems might utilize digital instruments and additional processors as well for actuating, signal-conditioning, or measuring functions. For example, a stepper motor that responds with incremental motion steps when driven by pulse signals can be considered a digital actuator. Furthermore, it usually contains digital logic circuitry in its drive system. Similarly, a two-position solenoid is a digital (binary) actuator. Digital flow control may be accomplished using a digital control valve. A typical digital valve consists of a bank of orifices, each sized in proportion to a place value of a binary word (2^i, $i=0, 1$, $2, ..., n$). Each orifice is actuated by a separate rapid-acting on/off solenoid. In this manner, many digital combinations of flow values can be obtained. Direct digital measurement of displacements and velocities can be made using shaft encoders. These are digital transducers that generate coded outputs (e.g., in binary or gray-scale representation) or pulse signals that can be coded using counting circuitry. Such outputs can be read in by the control computer with relative ease. Frequency counters also generate digital signals that can be fed directly into a digital controller. When measured signals are in the analog form, an analog front end is necessary to interface the transducer and the digital controller. Input/output interface cards that can take both analog and digital signals are available with digital controllers.

Analog measurements and reference signals have to be sampled and encoded prior to digital processing within the controller. Digital processing can be effectively used for signal conditioning as well. Alternatively, digital signal processing (DSP) chips can function as digital controllers. However, analog signals have to be *preconditioned* using analog circuitry prior to digitizing in order to eliminate or minimize problems due to *aliasing distortion* (high-frequency components above half the sampling frequency appearing as low-frequency components) and *leakage* (error due to signal truncation) as well as to improve the signal level and filter out extraneous noise. The drive system of a plant typically takes in analog signals. Often, the digital output from controller has to be converted into analog form for this reason. Both *analog-to-digital conversion* (ADC) and *digital-to-analog conversion* (DAC) can be interpreted as signal-conditioning (modification) procedures. If more than one output signal is measured, each signal will have to be conditioned and processed separately. Ideally, this will require separate conditioning and processing hardware for each signal channel. A less expensive (but slower) alternative would be to time-share this expensive equipment by using a *multiplexer*. This device will pick one channel of data from a bank of data channels in a sequential manner and connect it to a common input device.

The current practice of using dedicated, microprocessor-based, and often decentralized (distributed) digital control systems in industrial applications can be rationalized in terms of the major advantages of digital control.

10.1.3 Advantages of Digital Control

The following are some of the important advantages of digital control.

1. Digital control is less susceptible to noise or parameter variation in instrumentation because data can be represented, generated, transmitted, and processed as binary words, with bits possessing two identifiable states.

2. Very high accuracy and speed are possible through digital processing. Hardware implementation is usually faster than software implementation.

3. Digital control can handle repetitive tasks extremely well, through programming.

4. Complex control laws and signal-conditioning methods that might be impractical to implement using analog devices can be programmed.

5. High reliability in operation can be achieved by minimizing analog hardware components and through decentralization using dedicated microprocessors for various control tasks.

6. Large amounts of data can be stored using compact, high-density data storage methods.

7. Data can be stored or maintained for very long periods of time without drift and without being affected by adverse environmental conditions.

8. Fast data transmission is possible over long distances without introducing excessive dynamic delays, as in analog systems.

9. Digital control has easy and fast data retrieval capabilities.

10. Digital processing uses low operational voltages (e.g., 0–12 V dc).

11. Digital control is cost effective.

10.2 Signal Sampling and Control Bandwidth

Sampling of signals is needed in computer control systems. Aliasing distortion occurs when data are sampled from a continuous (analog) signal.

10.2.1 Sampling Theorem

If a time signal $x(t)$ is sampled at equal steps of ΔT, no information regarding its frequency spectrum $X(f)$ is obtained for frequencies higher than $f_c = 1/(2\Delta T)$. This fact is known as *Shannon's sampling theorem*, and the limiting (cut-off) frequency is called the *Nyquist frequency*.

It can be shown that the aliasing error is caused by "folding" of the high-frequency segment of the frequency spectrum beyond the Nyquist frequency into the low-frequency segment. This is illustrated in Figure 10.2. The aliasing error becomes more and more prominent for frequencies of the spectrum closer to the Nyquist frequency. In digital signal analysis and control, a sufficiently small sample step ΔT should be chosen in order to reduce aliasing distortion in the frequency domain, depending on the highest frequency of interest in the analyzed signal. This however, increases the signal processing time and the computer storage requirements, which is undesirable particularly in real-time analysis. It also can result in stability problems in numerical computations. The Nyquist sampling criterion requires that the sampling rate $(1/\Delta T)$ for a signal should be at least twice the highest frequency of interest. Instead of making the sampling rate very high, a moderate value that satisfies the Nyquist sampling criterion is used in practice, together with an *antialiasing filter* to remove the distorted frequency components.

10.2.2 Antialiasing Filter

It should be clear from Figure 10.2 that, if the original signal is low-pass filtered at a cut-off frequency equal to the Nyquist frequency, then the aliasing distortion due to sampling would not occur. A filter of this type is called an antialiasing filter. Analog hardware filters may be used for this purpose. In practice, it is not possible to achieve perfect filtering. Hence, some aliasing could remain even after using an antialiasing filter. Such residual errors may be reduced by using a filter cut-off frequency that is slightly less than the Nyquist frequency. Then the resulting spectrum would only be valid up to this filter cut-off frequency (and not up to the theoretical limit of Nyquist frequency). Aliasing reduces the valid frequency range in digital Fourier results. Typically, the useful frequency limit is $f_c/1.28$ so that the last 20% of the spectral points near the Nyquist frequency should be neglected. Note that sometimes $f_c/1.28 (\cong 0.8 f_c)$ is used as the filter cut-off frequency. In this case the computed spectrum is accurate up to $0.8 f_c$ and not up to f_c.

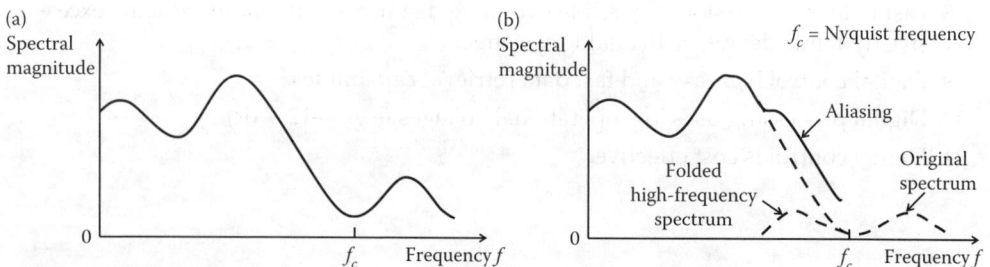

FIGURE 10.2
Aliasing distortion of a frequency spectrum. (a) Original spectrum. (b) Distorted spectrum due to aliasing.

10.2.3 Control Bandwidth

Control bandwidth represents the maximum possible speed of control. It is an important specification in both analog control and digital control. In digital control, the data sampling rate (in samples/second) has to be several times higher than the control bandwidth (in Hertz) so that sufficient data would be available to compute the control action. Also, from Shannon's sampling theorem, control bandwidth is given by half the rate at which the control action is computed. The control bandwidth provides the frequency range within which a system can be controlled (assuming that all the devices in the system can operate within this bandwidth).

Example 10.1

a. If a sensor signal is sampled at f_s Hz, suggest a suitable cut-off frequency for an antialiasing filter to be used in this application.
b. Suppose that a sinusoidal signal of frequency f_1 Hz is sampled at the rate of f_s samples/s. Another sinusoidal signal of the same amplitude, but of a higher frequency f_2 Hz was found to yield the same data when sampled at f_s. What is the likely analytical relationship between f_1, f_2, and f_s?
c. Consider a plant of transfer function: $G(s) = k/(1 + \tau s)$
   What is the static gain of this plant? Show that the magnitude of the transfer function reaches $1/\sqrt{2}$ of the static gain when the excitation frequency is $1/\tau$ rad/s. Note that the frequency, $\omega_b = 1/\tau$ rad/s, may be taken as the operating bandwidth of the plant.
d. Consider a chip refiner that is used in the pulp and paper industry. The machine is used for mechanical pulping of wood chips. It has a fixed plate and a rotating plate, driven by an induction motor. The gap between the plates is sensed and is adjustable as well. As the plate rotates, the chips are ground into a pulp within the gap. A block diagram of the plate-positioning control system is shown in Figure 10.3.

Suppose that the torque sensor signal and the gap sensor signal are sampled at 100 Hz and 200 Hz, respectively, into the digital controller, which takes 0.05 s to compute each positional command for the servovalve. The time constant of the servovalve is $(0.05/2\pi)$ s and that of the mechanical load is $(0.2/2\pi)$ s. Estimate the control bandwidth and the operating bandwidth of the positioning system.

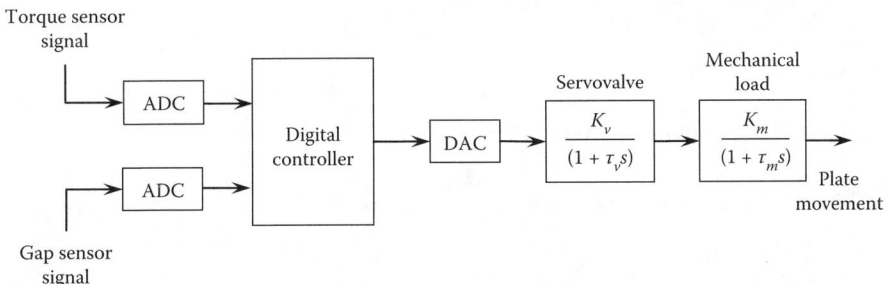

FIGURE 10.3
Block diagram of the plate positioning control system for a chip refiner.

Solution

a. In theory, the cut-off frequency of the antialiasing filter has to be $(1/2)f_s$, which is the Nyquist frequency. In practice, however, $0.4\,f_s$ would be desirable, providing a useful spectrum of only up to $0.4\,f_s$.

b.
$$f_2 = f_c + (f_c + f_1) = 2f_c - f_1 \;\rightarrow\; f_2 = f_s - f_1 \tag{10.1}$$

c. $G(j\omega) = k/(1 + \tau j\omega) =$ frequency transfer function where, ω is in rad/s.
Static gain is the transfer function magnitude at steady-state (i.e., at zero frequency).
Hence:

Static gain $= G(0) = k$

When $\omega = \dfrac{1}{\tau}$: $G(j\omega) = \dfrac{k}{(1+j)}$

Hence $|G(j\omega)| = \dfrac{k}{\sqrt{2}}$ at this frequency.

This is the half-power bandwidth.

d. Due to sampling, the torque signal has a bandwidth of $1/2 \times 100$ Hz $= 50$ Hz, and the gap sensor signal has a bandwidth of $1/2 \times 200$ Hz $= 100$ Hz. Control cycle time $= 0.05$ s, which provides control signals at a rate of $1/0.05$ Hz $= 20$ Hz.

Since: $20\,\mathrm{Hz} < \min\left(\dfrac{50}{2}\mathrm{Hz}, \dfrac{100}{2}\mathrm{Hz}\right)$

we have adequate bandwidth from the sampled sensor signals to compute the control signal. The control bandwidth from the digital controller

$= 1/2 \times 20$ Hz (from Shannon's sampling theoreom) $= 10$ Hz

But, the servovalve is also part of the controller. Its bandwidth

$= \dfrac{1}{\tau_v}\,\mathrm{rad/s} = \dfrac{1}{2\pi\tau_v}\mathrm{Hz} = \dfrac{2\pi}{2\pi \times 0.05}\mathrm{Hz} = 20\,\mathrm{Hz}$

Hence:

Control bandwidth $= \min\,(10\ \mathrm{Hz},\ 20\ \mathrm{Hz}) = 10$ Hz.

Bandwidth of the mechanical load

$= \dfrac{1}{\tau_m}\,\mathrm{rad/s} = \dfrac{1}{2\pi\tau_m}\mathrm{Hz} = \dfrac{2\pi}{2\pi \times 0.2}\mathrm{Hz} = 5\,\mathrm{Hz}$

Hence:

Operating bandwidth of the system $= \min\,(10\ \mathrm{Hz},\ 5\ \mathrm{Hz}) = 5$ Hz.

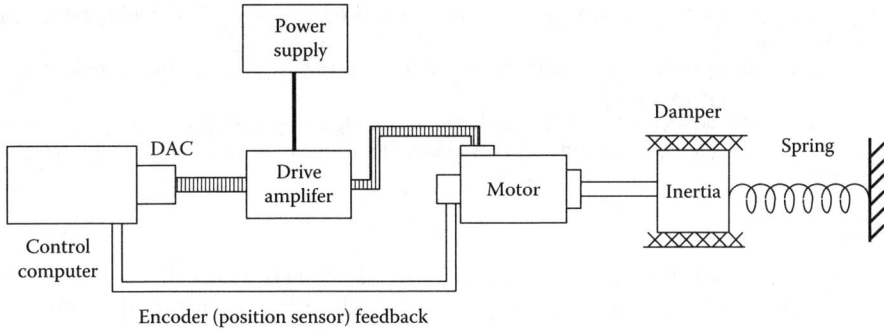

FIGURE 10.4
Digital control system for a mechanical positioning application.

Example 10.2

Consider the digital control system for a mechanical position application, as schematically shown in Figure 10.4. The control computer generates a control signal according to an algorithm, on the basis of the desired position and actual position, as measured by an optical encoder. This digital signal is converted into the analog form using a DAC and is supplied to the drive amplifier. Accordingly, the current signals needed to energize the motor windings are generated by the amplifier. The inertial element, which has to be positioned is directly (and rigidly) linked to the motor rotor and is resisted by a spring and a damper, as shown.

Suppose that the combined transfer function of the drive amplifier and the electromagnetic circuit (torque generator) of the motor is given by: $k_e/(S^2 + 2\zeta_e\omega_e s + \omega_e^2)$ and the transfer function of the mechanical system including the inertia of the motor rotor is given by: $k_m/(s^2 + 2\zeta_m\omega_m s + \omega_m^2)$
Here:

$$k = \text{equivalent gain}$$

$$\zeta = \text{damping ratio}$$

$$\omega = \text{natural frequency}$$

with the subscripts $()_e$ and $()_m$ denoting the electrical and mechanical components respectively.
Also:

$$\Delta T_c = \text{time taken to compute each control action}$$

$$\Delta T_p = \text{pulse period of the position sensing encoder.}$$

The following numerical values are given:

$$\omega_e = 1000\pi \text{ rad/s}, \quad \zeta_e = 0.5, \quad \omega_m = 100\pi \text{ rad/s}, \quad \text{and} \quad \zeta_m = 0.3$$

For the purpose of this example, you may neglect loading effects and coupling effects due to component cascading and signal feedback.

i. Explain why the control bandwidth of this system cannot be much larger than 50 Hz.
ii. If $\Delta T_c = 0.02$ s, estimate the control bandwidth of the system.

iii. Explain the significance of ΔT_p in this application. Why, typically, ΔT_p should not be greater than $0.5\Delta T_c$?

iv. Estimate the operating bandwidth of the positioning system, assuming that significant plant dynamics are to be avoided.

v. If $\omega_m = 500\pi$ rad/s and $\Delta T_c = 0.02$s, with the remaining parameters kept as specified above, estimate the operating bandwidth of the system, again not exciting significant plant dynamics.

Solution

i. The drive system has a resonant frequency less than 500 Hz. Hence the flat region of the spectrum of the drive system would be about 1/10th of this; i.e., 50 Hz. This would limit the maximum spectral component of the drive signal to about 50 Hz. Hence the control bandwidth would be limited by this value.

ii. Rate at which the digital control signal is generated $= (1/0.02)$ Hz $= 50$ Hz. By Shannon's sampling theorem, the effective (useful) spectrum of the control signal is limited to $(1/2)\times 50$ Hz $= 25$ Hz. Even though the drive system can accommodate a bandwidth of about 50 Hz, the control bandwidth would be limited to 25 Hz, due to digital control, in this case.

iii. Note that ΔT_p corresponds to the sampling period of the measurement signal (for feedback). Hence its useful spectrum would be limited to $1/2\Delta T_p$, by Shannon's sampling theorem. Consequently, the feedback signal will not be able to provide any useful information of the process beyond the frequency $1/2\Delta T_p$. To generate a control signal at the rate of $1/\Delta T_c$ samples/s, the process information has to be provided at least up to $(1/\Delta T_c)$ Hz. To provide this information we must have:

$$\frac{1}{2\Delta T_p} \geq \frac{1}{\Delta T_c} \quad \text{or} \quad \Delta T_p \leq 0.5\,\Delta T_c. \tag{10.2}$$

Note: This guarantees that at least two points of sampled data from the sensor are used for computing each control action.

iv. The resonant frequency of the plant (positioning system) is approximately (less than) $(100\pi/2\pi)$ Hz $\doteq 50$ Hz. At frequencies near this, the resonance will interfere with control, and should be avoided if possible, unless the resonances (or modes) of the plant themselves need to be modified through control. At frequencies much larger than this, the process will not significantly respond to the control action, and will not be of much use (the plant will be felt like a rigid wall). Hence, the operating bandwidth has to be sufficiently smaller than 50 Hz, say 25 Hz, in order to avoid plant dynamics.

Note: This is a matter of design judgment, based on the nature of the application (e.g., excavator, disk drive). Typically, however, one needs to control the plant dynamics. In that case it is necessary to use the entire control bandwidth (i.e., maximum possible control speed) as the operating bandwidth. In the present case, even if the entire control BW (i.e., 25 Hz) is used as the operating BW, it still avoids the plant resonance.

v. The plant resonance in this case is about $(500\pi/2\pi)$ Hz $\simeq 250$ Hz. This limits the operating bandwidth to about $(250\pi/2)$ Hz $\simeq 125$ Hz, so as to avoid plant dynamics. But, the control bandwidth is about 25 Hz because $\Delta T_c = 0.02$ s. The operating bandwidth cannot be greater than this value, and would be $\simeq 25$ Hz.

10.2.4 Bandwidth Design of a Control System

Based on the foregoing concepts, it is now possible to give a set of simple steps for designing a control system on the basis of bandwidth considerations.

Step 1: Decide on the maximum frequency of operation (BW_o) of the system based on the requirements of the particular application.

Step 2: Select process components (e.g., electro-mechanical components) that have the capacity to operate at BW_o and perform the required tasks.

Step 3: Select feedback sensors with a flat frequency spectrum (operating frequency range) greater than $4 \times BW_o$.

Step 4: Develop a digital controller with a sampling rate greater than $4 \times BW_o$ for the sensor feedback signals (keeping within the flat spectrum of the sensors) and a direct-digital control cycle time (period) of $1/(2 \times BW_o)$. *Note:* Digital control actions are generated at a rate of $2 \times BW_o$.

Step 5: Select the control drive system (interface analog hardware, filters, amplifiers, actuators, etc.) that have a flat frequency spectrum of at least BW_o.

Step 6: Integrate the system and test the performance. If the performance specifications are not satisfied, make necessary adjustments and test again.

10.2.5 Control Cycle Time

In the engineering literature it is often used that $\Delta T_c = \Delta T_p$, where $\Delta T_c =$ control cycle time (period at which the digital control actions are generated) and $\Delta T_p =$ period at which the feedback sensor signals are sampled (see Figure 10.5a). This acceptable in systems where the significant frequency range of the plant is sufficiently smaller than $1/\Delta T_p$ (and $1/\Delta T_c$). In that case the sampling rate $1/\Delta T_p$ of the feedback measurements (and the Nyquist frequency $0.5/\Delta T_p$) will still be sufficiently larger than the significant frequency range of

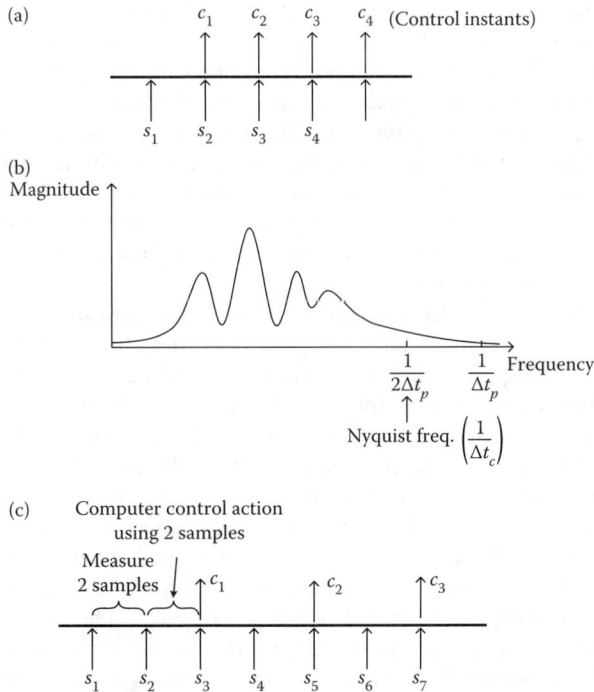

FIGURE 10.5
(a) Conventional sampling of feedback sensor signals for direct digital control. (b) Acceptable frequency characteristic of a plant for case (a). (c) Improved sampling criterion for feedback signals in direct digital control.

the plant (see Figure 10.5b) and hence the control system will function satisfactorily. But, the bandwidth criterion presented before satisfies $\Delta T_p \leq \Delta T_c$. This is a more desirable option. For example, in Figure 10.5c, two measurement samples are used in computing each control action. Here, the Nyquist frequency of the sampled feedback signals is double that of the previous case, and it will cover a larger (double) frequency range of the plant.

10.3 Digital Control Using z-Transform

In computer-based control systems, a suitable control algorithm has to be programmed into the memory of the control computer. A digital controller is functionally similar to its analog counterpart except that the input data to the controller and the output data from the controller are in the digital form (see Figure 10.1). The control law can be represented by a set of *difference equations*. These difference equations relate the discrete output signals from the controller and the discrete input signals into the controller. The problem of developing a digital controller can be interpreted as the formulation of appropriate difference equations that are able to generate the required control signals. Similarly, just the same way as an analog controller may be represented by a set of analog transfer functions, a digital controller may be represented by a set of *discrete transfer functions*. These discrete transfer functions, in turn, can be transformed into a set of difference equations.

Once a control law is available in the analog form, as a transfer function, the corresponding digital control law may be determined by obtaining the discrete transfer function that is equivalent to the analog transfer function. This approach is particularly useful when, for example, it is required to update (modernize) a well-established analog control system by replacing its analog compensator circuitry with a digital controller/compensator. Then the (Laplace) transfer function of the analog compensator can be obtained by testing or analysis (or both) of the compensator. The eventual objective would be to develop a difference equation to represent the analog compensator. This is a basic task in the development of a digital controller, and is conveniently handled by the z-transform method. This approach is developed in the present section.

A discrete transfer function necessarily depends on the sampling period T used to convert analog signals into discrete data (sampled data). Digital control action approaches the corresponding analog control action when T approaches zero. Faster sampling rates provide better accuracy and less aliasing error, but demand smaller processing cycle times, which in turn call for efficient processors and improved control algorithms for a given level of control complexity. Faster sampling rates are more demanding on the interface hardware as well. A large word size is needed to accurately represent data. By increasing the word size (number of bits per word), the *dynamic range* and the *resolution* of the represented data can be improved and the *quantization error* decreased. Even though the processing cycle time will generally increase by increasing the word size, on average there is also a speed advantage to increasing the word size of a computer. The larger the program size (number of instructions per program) the greater the memory requirements and, furthermore, the slower the associated control cycle for a given control computer. It follows that sampling rate, processing cycle time, data word size, and memory requirements are crucial parameters that are interrelated, in digital control.

10.3.1 z-Transform

Consider an infinite sequence of data:

$$\{x_k\}=\{\dots x_{-k}, x_{-k+1}, \dots, x_0, x_1, \dots, x_k, x_{k+1}, \dots\}. \tag{10.3}$$

This sequence can be represented by a polynomial function of the complex variable z:

$$X(z)=\sum_{k=-\infty}^{\infty} x_k z^{-k} \tag{10.4}$$

$X(z)$ is termed the z-transform of the sequence $\{x_k\}$. This relationship may be expressed using the z-transform operator "$\mathcal{Z}$" as:

$$\mathcal{Z}\{x_k\}=X(z) \tag{10.5}$$

Note from Equation 10.4 that $\{x_k\}$ uniquely determines $X(z)$ and vice versa. Since $X(z)$ is a continuous polynomial function of z it is convenient to use the z-transform instead of the sequence which it represents, in analyses that involve sequences of data. In digital control systems in particular, inputs and outputs of a digital controller are such data sequences, which are defined at discrete-time points. Hence z-transform techniques are quite useful in the design of digital controllers and compensators.

Note: Generally, for the summation in Equation 10.4 to converge, the magnitude of z has to be restricted to at least $|z|<1$.

In a digital control system, the controller reads sampled values of a continuous signal $x(t)$. Assuming that the sampling period T is constant, the corresponding discrete data values are given by:

$$x_k=x(k \cdot T). \tag{10.6}$$

Typically the signal is zero for negative values of time; hence, $x_k=0$ for $k<0$. But we will retain the full sequence including the negative portion of Equation 10.3 for the sake of analytical convenience.

Example 10.3

Consider a unit step signal given by:

$$\mathcal{U}(t)=1 \quad \text{for} \quad t \geq 0$$
$$=0 \quad \text{for} \quad t < 0$$

Suppose that this signal is sampled at sampling period T. The corresponding data sequence is:

$$\{\mathcal{U}_k\}=\{0,0,\dots,0,0,1,1,\dots,1,1,\dots\}$$

Determine the z-transform of this sequence.

Solution

By definition, the z-transform is given by:

$$\mathcal{U}(z) = \sum_0^\infty z^{-k}$$

By summation of series this can be expressed in the closed form:

$$\mathcal{U}(z) = \frac{1}{(1 - z^{-1})}$$

$$\text{Or: } \mathcal{U}(z) = \frac{z}{(z-1)}$$

Example 10.4

Consider the unit ramp signal given by:

$$x(t) = t \quad \text{for} \quad t \geq 0$$
$$= 0 \quad \text{for} \quad t < 0$$

What is the corresponding z-transform if the signal is sampled at period T?

Solution

Note that, by definition, the z-transform of the sampled data is expressed as:

$$X(z) = \sum_{k=0}^\infty kTz^{-k} = T\sum_{k=0}^\infty z^{-k}kz^{-k}$$

By using a well-known result in summation of series, this can be expressed in the closed form:

$$X(z) = \frac{Tz}{(z-1)^2}$$

A z-transform depends on the sampling period T in general. z-transforms corresponding to a selected set of time signals, sampled at T, are listed in Table 10.1.

10.3.2 Difference Equations

In the context of dynamic systems, difference equations are discrete-time models. Consider, in particular, a single-input single-output (SISO) system. The input to a discrete-time model of this system is the sequence $\{u_k\}$ and the output from the model is the sequence $\{y_k\}$ as represented in Figure 10.6.

An nth order linear dynamic system can be modeled in the continuous-time case by the nth order linear ordinary differential equation:

$$\bar{a}_n \frac{d^n y}{dt^n} + \bar{a}_{n-1} \frac{d^{n-1} y}{dt^{n-1}} + \cdots + \bar{a}_0 y = \bar{b}_m \frac{d^m u}{dt^m} + \bar{b}_{m-1} \frac{d^{m-1} u}{dt^{m-1}} + \cdots + \bar{b}_0 u \tag{10.7}$$

TABLE 10.1

Some useful z-Transforms

Time Signal $x(t)$	z-Transform $X(z)$
Unit impulse $\delta(t)$	$1/T$
Unit pulse	1
Unit step $u(t)$	$\dfrac{z}{z-1}$
Unit ramp t	$\dfrac{Tz}{(z-1)^2}$
$\exp(-at)$	$\dfrac{z}{z-\exp(-aT)}$
$\sin \omega t$	$\dfrac{z\sin\omega T}{z^2 - 2z\cos\omega T + 1}$
$\cos \omega t$	$\dfrac{z(z-\cos\omega T)}{z^2 - 2z\cos\omega T + 1}$
$\exp(-at)\sin \omega t$	$\dfrac{z\exp(-aT)\sin\omega T}{z^2 - 2z\exp(-aT)\cos\omega T + \exp(-2aT)}$
$\exp(-at)\cos \omega t$	$\dfrac{z^2 - z\exp(-aT)\cos\omega T}{z^2 - 2z\exp(-aT)\cos\omega T + \exp(-2aT)}$
$t \exp(-at)$	$\dfrac{Tz\exp(-aT)}{\left[z-\exp(-aT)\right]^2}$

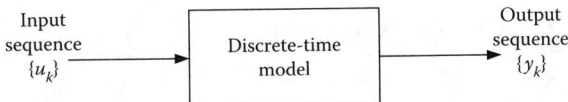

FIGURE 10.6
Block diagram representation of a single-input single-output discrete model.

in which:

$$u(t) = \text{Input}$$

$$y = \text{Output}$$

The corresponding discrete-time model may be expressed by the nth order linear difference equation:

$$a_0 y_k + a_1 y_{k-1} + \cdots + a_n y_{k-n} = b_0 u_k + b_1 u_{k-1} + \cdots + b_m u_{k-m} \qquad (10.8)$$

Note: The coefficients of the difference equation and the coefficients of the original differential equation are not identical. It is clear from Equation 10.8 that, provided the input sequence $\{u_k\}$ is known, the output sequence $\{y_k\}$ can be computed starting with the first n values of the sequence, which should be known. These initial n values are the *initial conditions(ICs)*, which are required to determine the complete solution of a difference equation. In general, the model parameters a_i and b_i in Equation 10.8 depend on the sampling period T.

10.3.3 Discrete Transfer Functions

For a time-invariant (i.e., constant-parameter) linear system, the coefficients $\bar{a}_i$ and $\bar{b}_i$ in Equation 10.7 are constants. Then, the system transfer function is given by

$$G(s) = \frac{\bar{b}_0 + \bar{b}_1 s + \cdots + \bar{b}_m s^m}{\bar{a}_0 + \bar{a}_1 s + \cdots + \bar{a}_n s^n} \tag{10.9}$$

This analog transfer function is obtained, in theory, by applying the Laplace transformation, with zero ICs, to Equation 10.7. In an analogous manner, by applying the z-transform to Equation 10.8, the corresponding discrete transfer function is obtained. To show this approach, multiply Equation 10.8 by z^{-k} and sum over $k(-\infty, \infty)$. This gives:

$$a_0 \sum y_k z^{-k} + a_1 \sum y_{k-1} z^{-k} + \cdots + a_n \sum y_{k-n} z^{-k} =$$
$$b_0 \sum u_k z^{-k} + b_1 \sum u_{k-1} z^{-k} + \cdots + b_m \sum u_{k-m} z^{-k}$$

which can be rewritten in the form:

$$a_0 \sum y_k z^{-k} + a_1 z^{-1} \sum y_{k-1} z^{-(k-1)} + \cdots + a_n z^{-n} \sum y_{k-n} z^{-(k-n)} =$$
$$b_0 \sum u_k z^{-k} + b_1 z^{-1} \sum u_{k-1} z^{-(k-1)} + \cdots + b_m z^{-m} \sum u_{k-m} z^{-(k-m)} \tag{i}$$

Since all summations on the left hand side of Equation (i) run through $-\infty$ to $+\infty$ each of them is equal to $Y(z)$ as evident from Equation 10.4. Similarly, each summation on the right hand side of Equation (i) is equal to $U(z)$. It follows that Equation (i) can be written as:

$$(a_0 + a_1 z^{-1} + \cdots + a_n z^{-n}) Y(z) = (b_0 + b_1 z^{-1} + \cdots + b_m z^{-m}) U(z)$$

Hence, the discrete transfer function is given by:

$$G(z) = \frac{Y(z)}{U(z)} = \frac{b_0 + b_1 z^{-1} + \cdots + b_m z^{-m}}{a_0 + a_1 z^{-1} + \cdots + a_n z^{-n}} \tag{10.10}$$

As mentioned earlier, the parameters a_i and bi depend on the sampling period T in general. Equations 10.8 and 10.10 represent *discrete models* and the Equations 10.7 and 10.9 represent continuous (analog) models of a dynamic system.

10.3.4 Time Delay

It should be intuitively clear from the preceding development that z^{-1} can be interpreted as an operator representing a time delay by one sampling period. Specifically:

$$\mathcal{Z}\{x_{k-1}\} = z^{-1}X(z) \tag{10.11}$$

This result can be verified directly from the definition of the z-transform—Equation 10.4. In general, for a delay of r sampling periods we have:

$$\mathcal{Z}\{x_{k-r}\} = z^{-r}X(z) \tag{10.12}$$

This should be compared with the property of Laplace transformation that allows us to interpret the Laplace variable s as the time-derivative operator.

A time delay by T in the continuous-time case can be represented by the transfer function exp $(-Ts)$, as can be verified using the definition of Laplace transform (see Appendix A). This establishes an equivalence between the z-transform and the Laplace transform, through the relation: $z^{-1} = \exp(Ts)$

Or:

$$z = \exp(Ts) \tag{10.13}$$

Caution should be exercised when using Equation 10.13. This equation provides a "mapping" between the s-domain and the z-domain. Specifically, suppose that we have the two functions $G_1(s)$ and $G_2(z)$. As s varies in some manner (say, along some contour) on the s-plane, there is a corresponding variation of $G_1(s)$ in the $G_1(s)$-plane. Then, z varies on the z-plane according to the mapping Equation 10.13 and $G_2(z)$ varies on the $G_2(z)$-plane according to this variation of z. Note carefully that, nowhere did we imply that G_1 and G_2 are the same functions. It is clear that $G_2(z)$ is "not" obtained by simply substituting Equation 10.13 in to $G_1(s)$. Hence, one should not attempt to obtain the discrete transfer function corresponding to a continuous (analog) transfer function by substituting Equation 10.13 into the analog transfer function.

Now we will establish another important property of the z-transform. Consider a signal $x(t)$. Its Laplace transform is denoted by $X(s)$. If we sample $x(t)$ at period T, we have the data sequence $\{x_k\}$. The corresponding z-transform is denoted by $X(z)$. *Note*: Here $X(s)$ and $X(z)$ are convenient notations. We should keep in mind that they represent two entirely different functions. $X(z)$ is not obtained from $X(s)$ by substituting z for s.

If we delay $x(t)$ by an integer multiple of T, say rT, the resulting signal is $x(t-rT)$. Its Laplace transform is given by $\exp(-rTs)X(s)$.

Now if we sample this delayed signal $x(t-rT)$ at sampling period T, we get the delayed sequence $\{x_{k-r}\}$. From Equation 10.12 we note that the corresponding z-transform is $z^{-r}X(z)$. Thus we can state the following general result.

General result: Consider a signal $x(t)$ whose Laplace transform is $X(s)$. Consider also a signal $y(t)$ whose Laplace transform can be expressed as:

$$Y(s) = f(\exp(Ts)) \cdot X(s)$$

TABLE 10.2

Some useful Properties of the z-transform

Item	z-Transform Results
$x(t)$	$X(z)$
$a_1 x_1(t) + a_2 x_2(t)$	$a_1 X_1(z) + a_2 X_2(z)$
$x(t - rT)$	$z^{-r} X(z)$
$Y(s) = f(\exp(Ts))X(s)$	$Y(z) = f(z)X(z)$
Final value theorem	$x_{ss} = \lim\limits_{z \to 1}(z - 1)X(z)$

in which $f(\ )$ is a polynomial function of $\exp(Ts)$. Then the z-transform of the sequence $\{y_k\}$, obtained by sampling $y(t)$ at period T, is given by:

$$Y(z) = f(z) \cdot X(z) \tag{10.14}$$

in which $X(z)$ is the z-transform of the sequence $\{x_k\}$, which is obtained by sampling $x(t)$ at period T. This result and several other properties of the z-transform are given in Table 10.2.

10.3.5 s–z Mapping

Equation 10.13 represents a mapping between the complex z-plane and the complex s-plane. This is one of the most important relationships in z-transform analysis. Let us further discuss the nature of this mapping.

Consider the case of complex poles:

$$s = -\zeta \omega_n \pm j\omega_d \tag{10.15}$$

These poles correspond to, for instance, a simple oscillator with undamped natural frequency ω_n, damped natural frequency ω_d, and damping ratio ζ. An s-plane representation of these complex conjugate poles is given in Figure 10.7. By substituting Equation 10.15 into Equation 10.13 we get the corresponding pole locations on the z-plane:

$$z = \exp(T(-\zeta \omega_n \pm j\omega_d)) \tag{10.16}$$

The magnitude of these two poles (on the z-plane) is:

$$|z| = \exp(-T\zeta \omega_n) \tag{10.17}$$

and the phase angle is:

$$\angle z = \pm T\omega_d \tag{10.18}$$

Let us discuss several special mappings given by these (magnitude and phase) relationships.

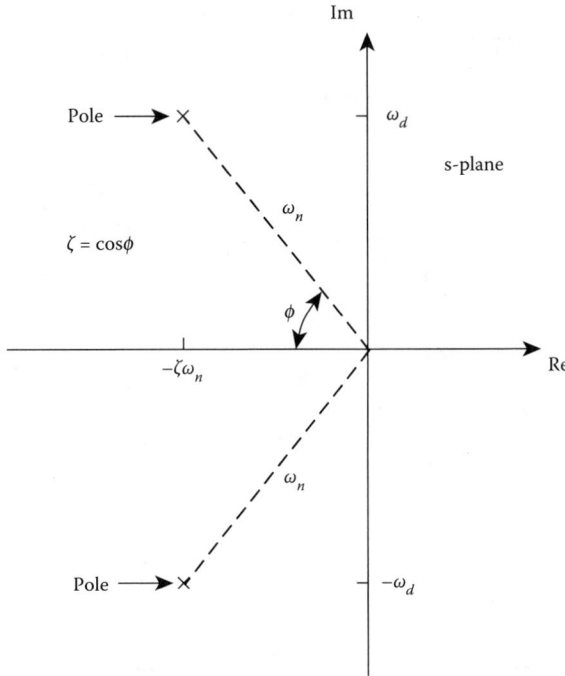

FIGURE 10.7
Representation of complex-conjugate poles on the s-plane.

1. **Constant $\zeta\omega_n$ Lines:** From Equation 10.17 it follows that when $\zeta\omega_n$ is a constant, $|z|$ is also a constant. Hence, constant $\zeta\omega_n$ lines on the s-plane (i.e., lines parallel to the imaginary axis) map onto circles centered at the origin on the z-plane, as shown in Figure 10.8.

   Note in particular that when $\zeta\omega_n=0$ (line A), we have $|z|=1$—a *unit circle*. The left hand side of the s-plane corresponds to the "inside" of the unit circle and the right hand side of the s-plane corresponds to the "outside" of the unit circle on the z-plane.

2. **Constant ω_d Lines:** It should be clear from Equation 10.18 that constant ω_d lines on the s-plane (i.e., lines parallel to the real axis) map onto constant phase angle lines on the z-plane. This is shown in Figure 10.9. Each line on the s-plane can be divided into a part on the left hand plane and a part on the right hand plane (RHP). Correspondingly, on the z-plane the line is divided into a part within the unit circle and a part outside the unit circle.

   The fact that a mapping from the s-plane to the z-plane is a *many-to-one mapping* should be clear from Equation 10.18. Note in particular that all the lines given by:

$$T\omega_d=2\pi r+c \tag{10.19}$$

on the s-plane, for integer r and constant c, are mapped onto the same line on the z-plane. This is because any integer change of r corresponds to a phase change by 2π on the z-plane, which returns the line to its original location.

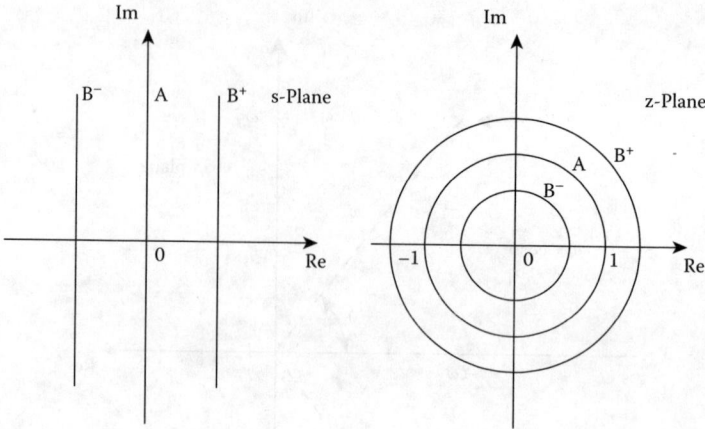

FIGURE 10.8
Constant $\zeta\omega_n$ lines.

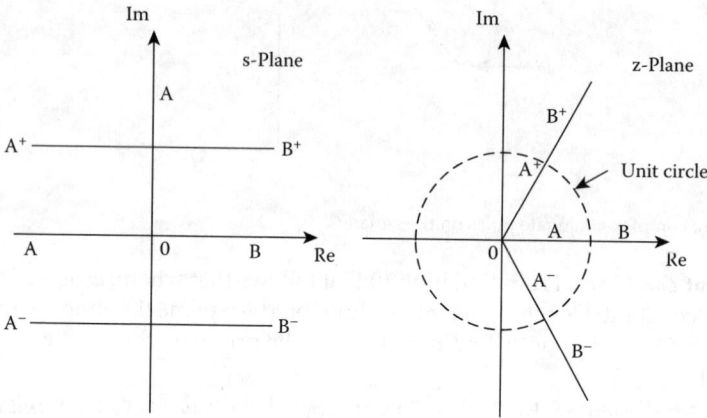

FIGURE 10.9
Constant ω_d lines.

3. **Constant ζ Lines:** It can be shown that constant ζ lines on the s-plane (i.e., straight lines through the origin) map onto spirals on the z-plane. This situation is sketched in Figure 10.10.

10.3.6 Stability of Discrete Models

Under the heading of constant $\zeta\omega_n$ lines, in the previous section, we noted that the left-half s-plane is mapped onto the inside of the unit circle on z-plane, as shown in Figure 10.11. It follows that stable poles in a discrete transfer function are those within the unit circle on the z-plane.

10.3.7 Discrete Final Value Theorem (FVT)

In the continuous-time case we have the familiar FVT:

$$x_{ss} = \lim_{s \to 0} sX(s) \tag{10.20}$$

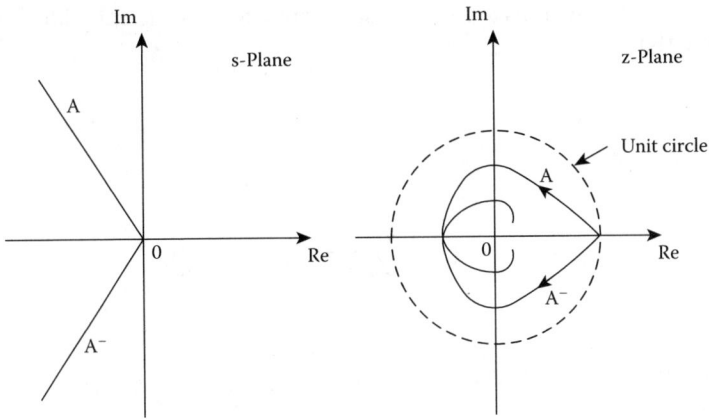

FIGURE 10.10
Constant ζ lines are spirals on the z-plane.

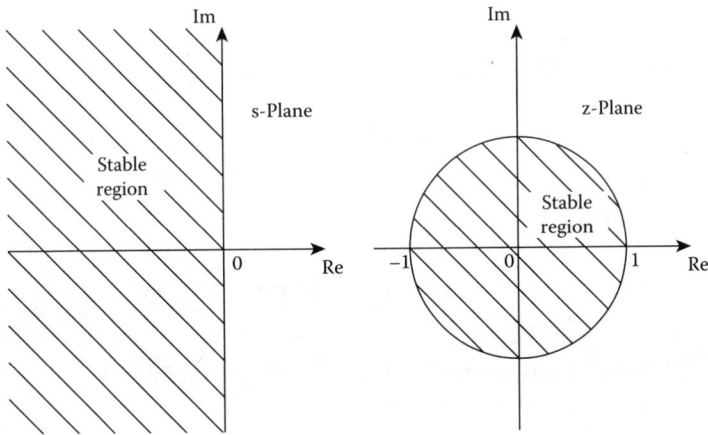

FIGURE 10.11
Stability region for discrete-time models.

in which x_{ss} is the steady-state value of the signal $x(t)$;

$$x_{ss} = \lim_{t \to \infty} x(t) \qquad (10.21)$$

and $X(s)$ is the Laplace transform of $x(t)$. It is assumed that a steady-state value exists for the considered signal.

To establish the discrete-time counterpart of the FVT, let us return to the definition of the z-transform—Equation 10.4. Assume that the steady-state value given by:

$$x_{ss} = \lim_{k \to \infty} x_k \qquad (10.22)$$

exists for the discrete-time signal (sequence) $\{x_k\}$. Now consider the product:

$$(z-1)X(z) = (z-1)\sum x_k z^{-k}$$

If the sequence $\{x_k\}$ converges to a steady-state value for a sufficiently large value of $k=N$, we can assume that:

$$x_N \approx x_{N+r} \quad \text{for all} \quad r > 0$$

Hence, we can write:

$$(z-1)X(z) \approx (z-1)\sum_{-\infty}^{N-1} x_k z^{-k} + (z-1)x_N \sum_{N}^{\infty} z^{-k}$$

Since the first summation on the right hand side of this relation is finite, its product with $(z-1)$ will vanish as $z \to 1$. The second summation on the right hand side can be written as:

$$\sum_{N}^{\infty} z^{-k} = z^{-N} \sum_{0}^{\infty} z^{-k} = \frac{z^{-N}}{1-z^{-1}} = \frac{z^{-N+1}}{z-1}$$

Its product with $(z-1)$ will approach unity as $z \to 1$. Hence:

$$\lim_{z \to 1}(z-1)X(z) \approx x_N$$

Exact equality is obtained as $N \to \infty$:

$$x_{ss} = \lim_{z \to 1}(z-1)X(z) \tag{10.23}$$

Equation 10.23 is the discrete-time FVT. This is listed in Table 10.2, along with other important properties of the z-transform.

10.3.8 Pulse Response Function

It is well known that if $g(t)$ is the impulse response function (i.e., response to a unit impulse input $\delta(t)$ of a system, then its Laplace transform $G(s)$ is the transfer function of the system. We write:

$$\mathcal{L}g(t) = G(s) \tag{10.24}$$

The response $y(t)$ to a general input $u(t)$ is given by the *convolution integral* (see Chapter 6 and Appendix A):

$$y(t) = \int g(t-\tau)u(\tau)d\tau = \int g(\tau)u(t-\tau)d\tau \tag{10.25}$$

The limits of integration may be chosen depending on the nonzero regions of the two functions $g(t)$ and $u(t)$.

Now consider the unit pulse input (given at $k=0$), which is defined by the sequence:

$$\{\ldots,0,0,\ldots,0,1,0,\ldots,0,0,\ldots\}$$

Note: The only nonzero sample value in this sequence is the "1" given at $k=0$.

The z-transform of this sequence is unity, as clear from Equation 10.4. This is given as the first entry in Table 10.1. From Equation 10.10 it follows that the z-transform of the pulse response sequence $\{g_k\}$ is the discrete transfer function $G(z)$:

$$\mathcal{Z}\{g_k\} = G(z) \tag{10.26}$$

The *discrete convolution* given below may be used to obtain the response to a general input, once the pulse response is known:

$$y_k = \sum_r g_{k-r} u_r = \sum_r g_r u_{k-r} \tag{10.27}$$

Equation 10.27 may be verified by direct substitution of Equation 10.4 into Equation 10.10.

10.3.8.1 Unit Pulse and Unit Impulse

The unit pulse is a pulse of unity height (magnitude) applied at $t=0$. Pulse width is the sampling period T. Note that the area of this pulse is T and not unity. Since the sample value of the unit pulse is 1 for the first sample and zero thereafter, as shown earlier, the z-transform of the unit pulse is 1.

A distinction between the unit pulse and the unit impulse should be recognized. The unit impulse $\delta(t)$ has an infinite height at $t=0$ and its area is unity. To sample such a signal, in practice, a very small sampling period T has to be used. Then, in the discrete approximation, the unit impulse is assumed to extend over the entire sample period of the first sample. The magnitude of the first sample has to be $1/T$ so that the area under the signal is unity, and this is inconsistent with the definition of the unit impulse. It follows that the z-transform of the unit impulse signal (whose Laplace transform is s) is given by $1/T$. These two important observations are listed as the first two entries of Table 10.1.

10.4 Digital Compensation

In Chapter 9, we noticed that analog compensator design could be interpreted as the development of a transfer function $G(s)$ that would modify the control signal so as to generate a desired system response. Signal modification by an analog transfer function is schematically shown in Figure 10.12a. Let us consider the possibility of using a digital device to accomplish the same task. This is termed *digital compensation*.

In a software-based digital compensator, a digital computer reads a sequence of data $\{u_k\}$ as obtained by sampling the true continuous-time signal $u(t)$, and produces a sequence of output data $\{y_k^*\}$ according to the *compensation algorithm*, which is stored inside the

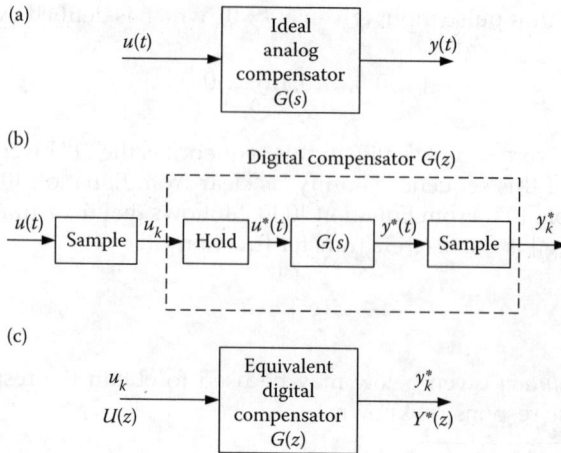

FIGURE 10.12
Digital compensation (a) An ideal analog compensator. (b) Schematic representation of digital compensation. (c) Equivalent digital compensator.

computer. The compensation algorithm may be expressed as an appropriate difference equation. Suppose that the discrete transfer function corresponding to this difference equation is $G(z)$. This process is schematically shown in Figure 10.12b or equivalently, in Figure 10.12c. Our objective in digital compensation is to establish a $G(z)$ that corresponds to an ideal analog compensator $G(s)$ such that the error between $\{y_k\}$ and $\{y_k^*\}$ is sufficiently small.

Note: $\{y_k\}$ is obtained by sampling the ideal (analog) compensator output $y(t)$ according to $y_k = y(k \cdot T)$.

It should be intuitively clear that if $T=0$ (i.e., infinite sampling rate) the output sequence $\{y_k^*\}$ would be identical to $\{y_k\}$, in theory. We know, however, that due to practical limitations it is impossible to achieve such a realization. The error $(y_k^* - y_k)$ depends primarily on the sampling period T and the holding method used for each input sample during this period.

The continuous and discrete equivalence is achieved by modeling the discrete compensation process as follows:

First sample the input signal $u(t)$ at sampling period T and then hold (or extrapolate) the sampled data to generate an equivalent continuous signal $u^*(t)$. Next, pass this through the ideal compensator $G(s)$. The resulting output is $y^*(t)$, which when sampled at period T, produces the sampled data sequence $\{y_k^*\}$. Clearly, in general $y^*(t)$ is different from the ideal $y(t)$, due to the error between $u(t)$ and $u^*(t)$ as a result of the "sample and hold" operation (with nonzero T). Hence, the error between the digital compensator and the analog compensator is caused essentially by this discrepancy between $u(t)$ and $u^*(t)$.

10.4.1 Hold Operation

The purpose of a hold operation is to extrapolate a data sequence (sampled data points) to obtain a piecewise continuous signal. The type of hold is determined by the extrapolation scheme that is used. If the data value is held constant until the next data value has arrived (i.e., extrapolation by a zeroth order polynomial) we have a *zero-order hold*. Only the current

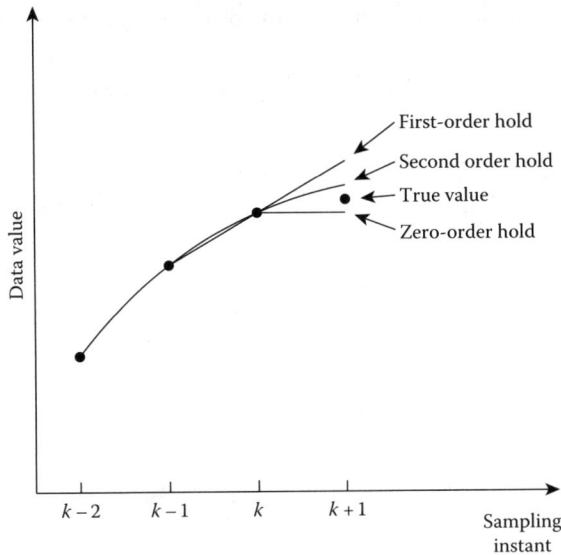

FIGURE 10.13
Several types of data hold.

data value is used in this extrapolation. This is typically the hold method used in ADC, since a constant data value is needed during the ADC process. If the current data value and the previous data value are used to extrapolate to the next data value using a straight line (i.e., extrapolation by a first-order polynomial), it is a *first-order hold*. If the current data sample and the two previous data samples are used to extrapolate to the next data sample by a quadratic curve (i.e., extrapolation by a second order polynomial), we have a *second order hold*, and so on. These various types of data hold are illustrated in Figure 10.13.

10.4.2 Discrete Compensator

We now return to the schematic diagram in Figure 10.12b. In the present analysis we will employ a zero-order hold, as this is what is commonly used in ADC. This may not be accurate enough, however, in some applications unless sufficiently small T is employed.

The zero-order hold converts a sampled data sequence $\{u_k\}$ into a continuous series of pulses. Consider, in general, the pulse corresponding to the data sample u_k. The Laplace transform of a pulse of unit height and width T that is originating at time $t = \tau$, is given by subtracting a unit step signal at $t = \tau + T$ from a unit step signal at $t = \tau$:

$$\Delta_\tau(s) = \mathcal{L}u(t - \tau) - \mathcal{L}u(t - \tau - T)$$

$$= \frac{1}{s}\exp(-\tau s) - \frac{1}{s}\exp(-(\tau + T)s)$$

Or:

$$\Delta_\tau(s) = \frac{1}{s}\exp(-\tau s)[1 - \exp(-Ts)] \tag{10.28}$$

Hence, the Laplace transform of a pulse of magnitude u_k given at time $t=kT$ is:

$$u_k \Delta_k(s) = \frac{u_k}{s} \exp(-kTs)[1 - \exp(-Ts)] \tag{10.29}$$

Note: This result is obtained by using $\tau=kT$ in Equation 10.28.

It follows that, in Figure 10.12b, the Laplace transform of the sampled and extrapolated (S/H) signal $u^*(t)$ is given by:

$$U^*(s) = \sum_k \frac{u_k}{s} \exp(-kTs)[1 - \exp(-Ts)]$$

Or:

$$U^*(s) = \frac{1}{s}[1 - \exp(-Ts)] \sum_k u_k \exp(-kTs) \tag{10.30}$$

The corresponding output signal $y^*(t)$ has the Laplace transform:

$$Y^*(s) = \{[1 - \exp(-Ts)] \sum u_k \exp(-kTs)\} \frac{G(s)}{s} \tag{10.31}$$

At this stage we use Equation 10.14 to obtain $Y^*(z)$:

$$Y^*(z) = \{[1 - z^{-1}] \sum u_k z^{-k}\} \mathcal{Z}\left[\frac{G(s)}{s}\right]$$

in which $\mathcal{Z}[\ ]$ denotes the z-transform of the sequence generated by a continuous signal whose Laplace transform is $[\ ]$.

Now using the definition of $U(z)$—Equation 10.4, we get:

$$Y^*(z) = (1 - z^{-1})U(z)\mathcal{Z}\left[\frac{G(s)}{s}\right] \tag{10.32}$$

It follows that the discrete transfer function $G(z)$ that is equivalent to the analog transfer function $G(s)$ is given by (see Figure 10.12c):

$$G(z) = (1 - z^{-1})\mathcal{Z}\left[\frac{G(s)}{s}\right] \tag{10.33}$$

Note: For the sake of emphasis, let us repeat what we have mentioned earlier. $G(z)$ is "not" obtained by replacing s in $G(s)$ according to the mapping equation (Equation 10.13). Furthermore, $G(z)$ is "not" obtained by replacing s by z in $G(s)$.

Equation 10.33 provides an important result, which may be directly employed to design digital compensators. The main steps of this procedure are given below.

Step 1: Design an analog compensator $G(s)$ to meet a given set of design specifications.

Step 2: Using Equation 10.33 obtain the discrete transfer function $G(z)$ of the corresponding digital compensator.

Step 3: Obtain the difference equation corresponding to $G(z)$.

The resulting difference equation can be programmed into a digital computer, forming the digital compensator.

Example 10.5

Consider the lead compensator: $G_c(s) = \left[\dfrac{0.051s + 1}{0.008s + 1} \right]$

as obtained in Chapter 9 for a motor speed control design problem. This is of the form:

$$G_c(s) = \left[\frac{bs + 1}{as + 1} \right]$$

Derive a digital compensator corresponding to this analog compensator.

Solution

To use Equation 10.33 we first determine the partial fractions:

$$\frac{(bs + 1)}{s(as + 1)} = \frac{A}{s} + \frac{B}{(as + 1)}$$

This gives the constants A and B. The time signals corresponding to these partial fractions are known from Laplace transform tables (see Appendix A). Specifically:

$$\mathcal{L}^{-1} \frac{1}{s} = \text{unit step function}$$

$$\mathcal{L}^{-1}\left[\frac{1}{s + 1/a} \right] = \exp(-t/a)$$

The corresponding discrete sequences have the z-transforms (see Table 10.1):

$$\frac{z}{z - 1} \quad \text{and} \quad \frac{z}{z - \exp(-T/a)}$$

Hence, we have from Equation 10.33:

$$G_c(z) = (1 - z^{-1})\left[\frac{Az}{(z - 1)} + \frac{B}{a} \frac{z}{(z - \exp(-T/a))} \right] = A + \frac{B}{a} \frac{(z - 1)}{(z - \exp(-T/a))}$$

This may be written as: $G_c(z) = K\dfrac{(z - \beta)}{(z - \alpha)} = K\dfrac{(1 - \beta z^{-1})}{(1 - \alpha z^{-1})}$

In view of Equations 10.8 and 10.10 we can write the corresponding difference equation as:

$$y_k - \alpha y_{k-1} = K(u_k - \beta u_{k-1})$$

This is the difference equation that should be programmed into the computer, for digital compensation.

Example 10.6

Using Equation 10.33 develop a discrete-time integrator and a discrete-time differentiator.
  Note: Since a zero-order hold is assumed, the results may not be very accurate.

Solution

For the continuous integrator we have: $G(s) = \dfrac{1}{s}$

To use Equation 10.33 we have to determine: $Z\left[\dfrac{G(s)}{s}\right] = Z\left[\dfrac{1}{s^2}\right]$

Note: $\mathcal{L}^{-1}\dfrac{1}{s^2} = t$

Hence, from Table 10.1: $Z\left[\dfrac{1}{s^2}\right] = \dfrac{Tz}{(z-1)^2}$

Then, from Equation 10.33, we get the discrete transfer function for the integrator as:

$$G(z) = (1 - z^{-1})\frac{Tz}{(z-1)^2} = \frac{T}{(z-1)} = \frac{Tz^{-1}}{(1-z^{-1})}$$

In view of Equations 10.8 and 10.10, we have the corresponding difference equation:

$$y_k - y_{k-1} = Tu_{k-1}$$

Or: $y_k = y_{k-1} + Tu_{k-1}$

This is the familiar *forward rectangular rule* of integration. This difference equation can be programmed into a control computer for use as a simple integration algorithm.

  Next, since the continuous differentiator is given by: $G(s) = s$ we have from Equation 10.33, the corresponding z-transform representation as:

$$G(z) = (1 - z^{-1})Z_1\left(\frac{s}{s}\right) = (1 - z^{-1})Z_1(1).$$

We know that 1 is the Laplace transform of the unit impulse $\delta(t)$. Hence, from Table 10.1, its z-transform is given by $1/T$. Accordingly, the discrete differentiator is given by:

$$G(z) = (1 - z^{-1})\frac{1}{T}$$

If the input to the differentiator is $u(t)$ and the differentiated output is $y(t)$, the difference equation of this discrete differentiator is given by:

$$y_k = \frac{1}{T}(u_k - u_{k-1})$$

This is the familiar *backward difference rule* for differentiation. The results of the present example may be used to develop an algorithm for a *digital PID controller*.

10.4.3 Direct Synthesis of Digital Compensators

The method of digital compensation (and control) as described in the previous sections is an indirect method in the sense that first an analog compensator or controller is developed and then it is approximated by a discrete transfer function using the z-transform method. Finally the corresponding difference equation is programmed into a digital device. An alternative method (a direct synthesis method) starts with a discrete transfer function $\tilde{G}(z)$ of a closed-loop system that responds with a desired response $\{y_k\}$ when a known input $\{u_k\}$ is applied, as indicated in Figure 10.14a. For example, $\tilde{G}(z)$ may be the discrete transfer function of a simple oscillator with specified values for damping ratio, natural frequency, and dc gain. Now, since the discrete transfer function $G_p(z)$ of the process (plant) to be controlled is assumed to be known, it is a straightforward algebraic exercise to determine a compensator transfer function $G_c(z)$ that will produce $\tilde{G}(z)$ for a given feedback structure. For example, if the unity feedback control structure, as shown in Figure 10.14b, is used we notice that:

$$\tilde{G}(z) = \frac{G_c(z)G_p(z)}{[1+G_c(z)G_p(z)]} \tag{10.34}$$

Now by straightforward algebraic manipulation, we get an expression for the desired compensator/controller:

$$G_c(z) = \frac{\tilde{G}(z)}{G_p(z)[1-\tilde{G}(z)]} \tag{10.35}$$

Since both $\tilde{G}(z)$ and $G_p(z)$ are known, we can directly determine $G_c(z)$.

10.4.4 Causality Requirement

Note that the discrete transfer function $G_c(z)$ that is obtained by direct synthesis, as described above, may not be physically realizable, and even when physically realizable it may not be well behaved. Hence, direct programming of the difference equation corresponding to this synthesized $G_c(z)$ into the control computer may not always produce the desired response. To illustrate this, note that when a discrete transfer function $G_c(z)$

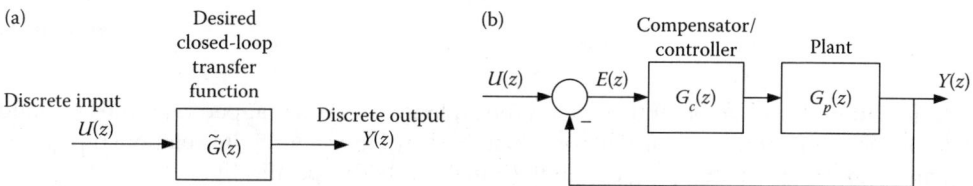

FIGURE 10.14
A direct synthesis method for digital compensators and controllers. (a) Desired system. (b) Unity feedback control structure.

is expanded as a series in powers of z^{-1}, we cannot have negative powers (i.e., we cannot have positive powers of z in the series expansion of $G_c(z)$). Otherwise, the corresponding difference equation will require future input values to determine the present output value, thereby violating the *causality* requirement for a "realizable" dynamic system. Hence, it is clear from Equation 10.35 that the specification $\tilde{G}(z)$ for a direct design will be limited by the nature of $G_p(z)$. In particular, it can be shown that for $G_c(z)$ to be realizable (i.e., for not to violate the causality requirement) it is required that the lowest power of z^{-1} in the series expansion of $\tilde{G}(z)$ be greater than or equal to the lowest power of z^{-1} in the series expansion of $G_p(z)$.

It should be understood that, in the discrete transfer function block diagrams shown in Figure 10.14a and b, the signal paths carry discrete sequences such as $\{u_k\}$ and $\{y_k\}$, and not continuous signals. In particular, in Figure 10.14b, the error (correction) signal is in fact $\{e_k\}$ and is given by:

$$e_k = u_k - y_k \tag{10.36}$$

and hence:

$$E(z) = U(z) - Y(z) \tag{10.37}$$

The closed-loop equation (Equation 10.34) originates from Equation 10.37. If some of the signal paths were analog, Equation 10.34 would not hold exactly. For example, suppose that $G_c(z)$ and $G_p(z)$ correspond to the continuous transfer functions $G_c(s)$ and $G_p(s)$, respectively. But, $\tilde{G}(z)$ will not be exactly equal to the discrete transfer function corresponding to the analog closed-loop transfer function:

$$\frac{G_c(s)G_p(s)}{\left[1 + G_c(s)G_p(s)\right]}$$

10.4.5 Stability Analysis Using Bilinear Transformation

Nyquist stability criterion and associated concepts of gain margin and phase margin cannot be directly extended to discrete transfer functions. The reason for this is straightforward: the stability region on the s-plane is the left hand plane whereas the stability region on the z-plane is the unit circle area. To overcome this difficulty, another transformation that maps a unit circle on to the left hand plane is used. One such transformation is the *bilinear transformation* given by:

$$w = \frac{z-1}{z+1} \tag{10.38}$$

Note that in this case the unit circle on the z-plane ($|z| \le 1$) is mapped onto the left hand side of the w-plane ($\mathrm{Re}(w) \le 0$). This is illustrated in Figure 10.15. Its relationship to the s-plane is given by the mapping equation (Equation 10.13). Specifically:

$$w = \frac{\exp(Ts)-1}{\exp(Ts)+1} \tag{10.39}$$

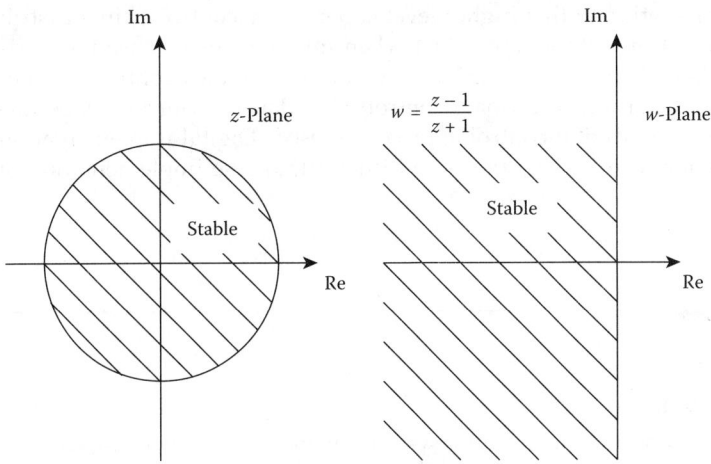

FIGURE 10.15
Bilinear transformation.

On the imaginary axis (frequency axis) of the s-plane we have $s=j\omega$. On the w-plane, this corresponds to the line:

$$w = \frac{\exp(Tj\omega)-1}{\exp(Tj\omega)+1} = j\tan\left(\frac{\omega T}{2}\right) \tag{10.40}$$

This is clearly the imaginary axis of the w-plane. It follows that the frequency axis of the s-plane corresponds to the imaginary axis of the w-plane. Consequently, we can define a frequency variable ω^* for the w-plane. In view of Equation 10.40, ω^* is related to the true frequency ω through:

$$\omega^* - \tan\left(\frac{\omega T}{2}\right) \tag{10.41}$$

This is a monotonic, but nonlinear, relationship. As a result of this monotonic relationship between the s-plane and the w-plane, we are able to use the same concepts of *relative stability* (gain margin, phase margin, etc.) for a discrete transfer function, when expressed in terms of the w variable (i.e., for a transfer function $G(w)$). Similarly, the well-known *Routh–Hurwitz stability criterion* (see Chapter 8) can be applied to the denominator polynomial of $G(w)$, in order to determine the stability of a discrete-time system.

10.4.6 Computer Implementation

Digital control is particularly preferred when the control algorithms are complex. The algorithm for a *three-point controller* (proportional integral derivative [PID] controller) for example, is quite simple and straightforward. Even though a PID controller can be easily implemented by analog means, or even by a *hardware digital controller*, one may decide to employ a simple microprocessor as the controller in each PID loop of a control system. The microprocessor approach has the advantages of low cost, small size, and flexibility. In particular, integration with a higher-level supervisory controller in a distributed-control environment will be rather convenient when microprocessor-based loop controllers are employed. Also integration of PID loops with more complex control schemes such as linearizing control (nonlinear feedback control) and adaptive control will be simplified when the software-based digital control approach is used. Digital implementation of lead and lag compensators can be slightly more difficult than the implementation of three-point controllers.

Problems

PROBLEM 10.1

Into what classification of control system components (actuators, signal modification devices, controllers, and measuring devices) would you put the following?

   a. Stepping motor
   b. Proportional-plus-integration circuit
   c. Power amplifier
   d. ADC
   e. DAC
   f. Optical increment encoder
   g. Process computer
   h. FFT analyzer
   i. Digital signal processor

PROBLEM 10.2

Compare analog control and direct digital control for motion control in high-speed applications of industrial manipulators. Give some advantages and disadvantages of each control method for this application.

PROBLEM 10.3

What is an antialiasing filter? In a particular application, the sensor signal is sampled at f_s Hz. Suggest a suitable cut-off frequency for an antialiasing filter to be used in this application.

PROBLEM 10.4

   a. Consider a multi degree-of-freedom robotic arm with flexible joints and links. The purpose of the manipulator is to accurately place a payload. Suppose that the second natural frequency (i.e., the natural frequency of the second flexible mode) of bending of the robot, in the plane of its motion, is more than four times the first natural frequency.
      Discuss pertinent issues of sensing and control (e.g., types and locations of the sensors, types of control, operating bandwidth, control bandwidth, sampling rate of sensing information) if the primary frequency of the payload motion is:
      (i)   one-tenth of the first natural frequency of the robot.
      (ii) very close to the first natural frequency of the robot.
      (iii) twice the first natural frequency of the robot.

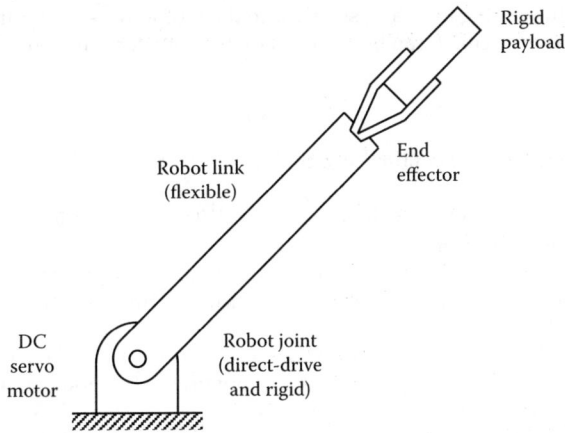

FIGURE P10.4
A single-link robotic manipulator.

    b. A single-link space robot is shown in Figure P10.4. The link is assumed to be uniform with length 10 m and mass 400 kg. The total mass of the end effector and the payload is also 400 kg. The robot link is assumed to be flexible while the other components are rigid. The modulus of rigidity of bending deflection of the link in the plane of robot motion is known to be $EI=8.25\times10^9$ N.m². The primary natural frequency of bending motion of a uniform cantilever beam with an end mass is given by: $\omega_1 = \lambda_1^2\sqrt{EI/m}$

where: m=mass per unit length
    λ_1=mode shape parameter for mode 1.
    For [beam mass/end mass]=1.0, it is known that $\lambda_1 l$=1.875 where l=beam length.
    Give a suitable operating bandwidth for the robot manipulation. Estimate a suitable sampling rate for response measurements, to be used in feedback control. What is the corresponding control bandwidth, assuming that the actuator and the signal conditioning hardware can accommodate this bandwidth?

PROBLEM 10.5

Suppose that the frequency range of interest in a particular signal is 0–200 Hz. Determine a suitable sampling rate (digitization speed) and the cut-off frequency for an appropriate antialiasing (low-pass) filter.

PROBLEM 10.6

A computer-based open-loop control system for a dc motor is shown by the block diagram in Figure P10.6. In response to an input command, the control computer generates digital values corresponding to a desired continuous input. These digital values are

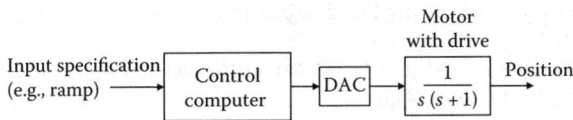

FIGURE P10.6
Open-loop digital control of a dc motor (unstable).

converted into the analog form and supplied to the motor drive circuit in real-time. The motor and its drive circuitry can be modeled by the transfer function:

$$G(s) = \frac{1}{s(s+1)}$$

Suppose that the conversion time T of the DAC is $T=1$ s.

a. Using the z-transform approach, obtain a difference equation that will approximate the motor response.
b. Assuming that the continuous input is a unit ramp, compute the first four discrete output values (at sampling period $T=1$ s) using the difference equations, with zero initial values.
c. For a unit ramp input, determine an expression for $Y(z)$—the z-transform of the discrete output values. By long division, obtain the first four discrete output values and compare them with the results in (b).
d. Suppose the input is a unit step. Using the difference equation, with zero ICs, compute the first few output samples. Show that the response behaves in an unstable manner. What is the main source of this instability (nature of the system or recursive algorithm)?

PROBLEM 10.7

Consider a continuous-time system given by the transfer function:

$$G(s) = \frac{1}{(s^2 + 1.2s + 1)}$$

a. Determine the undamped natural frequency and damping ratio of the system.
b. Write the input/output differential equation for this system.
c. Write an expression for the system response $y(t)$ for a unit step input. Plot this response.
d. From the plotted time response determine the % overshoot (P.O.) and the peak time (T_p) for the system.
e. Determine the discrete transfer function $G(z)$, which relates the discrete input and output data.
f. Write the difference equation corresponding to this discrete transfer function.
g. Using this difference equation, compute the time response of the system for a unit step input,
   (i) using the sampling period $T=1$ s
   (ii) using the sampling period $T=0.5$ s

Plot these two curves on the same paper (same scale) along with the exact response curve obtained in (c). Compare the three curves. Discuss any discrepancies. (Compare, in particular, P.O., and T_p for the three curves.)

PROBLEM 10.8

What are advantages of digital control over analog control? A lead compensator was designed for a control system using the classical analog approach. Its transfer function was found to be: $G_c(s) = (2s+1)/(s+1)$.

Assuming a sampling period of $T=1$, obtain a difference equation for digital implementation of this lead compensator.

PROBLEM 10.9

Consider a system whose transfer function is $G(s)$. The corresponding discrete transfer function $G(z)$ is obtained using:

$$G(z) = (1 - z^{-1})\mathcal{Z}\left[\frac{G(s)}{s}\right]$$

in which the operator $\mathcal{Z}$ denotes z-transformation. Note that the response of the system computed using the difference equation given by $G(z)$ is the same as the response of the original system $G(s)$, if the analog input was first sampled, then each sample is held using a zero-order hold circuit, then the resulting analog signal is applied to the original analog system $G(s)$ and, finally, the resulting response is sampled at the same sampling frequency as the input. Using this fact, explain why the discrete transfer function corresponding to $G_1(s)G_2(s)$ is not identical to the product $G_1(z)G_2(z)$ in which:

$G_1(z)$=discrete transfer function corresponding to $G_1(s)$
$G_2(z)$=discrete transfer function corresponding to $G_2(s)$.

If $G_1(s)$ and $G_2(s)$ are known, would you prefer the former process (i.e., convert the product $G_1(s)G_2(s)$) or the latter process (i.e., convert $G_1(s)$ and $G_2(s)$ separately and take the product) in order to obtain a difference equation for the product transfer function $G_1(s)G_2(s)$, for computer implementation?

PROBLEM 10.10

Schematic representation of signal processing associated with a vibration-test system is shown in Figure P10.10. An acceleration signal $x(t)$ of the test object is measured using an accelerometer/charge amplifier combination and sampled into the signal-processing computer at sampling period T. This discrete acceleration sequence $\{x_k\}$ is then integrated using parabolic integration, to form a velocity sequence $\{y_k\}$. Show that the parabolic integration algorithm is given by:

$$y_{k+1} = y_k + \frac{T}{12}\left[5x_{k+1} + 8x_k - x_{k-1}\right]$$

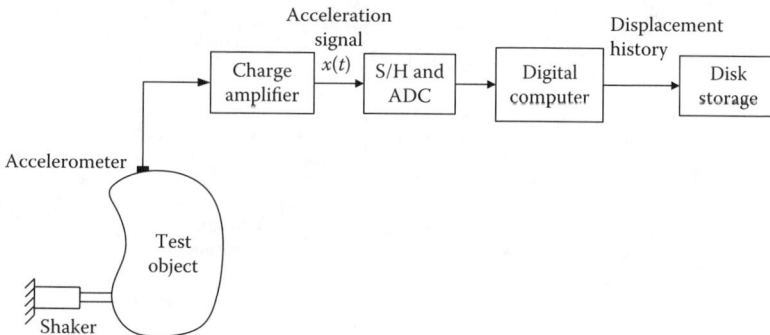

FIGURE P10.10
A vibration test system.

Note: In the parabolic integration algorithm, the discrete data are interpolated (every three points x_{k-1}, x_k, x_{k+1}) using a quadratic curve (a parabola).

What is the discrete transfer function $G(z)$ corresponding to this difference equation? Using this, derive an algorithm for double integration of $x(t)$, to obtain the displacement response. Show that the associated discrete transfer function is marginally stable. What is the practical implication of this observation? Suggest a way to improve this situation.

PROBLEM 10.11

Consider the analog feedback control system shown in Figure P10.11a. If we sample the error (correction) signal as shown in Figure P10.11b, this is equivalent to sampling both input signal and feedback signal. If the output signal is also sampled, then the signal paths can be completely represented by discrete sequences (only the holding operation is needed to convert the discrete sequences to sample and hold outputs). Hence, the discrete transfer function of the system shown in Figure P10.11c is identical to that of the system shown in Figure P10.11b. Note that $G(z)$ and $H(z)$ are, respectively, the discrete transfer functions corresponding to $G(s)$ and $H(s)$. What is the closed-loop discrete transfer function $\tilde{G}(z)$ corresponding to Figure P10.11b or c? Now suppose that the input and the output are sampled (and not the feedback signal). In this case we have the system shown in Figure P10.11d. What is the closed-loop discrete transfer function in this case? Verify that this is not identical to, but could be approximated by, the previous case.

Which discrete model (Figure P10.11c or Figure P10.11d) would be more accurate in the following two cases:

a. The forward path of the control loop has a digital controller, and a digital transducer (say, optical encoder) is used to measure the response signal for feedback.
b. The forward path of the control loop has a digital controller (computer) but the output sensor used to obtain the feedback signal is analog.

Is any one of the two discrete models shown in Figure P10.11 exactly equivalent to the case where the forward path and the feedback path are completely analog except that a digital transducer (with a zero-order hold) is used to measure the output signal for feedback?

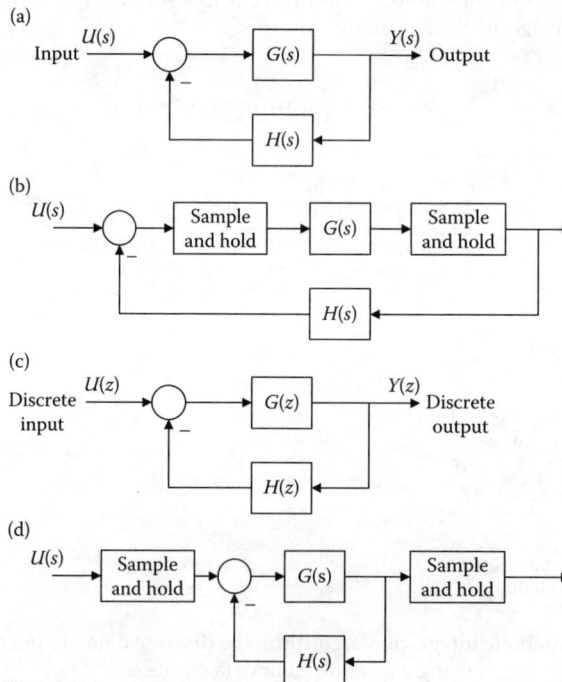

FIGURE P10.11
Discrete transfer function models of a closed-loop analog system. (a) Analog system. (b) Sampling the error and the output. (c) A discrete system equivalent to (b). (d) Sampling the input and the output.

PROBLEM 10.12

Consider the general control loop shown in Figure P10.12. The case of unity feedback (i.e., feedback transfer function $H=1$) is shown. Note that a loop with nonunity feedback $H(s)$ can be reduced to this equivalent form by moving $H(s)$ to the forward path of the loop and dividing the reference input by $H(s)$, since $U-HY=(U/H-Y)H$. Hence, the unity feedback case shown in Figure P10.12 can be considered the general case. The controller, with analog transfer function $G_c(s)$, converts the error (actually correction) signal $e(t)$ into the control signal $c(t)$. Consider the following three types of controllers, which are commonly used in industrial control systems:

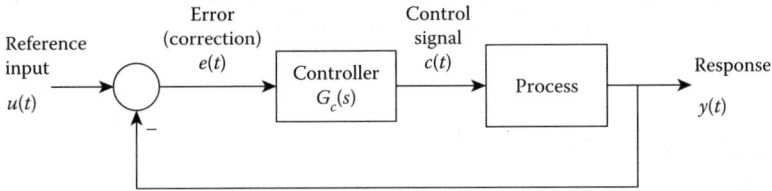

FIGURE P10.12
A general control loop.

a. Proportional plus integral (PI) control given by

$$G_c(s) = K_p\left(1 + \frac{1}{\tau_i s}\right)$$

b. Proportional plus derivative (PPD) control given by

$$G_c(s) = K_p(1 + \tau_d s)$$

c. Proportional plus integral plus derivative (PID) control given by

$$G_c(s) = K_p\left(1 + \frac{1}{\tau_i s} + \tau_d s\right)$$

By approximating the integration

$$i(t) = \int e(t)dt$$

by the forward rectangular rule:

$$i_k = i_{k-1} + Te_{k-1}$$

and the differentiation

$$d(t) = \frac{d}{dt}e(t)$$

by the backward difference rule:

$$d_k = \frac{1}{T}(e_k - e_{k-1})$$

obtain difference equations (digital control algorithms) for the three types of controllers PI, PPD, and PID. Note that the discrete control sequence $\{c_i\}$ can be computed from the discrete error (correction) sequence $\{e_i\}$ using these difference equations.

Give the discrete transfer functions (functions of z) corresponding to these three difference equations. Compare them with the functions obtained by converting the continuous transfer functions $G_c(s)$ into the corresponding discrete transfer functions $G_c(z)$ using the standard z-transform method.

PROBLEM 10.13

Consider a discrete transfer function $G_c(z)$. It may be expressed as a ratio of rational polynomials in z^{-1}. Suppose that the numerator polynomial is:

$$N(z^{-1}) = b_m z^{-m} + b_{m+1} z^{-(m+1)} + \cdots$$

and the denominator polynomial is:

$$D(z^{-1}) = a_n z^{-n} + a_{n+1} z^{-(n+1)} + \cdots$$

in which m and n are nonnegative integers. Show that for the causality requirement of $G_c(z)$ to be satisfied (i.e., for $G_c(z)$ to be physically realizable) we must have $m \geq n$.

Now consider the direct synthesis of digital controllers and compensators using the z-transform approach. Note that $G_p(z)$ is fixed by the given system and $\tilde{G}(z)$ is defined by the specified (required) performance of the closed-loop system. Of course, both $G_p(z)$ and $\tilde{G}(z)$ have to be realizable. Suppose that the series expansion (obtained, say, be long division) of $G_p(z)$ in powers of z^{-1} has n as the lowest power, and a similar series expansion of $\tilde{G}(z)$ has m as the lowest power. Show that for $G_c(z)$ to be physically realizable, we must have $m \geq n$.

PROBLEM 10.14

A *deadbeat controller* is a digital controller that provides a system response, which will settle down to the steady-state value in a finite (a very few) number of sample periods.

Note: The settling time of a system can be determined by the free response of the system to a unit pulse.

Consider a closed-loop system $\tilde{G}(z)$ that includes a deadbeat controller. Suppose that $\tilde{G}(z)$ is expanded as a series in powers of z^{-1}. Discuss characteristics of this series such that $\tilde{G}(z)$ represents a deadbeat control system.

11

Advanced Control

The emphasis of the present book has been on conventional control (also known as classical control) which is commonly used in academic curricula and engineering/industrial applications. It primarily deals with single-input–single-output (SISO) systems both in time domain and frequency domain. What are commonly identified as modern control techniques are time domain multivariable (multiinput–multioutput [MIMO]) techniques that use the state-space representation for the system. The present chapter will present some of these advanced control techniques, particularly in the categories of optimal control and modal control. In this context, linear quadratic regulator and pole-placement control will be studied as two particularly popular multivariable control techniques. Another control method that has been quite popular in engineering/industrial applications is fuzzy logic control, which is also presented in this chapter.

11.1 Modern Control

The class of multivariable (or MIMO) control that is identified as "modern control" depends primarily on the state-space representation the system to be controlled (see Chapter 2). The associated first order ordinary differential equations may be nonlinear (i.e., may contain nonlinear functions of the state variables), coupled (i.e., one equation may contain more than one state variable), and time-variant (i.e., may have the variable t explicitly in the equations, and the system parameters will vary with time). The so-called modern control techniques originated primarily in the United States and the former Soviet Union in the early 1950s. Hence the term "modern control" is a misnomer.

In the present chapter we will present in detail the following two "modern" control techniques:

a. Linear quadratic regulator (LQR)
b. Pole placement

In LQR, the objective is to minimize a cost function (maximize a performance index) and hence this technique falls under the general category of *optimal control*. In pole placement, the objective is to locate the poles (i.e., eigenvalues) of the system so that the modes (i.e., fundamental free natural responses) of the system behave in a desired manner (with respect to stability, speed of response, etc.) and hence this technique falls into the category of *modal control*. Both these techniques are state-space methods, and are specifically based on a linear representation (linear model) of the system (plant) that is controlled.

Several other modern control techniques will be outlined without presenting all the details. The chapter will be concluded with an introductory presentation of fuzzy logic control (FLC) which is a rule-based (or knowledge-based) control technique which falls into the class of *intelligent control*. FLC does not explicitly use a model of the system to be controlled (plant)—it is a model-free technique. It uses control knowledge that is expressed as a set of rules containing fuzzy terms such as "large" and "fast." Observations on the plant behavior (e.g., through response measurement) are matched with the rule base to generate control actions (which are the inferences of the decision-making system).

11.2 Time Response

The time response of a system describes how the system responds as a function of time (see Chapter 6). The frequency response describes how the system responds as a function of frequency (see Chapters 5, 6, and 8). Since the objective of control is to make the system behave in a desired manner, it is important to analyze how the system variables, the output variables in particular, vary with time. Both free response and forced response are important.

11.2.1 The Scalar Problem

Consider the first order system:

$$\dot{x} = a(t)x + b(t)u(t) \tag{11.1}$$

where x is the state (or response) and u is the input.
   Note: Time-varying parameters a and b.

11.2.1.1 Homogeneous Case (Input u=0)

$$\dot{x} = a(t)x \implies dx/x = a(t)dt$$

Integrate:

$$\ln\frac{x(t)}{x(t_o)} = \int_{t_o}^{t} a(\tau) \implies x(t) = x(t_o)e^{\int_{t_o}^{t} a(\tau)d\tau} \tag{11.2}$$

$$\text{If } a = \text{constant } x(t) = x(t_o)e^{a(t-t_o)} \tag{11.3}$$

   Note: The initial condition (IC) $x(t_o)$ must be specified for unique solution. Also $x(t)$ depends on the time difference $(t-t_o)$, not the absolute time, when a is constant.

11.2.1.2 Nonhomogeneous (Forced) Case

Multiply Equation 11.1 by $k(t)$:

$$k(t)\dot{x} - k(t)a(t)x = k(t)b(t)u(t) \tag{i}$$

Pick $k(t)$ such that:

$$\dot{k}(t) = -k(t)a(t) \tag{ii}$$

(to make the left hand side an exact differential). Then, the system equation (Equation 11.1) has a solution:

$$k(t) = k(t_o)e^{-\int_{t_o}^t a(\tau)d\tau} \tag{iii}$$

Hence, Equation (i) can be written as

$$\frac{d}{dt}[kx] = kb(t)u(t) \tag{iv}$$

Integrate:

$$k(t)x(t) = k(t_o)x(t_o) + \int_{t_o}^t k(\tau)b(\tau)u(\tau)d\tau$$

We get:

$$x(t) = \frac{k(t_o)}{k(t)}x(t_o) + \int_{t_o}^t \frac{k(\tau)}{k(t)}b(\tau)u(\tau)d\tau$$

$$= x(t_o)e^{\int_{t_o}^t a(\tau)d\tau} + \int_{t_o}^t e^{-\int_{t_o}^t a(\tau)d\tau}b(\tau)u(\tau)d\tau$$

Final result:

$$x(t) = x(t_o)e^{\int_{t_o}^t a(\tau)d\tau} + \int_{t_o}^t e^{-\int_{t_o}^t a(\tau)d\tau}b(\tau)u(\tau)d\tau \tag{11.4}$$

For constant a:

$$x(t) = e^{a(t-t_o)}x(t_o) + \int_{t_o}^t e^{a(t-\tau)}b(\tau)u(\tau)d\tau \tag{11.5}$$

Note: In general (see Chapter 6),
  zero-input response $\neq$ homogeneous solution (complementary solution)
  zero state (zero IC) response $\neq$ particular solution

(because the input affects the unknown coefficients in the homogeneous solution).

Example 11.1

Let, particular solution=zero state response+ $e^{a(t-t_o)}$
    This satisfies the system differential equation (Equation 11.1), with constant a. Then, homogeneous (complementary) solution= $e^{a(t-t_o)}[x(t_o)-1]$.

11.2.2 Time Response of a State-Space Model

The foregoing concepts of time response may be extended to the multivariable case of a state-space model.

11.2.2.1 Case of Constant System Matrix

Assume that A is a constant matrix.

$$\text{Homogeneous Case: } \dot{x} = Ax \tag{11.6}$$

$$\text{This satisfies: } x(t) = e^{A(t-t_o)}x(t_o) \tag{11.7}$$

$$\text{Nonhomogeneous (Forced) Case: } \dot{x} = Ax + Bu(t) \tag{11.8}$$

$$\text{Let } K(t) \text{ satisfy } \dot{K}(t) = -K(t)A \tag{i}$$

$$\text{For example } K(t) = e^{-A(t-t_o)} \tag{ii}$$

Multiply Equation 11.8 throughout by $K(t)$. Then, as for the scalar case:

$$\frac{d}{dt}[K(t)x] = K(t)B(t)u(t) \tag{iii}$$

Integrate Equation (iii):

$$x(t) = K^{-1}(t)K(t_o)x(t_o) + \int_{t_o}^{t} K^{-1}(t)K(\tau)u(\tau)d\tau \tag{iv}$$

With Equation (ii):

$$x(t) = e^{A(t-t_o)}x(t_o) + \int_{t_o}^{t} e^{A(t-\tau)}B(\tau)u(\tau)d\tau \tag{11.9}$$

11.2.2.2 Matrix Exponential

Note that the matrix exponential (which is called the *state-transition matrix*, because it changes the state vector over time) is needed for the response analysis of a constant-parameter state-space model. This matrix is given by:

$$e^{At} = I + At + \frac{1}{2!}A^2t^2 + ... \tag{11.10}$$

Some properties:

$$e^{-At} = I - At + \frac{1}{2!}A^2t^2 + \cdots$$

$$\frac{d}{dt}e^{At} = A\left[I + At + \cdots\right] = Ae^{At}$$

$$e^{At}\,e^{-At} = I$$

Hence

$$\left[e^{At}\right]^{-1} = e^{-At}$$

11.2.2.3 Methods of Computing e^{At}

Method 1 (Laplace Transform Method):
According to Equation 11.7: $\dot{x} = Ax \Rightarrow x(t) = e^{At}x(0)$
   Take Laplace Transform (see Appendix A):

$$sX(s) - x(0) = AX(s) \Rightarrow x(t) = L^{-1}(sI - A)^{-1}x(0)$$

Hence

$$e^{At} = L^{-1}(sI - A)^{-1} \tag{11.11}$$

Method 2 (Modal Transformation Method):
Determine the eigenvectors (see Appendix C) of A, and assemble them into the *modal matrix* M. Form the matrix J through the similarity transformation:

$$J = M^{-1}AM \tag{11.12}$$

At least in the case when A has distinct eigenvalues, it is known that J (the *Jordan matrix*) is diagonal, with eigenvalues as its diagonal elements. Then

$$e^{Jt} = \begin{bmatrix} e^{\lambda_1 t} & 0 & & 0 \\ 0 & e^{\lambda_2 t} & & 0 \\ & & & \\ 0 & 0 & 0 & e^{\lambda_n t} \end{bmatrix} \tag{11.13}$$

It is known that:

$$e^{At} = Me^{Jt}M^{-1} \tag{11.14}$$

Method 3 (Matrix Element Evaluation):
Consider the series expression of each matrix element in:

$$e^{At} = I + At + \frac{1}{2}A^2t^2 + \cdots$$

Then, see whether a closed-form expression can be written for each matrix element. Alternatively, for small t (e.g., for digital simulation over small time steps) truncate the series expression.

Method 4 (Use Cayley Hamilton Theorem):
See Appendix C. Determine coefficients $\alpha_0, \alpha_1, \ldots, \alpha_{n-1}$ by solving

$$e^{\lambda_1 t} = \alpha_0 + \alpha_1 \lambda_1 + \cdots + \alpha_{n-1} \lambda_1^{n-1}$$

$$\vdots$$

$$e^{\lambda_n t} = \alpha_0 + \alpha_1 \lambda_n + \cdots + \alpha_{n-1} \lambda_n^{n-1}$$

where λ_i are the eigenvalues of A. Then (see Appendix C),

$$e^{At} = \alpha_0 I + \alpha_1 A + \cdots + \alpha_{n-1} A^{n-1} \tag{11.15}$$

Example 11.2

Consider the system matrix:

$$A = \begin{bmatrix} 0 & 1 \\ 0 & -2 \end{bmatrix}$$

We will determine the matrix exponential of this matrix by each of the four methods that were outlined before.
Method 1:

$$sI - A = \begin{bmatrix} s & -1 \\ 0 & s+2 \end{bmatrix}$$

$$(sI - A)^{-1} = \begin{bmatrix} \dfrac{1}{s} & \dfrac{1}{s(s+2)} \\ 0 & \dfrac{1}{s+2} \end{bmatrix}$$

Take inverse Laplace of each term, using Laplace transform tables (see Table A.1). We get

$$e^{At} = L^{-1} \begin{bmatrix} \dfrac{1}{s} & \dfrac{1}{s(s+2)} \\ 0 & \dfrac{1}{s+2} \end{bmatrix} = \begin{bmatrix} 1 & \dfrac{1}{2}\left(1 - e^{-2t}\right) \\ 0 & e^{-2t} \end{bmatrix}$$

Method 2:
Determine eigenvalues of A: $\lambda_1 = 0$, $\lambda_2 = -2$
Determine eigenvectors by solving: $[A - \lambda I]\xi = 0$

We get

$$\text{For } \lambda_1: \xi_1 = \begin{bmatrix} 1 \\ 0 \end{bmatrix}$$

$$\text{For } \lambda_2: \begin{bmatrix} 2 & 1 \\ 0 & 0 \end{bmatrix}\begin{bmatrix} x \\ x \end{bmatrix} = \begin{bmatrix} 0 \\ 0 \end{bmatrix} \Rightarrow \xi_2 = \begin{bmatrix} 1 \\ -2 \end{bmatrix}$$

Then, form the modal matrix **M** and determine its inverse:

$$M = \begin{bmatrix} 1 & 1 \\ 0 & -2 \end{bmatrix}, M^{-1} = \begin{bmatrix} 1 & \dfrac{1}{2} \\ 0 & -\dfrac{1}{2} \end{bmatrix}.$$

Form the exponential of the Jordan matrix **J**.

$$e^{Jt} = \begin{bmatrix} e^0 & 0 \\ 0 & e^{-2t} \end{bmatrix}$$

Finally, use Equation 11.13 to determine the matrix exponential:

$$e^{At} = \begin{bmatrix} 1 & 1 \\ 0 & -2 \end{bmatrix}\begin{bmatrix} 1 & 0 \\ 0 & e^{-2t} \end{bmatrix}\begin{bmatrix} 1 & \dfrac{1}{2} \\ 0 & -\dfrac{1}{2} \end{bmatrix}$$

$$= \begin{bmatrix} 1 & \dfrac{1}{2}(1-e^{-2t}) \\ 0 & e^{-2t} \end{bmatrix}$$

Method 3:
Write the series expansion: $e^{At} = I + tA + t^2\dfrac{A^2}{2} + \cdots$

$$= \begin{bmatrix} 1 & 0 \\ 0 & 1 \end{bmatrix} + t\begin{bmatrix} 0 & 1 \\ 0 & -2 \end{bmatrix} + \dfrac{t^2}{2!}\begin{bmatrix} 0 & 1 \\ 0 & -2 \end{bmatrix}^2 + \cdots$$

$$= \begin{bmatrix} 1 & \dfrac{1}{2} - \dfrac{1}{2}\left(1 - 2t + \dfrac{(2t)^2}{2!} - \dfrac{(2t)^3}{3!} + \cdots\right) \\ 0 & 1 - 2t + \dfrac{(2t)^2}{2!} - \dfrac{(2t)^3}{2!} - \dfrac{(2t)^3}{3!} + \cdots \end{bmatrix}$$

$$= \begin{bmatrix} 1 & \dfrac{1}{2}(1-e^{-2t}) \\ 0 & e^{-2t} \end{bmatrix}$$

Method 4:

Form the two coefficient equations, using the eigenvalues,

$$\left.\begin{array}{l} \alpha_0 + \alpha_1\lambda_1 = e^{\lambda_1 t} \\ \alpha_0 + \alpha_1\lambda_2 = e^{\lambda_2 t} \end{array}\right] \quad \text{for} \quad \begin{array}{l} \lambda_1 = 0 \\ \lambda_2 = -2. \end{array}$$

and solve for the coefficients:

$$\Rightarrow \alpha_0 = 1, \alpha_1 = \frac{1}{2}\left(1 - e^{-2t}\right)$$

Then,

$$e^{At} = \alpha_0 I + \alpha_1 A = I + \frac{1}{2}\left(1 - e^{-2t}\right)A$$

$$= \begin{bmatrix} 1 & \frac{1}{2}\left(1 - e^{-2t}\right) \\ 0 & e^{-2t} \end{bmatrix}$$

Example 11.3

Find the response to initial state $x(0)$ of the system $\dot{x} = \begin{bmatrix} 0 & 1 \\ 0 & -2 \end{bmatrix} x$

Here we use the result (Equation 11.7) and the matrix exponential determined in the previous example. We have

$$x(t) = e^{At}x(0) = \begin{bmatrix} 1 & \frac{1}{2}\left(1 - e^{-2t}\right) \\ 0 & e^{-2t} \end{bmatrix} x(0)$$

$$= \begin{bmatrix} x_1(0) + \frac{1}{2}x_2(0) - \frac{1}{2}e^{-2t}x_2(0) \\ e^{-2t}x_2(0) \end{bmatrix}$$

Example 11.4

For the circuit shown in Figure 11.1a, the input voltage $e_i(t)$ is given in Figure 11.1b. The output is voltage e_0.

  i. Using current through inductor 2 H and voltage across the capacitor 1/2 F as the state variables, obtain a complete state model (including output equation) for the system.
 ii. Determine the value of the state vector at $t=0^+$.
iii. Find the eigenvalues and the corresponding eigenvectors of the system.
 iv. Using Equation (iii), obtain an expression for the state-transition matrix e^{At}.
  v. Obtain the response (output) $y(t)$ for the given input.

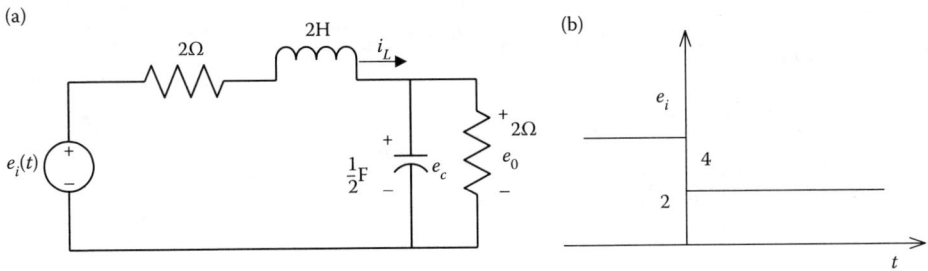

FIGURE 11.1
(a) An electrical circuit. (b) Input to the circuit.

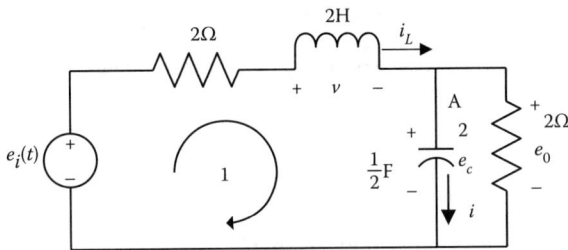

FIGURE 11.2
Writing equations for the electrical circuit.

Solution

See Figure 11.2.

The state-space shell:

$$2\frac{di_L}{dt} = v$$

$$\frac{1}{2}\frac{de_c}{dt} = i$$

Loop 1 (Compatibility equation): $e_i - 2i_L - v - e_c = 0$
Node A (Continuity equation): $i_L - i - (e_c)/2 = 0$
Eliminate the auxiliary variables v and i:

$$2\frac{di_c}{dt} = e_i - 2i_L - e_c$$

$$\frac{1}{2}\frac{de_c}{dt} = i_L - \frac{e_c}{2}$$

Hence, the state equations are:

$$\frac{di_L}{dt} = -i_L - \frac{1}{2}e_c + \frac{1}{2}e_i$$

$$\frac{de_c}{dt} = 2i_L - e_c$$

The output equation: $e_0 = e_c$

i. $A = \begin{bmatrix} -1 & -\dfrac{1}{2} \\ 2 & -1 \end{bmatrix}$ $B = \begin{bmatrix} \dfrac{1}{2} \\ 0 \end{bmatrix}$

$$C = [0\ 1]\ D = 0$$

ii. *Note*: State variables cannot undergo step changes. Hence

$$e_c(0^+) = e_c(0^-)\ i_L(0^+) = i_L(0^-)$$

$$\text{At } t=0^-:\ (di_L/dt) = (de_c/dt) = 0 \text{ (steady-state)}$$

$$e_i = 4$$

Hence:

$$0 = -i_L(0^-) - \frac{1}{2}e_c(0^-) + \frac{1}{2} \times 4$$

$$0 = 2i_L(0^-) - e_c(0^-)$$

$$\text{Solve: } i_L(0^-) = 1,\ e_c(0^-) = 2$$

$$\text{Hence: } x(0^+) = \begin{bmatrix} 1 \\ 2 \end{bmatrix}$$

iii. Find eigenvalues by solving the characteristic equation:

$$|\lambda I - A| = \begin{vmatrix} \lambda+1 & \dfrac{1}{2} \\ -2 & \lambda+1 \end{vmatrix} = (\lambda+1)^2 + 1 = 0$$

$$\text{Hence: } \lambda + 1 = \pm j$$

$$\text{Or: } \lambda_1 = -1 + j\ \lambda_2 = -1 - j$$

Determine the corresponding eigenvectors:

$$\xi: \begin{bmatrix} j & \dfrac{1}{2} \\ -2 & j \end{bmatrix} \begin{bmatrix} a+jb \\ p+jq \end{bmatrix} = \begin{bmatrix} 0 \\ 0 \end{bmatrix} \Rightarrow \begin{array}{l} ja - b + \dfrac{1}{2}p + \dfrac{j}{2}q = 0 \\ -2a - 2jb + jp - q = 0 \end{array}$$

$$\text{Hence: } a + (q/2) = 0,\ -b + (1/2)\,p = 0$$

$$\Rightarrow q = -2a,\ p = 2b$$

$$\xi_1 = \begin{bmatrix} a+jb \\ 2a-2ja \end{bmatrix} \text{Pick } a=1, b=0$$

$$\xi_1 = \begin{bmatrix} 1 \\ -2j \end{bmatrix}$$

Hence: $\xi_2 = \begin{bmatrix} 1 \\ 2j \end{bmatrix}$

iv. We will use Method 2 to determine the matrix exponential.

Form the modal matrix $M = \begin{bmatrix} 1 & 1 \\ -2j & 2j \end{bmatrix}$

Note: $|M| = 2j + 2j = 4j \neq 0$ (nonsingular matrix).

Form the matrix inverse: $M^{-1} = \dfrac{1}{4j}\begin{bmatrix} 2j & -1 \\ 2j & 1 \end{bmatrix} = \begin{bmatrix} \dfrac{1}{2} & \dfrac{j}{4} \\ \dfrac{1}{2} & -\dfrac{j}{4} \end{bmatrix}$

$$e^{Jt} = \begin{bmatrix} e^{(-1+j)t} & 0 \\ 0 & e^{(-1-j)t} \end{bmatrix}$$

By Method 2 of determining matrix exponential, we have:

$$e^{At} = Me^{Jt}M^{-1} = e^{-t}\begin{bmatrix} 1 & 1 \\ -2j & 2j \end{bmatrix}\begin{bmatrix} e^{jt} & 0 \\ 0 & e^{-jt} \end{bmatrix}\begin{bmatrix} \dfrac{1}{2} & \dfrac{j}{4} \\ \dfrac{1}{2} & \dfrac{-j}{4} \end{bmatrix}$$

$$= e^{-t}\begin{bmatrix} e^{jt} & e^{-jt} \\ -2je^{jt} & 2je^{-jt} \end{bmatrix}\begin{bmatrix} \dfrac{1}{2} & \dfrac{j}{4} \\ \dfrac{1}{2} & \dfrac{-j}{4} \end{bmatrix}$$

$$= e^{-t}\begin{bmatrix} \dfrac{1}{2}\left(e^{jt}+e^{-jt}\right) & \dfrac{j}{4}\left(e^{jt}-e^{-jt}\right) \\ -j\left(e^{jt}-e^{-jt}\right) & \dfrac{1}{2}\left(e^{jt}+e^{-jt}\right) \end{bmatrix} = e^{-t}\begin{bmatrix} \cos t & \dfrac{-1}{2}\sin t \\ 2\sin t & \cos t \end{bmatrix}$$

v. $x(t) = e^{At}x(0) + \displaystyle\int_0^t e^{A(t-\tau)}Bu(\tau)d\tau$

$$= e^{-t}\begin{bmatrix} \cos t & \dfrac{-1}{2}\sin t \\ 2\sin t & \cos t \end{bmatrix}\begin{bmatrix} 1 \\ 2 \end{bmatrix} + \int_0^t e^{-(t-\tau)}\begin{bmatrix} \cos(t-\tau) & \dfrac{-1}{2}\sin(t-\tau) \\ 2\sin(t-\tau) & \cos(t-\tau) \end{bmatrix}\begin{bmatrix} 1 \\ 0 \end{bmatrix}\times 2d\tau$$

$$= e^{-t} \begin{bmatrix} \cos t - \sin t \\ 2(\cos t + \sin t) \end{bmatrix} + \int_0^t e^{-(t-\tau)} \begin{bmatrix} \cos(t-\tau) \\ 2\sin(t-\tau) \end{bmatrix} d\tau$$

Now, direct integration gives:

$$\int_0^t e^{-(t-\tau)} \cos(t-\tau) d\tau = \frac{1}{2} - \frac{e^{-t}}{2}(\cos t - \sin t)$$

$$\int_0^t e^{-(t-\tau)} \sin(t-\tau) d\tau = \frac{1}{2} - \frac{e^{-t}}{2}(\cos t + \sin t)$$

Hence:

$$x(t) = e^{-t} \begin{bmatrix} \cos t - \sin t - \dfrac{1}{2}\cos t + \dfrac{1}{2}\sin t \\ 2\cos t + 2\sin t - \dfrac{1}{2}\cos t - \dfrac{1}{2}\sin t \end{bmatrix} + \begin{bmatrix} \dfrac{1}{2} \\ \dfrac{1}{2} \end{bmatrix}$$

$$= e^{-t} \begin{bmatrix} \dfrac{1}{2}\cos t - \dfrac{1}{2}\sin t \\ \dfrac{3}{2}\cos t + \dfrac{3}{2}\sin t \end{bmatrix} + \begin{bmatrix} \dfrac{1}{2} \\ \dfrac{1}{2} \end{bmatrix}$$

Note that the ICs are satisfied: $x(0) = \begin{bmatrix} 1 \\ 2 \end{bmatrix}$.

11.2.3 Time Response by Laplace Transform

Time variation of the state vector of a linear, constant-parameter dynamic system (Equation 11.8) can be obtained using the Laplace transform method (see Appendix A and Chapter 6). This idea was already used for the unforced case ($u = 0$), in the context of determining the matrix exponential. Now, let us address the general (forced) case more formally. The Laplace transform of the forced state-space equation:

$$\dot{x} = Ax + Bu(t) \tag{11.8}$$

is given by:

$$sX(s) - x(0) = AX(s) + BU(s) \tag{11.16}$$

Consequently:

$$x(t) = L^{-1} (sI - A)^{-1} x(0) + L^{-1} (sI - A)^{-1} BU(s) \tag{11.17}$$

in which I denotes the identity (unit) matrix. Note that L^{-1} denotes the inverse Laplace transform operator. The square matrix $(sI - A)^{-1}$ is known as the *resolvent matrix*. Its inverse

Laplace transform is the *state-transition matrix,* as clear from Equation 11.9 and already seen in Equation 11.11. Specifically:

$$\Phi(t) = L^{-1}(sI - A)^{-1} \tag{11.18}$$

We have already seen that $\Phi(t)$ is equal to the matrix exponential:

$$\Phi(t) = \exp(At) = I + At + \frac{1}{2!}A^2\,t^2 + \cdots \tag{11.19}$$

It follows from Equation 11.18 that the state-transition matrix maybe analytically determined as a closed-form matrix function by the direct use of inverse Laplace transformation on each term of the resolvent matrix. Since the product in the Laplace domain is a convolution integral in the time domain, and vice versa (see Appendix A) the second term on the right hand side of Equation 11.17 can be expressed as a matrix convolution integral. Hence, Equation 11.17 may be expressed as:

$$x(t) = \Phi(t)\,x(0) + \int_0^t \Phi(t-\tau)Bu(\tau)d\tau \tag{11.20}$$

The first part of this solution is the *zero-input response;* the second part is the *zero state response.*

11.2.4 Output Response

State variables are not necessarily measurable and generally are not system outputs. Linearized relationship between state variables and system output (response) variables $y(t)$ may be expressed as:

$$y(t) = Cx(t) \tag{11.21}$$

in which the output vector is:

$$y = [y_1,\, y_2,\, \ldots,\, y_p]^T \tag{11.22}$$

and C denotes the output (measurement) gain matrix. Hence, once x is known, y can be determined by using Equation 11.21.

11.2.4.1 Transfer Function Matrix

Suppose that the number of inputs $m > 1$ and the number of outputs $p > 1$. By substituting the Laplace transformed equation (Equation 11.21) into Equation 11.16, with zero IC $(x(0) = 0)$ we get the input-output relation

$$Y(s) = H(s)\,U(s) \tag{11.23}$$

where the transfer function matrix $H(s)$ of the system is given by

$$H(s) = C(sI - A)^{-1}\,B \tag{11.24}$$

11.2.5 Modal Response

A dynamic system has a set of natural, unforced responses that represent the preferred dynamics of the system. Such a preferred motion is called a "modal response" or a "mode of motion." Analytically, a mode of a linear, time-invariant system is represented by an eigenvalue and the corresponding eigenvector. In particular, with regard to a state-space model, in a given mode of response the state vector will remain proportional to the corresponding eigenvector, and the time variation of the associated response will be given by the time exponential function of the corresponding eigenvalue.

Since modal response corresponds to free (unforced) response of the system, we consider the corresponding state-space model

$$\dot{x} = Ax \tag{11.6}$$

Eigenvectors of A: nontrivial solutions ξ of

$$A\xi = \lambda\xi \text{ or } (A - \lambda I)\,\xi = 0 \tag{11.25}$$

For a nontrivial (i.e., nonzero) solution to be possible, the matrix $(A - \lambda I)$ should be singular (i.e., it should not have a finite inverse. Otherwise $\xi = 0$). Hence, we must have:

$$|A - \lambda I| = 0 \tag{11.26}$$

which is the characteristic equation of the system, whose roots are the eigenvalues (poles) of the system: $\lambda_1, \lambda_2, \ldots, \lambda_n$.

Note: If ξ is a solution of Equation 11.25 then any multiple $a\xi$ of it is also a solution. Hence, an eigenvector is arbitrary up to a multiplication factor.

Assume that A is a *normal matrix*; i.e.,

$$AA^H = A^H A \tag{11.27}$$

$\Rightarrow \xi_1, \xi_2, \ldots, \xi_n$ will be linearly independent eigenvectors, and will form a basis for the state-space Σ (see Appendix C).

Consider the forced state-space model

$$\dot{x} = Ax + B(t)u(t) \tag{11.28}$$

where B may be time-variant.

Note: $x \in \Sigma$ and $B(t)u(t) \in \Sigma$

Hence, the state vector can be expressed as a linear combination of the eigenvectors:

$$x(t) = q_1(t)\xi_1 + \cdots + q_n(t)\xi_n \tag{11.29}$$

Also:

$$B(t)u(t) = \beta_1(t)\,\xi_1 + \cdots + \beta_n(t)\xi_n \tag{11.30}$$

Substitute Equations 11.29 and 11.30 into Equation 11.28.

$$\Rightarrow \left(\dot{q}_1 - \lambda_1 q_1 - \beta_1\right)\xi_1 + \cdots + \left(\dot{q}_n - \lambda_n q_n - \beta_n\right)\xi_n = 0$$

Hence,

$$\dot{q}_i - \lambda_i q_i - \beta_i = 0 \quad [\because \xi_i \text{ are linearly independent}] \text{ for all } i \tag{11.31}$$

$$\text{Form the modal matrix:} \quad M = |\xi_1, \xi_2, \ldots, \xi_n| \tag{11.32}$$

$$\text{Equation 11.29 can be written as:} \quad x = Mq \tag{11.33}$$

This is a transformation between the state vector x and the modal coordinate vector q. Substitute Equation 11.33 into Equation 11.28:

$$\dot{q} = M^{-1}AMq + M^{-1}Bu \tag{11.34}$$

The Jordan matrix:

$$J = M^{-1}AM = \text{diag}(\lambda_1, \lambda_2, \lambda_3, \ldots, \lambda_n) \tag{11.35}$$

is a diagonal matrix of eigenvalues, in view of Equation 11.25. Hence, Equation 11.34 represents a set of uncoupled first order differential equations (see Equation 11.31), which can be integrated to obtain the modal responses q_i.

11.2.5.1 State Response through Modal Response

Step 1: Integrate the uncoupled Equation 11.34 to determine the responses q_i:

$$q_i(t) = e^{\lambda_i(t-t_o)}q_i(t_o) + \int_{t_o}^{t} e^{\lambda_i(t-\tau)} \beta_i(\tau)d\tau \tag{11.36}$$

Step 2: Transform the resulting response vector q back to x using Equation 11.33.

$$\text{Note:} \quad e^{At} = Me^{Jt} M^{-1} \text{ (this is a method to compute } e^{At}) \tag{11.37}$$

$$x(t) = e^{At} x(0) = Me^{Jt} M^{-1} x(0) = Me^{Jt} q(0) \tag{11.38}$$

11.2.5.2 Advantages of Modal Decomposition

1. Easier to determine the response.
2. Gives better insight about system and its response (e.g., stability, controability, observability).
3. Useful in system modeling (e.g., model reduction) and design (e.g., pole placement).

Example 11.5

Consider the mechanical system shown in Figure 11.3. This may be interpreted as a model of a motor and a rotor (of moments of inertia I_1 and I_2, respectively, which are connected through a fluid coupling (of damping constant b_1).

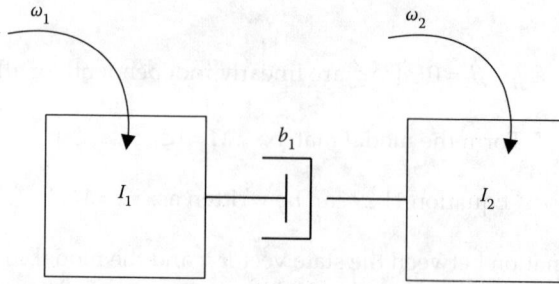

FIGURE 11.3
A rotor driven by a motor through fluid coupling.

The equations of motion are:

$$I_1\dot{\omega}_1 = -b_1(\omega_1 - \omega_2) \tag{i}$$

$$I_2\dot{\omega}_2 = b_1(\omega_1 - \omega_2) \tag{ii}$$

which form a second order state-space model.
Let $I_1 = I_2 = 1$, $b_1 = 1$

$$\text{The system matrix } A = \begin{bmatrix} -1 & 1 \\ 1 & -1 \end{bmatrix} \tag{iii}$$

By straightforward mathematics, the eigenvalues of A are:

$$\lambda_1 = 0, \ \lambda_2 = -2$$

and the corresponding eigenvectors are:

$$\xi_1 = \begin{bmatrix} 1 \\ 1 \end{bmatrix}, \ \xi_2 = \begin{bmatrix} 1 \\ -1 \end{bmatrix}$$

Hence the state response for free motion (input=0) may be expressed as:

$$\begin{bmatrix} \omega_1 \\ \omega_2 \end{bmatrix} = q_1(t)\begin{bmatrix} 1 \\ 1 \end{bmatrix} + q_2(t)\begin{bmatrix} 1 \\ -1 \end{bmatrix}$$

$$= p_1 e^{\lambda_1 t}\begin{bmatrix} 1 \\ 1 \end{bmatrix} + p_2 e^{\lambda_2 t}\begin{bmatrix} 1 \\ -1 \end{bmatrix} = p_1\begin{bmatrix} 1 \\ 1 \end{bmatrix} + p_2 e^{-2t}\begin{bmatrix} 1 \\ -1 \end{bmatrix} \tag{iv}$$

The constants p_1 and p_2 are determined by the ICs $\omega_1(0)$ and $\omega_2(0)$, using Equation (iv). The state-space of the system (which is a plane for this second order problem) is shown in Figure 11.4. The

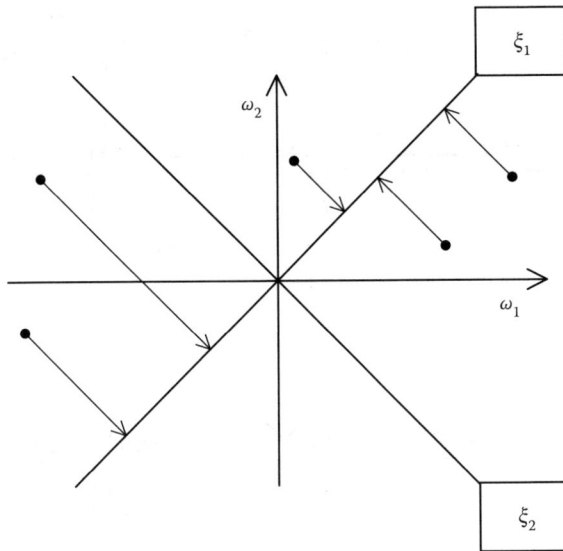

FIGURE 11.4
State-space and the eigenvectors of the example system.

two eigenvectors are indicated as two lines, the first one having a slope of 1 and the other having a slope of −1. We observe the following:

1. If the initial state vector falls on one of the eigenvectors, the subsequent response will remain on the same eigenvector.
2. If the ICs are such that $p_2=0$, then the second mode will not enter into the response. The response will remain on the line with slope 1, which corresponds to the first mode. (In fact, the system will be stationary in this case, because mode 1 has a zero eigenvalue.)
3. If the ICs are such that $p_1=0$, then mode 1 will not enter the response. Then, the response will remain on the line of slope −1, which corresponds to mode 2.
4. Since the second mode has a negative eigenvalue, the corresponding modal response will eventually decay to zero. Hence, the response for any arbitrary IC will eventually end up on the first eigenvector (line of slope 1).

Example 11.6

A torsional dynamic model of a pipeline segment is shown in Figure 11.5a. Free-body diagrams in Figure 11.5b show internal torques acting at sectioned inertia junctions, for free motion.
    The state vector is chosen as:

$$x=[\Omega_1, \Omega_2, T_1, T_2]^T \tag{i}$$

The corresponding system matrix may be determined as (see Chapter 2):

$$A = \begin{bmatrix} 0 & 0 & -\dfrac{1}{I_1} & \dfrac{1}{I_1} \\ 0 & 0 & -\dfrac{1}{I_2}\left(\dfrac{k_3}{k_1}\right) & -\dfrac{1}{I_2}\left(1+\dfrac{k_3}{k_2}\right) \\ k_1 & 0 & 0 & 0 \\ -k_2 & k_2 & 0 & 0 \end{bmatrix} \tag{ii}$$

(a)

(b)

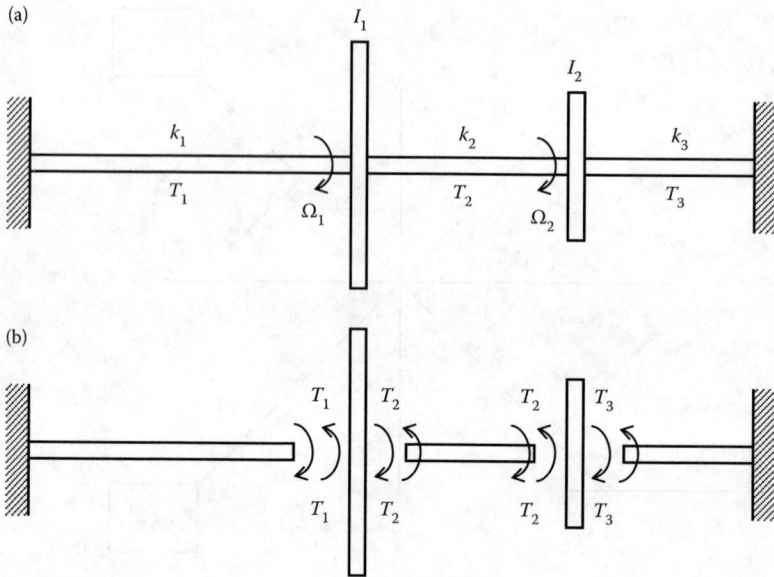

FIGURE 11.5
(a) Dynamic model of a pipeline segment. (b) Free-body diagrams.

The displacements are used as outputs:

$$y = \left[\frac{T_1}{k_1}, \frac{T_1}{k_1} + \frac{T_2}{k_2} \right]^T \tag{iii}$$

This output vector corresponds to the output-gain matrix:

$$C = \begin{bmatrix} 0 & 0 & \frac{1}{k_1} & 0 \\ 0 & 0 & \frac{1}{k_1} & \frac{1}{k_2} \end{bmatrix} \tag{iv}$$

For the special case given by $I_1 = I_2 = I$ and $k_1 = k_3 = k$, the system eigenvalues are

$$\lambda_1, \bar{\lambda}_1 = \pm j\omega_1 = \pm j\sqrt{\frac{k}{I}}$$

$$\lambda_2, \bar{\lambda}_2 = \pm j\omega_2 = \pm j\sqrt{\frac{k + 2k_2}{I}} \tag{v}$$

The magnitudes of these are in fact the two *natural frequencies* of oscillation of the system. The corresponding eigenvectors are:

$$X_1, \bar{X}_1 = R_1 \pm jI_1 = \frac{\alpha_1}{2} \left[\omega_1, \omega_1, \mp jk_1, 0 \right]^T$$

$$X_2, \bar{X}_2 = R_2 \pm jI_2 = \frac{\alpha_2}{2} \left[\omega_2, -\omega_2, \mp jk_1, \pm 2jk_2 \right]^T \tag{vi}$$

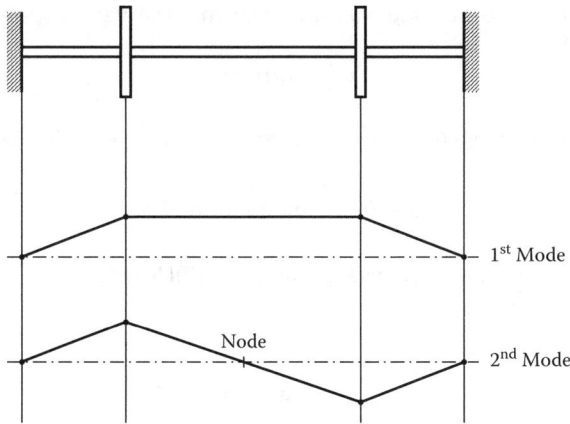

FIGURE 11.6
Mode shapes of the pipeline segment.

where α_1 and α_2 are arbitrary constants, because eigenvectors are arbitrary up to a multiplication factor.

Note that the first two elements of the state vector correspond to the angular velocities of the two inertia elements. In a modal motion of a particular natural frequency, the amplitude of velocity will be proportional to the amplitude of displacement. The modal contributions to the displacement vector (or velocity vector) are given by the first two elements of the eigenvectors:

$$Y_1 = \begin{bmatrix} 1 \\ 1 \end{bmatrix} \alpha_1 \sin \omega_1 t$$

and

$$Y_2 = \begin{bmatrix} 1 \\ -1 \end{bmatrix} \alpha_2 \sin \omega_2 t \qquad \text{(x)}$$

The mode shapes of motion are given by the amplitude vectors $S_1 = [1,1]^T$ and $S_2 = [1, -1]^T$, which are sketched in Figure 11.6. In general, each modal contribution introduces two unknown parameters α_i and Φ_i, into the free response (homogeneous solution), where Φ_i are the phase angles associated with the sinusoidal terms. For an n-degree-of-freedom (i.e., order-$2n$) system, this results in $2n$ unknowns, the determination of which requires the $2n$ ICs $x(0)$.

11.2.6 Time-Varying Systems

In this case the system matrix is time dependent, and is given by $A(t)$.

$$\text{Homogeneous (free) system:} \quad \dot{x} = A(t)x \qquad (11.39)$$

Let x_1 be the response to initial state $[1,0,...,0]^T$, and so on.

$$\text{Form the matrix: } U(t) = [x_1, x_2, ..., x_n] \text{ (Fundamental solution matrix)} \qquad (11.40)$$

Then, for any arbitrary initial state $x(t_o)$ it is clear that the state response is:

$$x(t) = U(t)x(t_o) \tag{11.41}$$

Nonhomogeneous (forced) system: $\dot{x} = A(t)x + B(t)u \tag{11.42}$

Differentiate Equation 11.41 and substitute Equation 11.39:

$$\dot{x}(t) = \dot{U}(t)x(t_o) = A(t)x(t) = A(t)U(t)x(t_o)$$

Hence:

$$\dot{U}(t) = A(t)U(t) \tag{11.43}$$

The fundamental solution matrix $U(t)$ is nonsingular in general. Hence:

$$U(t)U^{-1}(t) = I_n \tag{11.44}$$

Differentiating Equation 11.44 and using Equation 11.43 we can show that:

$$\frac{dU^{-1}}{dt} = -U^{-1}A(t) \tag{11.45}$$

We now use Equation 11.45 to get a perfect differential form for Equation 11.42:
$\dot{x} - A(t)x = B(t)u$
Premultiply Equation 11.42 by U^{-1}. This gives

$$\frac{d}{dt}\left[U^{-1}x(t)\right] = U^{-1}B(t)u$$

Integrate:

$$U^{-1}x(t) = U^{-1}(t_o)x(t_o) + \int_{t_o}^{t} U(\tau)B(\tau)u(\tau)d\tau$$

Premultiply by $U(t)$:

$$x(t) = U(t)U^{-1}(t_o)x(t_o) + \int_{t_o}^{t} U(t)U^{-1}(\tau)B(\tau)u(\tau)d\tau \tag{11.46}$$

Note: According to Equation 11.41 $U(t_0) = I$ (the identity matrix), but is not substituted in Equation 11.46 for the sake of mathematical clarity.

Define state-transition matrix: $\Phi(t, \tau) = U(t)U^{-1}(\tau) \tag{11.47}$

Then, Equation 11.46 may be written as:

$$x(t) = \Phi(t,t_o)x(t_o) + \int_{t_o}^{t} \Phi(t,\tau)B(\tau)u(\tau)d\tau \tag{11.48}$$

From Equation 11.48, for the homogeneous case ($u=0$) we have:

$$x(t) = \Phi(t, t_o)x(t_o) \tag{11.49}$$

which is identical to Equation 11.41, because $U(t_0)=I$.

11.2.6.1 Properties of the State-Transition Matrix

1. $\Phi(\tau, \tau)=I_n$ for any τ
2. $\dfrac{d}{dt}\Phi(t,\tau) = A(t)\Phi(t,\tau)$
3. $\Phi(t, \tau) = \Phi^{-1}(t, \tau)$
4. $\Phi(t_2, t_0) = \Phi(t_2,t_1)\Phi(t_1, t_0)$

Note: For constant A, we have:

$$\Phi(t, \tau) = e^{A(t-\tau)}$$

11.3 System Stability

As discussed in Chapter 8, if the response of a system grows beyond some acceptable level (say, goes to infinity in a mathematical sense) then the system is unstable. If the plant is unstable, the controller should make it stable, or it will not be possible to achieve the control objective. Now we will study some important concepts of stability in both linear and nonlinear systems. State-space models will be primarily used for this purpose.

11.3.1 Stability of Linear Systems

In classical control, stability of linear systems is determined using such approaches as the Routh–Hurwitz criterion, Bode diagrams, and root locus, as discussed in Chapter 8. In the present context of modern control, we will examine the stability of a linear, constant coefficient (time-invariant), and free (unforced) state-space model:

$$\dot{x} = Ax \tag{11.50}$$

Eigenvectors of A are the nontrivial solutions ξ of

$$A\xi = \lambda\xi \text{ or } (A - \lambda I)\xi = 0 \tag{11.51}$$

For a nontrivial (i.e., nonzero) solution to be possible, the matrix $(A - \lambda I)$ should be singular (i.e., it should not have a finite inverse. Otherwise $\xi = 0$). Hence, we must have:

$$|A - \lambda I| = 0 \tag{11.52}$$

which is the characteristic equation of the system, whose roots are the eigenvalues (poles) $\lambda_1, \lambda_2, ..., \lambda_n$ of the system. Assume distinct (i.e., unequal) eigenvalues. Then, the eigenvectors corresponding to these eigenvalues will be linearly independent. Accordingly, we can express the solution to Equation 11.50 as:

$$x(t) = q_1(t)\xi_1 + \cdots + q_n(t)\xi_n \tag{11.53}$$

which may be written as:

$$x = Mq \tag{11.54}$$

This is a transformation between the state vector x and the modal coordinate vector q:

$$q = [q_1, q_2, ..., q_n]^T \tag{11.55}$$

The modal matrix M is the matrix of eigenvectors, as given by:

$$M = [\xi_1, \xi_2, ..., \xi_n] \tag{11.56}$$

In view of Equation 11.51, the following relation is satisfied:

$$AM = MJ \text{ or } J = M^{-1} AM \tag{11.57}$$

where the Jordan matrix J is the diagonal matrix of eigenvalues:

$$J = \text{diag}(\lambda_1, \lambda_2, \lambda_3, ..., \lambda_n) \tag{11.58}$$

The substitution of the transformation (Equation 11.54) into the system equation (Equation 11.50) gives the transformed system:

$$\dot{q} = M^{-1} AMq = Jq \tag{11.59}$$

This is the set of uncoupled equations:

$$\dot{q}_i - \lambda_i q_i = 0 \quad \text{for } i = 1, 2, ..., n \tag{11.60}$$

The solution of Equation 11.60 is:

$$q_i(t) = e^{\lambda_i t} q_i(0) \tag{11.61}$$

Substitute Equation 11.61 into Equation 11.53. We get:

$$x(t) = q_1(0)e^{\lambda_1 t}\xi_1 + \cdots + q_n(0)e^{\lambda_n t}\xi_n \tag{11.62}$$

The IC $q_i(0)$ may be absorbed into the eigenvectors ξ_i, which are arbitrary up to a multiplication factor. Note that the above development is for the special case of free response, which is all what is needed with regard to system stability.

Equation 11.62 indicates that, if none of the eigenvalues of the system have a positive real part, then none of the terms on the right hand side of Equation 11.62 will grow. In that case, the free response of the system will not grow (will remain bounded), and the system is considered stable. In particular, if all the eigenvalues have negative real parts, the free response will decay to zero, and the system is said to be asymptotically stable.

11.3.1.1 General Case of Repeated Eigenvalues

The previous development assumed that the eigenvalues of the system are distinct; i.e., no repeated eigenvalues. Now let us relax that assumption and develop some concepts for the case of repeated eigenvalues. Still, the system stability will depend on the nature of the eigenvalues.

The system matrix A is of size $n \times n$ matrix (i.e., nth order system). Suppose that the eigenvalue λ_i of A has a multiplicity m_i and the remaining eigenvalues are distinct. Also, suppose that

$$\text{Rank}(A - \lambda_i I) = r_i \tag{11.63}$$

Then, from linear algebra, it is known that

1. There are $n - r_i$ independent eigenvectors corresponding to the eigenvalue λ_i.
2. There are $m_i - (n - r_i)$ generalized eigenvectors corresponding to λ_i
3. Together, 1 and 2 form a set of m_i independent vectors for the m_i repeated eigenvalues λ_i
4. Together with the $n - m_i$ independent eigenvectors corresponding to the remaining distinct eigenvalues, there are a total of n independent vectors, which may be assembled to form the generalized modal matrix M for A.

In the present case, the Jordan matrix:

$$J = M^{-1} A M \tag{11.64}$$

will be block diagonal (not strictly diagonal).

Note: Each independent eigenvector will have a Jordan block $\Rightarrow$ In total there will be $a_i = n - r_i$ Jordan blocks for the eigenvalue λ_i.

11.3.1.2 Generalized Eigenvectors

Eigenvector of λ_i is given by: $(A - \lambda_i I)\xi = 0$. Call this eigenvector ξ_1
Generalized eigenvectors of rank k are given by:

$$\xi_1 = (A - \lambda_i I)\xi_2, \ \xi_2 = (A - \lambda_i I)\xi_3, \ ..., \ \xi_k = (A - \lambda_i I)\xi_{k-1} \tag{11.65}$$

Note: $(A - \lambda_i I)^p \ \xi_p = 0$ for $p = 1, 2, ..., k$
We will illustrate these results by examples.

Example 11.7

Consider a system given by the system matrix

$$A = \begin{bmatrix} 5 & 4 & 3 \\ -1 & 0 & -3 \\ 1 & -2 & 1 \end{bmatrix}$$

The eigenvalues are: $\lambda_1 = -2, \lambda_2 = \lambda_3 = 4$.
Rank $(A - \lambda_2 I) = 2 = r_i$
So, there exists one eigenvector for $\lambda = -2$; and one eigenvector and one generalized eigenvector $\lambda = 4$.
Note: $n = 3$ and for $\lambda = 4$ we have $m_i = 2$; $r_i = 2$; $a_i = n - r_i = 1$ (number of Jordan blocks); $m_i - a_i = 2 - 1 = 1$ (number of generalized eigenvectors).
Independent eigenvectors are:

$$\xi_1 = \begin{bmatrix} 1 \\ -1 \\ -1 \end{bmatrix}, \quad \xi_2 = \begin{bmatrix} 1 \\ -1 \\ 1 \end{bmatrix}$$

Generalized eigenvector: $(A - 4I)\xi = \xi_2$

$$\begin{bmatrix} 1 & 4 & 3 \\ -1 & -4 & -3 \\ 1 & -2 & -3 \end{bmatrix} \begin{bmatrix} a \\ b \\ c \end{bmatrix} = \begin{bmatrix} 1 \\ -1 \\ 1 \end{bmatrix}$$

$$\text{A solution: } \begin{bmatrix} a \\ b \\ c \end{bmatrix} = \begin{bmatrix} 0 \\ 1 \\ -1 \end{bmatrix} = \xi_3$$

Form the modal matrix and determine its inverse:

$$M = \begin{bmatrix} 1 & 1 & 0 \\ -1 & -1 & 1 \\ -1 & 1 & -1 \end{bmatrix}, \quad M^{-1} = \frac{1}{2} \begin{bmatrix} 0 & -1 & -1 \\ 2 & 1 & 1 \\ 2 & 2 & 0 \end{bmatrix}$$

Note: $\det(M) = 2 \neq 0$. Hence M is nonsingular, as required.
The Jordan matrix is:

$$J = M^{-1}AM = \begin{bmatrix} -2 & 0 & 0 \\ 0 & 4 & 1 \\ 0 & 0 & 4 \end{bmatrix}$$

Note the Jordan block corresponding to the eigenvalue 4, in the bottom right hand corner of *J*.

Example 11.8

Consider the system with system matrix $A = \begin{bmatrix} 2 & 2 & -1 \\ -1 & -1 & 1 \\ -1 & 2 & 2 \end{bmatrix}$

System order $n = 3$; eigenvalues are: $\lambda_1 = \lambda_2 = \lambda_3 = 1$; multiplicity $m_i = 3$.
Rank $(A - \lambda_1 I) = 1 = r_i$; $a_i = n - r_i = 2 \Rightarrow$ two independent eigenvectors.
$m_i - a_i = 1 \Rightarrow$ one generalized eigenvector.
Eigenvectors: $(A - \lambda_1 I)\xi = 0 \Rightarrow$

$$\xi_1 = \begin{bmatrix} 1 \\ -1 \\ -1 \end{bmatrix} \quad \xi_2 = \begin{bmatrix} 1 \\ 0 \\ 1 \end{bmatrix}$$

Generalized eigenvector: $(A - \lambda_1 I)\xi = \xi_1$ or $\xi_2 \Rightarrow$

$$\begin{bmatrix} 1 & 2 & -1 \\ -1 & -2 & 1 \\ -1 & -2 & 1 \end{bmatrix} \begin{bmatrix} a \\ b \\ c \end{bmatrix} = \begin{bmatrix} 1 \\ -1 \\ -1 \end{bmatrix} \Rightarrow \begin{bmatrix} a \\ b \\ c \end{bmatrix} = \begin{bmatrix} 1 \\ 0 \\ 0 \end{bmatrix}$$

Form the modal matrix and the Jordan matrix:

$$M = \begin{bmatrix} 1 & 1 & 1 \\ -1 & 0 & 0 \\ -1 & 0 & 1 \end{bmatrix}, \quad J = M^{-1} A M = \begin{bmatrix} 1 & 1 & |0 \\ 0 & 1 & |0 \\ 0 & 0 & 1 \end{bmatrix}$$

Note the Jordan block in the top left hand corner of J.

11.3.2 Stability from Modal Response for Repeated Eigenvalues

Again consider the unforced case of constant A: $\dot{x} = Ax$
Form the generalized modal matrix M for the case with repeated eigenvalues (generalized eigenvectors). The modal transformation $x = Mq$ gives:

$$\dot{q} = M^{-1} A M q$$

$$q = e^{Jt} q(0) \tag{11.66}$$

$$x = M e^{Jt} M^{-1} x(0)$$

Where:

$$J = \begin{bmatrix} J_1 & & & 0 \\ & J_2 & & \\ & & 0 & \\ 0 & & & J_p \end{bmatrix} \tag{11.67}$$

The Jordan blocks are J_1, J_2, ..., J_p. Hence:

$$e^{Jt} = \begin{bmatrix} e^{J_1 t} & & 0 \\ & \ddots & \\ 0 & & e^{J_p t} \end{bmatrix} \quad \text{(System stability depends on this)} \tag{11.68}$$

11.3.2.1 Possibilities of Jordan Blocks and Modal Responses

1. $J_i = \lambda_i, \Rightarrow e^{J_i t} \equiv e^{\lambda_i t}$

2. $J_i = \begin{bmatrix} \lambda_i & 1 \\ 0 & \lambda_i \end{bmatrix} \Rightarrow e^{J_i t} = e^{\lambda_i t} \begin{bmatrix} 1 & t \\ 0 & 1 \end{bmatrix}$

3. $J_i = \begin{bmatrix} \lambda_i & 1 & 0 \\ 0 & \lambda_i & 1 \\ 0 & 0 & \lambda_i \end{bmatrix} \Rightarrow e^{J_i t} = e^{\lambda_i t} \begin{bmatrix} 1 & t & \dfrac{t^2}{2!} \\ 0 & 1 & t \\ 0 & 0 & 1 \end{bmatrix}$

4. $J_i = \begin{bmatrix} \lambda_i & 1 & & 0 \\ & \lambda_i & 1 & \\ & & \lambda_i & 1 \\ 0 & & & \lambda_i \end{bmatrix} \Rightarrow e^{J_i t} = e^{\lambda_i t} \begin{bmatrix} 1 & t & \dfrac{t^2}{2!} & \dfrac{t^3}{3!} \\ & 1 & t & \dfrac{t^2}{2!} \\ & & 1 & t \\ 0 & & & 1 \end{bmatrix}$

and so on. So, if there are generalized eigenvectors (i.e., if the number of linearly independent eigenvectors is less than the multiplicity of the corresponding eigenvalue) then, we get "t" terms in the modal respose (exponential of the Jordan matrix).

Conclusions:

1. $\text{Re}(\lambda_i) < 0 \Rightarrow$ Stable
2. $\text{Re}(\lambda_i) = 0$ and *all independent eigenvectors $\Rightarrow$ Stable *has one or more generalized eigenvectors $\Rightarrow$ Unstable
3. $\text{Re}(\lambda_i) > 0 \Rightarrow$ Unstable

11.3.3 Equilibrium

If a system remains in a particular state at all times, when not excited by an external force, then the system is said be in an equilibrium state. Mathematically,

If $x(t_0) = x_e$ and $x(t) = x_e$ for all $t \geq t_0$, with $u = 0 \Rightarrow x_e$ equilibrium point.

There are three types of equilibrium: stable, unstable, and neutral.

Definition 1: (Stability in the Sense of Lyapunov)
The origin ($x = 0$) of the state-space is a stable equilibrium point (in the sense of Lyapunov—i.s.L) if for any $\varepsilon > 0$ there exists $\delta(\varepsilon, t_0) > 0$ such that if $||x(t_0)|| < \delta$ then $||x(t)|| < \varepsilon$ for all $t > t_0$.

Note 1: If we make ε very small, unless the system is asymptotically stable, $\delta(\varepsilon, t_0)$ might not exist.

Note 2: In nonmathematical terms, "stability i.s.L" means that for any finite IC, the state response will remain finite (bounded).

Definition 2:
The origin is an asymptotically stable equilibrium point if:

1. Stable, and
2. There exists $\delta'(t_0)$ such that if $\|x(t_0)\| < \delta'$ then $\lim_{x \to \infty}\|x(t)\| = 0$

For time-variant systems:
Uniformly stable $\Rightarrow \delta$ is not a function of t_0
Uniformly asymptotically stable $\Rightarrow \delta'$ is not a function of t_0.

11.3.3.1 Bounded-Input Bounded-State (BIBS) Stability

For a bounded input and an arbitrary IC, the state $x(t)$ is bounded. Mathematically:

$$\text{If } \|u\| < K \text{ and for arbitrary } x(t_0) \text{ there exists } \delta(K, t_0, x(t_0)) > 0$$

$$\text{such that } \|x(t)\| < \delta \Rightarrow \text{BIBS stable.}$$

11.3.3.2 Bounded-Input Bounded-Output (BIBO) Stability

This is defined similar to BIBS stability, except using output y instead of state x.
Note: If δ and δ' can be made arbitrarily large $\Rightarrow$ globally stable.

11.3.4 Stability of Linear Systems

Consider the general time-variant case.

$$x(t) = \Phi(t, t_0)x(t_0) \tag{11.69}$$

Stable (i.s.L) iff $\|\Phi(t, t_0)\| \leq N(t_0)$ for all t (BIBS stability).
Note: "iff" means "if and only if"
Asymptotically stable iff $\|\Phi(t, t_0)\| \to 0$ as $t \to \infty$.

$$\text{Output } y(t) = \int_{-\infty}^{t} W(t, \tau)u(\tau)d\tau \tag{11.70}$$

$$\text{BIBO Stable iff } y(t) = \int_{-\infty}^{t} \|W(t, \tau)\|d\tau \leq M \text{ for all } t.$$

Note: We have already discussed linear constant (time-invariant) systems.

11.3.4.1 Frobenius' Theorem

If α_i are eigenvalues of $e^{A(t-t_0)}(= \Phi(t, t_0))$ and λ_i are eigenvalues of A
then, $\alpha_i = e^{\lambda_i(t-t_0)}$.
Note: This theorem leads to the same conclusions on stability as obtained above.

11.3.4.2 First Method of Lyapunov

A system is stable for small disturbances about an operating point if the linearized system at the operating point is asymptotically stable.

Note 1: Use Taylor series expansion to linearize the system (see Chapter 3).

Note 2: No conclusion if $\text{Re}(\lambda_i) = 0$ for the linearized system.

Example 11.9

Consider the second order nonlinear state-space model:

$$\dot{x}_1 = -2x_1 + x_2^3$$

$$\dot{x}_2 = x_1 + 4 + \cos t$$

Define small increments as:

$$x_1 = \bar{x}_1 + \hat{x}_1 \Rightarrow \dot{x}_1 = \dot{\hat{x}}$$

$$x_2 = \bar{x}_2 + \hat{x}_2 \Rightarrow \dot{x}_2 = \dot{\hat{x}}_2$$

with the input $u = \bar{u} + \hat{u} = \cos t$

Operating point (equilibrium point):

Note: Average value of the input is $\bar{u} = 0$ for the cosine function.

Then, at the equilibrium point:

$$0 = -2\bar{x}_1 + \bar{x}_2^3$$

$$0 = \bar{x}_1 + 4$$

$$\Rightarrow \bar{x}_1 = -4, \quad \bar{x}_2 = -2$$

The linearized state-space model is:

$$\dot{\hat{x}}_1 = -2\hat{x}_1 + 3\bar{x}_2\hat{x}_2 = -2\hat{x}_1 - 6\hat{x}_2$$

$$\dot{\hat{x}}_2 = \hat{x}_1 + \hat{u}$$

System matrix $A = \begin{bmatrix} -2 & -6 \\ 1 & 0 \end{bmatrix} \Rightarrow$ Characteristic polynomial $(\lambda + 2)\lambda + 6 = 0$.

or, $\lambda^2 + 2\lambda + 6 = 0 \Rightarrow$ eigenvalues $\lambda = \dfrac{-2 \pm \sqrt{4 - 24}}{2} = -1 \pm j\sqrt{5}$

$\Rightarrow$ Asymptotically stable

Eliminate $\hat{x}_2$ in the linearized state-space model. We get the system differential equation (input-output equation)

$$\ddot{\hat{x}}_1 + 2\dot{\hat{x}}_1 + 6\hat{x}_1 = -6\hat{u}$$

which confirms the above characteristic equation.

Example 11.10

Consider the mechanical system in Figure 11.7, which has a linear spring, a nonlinear spring, and a lumped mass. The state variables are defined as: $x_1 = x$ and $x_2 = \dot{x}$. Then, the state-space model is obtained (by writing Newton's second law) in the usual manner as

$$\dot{x}_1 = x_2$$

$$\dot{x}_2 = -\frac{3}{2}x_1 - \frac{1}{2}|x_1|x_1 + \frac{1}{2}f(t)$$

with the input $f(t) = 10 + \hat{f}(t)$ (see Figure 11.8).
  Operating point:
  Substitute $\bar{f} = 10$ and zero rates of changes. We get

$$0 = \hat{x}_2$$

$$0 = -\frac{3}{2}\bar{x}_1 - \frac{1}{2}|\bar{x}_1|\bar{x}_1 + \frac{1}{2} \times 10$$

For $\bar{x}_1 > 0$:

$$0 = -\frac{3}{2}\bar{x}_1 - \frac{1}{2}\bar{x}_1^2 + 5$$

$$\Rightarrow \bar{x}_1^2 + 3\bar{x}_1 - 10 = 0 \Rightarrow$$

$$\bar{x}_1 = 2 \text{ or } -5 \text{ (Reject } -5\text{)}.$$

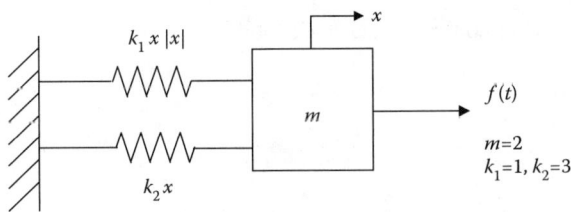

FIGURE 11.7
A Nonlinear mechanical system.

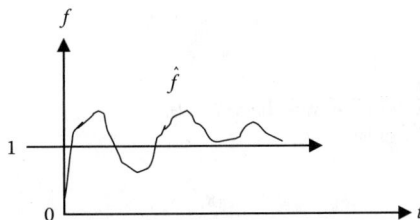

FIGURE 11.8
Input function.

For $\bar{x}_1 < 0$:

$$0 = -\frac{3}{2}\bar{x}_1 + \frac{1}{2}\bar{x}_1^2 + 5 \Rightarrow \bar{x}_1^2 - 3\bar{x}_1 + 10 = 0$$

$$\Rightarrow \bar{x}_1 = \frac{3 \pm \sqrt{9 - 40}}{2} = \text{complex} \quad (\text{Reject both roots}).$$

Linearization:

Since $\hat{x}_1 > 0$ we have $|x_1| = x_1$. Hence, the nonlinear term in the second state equation may be approximated as: $-\frac{1}{2}|x_1|x_1 = -\frac{1}{2}x_1^2 = -\frac{1}{2}[\bar{x}_1^2 + 2\bar{x}_1\hat{x}_1]$.

It follows that the linearized model is:

$$\dot{\hat{x}}_1 = \hat{x}_2,$$

$$\dot{\hat{x}}_2 = -\frac{3}{2}\hat{x}_1 - \bar{x}_1\hat{x}_1 + \frac{1}{2}\hat{f}(t)$$

with $|x_1| = 2$.

The system matrix is:

$$A = \begin{bmatrix} 0 & 1 \\ -\dfrac{7}{2} & 0 \end{bmatrix} \Rightarrow \lambda^2 + \frac{7}{2} = 0 \Rightarrow \lambda_1, \lambda_2 = \pm j\sqrt{\frac{7}{2}}$$

$\Rightarrow$ Both roots are imaginary $\Rightarrow$ Stable (i.s.L).

Example 11.11

Consider the second order nonlinear state-space model:

$$\dot{x}_1 = -x_1 + x_2$$

$$\dot{x}_2 = x_1 x_2 + u(t)$$

The operating point is obtained from:

$$0 = -\bar{x}_1 + \bar{x}_2$$

$$0 = \bar{x}_1\bar{x}_2 + \bar{u}$$

which gives: $\bar{x}_1 = \bar{x}_2 = \pm\sqrt{-\bar{u}}$

It follows that for a real system we must have $\bar{u} < 0$.

Linearized state-space model is:

$$\dot{\hat{x}}_1 = -\hat{x}_1 + \hat{x}_2$$

$$\dot{\hat{x}}_2 = \bar{x}_2\hat{x}_1 + \bar{x}_1\hat{x}_2 + \hat{u}(t)$$

The system matrix is:

$$A = \begin{bmatrix} -1 & 1 \\ \bar{x}_2 & \bar{x}_1 \end{bmatrix} \Rightarrow (\lambda + 1)(\lambda - \bar{x}_1) - \bar{x}_2 = 0$$

$$\lambda^2 + (1 - \bar{x}_1)\lambda - \bar{x}_1 - \bar{x}_2 = 0$$

Note: The product of the eigenvalues $\lambda_1 \lambda_2 = -\bar{x}_1 - \bar{x}_2$.

For $\bar{x}_1 = \bar{x}_2 = \sqrt{-\bar{u}}$: $\lambda_1 \lambda_2 < 0 \Rightarrow$ One eigenvalue will have a positive real part.
$\Rightarrow$ Unstable system.

For $\bar{x}_1 = \bar{x}_2 = -\sqrt{-\bar{u}}$: Both eigenvalues will have negative real parts.
$\Rightarrow$ Asymptotically stable system.

Example 11.12

Consider the Vander Pol equation (which, for example, represents the voltage build-up in a non-linear oscillator):

$$\ddot{V} - a(V^2 - 1)\dot{V} + kV = Q, a > 0, \quad k > 0$$

Here Q is the input.
Operating point:
Let us assume that Q=const.

$$\text{At the operating point } \left[\dot{V} = \ddot{V} = 0 \right] \Rightarrow k\bar{V} = Q \Rightarrow \bar{V} = V\frac{Q}{k}$$

Let $V = \bar{V} + \hat{v}$. The corresponding linearized equation is:

$$\ddot{\hat{v}} - a(\bar{V}^2 - 1)\dot{\hat{v}} + k\hat{v} = 0$$

It follows that we must have $a(\bar{V}^2 - 1) < 0$ for stability.
Hence, if $a > 0$ we must have $|V| < 1$ for stability.
If $|\bar{V}| = 0 \Rightarrow$ No conclusion about system stability.

11.3.5 Second Method (Direct Method) of Lyapunov

Stability of a system may be determined by the second method of Lyapunov, which is described now.

Theorem 1: If a positive definite function $V(\underline{x})$ can be determined such that $\dot{V}(\underline{x})$ is negative semidefinite then the origin is stable i.s.L.

Note: Such $V(\underline{x})$ is called a Lyapunov function (not necessarily a quadratic form). See Appendix C for the definitions of positive definite, etc. Also, $V(\underline{x})$ should be single-valued, continuous, and have continuous partial derivates.

Theorem 2: In Theorem 1, if $\dot{V}(\underline{x})$ is negative definite then the origin is asymptotically stable.

Note: Theorems 1 and 2 concern local stability at some ICs or operating points.

Theorem 3: If Theorem 2 is satisfied and, in addition, $V(\underline{x}) \to \infty$ as $\|\underline{x}\| \to \infty$, then the origin is globally asymptotically stable.

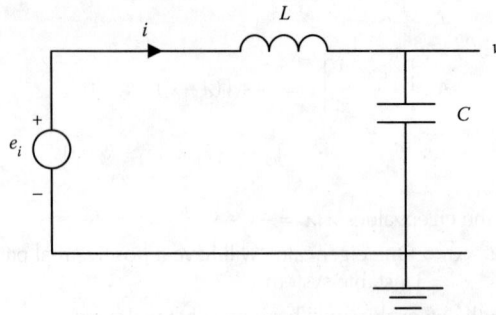

FIGURE 11.9
An electrical circuit.

Example 11.13

Consider the electrical circuit shown in Figure 11.9.
  Define the state variables (see Chapter 2): $x_1 = i$; $x_2 = V$
  We get the following state-space model:

$$\dot{x}_1 = -\frac{1}{L}x_2 + \frac{1}{L}e_i$$

$$\dot{x}_2 = \frac{1}{C}x_1$$

Consider the unforced (free) case: $e_i = 0$.

Try the positive definite function: $E = \frac{1}{2}Li^2 + \frac{1}{2}CV^2$

as a Lyapunov function. Differentiate and substitute the state equations:

$$\dot{E} = Li\frac{di}{dt} + CV\,\dot{V}$$

$$= Lx_1\left[-\frac{1}{L}x_2\right] + Cx_2\frac{1}{C}x_1 = 0 \;\Rightarrow\; \text{negative semidefinite.}$$

$\Rightarrow$ Stable i.s.L., but not asymptotically (globally) stable.
Note: $E \to \infty$ as $\|x\| \to \infty$.

Example 11.14

Consider the second order nonlinear state-space model

$$\dot{x}_1 = x_2 - ax_1(x_1^2 + x_2^2)$$

$$\dot{x}_2 = -x_1 - ax_2(x_1^2 + x_2^2)$$

Let us check stability of this system.
  Try the positive definite function: $V(\underline{x}) = x_1^2 + x_2^2$. Differentiate and substitute the state equations:

$$\dot{V}(\underline{x}) = 2x_1\dot{x}_1 + 2x_2\dot{x}_2 = -2a(x_1^2 + x_2^2)^2$$

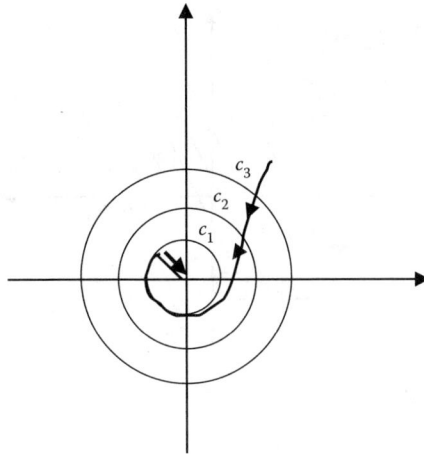

FIGURE 11.10
Free response of the asymptotically stable system.

This is negative definite. Also, $V(\underline{x}) \rightarrow \infty$ as $\|\underline{x}\| \rightarrow \infty$.

Hence, the system is globally asymptotically stable. See Figure 11.10 for a sketch of a free response of the system in the state-space (state plane, for this second-order system). The radii c_i are the constant values of $V(\underline{x})$.

Example 11.15

Let us examine the stability of the nonlinear, second order state-space model:

$$\dot{x}_1 = x_2$$

$$\dot{x}_2 = -ax_2 - bx_2{}^3 - x_1$$

As a Lyapunov function we try: $V(\underline{x}) = x_1^2 + x_2^2$

Differentiate and substitute the state equations:

$$\dot{V} = 2(x_1\dot{x}_1 + x_2\dot{x}_2) = -2\left[ax_2^2 + bx_2^4\right] = -2x_2^2\left[a + bx_2^2\right]$$

Case 1: $a > 0, b > 0$
  $\dot{V}$ is negative semidefinite $\Rightarrow$ Stable i.s.L
Case 2: $a < 0, b < 0$
  $\dot{V}$ is positive semidefinite $\Rightarrow$ Most likely unstable (or at best marginally stable).
Case 3: $a > 0, b > 0$
  Let $b = -\tilde{b}$.
  For stability we must have $a - \tilde{b}x_2^2 > 0 \Rightarrow x_2^2 < \dfrac{a}{b}$. This gives the stability region (i.s.L), as sketched in Figure 11.11.

11.3.5.1 Lyapunov Equation

Consider the free (unforced), linear, constant (time-invariant) system:

$$\dot{x} = Ax \tag{11.71}$$

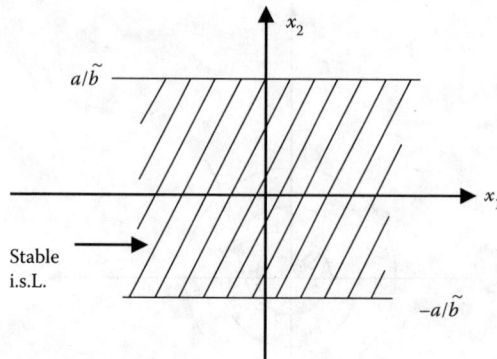

FIGURE 11.11
Stability region (shaded area).

and the quadratic form:

$$V(x) = x^T P x \tag{11.72}$$

where the matrix P is taken to be symmetric without any loss of generality. Assume that P is positive definite so that the quadratic form (Equation 11.72) is positive definite (see Appendix C). Differentiate Equation 11.72 and substitute Equation 11.71:

$$\dot{V}(x) = \dot{x}^T P x + x^T P \dot{x} = x^T A P x + x^T P A x = x^T (AP + PA) x \tag{11.73}$$

It follows that the system (Equation 11.71) is stable i.s.L if:

$$AP + PA = -N \tag{11.74}$$

where N is a positive semidefinite matrix. Equation 11.74 is called *Lyapunov equation*.

Theorem: A linear constant system is stable i.s.L if the Lyapunov equation $AP + PA = -N$ is satisfied where A is the system matrix, P is a positive definite matrix, and N is a positive semidefinite matrix.

Note: In the above, if N is positive definite, the system is asymptotically stable.

Procedure:

1. Pick N to be positive semidefinite (*Note*: N and P are symmetric)
2. Solve the $(1/2)n(n+1)$ equations in Equation 11.74 (*Note*: n is the system order)
3. Check conditions for the positive definiteness of P (for example, apply Sylvester's Theorem—see Appendix C).

Example 11.16

Consider the second-order system equation (unforced)

$$\ddot{x} + a_1 \dot{x} + a_0 x = 0$$

The system matrix in the companion form is: $A = \begin{bmatrix} 0 & 1 \\ -a_0 & -a_1 \end{bmatrix}$

Use: $P = \begin{bmatrix} P_{11} & P_{12} \\ P_{12} & P_{22} \end{bmatrix}$ and $P = \begin{bmatrix} 2 & 0 \\ 0 & 0 \end{bmatrix}$

Lyapunov equation gives:

$$\left(\begin{matrix} 2a_0 P_{12} = 2 \\ -P_{11} + a_1 P_{12} + a_0 P_{22} = 0 \\ -2P_{12} + 2a_1 P_{22} = 0 \end{matrix} \right) \quad Note: \frac{1}{2} n(n+1) = 3$$

Solution: $P_{12} = \dfrac{1}{a_0}$, $P_{22} = \dfrac{1}{a_0 a_1}$, $P_{11} = \dfrac{a_0 + a_1^2}{a_0 a_1}$

Apply Sylvester's theorem for positive definiteness of P:

$$P_{11} = \frac{a_0 + a_1^2}{a_0 a_1} > 0 \quad \text{and} \quad P_{11}P_{22} - P_{12}^2 = \frac{a_0}{a_0^2 a_1^2} > 0$$

From second condition: $a_0 > 0$
From first condition: $a_1 > 0$
Note: These are identical to the Routh–Hurwitz conditions of stability, in classical control.
$\Rightarrow$ eigenvalues of A will have negative real parts in this case.
$\Rightarrow$ asymptotically stable (not just stable i.s.L).
Theorem: The solution P of (the modified Lyapunov equation) $A^T P + PA = -I$ is positive definite iff the eigenvalues of A have negative real parts.
Note: This theorem will guarantee asymptotic stability for a linear constant system, through the solution of Lyapunov's equation.

Example 11.17

Re-do Example 11.16 with $N=I$.

11.4 Controllability and Observability

A controller that is commonly used in modern control engineering is state feedback (constant-gain). In this, the system states are measured and fed back through constant-gain amplifiers, into the system. Two main questions arise in this context:

1. Can the state (vector) of the system be changed from any arbitrary value to any other specified value in a finite time, through state feedback?

2. If not all the states in the system are available for measurement (observation) then, can the full state vector be completely determined from the outputs (or, measurable states) of the system over a finite duration?

The first question relates to controllability and the second question to observability of the system. In the present section we will discuss these two questions and some related issues. Note that since the origin of the state-space can be chosen arbitrarily, we can make the specified (second) state in Question 1 to be the origin.

Definition 1: A linear system is controllable at time t_0 if we can find an input u that transfers any arbitrary state $x(t_0)$ to the origin ($x=0$) in some finite time t_1. If this is true for any t_0 then the system is said to be completely controllable.

Definition 2: A linear system is observable at time t_0 if we can completely determine the state $x(t_0)$ from the output measurements y over the duration $[t_0, t_1]$ for finite t_1. If this is true for any t_0 then the system is said to be completely observable.

Now consider the time-invariant (i.e., constant-parameter) linear case only.

Criteria 1

Without giving a solid proof, we give some basic criteria for controllability now. These criteria should be intuitively clear.

Consider the state-space model:

$$\dot{x} = Ax + Bu \tag{11.75}$$

$$y = Cx + Du \tag{11.76}$$

Assume distinct eigenvalues $\Rightarrow$ the Jordan matrix J is diagonal. Apply the modal transformation:

$$x = Mq \tag{11.77}$$

where M is the matrix of independent eigenvectors (corresponding to the distinct eigenvalues) of the system. We get:

$$\dot{q} = Jq + M^{-1}Bu \tag{11.78}$$

$$y = CMq + Du \tag{11.79}$$

Then, the following are clear from Equations 11.78 and 11.79:

Controllable iff $M^{-1}B$ has all nonzero rows.
Observable iff CM has all nonzero columns.

To verify the first statement, note that if the ith row of $M^{-1}B$ is zero, then the input u has no effect on the corresponding modal response q_i. Then, that mode is not controllable.

To verify the second statement, note that if the ith column of CM is zero, then the modal response q_i will never be included in the output y. Then that mode is not observable.

Criteria 2

General criteria for controllability and observability of a linear constant system are as follows:

$$\text{Controllable iff Rank } [\, B \mid A\,B \mid \cdots \mid A^{n-1}B] = n \tag{11.80}$$

$$\text{Observable iff Rank } [C^T \mid A^T\, C^T \mid \cdots \mid A^{n-1}\, C^T] = n \tag{11.81}$$

Note: Iff means "If and only if."

Controllability Proof:

An indication of the procedure for the proof of Equation 11.80 is given now. The state response (forced) of system (Equation 11.75) is

$$x(t) = e^{A(t-t_o)}x(t_o) + \int_{t_o}^{t} e^{A(t-\tau)}B(\tau)u(\tau)d\tau \tag{11.82}$$

If controllable, from Definition 1, we must have (with $t_0 = 0$ and $t = t_1$)

$$0 = e^{At_1}x(0) + \int_0^{t_1} e^{A(t_1-\tau)}Bu(\tau)d\tau \Rightarrow x(0) = -\int_0^{t_1} e^{-A\tau}Bu(\tau)d\tau \tag{i}$$

$$= \int_0^{t_1} [\alpha_0(\tau)I + \alpha_1(\tau)A + \alpha_2(\tau)A^2 + ... + \alpha_{n-1}(\tau)A^{n-1}]Bu(\tau)d\tau$$

(from the finite series expansion of the matrix exponential)

$$= \sum_{j=0}^{n-1} A^j B \int_0^{t_1} \alpha_j(\tau)u(\tau)d\tau$$

$$= \sum_{j=0}^{n-1} A^j B v_j$$

Since the right hand side above should be able to form any arbitrary state vector on the left hand side, it is required that the terms $A^j B$ together should span the n-dimensional vector space (state-space); see Appendix C. In other words, the *controllability matrix*

$$P = [B|AB|\cdots|A^{n-1}B] \tag{11.83}$$

must have the rank n.

Note: What we proved is the "if" part. The "only if" part can be proved by starting with Rank $[B|AB|\cdots|A^{n-1}B] = n$ and the showing that the relation (i) can be made to hold for any arbitrary $x(t_1)$.

Observability Proof:

Consider the output equation (Equation 11.76) and substitute the state response (Equation 11.82):

$$y(t) = Cx(t) + Du(t)$$

$$= C[e^{At}x(0) + \int_0^{t} e^{A(t-\tau)}Bu(\tau)d\tau] + Du \tag{ii}$$

Here, without loss of generality, we have taken $t_0=0$. For observability, then, we need to determine a condition such that $x(0)$ can be completely determined with the output information from time 0 to finite t. Rearranging (ii) we get:

$$Ce^{At}x(0) = y(t) - C\int_0^t e^{A(t-\tau)}Bu(\tau)d\tau - Du$$

By using the finite series expansion for the matrix exponential we have:

$$C[\alpha_0(t)\,I + \alpha_1(t)\,A + \cdots + \alpha_{n-1}(t)A^{n-1}]\,x(0) = y(t) - f(u)$$

where the terms containing u are combined together. Note that the right hand side is completely known. Multiplying out the left hand side, the above result is written as:

$$[\alpha_0 C + \alpha_1 CA + \cdots + \alpha_{n-1}CA^{n-1}]\,x(0) = f(y,u) \qquad \text{(iii)}$$

where the right hand side is completely known. We require a condition such that Equation (iii) can be solved to obtain a unique solution for $x(0)$. In other words, Equation (iii) needs to form a set of n linearly independent equations. It is intuitively clear (see Appendix C) that the required condition is, the *observability matrix*:

$$Q = [C^T\,|A^T C^T\,|\cdots|A^{n-1}\,C^T] \qquad (11.84)$$

must have the rank n (i.e., the columns within [ ] should span the state-space. Here we have used the following theorem.

Theorem: If X is a real matrix, then X and $X^T X$ have the same rank.

Criteria 3 (Pole-Zero Cancellation)

Controllability and observability may be interpreted as well, in terms of pole-zero cancellation in the system transfer function. To illustrate this, consider a SISO system with pole-zero cancellation in its transfer function $G(s)$. We will show that a state variable realization of the system that includes the cancelled pole is either uncontrollable or unobservable or both.

The system transfer function (output-input in the Laplace domain, with zero ICs) is:

$$G(s) = \frac{Y(s)}{U(s)} \qquad (11.85)$$

Suppose that the cancelled pole (eigenvalue) is $s=p$. Expressing the system in the Jordan canonical form, with the cancelled pole p (assumed to be distinct), we have

$$\dot{q} = \begin{bmatrix} p & 0 \\ \hline 0 & \Delta \end{bmatrix} q + \begin{bmatrix} a_1 \\ \hline b \end{bmatrix} u \qquad (11.86)$$

$$y = [a_2\,|\,C^T]q \qquad (11.87)$$

Note that Δ in (11.86) is a block diagonal matrix (consisting of Jordan blocks). Also, for the SISO case, u and y in Equations 11.86 and 11.87 are scalars. Hence, we can write

$$\frac{Y(s)}{U(s)} = G(s) = [a_2 | C^T] \left[\begin{array}{c|c} s-p & 0 \\ \hline 0 & sI - \Delta \end{array} \right]^{-1} \left[\begin{array}{c} a_1 \\ b \end{array} \right]$$

$$= [a_2 | C^T] \left[\begin{array}{c|c} \dfrac{1}{s-p} & 0 \\ \hline 0 & | Q(s) \end{array} \right]^{-1} \left[\begin{array}{c} a_1 \\ b \end{array} \right] \tag{11.88}$$

$$\text{Where: } Q = (sI - \Delta)^{-1} \tag{11.89}$$

Multiply out the right hand side of Equation 11.88:

$$G(s) = [\dfrac{a_2}{s-p} | C^T Q(s)] \left[\begin{array}{c} a_1 \\ b \end{array} \right]$$

Further multiplication gives:

$$G(s) = \dfrac{a_1 a_2}{s-p} + C^T Q(s)b \tag{11.90}$$

Note from Equation 11.90 that three cases of pole-zero cancellation (i.e., the p term vanishes) are possible. These cases are considered below.

Case 1: $a_1 = 0, a_2 \neq 0$
It is clear from Equation 11.86 that in this case the input will not influence the mode corresponding to pole p. Hence the system is not controllable.

Case 2: $a_1 \neq 0, a_2 = 0$
It is clear from Equation 11.87 that in this case the modal response corresponding to pole p will not enter into the output y. Hence the system is not observable.

Case 3: $a_1 = 0, a_2 = 0$
It is clear from Equations 11.86 and 11.87 that in this case system is neither controllable nor observable.

In general, the following two theorems can be stated.

Theorem 1: The system (A, B) is controllable iff $(s\underline{I} - A)^{-1}B$ has no pole-zero cancellations.
Theorem 2: The system (A, C) is observable iff $C(sI - A)^{-1}$ has no pole-zero cancellations.

11.4.1 Minimum (Irreducible) Realizations

If a pole-zero cancellation is possible, what this means is that the system can be reduced to a lower order one. A state-space realization (A,B,C) is said to be a minimum realization if pole-zero cancellations are not possible, and hence the system order (dimension of the

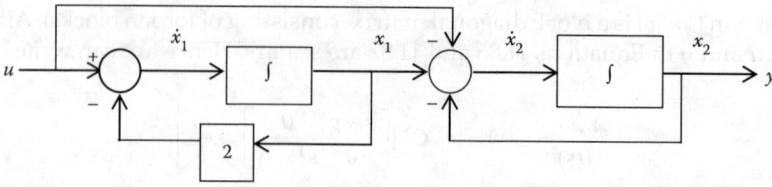

FIGURE 11.12
Simulation block diagram of a system.

state-space) cannot be reduced further. Equivalently, according to Criteria 3 presented above, a realization (A,B,C) is minimal iff (A,B) is controllable and (A,C) is observable).

Example 11.18

Consider the system represented by the simulation block diagram in Figure 11.12.
It is clear from Figure 11.12 that (see Chapter 5) the state-space model (the two state equations and the output equation) is:

$$\dot{x}_1 = -2x_1 + u$$

$$\dot{x}_2 = -x_1 - x_2 + u$$

$$y = x_2$$

It follows that the corresponding matrices are:

$$A = \begin{bmatrix} -2 & 0 \\ -1 & -1 \end{bmatrix}, \quad B = \begin{bmatrix} 1 \\ 1 \end{bmatrix}, \quad C = \begin{bmatrix} 0 & 1 \end{bmatrix}$$

Let us check controllability and observability of this system using the three groups of criteria that were presented before.
Criteria 2:

$$\text{Controllability matrix } P = \begin{bmatrix} 1 & -2 \\ 1 & -2 \end{bmatrix}.$$

Note that the the two columns of this matrix are not linearly independent. This is further confirmed since the determinant of P is zero. Hence the rank of P has to be less than 2. In fact it is clear that Rank $P=1 \Rightarrow$ uncontrollable.

Observability matrix $Q = \begin{bmatrix} 0 & -1 \\ 1 & -1 \end{bmatrix}$. Note that the Det $Q \neq 0$ (also, the two columns of Q are clearly independent). Hence, Rank $Q=2 \Rightarrow$ observable.
Criteria 3:
Controllability Check:
Straightforward linear algebra gives

$$sI - A = \begin{bmatrix} s+2 & 0 \\ 1 & s+1 \end{bmatrix} \Rightarrow (sI - A)^{-1} = \frac{1}{(s+1)(s+2)} \begin{bmatrix} s+1 & 0 \\ -1 & s+2 \end{bmatrix} \Rightarrow$$

$$(sI - A)^{-1}B = \frac{1}{(s+1)(s+2)} \begin{bmatrix} s+1 \\ s+1 \end{bmatrix} = \frac{(s+1)}{(s+1)(s+2)} \begin{bmatrix} 1 \\ 1 \end{bmatrix}$$

$\Rightarrow$ pole-zero cancellation $\Rightarrow$ system is uncontrollable.
Specifically, the mode with eigenvalue $\lambda_1 = -1$ is uncontrollabile.
Observability Check:
Straightforward linear algebra gives

$$C(sI - A)^{-1} = \frac{1}{(s+1)(s+2)} [-1 \quad (s+2)] \Rightarrow \text{no cancellation} \Rightarrow \text{observable.}$$

Criteria 1:
Eigenvalues of the system are obtained as follows:

$$\lambda I - A = \begin{bmatrix} \lambda + 2 & 0 \\ 1 & \lambda + 1 \end{bmatrix} \Rightarrow |\lambda I - A| = (\lambda + 2)(\lambda + 1) = 0$$

(for eigenvalues). Hence, the eigenvalues are: $\lambda_1 = -1$, $\lambda_2 = -2$.
The eigenvectors are obtained as follows:

$$\text{For } \lambda_1 = -1: \begin{bmatrix} 1 & 0 \\ 1 & 0 \end{bmatrix} \begin{bmatrix} a \\ b \end{bmatrix} = \begin{bmatrix} 0 \\ 0 \end{bmatrix} \Rightarrow \xi_1 = \begin{bmatrix} 0 \\ 1 \end{bmatrix}$$

$$\text{For } \lambda_2 = -2: \begin{bmatrix} 0 & 0 \\ 1 & -1 \end{bmatrix} \begin{bmatrix} a \\ b \end{bmatrix} = \begin{bmatrix} 0 \\ 0 \end{bmatrix} \Rightarrow a - b = 0 \Rightarrow \xi_2 = \begin{bmatrix} 1 \\ 1 \end{bmatrix}$$

The modal matrix (matrix of independent eigenvectors) $M = \begin{bmatrix} 0 & 1 \\ 1 & 1 \end{bmatrix}$
Controllability Check:

$$M^{-1} = \frac{1}{(-1)} \begin{bmatrix} 1 & -1 \\ -1 & 0 \end{bmatrix} = \begin{bmatrix} -1 & 1 \\ 1 & 0 \end{bmatrix} \Rightarrow M^{-1}B = \begin{bmatrix} -1 & 1 \\ 1 & 0 \end{bmatrix} \begin{bmatrix} 1 \\ 1 \end{bmatrix} = \begin{bmatrix} 0 \\ 1 \end{bmatrix} \Rightarrow \text{uncontrollable.}$$

Specifically, the mode with the eigenvalue λ_1 is uncontrollable.
Observability Check:

$$CM = [0 \quad 1] \begin{bmatrix} 0 & 1 \\ 1 & 1 \end{bmatrix} = [1 \quad 1] \Rightarrow \text{observable.}$$

Note: The modal equations of the system are:

$$\dot{q}_1 = -q_1$$
$$\dot{q}_2 = -2q_2 + u \Leftarrow \text{Jordan canonical form.}$$
$$y = q_1 + q_2$$

A simulation block diagram of this form is shown in Figure 11.13. Note that the input u does not reach the first mode (q_1), and hence, that mode is uncontrollable.

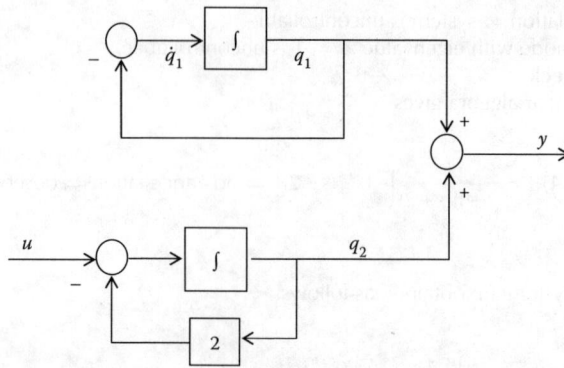

FIGURE 11.13
A system with an uncontrollable mode.

Example 11.19

What do you think of the system:

$$\dot{x} = \begin{bmatrix} 1 & 2 & 1 & -2 \\ -1 & 1 & -3 & 2 \\ 0 & 0 & 2 & 0 \\ 3 & 2 & 7 & 6 \end{bmatrix} x + \begin{bmatrix} -3 \\ 2 \\ 0 \\ 1 \end{bmatrix} u; \quad y = \begin{bmatrix} 1 & 2 & 3 & 4 \\ -4 & -3 & -2 & -1 \end{bmatrix} x$$

Solution

Here note that the third state equation is uncoupled from the rest and does not contain an input term. Specifically, we have:

$$\dot{x}_3 = 2x_3$$

Hence, the state x_3 is uncontrollable.

Example 11.20

What do you think of the system:

$$\dot{x} = \begin{bmatrix} 1 & 2 & 3 & 4 \\ 0 & 1 & 0 & 0 \\ 4 & 3 & 2 & 1 \\ 0 & 0 & 0 & -3 \end{bmatrix} x + \begin{bmatrix} 1 \\ 2 \\ 3 \\ 4 \end{bmatrix} u$$

$$y = [0 \ 1 \ 0 \ -3] x$$

Solution

Here note from the output equation that the states x_1 and x_3 are not "directly" present in the output variables. Furthermore, note from the state equations that the states x_2 and x_4 are uncoupled from the rest, and hence, these two states will not carry information about the states x_1 and x_3. It follows that the x_1 and x_3 are unobservable.

11.4.2 Companion Form and Controllability

Consider the input-output differential equation model of a system, as given by:

$$\frac{d^n y}{d_{t^n}} + a_{n+1}\frac{d^{n-1}y}{d_{t^{n-1}}} + \cdots + a_0 y = u \tag{11.91}$$

A state-space model $\dot{x} = Ax + Bu$ may be obtained by defining the state variables as $x_1 = y$, $x_2 = \dot{y}$, ..., $x_n = d^{n-1}y/dt^{n-1}$. This model is said to be in the *companion form*, and its matrices are:

$$A = \begin{bmatrix} 0 & 1 & 0 & . & . & . & 0 \\ 0 & 0 & 1 & . & . & . & 0 \\ 0 & 0 & 0 & . & . & . & 0 \\ . & . & . & . & . & . & . \\ . & . & . & . & . & . & . \\ 0 & 0 & 0 & . & . & . & 1 \\ -a_0 & -a_1 & -a_2 & . & . & . & -a_{n-1} \end{bmatrix}, \quad B = \begin{bmatrix} 0 \\ 0 \\ 0 \\ \vdots \\ 0 \\ 1 \end{bmatrix} \tag{11.92}$$

Using this model, according to Equation 11.83 we form the controllability matrix:

$$P = \begin{bmatrix} 0 & 0 & & & & 1 \\ 0 & 0 & & & \cdot\cdot & \\ \vdots & \vdots & 1 & & & \\ 0 & 1 & -a_0 & & & \\ 1 & -a_0 & -a_{n-2}+a_{n-1}a_0 & & {''} & {''} \end{bmatrix} \tag{11.93}$$

It can be shown that the determinant of the controllability matrix is nonzero. Specifically, Det $P = 1$. Hence, Rank $P = n \Rightarrow$ system is controllable.

Conclusion: If a system can be expressed in the companion form, then it is controllable.

11.4.3 Implication of Feedback Control

Now let us address the implication of feedback (constant-gain) on the controllability and observability of a system. First we will consider the general case of output feedback. The results can be used then for the special case of state feedback (where, $C = I$).

Consider the system:

$$\dot{x} = Ax + Bu \tag{11.94}$$

$$y = Cx \tag{11.95}$$

This system may be represented by the block diagram in Figure 11.14.

Suppose that (A, B) is uncontrollable and (A, C) is unobservable. Now apply the output feedback control:

$$u = u_{ref} - ky \tag{11.96}$$

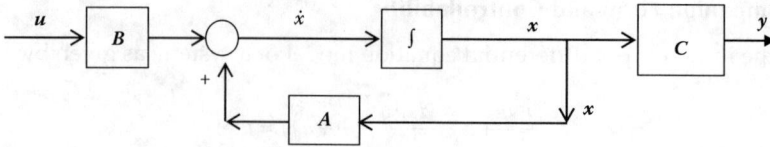

FIGURE 11.14
The open-loop system (plant).

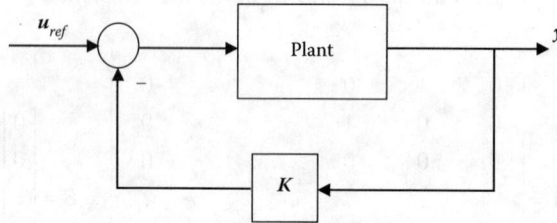

FIGURE 11.15
The closed-loop system.

where u_{ref} is some reference input. Substitute Equation 11.96 into Equations 11.94 and 11.95, to obtain the closed-loop system:

$$\dot{x} = (A - BKC)x + Bu_{ref} \tag{11.97}$$

This closed-loop system is shown in Figure 11.15. Note that the block denoted as the "Plant" is the open-loop system given in Figure 11.14.

The following two questions are posed:

Is the closed-loop system (*A-B K C, B*) controllable for some *K*?

Is the closed-loop system (*A-B K C, C*) observable for some *K*?

The answers to these questions are provided by the following two theorems.

Theorem 3: If (*A, B*) is uncontrollable, then for any compatible *K*, (*A-B K C, B*) is also uncontrollable.

Theorem 4: If (*A, C*) is unobservable, then for any compatible *K*, (*A-B K C, C*) is also unobservable.

Proof: What the controller (Equation 11.96) did was to change the nature of the input *u*. But, it is given that the system (Equation 11.94), Equation 11.95 is uncontrollable and unobservable for any *u*. Q.E.D.

11.4.4 State Feedback

1. Now consider the case of full state feedback, as given by:

$$u = u_{ref} - Kx \tag{11.98}$$

This is simply a special case of the output feedback (Equation 11.96), with *C=I*. Hence the Theorems 3 and 4 still hold. Specifically, an uncontrollable and an unobservable system will remain uncontrollable and unobservable with state feedback.

Example 11.21

Consider the open-loop system:

$$A = \begin{bmatrix} -2 & 0 \\ -1 & -1 \end{bmatrix}, \quad B = \begin{bmatrix} 1 & 2 \\ 1 & 2 \end{bmatrix}.$$

Controllability matrix: $P = [B, AB] = \begin{bmatrix} 1 & 2 & -2 & -4 \\ 1 & 2 & -2 & -4 \end{bmatrix}$

Note that all the columns in P are simple multiples of each other (i.e., they are linearly dependent). Hence:
Rank $P = 1 \Rightarrow$ uncontrollable.
Now apply a state feedback with the feedback gain matrix:

$$K = \begin{bmatrix} a & b \\ c & d \end{bmatrix}$$

Then we have $BK = \begin{bmatrix} a+2c & b+2d \\ a+2c & b+2d \end{bmatrix}$ and $A - BK = \begin{bmatrix} -2-a-2c & -b-2d \\ -1-a-2c & -1-b-2d \end{bmatrix}.$

The controllability matrix of the closed-loop system is

$$P = [B, (A - BK)B] = \begin{bmatrix} 1 & 2 & (-2-a-b-2c-2d) & 2(-2-a-b-2c-2d) \\ 1 & 2 & (-2-a-b-2c-2d) & 2(-2-a-b-2c-2d) \end{bmatrix}.$$

Again we note that all the columns of the controllability matrix are linearly dependent. Hence, Rank $P = 1 \Rightarrow$ uncontrollable.

11.4.5 Stabilizability

It is clear that (see Criterion 1 of controllability) if a system is controllable, then every one of its modes can be controlled by the input. In particular, any unstable modes (i.e., corresponding eigenvalues having positive real parts) can be stabilized through control. It follows that controllable systems are stabilizable. If a system is uncontrollable, still it is stabilizable provided that all its unstable modes are controllable.

Example 11.22

Consider the state-space realization:

$$A = \begin{bmatrix} 0 & 2 \\ 1 & 1 \end{bmatrix}, \quad B = \begin{bmatrix} 1 \\ 1 \end{bmatrix}, \quad C = \begin{bmatrix} 0 & 1 \end{bmatrix}.$$

It can be easily verified that:

Controllability matrix $P = [B, AB] = \begin{bmatrix} 1 & 2 \\ 1 & 2 \end{bmatrix}.$

The two columns of P are linearly dependent (also, Det $P=0$). Hence, Rank $P=1 \Rightarrow$ system is uncontrollable.

Note: It follows that what is given is not a minimum realization and that a pole-zero cancellation can be made.

$$\text{Observability matrix } Q = [C^T, A^T C^T] = \begin{bmatrix} 0 & 1 \\ 1 & 1 \end{bmatrix}$$

Note that Det $Q \neq 0$. Hence Rank $Q=2 \Rightarrow$ system is observable.

Now let us check whether the system is stabilizable. For this, we need to determine the modes of the system. If both modes of the system are unstable, then the system is not stabilizable, because it is uncontrollable. If both modes are stable, of course it is stabilizable (even without any feedback control). If one of the modes is unstable, then if that mode is controllable, the system is stabilizable. But, if the unstable mode is uncontrollable, then the system is not stabilizable. Let us check for these possibilities.

Eigenvalues:

$$\text{Det } (\lambda I - A) = \begin{vmatrix} \lambda & -2 \\ -1 & \lambda - 1 \end{vmatrix} = 0 \Rightarrow \lambda^2 - \lambda - 2 = 0 \Rightarrow \lambda_1 = -1, \ \lambda_2 = 2.$$

Note: First mode is stable and the second mode is unstable.

Eigenvectors: $(\lambda I - A)\xi = 0$

$$\text{For } \lambda_2: \begin{bmatrix} -1 & -2 \\ -1 & -2 \end{bmatrix} \begin{bmatrix} \xi_{11} \\ \xi_{12} \end{bmatrix} = \begin{bmatrix} 0 \\ 0 \end{bmatrix} \Rightarrow \xi_1 = \begin{bmatrix} -2 \\ 1 \end{bmatrix}$$

$$\text{For } \lambda_2: \begin{bmatrix} 2 & -2 \\ -1 & 1 \end{bmatrix} \begin{bmatrix} \xi_{21} \\ \xi_{22} \end{bmatrix} = \begin{bmatrix} 0 \\ 0 \end{bmatrix} \Rightarrow \xi_2 = \begin{bmatrix} 1 \\ 1 \end{bmatrix}$$

$$\text{The modal matrix } M = \begin{bmatrix} -2 & 1 \\ 1 & 1 \end{bmatrix} \Rightarrow M^{-1} = \frac{1}{(-3)} \begin{bmatrix} 1 & -1 \\ -1 & -2 \end{bmatrix} = \begin{bmatrix} -1/3 & 1/3 \\ 1/3 & 2/3 \end{bmatrix} \Rightarrow$$

$$M^{-1}B = \begin{bmatrix} -1/3 & 1/3 \\ 1/3 & 2/3 \end{bmatrix} \begin{bmatrix} 1 \\ 1 \end{bmatrix} = \begin{bmatrix} 0 \\ 1 \end{bmatrix}$$

It is seen that the first (stable) mode is uncontrollable and the second (unstable) mode is controllable (according to Criterion 1 of controllability). Hence the overall system is stabilizable.

11.5 Modal Control

Since the overall response of a system depends on the individual modes, it should be possible to control the system by controlling its modes. This is the basis of modal control. A mode is determined by the corresponding eigenvalue and eigenvector. In view of this, a popular approach of modal control is the *pole placement* or *pole assignment*. In this method of controller design, the objective is to select a feedback controller that will make the poles of the closed-loop system take up a set of desired values. The method of pole placement is addressed in this section.

Consider the uncontrolled (open-loop) system:

$$\dot{x} = Ax + Bu \qquad (11.99)$$

in the usual notation. Assume that all n states are available for feedback (in particular, assume that all n states are measurable). In general, we can implement a constant-gain feedback control law of the form:

$$u = -Kx \qquad (11.100)$$

in which K is the *feedback gain matrix*. By substituting Equation 11.100 into Equation 11.99 we get the closed-loop (controlled) system:

$$\dot{x} = (A - BK)u \qquad (11.101)$$

It follows that the closed-loop poles are the eigenvalues of the *closed-loop system matrix* $(A - BK)$. With this particular control structure, the problem of controller design reduces to the selection of a proper K that will assign desired values to the closed-loop poles.

11.5.1 Controller Design by Pole Placement

First we will study the case of a single input. Clearly, the results can be then extended to the case of multiple inputs. The system (Equation 11.99) for the case of a single input may be expressed as:

$$\dot{x} = Ax + bu \qquad (11.102)$$

where u is a scalar (single) input, and hence, b is a column vector of the same dimension (n) as the state vector. This open-loop system is shown in Figure 11.16. Next, using the complete state feedback control law:

$$u = u_{ref} - Kx \qquad (11.103)$$

one gets the closed-loop control system:

$$\dot{x} = (A - bK)x + bu_{ref} \qquad (11.104)$$

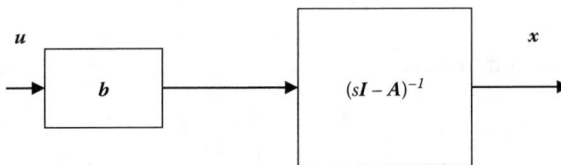

FIGURE 11.16
The open-loop system.

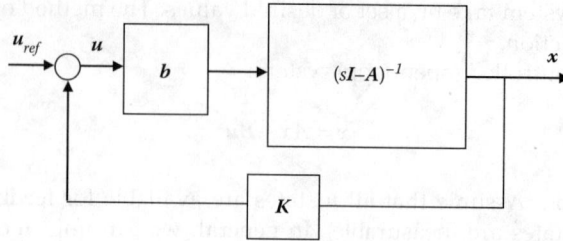

FIGURE 11.17
The closed-loop system.

where $A-bK$=closed-loop system matrix. The closed-loop system is shown in Figure 11.17.

Question: When can we arbitrarily assign (place) the poles of the closed-loop system matrix $\tilde{A}=A-bK$?

Theorem: Iff (A, b) is controllable, we can arbitrarily place the system poles using complete sate feedback for the corresponding single input.

Proof (By Construction): (If Part)

Suppose that the coefficients of the characteristic polynomial are a_i, and that the system is expressed in the companion form (*Note*: Any other form has to be transformed into this form) in which:

$$A = \begin{bmatrix} 0 & 1 & 0 & \cdots & 0 \\ 0 & 0 & 1 & \cdots & 0 \\ \cdots & \cdots & \cdots & \cdots & \cdots \\ -a_0 & -a_1 & \cdots & \cdots & -a_{n-1} \end{bmatrix}, \quad b = \begin{bmatrix} 0 \\ 0 \\ 0 \\ \vdots \\ 1 \end{bmatrix} \qquad (11.105)$$

It was shown that (A,b) is controllable in this case. Suppose that the feedback matrix is:

$$K = [k_1 \; k_2 \; \ldots \; k_n] \qquad (11.106)$$

Then:

$$bK = \begin{bmatrix} 0 & 0 & \cdots & 0 \\ 0 & 0 & \cdots & 0 \\ \vdots & \vdots & \ddots & \vdots \\ 0 & 0 & \cdots & 0 \\ k_1 & k_2 & \cdots & k_n \end{bmatrix}$$

and the closed-loop system matrix is:

$$\tilde{A} = A - bK = \begin{bmatrix} 0 & 1 & \cdots & & \cdots \\ 0 & 0 & \ddots & & \vdots \\ \vdots & \vdots & \vdots & & 1 \\ -a_0 - k_1 & \cdots & \cdots & & -a_{n-1} - k_n \end{bmatrix} \qquad (11.107)$$

The open-loop poles are: $(\lambda_1, \lambda_2, ..., \lambda_n)$ and the corresponding characteristic polynomial is:

$$s^n + a_{n-1}s^{n-1} + \cdots + a_0 = (s - \lambda_1)(s - \lambda_2) \cdots (s - \lambda_n) \tag{11.108}$$

Suppose that the closed-loop poles are to be located at the arbitrary values: $(\tilde{\lambda}_1, \tilde{\lambda}_2, \cdots, \tilde{\lambda}_n)$. According to Equation 11.107 it is clear that the desired poles can be assigned if the corresponding characteristic polynomial satisfies:

$$s^n + (a_{n-1} + k_n)s^{n-1} + \cdots + (a_0 + k_1) = (s - \tilde{\lambda}_1)(s - \tilde{\lambda}_2) \cdots (s - \tilde{\lambda}_n) \tag{11.109}$$

Clearly the gains k_i can be chosen to satisfy this equation.

According to this procedure, the following steps can be given for the controller design in pole placement with single input:

Step 1: If the system is not in companion form, transform into the companion form using $x = Rz$ (This is possible, if the system is controllable).

Step 2: Form the closed-loop characteristic polynomial using desired poles: $(\tilde{\lambda}_1, \tilde{\lambda}_2, \cdots, \tilde{\lambda}_n)$, resulting in the coefficients $\tilde{a}_i$, $i = 0,1,...,n-1$.

Step 3: By equating coefficients determine the feedback gains:

$$k_i = \tilde{a}_{i-1} - a_{i-2}, \ i = 1,2,...,n \tag{11.110}$$

Step 4: The corresponding feedback control law is

$$u_{ref} - u = Kz = KR^{-1}x \tag{11.111}$$

Example 11.23

A mechanical plant is given by the input-output differential equation $\ddot{x} + \dot{x} = u$, where u is the input and x is the output. Determine a feedback law that will yield approximately a simple oscillator response with a damped natural frequency of 1 unit and a damping ratio of $1/\sqrt{2}$.

Solution

It is well known that a stable complex conjugate pair of poles, representing damped oscillations, may be expressed as:

$$\lambda_1, \lambda_2 = -\zeta\omega_n \pm j\omega_d \tag{11.112}$$

where:
ω_n = undamped natural frequency
ζ = damping ratio,
and, the damped natural frequency is:

$$\omega_d = \sqrt{1 - \zeta^2}\omega_n \tag{11.113}$$

It is given that $\omega_d = 1$ and $\zeta = 1/\sqrt{2}$. Hence, from Equation 11.113 we get $\omega_n = \sqrt{2}$ and hence, $\zeta\omega_n = 1$. It follows that we need to place two poles at $-1 \pm j$. Also the third pole has to be far from these two on the left hand plane (LHP); say, at -10. We perform the following steps of pole placement design:

Step 1: Define the state variables as $x_1 = x$, $x_2 = \dot{x}_1$, and $x_3 = \dot{x}_2$. The corresponding state-space model is:

$$\dot{x}_1 = x_2$$

$$\dot{x}_2 = x_3$$

$$\dot{x}_3 = -x_3 + u$$

with the matrices:

$$A = \begin{bmatrix} 0 & 1 & 0 \\ 0 & 0 & 1 \\ 0 & 0 & -1 \end{bmatrix} \quad b = \begin{bmatrix} 0 \\ 0 \\ 1 \end{bmatrix}$$

Step 2: The characteristic equation of the closed-loop system should be:

$$\left(\lambda^2 + 2\tilde{\lambda} + 2\right)\left(\tilde{\lambda} + 10\right) = \lambda^3 + 12\tilde{\lambda}^2 + 22\tilde{\lambda} + 20$$

Step 3: The open-loop characteristic polynomial is $\lambda^3 + \lambda^2$
Hence, according to Equation 11.110 the control gains have to be chosen as:

$$k_1 = 20 - 0 = 20$$

$$k_2 = 22 - 0 = 22$$

$$k_3 = 12 - 1 = 11$$

Step 4: The corresponding feedback control law is:

$$u_{ref} - u = [20 \ 22 \ 11] \, x$$

Note: Transformation R is

$$R = [b \,|\, Ab \,|\, A^{n-1}b] \begin{bmatrix} a_1 & - & - & - & a_{n-1} & 1 \\ a_2 & - & - & a_{n-2} & 1 & 0 \\ - & - & - & - & - & - \\ a_{n-1} & 1 & - & - & 0 & 0 \\ 1 & 0 & 0 & - & 0 & 0 \end{bmatrix} \tag{11.114}$$

Question: Consider the multiinput case:

$$x = Ax + B^{n \times r} \, u^{r \times 1} \tag{11.115}$$

Can we convert this into the companion (rational) canonical form:

$$A^* = \begin{bmatrix} & 1 & & 0 \\ & & 1 & \\ 0 & & & 1 \\ -a_0 & -a_1 & \cdots & -a_{n-1} \end{bmatrix} \quad B^* = \begin{bmatrix} & 0 & & \\ & & 0 & \\ 1 & 1 & \cdots & 1 \end{bmatrix} \tag{11.116}$$

using a suitable transformation of the type:

$$x = Rz \tag{11.117}$$

so that:

$$A^* = R^{-1} AR \tag{11.118}$$

$$B^* = R^{-1} B \tag{11.119}$$

Answer: Not in general. Otherwise, the result:

$$\dot{z} = A^*z + B^*u \tag{11.120}$$

would imply that the system is controllable with just any one of the r inputs, which is a restrictive assumption, and cannot be true in general.

Note: If the system is controllable with any one input in the u vector, then we should be able to put (A^*, B^*) in the above form (Can you provide a proof for this?).

Observation 1: If a multiinput system can be put in the rational companion form, system is controllable with any u_i chosen from the u vector.

Observation 2: The control law $u = u_{ref} - Kx$ that can place the poles at arbitrary locations, is not unique in the multiinput case in general.

Proof: Consider the special case of rational companion form (A^*, B^*) in general (transformed from (A, B)), using R according to Equations 11.117 through 11.120).

First pick the feedback gain matrix:

$$K = \begin{bmatrix} k_1 & \cdots & \cdots & k_n \\ & & 0 & \end{bmatrix}$$

which gives:

$$A^* - B^*K = A^* - \begin{bmatrix} 0 & \cdots & \cdots & \cdots & 0 \\ \cdots & \cdots & \cdots & \cdots & \cdots \\ \cdots & \cdots & \cdots & \cdots & \cdots \\ 0 & \cdots & \cdots & \cdots & 0 \\ k_1 & \cdots & \cdots & \cdots & k_n \end{bmatrix}$$

From this K, we can assign poles arbitrarily.

Alternatively, pick:

$$K = \begin{bmatrix} 0 & \cdots & \cdots & 0 \\ k_1 & \cdots & \cdots & k_n \\ & & 0 & \end{bmatrix} \Rightarrow A^* - B^*K = A^* - \begin{bmatrix} 0 & \cdots & \cdots & \cdots & 0 \\ \cdots & \cdots & \cdots & \cdots & \cdots \\ \cdots & \cdots & \cdots & \cdots & \cdots \\ 0 & \cdots & \cdots & \cdots & 0 \\ k_1 & \cdots & \cdots & \cdots & k_n \end{bmatrix}$$

From this K as well, we can assign pole arbitrarily. So there are at least r possible K matrices that will do the job.

11.5.2 Pole Placement in the Multiinput Case

Consider the system:

$$\dot{x} = Ax + B_{n\times r}u_{r\times 1} \tag{11.121}$$

Use the feedback controller:

$$u = u_{ref} - Kx \tag{11.122}$$

$$\Rightarrow \dot{x} = (A - BK)x + Bu_{ref} \tag{11.123}$$

We want to assign the poles:

$$\Gamma = \left\{\tilde{\lambda}_1, \tilde{\lambda}_2, \cdots, \tilde{\lambda}_n\right\} \tag{11.124}$$

to $\tilde{A}$.

The corresponding closed-loop characteristic polynomial is:

$$\tilde{\Delta}(\lambda) = |\lambda I - A + BK| \tag{11.125}$$

$$= 0 \text{ if } \lambda = \tilde{\lambda}_i$$

$$= |(\lambda I - A)[I + (\lambda I - A)^{-1} BK]| = |\lambda I - A| \ |I + (\lambda I - A)^{-1} BK|$$

$$= \Delta(\lambda) \ |I + \Phi(\lambda) BK|$$

$$= \Delta(\lambda) \ |I_r + K\Phi(\lambda) B| \tag{11.126}$$

The last step (Equation 11.126) is obtained in view of the fact that the three matrices within the product of the right hand side expression may be rotated as indicated, which will be proved in the next section. Note that:

$$\Phi(\lambda) = (\lambda I - A)^{-1} \tag{11.127}$$

which is called the *resolvent matrix*, and is in fact the Laplace transform of the state-transition matrix (with $\lambda = s$):

$$\Phi(t) = L^{-1}\Phi(s) \tag{11.128}$$

Also, r is the number of inputs (dimension of the input vector) and I_r is the identity matrix of size r.

Proof:

Consider System 1 shown in Figure 11.18.

The following transfer relations are obtained:

Open-loop system: $\dot{x} = Ax + Bu$

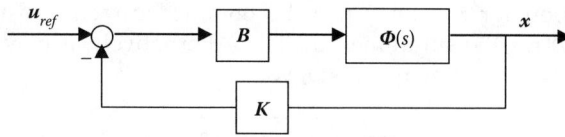

FIGURE 11.18
Control system 1.

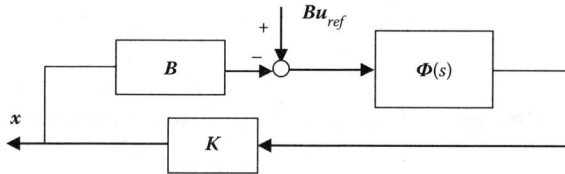

FIGURE 11.19
System 2, an equivalent version of system 1.

Laplace transform gives:

$$X = (sI - A)^{-1} BU = \Phi(s) BU \tag{i}$$

Closed-loop system: Use the feedback controller $u = u_{ref} - Kx$
    Laplace transform gives:

$$U = U_{ref} - KX \tag{ii}$$

Substitute Equation (ii) in Equation (i):

$$X = \Phi(s) B(U_{ref} - KX) \tag{iii}$$

Simple manipulation gives:

$$[I + \Phi(s) BK]^{-1} X = BU_{ref}$$

The corresponding transfer relation is:

$$X = [I + \Phi(s) BK]^{-1} BU_{ref} \tag{11.129}$$

Hence, the characteristic polynomial of the closed-loop system is:

$$\tilde{\Delta}(s) = |I + \Phi(s)BK| \tag{11.130}$$

Now consider System 2 shown in Figure 11.19.

Note that System 2 is obtained from System 1 simply by moving the reference input (u_{ref}) to the output side of block B. Accordingly, this input has been multiplied by B to retain the system equivalence. It follows that the two systems 1 and 2 should have the same characteristic polynomial.

The transfer relation for System 2 may be obtained directly by writing the signal equation (continuity) for the summing junction. Alternatively, it may be obtained from Equation (iii) by noting the following analogy:

System 1	System 2
u_{ref}	Bu_{ref}
x	$\tilde{x}$
B	$\Phi(s)$
$\Phi(s)$	K
K	B

By either approach, the continuity equation for the summing junction of System 2 is

$$K\Phi\left[BU_{ref} - B\tilde{X}\right] = \tilde{X} \tag{iv}$$

$$\text{Or: } \left[I_r + K\Phi B\right]\tilde{X} = K\Phi BU_{ref}$$

Hence, the transfer relation for System 2 is:

$$\tilde{X} = \left[I_r + K\Phi B\right]^{-1} K\Phi BU_{ref} \tag{11.131}$$

where I_r is the rth order identity matrix (r=number of inputs). It follows from Equation 11.131 that the characteristic polynomial of the closed-loop system (System 2) is

$$\tilde{\tilde{\Delta}}(s) = \left|I_r + K\Phi B\right| \tag{11.132}$$

Since Equations 11.130 and 11.132 should be identical in view of the system equivalence, the proof is complete.

11.5.3 Procedure of Pole Placement Design

For realizing the required set of closed-loop poles $\Gamma = \{\tilde{\lambda}_1, \tilde{\lambda}_2, \cdots, \tilde{\lambda}_n\}$ the characteristic equation $\tilde{\Delta}(\tilde{\lambda}_i) = 0$ has to be satisfied. Then, according to Equation 11.126 we need

$$\left|I_r + K\Phi(\tilde{\lambda}_i)B\right| = 0, \, i=1, 2, \ldots, n \tag{11.133}$$

One way of achieving this would be to make one column of the matrix $I_r + K\Phi(\tilde{\lambda}_i)B$ equal to the null column vector $\mathbf{0}$. To accomplish this, for a given pole $\tilde{\lambda}_i$ pick an appropriate column β_i in the matrix $\Phi(\tilde{\lambda}_i)B$, also pick the corresponding column α_i of I_r, and then set:

$$\alpha_i + K\beta_i = \mathbf{0}_{n \times 1} \tag{11.134}$$

Collect the n columns β_i, $i=1, 2, \ldots, n$ such that they form a linearly independent set of vectors. Assemble the n Equations 11.134 as:

$$[\alpha_1 \, \alpha_2 \, \ldots \, \alpha_n] + K[\beta_1 \, \beta_2 \, \ldots \, \beta_n] = \mathbf{0}_{n \times n} \tag{11.135}$$

Since the set of vectors β_i are linearly independent, the matrix assembled from them should be nonsingular, and can be inverted. Hence Equation 11.135 can be written as:

$$K = -[\alpha_1 \ \alpha_n] [\beta_1 \ \beta_n]^{-1} \tag{11.136}$$

The result (Equation 11.136) gives a feedback controller that will assign the closed-loop poles at the specified locations.

Note: The design (Equation 11.136) is not unique because more than one set of linearly independent vectors β_i would be possible.

Example 11.24

The open-loop system $A = \begin{bmatrix} 0 & 2 \\ 0 & 3 \end{bmatrix}$, $B = \begin{bmatrix} 0 \\ 1 \end{bmatrix}$ is given.

We will design a feedback controller that will place the closed-loop poles at $\Gamma = \{-3, -4\} = \{\tilde{\lambda}_1, \tilde{\lambda}_2\}$. From Equation 11.127, the resolvent matrix of the open-loop system is:

$$\Phi(\lambda) = \frac{1}{\lambda(\lambda - 3)} \begin{bmatrix} \lambda - 3 & 2 \\ 0 & \lambda \end{bmatrix}$$

The open-loop poles are directly obtained from this result as $\lambda_1 = 0$, $\lambda_2 = 3$. Now form the matrix product:

$$\Phi(\lambda)B = \frac{1}{\lambda(\lambda - 3)} \begin{bmatrix} 2 \\ \lambda \end{bmatrix} \tag{i}$$

Note that since the system has only one input, this product is a column vector, and hence there is only one choice of β_i for each closed-loop pole $\tilde{\lambda}_i$.

For $\tilde{\lambda}_1 = -3$: Substitute in Equation (i) to get:

$$\beta_1 = \frac{1}{18} \begin{bmatrix} 2 \\ -3 \end{bmatrix} = \begin{bmatrix} \frac{1}{9} \\ -\frac{1}{6} \end{bmatrix}$$

Also, since I_r in Equation 11.133 is a scalar for this single-input system, we have the corresponding $\alpha_1 = 1$ (scalar).

For $\tilde{\lambda}_2 = -4$: Substitute in Equation (i) to get

$$\beta_2 = \frac{1}{28} \begin{bmatrix} 2 \\ -4 \end{bmatrix} = \begin{bmatrix} \frac{1}{14} \\ -\frac{1}{7} \end{bmatrix}$$

and as before, the corresponding $\alpha_2 = 1$.

Finally Equation 11.136 gives the feedback gain matrix (row vector):

$$K = -[1 \ \ 1] \begin{bmatrix} \frac{1}{9} & \frac{1}{14} \\ -\frac{1}{6} & -\frac{1}{7} \end{bmatrix}^{-1} = -[1 \ \ 1] \begin{bmatrix} 36 & 18 \\ -47 & -28 \end{bmatrix} = [6 \ \ 10]$$

11.5.4 Placement of Repeated Poles

At a repeated root, the derivative of the characteristic polynomial also becomes zero. It follows that, for a repeated pole $\tilde{\lambda}_i$, we have the additional equation

$$\frac{d}{d\lambda}\tilde{\Delta}(\tilde{\lambda}_i) = 0 \qquad (11.137)$$

Note: If a pole is repeated three times, the second derivative of the characteristic polynomial also vanishes, giving the necessary two additional equations, and so on.

Now, according to Equation 11.137 the derivative of Equation 11.133 and also that of Equation 11.134 should vanish. But, since the terms α_i and K are independent of λ, the derivative of Equation 11.134) is:

$$K\frac{d}{d\lambda}\beta_i = 0 \qquad (11.138)$$

Again, it is understood that the chosen column $d\beta_i(\tilde{\lambda}_i)/d\lambda$ has to be independent with respect to the other chosen columns β_k. Then Equation 11.138 provides the additional equation that should be augmented with the $n-1$ independent equations of Equation 11.135, corresponding to the $n-1$ distinct poles, so as to be able to carry out the matrix inversion in the modified Equation 11.136 in obtaining the feedback gain matrix K. Specifically, if the pole $\tilde{\lambda}_i$ is repeated two give like poles, then Equation 11.136 has to be modified as follows:

$$K = -[\alpha_1, \cdots \alpha_i, 0, \cdots \alpha_{n-1}]\left[\beta_1 \cdots \beta_i, \frac{d\beta_i(\tilde{\lambda}_i)}{d\lambda}, \cdots \beta_{n-1}\right]^{-1} \qquad (11.139)$$

Example 11.25

If in the previous example we are required to place the closed-loop poles at $\tilde{\lambda}_1 = \tilde{\lambda}_2 = -3$, an appropriate feedback controller is designed now.

The additional Equation 11.138 is obtained by performing:

$$\frac{d}{d\lambda}\beta_1 = \frac{d}{d\lambda}\frac{1}{\lambda(\lambda-3)}\begin{bmatrix}2\\\lambda\end{bmatrix} = \frac{1}{(\lambda^2-3\lambda)^2}\begin{bmatrix}-4\lambda+6\\-\lambda^2\end{bmatrix} = \frac{1}{18\times18}\begin{bmatrix}18\\-9\end{bmatrix} = \begin{bmatrix}\dfrac{1}{18}\\-\dfrac{1}{36}\end{bmatrix}$$

Then, Equation 11.139 is written as:

$$K = -[1 \quad 0]\begin{bmatrix}\dfrac{1}{9} & \dfrac{1}{18}\\-\dfrac{1}{6} & -\dfrac{1}{36}\end{bmatrix}^{-1} = -36[1 \quad 0]\begin{bmatrix}4 & 2\\-6 & -1\end{bmatrix}^{-1}$$

11.5.5 Placement of Some Closed-Loop Poles at Open-Loop Poles

In some situations it may be required to retain some of the poles unchanged after feedback control. Mathematically, then, what is required would be place some of the closed-loop poles at the chosen open-loop poles. It is clear that, for such a pole, Equation 11.126 is satisfied for any control K:

$$\Delta(\lambda) \ |I_r + K\Phi(\lambda) \ B| = 0 \tag{11.140}$$

because the open-loop characteristic polynomial $\Delta(\lambda)$ also vanishes at the desired closed-loop pole. Then, we need to consider Equation 11.140 in its entirety in an expanded form, not just the second factor on the left hand side of Equation 11.140), as given in Equation 11.133. First, we write Equation 11.140 as: $|\Delta(\lambda) \ I_r + K\Delta(\lambda)\Phi(\lambda)B| = 0$.

Since the first term inside the determinant vanishes at the common pole, one has:

$$|K\Delta(\lambda)\Phi(\lambda)B| = 0 \tag{11.141}$$

As before, we make the determinant on the left hand side of Equation 11.141 vanish by setting a column within it be zero. Hence:

$$K\Delta(\lambda)\Phi(\lambda)B = 0 \tag{11.142}$$

In Equation 11.142, it should be clear that there is the same identical factor $\Delta(\lambda)$ in the denominator of the resolvent matrix $\Phi(\lambda)$, which will cancel out, leaving behind the Adjoint matrix of λI-A (see Appendix C). Hence, it will be finite for the value of the common pole. Specifically, we have:

$$K\text{Adj}(\lambda I - A)B = 0 \tag{11.143}$$

This result will provide the necessary equation for Equation 11.135 (and Equation 11.136), in determining the feedback controller K. The approach is illustrated now using an example.

Example 11.26

Consider the same open-loop system as in the two previous examples. Suppose that the closed-loop poles are to be placed at $\tilde{\lambda}_1 = 0, \tilde{\lambda}_2 = -3$, where 0 is also an open-loop pole. We now develop an appropriate feedback controller to achieve this.

Since $\tilde{\lambda}_1$ is identical to an open-loop pole, we have to use Equation 11.143 for this pole. Specifically: $\Delta(\lambda) = \lambda(\lambda - 3)$ and $\Phi(\lambda) = (\lambda I - A)^{-1} = \dfrac{1}{\lambda(\lambda - 3)}\begin{bmatrix} \lambda - 3 & 2 \\ 0 & \lambda \end{bmatrix}$

which gives: $\Delta(\lambda) \ \Phi(\lambda) = \text{Adj}(\lambda I - A) = \begin{bmatrix} \lambda - 3 & 2 \\ 0 & \lambda \end{bmatrix} \Rightarrow \text{Adj}(\lambda I - A)B = \begin{bmatrix} 2 \\ \lambda \end{bmatrix}$

The equation corresponding to Equation 11.143 for the common pole is:

$$\left\{ K \begin{bmatrix} 2 \\ \lambda \end{bmatrix} \right\}_{\lambda = \tilde{\lambda}_i} = 0 \Rightarrow K \begin{bmatrix} 2 \\ 0 \end{bmatrix} = 0 \tag{i}$$

This is one equation for **K**. We already had the other equation, from the original example as:

$$1 + K \begin{bmatrix} \dfrac{2}{18} \\[2mm] -\dfrac{3}{18} \end{bmatrix} = 0 \tag{ii}$$

By assembling Equations (i) and (ii) we get:

$$\begin{bmatrix} 1 & 0 \end{bmatrix} + K \begin{bmatrix} \dfrac{1}{9} & 2 \\[2mm] -\dfrac{1}{6} & 0 \end{bmatrix} = 0 \Rightarrow$$

$$K = -\begin{bmatrix} 1 & 0 \end{bmatrix} \begin{bmatrix} \dfrac{1}{9} & 2 \\[2mm] -\dfrac{1}{6} & 0 \end{bmatrix}^{-1} = -18\begin{bmatrix} 1 & 0 \end{bmatrix} \begin{bmatrix} 2 & 36 \\ -3 & 0 \end{bmatrix}^{-1}$$

$$= \dfrac{-18}{108}\begin{bmatrix} 1 & 0 \end{bmatrix} \begin{bmatrix} 0 & -36 \\ 3 & 2 \end{bmatrix} = -\dfrac{1}{6}\begin{bmatrix} 0 & -36 \end{bmatrix} = \begin{bmatrix} 0 & 6 \end{bmatrix}$$

$$\text{Check: } A - BK = \begin{bmatrix} 0 & 2 \\ 0 & 3 \end{bmatrix} - \begin{bmatrix} 0 & 0 \\ 0 & 6 \end{bmatrix} = \begin{bmatrix} 0 & 2 \\ 0 & -3 \end{bmatrix} \Rightarrow \tilde{\Delta}(\lambda) = \lambda(\lambda + 3)$$

Example 11.27

Figure 11.20 shows an inverted pendulum that is mounted on a mobile carriage. This model may be used for the analysis of the dynamic response and stability of systems such as rockets, space robots, and excavators. Suppose that the state variables are defined as: $x = \begin{bmatrix} z, & \dot{z}, & \theta, & \dot{\theta} \end{bmatrix}^T$

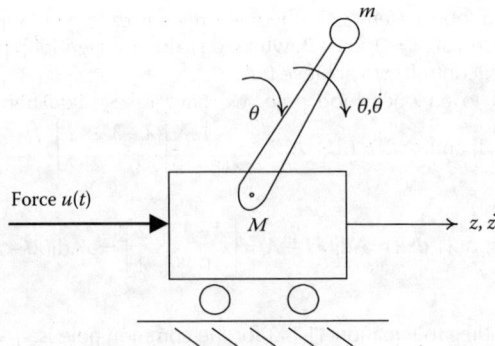

FIGURE 11.20
An inverted pendulum on a carriage.

and the equations of motion are linearized about an operating point, to obtain the linear state-space model:

$$\dot{x} = \begin{bmatrix} 0 & 1 & 0 & 0 \\ 0 & 0 & -1 & 0 \\ 0 & 0 & 0 & 1 \\ 0 & 0 & 11 & 0 \end{bmatrix} x + \begin{bmatrix} 0 \\ 1 \\ 0 \\ -1 \end{bmatrix} u \qquad (i)$$

It can be easily verified that the open-loop characteristic polynomial is:

$$\Delta(\lambda) = \lambda^4 - 11\lambda^2 \Rightarrow \lambda_1 = \lambda_2 = 0, \ \lambda_3 = \sqrt{11}, \ \lambda_4 = -\sqrt{11} \Rightarrow \text{unstable.}$$

Also, it can be verified (say, by Criterion 2) that the system is controllable.
⇒ Can assign poles arbitrarily using full state feedback.

Let the required closed-loop poles be $\Gamma = [\tilde{\lambda}_1, \tilde{\lambda}_2, \tilde{\lambda}_3, \tilde{\lambda}_4] = [-1, -2, -1-j, -1+j]$.

Use the feedback controller: $u = u_{ref} - Kx$ $\qquad (ii)$

where the feedback gain matrix (row vector) is $K = [k_1 \, k_2 \, k_3 \, k_4]$.
Substitute Equation (ii) in Equation (i) to obtain the closed-loop system:

$$\dot{x} = (A - BK)x + Bu_{ref}$$

$$= \underbrace{\begin{bmatrix} 0 & 1 & 0 & 0 \\ -k_1 & -k_2 & 1-k_3 & -k_4 \\ 0 & 0 & 0 & 1 \\ k_1 & k_2 & k_3-11 & k_4 \end{bmatrix}}_{A} x + \begin{bmatrix} 0 \\ 1 \\ 0 \\ -1 \end{bmatrix} u_{ref} \qquad (iii)$$

The characteristic polynomial of the closed-loop system is:

$$\tilde{\Delta}(\lambda) = \lambda^4 + (k_2 - k_4)\lambda^3 + (k_1 - k_3 + 11)\lambda^2 - 10k_2\lambda - 10k_1$$

$$= (\lambda+1)(\lambda+2)(\lambda+1+j)(\lambda+1-j) \text{ (as required)}$$

$$= \lambda^4 + 5\lambda^3 + 10\lambda^2 + 10\lambda + 4$$

By comparing the like coefficients in the characteristic polynomial, we require:

$$k_1 = -0.4$$

$$k_2 = -1$$

$$k_3 = -21.4$$

$$k_4 = -6$$

Note 1: With the assigned poles, the system will be asymptotically stable
⇒ Inverted pendulum balances, and the car returns to the initial position.
Note 2: See Appendix B for a MATLAB® treatment of this problem.

11.5.6 Pole Placement with Output Feedback

Often all the states may not be available for sensing and feedback in the control system. In that case a convenient option for control would be to use the system outputs (which are typically measurable) in place of the state variables, for feedback. Accurate placement of all the closed-loop poles at arbitrary locations may not be achievable with output feedback control. In this context the following theorem is useful.

Theorem: Consider the (controllable) system:

$$\dot{x} = Ax + u \tag{11.144}$$

and the output equation:

$$y = Cx \tag{11.145}$$

Note: $B = I_n = n$th order identity matrix ⇒ (A, B) controllable for any A.
There exists an output feedback:

$$u = u_{ref} - Ky \tag{11.146}$$

that will assign any set of poles (complex ones should be in conjugates) to the closed-loop system if the system (A, C) is observable.

Proof: Assume.

Note: The closed loop system is:

$$\dot{x} = Ax + [u_{ref} - KCx] = [A - KC]x + u_{ref} \tag{11.147}$$

Example 11.28

Consider the open-loop system:

$$\dot{x} = \begin{bmatrix} 0 & 0 & 5 \\ 1 & 0 & -1 \\ 0 & 1 & -3 \end{bmatrix} x + \begin{bmatrix} -2 & 0 \\ 1 & -2 \\ 0 & 1 \end{bmatrix} u$$

$$y = \begin{bmatrix} 0 & 0 & 1 \end{bmatrix} x$$

We can verify that:
(A, B) is controllable.
(A, C) is observable.
Note: $B \neq I_n$ in this system.
The characteristic polynomial of the open-loop system is shown to be:

$$\Delta(\lambda) = \lambda^3 + 3\lambda^2 + \lambda - 5 = [(\lambda + 2)^2 + 1][\lambda - 1] \Rightarrow \text{unstable}.$$

Now use the output feedback controller $u = u_{ref} - Ky$

with the control gain matrix:

$$K = \begin{bmatrix} k_1 \\ k_2 \end{bmatrix}$$

The closed-loop system matrix is:

$$A = A - BKC$$

$$= \begin{bmatrix} 0 & 0 & 5 \\ 1 & 0 & -1 \\ 0 & 1 & -3 \end{bmatrix} - \begin{bmatrix} -2 & 0 \\ 1 & -2 \\ 0 & 1 \end{bmatrix} \begin{bmatrix} k_1 \\ k_2 \end{bmatrix} [0 \ \ 0 \ \ 1] = \begin{bmatrix} 0 & 0 & 5 - 2k_1 \\ 1 & 0 & 2k_2 - k_1 + 1 \\ 0 & 1 & -k_2 - 3 \end{bmatrix}$$

The corresponding closed-loop characteristic polynomial is shown to be:

$$\tilde{\Delta}(\lambda) = \lambda^3 + (3 + k_2)\lambda^2 + (1 + k_1 - 2k_2)\lambda - (2k_1 + 5)$$

Pick: $k_1 = -3$, $k_2 = -2$.

The corresponding closed-loop characteristic polynomial is $\Delta(\lambda) = \lambda^3 + \lambda^2 + 2\lambda + 1$.
The corresponding poles are computed to be:

$$\tilde{\lambda}_1 = -0.57, \tilde{\lambda}_2 = -0.22 + j1.3, \tilde{\lambda}_3 = -0.22 - j1.3$$

Note: The closed-loop system is asymptotically stable.
If we use only one input (say u_1) (i.e., $k_2 = 0$), we have:

$$\tilde{\Delta}(\lambda) = \lambda^3 + 3\lambda^2 + (1 + k_1)\lambda - (2k_1 + 5)$$

Note: There is no choice of values for k_1 where $(1 + k_1)$ and $-(2k_1 + 5)$ are both positive. $\Rightarrow$ we cannot even make the system stable.
Similarly, if we set $k_1 = 0$ (only u_2 is used), we have

$$\tilde{\Delta}(\lambda) = \lambda^3 + (3 + k_2)\lambda^2 + (1 - 2k_2)\lambda - 5$$

This is unstable for any k_2.
Note: This example shows that leaving some of the inputs aside in a multivariable system can have disastrous consequences.
It has been mentioned that the assumption that all states are measurable and available for feedback is highly unrealistic. In most practical situations in order to implement a complete state feedback law, which may be developed as discussed previously, the complete state vector has to be estimated from the measurement (output) vector **y** or from a portion of the state vector **x**. A device (or a *filter*) that will generate the full state vector from output measurements or part of the state vector is known as a *state estimator* or an *observer*. Once a state estimator and a complete-state feedback control law are designed, the control scheme can be implemented by measuring

y or some of the state variables, supplying them to the state estimator (a digital computer) that will generate an estimate of *x* in real time and supplying the estimated *x* to the controller that will simply multiply it by the gain matrix **K** to generate the feedback control signal.

11.6 Optimal Control

A system can be controlled using a feedback control law so as to satisfy some performance requirements. In the previous section our objective was to place the system poles (eigenvalues) at suitable locations. Assuming that all states are available for feedback, we know that a constant-gain feedback of the state variables is able to accomplish this objective. Also, a constant-gain feedback of system outputs is able to achieve the same objective in a limited sense, under some special conditions. In the present section we will discuss optimal control, where the objective is to optimize a suitable *objective function* (e.g., maximize a *performance index* or minimize a *cost function*), by using an appropriate control law. A particularly favorite performance index is the infinite-time *quadratic integral* of the state variables and input variables, and a popular control law is again a linear constant-gain feedback of the system states. The associated controller is known as the LQR. Now, we will discuss this particular controller in some detail, after introducing the general optimal control problem, with sufficient analytical details.

11.6.1 Optimization through Calculus of Variations

The calculus of variations is a powerful approach that is used in the mathematics of optimization. In particular, the approach can be used to determine the optimal conditions of time integrals of functions, subject to various boundary conditions (BCs). In optimal control, the functions of state variables and input variables are particularly considered. The underlying approaches are first presented for the scalar (not vector) case of the system variables.

Consider the objective function:

$$J(x) = \int_{t_o}^{t_f} g(x(t), \dot{x}(t), t) dt \tag{11.148}$$

which needs to be optimized. Specifically, its relative extremum is to be determined. In Equation 11.148 g is interpreted as a function of a system variable (e.g., state variable) x and its time derivative $\dot{x}$. Note that the time variable t is explicitly present in g, thereby allowing the possibility of "time optimal control." Suppose that there are no constraint equations (i.e., consider the unconstrained optimization problem).

First assume that the starting time t_0 and the end time t_f of the process are fixed (specified). In applying the calculus of variations, we consider the variation in the cost function J as a result of a small virtual variation in the system response (trajectory) $x(t)$. For optimal conditions (i.e., the extremum of J), the variation δJ of J has to be set to zero. Specifically, we express the variation in J as:

$$\delta J = \int_{t_o}^{t_f} \delta g \, dt = \int_{t_o}^{t_f} \left[\frac{\partial g}{\partial x} \delta x + \frac{\partial g}{\partial \dot{x}} \delta \dot{x} \right] dt$$

$$= \int_{t_o}^{t_f} \frac{\partial g}{\partial x} \delta x \, dt + \left[\frac{\partial g}{\partial \dot{x}} \delta x \right]_{t_o}^{t_f} - \int_{t_o}^{t_f} \frac{d}{dt} \frac{\partial g}{\partial \dot{x}} \delta x \, dt$$

(integration by parts)　　　　　　(11.149)

$$= \int_{t_o}^{t_f} \left[\frac{\partial g}{\partial x} - \frac{d}{dt} \frac{\partial g}{\partial \dot{x}} \right] \delta x \, dt + \left[\frac{\partial g}{\partial \dot{x}} \delta x \right]_{t_o}^{t_f}$$

$$= 0 \quad \text{(for an optimum of } J)$$

Note that in Equation 11.149 we have not included changes in the end times, as they are taken to be fixed. First, we will consider below two cases under these conditions. Then we will allow for free end times and consider two further cases.

Case 1: End Times t_o, t_f Fixed; End States $x(t_0)$, $x(t_f)$ Specified
An example of a trajectory $x(t)$ and a small admissible variation (one that satisfies the BCs) for this case is sketched in Figure 11.21.

Since the end states $x(t_0)$ and $x(t_f)$ are also fixed in this case, we have the BCs $\delta x(t_0)=0$ and $\delta x(t_f)=0$. Accordingly, the second term on the right hand side of the final result in Equation 11.149 must be zero. Furthermore, since Equation 11.149 must vanish at the extremum conditions, for any arbitrary, small variation δx, it is required that the term within the square brackets representing the coefficient of δx must vanish as well. Specifically, then, the optimal condition is given by

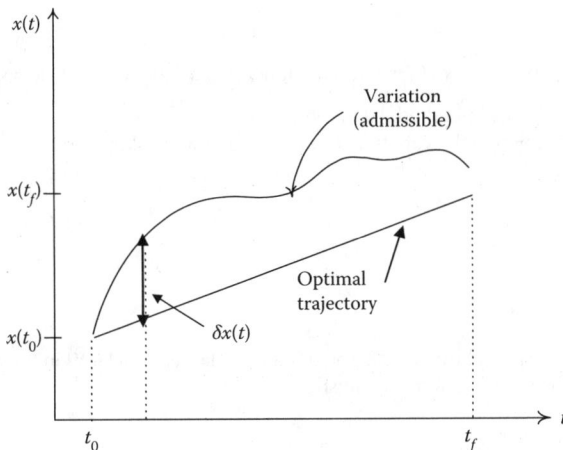

FIGURE 11.21
A trajectory and an admissible variation (fixed-fixed boundary conditions).

$$\frac{\partial g}{\partial x} - \frac{d}{dt}\frac{\partial g}{\partial \dot{x}} = 0 \tag{11.150}$$

This necessary condition for optimality is known as the *Euler equation.*

Example 11.29

Find the extremal trajectory $x^*(t)$ corresponding to the extremum of the cost function:

$$J = \int_0^1 [x^2(t) + \dot{x}^2(t)]dt$$

with the BCs: $x(0)=0$, $x(1)=1$.
   Apply Euler equation (Equation 11.150) with: $g(x,\dot{x}) = x^2(t) + \dot{x}^2(t)$

$$\text{We have: } 2x - \frac{d}{dt}(2\dot{x}) = 0$$

Hence: $\ddot{x} - x = 0$ for optimal trajectory.
   The solution of this differential equation is: $x = a_1 e_t + a_2 e^{-t}$
   Substitute the BCs:

$$0 = a_1 + a_2 \Rightarrow a_2 = -a_1$$

$$1 = a_1 e + \frac{a_2}{e} \Rightarrow a_1(e - 1/e) = 1$$

$$\text{We get: } x^* = \frac{e}{e^2 - 1}[e^t - e^{-t}]$$

(*Note*: Once we specify $x(t) \Rightarrow x^*(t)$ is known. Hence J is a function of x alone.)

Case 2: t_0 and t_f Fixed; $x(t_0)$ Specified, $x(t_f)$ Free
A trajectory and an admissible variation for this situation is sketched in Figure 11.22. In this case as well, the Euler equation has to be satisfied (using the same argument as before). Hence:

$$\frac{\partial g}{\partial x} - \frac{d}{dt}\frac{\partial g}{\partial \dot{x}} = 0 \tag{11.150}$$

But, the free BC needs special treatment. Specifically, in Equation 11.149 in the present case, $\delta x(t_f)$ is arbitrary. Hence, its coefficient must vanish:

$$\left.\frac{\partial g}{\partial \dot{x}}\right|_{t=t_f} = 0 \tag{11.151}$$

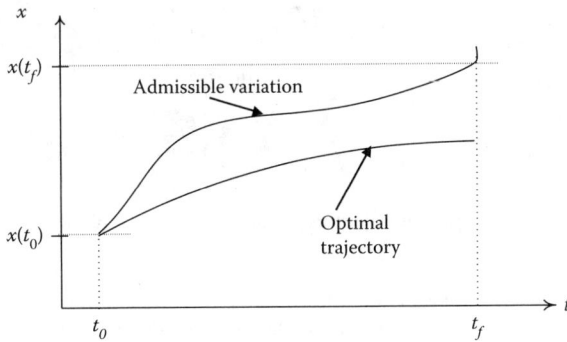

FIGURE 11.22
A trajectory and an admissible variation (Fixed-free BCs).

Example 11.30

Consider the same cost function as in the previous example:

$$J = \int_0^1 [x^2(t) + \dot{x}^2(t)]dt$$

Suppose that $x(1)$ is free. Consider the two cases:

a. $x(0)=0$
b. $x(0)=1$.

As before, from Euler equation we get: $\ddot{x} - x = 0$

$$\text{Its solution is known to be: } x = a_1 e^t + a_2 e^{-t} \qquad \text{(i)}$$

$$\text{Hence, by differentiating Equation (i) we get: } \dot{x} = a_1 e^t - a_2 e^{-t} \qquad \text{(ii)}$$

The free BC gives (from Equation 11.151):

$$2\dot{x}\big|_{t=1} = 0 \quad \Rightarrow \quad \dot{x}(1) = 0$$

Case (a): Given that $x(0)=0$, substitute in Equations (i) and (ii):

$$\left. \begin{array}{l} 0 = a_1 + a_2 \\[2mm] 0 = a_1 e - \dfrac{a_2}{e} \end{array} \right] \Rightarrow \begin{array}{l} a_2 = -a_1 \\[2mm] 0 = a_1 \left[e + \dfrac{1}{e} \right] \end{array}$$

Hence, $a_1 = 0$ and $a_2 = 0$, which correspond to the trivial solution $x(t)=0$.
   Case (b): Here $x(0)=1$:

$$1 = a_1 + a_2$$

$$0 = a_1 e - \dfrac{a_2}{e} \Rightarrow a_2 = a_1 e^2$$

Hence we get:

$$a_1(1+e^2)=1 \quad \Rightarrow \quad \begin{aligned} a_1 &= \frac{1}{1+e^2} \\ a_2 &= \frac{e^2}{1+e^2} \end{aligned}$$

The optimal trajectory is $x(t) = \frac{1}{(1+e^2)}(e^t + e^{2-t})$

Now we will consider two cases where the end time is not fixed. In this case $\delta t_f \neq 0$ and consequently, will introduce new terms into the variation of J, in addition to what is given in Equation 11.149.

Case 3: t_0 Fixed, t_f Free; $x(t_0)$ and $x(t_f)$ Specified
A sketch of this case is given in Figure 11.23.
Here there are two components that contribute to the variation in J.

$$\delta J = (\delta J)_{\text{Change in trajectory}} + (\delta J)_{\text{Change } \delta t_f} = 0 \text{ for extremum (as usual)}$$

$(\delta J)_{\text{Change in trajectory}} = $ same as given in Equation 11.149.

$$= \int_{t_0}^{t_f}\left[\frac{\partial g}{\partial x} - \frac{d}{dt}\frac{\partial g}{\partial \dot x}\right]\delta x\, dt + \left[\frac{\partial g}{\partial \dot x}\delta x\right]_{t_0}^{t_f}$$

The first term gives the Euler equation as before:

$$\frac{\partial g}{\partial x} - \frac{d}{dt}\frac{\partial g}{\partial \dot x} = 0 \tag{11.150}$$

(on optimal trajectory)

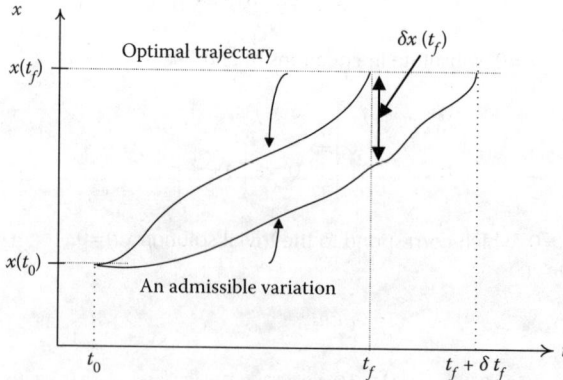

FIGURE 11.23
A trajectory and an admissible variation (fixed-fixed BCs; free end time).

$$\text{The second term} = \frac{\partial g}{\partial \dot{x}}\bigg|_{t_f} \delta x(t_f)$$

But, it is clear from Figure 11.23 that $\delta x(t_f) \cong -\dot{x}\big|_{t_f} \delta t_f \Rightarrow$

$$\text{Second term} = -\frac{\partial g}{\partial \dot{x}}\bigg|_{t_f} \dot{x}\,\delta t_f$$

$$\text{Next consider: } (\delta J)_{\text{Change } \delta t_f} = \int_{t_f}^{t_f+\delta t_f} g\,dt \cong g\big|_{t_f}\,\delta t_f$$

Hence the total variation corresponding to δt_f is $\left[g - \frac{\partial g}{\partial \dot{x}}\dot{x} \right]_{t_f} \delta t_f$

which also must vanish under optimal conditions. But, since t_f is free, δt_f must be arbitrary. Hence, its coefficient must vanish. This gives the BC:

$$\left[g - \frac{\partial g}{\partial \dot{x}}\dot{x} \right]_{t_f} = 0 \tag{11.152}$$

Example 11.31

Consider the same cost function as in the previous examples:

$$J = \int_0^{t_f} \left[x^2(t) + \dot{x}^2(t) \right] dt$$

Find the optimal solution for the BC

a. $x(0)=0$ and $x(t_f)=1$, with t_f free
b. $x(0)=1$ and $x(t_f)=2$, with t_f free

As before, from Euler equation we get: $\ddot{x} - x = 0$

$$\text{which has the solution: } x = a_1 e^t + a_2 e^{-t} \tag{i}$$

Case (a): The two BCs give:

$$0 = a_1 + a_2 \tag{ii}$$

$$1 = a_1 e^{t_f} + a_2 e^{-t_f} \tag{iii}$$

From Equation 11.152 we get

$$\left[g - \frac{\partial g}{\partial \dot{x}}\dot{x} \right]_{t_f} = \left[x^2 + \dot{x}^2 - 2\dot{x}^2 \right]_{t_f} = \left[x^2 - \dot{x}^2 \right]_{t_f} = \left[(a_1 e^t + a_2 e^{-t})^2 - (a_1 e^t - a_2 e^{-t})^2 \right]_{t_f} = \left[4a_1 a_2 \right]_{t_f} \tag{iv}$$

$$= 4a_1 a_2 = 0$$

From Equations (ii) and (iv) we get $a_1 = 0$ and $a_2 = 0$, which correspond to the trivial solution $x(t) = 0$, and furthermore Equation (iii) is not satisfied for finite t_f. Hence, this case is not feasible.

Case (b): The two BCs give:

$$1 = a_1 + a_2 \tag{v}$$

$$2 = a_1 e^{t_f} + a_2 e^{-t_f} \tag{vi}$$

and, as before, the free end time condition (Equation 11.152) gives:

$$a_1 \, a_2 = 0 \tag{iv}$$

From Equation (iv) and (v) it is clear that two solutions are possible:

1. $a_1 = 0$, $a_2 = 1$
2. $a_2 = 0$, $a_1 = 1$

Substitute in Equation (vi);

1. $2 = e^{-t_f} \Rightarrow$ no positive solution for t_f
2. $2 = e^{t_f} \quad \Rightarrow \quad t_f = \ln 2$

Case 4: t_0 Fixed; $x(t_0)$ Specified; t_f and x_f Free
In this case we assume that x_f and t_f are completely free and independent of each other so that t_f can take any value in a set and x_f can take any value in another set. A trajectory and an admissible variation thereof for this situation are sketched in Figure 11.24.

Note: In this case we can consider either δt_f and $\delta x(t_f)$ are arbitrary or δt_f and δx_f are arbitrary. Both will result in the same BC.

We proceed as in Case 3:

$$\delta J = (\delta J)_{change\text{-}trajectory} + (\delta J)_{change\text{-}t_f}$$

$$= \underbrace{\int_{t_0}^{t_f} \left[\frac{\partial g}{\partial x} - \frac{d}{dt} \frac{\partial g}{\partial \dot{x}} \right] \delta x \, dt + \frac{\partial g}{\partial \dot{x}} \bigg|_{t_f} \partial x_{(t_f)}}_{(\delta J)_{traject}} + \underbrace{g \big|_{t_f} \delta t_f}_{(\delta J)_{change\text{-}t_f}} \tag{11.153}$$

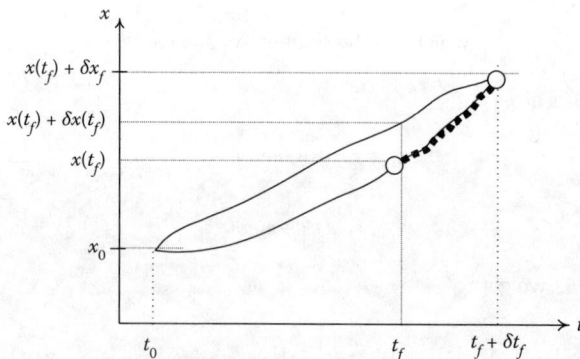

FIGURE 11.24
A trajectory and an admissible variation (starting conditions fixed; end conditions completely free).

The first term on the right hand side of Equation 11.153 gives the Euler equation, as before.

$$\frac{\partial g}{\partial x} - \frac{d}{dt}\frac{\partial g}{\partial \dot{x}} = 0 \qquad (11.150)$$

(on optimal trajectory)

Now, since δt_f and $\delta x(t_f)$ are arbitrary, their coefficients in Equation 11.153 must vanish. These result in the end conditions:

$$\left.\frac{\partial g}{\partial \dot{x}}\right|_{t_f} = 0 \qquad (11.154)$$

$$g\big|_{t_f} = 0 \qquad (11.155)$$

Alternatively, the same BCs (Equations 11.154 and 11.155) may be obtained by using the fact that δt_f and δx_f are arbitrary. We start by noting (see Figure 11.24) that:

$$\delta x_f = \delta x(t_f) + \dot{x}\delta t_f \qquad (11.156)$$

Substitute (Equation 11.156) in the boundary terms of (Equation 11.153) to give:

$$\left.\frac{\partial g}{\partial \dot{x}}\right|_{t_f}\left[\delta x_f - \dot{x}\delta t_f\right] + g\big|_{t_f}\,\delta t_f = 0$$

$$\text{Or: } \left.\frac{\partial g}{\partial \dot{x}}\right|_{t_f}\delta x_f + \left[g - \dot{x}\frac{\partial g}{\partial \dot{x}}\right]_{t_f}\delta t_f = 0 \qquad (11.157)$$

Since δx_f is arbitrary, from Equation (i) above we get the same result as above:

$$\left.\frac{\partial g}{\partial \dot{x}}\right|_{t_f} = 0 \qquad (11.154)$$

Since δt_f is arbitrary, from Equation (i) we have:

$$\left[g - \dot{x}\frac{\partial g}{\partial \dot{x}}\right]_{t_f} = 0 \qquad (11.158)$$

Substitute Equation 11.154 into Equation 11.157: $\Rightarrow g\big|_{t_f} = 0$, which is Equation 11.155 as obtained before.

Now we will consider three examples for the application of this case.

Example 11.32

Consider the usual cost function: $J = \int\limits_{0}^{t_f}\left[x^2(t) + \dot{x}^2(t)\right]dt$

with the initial state specified (completely) as (a): $x(0)=0$ and (b): $x(0)=1$, while the final time and state are free.

We proceed as before by applying Euler equation, which gives:

$$x = a_1 e^t + a_2 e^{-t}$$

Also, we have:

$$g = x^2 + \dot{x}^2 = 2[a_1^2 e^{2t} + a_2^2 e^{-2t}]$$

$$\frac{\partial g}{\partial \dot{x}} = 2\dot{x} = 2[a_1 e^t - a_2 e^{-t}]$$

The BCs (Equations 11.155 and 11.154) give:

$$0 = 2[a_1^2 e^{2t_f} + a_2^2 e^{-2t_f}] \qquad \text{(i)}$$

$$0 = 2[a_1 e^{t_f} - a_2 e^{-t_f}] \qquad \text{(ii)}$$

For finite t_f: multiply Equation (i) by $\exp(2t_f)$. We get:

$$a_2^2 = -[a_1^2 e^{4t_f}] \Rightarrow a_1 = 0, a_2 = 0 \qquad \text{(iii)}$$

Multiply Equation (ii) by $\exp(t_f)$. We get:

$$a_2 = [a_1 e^{2t_f}] \qquad \text{(iv)}$$

a. The IC gives: $0 = a_1 + a_2$.
This satisfies both Equations (iii) and (iv), but the solution is trivial and is not useful.
b. The IC gives $1 = a_1 + a_2$
This can satisfy Equation (iv) but not Equation (iii). Hence, a solution does not exist.

Example 11.33

Consider the cost function: $J = \int\limits_0^{t_f} [-2x(t) + \dot{x}^2(t)] dt$
with the IC $x(0) = 0$.

Euler equation gives: $-2 - \dfrac{d}{dt}(2\dot{x}) = 0 \Rightarrow \ddot{x} + 1 = 0$

$$\Rightarrow \dot{x} = -t + a_1 \Rightarrow x = -\frac{1}{2}t^2 + a_1 t + a_2$$

IC gives $0 = a_2$.
The final conditions give:

$$g\Big|_{t_f} = -2\left[-\frac{1}{2}t_f^2 + a_1 t_f\right] + [-t_f + a_1]^2 = 0 \qquad \text{(i)}$$

$$\frac{\partial g}{\partial \dot{x}}\Big|_{t_f} = 2\dot{x}\Big|_{t_f} = 2\left(-t_f + a_1\right) = 0 \tag{ii}$$

From Equation (ii) we get:

$$a_1 = t_f \tag{iii}$$

Substitute Equation (iii) in Equation (i):

$$-2\left[-\frac{1}{2}t_f^2 + t_f^2\right] = 0 \Rightarrow t_f = 0 \Rightarrow \quad \text{trivial solution.}$$

Example 11.34

Consider the cost function: $J = \int_{\pi/2}^{t_f} \left[\dot{x}^2(t) - x^2\right]dt$

with the IC $x\left(\dfrac{\pi}{2}\right) = 1$. The final time and state are free.

Euler equation gives:

$$-2x - \frac{d}{dt}(2\dot{x}) = 0 \;\Rightarrow\; \ddot{x} + x = 0 \;\Rightarrow$$

$$x = a_1 \sin(t) + a_2 \cos(t)$$

The IC gives $1 = a_1$

$$\Rightarrow x = \sin(t) + a_2 \cos(t)$$

$$\dot{x} = \cos(t) - a_2 \sin(t)$$

Hence we have:

$$g = \dot{x}^2 - x^2$$
$$= \cos^2(t) + a_2^2 \sin^2(t) - 2a_2 \sin(t)\cos(t) - \sin^2(t) - a_2^2 \cos^2 t - 2a_2 \sin(t)\cos(t)$$
$$= \cos(2t) - a_2^2 \cos(2t) - 2a_2 \sin(2t)$$

$$\frac{\partial g}{\partial \dot{x}} = 2\dot{x} = 2\cos(t) - 2a_2 \sin(t)$$

Now substitute in the final conditions (Equations 11.154 and 11.155):

$$0 = 2\cos(t_f) - 2a_2 \sin(t_f)$$
$$0 = \cos(2t_f) - a_2^2 \cos(2t_f) - 2a_2 \sin(2t_f)$$

Solve these two (nonlinear) simultaneous equations to determine t_f and a_2, and hence $x(t)$.

Case 5: t_o Fixed, $x(t_0)$ Specified; t_f and x_f are Related through $x_f = \theta(t_f)$

Here, as in Case 4, the initial state is completely specified. The final time and state are not completely free however, but related according to a specified function.

We can use the variation given by Equation 11.153 subject to:

$$x_f = \theta(t_f) \qquad (11.159)$$

The Euler equation holds as usual:

$$\frac{\partial g}{\partial x} - \frac{d}{dt}\frac{\partial g}{\partial \dot{x}} = 0 \qquad (11.150)$$

The final conditions are governed by Equation 11.157:

$$\frac{\partial g}{\partial \dot{x}}\Big|_{t_f} \delta x_f + \left[g - \dot{x}\frac{\partial g}{\partial \dot{x}}\right]_{t_f} \delta t_f = 0 \qquad (11.157)$$

subject to Equation 11.159, which should be expressed in the differential form:

$$\delta t_f = \dot{\theta}(t_f)\delta t_f \qquad (11.160)$$

Substitute Equation 11.160 in Equation 11.157 to get:

$$\left[\dot{\theta}\frac{\partial g}{\partial \dot{x}} + g - \dot{x}\frac{\partial g}{\partial \dot{x}}\right]_{t_f} \delta t_f = 0 \qquad (i)$$

Since δt_f in Equation (i) can be treated as arbitrary, its coefficient must vanish. Accordingly, we have the end condition:

$$\left[\dot{\theta}\frac{\partial g}{\partial \dot{x}} + g - \dot{x}\frac{\partial g}{\partial \dot{x}}\right]_{t_f} = 0 \qquad (11.161)$$

Example 11.35

Consider the cost function: $J = \int_0^{t_f} \left[x^2(t) + \dot{x}^2(t)\right]dt$

with the IC $x(0)=0$, and the final condition $\theta=t$.

Here, Euler equation, as before, gives:

$$x = a_1 e^t + a_2 e^{-t} \qquad (i)$$

$$\text{IC: } 0 = a_1 + a_2 \qquad (ii)$$

Also, from the given information we have:

$$g = x^2(t) + \dot{x}^2(t) = 2\left[a_1^2 e^{2t} + a_2^2 e^{-2t}\right]$$

$$\frac{\partial g}{\partial \dot{x}} = 2\dot{x}(t) = 2(a_1 e^t - a_2 e^{-t}) \qquad (iii)$$

$$\dot{\theta} = 1$$

Substitute Equation (ii) in the final condition (Equation 11.161):

$$\left[1 \times 2(a_1 e^{t_f} - a_2 e^{-t_f}) + 2(a_1^2 e^{2t_f} + a_2^2 e^{-2t_f}) - 2(a_1 e^{t_f} - a_2 e^{-t_f})^2\right] = 0 \tag{iv}$$

and since $x = t$ at the final state, we get from Equation (i):

$$t_f = a_1 e^{t_f} + a_2 e^{-t_f} \tag{v}$$

Now, Equation (iv) simplifies to:

$$2(a_1 e^{t_f} - a_2 e^{-t_f}) + 4 a_1 a_2 = 0 \tag{iv}*$$

Substitute Equation (ii) in Equations (iv)* and (v):

$$t_f = a_1(e^{t_f} - e^{-t_f})$$
$$2 a_1(e^{t_f} + e^{-t_f}) - 4 a_1^2 = 0 \tag{vi}$$

Solve the nonlinear simultaneous equations (vi) to determine t_f and a_1, and hence $x(t)$.

11.6.2 Cost Function having a Function of End State

Now consider a cost function that has an explicit term which is a function of the end state, as:

$$J = \int_{t_0}^{t_f} g(x, \dot{x}, t) dt + h(x(t_f), t_f) \tag{11.162}$$

where the IC is specified. We can write Equation 11.162 as:

$$J = \int_{t_0}^{t_f}\left(g + \frac{dh}{dt}\right) dt + h(x(t_0), t_0) \tag{11.162}*$$

which can be verified by integrating the second term of the integrand. Now the last term in Equation 11.162* is a constant, and may be dropped. Consequently, we have the equivalent cost function:

$$\tilde{J} = \int_{t_0}^{t_f}\left(g + \frac{dh}{dt}\right) dt = \int_{t_0}^{t_f}\left(g + \frac{\partial h}{\partial t} + \frac{\partial h}{\partial x}\dot{x}\right) dt \tag{11.163}$$

$$\text{Let: } g_a = g + \frac{\partial h}{\partial t} + \frac{\partial h}{\partial x}\dot{x} \tag{11.164}$$

As usual, then, Euler equation is:

$$\frac{\partial g_a}{\partial x} - \frac{d}{dt}\frac{\partial g_a}{\partial \dot{x}} = 0 \qquad (11.150)*$$

which, in view of Equation 11.164, may be written as:

$$\frac{\partial g}{\partial x} + \frac{\partial^2 h}{\partial x \partial t} + \frac{\partial^2 h}{\partial x^2}\dot{x} - \frac{d}{dt}\left(\frac{\partial g}{\partial \dot{x}} + \frac{\partial h}{\partial x}\right) = 0 \Rightarrow$$

$$\frac{\partial g}{\partial x} - \frac{d}{dt}\left(\frac{\partial g}{\partial \dot{x}}\right) + \frac{\partial^2 h}{\partial x \partial t} + \frac{\partial^2 h}{\partial x^2}\dot{x} - \frac{\partial}{\partial t}\frac{\partial h}{\partial x} - \frac{\partial}{\partial x}\frac{\partial h}{\partial x}\dot{x} = 0 \Rightarrow$$

$$\frac{\partial g}{\partial x} - \frac{d}{dt}\left(\frac{\partial g}{\partial \dot{x}}\right) = 0 \qquad (11.150)$$

It is seen that the Euler equation remains the same as before.

The end condition is obtained using Equation 11.157 as before, to get:

$$\frac{\partial g_a}{\partial \dot{x}}\Big|_{t_f}\delta x_f + \left(g_a - \frac{\partial g_a}{\partial \dot{x}}\dot{x}\right)\Big|_{t_f}\delta t_f = 0 \qquad (11.157)*$$

Substitute Equation 11.164 in Equation 11.157* $\Rightarrow$

$$\left[\frac{\partial g}{\partial \dot{x}} + \frac{\partial h}{\partial x}\right]_{t_f}\delta x_f + \left[g + \frac{\partial h}{\partial t} - \frac{\partial h}{\partial x}\dot{x} - \left\{\frac{\partial g}{\partial \dot{x}} + \frac{\partial h}{\partial x}\right\}\dot{x}\right]_{t_f}\delta t_f = 0 \Rightarrow$$

$$\left[\frac{\partial g}{\partial \dot{x}} + \frac{\partial h}{\partial x}\right]_{t_f}\delta x_f + \left[g - \frac{\partial g}{\partial \dot{x}}\dot{x} + \frac{\partial h}{\partial t}\right]_{t_f}\delta t_f = 0 \qquad (11.165)$$

This condition may be applied, as usual. For example, if the end time and state are free, then the two coefficients in Equation 11.165 should vanish. If there exits a relation between the end state and time, its differential form should be substituted in Equation 11.165, and the resulting coefficient should be set to zero.

11.6.3 Extension to the Vector Problem

Thus far, we treated the "scalar" problem of optimization where the system trajectory $x(t)$ is a scalar (or, the state-space is one-dimensional). Now let us extend the ideas to the "vector" case where $x(t)$ is an nth order vector. The corresponding cost function (which is a scalar) may be expressed as (see Equation 11.162)

$$J = \int_{t_0}^{t_f} g(x,\dot{x},t)dt + h(x(t_f),t_f) \qquad (11.166)$$

Then, we have the following optimal conditions:

$$\text{Euler Equation:} \quad \left[\frac{\partial g}{\partial x} - \frac{d}{dt}\frac{\partial g}{\partial \dot{x}} \right]^T = 0 \qquad (11.167)$$

Note that Equation 11.167 is given as a column vector equation.

$$\text{Boundary (Final) Condition:} \quad \left[\frac{\partial g}{\partial \dot{x}} + \frac{\partial h}{\partial x} \right]_{t_f} \delta x_f + \left[g - \frac{\partial g}{\partial \dot{x}} \dot{x} + \frac{\partial h}{\partial t} \right]_{t_f} \delta t_f = 0 \quad (11.168)$$

Note that Equation 11.168 is a scalar equation.

11.6.4 General Optimal Control Problem

The general problem of optimization, in control systems, should satisfy the system equations (state-space equations) and also a set of constraints. Note that the system equations themselves are constraints (dynamic constraints), and may be incorporated into an "augmented" cost function through the use of Lagrange multipliers. This general problem is presented now.

The objective function that is to be optimized is expressed as

$$J(u) = \int_{t_0}^{t_f} g(x, u, t)dt + h(x(t_f), t_f) \qquad (11.169)$$

where x is an n-vector (i.e., an nth order vector), u is an m-vector, subject to the n state equations (dynamic constraints):

$$\dot{x} = f(x, u, t) \qquad (11.170)$$

and r isoperimetric constraints:

$$\int_{t_0}^{t_f} e(x, t)dt = k \qquad (11.171)$$

This optimization problem is solved by first converting the constraints (Equation 11.171) into state-space equations and then incorporating the overall set of state equations into an augmented cost function, through Lagrange multipliers. The associated steps are given below.

Method:

1. Define r new state variables

$$x_{n+i} = \int_{t_0}^{t} e_i(x, \tau)d\tau \quad i = 1, 2, \ldots, r \qquad (11.172)^*$$

$$\text{with } x_{n+i}(t_0)=0 \text{ and } x_{n+i}(t_f)=k_i \tag{11.173}$$

$$\text{Then } \dot{x}_{n+i}=e_i(x,t)=f_{n+i}(x,t) \quad i=1,2,\dots,r \tag{11.172}$$

(say)

2. Use $n+r$ Lagrange multipliers λ.

The constrained optimization problem is

$$\bar{J}=\int_{t_0}^{t_f}\underbrace{\{g(x,u,t)+\lambda^T(f-\dot{x})\}}_{g_a}dt+h(x(t_f),t_f) \tag{11.174}$$

This gives the Euler equations:

$$\frac{d}{dt}\left[\frac{\partial g_a}{\partial \dot{x}}\right]^T-\left[\frac{\partial g_a}{\partial x}\right]^T=0 \;(n+r \text{ equations}) \Rightarrow$$

$$\frac{d}{dt}(-\lambda)-\left[\frac{\partial g}{\partial x}\right]^T-\frac{\partial f}{\partial x}\lambda=0 \;(n+r \text{ equations}) \Rightarrow$$

$$\dot{\lambda}=-\left[\frac{\partial}{\partial x}(g+\lambda^T f)\right]^T \;(n+r \text{ equations}) \tag{11.175}$$

Note: The last r equations give $\dot{\lambda}_i=0$ for $i=n+1,\dots,n+r$ because the right hand side is not a function $x_{n+1},\dots,x_{n+r}$

Also the original state equations (dynamic constraints) must be satisfied:

$$\dot{x}=f$$

which may be written as:

$$\dot{x}=\left[\frac{\partial}{\partial \lambda}(g+\lambda^T f)\right]^T \tag{11.176}$$

Note: The Equations 11.176 are equivalent to Euler equations with respect to λ.

Specifically $(d/dt)[\partial g_a/\partial \dot{\lambda}]^T-[\partial g_a/\partial \lambda]^T=0$; but since g_a does not have $\dot{\lambda}$ terms, the first term on the left hand side will vanish, giving $[\partial g_a/\partial \lambda]^T=0$, which then gives Equation 11.176.

Furthermore, if we write Euler equations with respect to u, we have $(d/dt)[\partial g_a/\partial \dot{u}]^T-[\partial g_a/\partial u]^T=0$. As before, the first term on the left hand side will vanish, giving $[\partial g_a/\partial u]^T=0$. Hence in view of g_a as given in Equation 11.174 we get:

$$\left[\frac{\partial(g+\lambda^T f)}{\partial u}\right]^T=0 \tag{11.177}$$

The result (Equation 11.177) corresponds to Pontryagin's minimum principle, as will be noted later.

11.6.5 Boundary Conditions

The boundary (final) conditions for the augmented (constrained) optimization problem may be obtained from Equation 11.168 by substituting g_a for g. Specifically:

$$\left[\frac{\partial g_a}{\partial \dot{x}}+\frac{\partial h}{\partial x}\right]_{t_f}\delta x_f+\left[g_a-\frac{\partial g_a}{\partial x}\dot{x}+\frac{\partial h}{\partial t}\right]_{t_f}\delta t_f=0\Rightarrow$$

$$\left[-\lambda^T+\frac{\partial h}{\partial x}\right]_{t_f}\delta x_f+\left[g+\lambda^T(f-\dot{x})+\lambda^T\dot{x}+\frac{\partial h}{\partial t}\right]_{t_f}\delta t_f=0\Rightarrow$$

$$\left[-\lambda^T+\frac{\partial h}{\partial x}\right]_{t_f}\delta x_f+\left[g+\partial^T f+\frac{\partial h}{\partial t}\right]_{t_f}\delta t_f=0 \tag{11.178}$$

Note: $\delta x_f=0$ for the states $x_{n+1}, \ldots, x_{n+r}$ because they are fixed at $k_1, \ldots, k_r$.

11.6.6 Hamiltonian Formulation

The same results as obtained before may be expressed in a convenient and compact form using the Hamiltonian formulation. Here, we define the Hamiltonian:

$$H(x, u, \lambda, t)=g(x, u, t)+\lambda^T f \tag{11.179}$$

It follows from the results (Equations 11.175 and 11.176) that:

$$\dot{\lambda}=-\frac{\partial^T H}{\partial x} \tag{11.180}$$

$$\dot{x}-\frac{\partial^T H}{\partial \lambda} \tag{11.181}$$

The boundary (final) conditions (Equation 11.178) may be written as:

$$\left[-\lambda^T+\frac{\partial h}{\partial x}\right]_{t_f}\delta x_f+\left[H+\frac{\partial h}{\partial t}\right]_{t_f}\delta t_f=0 \tag{11.182}$$

11.6.7 Pontryagin's Minimum Principle

The result (Equation 11.178) may be written as:

$$\frac{\partial^T H}{\partial u}=0 \tag{11.183}$$

This is known as Pontryagin's Minimum Principle, and is stated as follows:

An optimal control must minimize the Hamiltonian.

This is a necessary but not sufficient condition, and is true for both bounded and unbounded u. Note that if the Hamiltonian is expressed as the negative of what is given in Equation 11.179, we get Pontryagin's Maximun Principle.

11.7 Linear Quadratic Regulator (LQR)

Often, the system to be controlled is represented by a linear model, which will hold at least in a small neighborhood of the operating point. Furthermore, a cost function that is a time integral of a quadratic form in the state and input variables, is applicable in many situations; for example, in designing an asymptotically stable regulator that uses minimal control energy. Then, what results is an LQR. Specifically, we seek to minimize the cost function:

$$J = \frac{1}{2} \int_{t_0}^{t_f} [x^T Q(t)x + u^T R(t)u]dt + \frac{1}{2} x^T(t_f) Sx(t_f) \tag{11.184}$$

subject to the dynamic constraints given by a set of linear (but, can be time-varying) state equations:

$$\dot{x} = A(t)x + B(t)u \tag{11.185}$$

In Equation 11.184 it is assumed that Q is positive semidefinite, R is positive definite, and S is positive semidefinite (see Appendix C). Also, without loss of generality, these three *weighting matrices* are assumed symmetric. The final time t_f is assumed fixed and the final state $x(t_f)$ is assumed free.

The solution to the LQR optimal control problem is now determined using the Hamiltonian approach, as developed before. First we form the Hamiltonian:

$$H = \frac{1}{2} x^T Q x + \frac{1}{2} u^T R u + \lambda^T (Ax + Bu) \tag{11.186}$$

11.7.1 The Euler Equations

Equations 11.180 and 11.183 give:

$$\dot{\lambda} = -\frac{\partial^T H}{\partial x} = -[Qx + A^T \lambda]$$

$$\frac{\partial H^T}{\partial u} = 0 = Ru + B^T \lambda$$

Hence we have the optimality conditions:

$$\dot{\lambda} = -Qx - A^T\lambda \qquad (11.187)$$

$$u = -R^{-1}B^T\lambda \qquad (11.188)$$

11.7.2 Boundary Conditions

For the given problem $\delta t_f = 0$ and δx_f is arbitrary. Hence, in Equation 11.182 the coefficient of δx_f must vanish, and we get:

$$\left[-\lambda^T + \frac{\partial h}{\partial x} \right] = 0 \Rightarrow \left[-\lambda + \frac{\partial^T}{\partial x}\left[\frac{1}{2}x^T Sx \right] \right]_{t_f} = 0 \Rightarrow -\lambda(t_f) + Sx(t_f) = 0$$

$$\lambda(t_f) = Sx(t_f) \qquad (11.189)$$

Note: Equation 11.188 gives the feedback control law. But here we need to determine the Lagrange multiplier λ, by solving Equation 11.187 subject to the final condition (Equation 11.189).

Kalman has shown that the solution of Equation 11.187 this is of the form:

$$\lambda = K(t)x \qquad (11.190)$$

Proof: A proof of Equation 11.190 may be given by construction, as follows. Differentiate Equation 11.190 to get:

$$\dot{\lambda} = \dot{K}x + K\dot{x}$$

$$= \dot{K}x + K(Ax + Bu) \text{ (by substituting Equation 11.185)}$$

$$= \dot{K}x + KAx + KB(-R^{-1}B^T Kx) \text{ (by substituting Equations 11.188 and 11.190)}$$

$$= (\dot{K} + KA - KBR^{-1}B^T K)x \qquad \text{(i)}$$

$$= -(Q + A^T K)x \text{ (from Equations 11.187 and 11.190)} \qquad \text{(ii)}$$

Hence, from Equations (i) and (ii) we have

$$[\dot{K} + KA + A^T K - KBR^{-1}B^T K + Q]x = 0 \qquad \text{(iii)}$$

Now we can show that the expression within the brackets in Equation (iii) must vanish, as $x \neq 0$ in general. Specifically, if $x = 0$ then with $u = 0$, the system will stay in this equilibrium state for ever giving the minimal solution for J. Hence at least during an early time interval we must have $x \neq 0$, and Equation (iii) gives:

$$\dot{K} + KA + A^T K - KBR^{-1}B^T K + Q = 0 \qquad (11.191)$$

with the BC obtained from Equations 11.189 and 11.190 as:

$$K(t_f) = S \qquad (11.192)$$

Equation 11.191 is the nonlinear, time-varying (nonsteady) matrix *Riccati equation*, whose solution must be symmetric since $K(t_f) = S$ is symmetric . This completes the proof, since Equation 11.190 is shown to be a suitable solution. Equations 11.188 and 11.190 then provide the optimal controller:

$$u = -R^{-1} B^T K x \qquad (11.193)$$

Note: Integrate Equation 11.191 backwards from t_f to t_0 using the BC (Equation 11.192) to get the gain matrix $K(t)$. Store this and then compute the feedback control signal u using Equation 11.193. To perform the integration, set $t = -\tau$. Then, Equation 11.191 may be written as:

$$\frac{dK}{d\tau} = KA + A^T K - KBR^{-1}B^T K + Q \qquad (11.194)$$

Integrate Equation 11.194 from $-t_f$ to $-t_0$, with the modified IC:

$$K(-t_f) = S \qquad (11.195)$$

Note 1: Since K is symmetric, only $(1/2)n\,(n+1)$ equations need to be integrated.
    Note 2: If $x(t_f) = 0$, then from Equation (iii) above, it follows that the Riccati equation need not be satisfied at $t = t_f$ for optimality. Then $K(t_f)$ can become ambiguous.

11.7.3 Infinite-Time LQR

Now we will consider the special time-invariant (stationary) case of the problem given by Equations 11.184 and 11.185, with $S = 0$ and the final time is infinity:

$$J = \frac{1}{2} \int_0^\infty \left[x^T Q x + u^T R u \right] dt \qquad (11.196)$$

$$\dot{x} = Ax + Bu \qquad (11.197)$$

Note: The weighting matrices Q and R and the system matrices A and B are constant in this case.
    Kalman has shown that the solution to this problem can be obtained from the previous solution of Equation 11.191, at steady-state, by setting $t \to \infty$. Since K is a constant matrix, at steady-state, we have $\dot{K} = 0$. The optimal control law for this LQR problem is then

$$u = -R^{-1} B^T K x \qquad (11.198)$$

where K is the positive definite solution of the nonlinear, steady-state, matrix *Riccati equation*:

$$KA + A^T K - KBR^{-1} B^T K + Q = 0 \qquad (11.199)$$

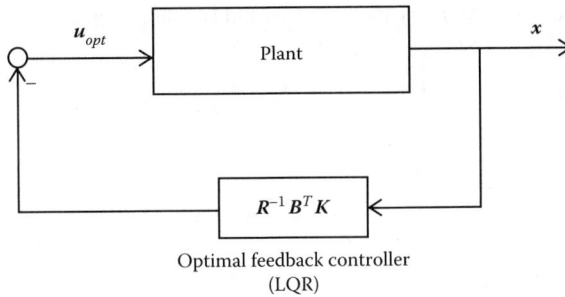

FIGURE 11.25
Linear quadratic regulator (LQR).

As before, since K is symmetric, only $(1/2)n(n+1)$ equations need to be solved. The optimal controller (Equation 11.198) is shown by the feedback control system in Figure 11.25.

Example 11.36

A plant is given by the state equation (simple integrator): $\dot{x} = u$
with the end conditions: $x(0)=1$, $x(1)$ free.
The system is to be controlled so as to minimize the cost function:

$$J = \int_0^1 (x^2 + u^2)dt$$

Determine the optimal control signal, optimal trajectory, and the control law.

Solution

Method 1 (General Solution Approach):
Hamiltonian: $H = x^2 + u^2 + \lambda u$
Hamilton (Euler) Equations:

$$\dot{\lambda} = -\frac{\partial H}{\partial x} = -2x \tag{i}$$

$$\frac{\partial H}{\partial u} = 2u + \lambda = 0 \quad \Rightarrow \quad u = -\frac{1}{2}\lambda \tag{ii}$$

BCs (Equation 11.182):

$$\left[-\lambda + \frac{\partial h}{\partial x}\right]_f \delta x_f + \left[H + \frac{\partial h}{\partial t}\right]_{t_f} \delta t_f$$

But, $h=0$ and hence $\partial h/\partial x=0$, $\partial h/\partial t=0$ for this case. Also, $\delta t_f=0$. Then, setting the coefficient of δx_f to zero, we get the BC

$$\lambda(1)=0 \tag{iii}$$

Substitute the state equation in Equation (ii). This and Equation (i) give

$$\left.\begin{array}{l} \dot{x} = -\dfrac{1}{2}\lambda \quad \text{(iv)} \\[2mm] \dot{\lambda} = -2x \quad \text{(i)} \end{array}\right\} \Rightarrow \ddot{x} = -\dfrac{1}{2}\dot{\lambda} = -\dfrac{1}{2}(-2x) = x \Rightarrow$$

$$\ddot{x} - x = 0 \Rightarrow x = c_1 e^t + c_2 e^{-t} \tag{v}$$

Substitute the given IC $x(0)=1$:

$$c_1 + c_2 = 1 \tag{vi}$$

From Equations (iv), (v) and (vi):

$$\lambda = -2\dot{x} = -2(c_1 e^t - c_2 e^{-t}) = -2c_1 e^t + 2c_2 e^{-t} \tag{vii}$$

Substitute Equation (vii) in Equation (iii):

$$\lambda(1) = 0 = -2c_1 e + 2c_2 e^{-1} \tag{viii}$$

Solve Equations (vi) and (viii):

$$-c_1 e + (1-c_1)\dfrac{1}{e} = 0 \Rightarrow$$

$$c_1 = \dfrac{1}{(e^2+1)}$$

$$c_2 = \dfrac{e^2}{e^2+1}$$

From Equations (vii) and (ii), the optimal control signal is $u = -\dfrac{1}{2}(-2c_1 e^t + 2c_2 e^{-t}) = c_1 e^t - c_2 e^{-t} \Rightarrow$

$$u_{opt} = \dfrac{1}{(e^2+1)} e^t - \dfrac{e^2}{(e^2+1)} e^{-t}$$

From Equation (v), the corresponding optimal trajectory is:

$$x_{opt} = \dfrac{1}{(e^2+1)} e^t + \dfrac{e^2}{(e^2+1)} e^{-t}$$

(Check: $\dot{x} = u$).

The optimal control law may be determined as the ratio u_{opt}/x_{opt}, and is left here as an exercise.

Method 2 (Riccati Equation, Linear Regulator, Approach):

For the given problem,

$$Q=2, \ R=2, \ S=0, \ A=0, \ B=1$$

Then, from Equation 11.193

$$u = -R^{-1}B^T Kx = -\frac{1}{2} \times 1 \times Kx = -\frac{1}{2}Kx \tag{i}$$

where K satisfies Equations 11.194 and 11.195: $\dot{K} + KA + A^T K - KBR^{-1}BK + Q = 0$; $K(t_f) = S \Rightarrow$

$$\dot{K} - \frac{1}{2}K^2 + 2 = 0; \ K(1) = 0 \Rightarrow 2\frac{dK}{dt} = -(4 - K^2) \Rightarrow \frac{dK}{4 - K^2} = -\frac{1}{2}dt \Rightarrow$$

$$\frac{1}{4}\log\frac{2+K}{2-K} = -\frac{1}{2}t + C$$

Substitute $K(1) = 0$: $-1/2 + C = 0 \Rightarrow$

$$\frac{1}{4}\log\frac{2+K}{2-K} = -\frac{1}{2}t + \frac{1}{2}$$

which simplifies to:

$$K = \frac{2[e^{2(1-t)} - 1]}{[e^{2(1-t)} - 1]} \tag{ii}$$

Substitute Equation (ii) in Equation (i) to obtain the optimal control law: $u = (-1/2)Kx \Rightarrow$

$$u_{opt} = -\frac{[e^{2(1-t)} - 1]}{[e^{2(1-t)} + 1]}x$$

Note: This result can be substituted in the state equation and integrated, to determine the optimal $x(t)$, and is left as an exercise.

Check:
Substitute the result for $x_{opt}(t)$ as obtained from Method 1, into the optimal control law obtained in Method 2:

$$u_{opt} = -\frac{[e^{2(1-t)} - 1]}{[e^{2(1-t)} + 1]}x = -\frac{[e^{-2(t-1)} - 1]}{[e^{-2(t-1)} + 1]}\left[\frac{1}{e^2 + 1}e^t + \frac{e^2}{e^2 + 1}e^{-t}\right]$$

$$= -\frac{[e^{-2(t-1)} - 1]}{[e^{-2(t-1)} + 1]}\frac{e^t}{e^2 + 1}[1 + e^2 e^{-2t}]$$

$$= \frac{1}{(e^2 + 1)}e^t - \frac{e^2 e^{-t}}{(e^2 + 1)}$$

This is identical to the optimal control signal as obtained in Method 1, and hence confirms that the results obtained from Method 1 and Method 2 are identical.

11.7.4 Control System Design

Assuming that all n states are available for feedback (in particular, assuming that all n states are measurable), we can implement a constant-gain feedback control law of the form:

$$u = -Kx \qquad\qquad (11.200)$$

in which K is the feedback gain matrix, which is determined by the optimal control approach as described in this section; for example, by minimizing the infinite-time quadratic integral. Standard software packages use efficient recursive algorithms to solve the matrix Riccati equation, which is needed in this approach. In particular, MATLAB® may be used (see Appendix B).

   The assumption that all states are measurable and available for feedback is somewhat unrealistic in general. In many practical situations in order to implement a complete state feedback law developed as above, the complete state vector has to be estimated from the measurement (output) vector y or from a portion of the state vector. A device (or a filter) that will generate the full state vector from output measurements or part of the state vector is known as a state estimator or an observer. Once a state estimator and a complete state feedback control law are designed, the control scheme can be implemented by measuring y or some of the state variables, supplying them to the state estimator (a digital computer) that will generate an estimate of x in real time, and supplying the estimated x to the controller that will simply multiply it by the control gain K to generate the feedback control signal. Furthermore, since the poles of the closed-loop system can be placed at arbitrary locations using just on input (assuming controllability), the remaining control inputs may be used to optimize to some extent an appropriate performance index. Such an approach will combine modal control and optimal control, and is beyond the scope of the present treatment.

Example 11.37

Consider the damped mechanical system shown in Figure 11.26a. The equations of motion of the system are:

$$\begin{bmatrix} m & 0 \\ 0 & m \end{bmatrix} \ddot{y} + \begin{bmatrix} c & 0 \\ 0 & c \end{bmatrix} \dot{y} + \begin{bmatrix} 3k & -2k \\ -2k & 5k \end{bmatrix} y = \begin{bmatrix} 0 \\ f(t) \end{bmatrix}$$

Assume the following parameters values: $m = 2$ kg, $k = 10$ N.m^{-1}, $c = 0.6$ N.s.m^{-1}.

   Suppose that the displacement and velocities of the masses are taken as the states. An LQR controller may be developed using MATLAB®, to generate the forcing inputs for the system. In this example, the following MATLAB code is used, which is self-explanatory, with the indicated parameter values:

```
clear;
global u A B
m = 2.0;
k = 10.0;
c = 0.6;
M = [m 0 ;
     0 m];
K = [3*k -2*k;
```

```
   -2*k 5*k];
C= [c 0;
    0 c];
u= [0;0];
%Construct A,B
A1 = -inv(M)*K;
A2 = -inv(M)*C;
B1 = inv(M)
A= [ 0 0 1 0;
     0 0 0 1;
  A1(1,1) A1(1,2) A2(1,1) A2(1,2);
  A1(2,1) A1(2,2) A2(2,1) A2(2,2)];
B= [ 0 0
     0 0
  B1(1,1) B1(1,2)
  B1(2,1) B1(2,2)];
%define the costfunction weighting matrix Q,R
Q= [1, 1, 1, 1];
Q=diag(Q,0);
%decrease element of R can have fast response
R= [6 0;
    0 6];
[Kp,S,E] =lqr(A, B, Q, R);
t0 = 0;
tf = 30;
x0 = [0.1 0 0 0];%Define initial condition
T= 0.003; % Define the sampling time
t_init=t0;
t_final=T;
x1_last =x0(1);
x2_last =x0(2);
x3_last =x0(3);
x4_last =x0(4);
for i =1:(tf/T)
err =odeset('RelTol', 1e-6,'AbsTol',1e-8 );
[t,w] =ode45('sysmodel',[t_init t_final],[x1_last x2_last x3_last
x4_last],err);
x1lst =w(:,1);
x2lst =w(:,2);
x3lst =w(:,3);
x4lst =w(:,4);
kq =size(w(:,1));
x1_last =x1lst(kq(1));
x2_last =x2lst(kq(1));
x3_last =x3lst(kq(1));
x4_last =x4lst(kq(1));
q1(i) =w(1,1);
q2(i) =w(1,2);
q3(i) =w(1,3);
q4(i) =w(1,4);
q= [q1(i);q2(i);q3(i);q4(i)];
u = -Kp*q;%Control feedback
tt(i) =t_init;
t_init=t_final;
t_final=t_final+T
end
figure(1);
```

```
subplot(2,2,1),plot(tt,q1);
subplot(2,2,2),plot(tt,q2);
subplot(2,2,3),plot(tt,q3);
subplot(2,2,4),plot(tt,q4);
function wd=sysmodel(t,w);
 global A B u
 wd=zeros(4,1);
 wd=A*w+B*u;
```

The system response under LQR control is shown in Figure 11.26b.

  Note: See Appendix B for further MATLAB control system examples.

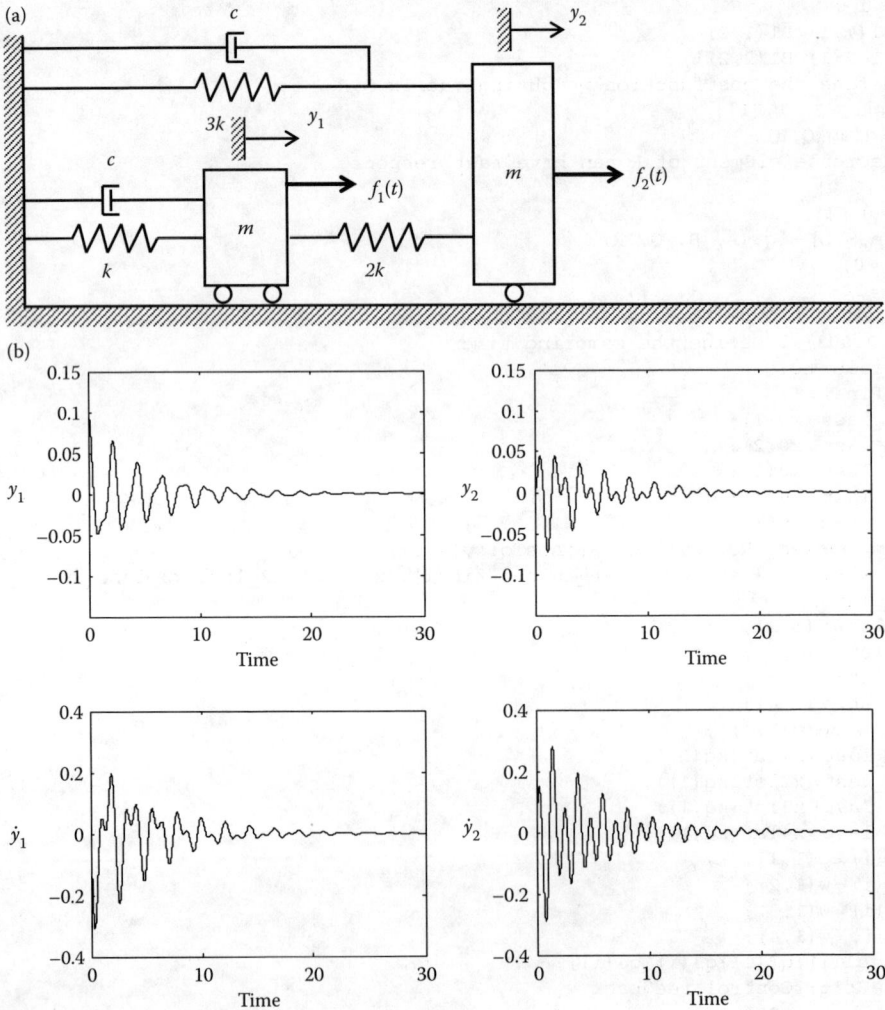

FIGURE 11.26
(a) A vibrating system; (b) the response under LQR control.

11.8 Other Advanced Control Techniques

Now we will outline several advanced control techniques, which may be classified into the same group as modern control techniques. It is beyond the scope of the present book give details of these techniques. The present section is intended to be an introduction to these techniques rather than a rigorous treatment.

11.8.1 Nonlinear Feedback Control

Simple, linear servo control is known to be inadequate for transient and high-speed operation of complex plants. Past experience of servo control in process applications is extensive, however, and servo control is extensively used in many commercial applications (e.g., robots). For this type of control to be effective, however, nonlinearities and dynamic coupling must be compensated faster than the control bandwidth at the servo level. One way of accomplishing this is by implementing a linearizing and decoupling controller inside the servo loops. This technique is termed *feedback linearization technique* (FLT). One such technique that is useful in controlling nonlinear and coupled dynamic systems such as robots is outlined here.

Consider a mechanical dynamic system (plant) given by:

$$M(q)\frac{d^2q}{dt^2} = n\left(q,\frac{dq}{dt}\right) + f(t) \tag{11.201}$$

in which:

$$f = \begin{bmatrix} f_1 \\ f_2 \\ \vdots \\ f_r \end{bmatrix} = \text{vector of input forces at various locations of the system}$$

$$q = \begin{bmatrix} q_1 \\ q_2 \\ \vdots \\ q_r \end{bmatrix} = \text{vector of response variables (e.g., positions)} \\ \text{at the forcing locations of the system}$$

$M(q)$ = inertia matrix (nonlinear)
$n(q,(dq/dt))$ = a vector of remaining nonlinear effects in the system (e.g., damping, backlash, gravitational effects)

Now suppose that we can model M by $\hat{M}$ and n by $\hat{n}$. Then, let us use the nonlinear (linearizing) feedback controller given by:

$$f = \hat{M}K\left[e + T_i^{-1}\int edt - T_d\frac{dq}{dt}\right] - \hat{n} \tag{11.202}$$

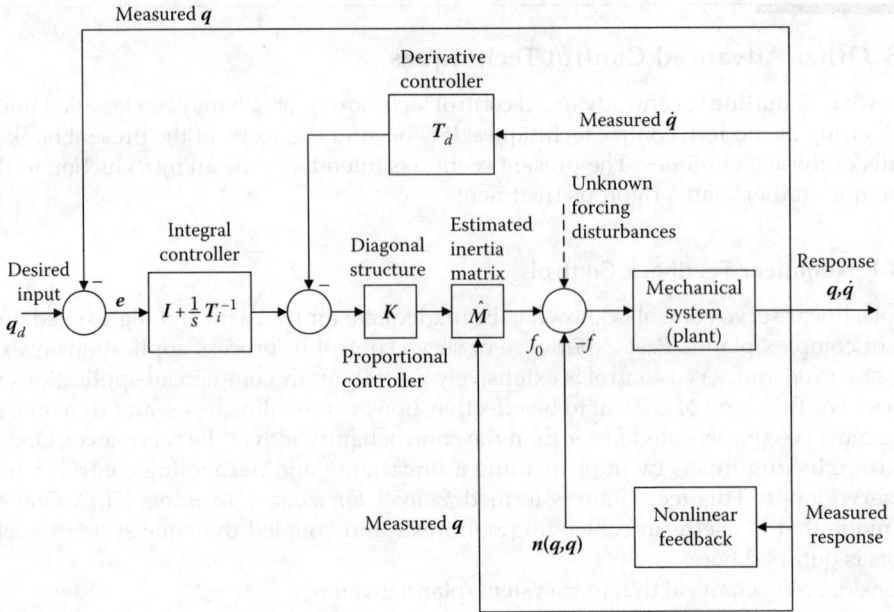

FIGURE 11.27
The structure of the model-based nonlinear feedback control system.

in which

$e = q_d - q$ = error (correction) vector

q_d = desired response

and K, T_i, and T_d are constant control parameter matrices. This control scheme is shown in Figure 11.27. By substituting the controller equation (Equation 11.202) into the plant equation (Equation 11.201) we get:

$$M\frac{d^2q}{dt^2} = n - \hat{n} + \hat{M}K\left[e + T_i^{-1}\int edt - T_d\frac{dq}{dt}\right] \qquad (11.203)$$

If our models are exact, we have $M = \hat{M}$ and $n = \hat{n}$. Then, because the inverse of matrix $\hat{M}$ exists in general (because the inertia matrix *is positive definite*), we get

$$\frac{d^2q}{dt^2} = K\left[e + T_i^{-1}\int edt - T_d\frac{dq}{dt}\right] \qquad (11.204)$$

Equation 11.204 represents a linear, constant parameter system with PID control. The proportional control parameters are given by the gain matrix K, the integral control parameters by T_i, and the derivative control parameters by T_d. It should be clear that we are free to select these parameters so as to achieve the desired response. In particular, if these three parameter matrices are chosen to be diagonal, then the control system, as given by Equation 11.204 and shown in Figure 11.27 will be uncoupled (i.e., one input affects only one output) and will not contain dynamic interactions. In summary, this controller has the advantages of linearizing and decoupling the system; its disadvantages are that accurate models will be needed and that the control algorithm is crisp and unable to handle qualitative or partially known information, learning, etc.

Instead of using analytical modeling, the parameters in $\hat{M}$ and $\hat{n}$ may be obtained through the measurement of various input-output pairs. This is called *model identification*, and can cause further complications in terms of instrumentation and data processing speed, particularly because some of the model parameters must be estimated in real time.

11.8.2 Adaptive Control

An adaptive control system is a feedback control system in which the values of some or all of the controller parameters are modified (adapted) during the system operation (in real time) on the basis of some performance measure, when the response (output) requirements are not satisfied. The techniques of adaptive control are numerous because many criteria can be employed for modifying the parameter values of a controller. According to the above definition, *self-tuning control* falls into the same category. In fact, the terms "adaptive control" and "self-tuning control" have been used interchangeably in the technical literature. Performance criteria used in self-tuning control may range from time-response or frequency-response specifications, parameters of "ideal" models, desired locations of poles and zeros, and cost functions. Generally, however, in self-tuning control of a system some form of parameter estimation or identification is performed on-line using input-output measurements from the system, and the controller parameters are modified using these estimated parameter values. A majority of the self-tuning controllers developed in the literature is based on the assumption that the plant (process) is linear and time-invariant. This assumption does not generally hold true for complex industrial processes. For this reason we shall restrict our discussion to an adaptive controller that has been developed for nonlinear and coupled plants.

On-line estimation or system identification, which may be required for adaptive control, may be considered to be a preliminary step of "learning." In this context, *learning control* and adaptive control are related, but learning is much more complex and sophisticated than a quantitative estimation of parameter values. In a learning system, control decisions are made using the cumulative experience and knowledge gained over a period of time. Furthermore, the definition of learning implies that a learning controller will "remember" and improve its performance with time. This is an evolutionary process that is true for intelligent controllers, but not generally for adaptive controllers.

Here, we briefly describe a *model-referenced adaptive control* (MRAC) technique. The general approach of MRAC is illustrated by the block diagram in Figure 11.28. In nonadaptive feedback control the response measurements are fed back into the drive controller through a feedback controller, but the values of the controller parameters (feedback gains) themselves are unchanged during operation. In adaptive control these parameter values are changed according to some criterion. In model-referenced adaptive control, in particular, the same reference input that is applied to the physical system is applied to a reference model as well. The difference between the response of the physical system and the output from the reference model is the error. The ideal objective is to make this error zero at all times. Then, the system will perform just like the reference model. The error signal is used by the adaptation mechanism to determine the necessary modifications to the values of the controller parameters in order to reach this objective. Note that the reference model is an idealized model which generates a desired response when subjected to the reference input, at least in an asymptotic manner (i.e., the error converges to zero). In this sense

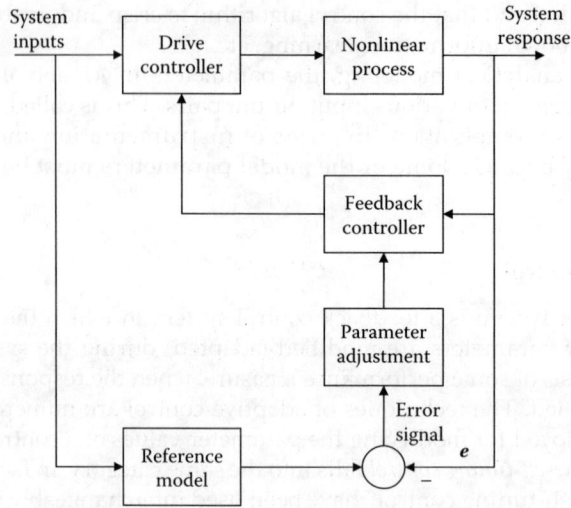

FIGURE 11.28
Model-referenced adaptive controller.

it is just a means of performance specification and may not possess any resemblance or analogy to an analytical model of the process itself. For example, the reference model may be chosen as a linear, uncoupled system with desired damping and bandwidth properties (i.e., damping ratios and natural frequencies).

A popular approach to derive the adaptive control algorithm (i.e., the equations express-ing how the controller parameters should be changed in real time) is through the use of the MIT rule. In this method, the controller parameters are changed in the direction opposite to the maximum slope of the quadratic error function. Specifically, the quadratic function

$$V(p) = e^T W e \qquad (11.205)$$

is formed, where e is the error signal vector shown in Figure 11.28. W is a diagonal and positive-definite weighting matrix, and p is a vector of control parameters which will be changed (adapted) during control. The function V is minimized numerically with respect to p during the controller operation, subject to some simplifying assumptions. The details of the algorithm are found in the literature.

The adaptive control algorithm described here has the advantage that it does not neces-sarily require a model of the plant itself. The reference model can be chosen to specify the required performance, the objective of MRAC being to drive the response of the system toward that of the reference model. Several drawbacks exist in this scheme, however. Because the reference model is quite independent of the plant model, the required control effort could be excessive and the computation itself could be slow. Furthermore, a new con-trol law must be derived for each reference model. Also, the control action must be gener-ated much faster than the speed at which the nonlinear terms of the plant change because the adaptation mechanism has been derived by assuming that some of the nonlinear terms remain more or less constant.

Many other adaptive control schemes depend on a reasonably accurate model of the plant, not just a reference model. The models may be obtained either analytically or through identification (experimental). Adaptive control has been successfully applied in complex, nonlinear, and coupled systems, even though it has several weaknesses, as mentioned previously.

11.8.3 Sliding Mode Control

Sliding mode control, variable structure control, and suction control fall within the same class of control techniques, and are somewhat synonymous. The control law in this class is generally a switching controller. A variety of switching criteria may be employed. Sliding mode control may be treated as an adaptive control technique. Because the switching surface is not fixed, its variability is somewhat analogous to an adaptation criterion. Specifically, the error of the plant response is zero when the control falls on the sliding surface.

Consider a plant that is modeled by the nth order nonlinear ordinary differential equation:

$$\frac{d^n y}{dt^n} = f(y,t) + u(t) + d(t) \tag{11.206}$$

where

y = response of interest

$u(t)$ = input variable

$d(t)$ = unknown disturbance input

$f(\bullet)$ = an unknown nonlinear model of the process which depends on the response

vector: $y = \left[y, \dot{y}, \cdots, \frac{d^{n-1} y}{dt^{n-1}} \right]^T$

A time-varying sliding surface is defined by the differential equation:

$$s(y,t) = \left(\frac{d}{dt} + \lambda \right)^{n-1} \tilde{y} = 0 \tag{11.207}$$

with $\lambda > 0$. Note the response error $\tilde{y} = y \quad y_d$, where y_d is the desired response. Similarly, $\tilde{y} = y - y_d$ may be defined. It should be clear from Equation 11.207 that if we start from rest with zero initial error ($\tilde{y}(0) = 0$ with all the derivative of $\tilde{y}$ up to the $n - 1$th derivative being zero at $t = 0$) then $s = 0$ corresponds to $\tilde{y}(t) = 0$ for all t. This will guarantee that the desired trajectory $y_d(t)$ is tracked accurately at all times. Hence, the control objective would be to force the error state vector $\tilde{y}$ onto the sliding surface $s = 0$. This control objective will be achieved if the control law satisfies

$$s \, \text{sgn}(s) \leq -\eta \quad \text{with} \quad \eta > 0 \tag{11.208}$$

where sgn(s) is the *signum function*.

The nonlinear process $f(y, t)$ is generally unknown. Suppose that $f(y,t) = \hat{f}(y,t) + \Delta f(y,t)$ where $\hat{f}(y,t)$ is a completely known function, and Δf represents modeling uncertainty. Specifically, consider the control equation:

$$u = -\hat{f}(y,t) - \sum_{p=1}^{n-1} \binom{n-1}{p} \lambda^p \frac{d^{n-p} \tilde{y}}{dt^{n-p}} - K(y,t) \text{sgn}(s) \tag{11.209}$$

where $K(y,t)$ is an upper bound for the total uncertainty in the system (i.e., disturbance, model error, speed of error reduction, etc.) and:

$$\binom{n-1}{p} = \frac{(n-1)!}{p!(n-1-p)!}$$

This sliding-mode controller satisfies Equation 11.208, but has drawbacks arising from the sgn(s) function. Specifically, very high switching frequencies can result when the control effort is significant. This is usually the case in the presence of large modeling errors and disturbances. High-frequency switching control can lead to the excitation of high-frequency modes in the plant. It can also lead to chattering problems. This problem can be reduced if the signum function in Equation 11.209 is replaced by a *saturation function*, with a boundary layer $\pm\Phi$, as shown in Figure 11.29. In this manner, any switching that would have occurred within the boundary layer would be filtered out. Furthermore, the switching transitions would be much less severe. Clearly, the advantages of sliding mode control include robustness against factors such as nonlinearity, model uncertainties, disturbances, and parameter variations.

11.8.4 Linear Quadratic Gaussian (LQG) Control

This is an *optimal control* technique that is intended for quite linear systems with random input disturbances and output (measurement) noise. Consider the *linear* system given by the set of first order differential equations (*state equations*):

$$\frac{dx}{dt} = Ax + Bu + Fv \tag{11.210}$$

and the output equations:

$$y = Cx + w \tag{11.211}$$

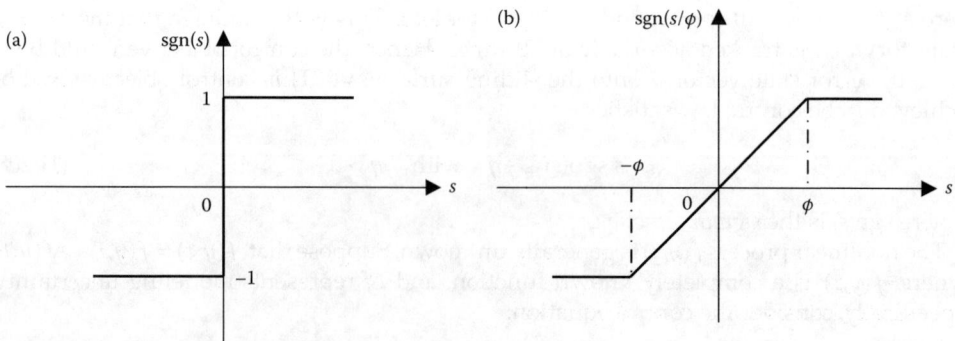

FIGURE 11.29
Switching functions used in sliding mode control. (a) Signum function. (b) Saturation function.

in which

$$x = \begin{bmatrix} x_1 \\ x_2 \\ \vdots \\ x_n \end{bmatrix}$$ is the state vector,

$$u = \begin{bmatrix} u_1 \\ u_2 \\ \vdots \\ u_r \end{bmatrix}$$ is the vector of system inputs, and

$$y = \begin{bmatrix} y_1 \\ y_2 \\ \vdots \\ y_m \end{bmatrix}$$ is the vector of system outputs

The vectors v and w represent input disturbances and output noise, respectively, which are assumed to be white noise (i.e., zero-mean random signals whose *power spectral density junction is* flat) with covariance matrices V and W. Also, A is called the system matrix, B the *input distribution matrix,* and C the *output formation matrix.* In LQG control the objective is to minimize the *performance index* (cost function):

$$J = E \left\{ \int_0^\infty (x^T Q x + u^T R u) dt \right\} \tag{11.212}$$

in which Q and R are diagonal matrices of weighting and E denotes the "expected value" (or mean value) of a random process. In the LQG method the controller is implemented as the two-step process:

1. Obtain the estimate $\hat{x}$ for the state vector x using a *Kalman filter* (with gain K_f).
2. Obtain the control signal as a product of $\hat{x}$ and a gain matrix K_0, by solving a noise-free linear quadratic optimal control problem.

This implementation is shown by the block diagram in Figure 11.30. As mentioned before, it can be analytically shown that the noise-free quadratic optimal controller is given by the gain matrix:

$$K_0 = R^{-1} B^T P_0 \tag{11.213}$$

where P_0 is the positive semidefinite solution of the *algebraic Riccati equation*:

$$A^T P_0 + P_0 A - P_0 B R^{-1} B^T P_0 + Q = 0 \tag{11.214}$$

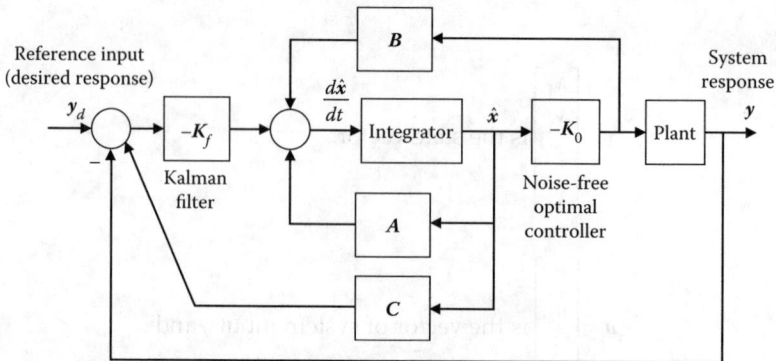

FIGURE 11.30
Linear quadratic Gaussian (LQG) control.

The Kalman filter is given by the gain matrix:

$$K_f = P_f\, C^T\, W^{-1} \tag{11.215}$$

where P_f is obtained as before by solving:

$$AP_f + P_f A^T - P_f\, C^T\, W^{-1}\, CP_f + FVF^T = 0 \tag{11.216}$$

An advantage of this controller is that the stability of the closed-loop control system is guaranteed as long as both the plant model and the Kalman filter are *stabilizable* and *detectable*. Note that if uncontrollable modes of a system are stable, the system is stabilizable. Similarly, if the unobservable modes of a system are stable, the system is detectable. Another advantage is the precision of the controller as long as the underlying assumptions are satisfied, but LQG control is also a model-based "crisp" scheme. Model errors and noise characteristics can significantly affect the performance. Also, even though stability is guaranteed, good stability margins and adequate robustness are not guaranteed in this method. Computational complexity (solution of two Riccati equations) is another drawback.

11.8.5 H_∞ Control

This is a relatively new optimal control approach which is quite different from the LQG method. However, this is a frequency-domain method. This technique assumes a linear plant with constant parameters, which may be modeled by a *transfer function* in the SISO case or by a *transfer matrix* in the MIMO case. Without going into the analytical details, let us outline the principle behind H_∞ control.

Consider the MIMO, linear, feedback control system shown by the block diagram in Figure 11.31, where I is an *identity matrix*. It satisfies the relation:

$$GG_c[y_d - y] = y$$

or

$$[I + GG_c]y = GG_c\, y_d \tag{11.217}$$

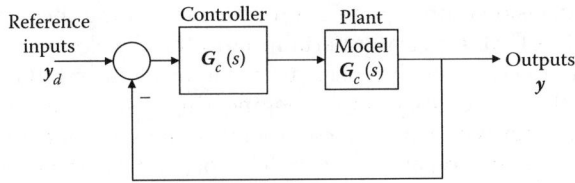

FIGURE 11.31
A linear multivariable feedback control system.

Because the plant G is fixed, the underlying design problem here is to select a suitable controller G_c that will result in a required performance of the system. In other words, the closed-loop transfer matrix:

$$H = \left[I + GG_c\right]^{-1} GG_c \tag{11.218}$$

must be properly "shaped" through an appropriate choice of G_c. The required shape of $H(s)$ may be consistent with the classical specifications such as

1. Unity $|H(j\omega)|$ or large $|GG_c(j\omega)|$ at low frequencies in order to obtain small steady-state error for step inputs
2. Small $|H(j\omega)|$ or small $|GG_c(j\omega)|$ at large frequencies so that high-frequency noise would not be amplified, and further, the controller would be physically realizable.
3. Adequately high gain and phase margins in order to achieve required stability levels.

Of course, in theory there is an "infinite" number of possible choices for $G_c(s)$ that will satisfy such specifications. The H_∞ method uses an optimal criterion to select one of these "infinite" choices. Specifically, the choice that minimizes the so-called "H_∞ norm" of the closed-loop transfer matrix $H(s)$, is chosen. The rationale is that this optimal solution is known to provide many desired characteristics (with respect to stability, robustness in the presence of model uncertainty and noise, sensitivity, etc.) in the control system.

The H_∞ norm of a transfer matrix H is the maximum value of the largest *singular value* of $H(j\omega)$, maximum being determined over the entire frequency range. A singular value of $H(j\omega)$ is the square root of an eigenvalue of the matrix $H(j\omega)H^T(j\omega)$.

The H_∞ control method has the advantages of stability and robustness. The disadvantages are that it is a "crisp" control method that is limited to linear, time-invariant systems, and that it is a model-based technique.

11.9 Fuzzy Logic Control

An intelligent controller may be interpreted as a computer-based controller that can somewhat "emulate" the reasoning procedures of a human expert in the specific area of control, to generate the necessary control actions. Here, techniques from the field of *artificial intelligence (AI)* are used for the purpose of acquiring and representing knowledge and for generating control decisions through an appropriate reasoning mechanism. With steady advances in the field of AI, especially pertaining to the development of practical *expert systems* or *knowledge systems*, there has been a considerable interest in using AI techniques for controlling complex processes. Complex engineering systems use intelligent control to cope with situations where conventional control techniques are not effective.

Intelligent control depends on efficient ways of representing and processing the control knowledge. Specifically, a knowledge base has to be developed and a technique of reasoning and making "inferences" has to be available. Knowledge-based intelligent control relies on knowledge that is gained by intelligently observing, studying, or understanding the behavior of a plant, rather than explicitly modeling the plant, to arrive at the control action. In this context, it also heavily relies on the knowledge of experts in the domain, and also on various forms of general knowledge. Modeling of the plant is implicit here. Soft computing is an important branch of study in the area of intelligent and knowledge-based systems. It has effectively complemented conventional AI in the area of machine intelligence (computational intelligence). Fuzzy logic, probability theory, neural networks, and genetic algorithms are cooperatively used in soft computing for knowledge representation and for mimicking the reasoning and decision-making processes of a human. Decision making with soft computing involves *approximate reasoning*, and is commonly used in intelligent control. This section presents an introduction to intelligent control, emphasizing FLC.

11.9.1 Fuzzy Logic

Fuzzy logic is useful in representing human knowledge in a specific domain of application and in reasoning with that knowledge to make useful inferences or actions. The conventional binary (bivalent) logic is crisp and allows for only two states. This logic cannot handle fuzzy descriptors, examples of which are "fast" which is a *fuzzy quantifier* and "weak" which is a *fuzzy predicate*. They are generally qualitative, descriptive, and subjective and may contain some overlapping degree of a neighboring quantity, for example, some degree of "slowness" in the case of the fuzzy quantity "fast." Fuzzy logic allows for a realistic extension of binary, crisp logic to qualitative, subjective, and approximate situations, which often exist in problems of intelligent machines where techniques of AI are appropriate.

In fuzzy logic, the knowledge base is represented by if-then rules of fuzzy descriptors. Consider the general problem of approximate reasoning. In this case the knowledge base K is represented in an "approximate" form, for example, by a set of if-then rules with *antecedent* and *consequent* variables that are fuzzy descriptors. First, the data D are preprocessed according to:

$$F_D = FP(D) \tag{11.219}$$

which, in a typical situation, corresponds to a data abstraction procedure called "fuzzification" and establishes the membership functions or membership grades that correspond to D. Then for a fuzzy knowledge base F_K, the fuzzy inference F_I is obtained through fuzzy-predicate approximate reasoning, as denoted by:

$$F_I = F_K \circ F_D \tag{11.220}$$

This uses a *composition* operator "$\circ$" for fuzzy matching of data (D) with the knowledge base (K) is carried out, and making inferences (I) on that basis.

Fuzzy logic is commonly used in "intelligent" control of processes and machinery. In this case the inferences of a fuzzy decision making system are the control inputs to the process. These inferences are arrived at by using the process responses as the inputs (context data) to the fuzzy decision-making system.

11.9.2 Fuzzy Sets and Membership Functions

A fuzzy set has a fuzzy boundary. The membership an element lying on the boundary is fuzzy: there is some possibility that the element is inside the set and a complementary possibility that it is outside the set. A fuzzy set may be represented by a membership function. This function gives the grade (degree) of membership within the set, of any element of the universe of discourse. The membership function maps the elements of the universe on to numerical values in the interval [0, 1]. Specifically:

$$\mu_A(x) \colon X \rightarrow [0, 1] \tag{11.221}$$

where $\mu_A(x)$ is the membership function of the fuzzy set A in the universe in X. Stated in another way, fuzzy set if A is a set of ordered pairs:

$$A = \{(x, \mu_A(x)); \; x \in X, \; \mu_A(x) \in [0,1]\} \tag{11.222}$$

The membership function $\mu_A(x)$ represents the grade of possibility that an element x belongs to the set A. It follows that a membership function is a *possibility function* and not a probability function. A membership function value of zero implies that the corresponding element is definitely not an element of the fuzzy set. A membership function value of unity means that the corresponding element is definitely an element of the fuzzy set. A grade of membership greater than 0 and less than 1 corresponds to a noncrisp (or fuzzy) membership, and the corresponding elements fall on the fuzzy boundary of the set. The closer the $\mu_A(x)$ is to 1 the more the x is considered to belong to A, and similarly the closer it is to 0 the less it is considered to belong to A. A typical fuzzy set is shown in Figure 11.32a and its membership function is shown in Figure 11.32b.

Note: A crisp set is a special case of fuzzy set, where the membership function can take the two values 1 (membership) and 0 (nonmembership) only. The membership function of a crisp set is given the special name *characteristic function*.

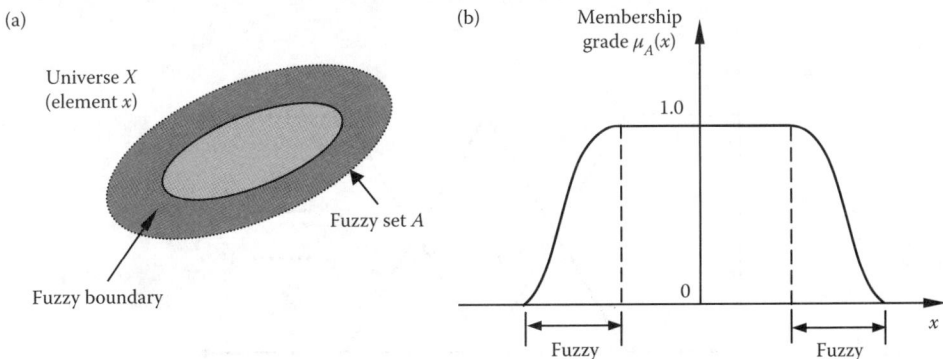

FIGURE 11.32
(a) A fuzzy set; (b) The membership function of a fuzzy set.

11.9.3 Fuzzy Logic Operations

It is well known that the "complement", "union", and "intersection" of crisp sets correspond to the logical operations NOT, OR, and AND, respectively, in the corresponding crisp, bivalent logic. Furthermore, it is known that, in the crisp bivalent logic, the union of a set with the complement of a second set represents an "implication" of the first set by the second set. Set inclusion (i.e., extracting a subset) is a special case of implication in which the two sets belong to the same universe. These operations (connectives) may be extended to fuzzy sets for corresponding use in fuzzy logic fuzzy reasoning. For fuzzy sets, the applicable connectives must be expressed in terms of the membership functions of the sets which are operated on. In view of the isomorphism between fuzzy sets and fuzzy logic, both the set operations and the logical connectives can be addressed together. Some basic operations that can be defined on fuzzy sets and the corresponding connectives of fuzzy logic are described now. Several methods are available to define the intersection and the union of fuzzy sets.

11.9.3.1 Complement (Negation, NOT)

Consider a fuzzy set A in a universe X. Its complement A' is a fuzzy set whose membership function is given by:

$$\mu_{A'}(x) = 1 - \mu_A(x) \quad \text{for all} \quad x \in X \tag{11.223}$$

The complement in fuzzy sets corresponds to the negation (NOT) operation in fuzzy logic, just as in crisp logic, and is denoted by $\bar{A}$ where A now is a fuzzy logic proposition (or a fuzzy state).

A graphic (membership function) representation of complement of a fuzzy set (or, negation of a fuzzy state) is given in Figure 11.33.

11.9.3.2 Union (Disjunction, OR)

Consider two fuzzy sets A and B in the same universe X. Their union is a fuzzy set containing all the elements from both sets, in a "fuzzy" sense. This set operation is denoted by $\cup$. The membership function of the resulting set $A \cup B$ is given by

$$\mu_{A \cup B}(x) = \max[(\mu_A(x), \mu_B(x)] \quad \forall x \in X \tag{11.224}$$

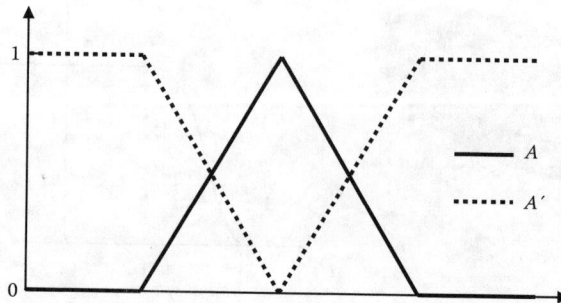

FIGURE 11.33
Fuzzy set complement or fuzzy logic NOT.

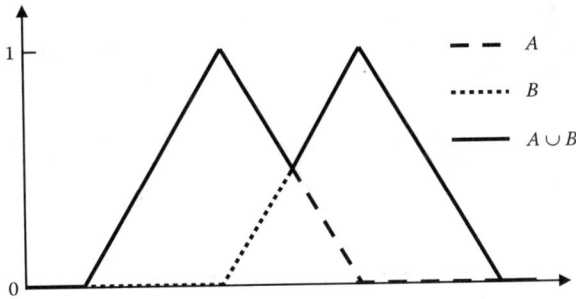

FIGURE 11.34
Fuzzy set union or fuzzy logic OR.

The union corresponds to a logical OR operation (called *Disjunction*), and is denoted by $A \cup B$, where A and B are fuzzy states or fuzzy propositions. The rationale for the use of *max* to represent fuzzy-set union is that, because element x may belong to one set or the other, the larger of the two membership grades should govern the outcome (union). Furthermore, this is consistent with the union of crisp sets. Similarly, the appropriateness of using *max* to represent fuzzy-logic operation "OR" should be clear. Specifically, since either of the two fuzzy states (or propositions) would be applicable, the larger of the corresponding two membership grades should be used to represent the outcome. A graphic (membership function) representation of union of two fuzzy sets (or, the logical combination OR of two fuzzy states in the same universe) is given in Figure 11.34.

Even though set intersection is applicable to sets in a common universe, a logical "OR" may be applied for concepts in different universes. In particular, when the operands belong to different universes, orthogonal axes have to be used to represent them in a common membership function.

11.9.3.3 Intersection (Conjunction, AND)

Again, consider two fuzzy set A and B in the same universe X. Their intersection is a fuzzy set containing all the elements that are common to both sets, in a "fuzzy" sense. This set operation is denoted by $\cap$. The membership function of the resulting set $A \cap B$ is given by:

$$\mu_{A \cap B}(x) = \min[(\mu_A(x), \mu_B(x)] \quad \forall x \in X \tag{11.225}$$

The union corresponds to a logical AND operation (called *Conjunction*), and is denoted by $A \cap B$, where A and B are fuzzy states or fuzzy propositions. The rationale for the use of *min* to represent fuzzy-set intersection is that, because the element x must simultaneously belong to both sets, the smaller of the two membership grades should govern the outcome (intersection).

Furthermore, this is consistent with the intersection of crisp sets. Similarly, the appropriateness of using *min* to represent fuzzy-logic operation "AND" should be clear. Specifically, since both fuzzy states (or propositions) should be simultaneously present, the smaller of the corresponding two membership grades should be used to represent the outcome. A graphic (membership function) representation of intersection of two fuzzy sets (or, the logical combination AND of two fuzzy states in the same universe) is given in Figure 11.35.

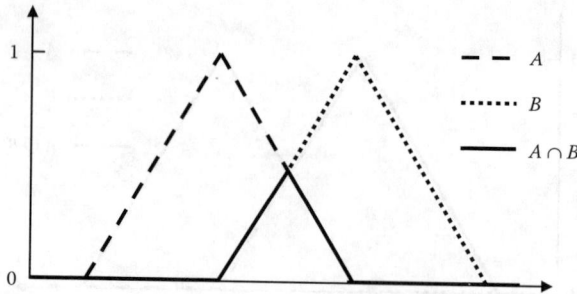

FIGURE 11.35
Fuzzy set intersection or fuzzy logic AND.

11.9.3.4 Implication (If-Then)

An if-then statement (a rule) is called an "implication." In a knowledge-based system, the knowledge base is commonly represented using if-then rules. In particular, a knowledge base in fuzzy logic may be expressed by a set of linguistic rules of the if-then type, containing fuzzy terms. In fact a fuzzy rule is a *fuzzy relation*. A knowledge base containing several fuzzy rules is also a relation, which is formed by combining (aggregating) the individual rules according to how they are interconnected.

Consider a fuzzy set A defined in a universe X and a second fuzzy set B defined in another universe Y. The fuzzy implication "If A then B," is denoted by $A{\to}B$. Note that in this fuzzy rule, A represents some "fuzzy" situation, and is the *condition* or the *antecedent* of the rule. Similarly, B represents another fuzzy situation, and is the *action* or the *consequent* of the fuzzy rule. The fuzzy rule $A{\to}B$ is a fuzzy relation. Since the elements of A are defined in X and the elements of B are defined in Y, the elements of $A{\to}B$ are defined in the *Cartesian product space* $X{\times}Y$. This is a two-dimensional space represented by two orthogonal axes (x-axis and y-axis), and gives the domain in which fuzzy rule (or fuzzy relation) is defined. Since A and B can be represented by membership functions, and additional orthogonal axis is needed to represent the membership grade.

Fuzzy implication may be defined (interpreted) in several ways. Two definitions of fuzzy implication are:

Method 1:

$$\mu_{A\to B}(x,y) = \min[(\mu_A(x), \mu_B(y)] \quad \forall x \in X, \; \forall y \in Y \tag{11.226}$$

Method 2:

$$\mu_{A\to B}(x,y) = \min[1, \{1 - \mu_A(x) + \mu_B(y)\}] \quad \forall x \in X, \forall y \in Y \tag{11.227}$$

These two methods are approaches for obtaining the membership function of the particular fuzzy relation given by an if-then rule (implication). Note that the first method gives an expression that is symmetric with respect to A and B. This is not intuitively satisfying because "implication" is not a commutative operation (specifically, $A{\to}B$ does not necessarily satisfy $B{\to}A$). In practice, however, this method provides a good, robust result. The second method has an intuitive appeal because in crisp bivalent logic, $A{\to}B$ has the same

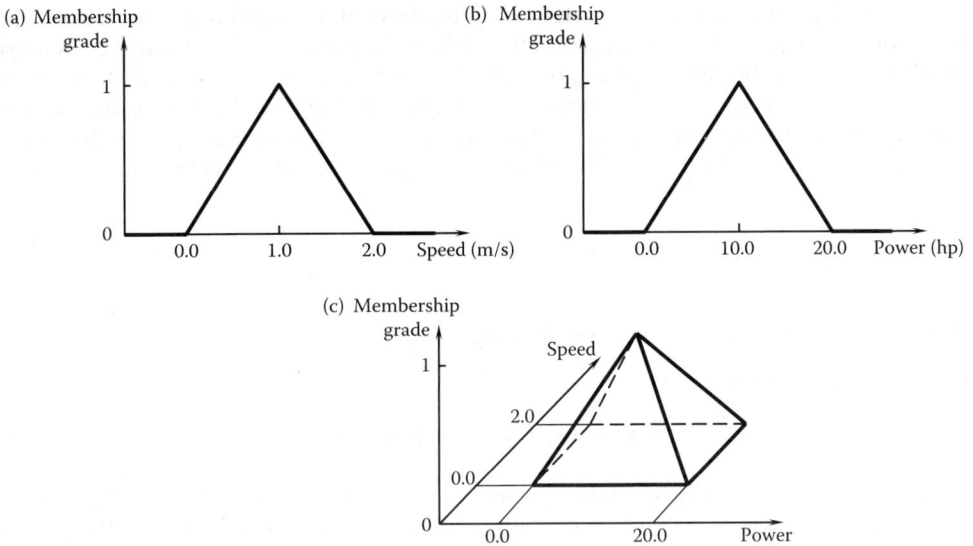

FIGURE 11.36
(a) Fuzzy set A; (b) Fuzzy set B; (c) Fuzzy implication $A \rightarrow B$.

truth table as [(NOT A) OR B] and hence are equivalent. Note that in Equation 11.227, the membership function is upper-bounded to 1 using the *bounded sum* operation, as required (A membership grade cannot be greater than 1). The first method is more commonly used because it is simpler to use and often provides quite accurate results.

An example of fuzzy implication using the first method is shown in Figure 11.36. Here the implication from a fuzzy set (Figure 11.36a to another fuzzy set (Figure 11.36b) is given by the fuzzy set in Figure 11.36c).

11.9.4 Compositional Rule of Inference

In knowledge-based systems, knowledge is often expressed as rules of the form:

"**IF** condition Y_1 is y_1 **AND IF** condition Y_2 is y_2 **THEN** action C is c."

In fuzzy knowledge-based systems (e.g., fuzzy control systems), rules of this type are linguistic statements of expert knowledge in which y_1, y_2, and c are fuzzy quantities (e.g., small negative, fast, large positive). These rules are fuzzy relations that employ the fuzzy implication (IF-THEN). The collective set of fuzzy relations forms the knowledge base of the fuzzy system. Let us denote the fuzzy relation formed by this collection of rules as the fuzzy set K. This relation is an aggregation of the individual rules, and may be represented by a multivariable membership function. In a fuzzy decision making process (e.g., in FLC), the rulebase (knowledge base) K is first collectively matched with the available data (context). Next, an inference is made on another fuzzy variable that is represented in the knowledge base, on this basis. The matching and inference making are done using the composition operation, as discussed previously. The application of composition to make inferences in this manner is known as the *compositional rule of inference* (CRI).

For example, consider a control system. Usually the context would be the measured outputs Y of the process. The control action that drives the process is C. Typically, both these variables are crisp, but let us ignore this fact for the time being and assume them to be fuzzy, for general consideration. Suppose that the control knowledge base is denoted by R, a fuzzy relation. The method of obtaining the rule-base R is analogous to model identification in conventional crisp control. Then, by applying the CRI we get the fuzzy control action as:

$$\mu_C = \max_Y \min(\mu_Y, \mu_R) \tag{11.228}$$

11.9.5 Extensions to Fuzzy Decision Making

Thus far we have considered fuzzy rules of the form:

$$\text{IF } x \text{ is } A_i \text{ AND IF } y \text{ is } B_i \text{ THEN } z \text{ is } C_i \tag{11.229}$$

where A_i, B_i, and C_i are fuzzy states governing the ith rule of the rulebase. This is the the Mamdani approach (Mamdani system or Mamdani model) named after the person who pioneered the application of this approach. Here, the knowledge base is represented as fuzzy protocols and represented by membership functions for A_i, B_i, and C_i, and the inference is obtained by applying the CRI. The result is a fuzzy membership function, which typically has to be *defuzzified* for use in practical tasks.

Several variations to this conventional method are available. One such version is the *Sugeno model* (or, *Takagi-Sugeno-Kang model* or *TSK model*). Here, the knowledge base has fuzzy rules with crisp functions as the consequent, of the form:

$$\text{IF } x \text{ is } A_i \text{ AND IF } y \text{ is } B_i \text{ THEN } c_i = f_i(x, y) \tag{11.230}$$

for Rule i, where, f_i is a crisp function of the condition variables (antecedent) x and y. Note that the condition part of this rule is the same as for the Mamdani model (Equation 11.229), where A_i and B_i are fuzzy sets whose membership functions are functions of x and y, respectively. The action part is a crisp function of the condition variables, however. The inference $\hat{c}(x,y)$ of the fuzzy knowledge-based system is obtained directly as a crisp function of the condition variables x and y, as follows:

For Rule i, a weighting parameter $w_i(x, y)$ is obtained corresponding to the condition membership functions, as for the Mamdani approach, by using either the "min" operation or the "product" operation. For example, using the "min" operation we form:

$$w_i(x,y) = \min[\mu_{A_i}(x), \mu_{B_i}(y)] \tag{11.231}$$

The crisp inference $\hat{c}(x,y)$ is determined as a weighted average of the individual rule inferences (crisp) $c_i = f_i(x, y)$ according to:

$$\hat{c}(x,y) = \frac{\sum_{i=1}^{r} w_i c_i}{\sum_{i=1}^{r} w_i} = \frac{\sum_{i=1}^{r} w_i(x,y) f_i(x,y)}{\sum_{i=1}^{r} w_i(x,y)} \tag{11.232}$$

where r is the total number of rules. For any data x and y, the knowledge-based action $\hat{c}(x,y)$ can be computed from Equation 11.232, without requiring any defuzzification. The Sugeno model is particularly useful when the actions are described analytically through crisp functions, as in conventional crisp control, rather than linguistically. The TSK approach is commonly used in the applications of direct control and in simplified fuzzy models. The Mamdani approach, even though popular in low-level direct control, is particularly appropriate for knowledge representation and processing in expert systems and in high-level (hierarchical) control systems.

11.9.6 Basics of Fuzzy Control

Fuzzy control uses the principles of fuzzy logic-based decision making to arrive at the control actions. The decision making approach is typically based on the CRI, as presented before. In essence, some information (e.g., output measurements) from the system to be controlled is matched with a knowledge base of control for the particular system, using CRI. A fuzzy rule in the knowledge base of control is generally a "linguistic relation" of the form:

$$\text{IF } A_i \text{ THEN IF } B_i \text{ THEN } C_i \tag{11.233}$$

where A_i and B_i are fuzzy quantities representing process measurements (e.g., process error and change in error) and C_i is a fuzzy quantity representing a control signal (e.g., change in process input). What we have is a rule-base with a set of (n) rules:

$$\text{Rule 1: } A_1 \text{ and } B_1 \Rightarrow C_1$$

$$\text{Rule 2: } A_2 \text{ and } B_2 \Rightarrow C_2$$

$$\vdots$$

$$\text{Rule } n: A_n \text{ and } B_n \Rightarrow C_n$$

Because these fuzzy sets are related through IF-THEN implications and because an implication operation for two fuzzy sets can be interpreted as a "minimum operation" on the corresponding membership functions, the membership function of this fuzzy relation may be expressed as:

$$\mu_{Ri}(a, b, c) = \min[\mu_{Ai}(a), \mu_{Bi}(b), \mu_{Ci}(c)] \tag{11.234}$$

The individual rules in the rule-base are joined through ELSE connectives, which are OR connectives ("unions" of membership functions). Hence, the overall membership function for the complete rule-base (relation R) is obtained using the "maximum" operation on the membership functions of the individual rules:

$$\mu_R(a, b, c) = \max_i \mu_{Ri}(a, b, c) = \max_i \min[\mu_{Ai}(a), \mu_{Bi}(b), \mu_{Ci}(c)] \tag{11.235}$$

In this manner the membership function of the entire rule-base can be determined (or, "identified" in the terminology of conventional control) using the membership functions of the response variables and control inputs. Note that a fuzzy knowledge base is a multivariable function—a multidimensional array (a three-variable function or a dimensional array in the case of Equation 11.235) of membership function values. This array corresponds to a fuzzy control algorithm in the sense of conventional control. The control rule-base may represent linguistic expressions of experience, expertise, or knowledge of the domain experts (control engineers, skilled operators, etc.). Alternatively, a control engineer may instruct an operator (or a control system) to carry out various process tasks in the usual manner; monitor and analyze the resulting data; and learn appropriate rules of control, say by using neural networks.

Once a fuzzy control knowledge base of the form given by Equation 11.235 is obtained, we need a procedure to infer control actions using process measurements, during control. Specifically, suppose that fuzzy process measurements A' and B' are available. The corresponding control inference C' is obtained using the *CRI* (i.e., inference using the composition relation). The applicable relation is:

$$m_{C'}(c) = \sup_{a,b} \min[m_{A'}(a), m_{B'}(b), m_R(a,b,c)] \tag{11.236}$$

Note that in fuzzy inference, the data fuzzy sets A' and B' are jointly matched with the knowledge-base fuzzy relation R. This is a "join" operation, which corresponds to an AND operation (an "intersection" of fuzzy sets), and hence the *min* operation applies for the membership functions. For a given value of control action c, the resulting fuzzy sets are then mapped (projected) from a three-dimensional space $X \times Y \times Z$ of knowledge onto a one-dimensional space Z of control actions. This mapping corresponds to a set of OR connectives, and hence the *sup* operation applies to the membership function values, as expressed in Equation 11.236.

Actual process measurements are crisp. Hence, they have to be *fuzzified* in order to apply the CRI. This is conveniently done by reading the grade values of the membership functions of the measurement at the specific measurement values. Typically, the control action must be a crisp value as well. Hence, each control inference C' must be *defuzzified* so that it can be used to control the process. Several methods are available to accomplish defuzzification. In the *mean of maxima* method the control element corresponding to the maximum grade of membership is used as the control action. If there is more than one element with a maximum (peak) membership value, the mean of these values is used. In the *center of gravity* (or *centroid*) *method* the centroid of the membership function of control decision is used as the value of crisp control action. This weighted control action is known to provide a somewhat sluggish, yet more robust control.

Because process measurements are crisp, one method of reducing the real-time computational overhead is to precompute a decision table relating quantized measurements to crisp control actions. The main disadvantage of this approach is that it does not allow for convenient modifications (e.g., rule changes and quantization resolution adjustments) during operation. Another practical consideration is the selection of a proper sampling period in view of the fact that process responses are generally analog signals. Factors such as process characteristics, required control bandwidth, and the processing time needed for one control cycle, must be taken into account in choosing a sampling period. Scaling or gain selection for various signals in a FLC system is another important consideration. For reasons of processing efficiency, it is customary to scale the process variables and control signals in a fuzzy control algorithm. Furthermore, adjustable gains can be cascaded with these system variables so that they may serve as tuning parameters for the controller. A proper tuning algorithm would be needed, however. A related consideration

Knowledge base

$$R: \underset{i}{ELSE} \, (IF \, Y_i \, THEN \, U_i)$$

⇓

Defuzzification	Inference engine (Compositional Rule of Inference)	Fuzzification
$\mu_{U'}(u) \to u'$	$\mu_{U'}(u) = \mu_{Y'}(y) \circ R$	$y' \to \mu_{Y'}(y)$

Control input u' Intelligent machine   Machine response y'

(motor currents, voltages, etc.)   (motions, etc.)

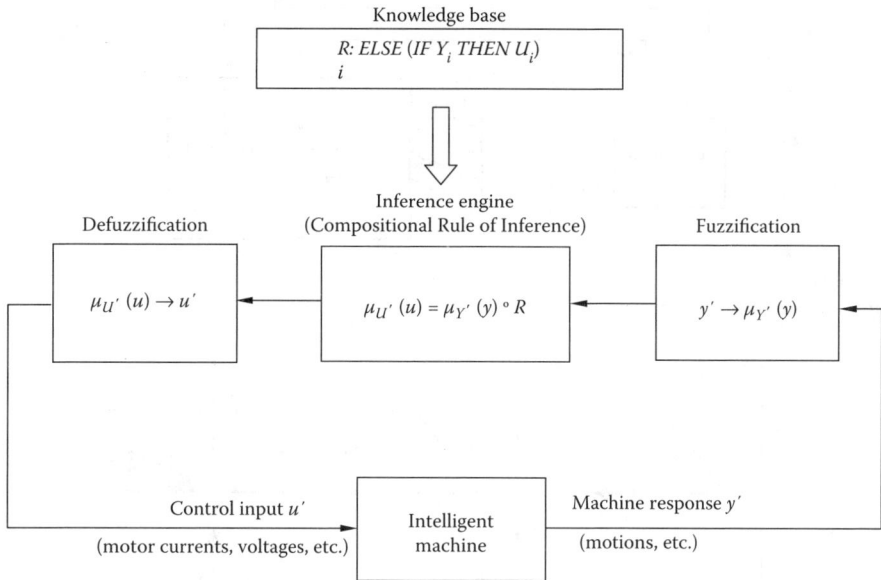

FIGURE 11.37
Structure of a direct fuzzy controller.

is real-time or on-line modification of a fuzzy rule-base. Specifically, rules may be added, deleted, or modified on the basis of some scheme of *learning and self-organization*. For example, using a model for the process and making assumptions such as input-output monotonicity, it is possible during control to trace and tag the rules in the rule-base that need attention. The control-decision table can be modified accordingly.

Hardware fuzzy processors (*fuzzy chips*) may be used to carry out the fuzzy inference at high speed. The rules, membership functions, and measured context data are generated as usual, through the use of a control "host" computer. The fuzzy processor is located in the same computer, which has appropriate interface (input-output) hardware and driver software. Regardless of all these, it is more convenient to apply the inference mechanism separately to each rule and then combine the result instead of applying it to the entire rule-base using the CRI.

Fuzzy logic is commonly used in direct control of processes and machinery. In this case the inferences of a fuzzy decision making system form the control inputs to the process. These inferences are arrived at by using the process responses as the inputs (context data) to the fuzzy system. The structure of a direct fuzzy controller is shown in Figure 11.37. Here, y represents the process output, u represents the control input, and R is the relation, which represents the fuzzy control knowledge base.

Example 11.38

Consider the room comfort control system schematically shown in Figure 11.38. The temperature (T) and humidity (H) are the process variables that are measured. These sensor signals are provided to the fuzzy logic controller, which determines the cooling rate (C) that should be generated by the air conditioning unit. The objective is to maintain a particular comfort level inside the room.

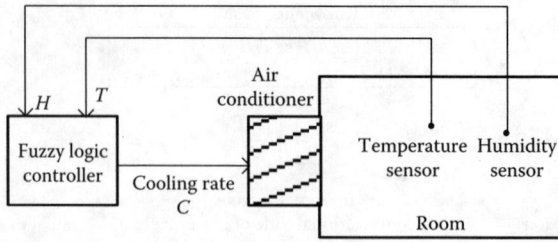

FIGURE 11.38
Comfort control system of a room.

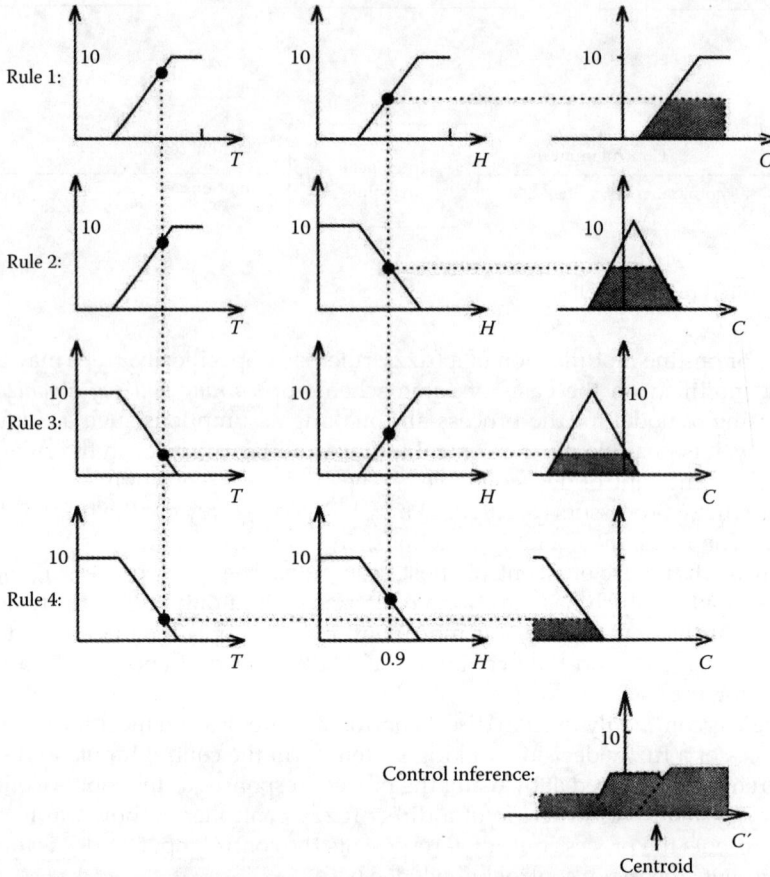

FIGURE 11.39
The fuzzy knowledge base of the comfort controller.

A simplified fuzzy rule-base of the comfort controller is graphically presented in Figure 11.39. The temperature level can assume one of two fuzzy states (*HG, LW*), which denote high and low, respectively, with the corresponding membership functions. Similarly, the humidity level can assume two other fuzzy states (*HG, LW*) with associated membership functions. Note that the membership functions of *T* are quite different from those of *H*, even though the same nomenclature is used. There are four rules, as given in Figure 11.39. The rule-base is:

Rule 1:		if	T	is	HG	and	H	is	HG	then	C	is	PH
Rule 2:	else	if	T	is	HG	and	H	is	LW	then	C	is	PL
Rule 3:	else	if	T	is	LW	and	H	is	HG	then	C	is	NL
Rule 4:	else	if	T	is	LW	and	H	is	LW	then	C	is	NH
	and	if											

The nomenclature used for the fuzzy states is as follows:

Temperature (T)	Humidity (H)	Change in Cooling Rate (C)
HG=High	HG=High	PH=Positive High
LW=Low	LW=Low	PL=Positive Low
		NH=Negative High
		NL=Negative Low

Application of the CRI is done here by using individual rule-based composition. Specifically, the measured information is composed with individual rules in the knowledge base and the results are aggregated to give the overall decision. For example, suppose that the room temperature is 30°C and the relative humidity is 0.9. Lines are drawn at these points, as shown in Figure 11.39, to determine the corresponding membership grades for the fuzzy states in the four rules. In each rule the lower value of the two grades of process response variables is then used to clip (or modulate) the corresponding membership function of C (a *min* operation). The resulting "clipped" membership functions of C for all four rules are superimposed (a *max* operation) to obtain the control inference C' as shown. This result is a fuzzy set, and it must be defuzzified to obtain a crisp control action $\hat{c}$ for changing the cooling rate. The centroid method may be used for defuzzification.

11.9.7 Fuzzy Control Surface

A fuzzy controller is a nonlinear controller. A well-defined problem of fuzzy control, with analytical membership functions and fuzzification and defuzzification methods, and well-defined fuzzy logic operators, may be expressed as a nonlinear control surface through the application of the CRI. The advantage then is that the generation of the control action becomes a simple and very fast step of reading the surface value (control action) for given values of crisp measurement (process variables). The main disadvantage is, the controller is fixed and cannot accommodate possible improvements to control rules and membership functions through successive learning and experience. Nevertheless, this approach to fuzzy control is quite popular. A useful software tool for developing fuzzy controllers is the MATLAB® Fuzzy Logic Toolbox.

Example 11.39

A schematic diagram of a simplified system for controlling the liquid level in a tank is shown in Figure 11.40a. In the control system, the error (actually, correction) is given by: e=Desired level−Actual level.

The change in error is denoted by Δe. The control action is denoted by u, where $u > 0$ corresponds to opening the inflow valve and $u < 0$ corresponds to opening the outflow valve. A low-level direct fuzzy controller is used in this control system, with the control rule-base as given in Figure 11.40b.

The membership functions for E, ΔE, and U are given in Figure 11.40c. Note that the error measurements are limited to the interval [−3a, 3a] and the Δerror measurements to [−3b, 3b]. The control actions are in the range [−4c, 4c].

(a)

Valve actuator

Inflow

Liquid level sensor

Level controller

Desired level

Valve actuator
Outflow

(b)

E \ ΔE	NL	NS	ZO	PS	PL
NL	NL	NL	NM	NS	ZO
NS	NL	NM	NS	ZO	PS
ZO	NM	NS	ZO	PS	PM
PS	NS	ZO	PS	PM	PL
PL	ZO	PS	PM	PL	PL

(c)

$\mu E(e)$

NL   NS   1.0 ZO   PS   PL

$-3a$ $-2a$ $-a$ 0 a $2a$ $3a$ e

$\mu \Delta E(\Delta e)$

NL   NS   1.0 ZO   PS   PL

$-3b$ $-2b$ $-b$ 0 b $2b$ $3b$ Δe

$\mu_U(U)$

NL   NM   NS   1.0 ZO   PS   PM   PL

$-4c$ $-3c$ $-2c$ $-c$ 0 c $2c$ $3c$ $4c$ u

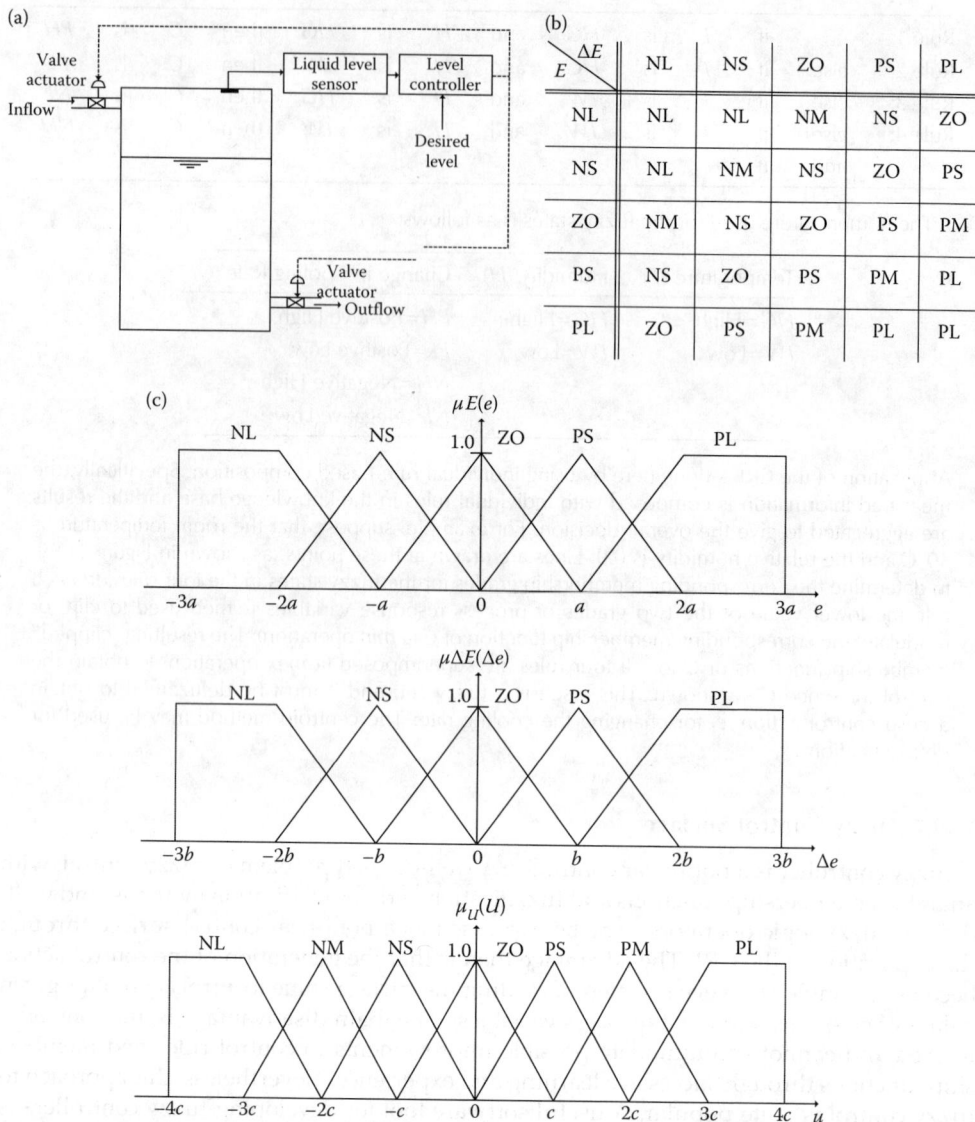

FIGURE 11.40
(a) Liquid level control system. (b) The control rule-base. (c) The membership functions of Error, Change in Error, and Control Action.

Following the usual steps of applying the CRI for this fuzzy logic controller, we can develop a crisp *control surface* $u(e, \Delta e)$ for the system, expressed in the three-dimensional coordinate system $(e, \Delta e, u)$, which then can be used as a simple and fast controller. This method is described next.

The crisp control surface is developed by carrying out the rule-based inference for each point: $(e, \Delta e)$ in the measurement space $E \times \Delta E$, using individual rule-based inference. Specifically:

$$\mu_{U'}(u) = \max_{i,j} \min[\mu_{Ei}(e_o), \mu_{\Delta Ej}(\Delta e_o), \mu_{Uk}(u)] \tag{11.237}$$

where:

μ_U=control inference membership function

e_o=crisp context variable "error" defined in $[-3a, 3a]$

Δe_o=crisp context variable "change in error" defined in $[-3b, -3b]$

E_i=fuzzy states of "error"

ΔE_j=fuzzy states of "change in error"

U_k=fuzzy states of "control action"

(i, j, k)=possible combinations of fuzzy states of error, change in error, and control action, within the rule-base.

To find the crisp control inference (u') for a set of crisp context data $(e, \Delta e)$, the fuzzy inference $\mu_{U'}(u)$ is defuzzified using the center of gravity (centroid) method, which for the continuous case, is:

$$u' = \frac{\int_{u \in U} u \mu_{U'}(u) du}{\int_{u \in U} \mu_{U'}(u) du} \tag{11.238a}$$

or, for the discrete case, it is:

$$u' = \frac{\sum_{u_i \in U} u_i \mu_{U'}(u_i) du}{\sum_{u_i \in U} \mu_{U'}(u_i) du} \tag{11.238b}$$

where $U=[-4c, 4c]$. Also, if the geometric shape of the inference is simple (e.g., piecewise linear), the centroid can be computed by the moment method:

$$u' = \frac{\sum_{i=1}^{n} area_i m_i}{\sum_{i=1}^{n} area_i} \tag{11.238c}$$

where:

$area_i$=area of the ith subregion

m_i=distance of the centroid of the ith subregion, on the control axis.

To demonstrate this procedure, consider a set of context data $(e_o, \Delta e_o)$, where e_o is in $[-3a, -2a]$ and Δe_o is in $[-b/2, 0]$. Then, from the membership functions and the rule-base, it should be clear that only two rules are valid in this region, as given below:

R_1: if e is *NL* and Δe is *NS* then u is *NL*

R_2: if e is *NL* and Δe is *ZO* then u is *NM*

Since, in the range $[-3a, -2a]$, the membership grade of singleton fuzzification of e_o is always 1, the lower grade of the two context values is the one corresponding to the singleton fuzzification of Δe_o for both rules. Then, in applying the individual rule-based inference, the lower grade value of the two context variables is used to clip off the corresponding membership function of the control action variable U in each rule (this is a *min* operation). The resulting membership functions of U for the two applicable rules are superimposed (this is a *max* operation) to obtain the control inference U', as shown in Figure 11.41.

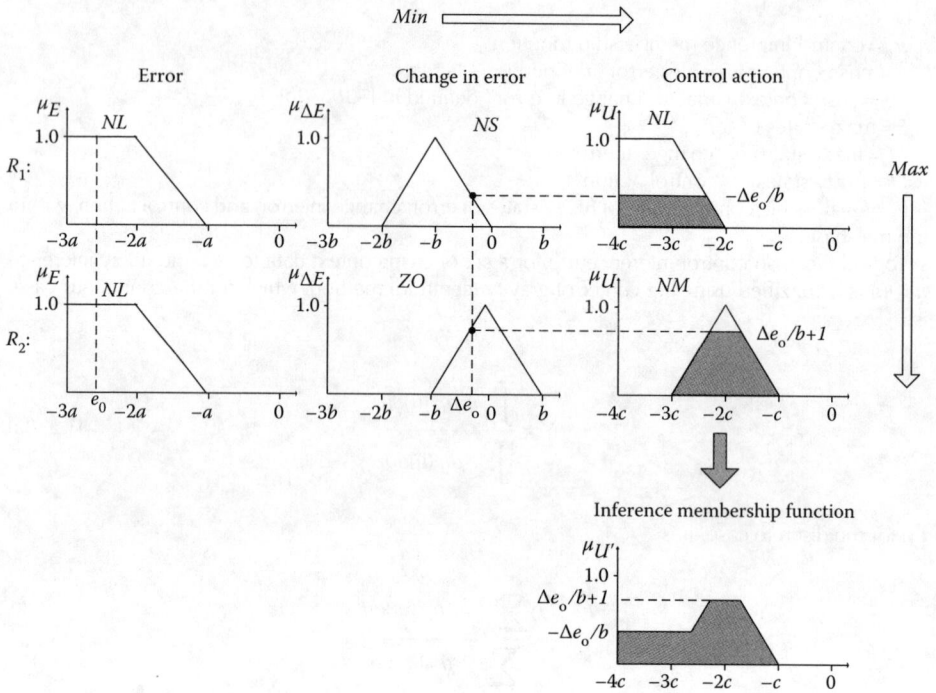

FIGURE 11.41
Individual rule-based inference for $e_o[-3a, -2a]$ and $\Delta e_o[-b/2, 0]$.

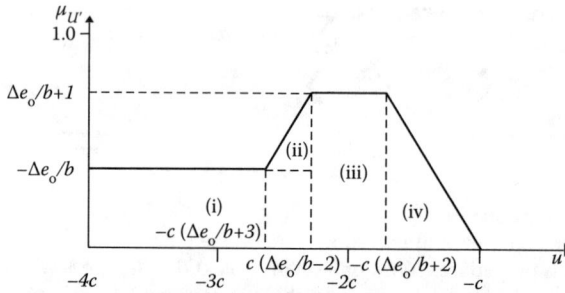

FIGURE 11.42
Subregions and critical points for calculation of the centroid.

For defuzzification, we apply the moment method to find the centroid of the resulting membership function of control inference, as shown in Figure 11.42. Note that the critical points in Figures 11.41 and 11.42 (e.g., $-\Delta e_o/b$, $-c(\Delta e_o/b+3)$, etc.) are found from the corresponding membership functions.

From the moment method, we obtain the crisp control action as a function of e and Δe. The above procedure is repeatedly applied to all possible ranges of e $[-3a, 3a]$ and Δe $[-3b, 3b]$, to

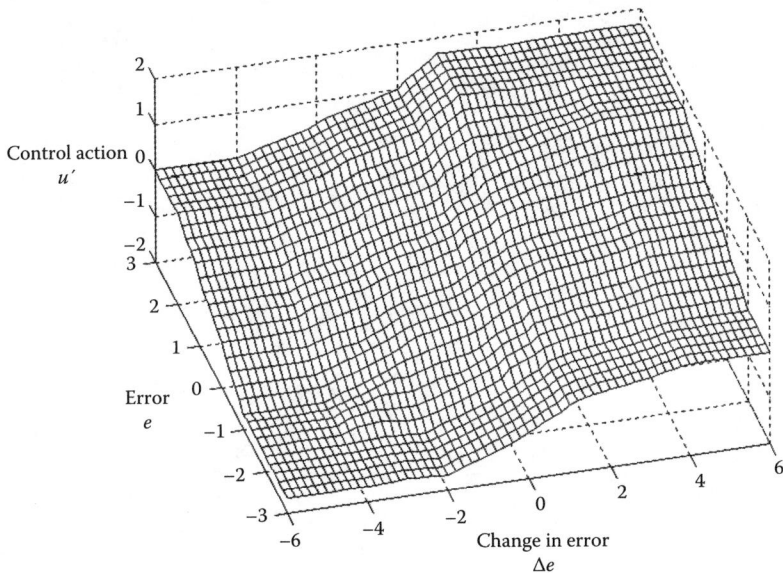

FIGURE 11.43
Control surface with $a=1$, $b=2$, and $c=0.5$.

obtain the complete control surface. Also, the procedure can be implemented in a computer program to generate a control surface. A control surface with $a=1$, $b=2$, and $c=0.5$ is shown in Figure 11.43.

In the present example what we have applied is in fact the Mamdani approach (Mamdani system or Mamdani model). Sugeno model (or TSK model) could have been used as well, thereby avoiding the defuzzification step.

Problems

PROBLEM 11.1

Consider the double pendulum (or a two-link robot with revolute joints) having arm lengths l_1 and l_2, and the end masses m_1 and m_2, as shown in Figure P11.1.

a. Obtain the equations of motion for the system in terms of the absolute angles of swing θ_1 and θ_2 about the vertical equilibrium configuration. Linearize the equations for small motions $\theta_1, \dot{\theta}_1, \theta_2$ and $\dot{\theta}_2$.
b. For the special case of $m_1=m_2=m$ and $l_1=l_2=l$, what are the natural frequencies of the system?
c. Express the free response of the system to an IC excitation of $\theta(0)$ and $\dot{\theta}(0)$.
d. Express the free response as obtained in (c) for the case $l=9.81\,\mathrm{m}$ with $\theta(0)=[1-(1/\sqrt{2})]^T$ and $\dot{\theta}(0)=0$. Sketch this response for a time period of 20 seconds.

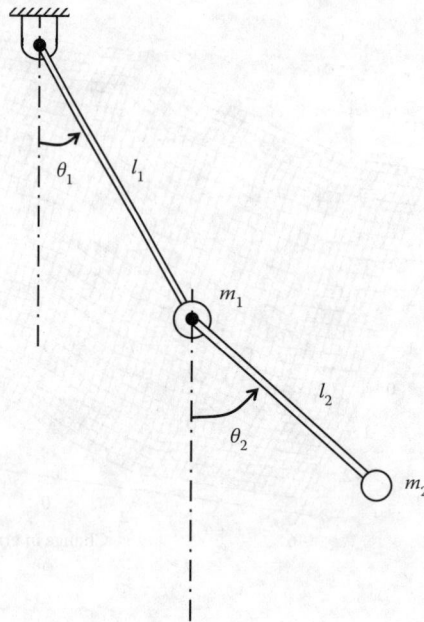

FIGURE P11.1
A double pendulum or a two-link robot arm with revolute joints.

PROBLEM 11.2

Prove the following results:

a. $\Phi(\tau, \tau) = I_n$ for any τ

b. $\dfrac{d}{dt}(t, \tau) = A(t)\,\Phi(t, \tau)$

c. $\Phi(\tau, t) = \Phi^{-1}(t, \tau)$

d. $\Phi(t_2, t_0) = \Phi(t_2, t_1)\,\Phi(t_1, t_0)$

where $\Phi(t_1, t_2)$ is the state-transition matrix of a time-varying system whose system matrix is $A(t)$.

PROBLEM 11.3

Consider a system whose system matrix is:

$$A = \begin{bmatrix} -2 & 4 & 1 \\ 0 & -2 & 3 \\ 0 & 0 & -2 \end{bmatrix}$$

a. Determine the eigenvalues of the system.

b. What are the corresponding eigenvectors (and, if necessary, generalized eigenvectors)?

c. Discuss stability of the system.

PROBLEM 11.4

A SISO dynamic system has the following transfer function:

$$G(s) = \frac{2(s+2)}{(s+1)(s+3)(s+4)}$$

a. What are the eigenvalues of the system?
b. What are the corresponding eigenvectors?
c. Discuss system stability.

PROBLEM 11.5

A system has matrix:

$$A = \begin{bmatrix} 1 & 2 & 3 \\ 0 & 1 & 4 \\ 0 & 0 & 1 \end{bmatrix}$$

Obtain the Jordan form J of this system.

PROBLEM 11.6

Consider the dynamic system represented by the input-output differential equation:

$$\dddot{y} + 15\ddot{y} + 74\dot{y} + 120y = \dddot{u} + 6\ddot{u} + 11\dot{u} + 6u$$

a. Express the system in a state-space form.
b. Determine the eigenvalues and eigenvectors of the system matrix.
c. What is the Jordan form of the system?

PROBLEM 11.7

Consider the second order nonlinear state-space model (Van der Pol's equation):

$$\dot{x}_1 = x_2$$

$$\dot{x}_2 = -x_1 + a(1 - x_1^2)x_2$$

Investigate the stability of the system with respect to the parameter a.

PROBLEM 11.8

A state-space realization of a dynamic system is given by:

$$A = \begin{bmatrix} -2 & 0 \\ -1 & 1 \end{bmatrix}, \quad B = \begin{bmatrix} 1 \\ 1 \end{bmatrix}, \quad C = \begin{bmatrix} 0 & 1 \end{bmatrix}$$

a. Is the system controllable?
b. Is the system observable?
c. Is the system stabilizable?
d. Is the system detectable?

PROBLEM 11.9

A state-space realization of a dynamic system is given by:

$$A = \begin{bmatrix} -2 & 0 \\ -1 & 1 \end{bmatrix}, \quad B = \begin{bmatrix} 0 \\ 1 \end{bmatrix}, \quad C = \begin{bmatrix} 1 & 1 \end{bmatrix}$$

Investigate controllability and observability of the system through pole-zero cancellation.

PROBLEM 11.10

Consider a system with the following matrices:

$$A = \begin{bmatrix} -2 & 0 \\ -1 & -1 \end{bmatrix}, \quad B = \begin{bmatrix} 0 & 2 \\ 1 & 0 \end{bmatrix}.$$

a. Is the system controllable with respect to the first input?
b. Is the system controllable with respect to the second input?
c. Is the system controllable with respect to both inputs?

PROBLEM 11.11

Without using the mathematical condition of controllability, investigate the controllability of the following system:

$$\dot{x} = \begin{bmatrix} 1 & 2 & 1 & -2 \\ 0 & -1 & 0 & 0 \\ 3 & -1 & 2 & -2 \\ 3 & 2 & 7 & 6 \end{bmatrix} x + \begin{bmatrix} -3 \\ 0 \\ -1 \\ 1 \end{bmatrix} u$$

Is the system stabilizable?

PROBLEM 11.12

Through direct observation and without performing any numerical calculations, establish the observability of the following state-space realization:

$$\dot{x} = \begin{bmatrix} 1 & 2 & 3 & -2 \\ 0 & 1 & 0 & 0 \\ 0 & 0 & 2 & 0 \\ 0 & -1 & 0 & -2 \end{bmatrix} x + \begin{bmatrix} 1 \\ 2 \\ -1 \\ 3 \end{bmatrix} u$$

$$y = \begin{bmatrix} 0 & -2 & 1 & 0 \end{bmatrix} x$$

PROBLEM 11.13

Consider the open-loop system given by:

$$A = \begin{bmatrix} 1 & -2 \\ 0 & -1 \end{bmatrix}, \quad B = \begin{bmatrix} 0 \\ 1 \end{bmatrix}$$

a. What are the eigenvalues of the system?
b. Design a complete state feedback law that will place the eigenvalues of the system at $\Gamma = \{-2, -3\}$.

PROBLEM 11.14

For the system in Problem 11.13, design a complete state feedback law that will place the eigenvalues at $\Gamma=\{-2,-1\}$.

PROBLEM 11.15

For the system in Problem 11.13, design a complete state feedback law that will place the eigenvalues at $\Gamma=\{-2,-2\}$.

PROBLEM 11.16

Consider an open-loop system whose transfer function is:

$$G(s) = \frac{2}{s(s+2)}$$

Design a complete state feedback law that will place the system eigenvalues at $\Gamma=\{-2,-2\}$.

PROBLEM 11.17

A third order open-loop system has the state-space realization:

$$A = \begin{bmatrix} 1 & 0 & 0 \\ -1 & -1 & 0 \\ 2 & 3 & -1 \end{bmatrix}, \quad B = \begin{bmatrix} 0 & 0 \\ 0 & 1 \\ -1 & 0 \end{bmatrix}, \quad C = \begin{bmatrix} 1 & -2 & 0 \end{bmatrix}$$

a. Check whether the system is controllable and observable.
b. Design a complete-state feedback law that will place the system poles at $\Gamma = \{-1,-1,-1\}$.
c. Design an output feedback law that will stabilize the open-loop system.

PROBLEM 11.18

Find the optimal trajectory $x^*(t)$ corresponding to the minimum of the cost function:

$$J = \int_0^2 [x^2(t) + 2\dot{x}^2(t)]dt$$

with the BC: $x(0)=0$, $x(2)=4$.

PROBLEM 11.19

Find the optimal trajectory $x^*(t)$ corresponding to the minimum of the cost function:

$$J = \int_0^2 [x^2(t) + 2\dot{x}^2(t)]dt$$

with the BC: $x(0)=2$, $x(2)$ is free.

PROBLEM 11.20

Consider the cost function:

$$J = \int_0^{t_f} [x^2(t) + 2\dot{x}^2(t)]dt$$

Find the optimal trajectory and the corresponding final time for the BC $x(0)=1$ and $x(t_f)=4$, with t_f free.

PROBLEM 11.21

Consider the cost function:

$$J = \int_0^{t_f} [x^2(t) + \dot{x}^2(t)]dt$$

Find the optimal trajectory and the corresponding final time for the following BC:
IC $x(0)=1$, and the final condition $\theta=2e^{-t}$.

PROBLEM 11.22

A plant is given by the state equation:

$$\dot{x} = 2x + u$$

with the end conditions: $x(0)=1$, $x(1)$ free. The system is to be controlled so as to
Minimize the cost function:

$$J = \int_0^1 (x^2 + u^2)dt$$

Determine the optimal control signal, optimal trajectory, and the control law.

PROBLEM 11.23

Discuss why nonlinear feedback control could be very useful in controlling complex
mechanical systems with nonlinear and coupled dynamics. What are the shortcomings of nonlinear feedback control? Consider the two-link manipulator that carries a
point load (weight W) at the end effector, as shown in Figure P11.23. Its dynamics can
be expressed as:

$$I\ddot{q} + b = \tau$$

where q=vector of (relative rotations) q_1 and q_2
  τ=vector of drive torques τ_1 and τ_2 at the two joints, corresponding to the coordinates
q_1 and q_2
  I=second order inertia matrix= $\begin{bmatrix} I_{11} & I_{12} \\ I_{21} & I_{22} \end{bmatrix}$

  b=vector of joint-friction, gravitational, centrifugal, and coriolis torques (components
are b_1 and b_2)
  Neglecting joint friction, and with zero payload ($W=0$), show that

$$I_{11} = m_1 d_1^2 + I_1 + I_2 + m_2(\ell_1^2 + d_2^2 + d\ell_1 d_2 \cos q_2)$$

$$I_{12} = I_{21} + I_2 + m_2 d_2^2 + m_2 \ell d_2 \cos q_2$$

$$I_{22} = I_2 + m_2 d_2^2$$

$$b_1 = m_1 g d_1 \cos q_1 + m_2 g [d_1 \cos q_1 + d_2 \cos(q_1 + q_2)]$$

$$- m_2 \ell_1 d_2 \dot{q}_2^2 \sin q_2 - 2 m_2 \ell_1 d_2 \dot{q}_1 \dot{q}_2 \sin q_2$$

$$b_2 = m_2 g d_2 \cos(q_1 + q_2) + m_2 \ell_1 d_2 \dot{q}_1^2 \sin q_2$$

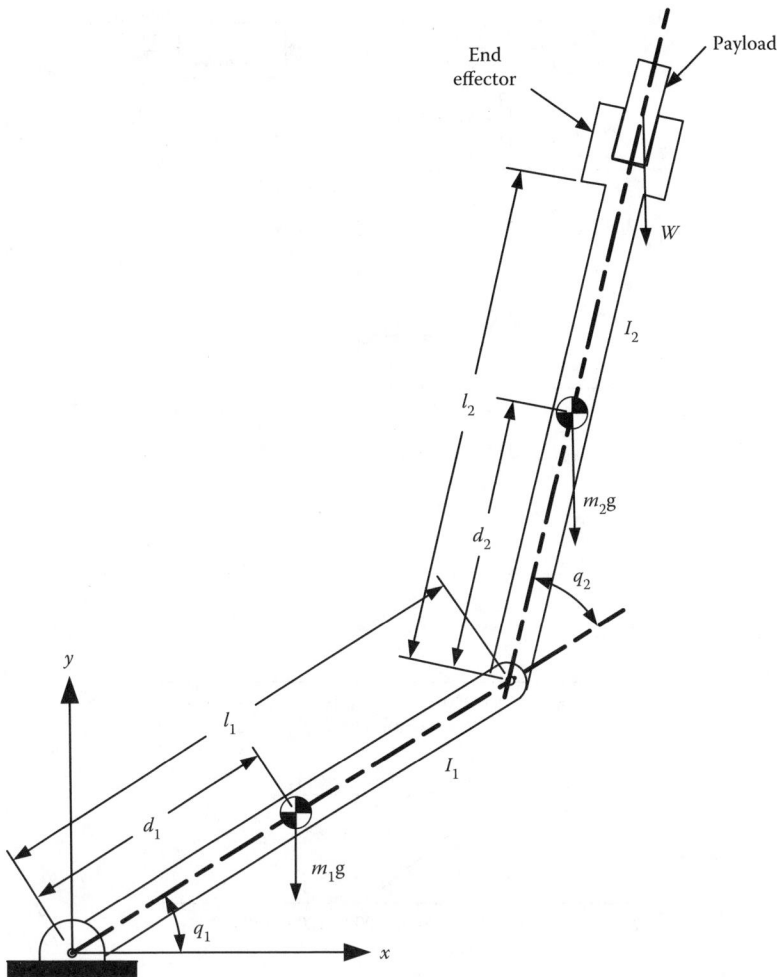

FIGURE P11.23
A two-link robotic manipulator.

The geometric parameters ℓ_1, ℓ_2, d_1, and d_2 are defined in Figure P11.23.

What variables have to be measured for nonlinear feedback control? Noting that the elements of b are more complex (even after neglecting joint friction, backlash, and payload) than the elements of I, justify using nonlinear feedback control for this system instead of using a control method based on an accurate dynamic model.

PROBLEM 11.24

  i. Which control method would you recommend for each of the following applications:
   (a) Servo control of a single-axis positioning table with a permanent-magnet dc motor (linear).
   (b) Active control of a vehicle suspension system (liner, multivariable).
   (c) Control of a rotary cement kiln (nonlinear, complex, difficult to model).
 ii. A metallurgical process consists of heat treatment of a bulk of material for a specified duration of time at a suitable temperature. The heater is controlled by its fuel supply rate. A schematic diagram of the system is shown in Figure P11.24a.

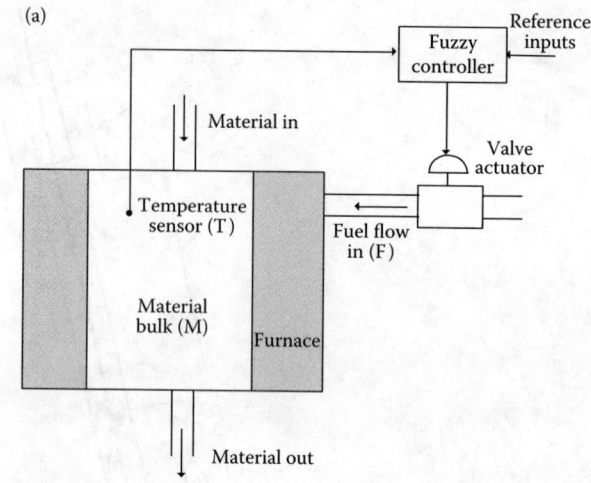

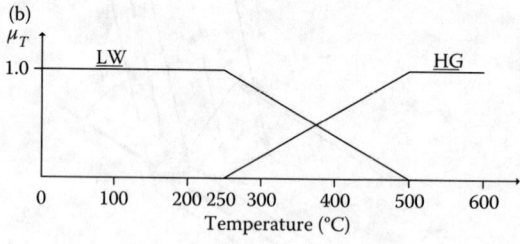

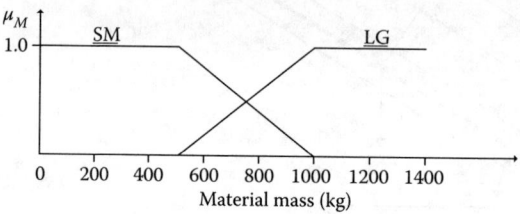

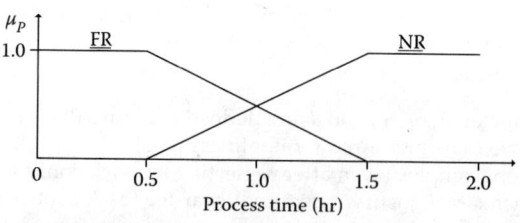

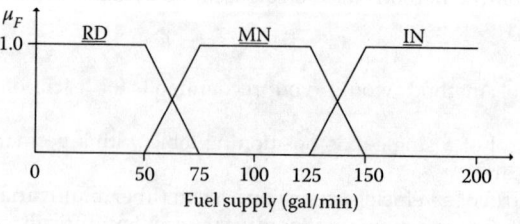

FIGURE P11.24
(a) A metallurgical heat treatment process. (b) Membership functions.

The following fuzzy quantities are defined, with the corresponding states:

T: Temperature of the material (*LW*=low; *HG*=high)
M: Mass of the material (*SM*=small; *LG*=large)
P: Process termination time (*FR*=far; *NR*=near)
F: Fuel supply rate (*RD*=reduce; *MN*=maintain; *IN*=increase)

The membership functions of these quantities are given in Figure P11.24b. A simple rule-base that is used in a fuzzy controller for the fuel supply unit is given below:

If *T* is *LW* and *P* is *FR* then *F* is *IN*
or if *T* is *HG* then *F* is *RD*
or if *M* is *SM* and *P* is *NR* then *F* is *MN*
or if *M* is *LG* and *P* is *FR* then *F* is *IN*
or if *P* is *NR* then *F* is *RD*
End if.

At a given instant, the following set of process data is available:

Temperature=300°C
Material mass=800 kg
Process operation time=1.3 hr

Determine the corresponding inference membership function for the fuel supply, and a crisp value for the control action. Comment on the suitability of this inference.

PROBLEM 11.25

Consider the experimental setup of an inverted pendulum shown in Figure P11.25.

Suppose that direct FLC is used to keep the inverted pendulum upright. The process measurements are the angular position, about the vertical (*ANG*) and the angular velocity (*VEL*) of the pendulum. The control action (*CNT*) is the current of the motor driving the positioning trolley. The variable *ANG* takes two fuzzy states: positive large (*PL*) and negative large (*NL*). Their memberships are defined in the support set $[-30°, 30°]$ and are trapezoidal. Specifically:

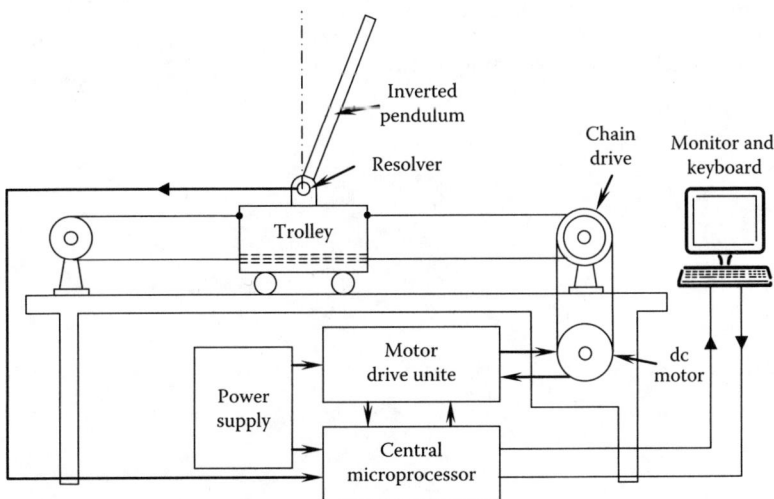

FIGURE P11.25
A computer-controlled inverted pendulum.

$\mu_{PL} = 0$ for $ANG = [-30°, -10°]$
  $= $linear $[0,1.0]$ for $ANG = [-10°, 20°]$
  $= 1.0$ for $ANG = [20°, 30°]$
$\mu_{NL} = 1.0$ for $ANG = [-30°, -20°]$
  $= $ linear $[1.0, 0]$ for $ANG = [-20°, 10°]$
  $= 0$ for $ANG = [10°, 30°]$

The variable *VEL* takes two fuzzy states *PL* and *NL*, which are quite similarly defined in the support set $[-60°/s, 60°/s]$. The control inference *CNT* can take three fuzzy states: Positive large (*PL*), no change (*NC*), and negative large (*NL*). Their membership functions are defined in the support set $[-3A, 3A]$ and are either trapezoidal or triangular. Specifically:

$\mu_{PL} = 0$ for $CNT = [-3A, 0]$
  $= $ linear $[0,1.0]$ for $CNT = [0.2A]$
  $= 1.0$ for $CNT = [2A, 3A]$
$\mu_{NC} = 0$ for $CNT = [-3A, -2A]$
  $= $ linear $[0, 1.0]$ for $CNT = [-2A, 0]$
  $= $ linear $[1.0, 0]$ for $CNT = [0, 2A]$
  $= 0$ for $CNT = [2A, 3A]$
$\mu_{NL} = 1.0$ for $CNT = [-3A, -2A]$
  $= $ linear $[1.0,0]$ for $CNT = [-2A, 0]$
  $= 0$ for $CNT = [0, 3A]$

The following four fuzzy rules are used in control:

	if	*ANG*	is	*PL*	and	*VEL*	is	*PL*	then	*CNT*	is	*NL*
Else	if	*ANG*	is	*PL*	and	*VEL*	is	*NL*	then	*CNT*	is	*NC*
Else	if	*ANG*	is	*NL*	and	*VEL*	is	*PL*	then	*CNT*	is	*NC*
Else	if	*NAG*	is	*NFL*	and	*EL*	is	*NFL*	then	*CT*	is	*LP*
End	if.											

a. Sketch the four rules in a membership diagram for the purpose of making control inferences using individual rule-based inference.
b. If the process measurements of $ANG = 5°$ and $VEL = 15°/s$ are made, indicate on your sketch the corresponding control inference.

12

Control System Instrumentation

This chapter introduces the subject of instrumentation, as related to control engineering. It considers the "instrumenting" of a control system with sensors, transducers, actuators, and associated hardware. The components have to be properly chosen and interconnected in order to achieve a specified level of performance. Relevant issues are addressed. A representative set of analog and digital sensors are presented. Stepper motor and dc motor are presented as popular actuators in control systems. Procedures of motor selection and control are addressed. The use of the computer software tool LabVIEW® for data acquisition and control, particularly in laboratory experimentation, is illustrated.

12.1 Control System Instrumentation

A control system contains a controller as an integral part. The purpose of the controller is to generate control signals, which will drive the process to be controlled (the plant) in the desired manner. Actuators are needed to perform the control actions as well as to drive the plant directly. Sensors and transducers are necessary to measure output signals (process responses) for feedback control; to measure input signals for feedforward control; to measure process variables for system monitoring, diagnosis and supervisory control; and for a variety of other purposes. Since many different types and levels of signals are present in a control system, signal modification (including signal conditioning and signal conversion) is indeed a crucial function associated with any control system. In particular, signal modification is an important consideration in component interfacing.

Potentiometers, differential transformers, resolvers, synchros, gyros, strain gauges, tachometers, piezoelectric devices, fluid flow sensors, pressure gauges, thermocouples, thermistors, and resistance temperature detectors (RTDs) are examples of sensors used to measure process response for monitoring its performance and possible feedback for control. Actuating devices (actuators) include stepper motors, dc motors, ac motors, solenoids, valves, and relays, which are also commercially available to various specifications. An actuator may be directly connected to the driven load, and this is known as the "direct-drive" arrangement. More commonly, however, a transmission device may needed to convert the actuator motion into a desired load motion and for proper matching of the actuator with the driven load. An important factor that we must consider in any practical control system is noise, including external disturbances. Noise may represent actual contamination of signals or the presence of other unknowns, uncertainties, and errors, such as parameter variations and modeling errors. Furthermore, weak signals will have to be amplified, and the form of a signal might have to be modified at various points of interaction. Charge amplifiers, lock-in amplifiers, power amplifiers, switching amplifiers, linear amplifiers, pulse-width-modulated (PWM) amplifiers, tracking filters, low-pass filters, high-pass filters, band-pass filters, and band-reject filters or notch filters are some of the signal-conditioning devices used in analog control systems. Additional components, such as power supplies and surge-protection units, are often needed in control, but they are only indirectly related to control functions. Relays and other switching and transmission devices, and modulators and demodulators may also be included.

The subject of control system instrumentation deals with "instrumenting" a control system through the incorporation of suitable sensors, actuators, and associated hardware. It is clear that the subject should deal with sensors and transducers, actuators, signal modification, and component interconnection. Several applications and their use of sensors and actuators are noted in Table 12.1. A simplified schematic example of an "instrumented" control system is shown in Figure 12.1.

TABLE 12.1

Sensors and Actuators Used in Some Common Engineering Applications

Process	Typical Sensors	Typical Actuators
Aircraft	Displacement, speed, acceleration, elevation, heading, force pressure, temperature, fluid flow, voltage, current, global positioning system (GPS)	dc motors, stepper motors, relays, valve actuators, pumps, heat sources, jet engines
Automobile	Displacement, speed, force, pressure, temperature, fluid flow, fluid level, voltage, current	dc motors, stepper motors, valve actuators, pumps, heat sources
Home heating system	Temperature, pressure, fluid flow	Motors, pumps, heat sources
Milling machine	Displacement, speed, force, acoustics, temperature, voltage, current	dc motors, ac motors
Robot	Optical image, displacement, speed, force, torque, voltage, current	dc motors, stepper motors, ac motors, hydraulic actuators
Wood drying kiln	Temperature, relative humidity, moisture content, air flow	ac motors, dc motors, pumps, heat sources

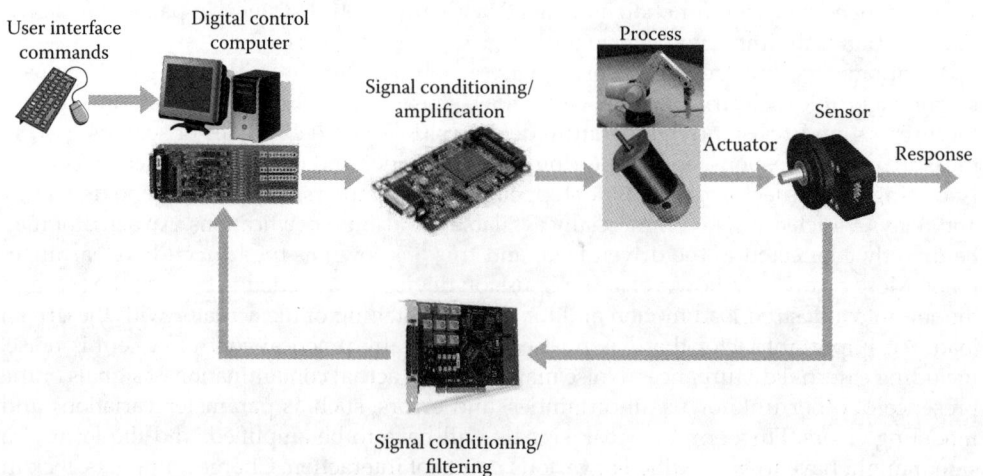

FIGURE 12.1
An instrumented feedback control system.

12.2 Component Interconnection

When components such as sensors and transducers, control boards, process (plant) equipment, and signal-conditioning hardware are interconnected, it is necessary to *match* impedances properly at each interface in order to realize their rated performance level. One adverse effect of improper impedance matching is the *loading effect*. For example, in a measuring system, the measuring instrument can distort the signal that is being measured. The resulting error can far exceed other types of measurement error. Both electrical and mechanical loading are possible. Electrical loading errors result from connecting and output unit such as a measuring device that has a low input impedance to an input device such as a signal source. Mechanical loading errors can result in an input device due to inertia, friction, and other resistive forces generated by an interconnected output component.

Impedance can be interpreted either in the traditional electrical sense or in the mechanical sense, depending on the type of signals that are involved. For example, a heavy accelerometer can introduce an additional dynamic load, which will modify the actual acceleration at the monitoring location. Similarly, a voltmeter can modify the currents (and voltages) in a circuit, and a thermocouple junction can modify the temperature that is being measured as a result of the heat transfer into the junction. In mechanical and electrical systems, loading errors can appear as phase distortions as well. Digital hardware also can produce loading errors. For example, an analog-to-digital conversion (ADC) board can load the amplifier output from a strain gage bridge circuit, thereby affecting digitized data.

Another adverse effect of improper impedance consideration is inadequate output signal levels, which make the output functions such as signal processing and transmission, component driving, and actuation of a final control element or plant very difficult. In context of sensor-transducer technology it should be noted here that many types of transducers (e.g., piezoelectric accelerometers, impedance heads, and microphones) have high output impedances on the order of a thousand megohms (1 megohm or $1\ M\Omega = 1 \times 10^6\ \Omega$). These devices generate low output signals, and they would require conditioning to step up the signal level. *Impedance-matching amplifiers*, which have high input impedances and low output impedances (a few ohms), are used for this purpose (e.g., charge amplifiers are used in conjunction with piezoelectric sensors). A device with a high input impedance has the further advantage that it usually consumes less power (v^2/R is low) for a given input voltage. The fact that a low input impedance device extracts a high level of power from the preceding output device may be interpreted as the reason for loading error.

12.2.1 Cascade Connection of Devices

Consider a standard two-port electrical device. The *output impedance* Z_o of such a device is defined as the ratio of the open-circuit (i.e., no-load) voltage at the output port to the short-circuit current at the output port.

Open-circuit voltage at output is the output voltage present when there is no current flowing at the output port. This is the case if the output port is not connected to a load (impedance). As soon as a load is connected at the output of the device, a current will flow through it, and the output voltage will drop to a value less than that of the open-circuit voltage. To measure the open-circuit voltage, the rated input voltage is applied at the input port and maintained constant, and the output voltage is measured using a voltmeter

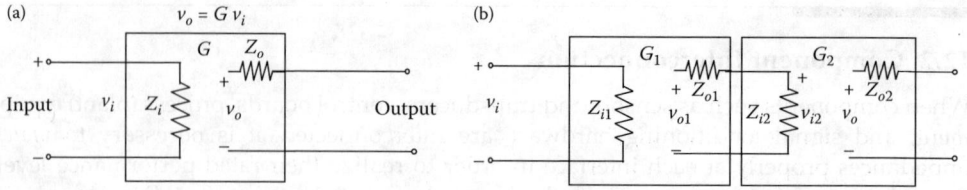

FIGURE 12.2
(a) Schematic representation of input impedance and output impedance. (b) Cascade connection of two two-port devices.

that has a very high (input) impedance. To measure the short-circuit current, a very low-impedance ammeter is connected at the output port.

The *input impedance* Z_i is defined as the ratio of the rated input voltage to the corresponding current through the input terminals while the output terminals are maintained as an open circuit.

Note that these definitions are associated with electrical devices. A generalization is possible by interpreting voltage and velocity as *across variables*, and current and force as *through variables*. Then mechanical *mobility* should be used in place of electrical impedance, in the associated analysis.

Using these definitions, input impedance Z_i and output impedance Z_o can be represented schematically as in Figure 12.2a. Note that v_o is the open-circuit output voltage. When a load is connected at the output port, the voltage across the load will be different from v_o. This is caused by the presence of a current through Z_o. In the frequency domain, v_i and v_o are represented by their respective *Fourier spectra*. The corresponding transfer relation can be expressed in terms of the complex frequency response (transfer) function $G(j\omega)$ under open-circuit (no-load) conditions:

$$v_o = Gv_i \tag{12.1}$$

Now consider two devices connected in cascade, as shown in Figure 12.2b. It can be easily verified that the following relations apply:

$$v_{o1} = G_1 v_i \tag{12.2}$$

$$v_{i2} = \frac{Z_{i2}}{Z_{o1} + Z_{i2}} v_{o1} \tag{12.3}$$

$$v_o = G_2 v_{i2} \tag{12.4}$$

These relations can be combined to give the overall input–output relation:

$$v_o = \frac{Z_{i2}}{Z_{o1} + Z_{i2}} G_2 G_1 v_i \tag{12.5}$$

We see from Equation 12.5 that the overall frequency transfer function differs from the ideally expected product ($G_2 G_1$) by the factor:

$$\frac{Z_{i2}}{Z_{o1} + Z_{i2}} = \frac{1}{Z_{o1} / Z_{i2} + 1} \tag{12.6}$$

Note that cascading has "distorted" the frequency response characteristics of the two devices. If $Z_{o1}/Z_{i2} \ll 1$, this deviation becomes insignificant. From this observation, it can be concluded that when frequency response characteristics (i.e., dynamic characteristics) are important in a cascaded device, cascading should be done such that the output impedance of the first device is much smaller than the input impedance of the second device.

12.2.2 Impedance Matching Amplifiers

When two electrical components are interconnected, current (and energy) will flow between the two components. This will change the original (unconnected) conditions. This is known as the (electrical) loading effect, and it has to be minimized. At the same time, adequate power and current would be needed for signal communication, conditioning, display, etc. Both situations can be accommodated through proper matching of impedances when the two components are connected. Usually an impedance matching amplifier (impedance transformer) would be needed between the two components.

From the analysis given in the preceding section, it is clear that the signal-conditioning circuitry should have a considerably large input impedance in comparison to the output impedance of the sensor-transducer unit in order to reduce loading errors. The problem is quite serious in measuring devices such as piezoelectric sensors, which have very high output impedances. In such cases, the input impedance of the signal-conditioning unit might be inadequate to reduce loading effects; also, the output signal level of these high-impedance sensors is quite low for signal transmission, processing, actuation, and control. The solution for this problem is to introduce several stages of amplifier circuitry between the output of the first hardware unit (e.g., sensor) and the input of the second hardware unit (e.g., data acquisition unit). The first stage of such an interfacing device is typically an *impedance-matching amplifier* that has very high input impedance, very low output impedance, and almost unity gain. The last stage is typically a stable high-gain amplifier stage to step up the signal level. Impedance-matching amplifiers are, in fact, *operational amplifiers* with feedback.

When connecting a device to a signal source, loading problems can be reduced by making sure that the device has a high input impedance. Unfortunately, this will also reduce the level (amplitude, power) of the signal received by the device. In fact, a high-impedance device may reflect back some harmonics of the source signal. A termination resistance may be connected in parallel with the device in order to reduce this problem.

In many data acquisition systems, output impedance of the output amplifier is made equal to the transmission line impedance. When maximum power amplification is desired, *conjugate matching* is recommended. In this case, input impedance and output impedance of the matching amplifier are made equal to the complex conjugates of the source impedance and the load impedance, respectively.

12.2.3 Operational Amplifier

Operational amplifier is a very versatile device, primarily due to its very high input impedance, low output impedance, and very high gain. An op-amp could be manufactured in the discrete-element form using, say, ten bipolar junction transistors and as many discrete resistors or alternatively (and preferably) in the modern monolithic form as an IC chip that may be equivalent to over 100 discrete elements. In any form, the device has an *input impedance* Z_i, an *output impedance* Z_o and a gain K. Hence, a schematic model for an op-amp can be given as in Figure 12.3a. Op-amp packages are available in several forms. Very

(a)

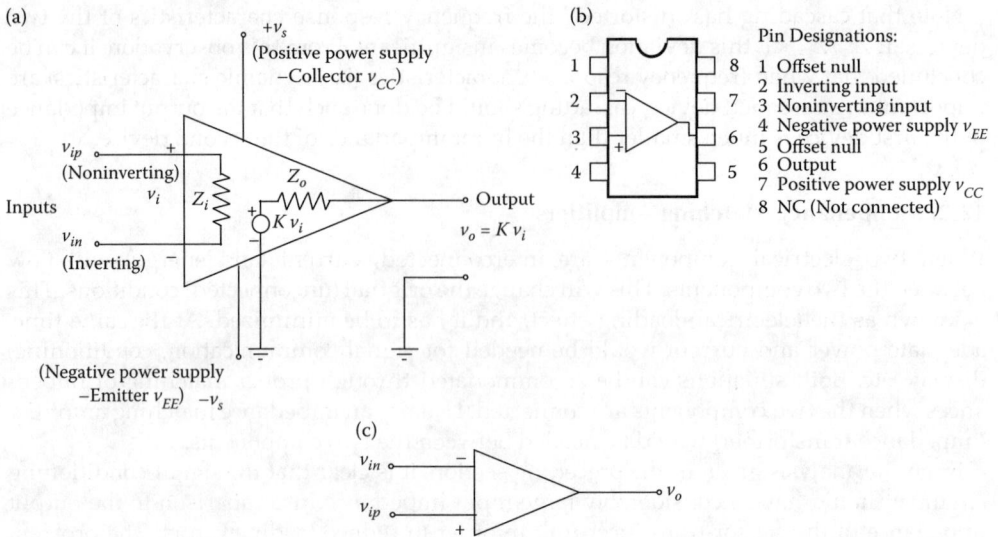

(b)

Pin Designations:

1 Offset null
2 Inverting input
3 Noninverting input
4 Negative power supply v_{EE}
5 Offset null
6 Output
7 Positive power supply v_{CC}
8 NC (Not connected)

(c)

FIGURE 12.3
Operational amplifier. (a) A schematic model. (b) Eight-pin dual in-line package (DIP). (c) Conventional circuit symbol.

common is the eight-pin dual in-line package (DIP) or V package, as shown in Figure 12.3b. The assignment of the pins (pin configuration or pin-out) is as shown in the figure, which should be compared with Figure 12.7a. Note the counter clockwise numbering sequence starting with the top left pin next to the semi circular notch (or, dot). This convention of numbering is standard for any type of IC package, not just op-amp packages. Other packages include eight-pin metal-can package or T package, which has a circular shape instead of the rectangular shape of the previous package, and the 14-pin rectangular "Quad" package which contains four op-amps (with a total of eight input pins, four output pins, and two power supply pins). The conventional symbol of an op-amp is shown in Figure 12.3c. Typically, there are five terminals (pins or lead connections) to an op-amp. Specifically, there are two input leads (a positive or noninverting lead with voltage v_{ip} and a negative or inverting lead with voltage v_{in}), an output lead (voltage v_o), and two bipolar power supply leads ($+v_s$ or v_{CC} or collector supply and $-v_s$ or v_{EE} or emitter supply). The typical supply voltage is ±22 V. Some of the pins may not be normally connected; for example, pins 1, 5, and 8 in Figure 12.3b.

The open loop voltage gain K is very high (10^5–10^9) for a typical op-amp. Furthermore, the input impedance Z_i could be as high as 10 MΩ (typical is 2 MΩ) and the output impedance is low, of the order of 10 Ω and may reach about 75 Ω for some op-amps.

In analyzing operational amplifier circuits under unsaturated conditions, we use the following two characteristics of an op-amp:

1. Voltages of the two input leads should be (almost) equal.
2. Currents through each of the two input leads should be (almost) zero.

The first property is credited to high open-loop gain, and the second property to high input impedance in an operational amplifier.

12.2.3.1 Use of Feedback in Op-Amps

An op-amp cannot be used without modification as an amplifier because it is not very stable in the open-loop form. The two main factors which contribute to this problem are: frequency response and drift. In other words, op-amp gain K does not remain constant; it can vary with frequency of the input signal (i.e., frequency response function is not flat in the operating range); and, also it can vary with time (i.e., drift). Since gain K is very large, by using feedback we can virtually eliminate its effect at the amplifier output. This *closed-loop* form of an op-amp has the advantage that the characteristics and the accuracy of the output of the overall circuit depends on the passive components (e.g., resistors and capacitors) in it, which can be provide at high precision, and not the parameters of the op amp itself.

12.2.4 Instrumentation Amplifiers

An instrumentation amplifier is typically a special-purpose voltage amplifier dedicated to instrumentation applications. Examples include amplifiers used for producing the output from a bridge circuit (bridge amplifier) and amplifiers used with various sensors and transducers. An important characteristic of an instrumentation amplifier is the adjustable-gain capability. The gain value can be adjusted manually in most instrumentation amplifiers. In more sophisticated instrumentation amplifiers the gain is *programmable* and can be set by means of digital logic. Instrumentation amplifiers are normally used with low-voltage signals.

12.2.4.1 Differential Amplifier

Usually, an instrumentation amplifier is also a *differential amplifier* (sometimes termed *difference amplifier*). In a differential amplifier both input leads are used for signal input, whereas in a single-ended amplifier one of the leads is grounded and only one lead is used for signal input. Ground-loop noise can be a serious problem in single-ended amplifiers. Ground-loop noise can be effectively eliminated using a differential amplifier because noise loops are formed with both inputs of the amplifier and, hence, these noise signals are subtracted at the amplifier output. Since the noise level is almost the same for both inputs, it is canceled out. Any other noise (e.g., 60 Hz line noise) that might enter both inputs with the same intensity will also be canceled out at the output of a differential amplifier.

A basic differential amplifier that uses a single op-amp is shown in Figure 12.4a. The input–output equation for this amplifier can be obtained as:

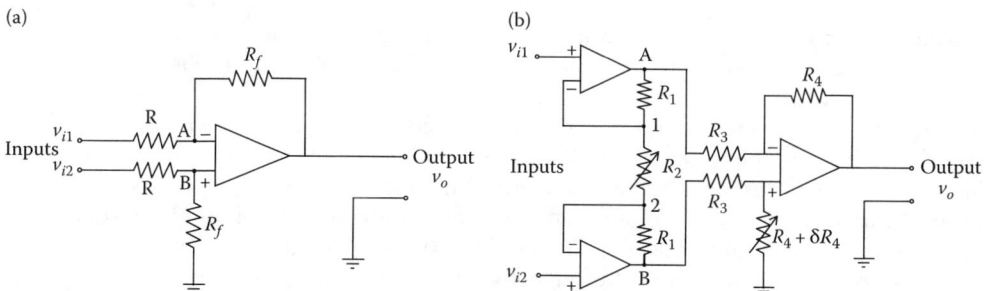

FIGURE 12.4
(a) A basic differential amplifier. (b) A basic instrumentation amplifier.

$$v_o = \frac{R_f}{R}(v_{i2} - v_{i1})$$ (12.7)

Two things are clear from Equation 12.7. First, the amplifier output is proportional to the "difference" and not the absolute value of the two inputs v_{i1} and v_{i2}. Second, voltage gain of the amplifier is R_f/R. This is known as the *differential gain*. It is clear that the differential gain can be accurately set by using high-precision resistors R and R_f.

The basic differential amplifier, shown in Figure 12.4a and discussed above, is an important component of an instrumentation amplifier. In addition, an instrumentation amplifier should possess the capability of adjustable gain. Furthermore, it is desirable to have a very high input impedance and very low output impedance at teach input lead. It is desirable for an instrumentation amplifier to possess a higher and more stable gain, and also a higher input impedance than a basic differential amplifier. An instrumentation amplifier that possesses these basic requirements may be fabricated in the monolithic IC form as a single package. Alternatively, in may be built using three differential amplifiers and high precision resistors, as shown in Figure 12.4b. The amplifier gain can be adjusted using the fine-tunable resistor R_2. Impedance requirements are provided by two voltage-follower type amplifiers, one for each input, as shown. The variable resistance δR_4 is necessary to compensate for errors due to unequal common-mode gain. The equation for the instrumentation amplifier (with $\delta R4 = 0$) is:

$$v_o = \frac{R_4}{R_3}\left(1 + \frac{2R_1}{R_2}\right)(v_{i2} - v_{i1})$$ (12.8)

12.3 Motion Sensors

By motion, we mean the four kinematic variables:

- Displacement (including position, distance, proximity, and size or gage)
- Velocity
- Acceleration
- Jerk

Note that each variable is the time derivative of the preceding one. Motion measurements are extremely useful in controlling mechanical responses and interactions in control systems. Numerous examples can be cited: the rotating speed of a work piece and the feed rate of a tool are measured in controlling machining operations. Displacements and speeds (both angular and translatory) at joints (revolute and prismatic) of robotic manipulators or kinematic linkages are used in controlling manipulator trajectory. In high-speed ground transit vehicles, acceleration and jerk measurements can be used for active suspension control to obtain improved ride quality. Angular speed is a crucial measurement that is used in the control of rotating machinery, such as turbines, pumps, compressors, motors, and generators in power-generating plants. Proximity sensors (to measure displacement) and accelerometers (to measure acceleration) are the two most common types of measuring devices used in machine protection systems for condition monitoring, fault detection, diagnostic, and on-line (often real-time) control of large and complex machinery. The accelerometer is often the only measuring device used in controlling dynamic test rigs. Displacement measurements are used for valve control in process applications. Plate thickness (or gage) is continuously monitored by the automatic gage control (AGC) system in steel rolling mills.

12.3.1 Linear-Variable Differential Transformer (LVDT)

Differential transformer is a noncontact displacement sensor, which does not possess many of the shortcomings of the potentiometer. It is a variable-inductance transducer, and is also a variable-reluctance transducer and a mutual-induction transducer. Furthermore, unlike the potentiometer, the differential transformer is a passive device.

In its simplest form (see Figure 12.5), the LVDT consists of an insulating, nonmagnetic "form" (a cylindrical structure on which a coil is wound, and is integral with the housing), which has a primary coil in the mid-segment and a secondary coil symmetrically wound in the two end segments, as depicted schematically in Figure 12.5b. The housing is made of magnetized stainless steel in order to shield the sensor from outside fields. The primary coil is energized by an ac supply of voltage v_{ref}. This will generate, by mutual induction, an ac of the same frequency in the secondary coil. A core made of ferromagnetic material is inserted coaxially through the cylindrical form without actually touching it, as shown. As the core moves, the reluctance of the flux path changes. The degree of flux linkage depends on the axial position of the core. Since the two secondary coils are connected in series opposition, so that the potentials induced in the two secondary coil segments

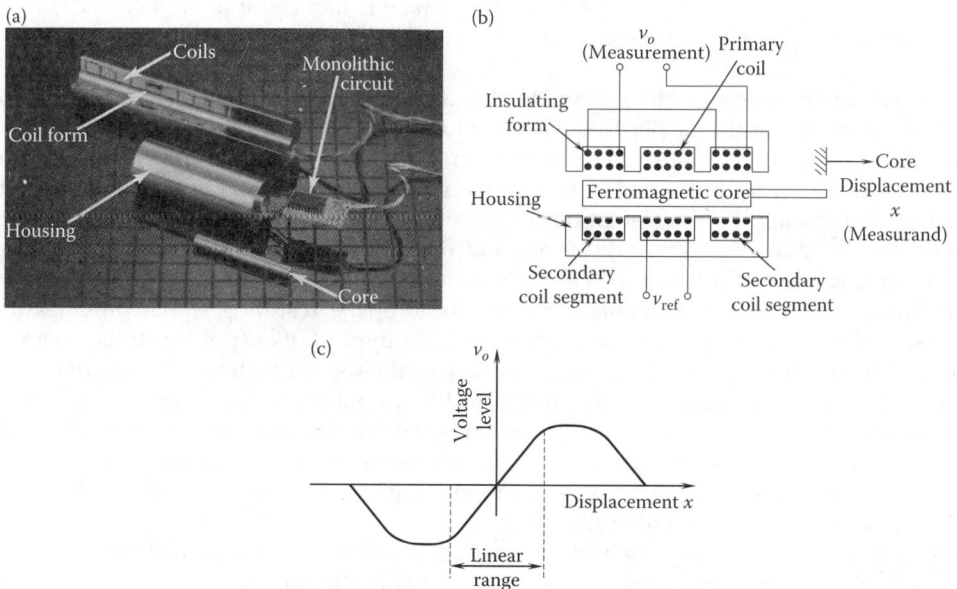

FIGURE 12.5
LVDT. (a) A commercial unit. (From: Scheavitz Sensors, Measurement Specialties, Inc. With permission.) (b) Schematic diagram. (c) A typical operating curve.

oppose each other, it is seen that the net induced voltage is zero when the core is centered between the two secondary winding segments. This is known as the *null position*. When the core is displaced from this position, a nonzero induced voltage will be generated. At steady-state, the amplitude v_o of this induced voltage is proportional to the core displacement x in the linear (operating) region (see Figure 12.5c). Consequently, v_o may be used as a measure of the displacement. Note that because of opposed secondary windings, the LVDT provides the direction as well as the magnitude of displacement. If the output signal is not demodulated, the direction is determined by the phase angle between the primary (reference) voltage and the secondary (output) voltage, which includes the carrier signal.

For an LVDT to measure transient motions accurately, the frequency of the reference voltage (the carrier frequency) has to be at least ten times larger than the largest significant frequency component in the measured motion, and typically can be as high as 20 kHz. For quasi-dynamic displacements and slow transients on the order of a few Hertz, a standard ac supply (at 60 Hz line frequency) is adequate. The performance (particularly sensitivity and accuracy) is known to improve with the excitation frequency, however. Since the amplitude of the output signal is proportional to the amplitude of the primary signal, the reference voltage should be regulated to get accurate results. In particular, the power source should have a low output impedance.

12.3.2 Signal Conditioning

Signal conditioning associated with differential transformers includes filtering and amplification. Filtering is needed to improve the signal-to-noise ratio of the output signal. Amplification is necessary to increase the signal strength for data acquisition and processing. Since the reference frequency (carrier frequency) is induced into (and embedded in) the output signal, it is also necessary to interpret the output signal properly, particularly for transient motions.

The secondary (output) signal of an LVDT is an amplitude-modulated signal where the signal component at the carrier frequency is modulated by the lower-frequency transient signal produced as a result of the core motion (x). Two methods are commonly used to interpret the crude output signal from a differential transformer: rectification and demodulation. Block diagram representations of these two procedures are given in Figure 12.6. In the first method (*rectification*) the ac output from the differential transformer is rectified to obtain a dc signal. This signal is amplified and then low-pass filtered to eliminate any high-frequency noise components. The amplitude of the resulting signal provides the transducer reading. In this method, phase shift in the LVDT output has to be checked separately to determine the direction of motion. In the second method (*demodulation*), the carrier frequency component is rejected from the output signal by comparing it with a phase-shifted and amplitude-adjusted version of the primary (reference) signal. Note that phase shifting is necessary because, as discussed before, the output signal is not in phase with the reference signal. The result is the modulating signal (proportional to x), which is subsequently amplified and filtered.

As a result of advances in miniature integrated circuit technology, differential transformers with built-in microelectronics for signal conditioning are commonly available today. A dc differential transformer uses a dc power supply (typically, ±15 V) to activate it. A built-in oscillator circuit generates the carrier signal. The rest of the device is identical to an ac differential transformer. The amplified full-scale output voltage can be as high as ±10 V. Advantages of the LVDT include the following.

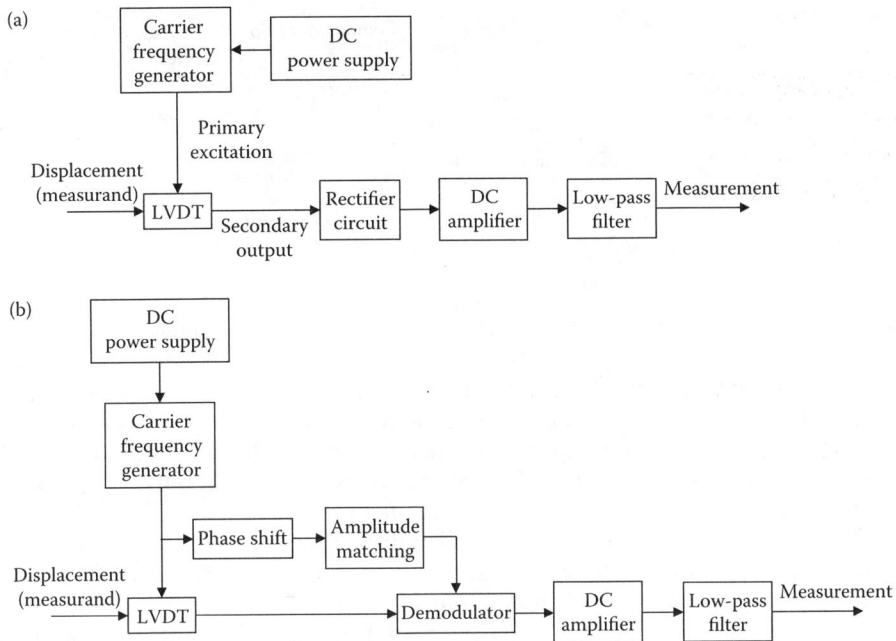

FIGURE 12.6
Signal-conditioning methods for a differential transformer. (a) Rectification. (b) Demodulation.

1. It is essentially a noncontacting device with no frictional resistance. Near-ideal electromechanical energy conversion and light-weight core will result in very small resistive forces. Hysteresis (both magnetic hysteresis and mechanical backlash) is negligible.

2. It has low output impedance, typically on the order of 100 Ω (Signal amplification is usually not needed beyond what is provided by the conditioning circuit.)

3. Directional measurements (positive/negative) are obtained.

4. It is available in small sizes (e.g., 1 cm long with maximum travel of 2 mm).

5. It has a simple and robust construction (inexpensive and durable).

6. Fine resolutions are possible (theoretically, infinitesimal resolution; practically, much better than a coil potentiometer).

12.3.3 DC Tachometer

This is a permanent-magnet (PM) dc velocity sensor in which the principle of electromagnetic induction between a permanent magnet and a conducting coil is used. Depending on the configuration, either rectilinear speeds or angular speeds can be measured. Schematic diagrams of the two configurations are shown in Figure 12.7. These are passive transducers, because the energy for the output signal v_o is derived from the motion (measured signal) itself. The entire device is usually enclosed in a steel casing to shield (isolate) it from ambient magnetic fields.

In the rectilinear velocity transducer (Figure 12.7a), the conductor coil is wound on a core and placed centrally between two magnetic poles, which produce a cross-magnetic field.

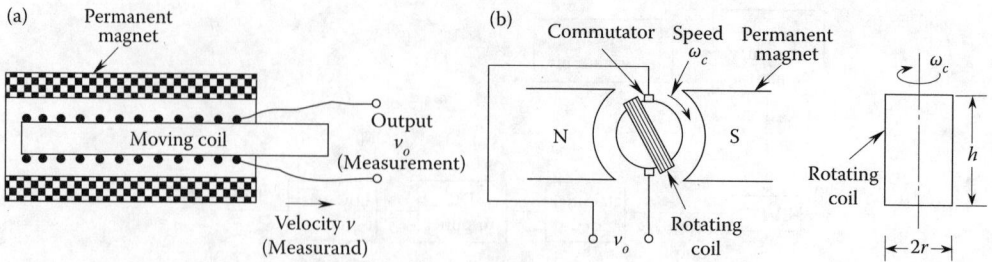

FIGURE 12.7
Permanent-magnet dc transducers. (a) Rectilinear velocity transducer. (b) DC tachometer.

The core is attached to the moving object whose velocity v must be measured. This velocity is proportional to the induced voltage v_o. Alternatively, a moving magnet and a fixed coil may be used as a dc tachometer. This arrangement is perhaps more desirable since it eliminates the need for any sliding contacts (slip rings and brushes) for the output leads, thereby reducing mechanical loading error, wear, and related problems.

The dc tachometer (or, tachogenerator) is a common transducer for measuring angular velocities. Its principle of operation is the same as that for a dc generator (or, back-driving of a dc motor). This principle of operation is illustrated in Figure 12.7b. The rotor is directly connected to the rotating object. The output signal that is induced in the rotating coil is picked up as dc voltage v_o using a suitable *commutator* device—typically consisting of a pair of low-resistance carbon brushes—that is stationary but makes contact with the rotating coil through split slip rings so as to maintain the direction of the induced voltage the same throughout each revolution. According to Faraday's law, the induced voltage is proportional to the rate of change of magnetic flux linkage. For a coil of height h and width $2r$ that has n turns, moving at an angular speed ω_c in a uniform magnetic field of flux density β, this is given by:

$$v_o = (2nhr\beta)\omega_c = k\omega_c \tag{12.9}$$

This proportionality between v_o and ω_c is used to measure the angular speed ω_c. The proportionality constant k is known as the *back-e.m.f. constant* or the *voltage constant*.

12.3.3.1 Electronic Commutation

Slip rings and brushes and associated drawbacks can be eliminated in a dc tachometer by using electronic commutation. In this case a PM rotor together with a set of stator windings are used. The output of the tachometer is drawn from the stationary (stator) coil. It has to be converted to a dc signal using an electronic switching mechanism, which has to be synchronized with the rotation of the tachometer. As a result of switching and associated changes in the magnetic field of the output signal, induced voltages known as *switching transients* will result. This is a drawback in electronic commutation.

12.3.4 Piezoelectric Accelerometer

The piezoelectric accelerometer (or, *crystal accelerometer*) is an acceleration sensor, which uses a piezoelectric element to measure the inertia force caused by acceleration.

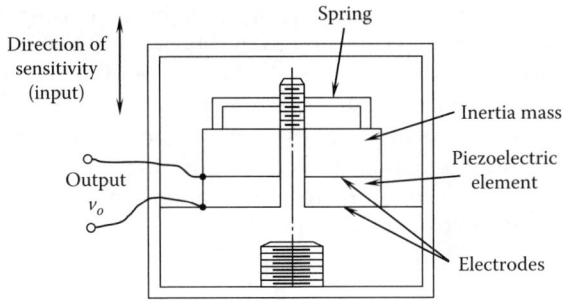

FIGURE 12.8
A compression-type piezoelectric accelerometer.

A piezoelectric velocity transducer is simply a piezoelectric accelerometer with a built-in integrating amplifier in the form of a miniature integrated circuit.

The advantages of piezoelectric accelerometers over other types of accelerometers are their light weight and high-frequency response (up to about 1 MHz). However, piezoelectric transducers are inherently high output impedance devices, which generate small voltages (on the order of 1 mV). For this reason, special impedance-transforming amplifiers (e.g., charge amplifiers) have to be employed to condition the output signal and to reduce loading error.

A schematic diagram for a compression-type piezoelectric accelerometer is shown in Figure 12.8. The crystal and the inertia mass are restrained by a spring of very high stiffness. Consequently, the fundamental natural frequency or resonant frequency of the device becomes high (typically 20 kHz). This gives a reasonably wide useful range (typically up to 5 kHz). The lower limit of the useful range (typically 1 Hz) is set by factors such as the limitations of the signal-conditioning system, the mounting methods, the charge leakage in the piezoelectric element, the time constant of the charge-generating dynamics, and the signal-to-noise ratio.

In a compression-type crystal accelerometer, the inertia force is sensed as a compressive normal stress in the piezoelectric element. There are also piezoelectric accelerometers where the inertia force is applied to the piezoelectric element as a shear strain or as a tensile strain.

For an accelerometer, acceleration is the signal that is being measured (the measurand). Hence, accelerometer sensitivity is commonly expressed in terms of electrical charge per unit acceleration or voltage per unit acceleration. Sensitivity depends on the piezoelectric properties, the way in which the inertia force is applied to the piezoelectric element (e.g., compressive, tensile, shear), and the mass of the inertia element. If a large mass is used, the reaction inertia force on the crystal will be large for a given acceleration, thus generating a relatively large output signal. Large accelerometer mass results in several disadvantages, however. In particular:

1. The accelerometer mass distorts the measured motion variable (mechanical loading effect).
2. A heavy accelerometer has a lower resonant frequency and hence a lower useful frequency range.

For a given accelerometer size, improved sensitivity can be obtained by using the shear-strain configuration. In this configuration, several shear layers can be used (e.g., in a *delta arrangement*) within the accelerometer housing, thereby increasing the effective shear area and hence the sensitivity in proportion to the shear area. Another factor that should be considered in selecting an accelerometer is its *cross-sensitivity* or transverse sensitivity. Cross-sensitivity is present because a piezoelectric element can generate a charge in response to forces and moments (or, torques) in orthogonal directions as well. The problem can be aggravated due to manufacturing irregularities of the piezoelectric element, including material unevenness and incorrect orientation of the sensing element, and due to poor design. Cross-sensitivity should be less than the maximum error (percentage) that is allowed for the device (typically 1%).

12.3.4.1 Charge Amplifier

Piezoelectric signals cannot be read using low-impedance devices. The two primary reasons for this are:

1. High output impedance in the sensor results in small output signal levels and large loading errors.
2. The charge can quickly leak out through the load.

A charge amplifier is commonly used as the signal-conditioning device for piezoelectric sensors, in order to overcome these problems to a great extent. Because of impedance transformation, the impedance at the output of the charge amplifier becomes much smaller than the output impedance of the piezoelectric sensor. This virtually eliminates loading error and provides a low-impedance output for purposes such as signal communication, acquisition, recording, processing, and control. Also, by using a charge amplifier circuit with a relatively large time constant, speed of charge leakage can be decreased.

12.3.5 Digital Transducers

Any measuring device that presents information as discrete samples and that does not introduce a *quantization error* when the reading is represented in the digital form may be classified as a digital transducer. Digital measuring devices (or digital transducers, as they are commonly known) generate discrete output signals such as pulse trains or encoded data that can be directly read by a digital controller. Nevertheless, the sensor stage of a digital measuring device is usually quite similar to that of an analog counterpart. There are digital measuring devices that incorporate microprocessors to perform numerical manipulations and conditioning locally and provide output signals in either digital form or analog form. These measuring systems are particularly useful when the required variable is not directly measurable but could be computed using one or more measured outputs (e.g., power=force×speed). Although a microprocessor is an integral part of the measuring device in this case, it performs not a measuring task but, rather, a conditioning task.

When the output of a digital transducer is a pulse signal, a common of reading the signal is by using a counter, either to count the pulses (for high-frequency pulses) or to count clock cycles over one pulse duration (for low-frequency pulses). The count is placed as a digital word in a buffer, which can be accessed by the host (control) computer, typically at a constant frequency (sampling rate). On the other hand, if the output of a digital transducer is automatically available in a coded form (e.g., natural binary code or gray code) it can be directly read by a computer. In the latter case, the coded signal is normally generated by a parallel set of pulse signals; each pulse transition generates one bit of the digital word, and the numerical value of the word is determined by the pattern of the generated pulses. Data acquisition from (i.e., computer interfacing) a digital transducer is commonly done using a general-purpose input–output (I/O) card; for example, a motion control (servo) card, which may be able to accommodate multiple transducers (e.g., eight channels of encoder inputs with 24-bit counters), or using a data acquisition card specific to the particular transducer.

There are several advantages of digital signals (or, digital representation of information) in comparison to analog signals. Notably:

1. Digital signals are less susceptible to noise, disturbances, or parameter variation in instruments because data can be generated, represented, transmitted, and processed as binary words consisting of bits, which possess two identifiable states.

2. Complex signal processing with very high accuracy and speed are possible through digital means (hardware implementation is faster than software implementation).

3. High reliability in a system can be achieved by minimizing analog hardware components.

4. Large amounts of data can be stored using compact, high-density, data storage methods.

5. Data can be stored or maintained for very long periods of time without any drift or being affected by adverse environmental conditions.

6. Fast data transmission is possible over long distances without introducing significant dynamic delays, as in analog systems.

7. Digital signals use low voltages (e.g., 0–12 V dc) and low power.

8. Digital devices typically have low overall cost.

12.3.6 Shaft Encoders

Any transducer that generates a coded (digital) reading of a measurement can be termed an encoder. Shaft encoders are digital transducers that are used for measuring angular displacements and angular velocities. Shaft encoders can be classified into two categories, depending on the nature and the method of interpretation of the transducer output: incremental encoders, absolute encoders.

The output of an incremental encoder is a pulse signal, which is generated when the transducer disk rotates as a result of the motion that is being measured. By counting the pulses or by timing the pulse width using a clock signal, both angular displacement and angular velocity can be determined. With an incremental encoder, displacement is obtained with respect to some reference point. The reference point can be the home position of the moving component (say, determined by a limit switch); or a reference point on the encoder disk, as indicated by a reference pulse (index pulse) generated at that location on the disk. Furthermore, the index pulse count determines the number of full revolutions.

An absolute encoder (or, whole-word encoder) has many pulse tracks on its transducer disk. When the disk of an absolute encoder rotates, several pulse trains—equal in number to the tracks on the disk—are generated simultaneously. At a given instant, the magnitude of each pulse signal will have one of two signal levels (i.e., a binary state), as determined by a level detector (or, edge detector). This signal level corresponds to a binary digit (0 or 1).

Hence, the set of pulse trains gives an encoded binary number at any instant. The pulse windows on the tracks can be organized into some pattern (code) so that the generated binary number at a particular instant corresponds to the specific angular position of the encoder disk at that time. The pulse voltage can be made compatible with some digital interface logic (e.g., transistor-to-transistor logic, or TTL). Consequently, the direct digital readout of an angular position is possible with an absolute encoder, thereby expediting digital data acquisition and processing. Absolute encoders are commonly used to measure fractions of a revolution. However, complete revolutions can be measured using an additional track, which generates an index pulse, as in the case of incremental encoder.

12.3.7 Optical Encoder

By far, the optical encoder is most popular and cost effective encoder used in motion sensing. The optical encoder uses an opaque disk (code disk) that has one or more circular tracks, with some arrangement of identical transparent windows (slits) in each track. A parallel beam of light (e.g., from a set of light-emitting diodes (LEDs)) is projected to all tracks from one side of the disk. The transmitted light is picked off using a bank of photosensors on the other side of the disk, which typically has one sensor for each track. This arrangement is shown in Figure 12.9a, which indicates just one track and one pick-off sensor. The light sensor could be a silicon photodiode or a phototransistor. Since the light from the source is interrupted by the opaque regions of the track, the output signal from the photosensor is a series of voltage pulses. This signal can be interpreted (e.g., through edge detection or level detection) to obtain the increments in the angular position. The resulting pulse count gives the angular position and the pulse frequency gives the angular velocity of the disk.

The opaque background of transparent windows (the window pattern) on an encoder disk may be produced by contact printing techniques. The precision of this production procedure is a major factor that determines the accuracy of optical encoders. Note that a transparent disk with a track of opaque spots will work equally well as the encoder

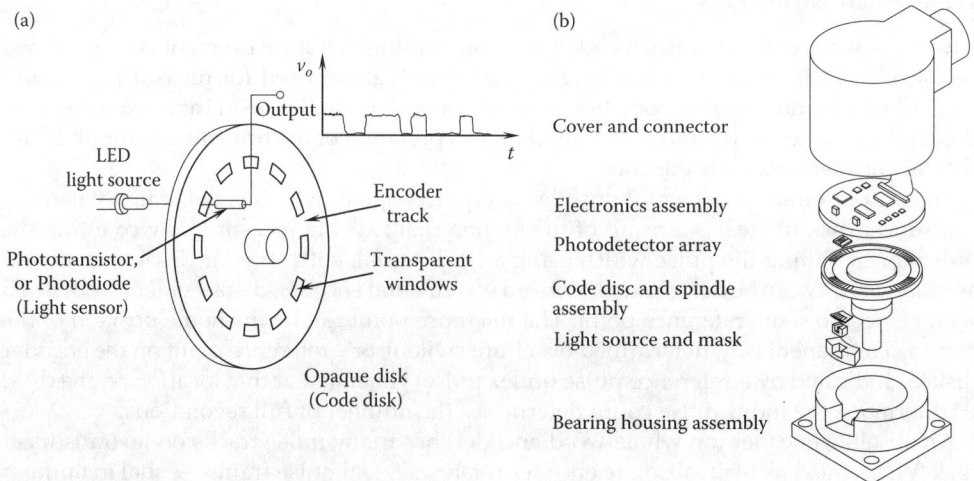

FIGURE 12.9
(a) Schematic representation of an (incremental) optical encoder. (b) Components of a commercial incremental encoder. (From: BEI Electronics, Inc. With permission.)

disk of an optical encoder. In either form, the track has a 50% duty cycle (i.e., length of the transparent region=length of the opaque region). A commercially available optical encoder is shown in Figure 12.9b.

An incremental encoder disk requires only one primary track that has equally spaced and identical window (pick-off) regions. The window area is equal to the area of the inter-window gap (i.e., 50% duty cycle). Usually, a reference track that has just one window is also present in order to generate a pulse (known as the index pulse) to initiate pulse counting for angular position measurement and to detect complete revolutions. It will also need a sensor at a quarter-pitch separation (pitch=center-to-center distance between adjacent windows) to generate a *quadrature signal*, which will identify the direction of rotation. Some designs of incremental encoders have two identical tracks, one at a quarter-pitch offset from the other, and the two pick-off sensors are placed radially without offset. The two (quadrature) signals obtained with this arrangement will be similar to those with the previous arrangement. A pick-off sensor for receiving a reference pulse is also used in some designs of incremental encoders (three-track incremental encoders).

In many control applications, encoders are built into the plant itself, rather than being externally fitted onto a rotating shaft. For instance, in a robot arm, the encoder might be an integral part of the joint motor and may be located within its housing. This reduces coupling errors (e.g., errors due to backlash, shaft flexibility, and resonances added by the transducer and fixtures), installation errors (e.g., misalignment and eccentricity), and overall cost. Encoders are available in sizes as small as 2 cm and as large as 15 cm in diameter.

12.4 Stepper Motors

Stepper motors are a popular type of actuators. They are driven in fixed angular steps (increments). Each step of rotation is the response of the motor rotor to an input pulse (or a digital command). In this manner, the stepwise rotation of the rotor can be synchronized with pulses in a command-pulse train, assuming of course that no steps are missed, thereby making the motor respond faithfully to the input signal (pulse sequence) in an open-loop manner. Like a conventional continuous-drive motor, a stepper motor is also an electromagnetic actuator, in that it converts electromagnetic energy into mechanical energy to perform mechanical work. The terms *stepper motor, stepping motor,* and *step motor* are synonymous and are often used interchangeably.

One common feature in any stepper motor is that the stator of the motor contains several pairs of field windings (or phase windings) that can be switched on to produce electromagnetic pole pairs (N and S). These pole pairs effectively pull the motor rotor in sequence so as to generate the torque for motor rotation. By switching the currents in the phases in an appropriate sequence, either a clockwise (CW) rotation or a counterclockwise (CCW) rotation can be produced. The polarities of a stator pole may have to be reversed in some types of stepper motors in order to carry out a stepping sequence. Although the commands that generate the switching sequence for a phase winding could be supplied by a microprocessor or a personal computer (a software approach) it is customary to generate it through hardware logic in a device called a *translator* or an *indexer*. This approach is more effective because the switching logic for a stepper motor is fixed, as noted in the foregoing discussion. *Microstepping* provides much smaller step angles. This is achieved by changing the phase currents by small increments (rather than on, off, and reversal) so that the detent (equilibrium) position of the rotor shifts in correspondingly small angular increments.

12.4.1 Stepper Motor Classification

Most classifications of stepper motors are based on the nature of the motor rotor. One such classification considers the magnetic character of the rotor. Specifically, a variable-reluctance (VR) stepper motor has a soft-iron rotor while a PM stepper motor has a magnetized rotor. The two types of motors operate in a somewhat similar manner. Specifically the stator magnetic field (polarity) is stepped so as to change the minimum reluctance (or detent) position of the rotor in increments. Hence both types of motors undergo similar changes in reluctance (magnetic resistance) during operation. A disadvantage of VR stepper motors is that since the rotor is not magnetized, the holding torque is zero when the stator windings are not energized (power-off). Hence, there is no capability to hold the load at a given position under power-off conditions unless mechanical brakes are employed. A hybrid stepper motor possesses characteristics of both VR steppers and PM steppers. The rotor of a hybrid stepper motor consists of two rotor segments connected by a shaft. Each rotor segment is a toothed wheel and is called a *stack*. The two rotor stacks form the two poles of a permanent magnet located along the rotor axis. Hence an entire stack of rotor teeth is magnetized to be a single pole (which is different from the case of a PM stepper where the rotor has multiple poles). The rotor polarity of a hybrid stepper can be provided either by a permanent magnet, or by an electromagnet using a coil activated by a unidirectional dc source and placed on the stator to generate a magnetic field along the rotor axis.

Another practical classification that is used in this book is based on the number of "stacks" of teeth (or rotor segments) present on the rotor shaft. In particular, a hybrid stepper motor has two stacks of teeth. Further sub classifications are possible, depending on the tooth pitch (angle between adjacent teeth) of the stator and tooth pitch of the rotor. In a *single-stack stepper motor*, the rotor tooth pitch and the stator tooth pitch generally have to be unequal so that not all teeth in the stator are ever aligned with the rotor teeth at any instant. It is the misaligned teeth that exert the magnetic pull, generating the driving torque. In each motion increment, the rotor turns to the minimum reluctance (stable equilibrium) position corresponding to that particular polarity distribution of the stator. In *multiple-stack stepper motors*, operation is possible even when the rotor tooth pitch is equal to the stator tooth pitch, provided that at least one stack of rotor teeth is rotationally shifted (misaligned) from the other stacks by a fraction of the rotor tooth pitch. In this design, it is this *inter-stack misalignment* that generates the drive torque for each motion step. It should be obvious that unequal-pitch multiple stack steppers are also a practical possibility. In this design, each rotor stack operates as a separate single-stack stepper motor. A photograph of the internal components of a two-stack stepper motor is given in Figure 12.10.

12.4.2 Driver and Controller

In principle, the stepper motor is an open-loop actuator. In its normal operating mode, the stepwise rotation of the motor is synchronized with the command pulse train. Under highly transient conditions near rated torque, "pulse missing" can be a problem.

A stepper needs a "control computer" or at least a hardware "indexer" to generate the pulse commands and a "driver" to interpret the commands and correspondingly generate the proper currents for the phase windings of the motor. This basic arrangement is shown

FIGURE 12.10
A commercial two-stack stepper motor (From: Danaher Motion. With permission.)

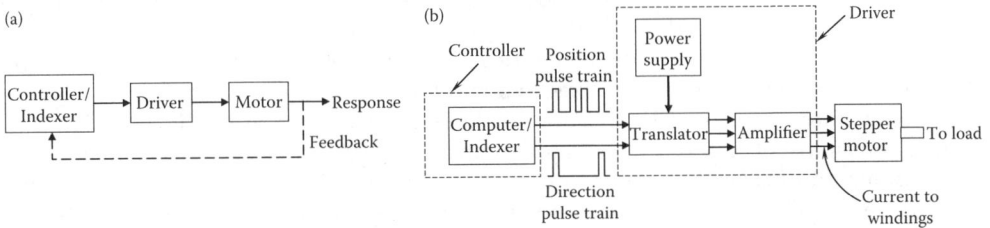

FIGURE 12.11
(a) The basic control system of a stepper motor. (b) The basic components of a driver.

in Figure 12.11a. For feedback control, the response of the motor has to be sensed (say, using an optical encoder) and fed back into the controller (see the dotted line in Figure 12.11a) for taking the necessary corrective action to the pulse command, when an error is present. The basic components of the driver for a stepper motor are identified in Figure 12.11b. It consists of a logic circuit called "translator" to interpret the command pulses and switch the appropriate analog circuits to generate the phase currents. Since sufficiently high current levels are needed for the phase windings, depending on the motor capacity, the drive system includes amplifiers powered by a power supply.

The command pulses are generated either by a control computer (a desktop computer or a microprocessor)—the software approach or by a variable-frequency oscillator (or, an indexer)—the basic hardware approach. For bidirectional motion, two pulse trains are necessary—the position-pulse train and the direction-pulse train, which are determined by the required motion trajectory. The position pulses identify the exact times at which angular steps should be initiated. The direction pulses identify the instants at which the direction of rotation should be reversed. Only a position pulse train is needed for unidirectional operation. Generation of the position pulse train for steady-state operation at a constant speed is a relatively simple task. In this case, a single command identifying the stepping rate (pulse rate), corresponding to the specified speed, would suffice. The logic circuitry within the translator will latch onto a constant-frequency oscillator, with the frequency determined by the required speed (stepping rate), and continuously cycle the switching sequence at this frequency. This is a hardware approach to open-loop control of a stepping motor. For steady-state operation, the stepping rate can be set by manually adjusting the knob of a potentiometer connected to the translator. For simple motions (e.g., starting from rest and stopping after reaching a certain angular position), the commands that generate the pulse train (commands to the oscillator) can be set manually. Under the more complex and transient operating conditions that are present when following intricate motion trajectories, however, a computer-based (or, microprocessor-based) generation of the pulse commands, using programmed logic, would be necessary. This is a software approach, which is usually slower than the hardware approach. Sophisticated feedback control schemes can be implemented as well through such a computer-based controller.

The *translator* module has logic circuitry to interpret a pulse train and "translate" it into the corresponding switching sequence for stator field windings (on/off/reverse state for each phase of the stator). The translator also has solid-state switching circuitry (using gates, latches, triggers, etc.), to direct the field currents to the appropriate phase windings according to the particular switching state. A "packaged" system typically includes both indexer (or, controller) functions and driver functions. As a minimum, it possesses the capability to generate command pulses at a steady rate, thus assuming the role of the pulse generator (or, indexer) as well as the translator and switching amplifier functions. The stepping rate or direction may be changed manually using knobs or through a user interface.

The translator may not have the capability to keep track of the number of steps taken by the motor (i.e., a step counter). A packaged device that has all these capabilities, including pulse generation, the standard translator functions, and drive amplifiers, is termed a *preset indexer*. It usually consists of an oscillator, digital microcircuitry (integrated-circuit chips) for counting and for various control functions, and a translator, and drive circuitry in a single package. The required angle of rotation, stepping rate, and direction are set either manually, by turning the corresponding knobs. With a more sophisticated programmable preset indexer, these settings can be programmed through computer commands from a standard interface. An external pulse source is not needed in this case. A programmable indexer—consisting of a microprocessor and microelectronic circuitry for the control of position and speed and for other programmable functions, memory, a pulse source (an oscillator), a translator, drive amplifiers with switching circuitry, and a power supply—represents a "programmable" controller for a stepping motor. A programmable indexer can be programmed using a personal computer or a hand-held programmer (provided with the indexer) through a standard interface (e.g., RS232 serial interface). Control signals within the translator are on the order of 10 mA, whereas the phase windings of a stepper motor require large currents on the order of several amperes. Control signals from the translator have to be properly amplified and directed to the motor windings by means of "switching amplifiers" for activating the required phase sequence.

Power to operate the translator (for logic circuitry, switching circuitry, etc.), and to operate phase excitation amplifiers comes from a dc *power supply* (typically 24 V dc). A regulated (i.e., voltage maintained constant irrespective of the load) power supply is preferred. A packaged unit that consists of the translator (or preset indexer), the switching amplifiers, and the power supply, is what is normally termed a *motor-drive system*. The leads of the output amplifiers of the drive system carry currents to the phase windings on the stator (and to the rotor magnetizing coils located on the stator in the case of an electromagnetic rotor) of the stepping motor. The *load* may be connected to the motor shaft directly or through some form of mechanical coupling device (e.g., harmonic drive, tooth-timing belt drive, hydraulic amplifier, rack and pinion).

12.4.3 Stepper Motor Selection

Selection of a stepper motor for a specific application cannot be made on the basis of geometric parameters alone. Torque and speed considerations are often more crucial in the selection process. For example, a faster speed of response is possible if a motor with a larger torque-to-inertia ratio is used.

12.4.3.1 Torque Characteristics and Terminology

The torque that can be provided to a load by a stepper motor depends on several factors. For example, the motor torque at constant speed is not the same as that when the motor "passes through" that speed (i.e., under acceleration, deceleration, or general transient conditions). In particular, at constant speed there is no inertia torque. Also, the torque losses due to magnetic induction are lower at constant stepping rates in comparison to variable stepping rates. It follows that the available torque is larger under steady (constant-speed) conditions. Another factor of influence is the magnitude of the speed. At low speeds (i.e., when the step period is considerably larger than the electrical time constant), the time taken for the phase current to build up or turned off is insignificant compared to the step time. Then the phase current waveform can be assumed rectangular. At high stepping rates, the induction effects dominate and as a result a phase may not reach its rated current during the duration of a step. As a result, the generated torque will be degraded. Furthermore, since the power provided by the power supply is limited, the torque×speed product of the motor is limited as well. Consequently, as the motor speed increases, the available torque must decrease in general. These two are the primary reasons for the characteristic shape of a speed-torque curve of a stepper motor where the peak torque occurs at a very low (typically zero) speed, and as the speed increases the available torque decreases. Eventually, at a particular limiting speed (known as the no-load speed) the available torque becomes zero.

The characteristic shape of the speed-torque curve of a stepper motor is shown in Figure 12.12. Some terminology is also given. What is given may be interpreted as experimental data measured under steady operating conditions (and averaged and interpolated). The given torque is called the "pull-out torque" and the corresponding speed is the "pull-out speed." In industry, this curve is known as the "pull-out curve."

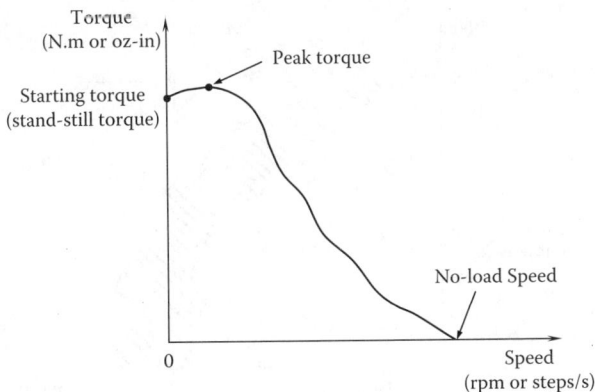

FIGURE 12.12
The speed-torque characteristics of a stepper motor.

Holding torque is the maximum static torque (see e.g. Equation 6.36), and is different from the maximum (pull-out) torque defined in Figure 12.12. In particular, the holding torque can be about 40% greater than the maximum pull-out torque, which is typically equal to the starting torque (or, stand-still torque). Furthermore, the static torque becomes higher if the motor has more than one stator pole per phase and if all these poles are excited at a time. The *residual torque* is the maximum static torque that is present when the motor phases are not energized. This torque is practically zero for a VR motor, but is not negligible for a PM motor. In some industrial literature, *detent torque* takes the same meaning as the residual torque. In this context, detent torque is defined as the torque ripple that is present under power-off conditions. A more appropriate definition for detent torque is the static torque at the present detention position (equilibrium position) of the motor, when the next phase is energized. According to this definition, detent toque is equal to $T_{max} \sin 2\pi/p$ where T_{max} is the holding torque, and p is the number of phases.

Some further definitions of speed-torque characteristics of a stepper motor are given in Figure 12.13. The pull-out curve or the *slew curve* here takes the same meaning as what is given in Figure 12.12. Another curve known as the *start-stop curve* or *pull-in curve* is also given.

The pull-out curve (or, slew curve) gives the speed at which motor can run under steady (constant-speed) conditions, under rated current and using appropriate drive circuitry. But, the motor is unable to steadily accelerate to the slew speed, starting from rest and applying a pulse sequence at constant rate corresponding to the slew speed. Instead, it should be accelerated first up to the pull-in speed by applying a pulse sequence corresponding to this speed. After reaching the start–stop region (pull-in region) in this manner, the motor can be accelerated to the pull-out speed (or to a speed lower than this, within the slew region). Similarly, when stopping the motor from a slew speed, it should be first decelerated (by down-ramping) to a speed in the start-stop region (pull-in region) and only when this region is reached satisfactorily, the stepping sequence should be turned off.

Since the drive system determines the current and the switching sequence of the motor phases and the rate at which the switching pulses are applied, it directly affects the speed-torque curve of a motor. Accordingly, what is given in a product data sheet should be interpreted as the speed-torque curve of the particular motor when used with a specified drive system and a matching power supply, and operating at rated values.

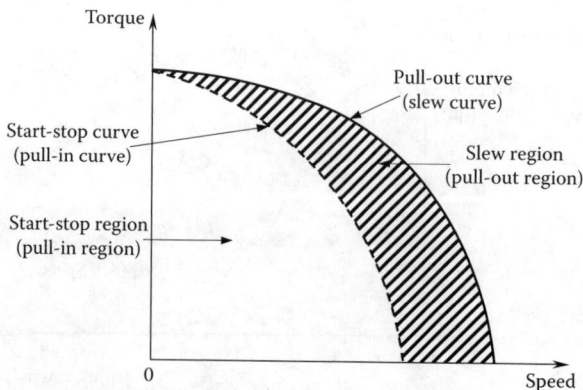

FIGURE 12.13
Further speed-torque characteristics and terminology.

12.4.3.2 Stepper Motor Selection Process

The effort required in selecting a stepper motor for a particular application can be reduced if the selection is done in a systematic manner. The following steps provide some guidelines for the selection process:

Step 1: List the main requirements for the particular application, according to the conditions and specifications for the particular application. These include operational requirements such as speeds, accelerations, and required accuracy and resolution, and load characteristics, such as size, inertia, fundamental natural frequencies, and resistance torques.

Step 2: Compute the operating torque and stepping rate requirements for the particular application.

Newton's second law is the basic equation employed in this step. Specifically, the required torque rating is given by:

$$T = T_R + J_{eq} \frac{\omega_{max}}{\Delta t} \tag{12.10}$$

where

T_R=net resistance torque

J_{eq}=equivalent moment of inertia (including rotor, load, gearing, dampers, etc.)

ω_{max}=maximum operating speed

Δt = time taken to accelerate the load to the maximum speed, starting from rest.

Step 3: Using the torque versus stepping rate curves pull-out curves) for a group of commercially available stepper motors, select a suitable stepper motor.

The torque and speed requirements determined in Step 2 and the accuracy and resolution requirements specified in Step 1 should be used in this step.

Step 4: If a stepper motor that meets the requirements is not available, modify the basic design.

This may be accomplished by changing the speed and torque requirements by adding devices such as gear systems (e.g., harmonic drive) and amplifiers (e.g., hydraulic amplifiers).

Step 5: Select a drive system that is compatible with the motor and that meets the operational requirements in Step 1.

Motors and appropriate drive systems are prescribed in product manuals and catalogs available from the vendors. For relatively simple applications, a manually controlled preset indexer or an open-loop system consisting of a pulse source (oscillator) and a translator could be used to generate the pulse signal to the translator in the drive unit. For more complex transient tasks, a software controller (a microprocessor or a personal computer), or a customized hardware controller may be used to generate the desired pulse command in open-loop operation. Further sophistication may be incorporated by using digital processor-based closed-loop control with encoder feedback, for tasks that require very high accuracy under transient conditions and for operation near the rated capacity of the motor.

The single most useful piece of information in selecting a stepper motor is the torque versus stepping rate curve (the pull-out curve). Other parameters that are valuable in the selection process include:

1. The step angle or the number of steps per revolution.

2. The static holding torque (maximum static torque of motor when powered at rated voltage).

3. The maximum slew rate (maximum steady-state stepping rate possible at rated load).

4. The motor torque at the required slew rate (pull-out torque, available from the pull-out curve).

5. The maximum ramping slope (maximum acceleration and deceleration possible at rated load).

6. The motor time constants (no-load electrical time constant and mechanical time constant).

7. The motor natural frequency (without an external load and near detent position).

8. The motor size (dimensions of poles, stator and rotor teeth, air gap and housing, weight, rotor moment of inertia).

9. The power supply ratings (voltage, current, and power).

Example 12.1

A schematic diagram of an industrial conveyor unit is shown in Figure 12.14. In this application, the conveyor moves intermittently at a fixed rate, thereby indexing the objects on the conveyor through a fixed distance d in each time period T. A triangular speed profile is used for each motion interval, having an acceleration and a deceleration that are equal in magnitude (see Figure 12.15).

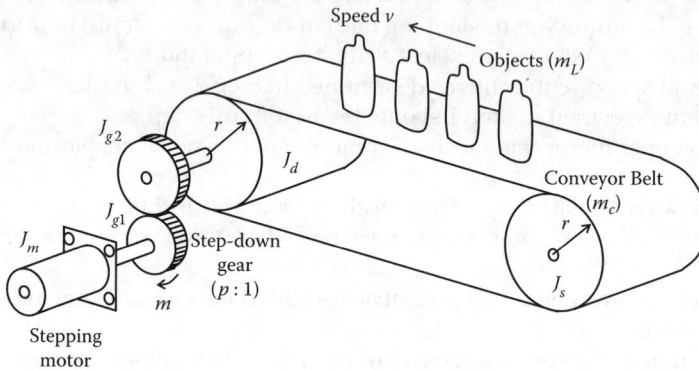

FIGURE 12.14
Conveyor unit with intermittent motion.

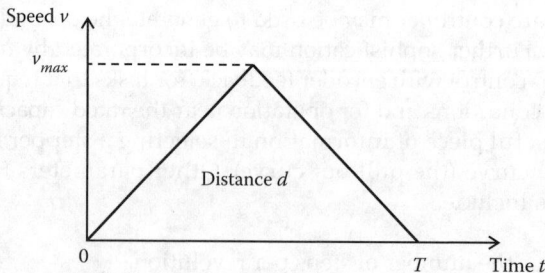

FIGURE 12.15
Speed profile for a motion period of the conveyor.

The conveyor is driven by a stepper motor. A gear unit with step down speed ratio p:1, where $p>1$, may be used if necessary, as shown in Figure 12.15.

a. Explain why the equivalent moment of inertia J_e at the motor shaft, for the overall system, is given by:

$$J_e = J_m + J_{g1} + (1/p^2)(J_{g2} + J_d + J_s) + (r^2/p^2)(m_c + m_L)$$

where J_m, J_{g1}, J_{g2}, J_d, and J_s are the moments of inertia of the motor rotor, drive gear, driven gear, drive cylinder of the conveyor, and the driven cylinder of the conveyor, respectively; m_c and m_L are the overall masses of the conveyor belt and the moved objects (load), respectively; and r is the radius of each of the two conveyor cylinders.

b. Four models of stepping motor are available for the application. Their specifications are given in Table 12.2 and the corresponding performance curves are given in Figure 12.16. The following values are known for the system:

$d=10$ cm, $T=0.2$ seconds, $r=10$ cm, $m_c=5$ kg, $m_L=5$ kg, $J_d=J_s=2.0\times10^{-3}$ kg m^2.

Also two gear units with $p=2$ and 3 are available, and for each unit $J_{g1}=50\times10^{-6}$ kg m^2 and $J_{g2}=200\times10^{-6}$ kg m^2.

Indicating all calculations and procedures, select a suitable motor unit for this application. You must not use a gear unit unless it is necessary to have one with the available motors.

What is the positioning resolution of the conveyor (rectilinear) for the final system?

Note: Assume an overall system efficiency of 80% regardless of whether a gear unit is used.

Solution

a. Angular speed of the motor and drive gear $=\omega_m$.
   Angular speed of the driven gear and conveyor cylinders $=(\omega_m/p)$.
   Rectilinear speed of the conveyor and objects $v=(r\omega_m/p)$.

TABLE 12.2

Stepper Motor Data

Model		50SM	101SM	310SM	1010SM
NEMA Motor frame size			23	34	42
Full step angle	degrees		1.8		
Accuracy	percent		±3 (noncumulative)		
Holding torque	oz-in	38	90	370	1050
	N-m	0.27	0.64	2.61	7.42
Detent torque	oz-in	6	18	25	20
	N-m	0.04	0.13	0.18	0.14
Rated phase current	Amps	1	5	6	8.6
Rotor inertia	oz-in-sec^2	1.66×10^{-3}	5×10^{-3}	26.5×10^{-3}	114×10^{-3}
	kg-m^2	11.8×10^{-6}	35×10^{-6}	187×10^{-6}	805×10^{-6}
Maximum radial load	lb		15	35	40
	N		67	156	178
Maximum thrust load	lb		25	60	125
	N		111	267	556
Weight	lb	1.4	2.8	7.8	20
	kg	0.6	1.3	3.5	9.1
Operating temperature	°C		−55 to +50		
Storage temperature	°C		−55 to +130		

Source: Aerotech, Inc. With permission.

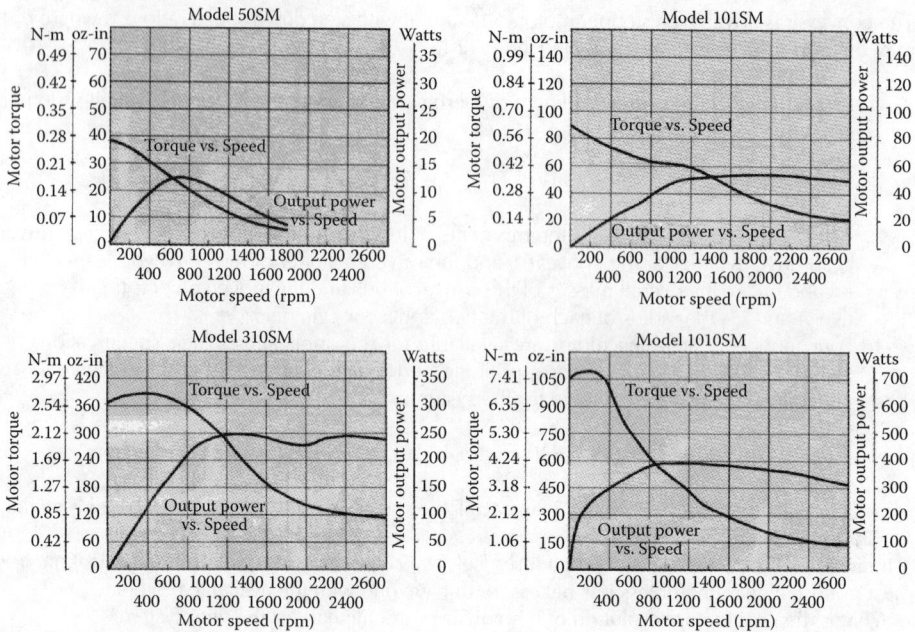

FIGURE 12.16
Stepper motor performance curves. (From: Aerotech, Inc. With permission.)

Determination of the Equivalent Inertia:

Determination of the equivalent moment of inertia of the system, as referred to the motor rotor, is an important step of the motor selection. This done by determining the kinetic energy of the overall system and equating it to the kinetic energy of the equivalent system as follows:

$$KE = \frac{1}{2}(J_m + J_{g1})\omega_m^2 + \frac{1}{2}(J_{g2} + J_d + J_s)\left(\frac{\omega_p}{p}\right)^2 + \frac{1}{2}(m_c + m_L)\left(\frac{r\omega_m}{p}\right)^2$$

$$= \frac{1}{2}[J_m + J_{g1} + \frac{1}{p^2}(J_{g2} + J_d + J_s) + \frac{r^2}{p^2}(m_c + m_L)]\omega_m^2$$

$$= \frac{1}{2}J_e\omega_m^2$$

Hence, the equivalent moment of inertia as felt at the motor rotor, is:

$$J_e = J_m + J_{g1} + \frac{1}{p^2}(J_{g2} + J_d + J_s) + \frac{r^2}{p^2}(m_c + m_L) \qquad \text{(i)}$$

b. Area of the speed profile is equal to the distance travelled. Hence:

$$d = \frac{1}{2}v_{max}T \qquad \text{(ii)}$$

Substitute numerical values: $0.1 = \frac{1}{2} v_{max} 0.2$ ➔ $v_{max} = 1.0$ m/s

The acceleration/deceleration of the system: $a = \frac{v_{max}}{T/2} = \frac{1.0}{0.2/2}$ m/s^2 = 10.0 m/s^2

Corresponding angular acceleration/deceleration of the motor:

$$\alpha = \frac{pa}{r} \tag{iii}$$

With an overall system efficiency of η, the motor torque T_m that is needed to accelerate/decelerate the system is given by:

$$\eta T_m = J_e \alpha = J_e \frac{pa}{r} = [J_m + J_{g1} + \frac{1}{p^2}(J_{g2} + J_d + J_s) + \frac{r^2}{p^2}(m_c + m_L)]\frac{pa}{r} \tag{iv}$$

Note: An alternative way to include energy dissipation into this equation is by using two separate terms: frictional torque referred to the motor rotor and gear efficiency. In the present problem, for simplicity, we use a single efficiency term whether a gear is present or not. In practice, however, it should be clear that the overall efficiency drops when a gear transmission is added.

$$\text{Maximum speed of the motor: } \omega_{max} = \frac{pv_{max}}{r} \tag{v}$$

Without gears ($p=1$) we have from Equation (iv):

$$\eta T_m = [J_m + J_d + J_s + r^2(m_c + m_L)]\frac{a}{r} \tag{vi}$$

$$\text{From Equation (v): } \omega_{max} = \frac{v_{max}}{r} \tag{vii}$$

Substitute numerical values.

Case 1: Without Gears
For an efficiency value $\eta=0.8$ (i.e., 80% efficient), we have from Equation (vi):

$$0.8 T_m = [J_m + 2 \times 10^{-3} + 2 \times 10^{-3} + 0.1^2(5+5)]\frac{10}{0.1} \text{ N.m}$$

$$\text{Or: } T_m = 125.0 [J_m + 0.104] \text{ N.m}$$

$$\text{From Equation (vii): } \omega_{max} = \frac{1.0}{0.1} \text{ rad/s} = 10 \times \frac{60}{2\pi} \text{ rpm} = 95.5 \text{ rpm}$$

The operating speed range is 0–95.5 rpm.
Note: The torque at 95.5 rpm is less than the starting torque for the first two motor models, and not so for the second two models (see the speed-torque curves in Figure 12.16). We must use the weakest point (i.e., lowest torque) in the operating speed range, in the motor selection process. Allowing for this requirement, Table 12.3 is formed for the four motor models.
It is seen that without a gear unit, the available motors cannot meet the system requirements.

TABLE 12.3

Data for Selecting a Motor without a Gear Unit

Motor Model	Available Torque at ω_{max} (N.m)	Motor Rotor Inertia (kg.m²)	Required Torque (N.m)
50 SM	0.26	11.8×10^{-6}	13.0
101 SM	0.60	35.0×10^{-6}	13.0
310 SM	2.58	187.0×10^{-6}	13.0
1010 SM	7.41	805.0×10^{-6}	13.1

TABLE 12.4

Data for Selecting a Motor with a Gear Unit

Motor Model	Available Torque at ω_{max} (N.m)	Motor Rotor Inertia (kg.m²)	Required Torque (N.m)
50 SM	0.25	11.8×10^{-6}	6.53
101 SM	0.58	35.0×10^{-6}	6.53
310 SM	2.63	187.0×10^{-6}	6.57
1010 SM	7.41	805.0×10^{-6}	6.73

Case 2: With Gears

Note: Usually the system efficiency drops when a gear unit is introduced. In the present exercise we use the same efficiency for reasons of simplicity.

With an efficiency of 80%, we have $\eta=0.8$. Then from Equation (iv):

$$0.8T_m = \left[J_m + 50\times10^{-6} + \frac{1}{p^2}(200\times10^{-6} + 2\times10^{-3} + 2\times10^{-3}) + \frac{0.1^2}{p^2}(5+5) \right] p \times \frac{10}{0.1} \text{ N.m}$$

$$\rightarrow T_m = 125.0 \left[J_m + 50\times10^{-6} + \frac{1}{p^2}\times104.2\times10^{-3} \right] p \text{ N.m}$$

From Equation (v): $\omega_{max} = \dfrac{1.0p}{0.1}$ rad/s $= 10p\times\dfrac{60}{2\pi}$ rpm $\rightarrow \omega_{max} = 95.5p$ rpm

First try the case of $p=2$ we have $\omega_{max}=191.0$ rpm. Table 12.4 is formed for the present case. It is seen that with a gear of speed ratio $p=2$, motor model 1010 SM satisfies requirement.

With full stepping, step angle of the rotor$=1.8°$. Corresponding step in the conveyor motion is the positioning resolution.

With $p=2$ and $r=0.1$ m, the position resolution is $\dfrac{1.8°}{2} \times \dfrac{\pi}{180°} \times 0.1 = 1.57\times10^{-3}$ m.

12.5 dc Motors

A dc motor converts direct current electrical energy into rotational mechanical energy. A major part of the torque generated in the rotor (armature) of the motor is available to drive an external load. The dc motor is probably the earliest form of electric motor. Because of features such as high torque, speed controllability over a wide range, portability, well-behaved

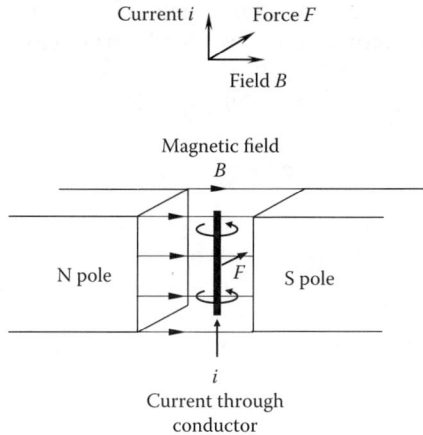

FIGURE 12.17
Operating principle of a dc motor.

speed-torque characteristics, easier and accurate modeling, and adaptability to various types of control methods, dc motors are still widely used in numerous control applications including robotic manipulators, transport mechanisms, disk drives, positioning tables, machine tools, and servovalve actuators.

The principle of operation of a dc motor is illustrated in Figure 12.17. Consider an electric conductor placed in a steady magnetic field at right angles to the direction of the field. Flux density B is assumed constant. If a dc current is passed through the conductor, the magnetic flux due to the current will loop around the conductor, as shown in the figure. Consider a plane through the conductor, parallel to the direction of flux of the magnet. On one side of this plane, the current flux and the field flux are additive; on the opposite side, the two magnetic fluxes oppose each other. As a result, an imbalance magnetic force F is generated on the conductor, normal to the plane. This force is given by (*Lorentz's law*):

$$F = Bil \tag{12.11}$$

where
  B = flux density of the original field
  i = current through the conductor
  l = length of the conductor
  Note: If the field flux is not perpendicular to the length of the conductor, it can be resolved into a perpendicular component that generates the force and to a parallel component that has no effect.

The active components of i, B, and F are mutually perpendicular and form a right-hand triad, as shown in Figure 12.17. Alternatively, in the vector representation of these three quantities, the vector F can be interpreted as the **cross product** of the vectors i and B. Specifically, $F = i \times B$.

If the conductor is free to move, the generated force will move it at some velocity v in the direction of the force. As a result of this motion in the magnetic field B, a voltage is induced in the conductor. This is known as the back electromotive force, or *back e.m.f.*, and is given by:

$$v_b = Blv \tag{12.12}$$

According to *Lenz's law*, the flux due to the back e.m.f. v_b will be opposing the flux due to the original current through the conductor, thereby trying to stop the motion. This is the cause of *electrical damping* in motors. Equation 12.11 determines the armature torque (motor torque), and Equation 12.12 governs the motor speed.

12.5.1 Rotor and Stator

A dc motor has a rotating element called rotor or armature. The rotor shaft is supported on two bearings in the motor housing. The rotor has many closely spaced slots on its periphery. These slots carry the rotor windings, as shown in Figure 12.18a. Assuming the field flux is in the radial direction of the rotor, the force generated in each conductor will be in the tangential direction, thereby generating a torque (force×radius), which drives the rotor. The rotor is typically a laminated cylinder made from a ferromagnetic material. A ferromagnetic core helps concentrate the magnetic flux toward the rotor. The lamination reduces the problem of magnetic hysteresis and limits the generation of eddy currents and associated dissipation (energy loss by heat generation) within the ferromagnetic material. More advanced dc motors use powdered-iron-core rotors rather than the laminated-iron-core variety, thereby further restricting the generation and conduction/dissipation of eddy currents and reducing various nonlinearities such as hysteresis. The rotor windings (armature windings) are powered by the supply voltage v_a.

The fixed magnetic field (which interacts with the rotor coil and generates the motor torque) is provided by a set of fixed magnetic poles around the rotor. These poles form the stator of the motor. The stator may consist of two opposing pole of a permanent magnet. In industrial dc motors, however, the field flux is usually generated not by a permanent magnet but electrically in the stator windings, by an electromagnet, as schematically shown in Figure 12.18a. Stator poles are constructed from ferromagnetic sheets (i.e., a *laminated construction*). The stator windings are powered by supply voltage v_f, as shown in Figure 12.18a. Furthermore, note that in Figure 12.18a, the net stator magnetic field is perpendicular to the net rotor magnetic field, which is along the commutation plane. The resulting forces that attempt to pull the rotor field toward the stator field may be interpreted as the cause of the motor torque (which is maximum when the two fields are at right angles).

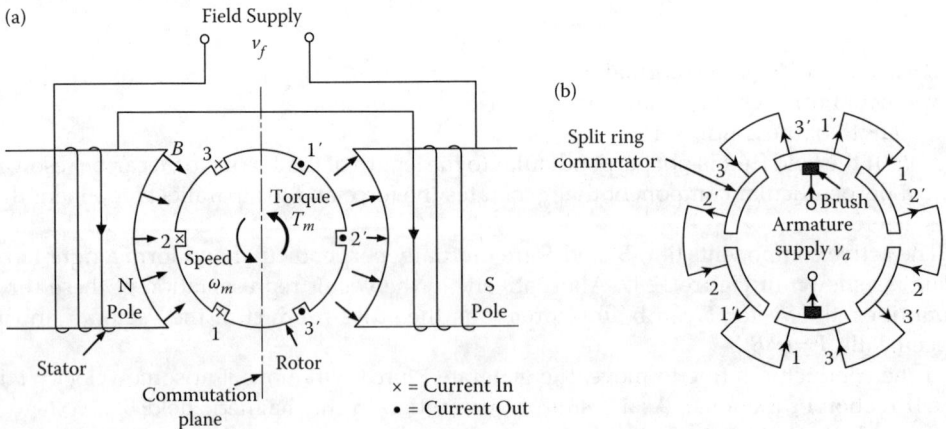

FIGURE 12.18
(a) Schematic diagram of a dc motor. (b) Commutator wiring.

12.5.2 Commutation

A plane known as the "commutation plane" symmetrically divides two adjacent stator poles of opposite polarity. In the two-pole stator shown in Figure 12.18a, the commutation plane is at right angles to the common axis of the two stator poles, which is the direction of the stator magnetic field. It is noted that on one side of the plane, the field is directed toward the rotor, while on the other side the field is directed away from the rotor. Accordingly, when a rotor conductor rotates from one side of the plane to the other side, the direction of the generated torque will be reversed. Such a scenario is not useful since the average torque will be zero in that case.

In order to maintain the direction of torque in each conductor group (one group is numbered 1, 2, 3, and the other group is numbered 1', 2', and 3', in Figure 12.18a), the direction of current in a conductor has to change as the conductor crosses the commutation plane. Physically, this may be accomplished by using a split ring and brush commutator, shown schematically in Figure 12.18b. The armature voltage is applied to the rotor windings through a pair of stationary conducting blocks made of graphite (conducting soft carbon), which maintain sliding contact with the split ring. These contacts are called "brushes" because historically, they were made of bristles of copper wire in the form of a brush. The graphite contacts are cheaper, more durable primarily due to reduced sparking (arcing) problems, and provide more contact area (less electrical contact resistance). Also the contact friction is lower. The split ring segments, equal in number to the conductor slots in the rotor, are electrically insulated from one another, but the adjacent segments are connected by the armature windings in each opposite pair of rotor slots, as shown in Figure 12.18b. For the rotor position shown in Figure 12.18, note that when the split ring rotates in the counter-clockwise direction through 30°, the current paths in conductors 1 and 1' reverse but the remaining current paths are unchanged, thus achieving the required commutation.

12.5.3 Brushless dc Motors

There are several shortcomings of the slip ring and brush mechanisms, which are used for current transmission through moving members, even with the advances from the historical copper brushes to modern graphite contacts. The main disadvantages include rapid wearout, mechanical loading, wear and heating due to sliding friction, contact bounce, excessive noise, and electrical sparks (arcing) with the associated dangers in hazardous (e.g., chemical) environments, problems of oxidation, problems in applications that require wash-down (e.g., in food processing), and voltage ripples at switching points.

Conventional remedies to these problems—such as the use of improved brush designs and modified brush positions to reduce arcing—are inadequate in sophisticated applications. Also, the required maintenance (to replace brushes and resurface the split-ring commutator) can be rather costly.

Brushless dc motors have PM rotors. Since in this case the polarities of the rotor cannot be switched as the rotor crosses a commutation plane, commutation is accomplished by electronically switching the current in the stator winding segments. Note that this is the reverse of what is done in brushed commutation, where the stator polarities are fixed and the rotor polarities are switched when crossing a commutation plane. The stator windings of a brushless dc motor can be considered the armature windings whereas for a brushed dc motor, rotor is the armature.

PM motors are less nonlinear than the electro-magnet motors because the field strength generated by a permanent magnet is rather constant and independent of the current through a coil. This is true whether the permanent magnet is in the stator (i.e., a brushed motor) or in the rotor (i.e., a brushless dc motor or a PM stepper motor).

12.5.4 DC Motor Equations

Consider a dc motor with separate windings in the stator and the rotor. Each coil has a resistance (R) and an inductance (L). When a voltage (v) is applied to the coil, a current (i) flows through the circuit, thereby generating a magnetic field. As discussed before, a force is produced in the rotor windings, and an associated torque (T_m), which turns the rotor. The rotor speed (ω_m) causes the flux linkage of the rotor coil with the stator field to change at a corresponding rate, thereby generating a voltage (back e.m.f.) in the rotor coil.

Equivalent circuits for the stator and the rotor of a conventional dc motor are shown in Figure 12.19a. Since the field flux is proportional to field current i_f, we can express the magnetic torque of the motor as:

$$T_m = k i_f i_a \tag{12.13}$$

This directly follows from Equation 12.11. Next, in view of Equation 12.12, the back e.m.f. generated in the armature of the motor is given by:

$$v_b = k' i_f \omega_m \tag{12.14}$$

The following notation has been used:

i_f = field current

i_a = armature current

ω_m = angular speed of the motor.

and k and k' are motor constants, which depend on factors such as the rotor dimensions, the number of turns in the armature winding, and the permeability (inverse of reluctance) of the magnetic medium. In the case of ideal electrical-to-mechanical energy conversion at the rotor (where the rotor coil links with the stator field), we have $T_m \omega_m = v_b i_a$ with consistent units (e.g., torque in Newton-meters, speed in radians per second, voltage in volts, and current in amperes). Then we observe that

$$k = k' \tag{12.15}$$

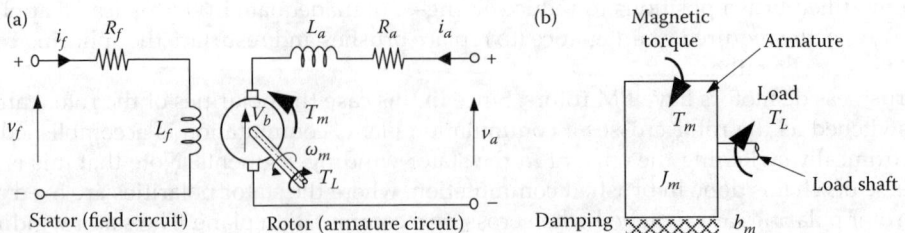

FIGURE 12.19
(a) The equivalent circuit of a conventional dc motor (separately excited). (b) Armature mechanical loading diagram.

The field circuit equation is obtained by assuming that the stator magnetic field is not affected by the rotor magnetic field (i.e., the stator inductance is not affected by the rotor) and that there are no eddy current effects in the stator. Then, from Figure 12.19a:

$$v_f = R_f i_f + L_f \frac{di_f}{dt} \qquad (12.16)$$

where v_f = supply voltage to the stator; R_f = resistance of the field winding; L_f = inductance of the field winding.

The equation for the armature rotor circuit is written as (see Figure 12.19a).

$$v_a = R_a i_a + L_a \frac{di_a}{dt} + v_b \qquad (12.17)$$

where v_a = supply voltage to the armature; R_a = resistance of the armature winding; L_a = leakage inductance in the armature winding.

It should be emphasized here that the primary inductance or *mutual inductance* in the armature winding is represented in the back e.m.f. term v_b. The leakage inductance, which is usually neglected, represents the fraction of the armature flux that is not linked with the stator and is not used in the generation of useful torque. This includes self-inductance in the armature.

The mechanical equation of the motor is obtained by applying Newton's second law to the rotor. Assuming that the motor drives some load, which requires a load torque T_L to operate, and that the frictional resistance in the armature can be modeled by a linear viscous term, we have (see Figure 12.19b):

$$J_m \frac{d\omega_m}{dt} = T_m - T_L - b_m \omega_m \qquad (12.18)$$

where J_m = moment of inertia of the rotor; b_m = equivalent (mechanical) damping constant for the rotor.

Note: The load torque may be due, in part, to the inertia of the external load that is coupled to the motor shaft. If the coupling flexibility is neglected, the load inertia may be directly added to (i.e., lumped with) the rotor inertia after accounting for the possible existence of a speed reducer (gear, harmonic drive, etc.). In general, a separate set of equations is necessary to represent the dynamics of the external load. Equations 12.13 through 12.18 form the dynamic model for a dc motor.

12.5.4.1 Steady-State Characteristics

In selecting a motor for a given application, its steady-state characteristics are a major determining factor. In particular, steady-state torque-speed curves are employed for this purpose. The rationale is that, if the motor is able to meet the steady-state operating requirements, with some design conservatism, it should be able to tolerate small deviations under transient conditions of short duration. In the separately excited case shown in Figure 12.19a, where the armature circuit and field circuit are excited by separate and independent voltage sources, it can be shown that the steady-state torque-speed curve is a straight line.

The shape of the steady-state speed-torque curve will be modified if a common voltage supply is used to excite both the field winding and the armature winding. Here, the two windings have to be connected together. There are three common ways the windings of the rotor and the stator are connected. They are known as: Shunt-wound motor; Series-wound motor; and Compound-wound motor. In a shunt-wound motor, the armature windings and the field windings are connected in parallel. In the series-wound motor, they are connected in series. In the compound-wound motor, part of the field windings is connected with the armature windings in series and the other part is connected in parallel. In a shunt-wound motor at steady-state, the back e.m.f. v_b depends directly on the supply voltage. Since the back e.m.f. is proportional to the speed, it follows that speed controllability is good with the shunt-wound configuration. In a series-wound motor, the relation between v_b and the supply voltage is coupled through both the armature windings and the field windings. Hence its speed controllability is relatively poor. But in this case, a relatively large current flows through both windings at low speeds of the motor, giving a higher starting torque. Also, the operation is approximately at constant power in this case. Since both speed controllability and higher starting torque are desirable characteristics, compound-wound motors are used to obtain a performance in between the two extremes. The torque-speed characteristics for the three types of winding connections are shown in Figure 12.20.

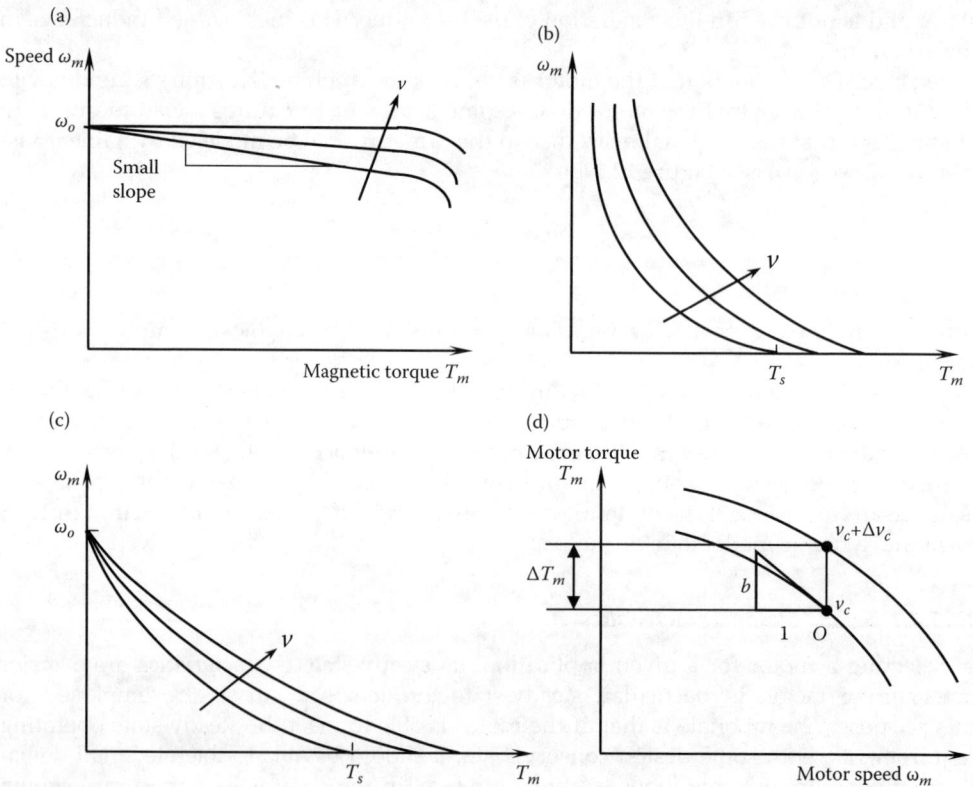

FIGURE 12.20
Torque-speed characteristic curves for dc motors (a) Shunt-wound. (b) Series-wound. (c) Compound-wound. (d) General case.

12.5.5 Experimental Model for dc Motor

In general, the speed-torque characteristic of a dc motor is nonlinear. A linearized dynamic model can be extracted from the speed-torque curves. One of the parameters of the model is the damping constant. First we will examine this.

12.5.5.1 Electrical Damping Constant

Newton's second law governs the dynamic response of a motor. In Equation 12.18, for example, b_m denotes the mechanical (viscous) damping constant and represents mechanical dissipation of energy. As is intuitively clear, mechanical damping torque opposes motion—hence the negative sign in the $b_m \omega_m$ term in Equation 12.18. Note, further, that the magnetic torque T_m of the motor is also dependent on speed ω_m. In particular, the back e.m.f., which is governed by ω_m, produces a magnetic field, which tends to oppose the motion of the motor rotor. This acts as a damper, and the corresponding damping constant is given by:

$$b_e = -\frac{\partial T_m}{\partial \omega_m} \tag{12.19}$$

This parameter is termed the *electrical damping constant*. Caution should be exercised when experimentally measuring b_e. Note that in constant speed tests, the inertia torque of the rotor will be zero; there is no torque loss due to inertia. Torque measured at the motor shaft includes as well the torque reduction due to mechanical dissipation (mechanical damping) within the rotor, however. Hence the magnitude b of the slope of the speed-torque curve as obtained by a steady-state test is equal to $b_e + b_m$, where b_m is the equivalent viscous damping constant representing mechanical dissipation at the rotor.

12.5.5.2 Linearized Experimental Model

To extract a linearized experimental model for a dc motor, consider the speed-torque curves shown in Figure 12.20d. For each curve, the excitation voltage v_c is maintained constant. This is the voltage that is used in controlling the motor, and is termed control voltage. It can be, for example, the armature voltage, the field voltage, or the voltage that excites both armature and field windings in the case of combined excitation (e.g., shunt-wound motor). One curve in Figure 12.20d is obtained at control voltage v_c and the other curve is obtained at $v_c + \Delta v_c$. Note also that a tangent can be drawn at a selected point (operating point O) of a speed-torque curve. The magnitude b of the slope (which is negative) corresponds to a damping constant, which includes both electrical and mechanical damping effects). What mechanical damping effects are included in this parameter depends entirely on the nature of mechanical damping that was present during the test (primarily bearing friction). We have the *damping constant* as the magnitude of the slope at the operating point:

$$b = -\frac{\partial T_m}{\partial \omega_m}\bigg|_{v_c = \text{constant}} \tag{12.20}$$

Next draw a vertical line through the operating point O. The torque intercept ΔT_m between the two curves can be determined in this manner. Since a vertical line is a constant speed line, we have the *voltage gain*:

$$k_v = \left.\frac{\partial T_m}{\partial v_c}\right|_{\omega_m = \text{constant}} = \frac{\Delta T_m}{\Delta v_c} \tag{12.21}$$

Now, using the well-known relation for total differential we have:

$$\delta T_m = \left.\frac{\partial T_m}{\partial \omega_m}\right|_{v_c} \delta\omega_m + \left.\frac{\partial T_m}{\partial v_c}\right|_{\omega_m} \delta v_c = -b\delta\omega_m + k_v\delta v_c \tag{12.22}$$

Equation 12.22 is the linearized model of the motor. This may be used in conjunction with the mechanical equation of the motor rotor, for the incremental motion about the operating point:

$$J_m \frac{d\delta\omega_m}{dt} = \delta T_m - \delta T_L \tag{12.23}$$

Note that Equation 12.23 is the incremental version of Equation 12.18 except that the overall damping constant of the motor (including mechanical damping) is included in Equation 12.22. The torque needed to drive the rotor inertia, however, is not included in Equation 12.22 because the steady-state curves are used in determining the parameters for this equation. The inertia term is explicitly present in Equation 12.23.

12.5.6 Control of dc Motors

Both speed and torque of a dc motor may have to be controlled for proper performance in a given application of a dc motor. By using proper winding arrangements, dc motors can be operated over a wide range of speeds and torques. Because of this adaptability, dc motors are particularly suitable as variable-drive actuators. Historically, ac motors were employed almost exclusively in constant-speed applications, but their use in variable-speed applications was greatly limited because speed control of ac motors was found to be quite difficult, by conventional means. Since variable-speed control of a dc motor is quite convenient and straightforward, dc motors have dominated in industrial control applications for many decades.

Following a specified motion trajectory is called servoing, and servomotors (or servo-actuators) are used for this purpose. The vast majority of servomotors are dc motors with feedback control of motion. Servo control is essentially a motion control problem, which involves the control of position and speed. There are applications, however, that require torque control, directly or indirectly, but they usually require more sophisticated sensing and control techniques. Control of a dc motor is accomplished by controlling either the stator field flux or the armature flux. If the armature and field windings are connected through the same circuit, both techniques are incorporated simultaneously. Specifically, the two methods of control are: Armature control and field control.

12.5.6.1 Armature Control

In an armature-controlled dc motor, the armature voltage v_a is used as the control input, while keeping the conditions in the field circuit constant. In particular, the field current i_f is assumed constant. Consequently, Equations 12.13 and 12.14 can be written as:

$$T_m = k_m i_a \tag{12.24}$$

$$v_b = k'_m \omega_m \tag{12.25}$$

The parameters k_m and k'_m are termed the *torque constant* and the *back e.m.f. constant*, respectively.

Note: With consistent units, $k_m = k'_m$ in the case of ideal electrical-to-mechanical energy conversion at the motor rotor.

In the Laplace domain, Equation 12.17 becomes:

$$v_a - v_b = (L_a s + R_a) i_a \tag{12.26}$$

Note: For convenience, time domain variables (functions of t) are used to denote their Laplace transforms (functions of s). It is understood, however, that the time functions are not identical to the Laplace functions.

In the Laplace domain, the mechanical equation (Equation 12.18) becomes:

$$T_m - T_L = (J_m s + b_m) \omega_m \tag{12.27}$$

where J_m and b_m denote the moment of inertia and the rotary viscous damping constant, respectively, of the motor rotor. Equations 12.22 through 12.27 are represented in the block diagram form, in Figure 12.21. Note that the speed ω_m is taken as the motor output. If the motor position θ_m is considered the output, it is obtained by passing ω_m through an integration block $1/s$. Note, further, that the load torque T_L, which is the useful (effective) torque transmitted to the load that is being driven, is an (unknown) input to the system. Usually, T_L increases with ω_m because a larger torque is necessary to drive a load at a higher speed. If a linear (and dynamic) relationship exists between T_L and ω_m at the load, a feedback path can be completed from the output speed to the input load torque

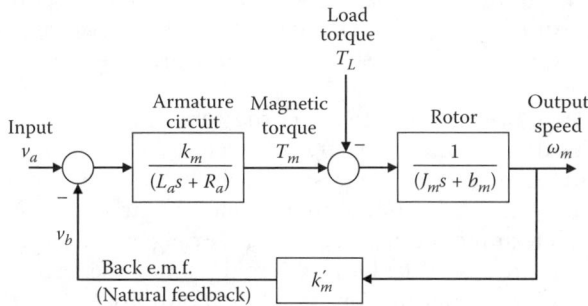

FIGURE 12.21
Open-loop block diagram for an armature-controlled dc motor.

through a proper load transfer function (load block). The system shown in Figure 12.21 is not a feedback control system. The feedback path, which represents the back e.m.f., is a "natural feedback" and is characteristic of the process (dc motor); it is not an external control feedback loop.

The overall transfer relation for the system is obtained by first determining the output for one of the inputs with the other input removed, and then adding the two output components obtained in this manner, in view of the *principle of superposition*, which holds for a linear system. We get:

$$\omega_m = \frac{k_m}{\Delta(s)} v_a - \frac{(L_a s + R_a)}{\Delta(s)} T_L \tag{12.28}$$

where $\Delta(s)$ is the *characteristic polynomial* of the system, given by

$$\Delta(s) = (L_a s + R_a)(J_m s + b_m) + k_m k_m' \tag{12.29}$$

This is a second order polynomial in the Laplace variable s.

12.5.6.2 Motor Time Constants

The *electrical time constant* of the armature is:

$$\tau_a = \frac{L_a}{R_a} \tag{12.30}$$

which is obtained from Equation 12.17 or 12.26. The mechanical response of the rotor is governed by the *mechanical time constant*:

$$\tau_m = \frac{J_m}{b_m} \tag{12.31}$$

which is obtained from Equation 12.18 or 12.27. Usually, τ_m is several times larger than τ_a, because the leakage inductance L_a is quite small (leakage of the flux linkage is negligible for high-quality dc motors). Hence, τ_a can be neglected in comparison to τ_m for most practical purposes. In that case, the transfer functions in Equation 12.28 become first order.

Note that the characteristic polynomial is the same for both transfer functions in Equation 12.28, regardless of the input (v_a or T_L). This should be the case because, $\Delta(s)$ determines the natural response of the system and does not depend on the system input. True time constants of the motor are obtained by first solving the characteristic equation $\Delta(s) = 0$ to determine the two roots (poles or eigenvalues), and then taking the reciprocal of the magnitudes (*Note:* Only the real part of the two roots is used if the roots are complex). For an armature-controlled dc motor, these true time constants are not the same as τ_a and τ_m because of the presence of the coupling term $k_m k_m'$ in $\Delta(s)$ (see Equation 12.29). This also follows from the presence of the natural feedback path (back e.m.f.) in Figure 12.21.

Example 12.2

Determine an expression for the dominant time constant of an armature-controlled dc motor.

Solution

By neglecting the electrical time constant in Equation 12.29, we have the approximate characteristic polynomial:

$$\Delta(s) = R_a(J_m s + b_m) + k_m k_m' \rightarrow \Delta(s) = k'(\tau s + 1)$$

where τ=overall dominant time constant of the system.
It follows that the dominant time constant is given by:

$$\tau = \frac{R_a J_m}{(R_a b_m + k_m k_m')} \tag{12.32}$$

12.5.6.3 Field Control

In field-controlled dc motors, the armature current is assumed to be kept constant, and the field voltage is used as the control input. Since i_a is assumed constant, Equation 12.13 can be written as:

$$T_m = k_a i_f \tag{12.33}$$

where k_a is the electromechanical torque constant for the motor. The back e.m.f. relation and the armature circuit equation are not used in this case. Equations 12.16 and 12.18 are written in the Laplace form as:

$$v_f = (L_f s + R_f) i_f \tag{12.34}$$

$$T_m - T_L = (J_m s + b_m)\omega_m \tag{12.35}$$

Equations 12.33 through 12.35 can be represented by the open-loop block diagram given in Figure 12.22.

Note that even though i_a is assumed constant, this is not strictly true. This should be clear from the armature circuit equation (Equation 12.17). It is the armature supply voltage v_a that is kept constant. Even though L_a can be neglected, then, i_a depends on the back e.m.f. v_b, which changes with the motor speed as well as the field current i_f. Under these

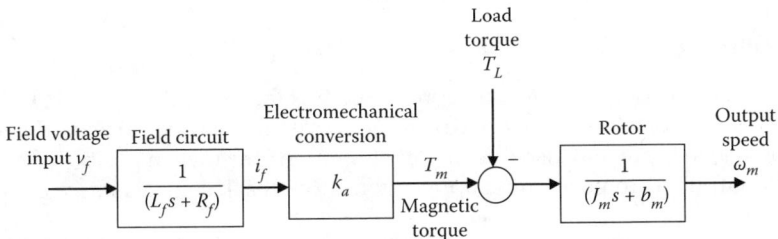

FIGURE 12.22
Open-loop block diagram for a field-controlled dc motor.

conditions, the block representing k_a in Figure 12.22 is not a constant gain, and in fact it is not linear. At least, a feedback will be needed into this block from output speed. This will also add another electrical time constant, which depends on the dynamics of the armature circuit. It will also introduce a coupling effect between the mechanical dynamics (of the rotor) and the armature circuit electronics. For the present purposes, however, we assume that k_a is a constant gain.

Now, we return to Figure 12.22. Since the system is linear, the principle of superposition holds. According to this, the overall output ω_m is equal to the sum of the individual outputs due to the two inputs v_f and T_L, taken separately. It follows that transfer relationship is given by:

$$\omega_m = \frac{k_a}{(L_f s + R_f)(J_m s + b_m)} v_f - \frac{1}{(J_m s + b_m)} T_L \tag{12.36}$$

In this case, the electrical time constant originates from the field circuit and is given by:

$$\tau_f = \frac{L_f}{R_f} \tag{12.37}$$

The mechanical time constant τ_m of the field-controlled motor is the same as that for the armature-controlled motor, and can be defined by Equation 12.31:

$$\tau_m = \frac{J_m}{b_m} \tag{12.31}$$

The characteristic polynomial of the open-loop field-controlled motor is:

$$\Delta(s) = (L_f s + R_f)(J_m s + b_m) \tag{12.38}$$

It follows that τ_f and τ_m are the true time constants of the system, unlike in an armature controlled motor. As in an armature-controlled dc motor, however, the electrical time constant is several times smaller and can be neglected in comparison to the mechanical time constant. Furthermore, as for an armature-controlled motor, the speed and the angular position of a field-controlled motor have to be measured and fed back for accurate motion control.

12.5.7 Feedback Control of dc Motors

Open-loop operation of a dc motor, as represented by Figures 12.21 (armature control) and 12.22 (field control), can lead to excessive error and even instability, particularly because of the unknown load input and also due to the integration effect when position (not speed) is the desired output (as in positioning applications). Feedback control is necessary in these circumstances.

In feedback control, the motor response (position, speed, or both) is measured using an appropriate sensor and fed back into the controller, which generates the control signal for the drive hardware of the motor. An optical encoder can be used to sense both position and speed and a tachometer may be used to measure the speed alone. The following three types of feedback control are important:

1. Velocity feedback
2. Position plus velocity feedback
3. Position feedback with a multi term controller

12.5.7.1 Velocity Feedback Control

Velocity feedback is particularly useful in controlling the motor speed. In velocity feedback, motor speed is sensed using a device such as a tachometer or an optical encoder, and is fed back to the controller, which compares it with the desired speed, and the error is used to correct the deviation. Additional dynamic compensation (e.g., lead or lag compensation) may be needed to improve the accuracy and the effectiveness of the controller, and can be provided using either analog circuits or digital processing. The error signal is passed through the compensator in order to improve the performance of the control system.

12.5.7.2 Position Plus Velocity Feedback Control

In position control, the motor angle θ_m is the output. In this case, the open-loop system has a free integrator, and the characteristic polynomial is $s(\tau s + 1)$. This is a marginally stable system. In particular, if a slight disturbance or model error is present, it will be integrated out, which can lead to a diverging error in the motor angle. In particular, the load torque T_L is an input to the system, and is not completely known. In control systems terminology, this is a disturbance (an unknown input), which can cause unstable behavior in the open-loop system. In view of the free integrator at the position output, the resulting unstable behavior cannot be corrected using velocity feedback alone. Position feedback is needed to remedy the problem. Both position and velocity feedback are needed. The feedback gains for the position and velocity signals can be chosen so as to obtain the desired response (speed of response, overshoot limit, steady-state accuracy, etc.). Block diagram of a position plus velocity feedback control system for a dc motor is shown in Figure 12.23. The motor block is given by Figure 12.21 for an armature-controlled motor, and by Figure 12.22 for a field-control motor (*Note*: Load torque input is integral in each of these two models). The

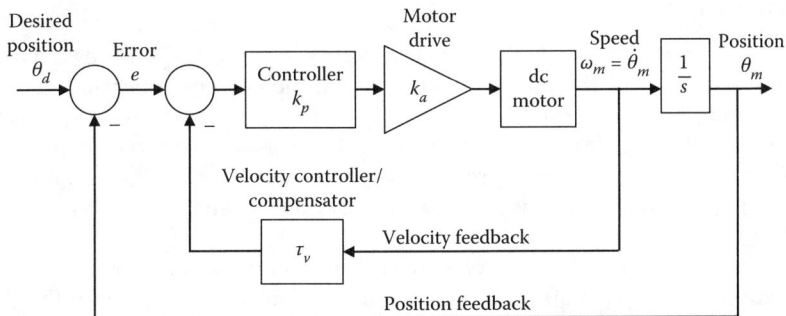

FIGURE 12.23
Position plus velocity feedback control of a dc motor.

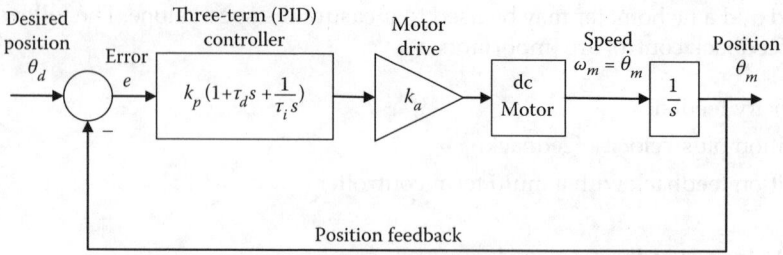

FIGURE 12.24
PID control of the position response of a dc motor.

drive unit of the motor is represented by an amplifier of gain k_a. Control system design involves selection of proper parameter values for sensors and other components in the control system.

12.5.7.3 Position Feedback with PID Control

A popular method of controlling a dc motor is to use just position feedback, and then compensate for the error using a three-term controller having the proportional, integral and derivative (PID) actions. A block diagram for this control system is shown in Figure 12.24.

In the control system of a dc motor (Figure 12.23 or 12.24), the desired position command may be provided by a potentiometer, as a voltage signal. The measurements of position and speed also are provided as voltage signals. Specifically, in the case of an optical encoder, the pulses are detected by a digital pulse counter, and read into the digital controller. This reading has to be calibrated to be consistent with the desired position command. In the case of a tachometer, the velocity reading is generated as a voltage, which has to be calibrated then, to be consistent with the desired position signal.

It is noted that proportional plus derivative control (PPD control or PD control) with position feedback, has a similar effect as position plus velocity (speed) feedback control. But, the two are not identical because the latter adds a zero to the system transfer function, requiring further considerations in the controller design, and affecting the motor response. In particular, the zero modifies the sign and the ratio in which the two response components corresponding to the two poles contribute to the overall response.

12.5.8 Motor Driver

The driver of a dc motor is a hardware unit, which generates the necessary current to energize the windings of the motor. By controlling the current generated by the driver, the motor torque can be controlled. By receiving feedback from a motion sensor (encoder, tachometer, etc.), the angular position and the speed of the motor can be controlled.

Note: When an optical encoder is provided with the motor—a typical situation—it is not necessary to use a tachometer as well, because the encoder can generate both position and speed measurements.

The drive unit primarily consists of a drive amplifier, with additional circuitry and a dc power supply. In typical applications of motion control and servoing, the drive unit is a *servoamplifier* with auxiliary hardware. The driver is commanded by a control input provided by a host computer (personal computer or PC) through an interface (I/O) card.

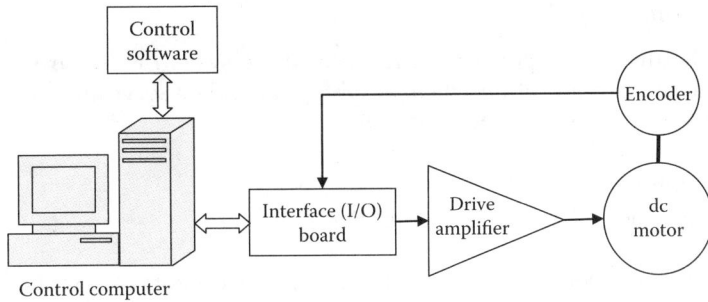

FIGURE 12.25
Components of a dc motor control system.

A suitable arrangement is shown in Figure 12.25. Also, the driver parameters (e.g., amplifier gains) are software programmable and can be set by the host computer.

The control computer receives a feedback signal of the motor motion, through the interface board, and generates a control signal, which is provided to the drive amplifier, again through the interface board. Any control scheme can be programmed (say, in C language) and implemented in the control computer. In addition to typical servo control schemes such as PID and position-plus-velocity feedback, other advanced control algorithms (e.g., optimal control techniques such as linear quadratic regulator (LQR) and linear quadratic Gaussian (LQG), adaptive control techniques such as model-referenced adaptive control, switching control techniques such as sliding-mode control, nonlinear control schemes such as feedback linearization technique (FLT), and intelligent control techniques such as fuzzy logic control) may be applied in this manner. If the computer does not has the processing power to carry out the control computations at the required speed (i.e., control bandwidth), a digital signal processor (DSP) may be incorporated into the computer. But, with modern computers, which can provide substantial computing power at low cost, DSPs are not needed in most applications.

12.5.8.1 Interface Board

The I/O card is a hardware module with associated driver software, based in a host computer (PC), and connected through its bus (ISA bus). It forms the input–output link between the motor and the controller. It can provide many (say, eight) analog signals to drive many (eight) motors, and hence termed a *multi axis* card. It follows that the digital-to-analog conversion (DAC) capability is built into the I/O card (e.g., 16 bit DAC including a sign bit, ±10 V output voltage range). Similarly, the ADC function is included in the I/O card (e.g., eight analog input channels with 16 bit ADC including a sign bit, ±10 V output voltage range). These input channels can be used for analog sensors such as tachometers, potentiometers, and strain gauges. Equally important are the encoder channels to read the pulse signals from the optical encoders mounted on the dc servomotors. Typically the encoder input channels are equal in number to the analog output channels (and the number of axes; e.g., eight). The position pulses are read using counters (e.g., 24-bit counters), and the speed is determined by the pulse rate. The rate at which the encoder pulses are counted can be quite high (e.g., 10 MHz). In addition a number of bits (e.g., 32) of digital input and output may be available through the I/O card, for use in simple digital sensing, control, and switching functions.

12.5.8.2 Drive Unit

The primary hardware component of the motor drive system is the drive amplifier. In typical motion control applications, these amplifiers are called servo amplifiers. Two types of drive amplifiers are commercially available:

1. Linear amplifier.
2. Pulse-width-modulation (PWM) amplifier.

A linear amplifier generates a voltage output, which is proportional to the control input provided to it. Since the output voltage is proportioned by dissipative means (using resistor circuitry), this is a wasteful and in efficient approach. Furthermore, fans and heat sinks have to be provided to remove the generated heat, particularly in continuous operation. To understand the inefficiency associated with a linear amplifier, suppose that the operating output range of the amplifier is 0–20 V, and that the amplifier is powered by a 20 V power supply. Under a particular operating condition, suppose that the motor is applied 10 V and draws a current of 4 A. The power used by the motor then is 10×4 W $= 40$ W. Still, the power supply provides 20 V at 5 A, thereby consuming 100 W. This means, 60 W of power is dissipated, and the efficiency is only 40%. The efficiency can be made close to 100% using modern PWM amplifiers, which are nondissipative devices depending on high-speed switching at constant voltage to control the power supplied to the motor, as discussed next.

Modern servo amplifiers use PWM to drive servomotors efficiently under variable-speed conditions, without incurring excessive power losses. Integrated microelectronic design makes them compact, accurate, and inexpensive. The components of a typical PWM drive system are shown in Figure 12.26. Other signal conditioning hardware (e.g., filters) and auxiliary components such as isolation hardware, safety devices including tripping hardware, and cooling fan are not shown in the figure. In particular, note the following components, connected in series:

1. A velocity amplifier (a differential amplifier).
2. A torque amplifier.
3. A PWM amplifier.

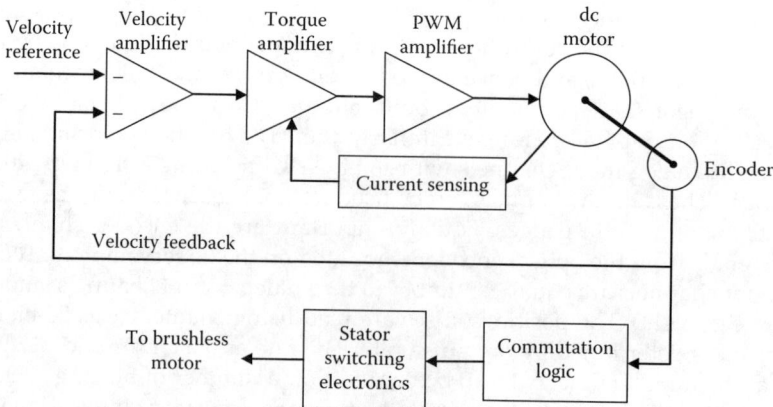

FIGURE 12.26
The main components of a PWM drive system for a dc motor.

The power can come from an ac line supply, which is rectified in the drive unit to provide the necessary dc power for the electronics. Alternatively, leads may be provided for an external power supply (e.g., 15 V dc). The reference velocity signal and the feedback signal (from an encoder or a tachometer) are connected to the input leads of the velocity amplifier. The resulting difference (error signal) is conditioned and amplified by the torque amplifier to generate a current corresponding to the required torque (corresponding to the driving speed). The motor current is sensed and fed back to this amplifier, to improve the torque performance of the motor. The output from the torque amplifier is used as the modulating signal to the PWM amplifier. The reference switching frequency of a PWM amplifier is high (on the order of 25 kHz). PWM is accomplished by varying the duty cycle of the generated pulse signal, through switching control, as explained next. The PWM signal from the amplifier (e.g., at 10 V) is used to energize the field windings of a dc motor. A brushless dc motor needs electronic commutation. This may be accomplished using the encoder signal to time the switching of the current through the stator windings.

Consider the voltage pulse signal shown in Figure 12.27. The following notation is used:

T = pulse period (i.e., interval between the successive on times)

T_o = on period (i.e., interval between on time to the next off time).

Then, the *duty cycle* is given by the percentage:

$$d = \frac{T_o}{T} \times 100\% \tag{12.39}$$

Note: In PWM, voltage level v_{ref} and the pulse frequency $1/T$ are kept fixed, and what is varied is T_o.

PWM is achieved by "chopping" the reference voltage so that the average voltage is varied. It is easy to see that, with respect to an output pulse signal, the duty cycle is given by the ratio of average output to the peak output:

$$\text{Duty cycle} = \frac{\text{Average output}}{\text{Peak output}} \times 100\% \tag{12.40}$$

Equation 12.39 or 12.40 also verifies that the average level of a PWM signal is proportional to the duty cycle (or the on time period T_o) of the signal. It follows that the output level (i.e., the average value) of a PWM signal can be varied simply by changing the signal-on time period (in the range 0 to T) or equivalently by changing the duty cycle (in the range 0–100%). This relationship between the average output and the duty cycle is linear. Hence

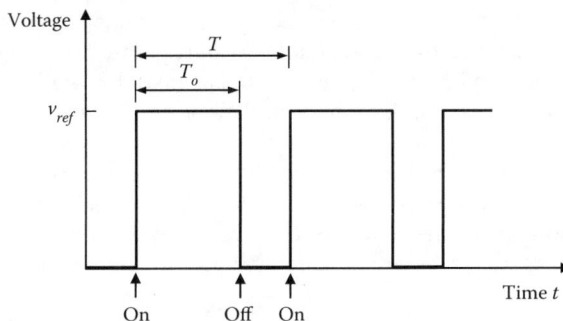

FIGURE 12.27
Duty cycle of a PWM signal.

a digital or software means of generating a PWM signal would be to use a straight line from 0 to the maximum signal level, spanning the period (T) of the signal. For a given output level, the straight line segment at this height, when projected on the time axis, gives the required on-time interval (T_o).

12.5.9 dc Motor Selection

DC motors, dc servomotors in particular, are suitable for applications requiring continuous operation (continuous duty) at high levels of torque and speed. Brushless permanent-magnet motors with advanced magnetic material provide high torque/mass ratio, and are preferred for continuous operation at high throughput (e.g., component insertion machines in the manufacture of printed-circuit boards, portioning and packaging machines, printing machines) and high speeds (e.g., conveyors, robotic arms), in hazardous environments (where spark generation from brushes would be dangerous), and in applications that need minimal maintenance and regular wash-down (e.g., in food processing applications). For applications that call for high torques and low speeds at high precision (e.g., inspection, sensing, product assembly), torque motors or regular motors with suitable speed reducers (e.g., harmonic drive, gear unit commonly using worm gears, etc.) may be employed.

A typical application involves a "rotation stage" producing rotary motion for the load. If an application requires linear (rectilinear) motions, a "linear stage" has to be used. One option is to use a rotary motor with a rotatory-to-linear motion transmission device such a lead screw or ball screw and nut, rack and pinion, or conveyor belt. This approach introduces some degree of nonlinearity and other errors (e.g., friction, backlash). For high-precision applications, linear motor provides a better alternative. The operating principle of a linear motor is similar to that of a rotary motor, except linearly moving armatures on linear bearing or guideways are used instead of rotors mounted on rotary bearings.

When selecting a dc motor for a particular application, a matching drive unit has to be chosen as well. Due consideration must be given to the requirements (specifications) of power, speed, accuracy, resolution, size, weight, and cost, when selecting a motor and a drive system. In fact vendor catalogs give the necessary information for motors and matching drive units, thereby making the selection far more convenient. Also, a suitable speed transmission device (harmonic drive, gear unit, lead screw and nut, etc.) may have to be chosen as well, depending on the application.

12.5.9.1 Motor Data and Specifications

Torque and speed are the two primary considerations in choosing a motor for a particular application. Speed-torque curves are available, in particular. The torques given in these curves are typically the maximum torques (known as peak torques), which the motor can generate at the indicated speeds. A motor should not be operated continuously at these torques (and current levels) because of the dangers of overloading, wear, and malfunction. The peak values have to be reduced (say, by 50%) in selecting a motor to match the torque requirement for continuous operation. Alternatively, the continuous torque values as given by the manufacturer should be used in the motor selection.

Motor manufacturers' data that are usually available to users include the following:

1. Mechanical data.
   - Peak torque (e.g., 65 N.m).
   - Continuous torque at zero speed or continuous stall torque (e.g., 25 N.m).

- Frictional torque (e.g., 0.4 N.m).
- Maximum acceleration at peak torque (e.g., 33×10^3 rad/s^2).
- Maximum speed or no-load speed (e.g., 3,000 rpm).
- Rated speed or speed at rated load (e.g., 2,400 rpm).
- Rated output power (e.g., 5,100 W).
- Rotor moment of inertia (e.g.,0.002 kg.m^2).
- Dimensions and weight (e.g., 14 cm diameter, 30 cm length, 20 kg).
- Allowable axial load or thrust (e.g., 230 N).
- Allowable radial load (e.g., 700 N).
- Mechanical (viscous) damping constant (e.g., 0.12 N.m/krpm).
- Mechanical time constant (e.g., 10 ms).

2. Electrical data.
   - Electrical time constant (e.g., 2 ms).
   - Torque constant (e.g., 0.9 N.m/A for peak current or 1.2 N.m/A rms current).
   - Back e.m.f. constant (e.g., 0.95 V/rad/s for peak voltage).
   - Armature/field resistance and inductance (e.g., 1.0 Ω, 2 mH).
   - Compatible drive unit data (voltage, current, etc.).

3. General data.
   - Brush life and motor life (e.g., 5×10^8 revolutions at maximum speed).
   - Operating temperature and other environmental conditions (e.g., 0–40°C).
   - Thermal resistance (e.g., 1.5°C/W).
   - Thermal time constant (e.g., 70 minutes).
   - Mounting configuration.

Quite commonly, motors and drive systems are chosen from what is commercially available. Customized production may be required, however, in highly specialized and research and development applications where the cost may not be a primary consideration. The selection process involves matching the engineering specifications for a given application with the data of commercially available motor systems.

12.5.9.2 Selection Considerations

When a specific application calls for large speed variations (e.g., speed tracking over a range of 10 dB or more), armature control is preferred. Note, however, that at low speeds (typically, half the rated speed), poor ventilation and associated temperature buildup can cause problems. At very high speeds, mechanical limitations and heating due to frictional dissipation become determining factors. For constant-speed applications, shunt-wound motors are preferred. Finer speed regulation may be achieved using a servo system with encoder or tachometer feedback or with phase-locked operation. For constant power applications, the series-wound or compound-wound motors are preferable over shunt-wound units. If the shortcomings of mechanical commutation and limited brush life are critical, brushless dc motors should be used.

A simple way to determine the operating conditions of a motor is by using its torque-speed curve, as illustrated in Figure 12.28. What is normally provided by the manufacturer

(b)

(a)

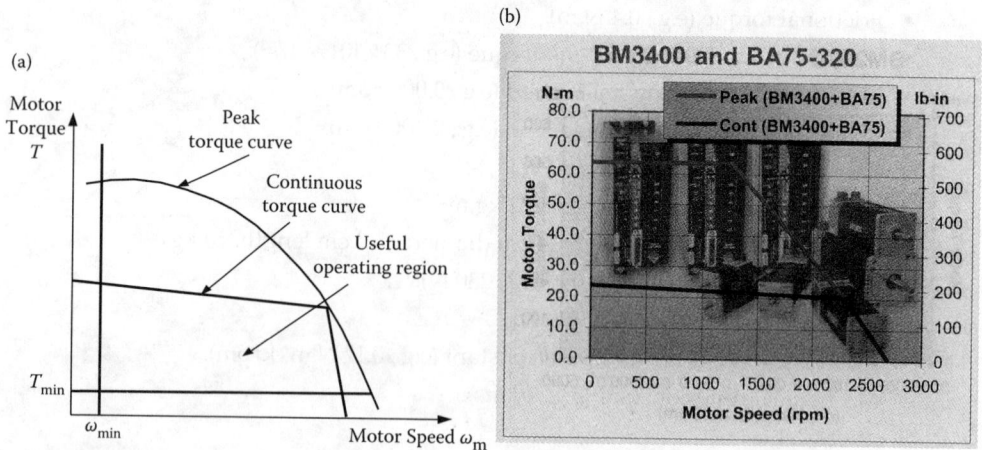

BM3400 and BA75-320

FIGURE 12.28

(a) Representation of the useful operating region for a dc motor. (b) Speed-torque characteristics of a commercial brushless dc servomotor with a matching amplifier. (From: Aerotech, Inc. With permission.)

is the peak torque curve, which gives the maximum torque the motor (with a matching drive system) can provide at a given speed, for short periods (say, 30% duty cycle). The actual selection of a motor should be based on its continuous torque, which is the torque that the motor is able to provide continuously at a given speed, for long periods without overheating or damaging the unit. If the continuous torque curve is not provided by the manufacturer, the peak torque curve should be reduced by about 50% (or even by 70%) for matching with the specified operating requirements. The minimum operating torque T_{min} is limited mainly by loading considerations. The minimum speed ω_{min} is determined primarily by operating temperature. These boundaries along with the continuous torque curve define the useful operating region of the particular motor (and its drive system), as indicated in Figure 12.28a. The optimal operating points are those that fall within this segment on the continuous torque-speed curve. The upper limit on speed may be imposed by taking into account transmission limitations in addition to the continuous torque-speed capability of the motor system.

12.5.9.3 *Motor Sizing Procedure*

Motor sizing is the term used to denote the procedure of matching a motor (and its drive system) to a load (demand of the specific application). The load may be given by a load curve, which is the speed-torque curve representing the torque requirements for operating the load at various speeds (see Figure 12.29). Clearly, greater torques are needed to drive a load at higher speeds. For a motor and a load, the acceptable operating range is the interval where the load curve overlaps with the operating region of the motor (segment AB in Figure 12.29). The optimal operating point is the point where the load curve intersects with the speed-torque curve of the motor (point A in Figure 12.29).

Sizing a dc motor is similar to sizing a stepper motor, as studied before. The same equations may be used for computing the load torque (demand). The motor characteristic (speed-torque curve) gives the available torque, as in the case of a stepper motor. The main difference is: a stepper motor is not suitable for continuous operation for long periods and

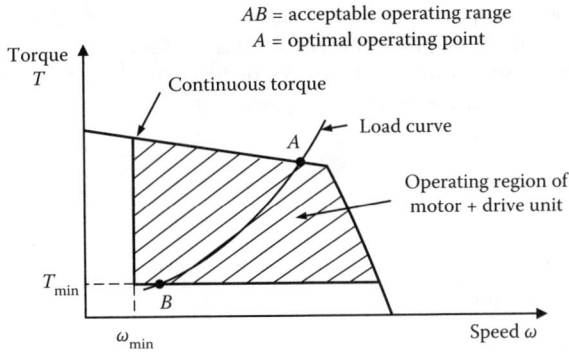

FIGURE 12.29
Sizing a motor for a given load.

at high speeds, whereas a dc motor can perform well in such situations. In this context, a dc motor can provide high torques, as given by its "peak torque curve," for short periods, and reduced torques, as given by its "continuous torque curve" for long periods of operation. In the motor sizing procedure, then, the peak torque curve may be used for short periods of acceleration and deceleration, but the continuous torque curve (or, the peak torque curve reduced by about 50%) must be used for continuous operation for long periods.

12.5.9.4 Inertia Matching

The motor rotor inertia (J_m) should not be very small compared to the load inertia (J_L). This is particularly critical in high speed and highly repetitive (high throughput) applications. Typically, for high speed applications, the value of J_L/J_m may lie in the range 5–20. For low speed applications, J_L/J_m can be as high as 100. This assumes direct drive applications.

A gear transmission may be needed between the motor and the load in order to amplify the torque available from the motor, which also reduces the speed at which the load is driven. Then, further considerations have to be made in inertia matching. In particular, neglecting the inertial and frictional loads due to gear transmission, it can be shown that best acceleration conditions for the load are possible if:

$$\frac{J_L}{J_m} = r^2 \tag{12.41}$$

where r is the step down gear ratio (i.e., motor speed/load speed). Since J_L/r^2 is the load inertia as felt at the motor rotor, the optimal condition (Equation 12.41) is when this equivalent inertia (which moves at the same acceleration as the rotor) is equal to the rotor inertia (J_m).

12.5.9.5 Drive Amplifier Selection

Usually, the commercial motors come with matching drive systems. If this is not the case, some useful sizing computations can be done to assist the process of selecting a drive amplifier. As noted before, even though the control procedure becomes linear and convenient when linear amplifiers are used, it is desirable to use PWM amplifiers in view of their high efficiency (and associated low thermal dissipation).

The required current and voltage ratings of the amplifier, for a given motor and a load, may be computed rather conveniently. The required motor torque is given by:

$$T_m = J_m\alpha + T_L + T_f \tag{12.42a}$$

where
α = highest angular acceleration needed from the motor
T_L = worst-case load torque
T_f = frictional torque on the motor
If the load is a pure inertia (J_L), Equation 12.42a becomes:

$$T_m = (J_m + J_L)\alpha + T_f \tag{12.42b}$$

The current required to generate this torque in the motor is given by:

$$i = \frac{T_m}{k_m} \tag{12.43}$$

where k_m is the torque constant of the motor.
The voltage required to drive the motor is given by:

$$v = k'_m\omega_m + Ri \tag{12.44}$$

where $k'_m = k_m$ is the back e.m.f. constant, R is the winding resistance, and ω_m is the highest operating speed of the motor in driving the load. For a PWM amplifier, the supply voltage (from a dc power supply) is computed by dividing the voltage in Equation 12.44 by the lowest duty cycle of operation.

Note 1: For "peak curve" operation, pick amplifier and power supply with these ratings (voltage, current, power).

Note 2: For "continuous curve" operation, increase the current rating proportionately.

Note 3: If several amplifiers use the same single power supply: Increase the power rating of power supply in proportion.

12.5.10 Summary of Motor Selection

- It is a component matching problem.
- Components: Load, Motor, Sensors, Drive Systems, Transmission (Gear), etc.
- Typically: Motor and sensor (e.g., encoder) come together; Motor may include a harmonic drive (Transmission); Drive system (PWM amplifier, power supply, etc.) commercially come matched to the motor.
- Typically: Match the motor to the load; Select a gear unit if necessary.
- Continuous operation has more stringent performance requirements (due to thermal problems) than peak (intermittent) operation.

Example 12.3

A load of moment of inertia J_L=0.5 kg.m² is ramped up from rest to a steady speed of 200 rpm in 0.5 s using a dc motor and a gear unit of step down speed ratio r=5. A schematic representation of the system is shown in Figure 12.30a and the speed profile of the load is shown in Figure 12.30b. The load exerts a constant resistance of T_R=55N.m throughout the operation. The efficiency of the gear unit is e=0.7. Check whether the commercial brushless dc motor with its drive unit, whose characteristics are shown in Figure 12.28b, is suitable for this application. The moment of inertia of the motor rotor is J_m=0.002kg.m²

Solution

The load equation to compute the torque required from the motor is given by:

$$T_m = \left(J_m + \frac{J_L}{er^2} \right) r\alpha + \frac{T_R}{er} \qquad (12.45)$$

where α=load acceleration, and the remaining parameters are as defined in the example. The derivation of Equation 12.45 is straightforward. From the given speed profile we have:
    Maximum load speed=200 rpm=20.94 rad/s

$$\text{Load acceleration} = \frac{20.94}{0.5} \text{rad/s}^2 = 42 \text{ rad/s}^2.$$

Substitute the numerical values in Equation 12.44, under worst-case conditions, to compute the required torque from the motor. We have:

$$T_m = \left(0.002 + \frac{0.05}{0.7 \times 5^2} \right) 5 \times 42 + \frac{55.0}{0.7 \times 5} \text{ N.m} = 1.02 + 15.71 \text{ N.m} = 16.73 \text{ N.m}$$

Under worst-case conditions, at least this much of torque would be required from the motor, operating at a speed of 200×5=1000 rpm. Note from Figure 12.28b that the load point (1000 rpm, 16.73 N.m) is sufficiently below even the continuous torque curve of the given motor (with its drive unit). Hence this motor is adequate for the task.

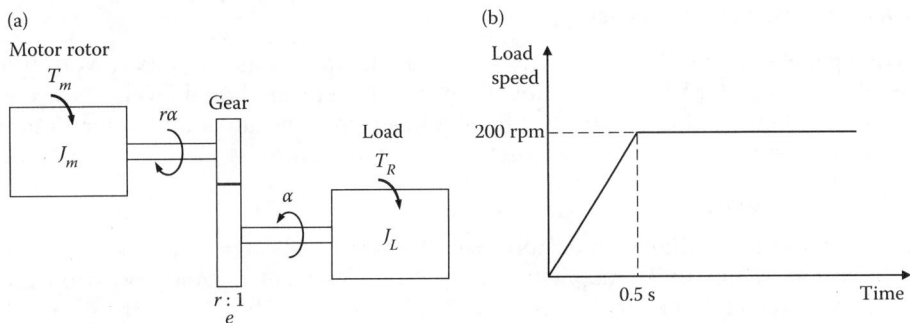

(a)   (b)

FIGURE 12.30
(a) A load driven by a dc motor through a gear transmission. (b) Speed profile of the load.

12.6 Control Experiments Using LabVIEW®

In this section two experiments in control which are used in the mandatory undergraduate course in feedback control in the Department of Mechanical Engineering, University of British Columbia, are outlined. These two experiments use the software tool LabVIEW or Laboratory Virtual Engineering Workbench, which is described in Appendix B.

12.6.1 Experiment 1: Tank Level Display

The main objective of this laboratory experiment is to employ LabVIEW®'s graphical programming tool for monitoring, data acquisition and display in a two-tank fluid flow system. Specific objectives are:

- Build a LabVIEW G program that acquires voltage signals corresponding to the liquid levels of the tanks.
- Calibrate the LabVIEW program to display the liquid levels.
- Determine the valve resistance using the experimental data acquired through the developed LabVIEW program and fundamentals of fluid flow through valves.

12.6.1.1 Procedure

The experimental setup is schematically shown in Figure 12.31. It contains two interconnected tanks that can continuously supply fluid at two different flow rates.

In the first part of experiment, a virtual instrument VI is built whose front panel and block diagram are as shown in Figures 12.32 and 12.33, respectively. The purpose of the VI is to perform the following functions:

- To continuously acquire signals from the two pressure transducers placed in the base of two tanks.
- To convert the acquired signal counts into voltages and display them.
- To convert the voltages into liquid levels, display them and monitor their time response.

12.6.1.2 Creating the Front Panel

The front panel shown in Figure 12.32 has three digital indicators for Volts A, Volts B, and Time (seconds), two tank level indicators to display the current liquid levels, a waveform chart that displays the time response of the liquid levels in the tanks, and a stop button to stop data acquisition. The steps of building the front panel are:

1. Select New VI from the LabVIEW® startup screen.
2. Create the three digital indicators: Select Digital Indicator from the Numeric Indicators (Num Inds) subpalette of the Controls palette. Drag and drop the indicator on the front panel. Type Volts A inside the label box. Using the positioning tool from the Tools palette, drag the indicator to the desired location on the front panel. Repeat the process to create the other two indicators, and label them. *Note*: Each time you create a new control or indicator, LabVIEW automatically creates the corresponding terminals in the block diagram.

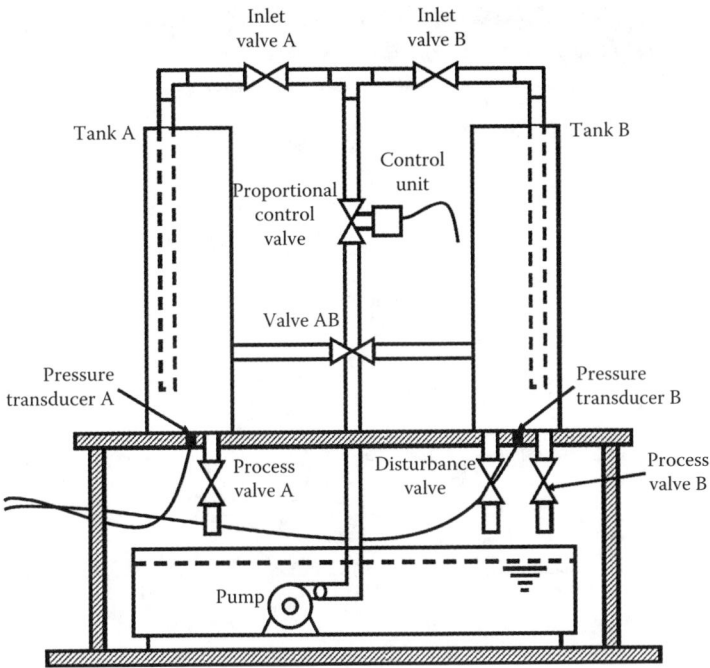

FIGURE 12.31
Schematic experimental setup for the two-tank fluid flow system.

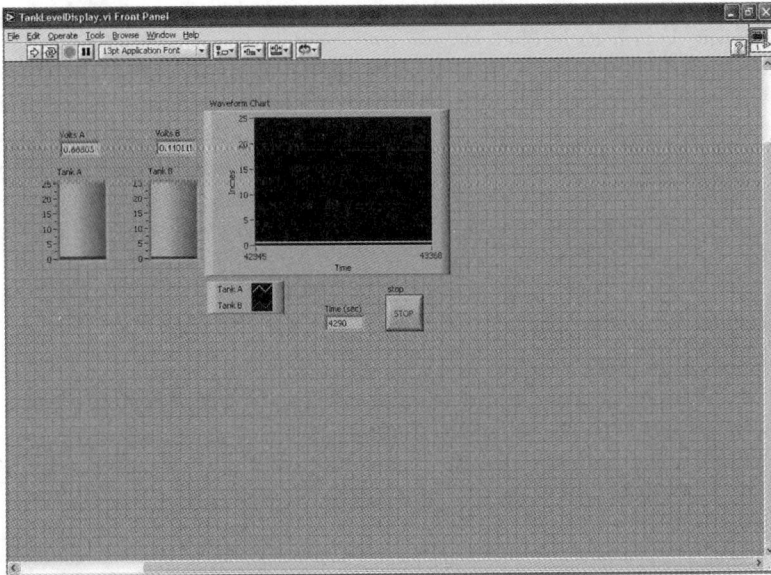

FIGURE 12.32
The front panel.

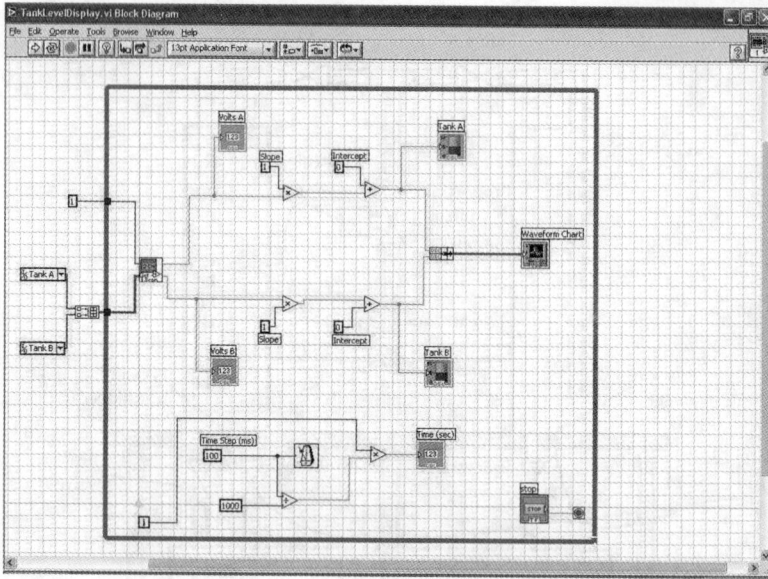

FIGURE 12.33
Block diagram.

3. Similarly, create the two tank level indicators by dragging them from the Numeric subpalette of the Controls palette. Add labels and position the tanks. You can set the upper limit of the scale of each tank to read 25.0 (inches) (*Note*: 1 inch=2.54 cm) by directly typing over the initial 10.0 upper limit.

4. Create the waveform chart by selecting it from the Graph Indicators (Graph Inds) subpalette of the Controls palette. Add the chart label, position it and set its *y*-axis upper limit. If you right-click on the graph using the mouse, a pop menu will appear, which will allow you to control the functionality and appearance; for example, showing or hiding the graph palette and the *x*-axis scale. Set the *x*-axis scale to *AutoScale X*, also select the *y*-axis scale, and make sure that there is no check mark beside *AutoScale Y*. You can also resize the legend of the chart by dragging its corners to allow for the two variables.

5. Create the stop button by selecting it from the Boolean subpalette of the Controls palette.

12.6.1.3 Creating the Block Diagram

Switch to the block diagram window and note the different terminals that LabVIEW® automatically created corresponding to the front panel objects. Use Figure 12.31 as a guide for positioning the different terminals in the diagram.

The subVI *Get Single Scan.vi* is set up to communicate with the data acquisition card installed in the computer and to perform a single scan on two specified channels to which the pressure transducers are connected. The *Get Single Scan.vi* SubVI takes the device number and the channels to scan the input. The SubVI also returns two double values corresponding to the pressure signals. The output from *Two Ch Single Scan.vi* subVI are 12-bit unsigned integers ranging from zero to 4,095 (largest 12-bit integer), which are internally converted to value in Volts.

The voltages are then converted into liquid heights using two sets of linear conversion factors each consisting of a slope and an intercept. The following steps are used to build the block diagram:

1. Insert the subVI *Get Single Scan.vi* into the block diagram by choosing *Select a VI* from the Functions palette. This subVI has two output terminals and two input terminals.

2. Create a new constant for the device input terminal and set the value to 1 by selecting the *Numeric Constant* from the Numeric subpalette of the All Functions palette.

3. Create two measurement channels by selecting the *NI Measurement* and then selecting the *Data Acquisition* subpalettes from the All Functions palette.

4. Right click on a channel and select "New Channel." Make sure that the drop down list box has Analog Input selected, and then press the next button. Type "Tank A" for channel name and leave the description blank, and then press the next button (if Tank A is already made create a Tank C for practice). Make sure that Voltage is selected on the drop down menu, and then press the next button. This ensures that the readings are Voltage measurements. It will now ask you to name the units for this measurement (optional). Insert Volts in the Units box, set the minimum and maximum values to –5 V and +5 V, respectively, and press the next button. Select "No scaling" and press the next button. It will ask what device you want to use. This allows you to select between multiple devices in systems that have more than one. Select Dev1, set "Which channel…" to 0, select the analog input mode of "Referenced single ended", and click the finish button. Repeat for "Tank B" but set the analog channel ("Which channel…") to 1.

5. Left click on the first channel and set it to "Tank A." Left click on the second channel and set it to "Tank B." You may select "Tank A" or "Tank B" on both channel tools of the block diagram even though you setup each only once. This enables the user to setup the channels just once at the beginning and have them available anywhere in the program.

6. To combine the signals from "Tank A" and "Tank B", place *Build Array* from the *Array* subpalette from the *Functions* palette. To make two input channels, right click the placed *Build Array* block and select *Add Input.*

7. Add the two sets of Volts to Inches linear conversion factors with preliminary values of 1.0, 1.0, 0.0 and 0.0 to the block diagram by selecting *Numeric Constant* from the Numeric subpalette of the All Functions palette. Label these constants as slope and intercept as shown. The values of these conversion factors may be adjusted later.

8. The waveform chart has a single input terminal. To plot two variables, they must be "bundeled" together into a cluster using the *Bundle* function available in the Cluster subpalette of the All Functions palette. Add a Bundle function to the block diagram. The Bundle function has two input terminals and one output terminal.

9. Using the wiring tool found on the Tools palette, wire the terminals of the objects on the block diagram as shown in Figure 12.31. To wire from one terminal to another, click on the first terminal, move the tool to the second terminal, and click on the second terminal. It does not matter on which terminal you initiate the wiring.

10. The current status of the block diagram may allow the VI to perform a single data acquisition operation for the tank level. To perform this operation repeatedly, the objects in the block diagram must be placed inside a While Loop found in the Structures subpalette of the Functions palette. The While Loop is a resizable box. The While Loop is placed in the block diagram by clicking with the While Loop icon in an area above and to the left of all the objects that need to be executed within the loop, and then draging out a rectangle while holding down the mouse button. The While Loop has two terminals: a Boolean input terminal (if symbol is loop executes while input is TRUE) and an optional numeric output terminal that outputs the number of times the loop has executed. *Note*: In the diagram the constants for the device and for the channels are placed outside of the while loop. This has no bearing on how the program operates. But if these are to be controls so that you can configure which channels are which only on program start, then it is vital that the controls be outside of the while loop. Otherwise during each loop the program will recheck these controls and set them to the values present at the time of activation. The stop button must to be in the while loop. Otherwise it only checks the status of the button once at startup and does not change the value during execution of the program.

11. To control the loop timing, place the Wait Until Next ms Multiple function located in the Time & Dialog subpalette of the Functions palette inside the loop. This timing control function has a single input terminal and a single optional output terminal. It waits until the millisecond timer is a multiple of the specified input value (in ms) before returning the optional millisecond timer value. Wire a Numeric Constant of 100.0 (ms) to the input terminal of the timing function.

12. To display the elapsed time since the start of the run, first convert the time step of the While Loop execution in milliseconds to seconds (i.e., divide by 1000.0), then multiply it by the number of times the loop has executed (i), and wire the result to the Time (seconds) digital indicator.

13. To stop data acquisition and exit the While loop, the stop button terminal should be wired to the Boolean input terminal of the loop. Since the default output value of the stop button is FALSE and clicking the button will change it into TRUE, the while loop by default stops on a TRUE condition. Hence wiring the button directly to the Boolean input terminal will stop the program when the stop button is pressed.

14. Select Save from the File menu to save the work on a disk or flash memory. You can run the VI by clicking on the Run button on either the front panel or block diagram Toolbars.

12.6.1.4 Calibrating the VI

If you run the present VI, the displayed tank levels will be the voltage readings and not the actual heights. Selection of the appropriate values for linear conversion factors (slope and intercept) from volts to inches, for each pressure transducer, can be accomplished as follows:

1. Run the Tank Level Display VI.
2. While Inlet Valve A is open and all other valves are closed, turn on the pump to fill Tank A with water to a known height, say 6 inches. Turn the pump off, wait

for one second, and record the corresponding voltage. (Turning the pump off also reduces the noise in the voltage reading).

3. Repeat Step 2 for water heights of 8, 10, 12, ... , 20 inches.

4. Use the eight data points to plot the height against the voltage for Tank A. The slope and intercept can then be measured graphically.

5. Close Inlet Valve A, open Inlet valve B, and repeat Steps 2–4 for Tank B.

6. Stop the Tank Level Display VI.

7. Insert the slope and intercept values into the block diagram by typing them over the preliminary values 1.0, 1.0, 0.0 and 0.0 for the conversion factors.

8. Now the Tank Level Display VI is ready to display actual tank levels.

12.6.1.5 Finding the Resistances of the Process Valves

The inside diameter of each tank is 5 inches and the outside diameters of the two pipes that run inside each tank are 1.25 inches and 0.875 inches. Use the following steps to determine the resistance of Process Valves A and B:

1. Run the Tank Level Display VI.

2. While the Inlet Valve A is open and all other valves are closed, turn on the pump to fill Tank A with water. Turn the pump off.

3. Open the Process Valve A and let the tank drain.

4. Using the time indicator on the front panel of the Tank Level Display VI, monitor the height drop from about 20 inches to about 5 inches by recording the time and the corresponding tank height using 10 s time intervals.

5. Use the linear equation for the flow rate through a constriction ($\alpha=1$) to graph the resistance of Process Valve A versus the water height in Tank A.

6. Use a similar procedure to graph the resistance of Process Valve B versus the water height in Tank B.

The graphs obtained in this manner are based on the assumption of slow laminar flow through the valve ($\alpha=1$). However, a value of $\alpha=2$ is more suitable for high flow rates and flows through short constrictions such as valves.

Assuming the operating height of water in Tank A to be $h_{A_o} = 10$ inches, use the time interval that contains this height to find the resistance of Process Valve A based on the linearized equation at this operating point. Also find a linearized value for the resistance of Process Valve B at an operating height of $h_{B_o} = 7$ inches.

To find the resistance of Valve AB at the same operating point $h_{A_o} = 10$ inches and $h_{B_o} = 7$ inches, fill Tank A to an initial height of 12 inches and Tank B to 5 inches. Open Valve AB and allow Tank A to drain into Tank B. Record the time it takes for Tank A level to drop from 10.5 inches to 9.5 inches (or Tank B height to rise from 6.5 inches to 7.5 inches). In this case, assuming the water level in Tank A is higher than that in B, the flow through Valve AB, which connects the two tanks A and B, can be written as

$$\rho A_A \frac{\Delta h_A}{\Delta t} = -\rho A_B \frac{\Delta h_B}{\Delta t} = -\frac{1}{R_{AB}} \sqrt{\rho g(h_A - h_B)} \qquad (12.46)$$

which can be linearized about the operating point $h_A = h_{A_0} + \Delta h_A$ and $h_B = h_{B_0} + \Delta h_B$ to give (see Chapter 3):

$$R_{AB} = -\frac{1}{A_A}\sqrt{\frac{g(h_{A_0} - h_{B_0})}{\rho}}\left(1 + \frac{1}{2}\frac{\Delta h_A - \Delta h_B}{h_{A_0} - h_{B_0}}\right)\frac{\Delta t}{\Delta h_A} \tag{12.47}$$

12.6.2 Experiment 2: Process Control Using LabVIEW®

The objectives of this experiment are to use LabVIEW VI to: implement and compare ON/OFF control and proportional control strategies; use these controllers to control the; and examine the time responses of the system (liquid levels in the tanks of Figure 12.31) under disturbance and step inputs.

12.6.2.1 Two-Tank System

The experimental system contains two tanks of 5 inches in diameter which are capable of holding up to 24 inches of water (Figure 12.31). The pump may be turned on or off and the control valve may be adjusted to control the flow rates.

The front panel and the block diagram of the Tank Level Control VI are shown in Figures 12.34 and 12.35, respectively. *Note*: In the block diagram the True/False case structure corresponding to the selected controller is left blank, which must be completed by you. The purpose of this VI is to continuously acquire signals from the pressure transducers at the base of the two liquid tanks and: convert the acquired readings into voltages and then into liquid heights; Select either ON/OFF control or proportional control as needed; Select an appropriate voltage (0–10 V) to control the input flow through the proportional control valve according to the desired control strategy; Convert the voltage into integer counts; Send the selected control signal to the control valve.

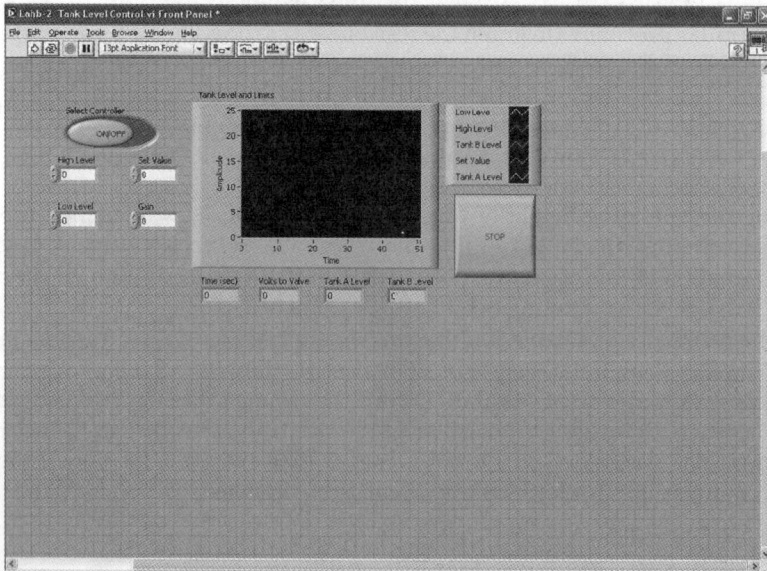

FIGURE 12.34
Front panel of the tank level control VI.

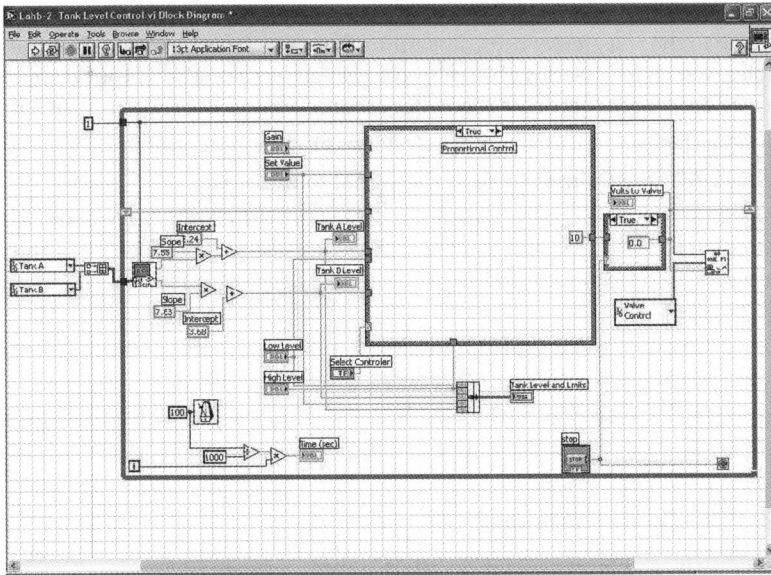

FIGURE 12.35
Block diagram of the tank level control VI.

12.6.2.2 Description of the Front Panel

The front panel of the Tank Level Control VI (see Figure 12.34) consists of the following controls and indicators:

1. A control switch to shift between ON/OFF control and proportional control.
2. Two digital controls to set the high and low limits for the ON/OFF control.
3. Two digital controls to define the set value and gain for the proportional control.
4. A waveform chart to plot the acquired tank level as well as the high limit, the low limit, and the set value.
5. Four digital indicators to show Tank A level, Tank B level, the voltage signal to the proportional control valve, and the elapsed time.
6. A stop button.

Figure 12.35 shows the block diagram with the proportional controller selected. It can switch to the ON/OFF block diagram by clicking the "True/False" switch. *Note*: The necessary instructions must be provided within the Case Structure box to implement the two control algorithms. The block diagram consists of the following functions:

1. Terminals corresponding to the control switch, digital controls, digital indicators, waveform chart and stop button on the front panel.
2. The subVI Get Single Scan.vi to acquire the signal reading from the pressure transducer.
3. Conversion factors to transform the acquired volts into liquid level units. Use the values obtained from Experiment 1.

4. A Case Structure that can allow for two different sets of instructions to be executed according to the selected control algorithm. The case structure has all the necessary variables, which are used within, wired to its left border. The output from the right border of the case structure is the voltage signal sent to the flow control valve.

5. Conversion factors to transform the voltage into integer counts.

6. The subVI A0 One Point.vi to send the signal to the valve.

7. A while loop to continuously repeat the data acquisition and control process. It has Shift Registers attached to its border to allow for the previous value of the control voltage signal to be stored and made available for use in the next iteration of the while loop.

8. A timing function to control the time between two successive While loop iterations and to calculate the elapsed time.

9. A True/False case structure that sends a 0 V signal to the valve to terminate the control process once the stop button is hit.

12.6.2.3 ON/OFF Control Algorithm

In ON/OFF control, the level should fluctuate within a range of high and low level limits supplied by the user, according to the following algorithm:

1. If the acquired tank level is between the two limits, the last valve setting should be preserved.

2. If the tank level exceeds the high limit, the valve must be completely shut off by sending a signal of 0 V.

3. If the tank level drops below the low limit, the valve is turned fully open by sending a signal of 10 V.

12.6.2.4 Proportional Control Algorithm

For proportional control, the user provides a set value for the desired tank level and an appropriate gain. The proportional control scheme is as follows:

1. The acquired tank level is first subtracted from the set value to get the level error signal.

2. The level error is multiplied by the controller gain, giving the product in units of volts.

3. The resulting voltage signal is limited to be within 0 and 10 volts. If it is less than 0, send a 0 V, and if it is higher than 10, use 10 V.

The block diagrams of the ON/OFF Control and the Proportional Control are shown in Figures 12.36 and 12.37, respectively.

12.6.2.5 Single-Tank Process Control

Water level in Tank B is controlled by controlling the flow through the proportional control valve. The water flows into Tank B through Inlet Valve B. Process Valve B should be fully

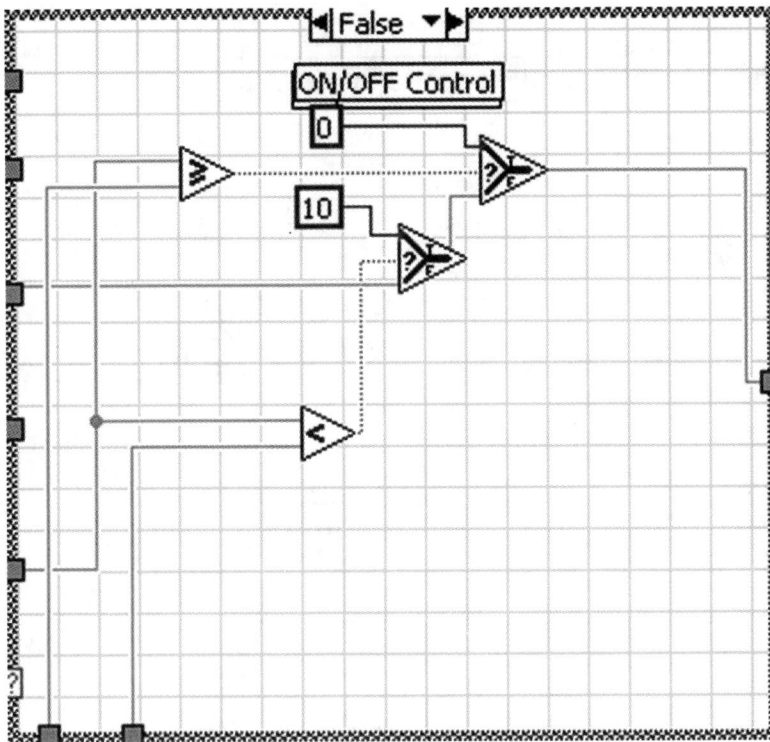

FIGURE 12.36
Block diagram of ON/OFF control.

open. Inlet Valve A and Valve AB should be closed. The operating liquid height in Tank B is to be maintained at 7 inches. Initially the Disturbance Valve should be closed.

ON/OFF Control Procedure:

1. Start with Tank B empty.
2. Turn on the pump.
3. Select ON/OFF control and set the High and Low Limits to 8 and 6 inches, respectively.
4. Run the VI.
5. Test the operation of the implemented ON/OFF algorithm (observe the voltage signal).
6. Examine and sketch the time response of the system.
7. Record the filling and draining times between the two limits.
8. Calculate the filling and draining rates (slope of the level-time curve).
9. Open the Disturbance Valve.
10. Examine and sketch the time response of the system to the disturbance.
11. Record the filling and draining times between the two limits.

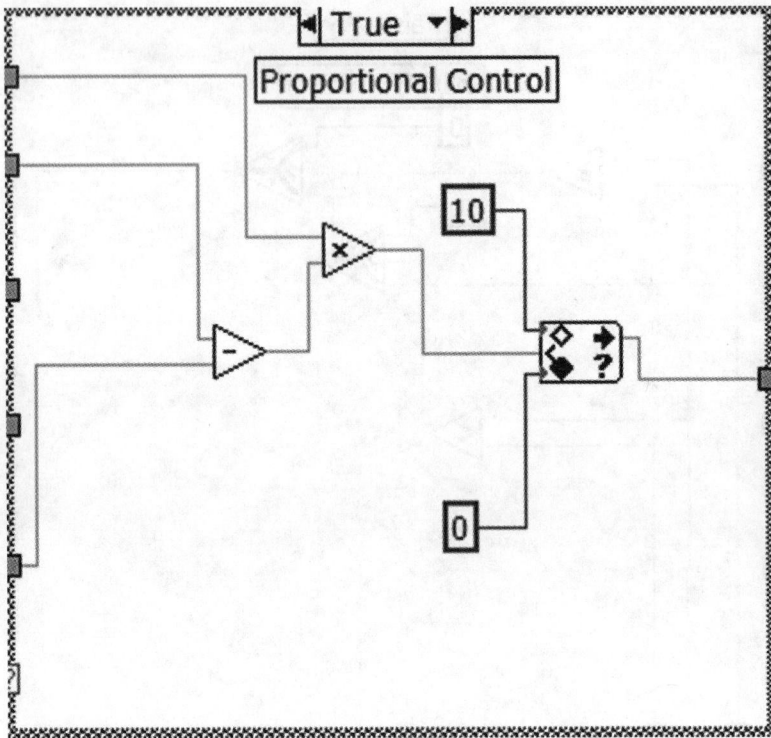

FIGURE 12.37
Block diagram of proportional control.

Proportional Control Procedure:

1. Start with Tank B empty.
2. Turn on the pump.
3. Select Proportional control with a set value of 7 inches and a gain of 1.0.
4. Run the VI.
5. Test the operation of the implemented proportional control algorithm.
6. Examine and sketch the time response of the system.
7. Record the steady-state level of the response and the corresponding voltage signal.
8. Calculate the steady-state error.
9. Open the Disturbance Valve.
10. Examine and sketch the time response of the system to the disturbance.
11. Find the steady-state error and the corresponding voltage signal under disturbance input.
12. Close the Disturbance Valve.
13. Set the gain to 3.0 and repeat Steps 6–12.

14. Set the gain to 5.0 and repeat Steps 6–12.

15. With a gain of 5.0 and the Disturbance Valve closed, increase the set value to 9 inches. Examine the response of the system to this step input.

For the ON/OFF control of the single-tank system with and without disturbance, the corresponding system time responses are shown in Figure 12.38.

For the proportional control of the single-tank system with and without disturbance, the corresponding system time responses are shown in Figures 12.39 through 12.41.

From the responses it is seen that as the gain is increased the system steady-state error decreases. Similarly, opening the disturbance valve increases the steady-state error.

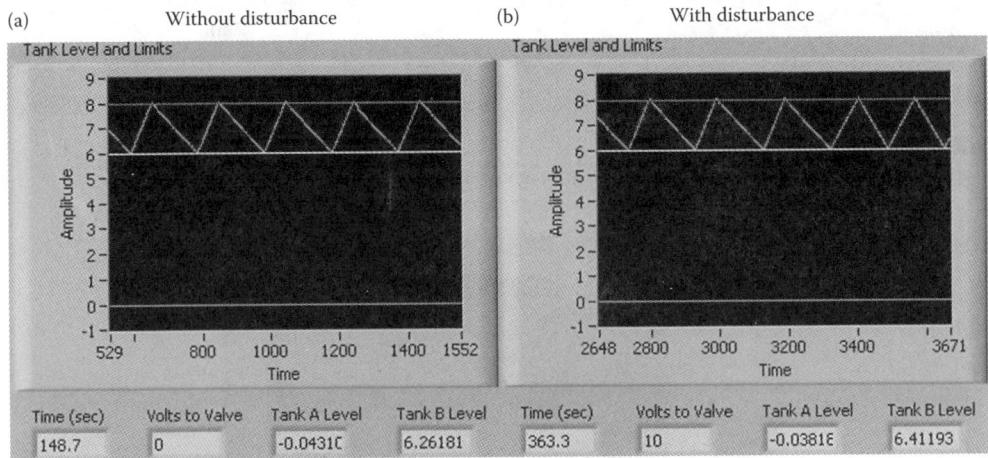

FIGURE 12.38
ON/OFF control with and without disturbance.

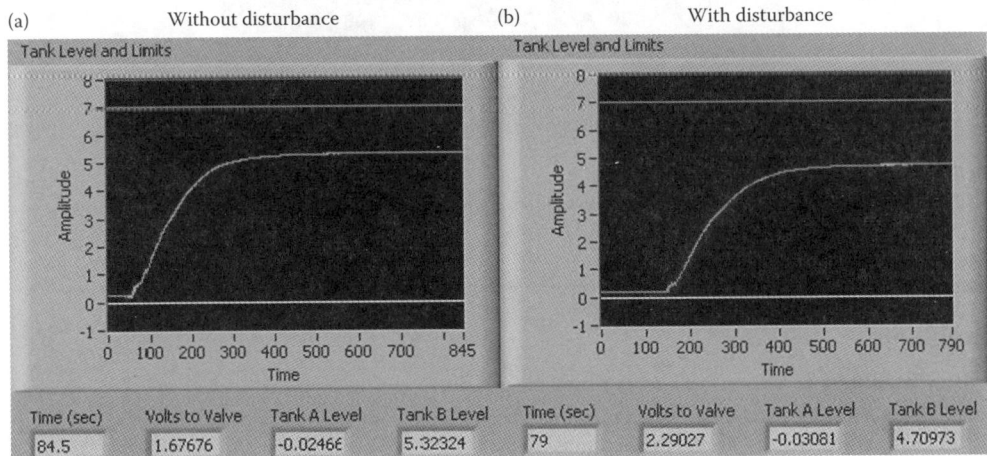

FIGURE 12.39
Proportional control with and without disturbance (gain=1.0).

(a)   Without disturbance   (b)   With disturbance

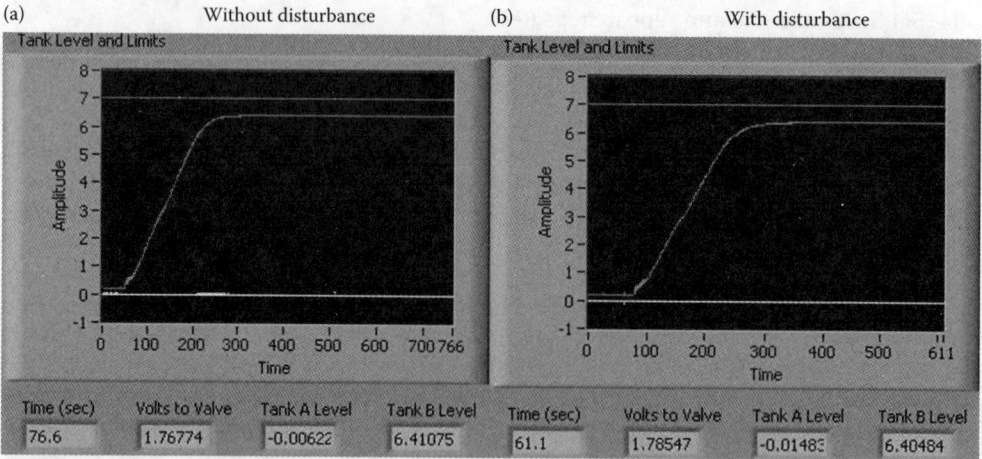

Time (sec)	Volts to Valve	Tank A Level	Tank B Level	Time (sec)	Volts to Valve	Tank A Level	Tank B Level
76.6	1.76774	−0.00622	6.41075	61.1	1.78547	−0.01483	6.40484

FIGURE 12.40
Proportional control with and without disturbance (gain=3.0).

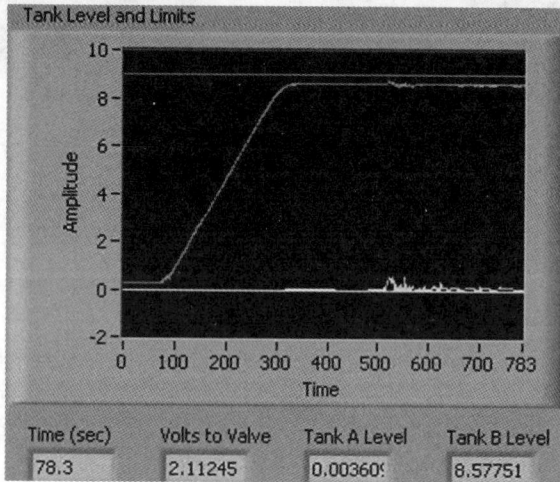

Time (sec)	Volts to Valve	Tank A Level	Tank B Level
78.3	2.11245	0.003609	8.57751

FIGURE 12.41
Proportional control without disturbance (set value=9, gain=5.0).

Problems

PROBLEM 12.1

A lag network used as the compensatory element of a control system is shown in Figure P12.1. Show that its transfer function is given by $(v_o/v_i)=Z_2/(R_1+Z_2)$ where $Z_2=R_2+(1/Cs)$.
What is the input impedance and what is the output impedance for this circuit?
Also, if two such lag circuits are cascaded as shown in Figure 2.2b, what is the overall transfer function?

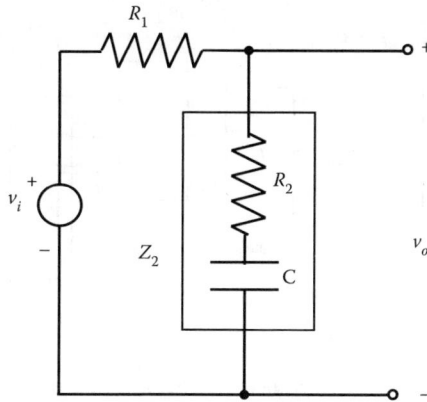

FIGURE P12.1
A single circuit module.

How would you make this transfer function become close to the ideal result: $\{Z_2/(R_1+Z_2)\}^2$

PROBLEM 12.2

Consider a dc power supply of voltage v_s and output impedance (resistance) R_s. It is used to power a load of resistance R_l, as shown in Figure P12.2. What should be the relationship between R_s and R_l if the objective is to maximize the power absorbed by the load?

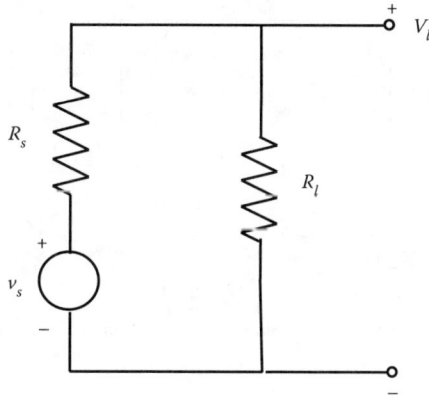

FIGURE P12.2
A load powered by a dc power supply.

PROBLEM 12.3

Consider the mechanical system where a torque source (motor) of torque T and moment of inertia J_m is used to drive a purely inertial load of moment of inertia J_L as shown in Figure P12.3a. What is the resulting angular acceleration $\ddot{\theta}$ of the system? Neglect the flexibility of the connecting shaft.

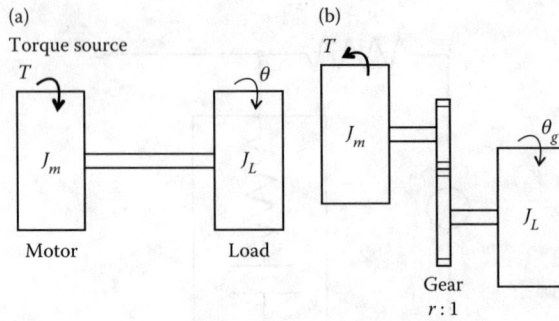

FIGURE P12.3
An inertial load driven by a motor. (a) Without gear transmission. (b) With a gear transmission.

Now suppose that the load is connected to the same torque source through an ideal (loss free) gear of motor-to-load speed ratio $r : 1$, as shown in Figure P12.3b. What is the resulting acceleration $\ddot{\theta}_g$ of the load?

Obtain an expression for the normalized load acceleration $a = \ddot{\theta}_g / \ddot{\theta}$ in terms of r and $p = J_L / J_m$. Sketch a versus r for $p = 0.1$, 1.0, and 10.0. Determine the value of r in terms of p that will maximize the load acceleration a.

Comment on the results obtained in this problem.

PROBLEM 12.4

Figure P12.4 shows a schematic diagram of a simplified signal conditioning system for an LVDT. The system variables and parameters are as indicated in the figure.

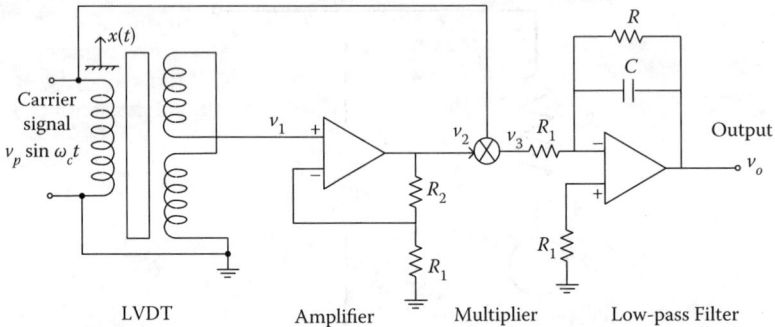

FIGURE P12.4
Signal conditioning system for an LVDT.

In particular:

$x(t)$ = displacement of the LVDT core (measurand, to be measured)

ω_c = frequency of the carrier voltage

v_o = output signal of the system (measurement).

The resistances R_1, R_2, and R, and the capacitance C are as marked. In addition, we may introduce a transformer parameter r for the LVDT, as required.

  i. Explain the functions of the various components of the system shown in Figure P12.4.
 ii. Write equations for the amplifier and filter circuits and, using them, give expressions for the voltage signals v_1, v_2, v_3, and v_o marked in Figure P12.4. *Note*: the excitation in the primary coil is $v_p \sin \omega_c t$.
iii. Suppose that the carrier frequency is $\omega_c = 500$ rad/s and the filter resistance $R = 100\text{k}\copyright$. If no more than 5% of the carrier component should pass through the filter, estimate the required value of the filter capacitance C. Also, what is the useful frequency range (measurement bandwidth) of the system in rad/s, with these parameter values?
 iv. If the displacement $x(t)$ is linearly increasing (i.e., speed is constant), sketch the signals $u(t)$, v_1, v_2, v_3, and v_o as functions of time.

PROBLEM 12.5

A dc tachometer is shown schematically in Figure P12.5a. The field windings are powered by dc voltage v_f. The across variable at the input port is the measured angular speed ω_i. The corresponding torque T_i is the through variable at the input port. The output voltage v_o of the armature circuit is the across variable at the output port. The corresponding current i_o is the through variable at the output port. The free-body diagram of the armature is shown in Figure P12.5b. Obtain a transfer-function model for

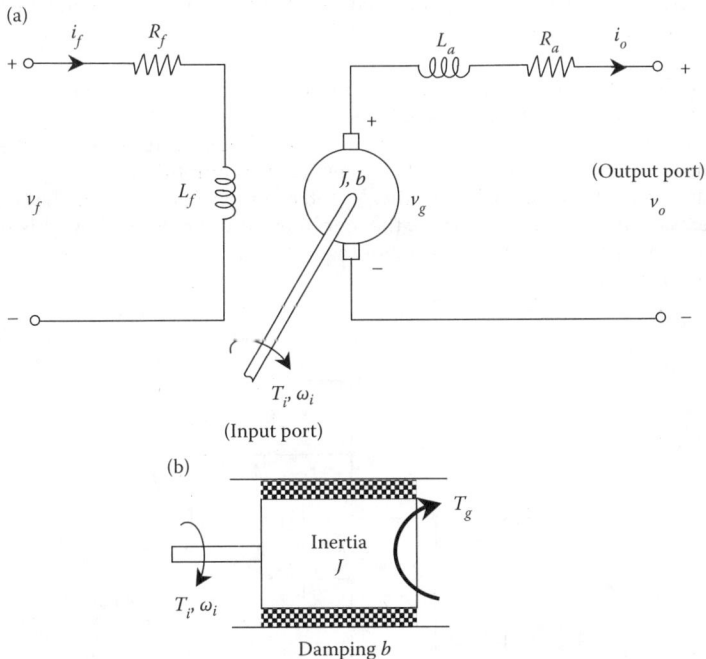

FIGURE P12.5
A dc tachometer example. (a) Equivalent circuit. (b) Armature free-body diagram.

this device. Discuss the assumptions needed to "decouple" this result into a practical input–output model for a tachometer. What are the corresponding design implications? In particular discuss the significance of the mechanical time constant and the electrical time constant of the tachometer.

PROBLEM 12.6

Applications of piezoelectric sensors are numerous; push-button devices and switches, airbag micro-electromechanical (MEMS) sensors in vehicles, pressure and force sensing, robotic tactile sensing, accelerometers, glide testing of computer disk-drive heads, excitation sensing in dynamic testing, respiration sensing in medical diagnostics, and graphics input devices for computers. Discuss advantages and disadvantages of piezoelectric sensors.

What is cross-sensitivity of a sensor? Indicate how the anisotropy of piezoelectric crystals (i.e., charge sensitivity quite large along one particular crystal axis) is useful in reducing cross-sensitivity problems in a piezoelectric sensor.

PROBLEM 12.7

As a result of advances in microelectronics, piezoelectric sensors (such as accelerometers and impedance heads) are now available in miniature form with built-in charge amplifiers in a single integral package. When such units are employed, additional signal conditioning is usually not necessary. An external power supply unit is needed, however, to provide power for the amplifier circuitry. Discuss the advantages and disadvantages of a piezoelectric sensor with built-in microelectronics for signal conditioning.

A piezoelectric accelerometer is connected to a charge amplifier. An equivalent circuit for this arrangement is shown in Figure P12.7.

a. Obtain a differential equation for the output v_o of the charge amplifier, with acceleration a as the input, in terms of the following parameters: S_a=charge sensitivity of the accelerometer (charge/acceleration); R_f=feedback resistance of the charge amplifier; τ_c= time constant of the system (charge amplifier).

b. If an acceleration pulse of magnitude a_o and duration T is applied to the accelerometer, sketch the time response of the amplifier output v_o. Show how this response varies with τ_c. Using this result, show that the larger the τ_c the more accurate the measurement.

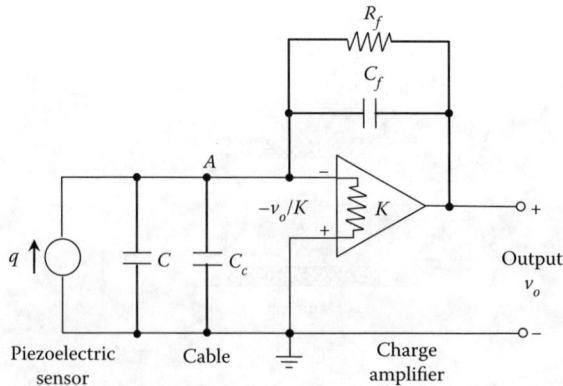

FIGURE P12.7
A piezoelectric sensor and charge amplifier combination.

PROBLEM 12.8

A positioning table uses a backlash-free high precision lead screw of lead 2 cm/rev, which is driven by a servo motor with a built-in optical encoder for feedback control. If the required positioning accuracy is ±10 μm determine the number of windows required in the encoder track. Also, what is the minimum bit size required for the digital data register/buffer of the encoder count?

PROBLEM 12.9

The rotating speed can be determined using an incremental optical encoder by two ways: For high speeds, pulse are counted over a sample period and the pulse rate is computed; For low speeds, the duration of a pulse is timed and since the angle of rotation corresponding to a pulse is a known fixed value, the speed is counted from the information.

An incremental encoder with 500 windows in its track is used for speed measurement. Suppose that:

a. in the pulse-counting method, the count (buffer) is read at the rate of 10 Hz.
b. in the pulse-timing method, a clock of frequency 10 MHz is used.

Determine the percentage resolution for each of these two methods when measuring a speed of:

i. 1 rev/s
ii. 100 rev/s.

PROBLEM 12.10

Consider the two quadrature pulse signals (say, A and B) from an incremental encoder. Using sketches of these signals, show that in one direction of rotation, signal B is at a high level during the up-transition of signal A, and in the opposite direction of rotation, signal B is at a low level during the up-transition of signal A. Note that the direction of motion can be determined in this manner, by using level detection of one signal during the up-transition of the other signal.

PROBLEM 12.11

Suppose that a feedback control system (Figure P12.11) is expected to provide an accuracy within $\pm \Delta y$ for a response variable y. Explain why the sensor that measures y should have a resolution of $\pm(\Delta y/2)$ or better for this accuracy to be possible. An x–y table has a travel of 2 m. The feedback control system is expected to provide an accuracy of ±1 mm. An optical encoder is used to measure the position for feedback in each direction (x and y). What is the minimum bit size that is required for each encoder output buffer? If the motion sensor used is an absolute encoder, how many tracks and how many sectors should be present on the encoder disk?

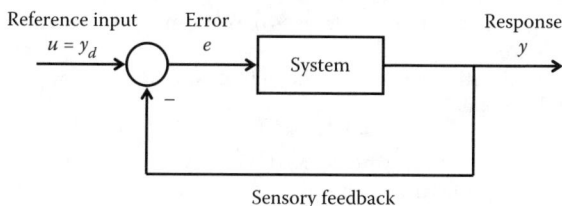

FIGURE P12.11
A feedback control loop.

PROBLEM 12.12

Compare and contrast an optical incremental encoder against a potentiometer, by giving advantages and disadvantages, for an application involving the sensing of a rotatory motion.

A schematic diagram for the servo control loop of one joint of a robotic manipulator is given in Figure P12.12.

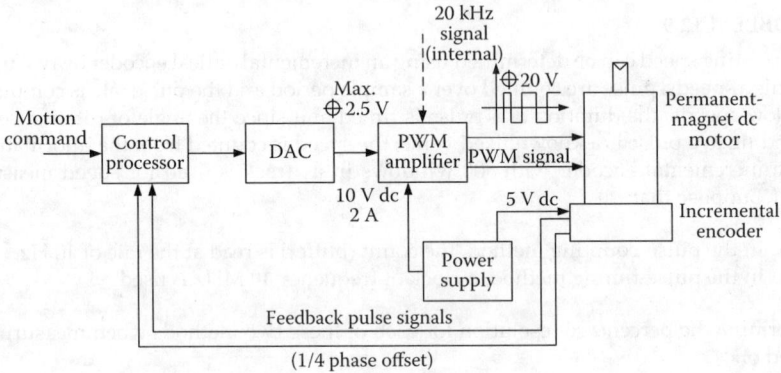

FIGURE P12.12
A servo loop of a robot.

The motion command for each joint of the robot is generated by the robot controller, in accordance with the required trajectory. An optical incremental encoder is used for both position and velocity feedback in each servo loop. Note that for a six-degree-of-freedom robot there will be six such servo loops. Describe the function of each hardware component shown in the figure and explain the operation of the servo loop.

After several months of operation the motor of one joint of the robot was found to be faulty. An enthusiastic engineer quickly replaced the motor with an identical one without realizing that the encoder of the new motor was different. In particular, the original encoder generated 200 pulses/rev whereas the new encoder generated 720 pulses/rev. When the robot was operated the engineer noticed an erratic and unstable behavior at the repaired joint. Discuss reasons for this malfunction and suggest a way to correct the situation.

PROBLEM 12.13

a. A position sensor is used in a microprocessor-based feedback control system for accurately moving the cutter blades of an automated meat-cutting machine. The machine is an integral part of the production line of a meat processing plant. What are primary considerations in selecting the position sensor for this application? Discuss advantages and disadvantages of using an optical encoder in comparison to a LVDT in this context.

b. Figure P12.13 illustrates one arrangement of the optical components in a linear incremental encoder.

The moving code plate has uniformly spaced windows as usual, and the fixed masking plate has two groups of identical windows, one above each of the two photodetectors. These two groups of fixed windows are positioned in half-pitch out of phase so that when one detector receives light from its source directly through the aligned windows

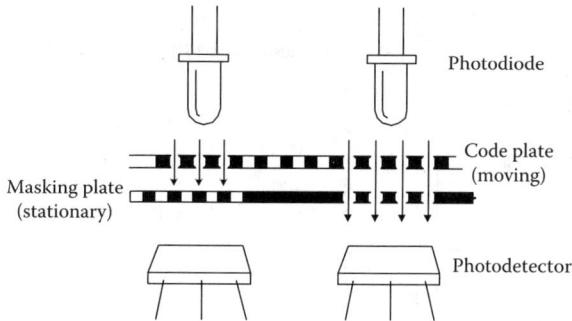

FIGURE P12.13
Photodiode-detector arrangement of a linear optical encoder.

of the two plates, the other detector has the light from its source virtually obstructed by the masking plate.

Explain the purpose of the two sets of photodiode-detector units, giving a schematic diagram of the necessary electronics. Can the direction of motion be determined with the arrangement shown in Figure P12.13? If so, explain how this could be done. If not, describe a suitable arrangement for detecting the direction of motion.

PROBLEM 12.14

Typically, when a digital transducer is employed to generate the feedback signal for an analog controller, a digital-to-analog converter (DAC) would be needed to convert the digital output from the transducer into a continuous (analog) signal. Similarly, when a digital controller is used to drive an analog process, a DAC has to be used to convert the digital output from the controller into the analog drive signal. There exist ways, however, to eliminate the need for a DAC in these two types of situations.

  a. Show how a shaft encoder and a frequency-to-voltage converter can replace an analog tachometer in an analog speed-control loop.
  b. Show how a digital controller with PWM can be employed to drive a dc motor without the use of a DAC.

PROBLEM 12.15

  a. What parameters or features determine the step angle of a stepper motor? What is microstepping? Briefly explain how microstepping is achieved.
  b. A stepper-motor-driven positioning platform is schematically shown in Figure P12.15. Suppose that the maximum travel of the platform is L and this is accomplished in a time period of Δt. A trapezoidal velocity profile is used with a region of constant speed V in between an initial region of constant acceleration from rest and a final region of constant deceleration to rest, in a symmetric manner.
  (i) Show that the acceleration is given by:

$$a = V^2/(V \cdot \Delta t - L)$$

The platform is moved using a mechanism of light, inextensible cable and a pulley, which is directly (without gears) driven by a stepper motor. The platform moves on a pair of vertical guideways that use linear bearings and, for design purposes, the associated frictional resistance to platform motion

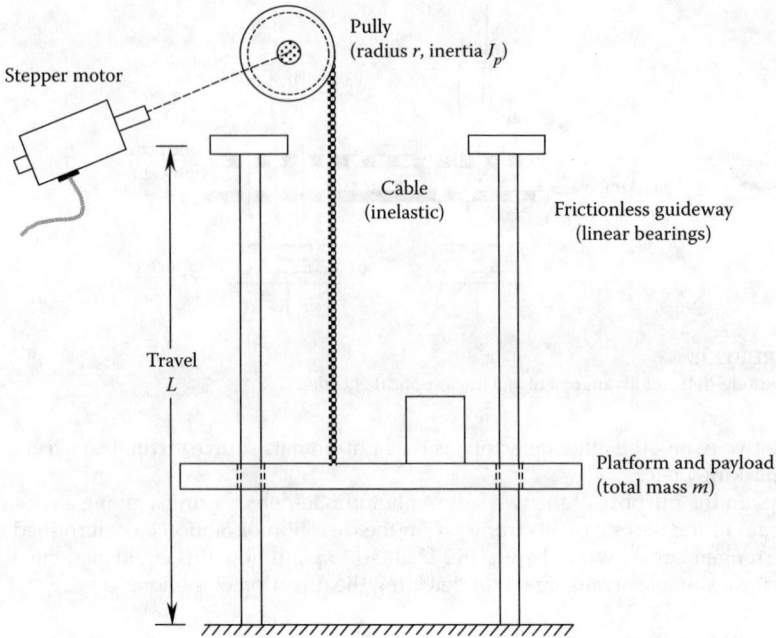

FIGURE P12.15
An automated positioning platform.

may be neglected. The frictional torque at the bearings of the pulley is not negligible, however. Suppose that:

$$\frac{\text{Frictional torque of the pulley}}{\text{Load torque on the pulley from the cable}} = e$$

Also, the following parameters are known:
J_p=moment of inertia of the pulley about the axis of rotation
r=radius of the pulley
m=equivalent mass of the platform and its payload.

(ii)   Show that the maximum operating torque that is required from the stepper motor is given by:

$$T = \left[J_m + J_p + (1+e)mr^2 \right] \frac{a}{r} + (1+e)rmg$$

in which J_m=moment of inertia of the motor rotor.

(iii)   Suppose that V=8.0 m/s, L=1.0 m, Δt =1.0 s, m=1.0 kg, J_p=3.0×10⁻⁴ kg.m², r=0.1 m, and e=0.1.

Four models of stepper motor are available, and their specifications given in Table 12.2 and Figure 12.16. Select the most appropriate motor (with the corresponding drive system) for this application. Clearly indicate all your computations and justify your choice.

(iv)   What is the position resolution of the platform, as determined by the chosen motor?

PROBLEM 12.16

An armature-controlled dc motor uses 2 hp under no-load conditions to maintain a constant speed of 600 rpm. The motor torque constant $k_m=1$ V.s, the rotor moment of inertia $J_m=0.1$ kg.m², and the armature circuit parameters are $R_a=10\ \Omega$ and $L_a=0.01$ H. Determine the electrical damping constant, the mechanical damping constant, the electrical time constant of the armature circuit, the mechanical time constant of the rotor, and the true time constants of the motor.

PROBLEM 12.17

Consider the position and velocity control system of Figure 12.23. Suppose that the motor model is given by the transfer function $k_m/(\tau_m s+1)$. Determine the closed-loop transfer function θ_m/θ_d. Next consider proportional plus derivative control system (i.e., Figure 12.24, with the integral controller removed) and the same motor model. What is the corresponding closed-loop transfer function (θ_m/θ_d)? Compare these two types of control, particularly with respect to speed of response, stability (percentage overshoot), and steady-state error.

PROBLEM 12.18

Figure P12.18 shows a schematic arrangement for driving a dc motor using a linear amplifier. The amplifier is powered by a dc power supply of regulated voltage v_s. Under a particular condition suppose that the linear amplifier drives the motor at voltage v_m and current i. Assume that the current drawn from the power supply is also i. Give an expression for the efficiency at which the linear amplifier is operating under these conditions. If $v_s=50$ V, $v_m=20$ V and $i=5$ A, estimate the efficiency of operation of the linear amplifier.

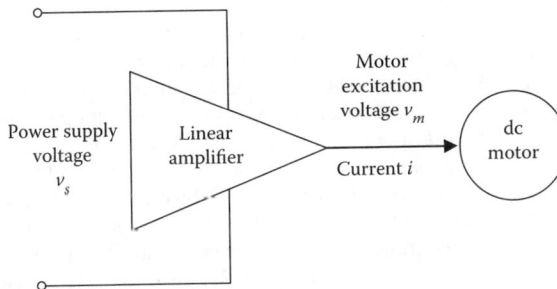

FIGURE P12.18
A linear amplifier for a dc motor.

PROBLEM 12.19

A brushless dc motor and a suitable drive unit are to be chosen for a continuous drive application. The load has a moment of inertia 0.016 kg.m², and faces a constant resisting torque of 35.0 N.m (excluding the inertia torque) throughout the operation. The application involves accelerating the load from rest to a speed of 250 rpm in 0.2 s, maintaining it at this period for extended periods, and then decelerating to rest in 0.2 s. A gear unit with step down gear ratio 4 is to be used with the motor. Estimate a suitable value for the moment of inertia of the motor rotor, for a fairly optimal design. Gear efficiency is known to be 0.8. Determine a value for continuous torque and a corresponding value for operating speed on which a selection of a motor and a drive unit can be made.

PROBLEM 12.20

You are a control engineer who has been assigned the task of designing and instrumenting a control system. In the final project report you will have to describe the steps of establishing the design/performance specifications for the system, selecting and sizing sensors, transducers, actuators, drive systems, controllers, signal conditioning and interface hardware, and software for the instrumentation and component integration of this system. Keeping this in mind, write a project proposal giving the following information:

1. Select a process (plant) as the system to be developed. Describe the plant indicating the purpose of the plant, how the plant operates, what is the system boundary (physical or imaginary), what are important inputs (e.g., voltages, torques, heat transfer rates, flow rates), response variables (e.g., displacements, velocities, temperatures, pressures, currents, voltages), and what are important plant parameters (e.g., mass, stiffness, resistance, inductance, conductivity, fluid capacity). You may use sketches.
2. Indicate the performance requirements (or, operating specifications) for the plant (i.e., how the plant should behave under control). You may use any available information on such requirements as accuracy, resolution, speed, linearity, stability, and operating bandwidth.
3. Give any constraints related to cost, size, weight, environment (e.g., operating temperature, humidity, dust-free or clean room conditions, lighting, wash-down needs), etc.
4. Indicate the type and the nature of the sensors and transducers present in the plant and what additional sensors and transducers might be needed for properly operating and controlling the system.
5. Indicate the type and the nature of the actuators and drive systems present in the plant and which of these actuators have to be controlled. If you need to add new actuators (including control actuators) and drive systems, indicate such requirements in sufficient detail.
6. Mention what types of signal modification and interfacing hardware would be needed (i.e., filters, amplifiers, modulators, demodulators, ADC, DAC, and other data acquisition and control needs). Describe the purpose of these devices. Indicate any software (e.g., driver software) that may be needed along with this hardware.
7. Indicate the nature and operation of the controllers in the system. State whether these controllers are adequate for your system. If you intend to add new controllers briefly give their nature, characteristics, objectives, etc. (e.g., analog, digital, linear, nonlinear, hardware, software, control bandwidth).
8. Describe how the users and/or operators interact with the system, and the nature of the user interface requirements (e.g., graphic user interface or GUI).

The following plants/systems may be considered:

1. A hybrid electric vehicle
2. A household robot
3. A smart camera
4. A smart airbag system for an automobile
5. Rover mobile robot for Mars exploration, developed by NASA.
6. An automated guided vehicle (AGV) for a manufacturing plant
7. A flight simulator
8. A hard disk drive for a personal computer
9. A packaging and labeling system for a grocery item
10. A vibration testing system (electrodynamic or hydraulic)
11. An active orthotic device to be worn by a person to assist a disabled or weak hand (which has some sensation, but not fully functional)

Appendix A: Transform Techniques

Many people use "transforms" without even knowing it. A "transform" is simply a number, variable, or function in a different form. For example, since $10^2 = 100$, you can use the exponent (2) to represent the number 100. Doing this for all numbers (i.e., using their exponent to the base 10), results in a "table of logarithms." One can perform mathematical computations using only "logarithms." The logarithm transforms all numbers into their exponential equivalents; a table of such transforms (i.e., a log table) enables a user to quickly transform any number into its exponent, do the computations using exponents (where, a product becomes an addition and a division becomes a subtraction), and transform the result back (i.e., inverse logarithm) into the original form. It is seen that the computations have become simpler by using logarithms, but at the cost of the time and effort for transformation and inverse transformation.

Other common transforms include the Laplace Transform, Fourier Transform, and Z-transform. In particular, the Laplace Transform provides a simple, algebraic way to solve (i.e., integrate) a linear differential equation. Most functions that we use are of the form t^n, sin ωt, or e^t, or some combination of them. Thus, in the expression

$$y = f(t)$$

the function y is quite likely a power, a sine, or an exponential function. Also, often, we have to work with derivatives and integrals of these functions, and differential equations containing these functions. These tasks can be greatly simplified by the use of the Laplace transform.

Concepts of frequency-response analysis originate from the nature of the response of a dynamic system to a sinusoidal (i.e., harmonic) excitation. These concepts can be generalized because the time-domain analysis, where the independent variable is time (t) and the frequency-domain analysis, where the independent variable is frequency (ω) are linked through the Fourier transformation. Analytically, it is more general and versatile to use the Laplace transformation, where the independent variable is the Laplace variable (s) which is complex (nonreal). This is true because analytical Laplace transforms may exist even for time functions that do not have "analytical" Fourier transforms. But with compatible definitions, the Fourier transform results can be obtained form the Laplace transform results simply by setting $s = j\omega$. In the present appendix we will formally introduce the Laplace transformation and the Fourier transformation, and will illustrate how these techniques are useful in the analysis of dynamic systems. The preference of one domain over another will depend on such factors as the nature of the excitation input, the type of the analytical model available, the time duration of interest, and the quantities that need to be determined.

A.1 Laplace Transform

The Laplace transformation relates the time domain to the *Laplace domain* (also called *s-domain* or complex frequency domain). The Laplace transform $Y(s)$ of a piecewise-continuous function or signal $y(t)$ is given, by definition, as

$$Y(s) = \int_0^\infty y(t)\exp(-st)\,dt \tag{A.1}$$

and is denoted using the Laplace operator $\mathcal{L}$, as

$$Y(s) = \mathcal{L}y(t) \tag{A.1*}$$

Here, s is a complex independent variable known as the Laplace variable, defined by

$$s = \sigma + j\omega \tag{A.2}$$

where σ is a real-valued constant that will make the transform (Equation A.1) finite, ω is simply frequency, and $j=\sqrt{-1}$. The real value (a) can be chosen sufficiently large so that the integral in Equation A.1 is finite even when the integral of the signal itself (i.e., $\int y(t)dt$) is not finite. This is the reason why, for example, Laplace transform is better behaved than Fourier transform, which will be defined later, from the analytical point of view. The symbol s can be considered to be a constant, when integrating with respect to t, in Equation A.1.

The inverse Laplace transform (i.e., obtaining y from its Laplace transform) is

$$y(t) = \frac{1}{2\pi j} \int_{\sigma - j\omega}^{\sigma + j\omega} Y(s)\exp(st)\,ds \tag{A.3}$$

and is denoted using the inverse Laplace operator $\mathcal{L}^{-1}$, as

$$y(t) = \mathcal{L}^{-1}Y(s) \tag{A.3*}$$

The integration in Equation A.3 is performed along a vertical line parallel to the imaginary (vertical) axis, located at σ from the origin in the complex Laplace plane (s-plane). For a given piecewise-continuous function $y(t)$, the Laplace transform exists if the integral in Equation A.1 converges. A sufficient condition for this is

$$\int_0^\infty |y(t)|\exp(-\sigma t)\,dt < \infty \tag{A.4}$$

Convergence is guaranteed by choosing a sufficiently large and positive σ. This property is an advantage of the Laplace transformation over the Fourier transformation.

A.1.1 Laplace Transforms of Some Common Functions

Now we determine the Laplace transform of some useful functions using Equation A.1. Usually, however, we use Laplace transform tables to obtain these results.

A.1.1.1 Laplace Transform of a Constant

Suppose our function $y(t)$ is a constant, B. Then the Laplace transform is:

$$\mathcal{L}(B) = Y(s) = \int_0^\infty Be^{-st}dt$$

$$= B\frac{e^{-st}}{-s}\bigg|_0^\infty = \frac{B}{s}$$

A.1.1.2 Laplace Transform of the Exponential

If $y(t)$ is e^{at}, its Laplace transform is

$$\mathcal{L}(e^{at}) = \int_0^\infty e^{-st}e^{at}dt$$

$$= \int_0^\infty e^{(a-s)t}dt$$

$$= \frac{1}{(a-s)}e^{(a-s)t}\bigg|_0^\infty = \frac{1}{s-a}$$

Note: If $y(t)$ is e^{-at}, it is obvious that the Laplace transform is

$$\mathcal{L}(e^{-at}) = \int_0^\infty e^{-st}e^{-at}dt$$

$$= \int_0^\infty e^{-(a+s)t}dt$$

$$= \frac{-1}{(a-s)}e^{-(a+s)t}\bigg|_0^\infty = \frac{1}{s+a}$$

This result can be obtained from the previous result simply by replacing a with $-a$.

A.1.1.3 Laplace Transform of Sine and Cosine

In the following, the letter $j = \sqrt{-1}$. If $y(t)$ is $\sin \omega t$, the Laplace transform is

$$\mathcal{L}(\sin \omega t) = \int_0^\infty e^{-st}(\sin \omega t)dt$$

Consider the identities:

$$e^{j\omega t} = \cos \omega t + j \sin \omega t$$

$$e^{-j\omega t} = \cos \omega t - j \sin \omega t$$

If we add and subtract these two equations, respectively, we obtain the expressions for the sine and the cosine in terms of $e^{j\omega t}$ and $e^{-j\omega t}$:

$$\cos \omega t = \frac{1}{2}\left(e^{j\omega t} + e^{-j\omega t}\right)$$

$$\sin \omega t = \frac{1}{2i}\left(e^{j\omega t} - e^{-j\omega t}\right)$$

$$\mathcal{L}(\cos \omega t) = \frac{1}{2}L\left(e^{j\omega t}\right) + \frac{1}{2}L\left(e^{-j\omega t}\right)$$

$$\mathcal{L}(\sin \omega t) = \frac{1}{2}L\left(e^{j\omega t}\right) - \frac{1}{2}L\left(e^{-j\omega t}\right)$$

We have just seen that

$$\mathcal{L}\left(e^{at}\right) = \frac{1}{s-a} ; \quad \mathcal{L}\left(e^{-at}\right) = \frac{1}{s+a}$$

Hence,

$$\mathcal{L}\left(e^{j\omega t}\right) = \frac{1}{s - j\omega t}; \quad \mathcal{L}\left(e^{-j\omega t}\right) = \frac{1}{s + j\omega t}$$

Substituting these expressions, we get

$$\mathcal{L}(\cos \omega t) = \frac{1}{2}\left[\frac{1}{s - j\omega}\right] + \frac{1}{2}\left[\frac{1}{s + j\omega}\right]$$

$$= \frac{1}{2}\left[\frac{s + j\omega}{s^2 - (j\omega)^2} + \frac{s - j\omega}{s^2 - (j\omega)^2}\right]$$

$$= \frac{s}{s^2 + \omega^2}$$

$$\mathcal{L}(\sin\omega t) = \frac{1}{2j}L\left(e^{j\omega t} - e^{-j\omega t}\right)$$

$$= \frac{1}{2j}\left[\frac{1}{s - j\omega}\right] - \frac{1}{2j}\left[\frac{1}{s + j\omega}\right]$$

$$= \frac{1}{2j}\left[\frac{s + j\omega}{s^2 - (j\omega)^2} + \frac{s - j\omega}{s^2 - (j\omega)^2}\right]$$

$$= \frac{1}{2j}\left[\frac{2j\omega}{s^2 + \omega^2}\right]$$

$$= \frac{\omega}{s^2 + \omega^2}$$

A.1.1.4 Laplace Transform of a Derivative

Let us transform a derivative of a function. Specifically, the derivative of a function y of t is denoted by $\dot{y} = (dy/dt)$. Its Laplace transform is given by

$$\mathcal{L}(\dot{y}) = \int_0^\infty e^{-st}\dot{y}dt = \int_0^\infty e^{-st}\frac{dy}{dt}dt \tag{A.5}$$

Now we integrate by parts, to eliminate the derivative within the integrand.
Integration by Parts: From calculus we know that $d(uv) = udv + vdu$

By integrating we get $uv = \int udv + \int vdu$

Hence,

$$\int udv = uv - vdu \tag{A.6}$$

This is known as integration by parts.
In Equation A.5, let

$$u = e^{-st} \text{ and } v = y$$

Then, $dv = dy = \dfrac{dy}{dt}dt = \dot{y}dt$

$$du = \frac{du}{dt}dt = -se^{-st}dt.$$

Substitute in Equation A.5 to integrate by parts:

$$\mathcal{L}(\dot{y}) = \int\limits_0^\infty e^{-st}dy$$

$$= \int udv = uv - \int vdu$$

$$= e^{-st}y(t)\Big|_0^\infty - \int\limits_0^\infty -se^{-st}y(t)dt$$

$$= -y(0) + s\mathcal{L}[y(t)]$$

$$= s\mathcal{L}(y) - y(0)$$

where $y(0)$=initial value of y. This says that the Laplace transform of a first derivative $\dot{y}$, equals s times the Laplace transform of the function y minus the initial value of the function (the initial condition).

Note: We can determine the Laplace transforms of the second and higher derivatives by repeated application this result, for the first derivative. For example, the transform of the second derivative is given by

$$\mathcal{L}[\ddot{y}(t)] = \mathcal{L}\left[\frac{d\dot{y}(t)}{dt}\right] = s\mathcal{L}[\dot{y}(t)] - \dot{y}(0) = s\{s\mathcal{L}[y(t)] - y(0)\} - \dot{y}(0)$$

or, $\mathcal{L}[\ddot{y}(t)] = s^2\mathcal{L}[y(t)] - sy(0) - \dot{y}(0)$

A.1.2 Table of Laplace Transforms

Table A.1 shows the Laplace Transforms of some common functions. Specifically, the table lists functions as $y(t)$, and their Laplace transforms (on the right) as $Y(s)$ or $\mathcal{L}y(t)$. If one is given a function, one can get its Laplace transform from the table. Conversely, if one is given the transform, one can get the function from the table.

Some general properties and results of the Laplace transform are given in Table A.2.

In particular, note that, with zero initial conditions, differentiation can be interpreted as multiplication by s. Also, integration can be interpreted as division by s.

A.2 Response Analysis

The Laplace transform method can be used in the response analysis of dynamic systems, mechatronic and control systems in particular. We will give examples for the approach.

TABLE A.1

Laplace Transform Pairs

$y(t) = \mathcal{L}^{-1}[Y(s)]$	$\mathcal{L}[y(t)] = Y(s)$
B	B/s
e^{-at}	$1/(s+a)$
e^{at}	$1/(s-a)$
Sinh at	$a/(s^2 - a^2)$
cosh at	$s/(s^2 - a^2)$
sin ωt	$\omega/(s^2 + \omega^2)$
cos ωt	$s/(s^2 + \omega^2)$
e^{-at} sin ωt	$\omega/((s+a)^2 + \omega^2)$
e^{-at} cos ωt	$s+a/((s+a)^2 + \omega^2)$
Ramp t	$1/s^2$
$e^{-at}(1 - at)$	$s/(s+a)^2$
$y(t)$	$Y(s)$
$(dy/dt) = \dot{y}$	$sY(s) - y(0)$
$(d^2y/dt^2) = \ddot{y}$	$s^2Y(s) - sy(0) - \dot{y}(0)$
$(d^3y/dt^3) = \dddot{y}$	$s^3Y(s) - s^2y(0) - s\dot{y}(0) - \ddot{y}(0)$
$\int_a^t y(t)dt$	$\dfrac{1}{s}Y(s) - \dfrac{1}{s}\int_0^a y(t)dt$
$af(t) + bg(t)$	$aF(s) + bG(s)$
Unit step $U(t) = 1$ for $t \geq 0$ $= 0$ otherwise	$1/s$
Delayed step $cU(t-b)$	$\dfrac{c}{s}e^{-bs}$

| Pulse
$c[U(t) - U(t-b)]$ | $c\left(\dfrac{1 - e^{-bs}}{s}\right)$ |

| Impulse function $\delta(t)$ | 1 |
| Delayed impulse
$\delta(t-b) = \dot{U}(t-b)$ | e^{-bs} |

| Sine pulse | $\left(\dfrac{\omega}{s^2 + \omega^2}\right)(1 + e^{-(\pi s/\omega)})$ |

TABLE A.2

Important Laplace Transform Relations

$\mathcal{L}^{-1}F(s) = f(t)$	$\mathcal{L}f(t) = F(s)$
$\dfrac{1}{2\pi j}\displaystyle\int_{\sigma-j\infty}^{\sigma+j\infty} F(s)\exp(st)ds$	$\displaystyle\int_{0}^{\infty} f(t)\exp(-st)dt$
$k_1 f_1(t) + k_2 f_2(t)$	$k_1 F_1(s) + k_2 F_2(s)$
$\exp(-at)\, f(t)$	$F(s+a)$
$f(t-\tau)$	$\exp(-\tau s)F(s)$
$f^{(n)}(t) = \dfrac{d^n f(t)}{dt^n}$	$s^n F(s) - s^{n-1}f(0^+) - s^{n-2}f^1(0^+)$
	$\qquad -\cdots - f^{n-1}(0^+)$
$\displaystyle\int_{-\infty}^{t} f(t)dt$	$\dfrac{F(s)}{s} + \dfrac{\displaystyle\int_{-\infty}^{0} f(t)dt}{s}$
t^n	$\dfrac{n!}{s^{n+1}}$
$t^n e^{-at}$	$\dfrac{n!}{(s+a)^{n+1}}$

Example A.1

The capacitor-charge equation of the RC circuit shown in Figure A.1 is

$$e = iR + v \tag{i}$$

$$\text{For the capacitor, } i = C\frac{dv}{dt} \tag{ii}$$

Substitute Equation (ii) in Equation (i) to get the circuit equation:

$$e = RC\frac{dv}{dt} + v \tag{iii}$$

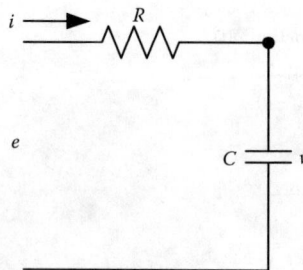

FIGURE A.1
An RC circuit with applied voltage e and voltage v across capacitor.

Take the Laplace transform of each term in Equation (iii), with all initial conditions$=0$:

$$E(s) = RCsV(s) + V(s)$$

The transfer function expressed as the output–input ratio (in the transform form) is:

$$\frac{V(s)}{E(s)} = \frac{V(s)}{sRCV(s) + V(s)} = \frac{1}{sRC + 1} = \frac{1}{\tau s + 1} \qquad \text{(iv)}$$

where $\tau = RC$.

The actual response can now be found from Table A.1 for a given input E. The first step is to get the transform into proper form (like Line 2):

$$\frac{1}{\tau s + 1} = \frac{1/\tau}{s + (1/\tau)} = \frac{a}{s + a} = a\left(\frac{1}{s + a}\right)$$

where $a = 1/\tau$. Suppose that input (excitation) e is a unit impulse. Its Laplace transform (see Table A.1) is $E = 1$. Then from Equation (iv),

$$V(s) = \frac{1}{\tau s + 1}$$

From Line 2 of Table A.1, the response is

$$v = ae^{-at} = \frac{1}{\tau}e^{-t/\tau} = \frac{1}{RC}e^{-t/RC}$$

A common transfer function for an overdamped second-order system (e.g., one with two RC circuit components of Figure A.1) would be

$$\frac{V(s)}{E(s)} = \frac{1}{(1 + \tau_1 s)(1 + \tau_2 s)}$$

This can be expressed as "partial fractions" in the from

$$\frac{A}{1 + \tau_1 s} + \frac{B}{1 + \tau_2 s}$$

and solved in the usual manner.

Example A.2

The transfer function of a thermal system is given by

$$G(s) = \frac{2}{(s + 1)(s + 3)}$$

If a unit step input is applied to the system, with zero initial conditions, what is the resulting response?

Solution

$$\text{Input } U(s) = \frac{1}{s} \text{ (for a unit step)}$$

$$\text{Since } \frac{Y(s)}{U(s)} = \frac{2}{(s+1)(s+3)}$$

the output (response)

$$Y(s) = \frac{2}{s(s+1)(s+3)}$$

Its inverse Laplace transform gives the time response. For this, first convert the expression into partial fractions as

$$\frac{2}{s(s+1)(s+3)} = \frac{A}{s} + \frac{B}{(s+1)} + \frac{C}{(s+3)} \qquad (i)$$

The unknown A is determined by multiplying Equation (i) throughout by s and then setting $s=0$. We get

$$A = \frac{2}{(0+1)(0+3)} = \frac{2}{3}$$

Similarly, B is obtained by multiplying Equation (i) throughout by $(s+1)$ and then setting $s=-1$. We get

$$B = \frac{2}{(-1)(-1+3)} = -1$$

Next, C is obtained by multiplying Equation (i) throughout by $(s+3)$ and then setting $s=-3$. We get

$$C = \frac{2}{(-3)(-3+1)} = \frac{1}{3}$$

Hence,

$$Y(s) = \frac{2}{3s} - \frac{1}{(s+1)} + \frac{1}{3(s+3)}$$

Take the inverse transform using Line 2 of Table A.1.

$$y(t) = \frac{2}{3} - e^{-t} + \frac{1}{3}e^{-3t}$$

Example A.3

The transfer function of a damped simple oscillator is known to be of the form

$$\frac{Y(s)}{U(s)} = \frac{\omega_n^2}{\left(s^2 + 2\zeta\omega_n s + \omega_n^2\right)}$$

where ω_n=undamped natural frequency; ζ=damping ratio.
   Suppose that a unit step input (i.e., $U(s) = (1/s)$) is applied to the system. Using Laplace transform tables determine the resulting response, with zero initial conditions.

Solution

$$Y(s) = \frac{1}{s} \cdot \frac{\omega_n^2}{\left(s^2 + 2\zeta\omega_n s + \omega_n^2\right)}$$

The corresponding partial fractions are of the form

$$Y(s) = \frac{A}{s} + \frac{Bs + C}{\left(s^2 + 2\zeta\omega_n s + \omega_n^2\right)} = \frac{\omega_n^2}{s\left(s^2 + 2\zeta\omega_n s + \omega_n^2\right)} \tag{i}$$

We need to determine A, B, and C.
Multiply Equation (i) throughout by s and set $s=0$. We get

$$A = 1$$

Next note that the roots of the characteristic equation

$$s^2 + 2\zeta\omega_n s + \omega_n^2 = 0$$

are

$$s = -\zeta\omega_n \pm \sqrt{\zeta^2 - 1}\,\omega_n = -\zeta\omega_n \pm j\omega_d$$

These are the poles of the system and are complex conjugates. Two equations for B and C are obtained by multiplying Equation (i) by $s + \zeta\omega_n - \sqrt{\zeta^2 - 1}\,\omega_n$ and setting $s = -\zeta\omega_n + \sqrt{\zeta^2 - 1}\,\omega_n$, and by multiplying Equation (i) by $s + \zeta\omega_n + \sqrt{\zeta^2 - 1}\,\omega_n$ and setting $s = -\zeta\omega_n - \sqrt{\zeta^2 - 1}\,\omega_n$. We obtain $B=-1$ and $C=-2\zeta\omega_n$. Consequently,

$$Y(s) = \frac{1}{s} - \frac{s + 2\zeta\omega_n}{\left(s^2 + 2\zeta\omega_n s + \omega_n^2\right)}$$

$$= \frac{1}{s} - \frac{s + \zeta\omega_n}{\left[\left(s + \zeta\omega_n\right)^2 + \omega_d^2\right]} - \frac{\zeta}{\sqrt{1 - \zeta^2}} \cdot \frac{\omega_d}{\left[\left(s + \zeta\omega_n\right)^2 + \omega_d^2\right]}$$

where $\omega_d = \sqrt{1-\zeta^2}\,\omega_n = $ damped natural frequency.

Now use Table A.1 to obtain the inverse Laplace transform:

$$y_{step}(t) = 1 - e^{-\zeta\omega_n t}\cos\omega_d t - \frac{\zeta}{\sqrt{1-\zeta^2}}e^{-\zeta\omega_n t}\sin\omega_d t$$

$$= 1 - \frac{e^{-\zeta\omega_n t}}{\sqrt{1-\zeta^2}}\left[\sin\phi\cos\omega_d t + \cos\phi\sin\omega_d t\right]$$

$$= 1 - \frac{e^{-\zeta\omega_n t}}{\sqrt{1-\zeta^2}}\sin(\omega_d t + \phi)$$

where $\cos\phi = \zeta = $ damping ratio; $\sin\phi = \sqrt{1-\zeta^2}$.

Example A.4

The open-loop response of a plant to a unit impulse input, with zero initial conditions, was found to be $2e^{-t}\sin t$. What is the transfer function of the plant?

Solution

By linearity, since a unit impulse is the derivative of a unit step, the response to a unit impulse is given by the derivative of the result given in the previous example; thus

$$y_{impulse}(t) = \frac{\zeta\omega_n}{\sqrt{1-\zeta^2}}e^{-\zeta\omega_n t}\sin(\omega_d t + \phi) - \frac{\omega_d}{\sqrt{1-\zeta^2}}e^{-\zeta\omega_n t}\cos(\omega_d t + \phi)$$

$$= \frac{\omega_n}{\sqrt{1-\zeta^2}}e^{-\zeta\omega_n t}\left[\cos\phi\sin(\omega_d t + \phi) - \sin\phi\cos(\omega_d t + \phi)\right]$$

or

$$y_{impulse}(t) = \frac{\omega_n}{\sqrt{1-\zeta^2}}e^{-\zeta\omega_n t}\sin\omega_d t$$

Compare this with the given expression. We have

$$\frac{\omega_n}{\sqrt{1-\zeta^2}} = 2;\ \zeta\omega_n = 1;\ \omega_d = 1$$

But,

$$\omega_n^2 = (\zeta\omega_n)^2 + \omega_d^2 = 1+1 = 2$$

Hence

$$\omega_n = \sqrt{2}$$

Hence

$$\zeta = \frac{1}{\sqrt{2}}$$

The system transfer function is:

$$\frac{\omega_n^2}{\left(s^2 + 2\zeta\omega_n s + \omega_n^2\right)} = \frac{2}{s^2 + 2s + 2}$$

Example A.5

Express the Laplace transformed expression

$$X(s) = \frac{s^3 + 5s^2 + 9s + 7}{(s+1)(s+2)}$$

as partial fractions. From the result, determine the inverse Laplace function $x(t)$.

Solution

$$X(s) = s + 2 + \frac{2}{s+1} - \frac{1}{s+2}$$

From Table A.1, we get the inverse Laplace transform

$$x(t) = \frac{d}{dt}\delta(t) + 2\delta(t) + 2e^{-t} - e^{-2t}$$

where $\delta(t)$ = unit impulse function.

A.3 Transfer Function

By the use of Laplace transformation, a *convolution integral* equation can be converted into an algebraic relationship. To illustrate this, consider the convolution integral which gives the response $y(t)$ of a dynamic system to an excitation input $u(t)$, with zero initial conditions. By definition Equation A.1, its Laplace transform, is written as

$$Y(s) = \int_0^\infty \int_0^\infty h(\tau)u(t-\tau)d\tau \exp(-st)dt \tag{A.7}$$

Note that $h(t)$ is the *impulse-response function* of the system. Since the integration with respect to t is performed while keeping τ constant, we have $dt = d(t - \tau)$. Consequently,

$$Y(s) = \int_{-\tau}^{\infty} u(t - \tau) \exp[-s(t - \tau)] \, d(t - \tau) \int_{0}^{\infty} h(\tau) \exp(-s\tau) d\tau$$

The lower limit of the first integration can be made equal to zero, in view of the fact that $u(t) = 0$ for $t < 0$. Again, by using the definition of Laplace transformation, the foregoing relation can be expressed as

$$Y(s) = H(s)U(s) \tag{A.8}$$

in which

$$H(s) = \mathcal{L}h(t) = \int_{0}^{\infty} h(t) \exp(-st) dt \tag{A.9}$$

Note that, by definition, the transfer function of a system, denoted by $H(s)$, is given by Equation A.8. More specifically, system transfer function is given by the ratio of the Laplace-transformed output and the Laplace-transformed input, with zero initial conditions. In view of Equation A.9, it is clear that the system transfer function can be expressed as the Laplace transform of the impulse-response function of the system. Transfer function of a linear and constant-parameter system is a unique function that completely represents the system. A physically realizable, linear, constant-parameter system possesses a unique transfer function, even if the Laplace transforms of a particular input and the corresponding output do not exist. This is clear from the fact that the transfer function is a system model and does not depend on the system input itself.

Note: The transfer function is also commonly denoted by $G(s)$. But in the present context we use $H(s)$ in view of its relation to $h(t)$.

Consider the nth order linear, constant-parameter dynamic system given by

$$a_n \frac{d^n y}{dt^n} + a_{n-1} \frac{d^{n-1} y}{dt^{n-1}} + \cdots + a_0 y = b_0 u + b_1 \frac{du(t)}{dt} + \cdots + b_m \frac{d^m u(t)}{dt^m} \tag{A.10}$$

For a physically realizable system, $m \leq n$. By applying Laplace transformation and then integrating by parts, it may be verified that

$$L \frac{d^k f(t)}{dt^k} = s^k \hat{F}(s) - s^{k-1} f(0) - s^{k-2} \frac{df(0)}{dt} - \cdots + \frac{d^{k-1} f(0)}{dt^{k-1}} \tag{A.11}$$

By definition, the initial conditions are set to zero in obtaining the transfer function. This results in

$$H(s) = \frac{b_0 + b_1 s + \cdots + b_m s^m}{a_0 + a_1 s + \cdots a_n s^n} \tag{A.12}$$

for $m \le n$. Note that Equation A.12 contains all the information that is contained in Equation A.10. Consequently, transfer function is an analytical model of a system. The transfer function may be employed to determine the total response of a system for a given input, even though it is defined in terms of the response under zero initial conditions. This is quite logical because the analytical model of a system is independent of the initial conditions of the system.

A.4 Fourier Transform

The Fourier transform $Y(f)$ of a signal $y(t)$ relates the time domain to the frequency domain. Specifically,

$$Y(f) = \int_{-\infty}^{+\infty} y(t)\exp(-j2\pi ft)dt$$

(A.13)

$$= \int_{-\infty}^{+\infty} y(t)e^{-\omega t}dt$$

Using the Fourier operator "$\mathcal{F}$" terminology:

$$Y(f) = \mathcal{F} \, y(t)$$

(A.14)

Note that if $y(t)=0$ for $t < 0$, as in the conventional definition of system excitations and responses, the Fourier transform is obtained from the Laplace transform by simply changing the variable according to $s=j2\pi f$ or $s=j\omega$. The Fourier is a special case of the Laplace, where, in Equation A.2, $\sigma=0$:

$$Y(f) = Y(s)\big|_{s=j2\pi f}$$

(A.15)

or

$$Y(\omega) = Y(s)\big|_{s=j\omega}$$

(A.16)

The (complex) function $Y(f)$ is also termed the (continuous) *Fourier spectrum* of the (real) signal $y(t)$. The inverse transform is given by:

$$y(t) = \int_{-\infty}^{+\infty} Y(f)\exp(j2\pi ft)df$$

(A.17)

or, $y(t) = \mathcal{F}^{-1}Y(f)$

Note that according to the definition given by Equation A.13, the Fourier spectrum $Y(f)$ is defined for the entire frequency range $f(-\infty, +\infty)$ which includes negative values. This is termed the *two-sided spectrum*. Since, in practical applications it is not possible to have "negative frequencies," the *one-sided spectrum* is usually defined only for the frequency range $f(0, \infty)$.

In order that a two-sided spectrum have the same amount of *power* as a one-sided spectrum, it is necessary to make the one-sided spectrum double the two-sided spectrum for $f>0$.

If the signal is not sufficiently transient (fast-decaying or damped), the infinite integral given by Equation A.13 might not exist, but the corresponding Laplace transform might still exist.

A.4.1 Frequency-Response Function (Frequency Transfer Function)

The Fourier integral transform of the impulse-response function is given by

$$H(f) = \int_{-\infty}^{\infty} h(t)\exp(-j2\pi ft)dt \tag{A.18}$$

where f is the *cyclic frequency* (measured in cycles/s or Hertz). This is known as the frequency-response function (or, frequency transfer function) of a system. Fourier transform operation is denoted as $\mathcal{F}h(t)=H(f)$. In view of the fact that $h(t)=0$ for $t<0$, the lower limit of integration in Equation A.18 could be made zero. Then, from Equation A.9, it is clear that $H(f)$ is obtained simply by setting $s=j2\pi f$ in $H(s)$. Hence, strictly speaking, we should use the notation $H(j2\pi f)$ and not $H(f)$. But for the notational simplicity we denote $H(j2\pi f)$ by $H(f)$. Furthermore, since the angular frequency $\omega=2\pi f$, we can express the frequency response function by $H(j\omega)$, or simply by $H(\omega)$ for the notational convenience. It should be noted that the frequency-response function, like the (Laplace) transfer function, is a complete representation of a linear, constant-parameter system. In view of the fact that both $u(t)=0$ and $y(t)=0$ for $t<0$, we can write the Fourier transforms of the input and the output of a system directly by setting $s=j2\pi f=j\omega$ in the corresponding Laplace transforms.

Then, from Equation A.8, we have

$$Y(f)=H(f)U(f) \tag{A.19}$$

Note: Sometimes for notational convenience, the same lowercase letters are used to represent the Laplace and Fourier transforms as well as the original time-domain variables.

If the Fourier integral transform of a function exists, then its Laplace transform also exists. The converse is not generally true, however, because of poor convergence of the Fourier integral in comparison to the Laplace integral. This arises from the fact that the factor $\exp(-\sigma t)$ is not present in the Fourier integral. For a physically realizable, linear, constant-parameter system, $H(f)$ exists even if $U(f)$ and $Y(f)$ do not exist for a particular input. The experimental determination of $H(f)$, however, requires system stability. For the nth order system given by Equation A.10, the frequency-response function is determined by setting $s=j2\pi f$ in Equation A.12 as

$$H(f) = \frac{b_0 + b_1 j2\pi f + \cdots + b_m(j2\pi f)^m}{a_0 + a_1 j2\pi f + \cdots + a_n(j2\pi f)^n} \tag{A.20}$$

This, generally, is a complex function of f, which has a magnitude denoted by $|H(f)|$ and a phase angle denoted by $\angle H(f)$.

A.5 The s-Plane

We have noted that the Laplace variable s is a complex variable, with a real part and an imaginary part. Hence, to represent it we will need two axes at right angles to each other—the real axis and the imaginary axis. These two axes from a plane, which is called the s-plane. Any general value of s (or, any variation or trace of s) may be marked on the s-plane.

A.5.1 An Interpretation of Laplace and Fourier Transforms

In the Laplace transformation of a function $f(t)$ we multiply the function by e^{-st} and integrate with respect to t. This process may be interpreted as determining the "components" $F(s)$ of $f(t)$ in the "direction" e^{-st} where s is a complex variable. All such components $F(s)$ should be equivalent to the original function $f(t)$.

In the Fourier transformation of $f(t)$ we multiply it by $e^{-j\omega t}$ and integrate with respect to t. This is the same as setting $s=j\omega$. Hence, the Fourier transform of $f(t)$ is $F(j\omega)$. Furthermore, $F(j\omega)$ represents the components of $f(t)$ that are in the direction of $e^{-j\omega t}$. Since $e^{-j\omega t} = \cos \omega t - j \sin \omega t$, in the Fourier transformation what we do is to determine the sinusoidal components of frequency ω, of a time function $f(t)$. Since s is complex $F(s)$ is also complex and so is $F(j\omega)$. Hence they all will have a real part and an imaginary part.

A.5.2 Application in Circuit Analysis

The fact that $\sin \omega t$ and $\cos \omega t$ are 90° out of phase is further confirmed in view of

$$e^{j\omega t} = \cos \omega t + j \sin \omega t \tag{A.21}$$

Consider the RLC circuit shown in Figure A.2. For the capacitor, the current (i) and the voltage (v) are related through

$$i = C \frac{dv}{dt} \tag{A.22}$$

If the voltage $v=v_0 \sin \omega t$, the current $i=v_0\omega C \cos \omega t$. Note that the magnitude of v/i is $1/\omega C$ (or, $1/2\pi fC$ where $\omega=2\pi f$; f is the cyclic frequency and ω is the angular frequency). But v and i are out of phase by 90°. In fact, in the case of a capacitor, i leads v by 90°. The equivalent circuit resistance of a capacitance is called *reactance*, and is given by

$$X_C = \frac{1}{2\pi fC} \tag{A.23}$$

$$= \frac{1}{\omega C} \tag{A.24}$$

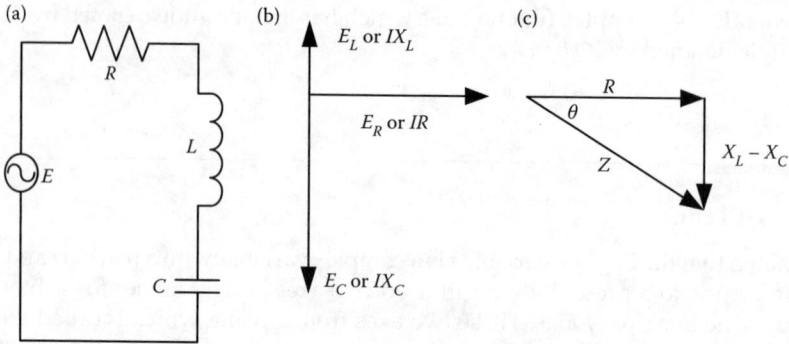

FIGURE A.2
(a) Series *RLC* circuit. (b) Phases of voltage drops. (c) Impedance triangle.

Note that this parameter changes with the frequency.

We cannot add the reactance of the capacitor and the resistance of the resistor algebraically; we must add them vectorialy because the voltages across a capacitor and resistor in series are not in phase, unlike in the case of a resistor. Also, the resistance in a resistor does not change with frequency. In a series circuit, as in Figure A.2, the current is identical in each element, but the voltages differ in both amplitude and phase; in a parallel circuit, the voltages are identical, but the currents differ in amplitude and phase.

Similarly, for an inductor

$$v = L\frac{di}{dt}$$ (A.25)

The corresponding reactance is

$$X_L = \omega L = 2\pi f L$$ (A.26)

If the voltage (*E*) across *R* in Figure A.2a is in the direction shown in Figure A.2b (i.e., pointing to the right), then the voltage across the inductor *L* must point upwards (90° leading) and the voltage across the capacitor *C* must point down (90° lagging). Since the current (*I*) is identical in each component of a series circuit, we see the directions of *IR*, *IX_L* and *IX_C* as in Figure A.2b, giving the impedance triangle shown in Figure A.2c.

To express these reactances in the *s* domain, we simply substitute *s* for *jω*:

$$-jX_C = \frac{1}{sC}$$

$$jX_L = s_L$$

The series impedance of the *RLC* circuit can be expressed as

$$Z = R + jX_L - jX_C = R + sL + \frac{1}{sC}$$

In this discussion, note the use of $\sqrt{-1}$ or *j*, to indicate a 90° phase change.

Appendix B: Software Tools

Modeling, analysis, design, data acquisition, and control are important activities within the field of Control Engineering. Computer software tools and environments are needed for effectively carrying out these, both at the learning level and at the professional application level. Several such environments and tools are commercially available. A selected few, which are particular useful for the tasks related to the present book are outlined here.

MATLAB®* is an interactive computer environment with a high-level language and tools for scientific and technical computation, modeling and simulation, design, and control of dynamic systems. Simulink® is a graphical environment for modeling, simulation, and analysis of dynamic systems, and is available as an extension to MATLAB. LabVIEW is a graphical programming language and a program development environment for data acquisition, processing, display, and instrument control.

B.1 Simulink®

Computer simulation of a dynamic model by using Simulink is outlined in Chapter 6. Simulink is a graphic environment that uses block diagrams. It is an extension to MATLAB.

B.2 MATLAB®

MATLAB interactive computer environment is very useful in computational activities in mechatronics. Computations involving scalars, vectors, and matrices can be carried out and the results can be graphically displayed and printed. MATLAB toolboxes are available for performing specific tasks in a particular area of study such as control systems, fuzzy logic, neural network, data acquisition, image processing, signal processing, system identification, optimization, model predictive control, robust control, and statistics. User guides, Web-based help, and on-line help from the parent company, The MathWorks, Inc., and various other sources. What is given here is a brief introduction to get started in MATLAB for tasks that are particularly related to control systems and mechatronics.

B.2.1 Computations

Mathematical computations can be done by using the MATLAB® command window. Simply type in the computations against the MATLAB prompt ">>" as illustrated next.

---

* MATLAB® and Simulink® are registered trademarks and products of The MathWorks, Inc. LabVIEW is a product of National Instruments, Inc.

B.2.2 Arithmetic

An example of a simple computation using MALAB® is given below.
```
>> x=2; y=-3;
>> z=x^2-x*y+4
z=14
```
In the first line we have assigned values 2 and 3 to two variables x and y. In the next line, the value of an algebraic function of these two variables is indicated. Then, MATLAB provides the answer as 14. Note that if you place a ";" at the end of the line, the answer will not be printed/displayed.

Table B.1 gives the symbols for common arithmetic operations used in MATLAB.

Following example shows the solution of the quadratic equation $ax^2+bx+c=0$:
```
>> a = 2;b=3;c=4;
>> x=(-b+sqrt(b^2-4*a*c))/(2*a)
x =
    -0.7500+1.1990i
```
The answer is complex, where i denotes $\sqrt{-1}$. Note that the function sqrt( ) is used, which provides the positive root only. Some useful mathematical functions are given in Table B.2.

Note: MATLAB is case sensitive.

TABLE B.1

MATLAB® Arithmetic Operations

Symbol	Operation
+	Addition
−	Subtraction
*	Multiplication
/	Division
^	Power

TABLE B.2

Useful Mathematical Functions in MATLAB

Function	Description
abs( )	Absolute value/magnitude
acos( )	Arc-cosine (inverse cosine)
acosh( )	Arc-hyperbolic-cosine
asin( )	Arc-sine
atan( )	Arc-tan
cos( )	Cosine
cosh( )	Hyperbolic cosine
exp( )	Exponential function
imag( )	Imaginary part of a complex number
log( )	Natural logarithm
log10( )	Log to base 10 (common log)
real( )	Real part of a complex number
sign( )	Signum function
sin( )	Sine
sqrt( )	Positive square root
tan( )	Tan function

B.2.3 Arrays

An array may be specified by giving the start value, increment, and the end value limit. An example is given below.

>> x=(.9:−.1:0.42)

x=

　0.9000　0.8000　0.7000　0.6000　0.5000

The entire array may be manipulated. For example, all the elements are multiplied by π as below:

>> x=x*pi

x=

　2.8274　　2.5133　　2.1991　　1.8850　　1.5708

The second and the fifth elements are obtained by:

>> x([2 5])

ans =

　　2.5133　1.5708

Next we form a new array y using x, and then plot the two arrays, as shown in Figure B.1.

>> y=sin(x);

>> plot(x,y)

A polynomial may be represented as an array of its coefficients. For example, the quadratic equation $ax^2+bx+c=0$ as given before, with $a = 2$, $b = 3$, and $c =$ M4, may be solved using the function "roots" as below.

>> p=[2 3 4];

>> roots(p)

ans =

　　−0.7500+1.1990i

　　−0.7500−1.1990i

The answer is the same as what we obtained before.

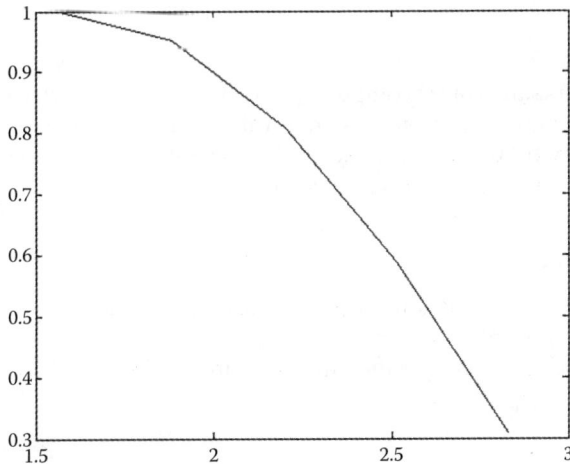

FIGURE B.1
A plot using MATLAB®.

TABLE B.3

Some Relational Operations

Operator	Description
<	Less than
<=	Less than or equal to
>	Greater than
>=	Greater than or equal to
= =	Equal to
~=	Not equal to

TABLE B.4

Basic Logical Operations

Operator	Description
&	AND
		OR
~	NOT

B.2.4 Relational and Logical Operations

Useful relational operations in MATLAB® are given in Table B.3. Basic logical operations are given in Table B.4.

Consider the following example.

```
>> x=(0:0.25:1)*pi
x =
   0    0.7854    1.5708    2.3562    3.1416
>> cos(x)>0
ans =
    1  1  1  0  0
>> (cos(x)>0)&(sin(x)>0)
ans =
    0  1  1  0  0
```

In this example, first an array is computed. Then the cosine of each element is computed. Next it is checked whether the elements are positive. (A truth value of 1 is sent out if true and a truth value of 0 if false.) Finally the "AND" operation is used to check whether both corresponding elements of two arrays are positive.

B.2.5 Linear Algebra

MATLAB® can perform various computations with vectors and matrices (see Appendix C). Some basic illustrations are given here.

A vector or a matrix may be specified by assigning values to its elements. Consider the following example.

```
>> b=[1.5 -2];
>> A=[2 1;-1 1];
>> b=b'
```

b =
    1.5000
  −2.0000
>> x=inv(A)*b
x =
    1.1667
  −0.8333

In this example, first a second order row vector and 2×2 matrix are defined. The row vector is transposed to get a column vector. Finally the matrix–vector equation $Ax=b$ is solved according to $x=A^{-1}b$. The determinant and the eigenvalues of A are determined by:

>> det(A)
ans =
    3
>> eig(A)
ans =
    1.5000+0.8660i
    1.5000−0.8660i

Both eigenvectors and eigenvalues of A computed as:

>> [V,P]=eig(A)
V =
  0.7071   0.7071
 −0.3536+0.6124i −0.3536−0.6124i
P =
  1.5000+0.8660i   0
    0   1.5000−0.8660i

Here the symbol V is used to denote the matrix of eigenvectors. The symbol P is used to denote the diagonal matrix whose diagonal elements are the eigenvalues.

Useful matrix operations in MATLAB are given in Table B.5 and several matrix functions are given in Table B.6.

B.2.6 M-Files

The MATLAB® commands have to be keyed in on the command window, one by one. When several commands are needed to carry out a task, the required effort can be tedious. Instead, the necessary commands can be placed in a text file, edited as appropriate (using

TABLE B.5

Some Matrix Operations in MATLAB

Operation	Description
+	Addition
−	Subtraction
*	Multiplication
/	Division
^	Power
'	Transpose

TABLE B.6

Useful Matrix Functions in MATLAB

Function	Description
det( )	Determinant
inv( )	Inverse
eig( )	Eigenvalues
[,]=eig( )	Eigenvectors and eigenvalues

text editor), which MATLAB can use to execute the complete task. Such a file is called an M-file. The file name must have the extension "m" in the form *filename.m*. A toolbox is a collection of such files, for use in a particular application area (e.g., control systems, fuzzy logic). Then, by keying in the M-file name at the MATLAB command prompt, the file will be executed. The necessary data values for executing the file have to be assigned beforehand.

B.3 Control Systems Toolbox

There are several toolboxes with MATLAB®, which can be used to analyze, compute, simulate, and design control problems. Both time-domain representations and frequency-domain representations can be used. Also, both classical and modern control problems can be handled. The application is illustrated here through several conventional control problems discussed in Chapters 7 through 11.

B.3.1 Compensator Design Example

Consider again the design problem given in Chapter 9, Example 9.2 (Figure 9.5). The MATLAB® single-input–single-output (SISO) Design Tool is used here to solve this problem.

B.3.1.1 Building the System Model

Build the transfer function model of the Motor and Filter, in the MATLAB® workspace, as follows:

```
Motor_G=tf([999], [10 1]);
Filter_H=tf([1], [0.1 1]);
```

To Open the SISO Design Tool, type
```
sisotool
```
at the MATLAB prompt (>>).

B.3.1.2 Importing Model into SISO Design Tool

Select *Import Model* under the *File* menu. This opens the *Import System Data* dialog box, as shown in Figure B.2a.

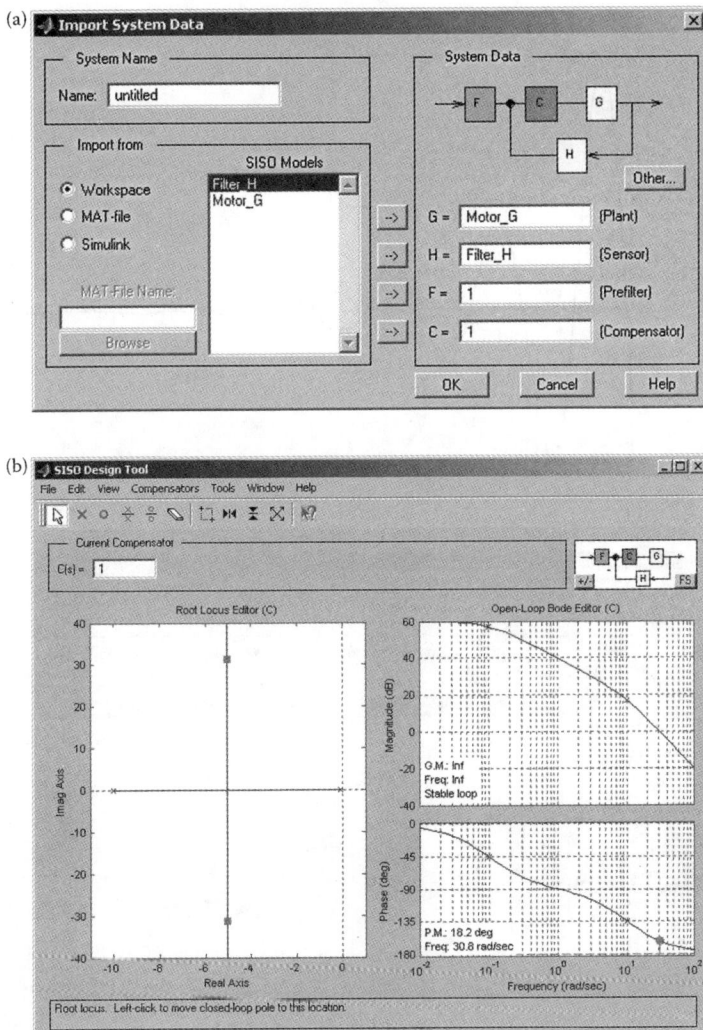

FIGURE B.2
(a) Importing the model into the SISO Design Tool. (b) Root locus and Bode plots for the motor model. (c) Closed-loop step response of the motor system without compensation. (d) Root locus and Bode plots of the compensated system. (e) Closed-loop step response of the compensated system.

Use the following steps to import the motor and filter models:

1. Select Motor_G under *SISO Models*.
2. Place it into the *G* Field under *Design Model* by pressing the right arrow button to the left of *G*.
3. Similarly import the filter model.
4. Press **OK**

(c)

(d)

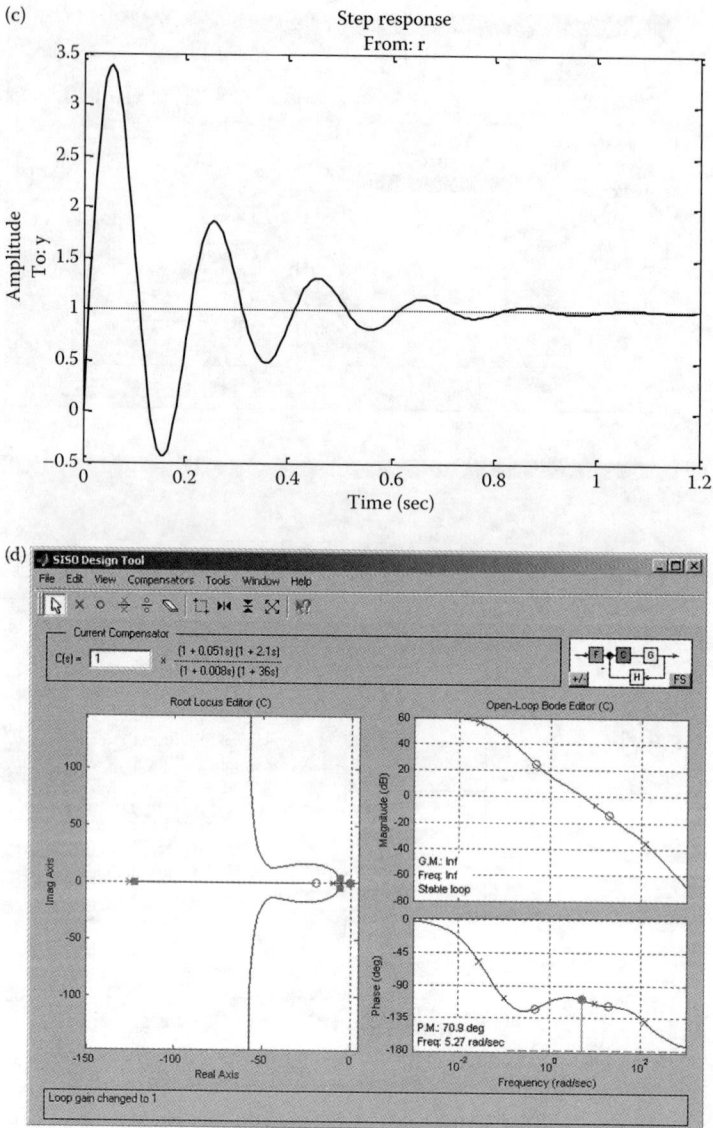

FIGURE B.2 (continued)

Now in the main window of the SISO Design Tool, will show the root locus and Bode plots of open-loop transfer function *GH* (see Figure B.2b). As given in the figure, the phase margin is 18.2°, which occurs at 30.8 rad/s (4.9 Hz).

The closed-loop step response, without compensation, is obtained by selecting ***Tools*** → ***Loop responses*** → ***closed-loop step*** from the main menu. The response is shown in Figure B.2c. It is noted that the phase margin is not adequate, which explains the oscillations and the long settling time. Also the *P.O.* is about 140%, which is considerably higher than the desired one (10%) and is not acceptable.

(e)

Step response
From: r

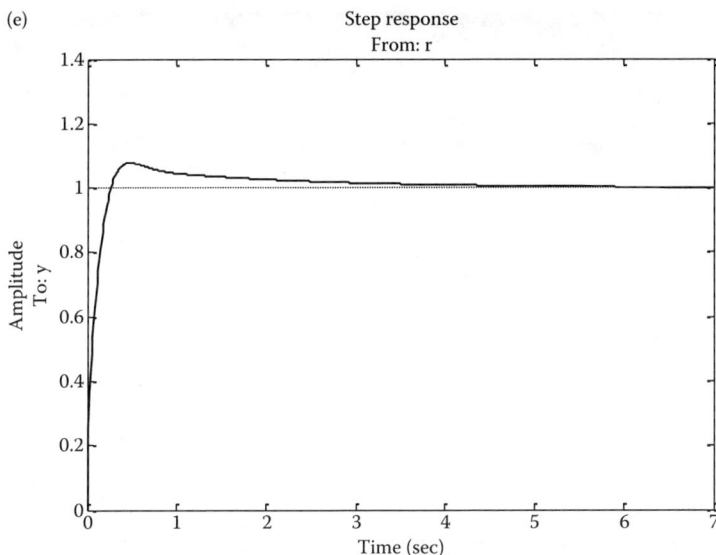

FIGURE B.2   (continued)

B.3.1.3 Adding Lead and Lag Compensators

To add a lead compensator, right-click the mouse in the white space of the Bode magnitude plot, choose *Add Pole/Zero* and then *lead* in the right-click menu for the open-loop Bode diagram. Move the zero and the pole of the lead compensator to get a desired phase margin about 60°.

To add a lag compensator, choose *Add Pole/Zero* and then *lag* in the right-click menu for the open-loop Bode diagram. Move the zero and the pole of the lag compensator to get a desired phase angle of about –115° at the crossing frequency, which corresponds to a phase margin of $180° - 115° = 65°$.

With the added lead and lag compensators, the root locus and Bode plots of the system are shown in Figure B.2d. The closed-loop step response of the system is shown in Figure B.2e.

B.3.2 PID Control with Ziegler–Nichols Tuning

Consider Example 9.5 given in Chapter 9, Figure 9.14. The SISO Design Tool is used. First build the transfer function model of the given system (call it Mill).

>   Mill_G=tf([1], [1 1 4 0]);
>   Filter_H=tf([1], [1]);

As before, import the system model into the SISO Design Tool.

B.3.2.1 Proportional Control

Even without using the Routh–Hurwitz method, we can change the gain setting by trial and error to obtain the proportional gain that will make the system marginally stable. As seen in Figure B.3a, when $K = 4$, the gain margin is just below 0 dB, which makes the system unstable. The response of the system is shown in Figure B.3b.

(a)

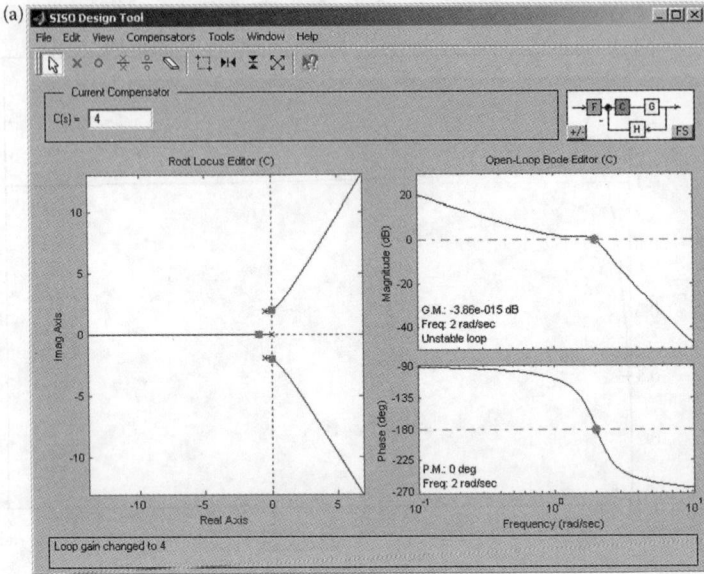

(b)

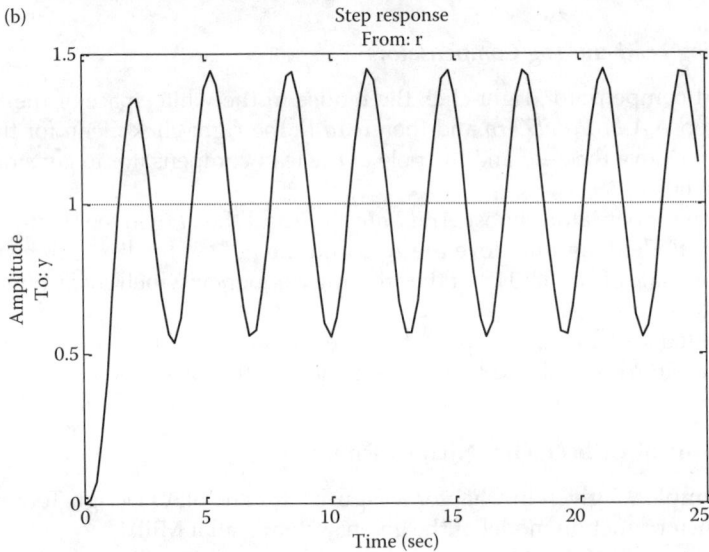

FIGURE B.3

(a) Root locus and Bode plots of the system with proportional gain $K_p=4$. (b) Step response of the closed-loop system with $K_p=4$. (c) Step response of the closed-loop system with $K_p=2$. (d) Bode plot of the system with PI control. (e) Step response of the system with PI control. (f) Bode plot of the system with PID control. (g) Step response of the system with PID control.

Referring to the Ziegler–Nichols controller settings, as given in Table 9.1, we can obtain the proper proportional gain as $K_p = 0.5 \times 4 = 2$. The corresponding system response is shown in Figure B.3c.

(c)

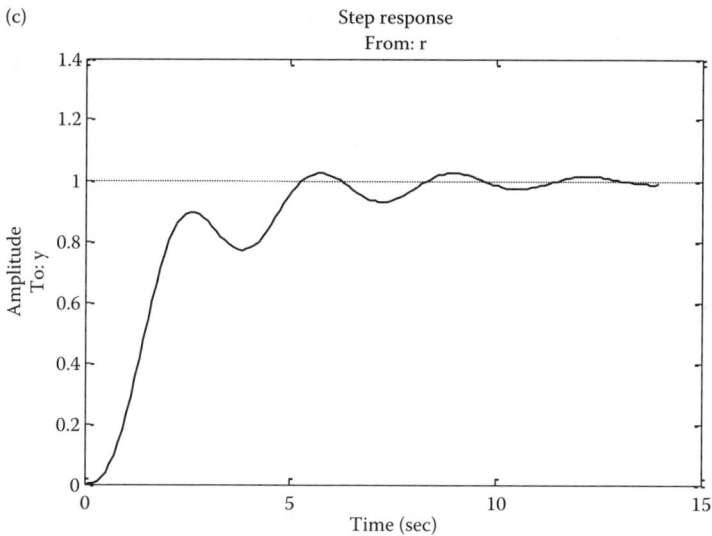

(d)

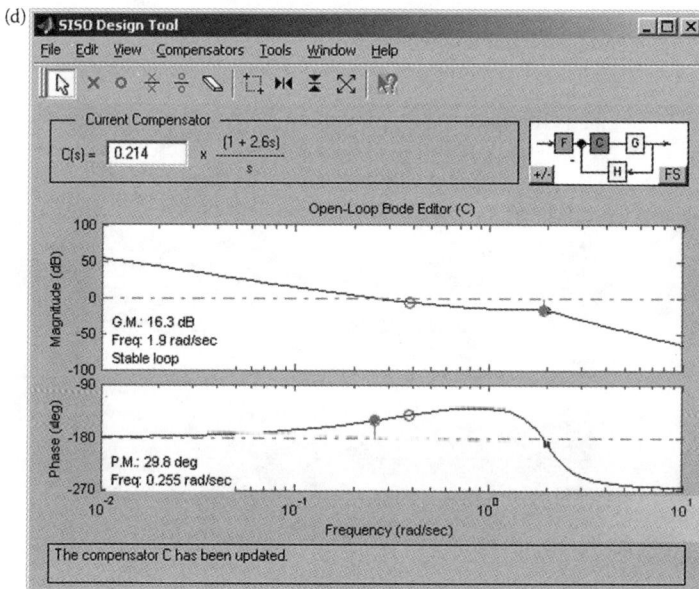

FIGURE B.3 (continued)

B.3.2.2 PI Control

Note that the period of oscillations (ultimate period) is

$$P_u = \frac{2\pi}{\omega_n} = \frac{2\pi}{2} = \pi \, \text{seconds}$$

(e)

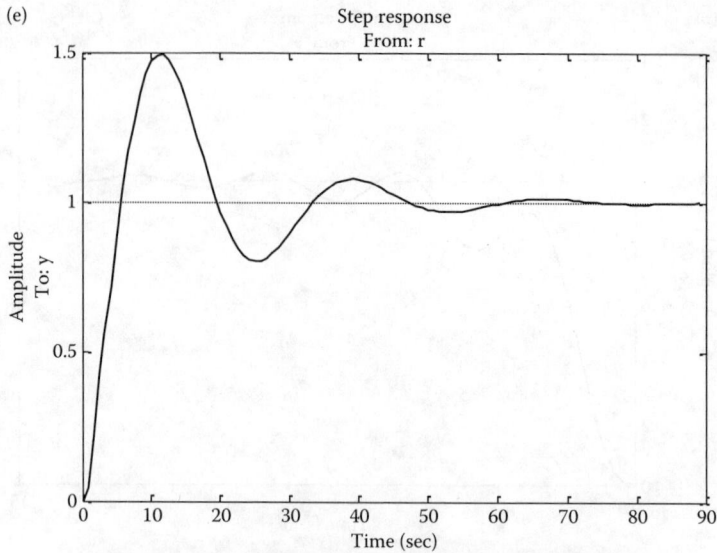

Step response
From: r

(f)

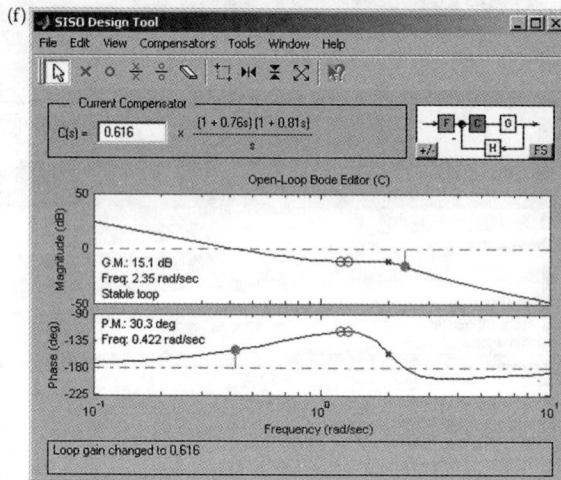

FIGURE B.3　(continued)

Hence, from the Ziegler–Nichols settings given in Table 9.1, we have for a PI controller,

$$K_p = 0.45 \times 4 = 1.8$$
$$\tau_i = 0.83\pi = 2.61 \text{ seconds.}$$

Hence, the PI controller transfer function is

$$K_p\left(1 + \frac{1}{\tau_i s}\right) = \frac{K_p \tau_i s + K_p}{\tau_i s} = \frac{4.68s + 1.8}{2.61s} = 0.214\frac{2.6s + 1}{s}$$

Insert this controller into C in the SISO Design Tool.

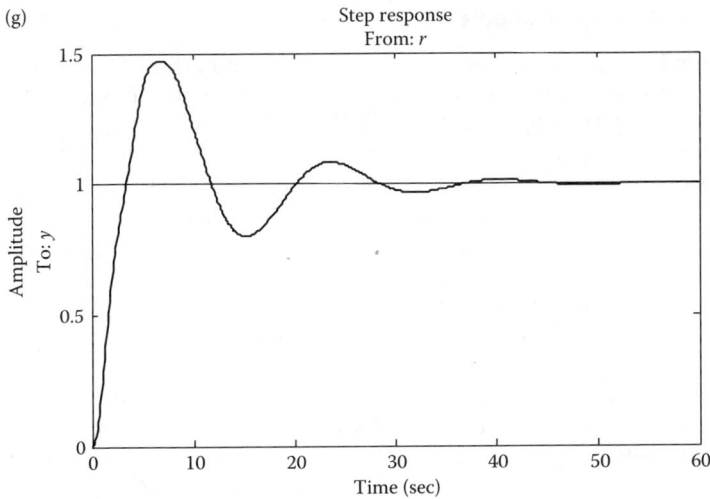

(g)

Step response
From: r

FIGURE B.3 (continued)

The corresponding system Bode plot and the step response are shown in Figures B.6d and e, respectively.

B.3.2.3 PID Control

From the Ziegler–Nichols settings given in Table 9.1, we have for a PID controller,

$$K_p = 0.6 \times 4 = 2.4$$
$$\tau_i = 0.5\pi = 1.57 \text{ seconds}$$
$$\tau_d = 0.125\pi = 0.393 \text{ seconds}.$$

The corresponding transfer function of the PID controller is

$$K_p \left(1 + \frac{1}{\tau_i s} + \tau_d s\right) = \frac{K_p \tau_i \tau_d s^2 + K_p \tau_i s + K_p}{\tau_i s} = \frac{1.48s^2 + 3.768s + 2.4}{1.57s} = \frac{0.94s^2 + 2.4s + 1.53}{s}$$

Use the MATLAB® function **roots** to calculate the roots of the numerator polynomial.
R=roots([0.94 2.4 1.53]);

R=−1.3217
−1.2315

Hence, the transfer function of the PID controller is

$$\frac{(s+1.32)(s+1.23)}{s} = 0.616 \frac{(0.76s + 1)(0.81s + 1)}{s}$$

Insert this controller into C of the SISO Design Tool. The corresponding Bode plot and the step response of the controlled system are shown in Figure B.6f and g.

B.3.3 Root Locus Design Example

Consider Example 9.3 given in Chapter 9, Figure 9.7. Again, the SISO Design Tool is used. First build the transfer function model for the rolling mill with no filter:

    Mill_G=tf([1], [1 5 0]);
    Filter_H=tf([1], [1]);

Then, as before, import the system model into the SISO Design Tool. The root locus and the step response of the closed-loop system are shown in Figures B.7a and b. From Figure B.4b, it is seen that the peak time and the 2% settling time do not meet the design specifications.

To add a lead compensator, right-click in the white space of the root locus plot, choose *Add Pole/Zero* and then **lead** in the right-click menu. Left-click on the root locus plot where we want to add the lead compensator.

Now we have to adjust the pole and zero of the lead compensator and the loop gain so that the root locus passes through the design region. To speed up the design process, turn on the grid setting for the root locus plot. The radial lines are constant damping ratio lines and the semicircular curves are constant undamped natural frequency lines (see Chapter 7).

On the root locus plot, drag the pole and zero of the lead compensator (pink cross or circle symbol on the plot) so that the root locus moves toward the design region. Left-click and move the closed-loop pole (small pink-color square box) to adjust the loop gain. As you drag the closed-loop pole along the locus, the current location of that pole, and the system damping ratio and natural frequency will be shown at the bottom of the graph.

Drag the closed-loop pole into the design region. The resulting lead compensator, the loop gain and the corresponding root locus are shown in Figure B.4c. The step response of the compensated closed-loop system is shown in Figure B.4d.

B.3.4 MATLAB® Modern Control Examples

Several examples in modern control engineering are given now to illustrate the use of MATLAB in control. The background theory is found in Chapter 11.

B.3.4.1 Pole Placement of a Third Order Plant

A mechanical plant is given by the input-output differential equation $\dddot{x} + \ddot{x} = u$, where u is the input and x is the output. Determine a feedback law that will yield approximately a simple oscillator with a damped natural frequency of 1 unit and a damping ratio of $1/\sqrt{2}$.

To solve this problem, first we define the state variables as $x_1 = x$, $x_2 = \dot{x}_1$, and $x_3 = \dot{x}_2$. The corresponding state-space model is

$$\dot{x} = \begin{bmatrix} \dot{x}_1 \\ \dot{x}_2 \\ \dot{x}_3 \end{bmatrix} = \underbrace{\begin{bmatrix} 0 & 1 & 0 \\ 0 & 0 & 1 \\ 0 & 0 & -1 \end{bmatrix}}_{A} \begin{bmatrix} x_1 \\ x_2 \\ x_3 \end{bmatrix} + \underbrace{\begin{bmatrix} 0 \\ 0 \\ 1 \end{bmatrix}}_{B} u$$

$$y = \underbrace{\begin{bmatrix} 1 & 0 & 0 \end{bmatrix}}_{C} x$$

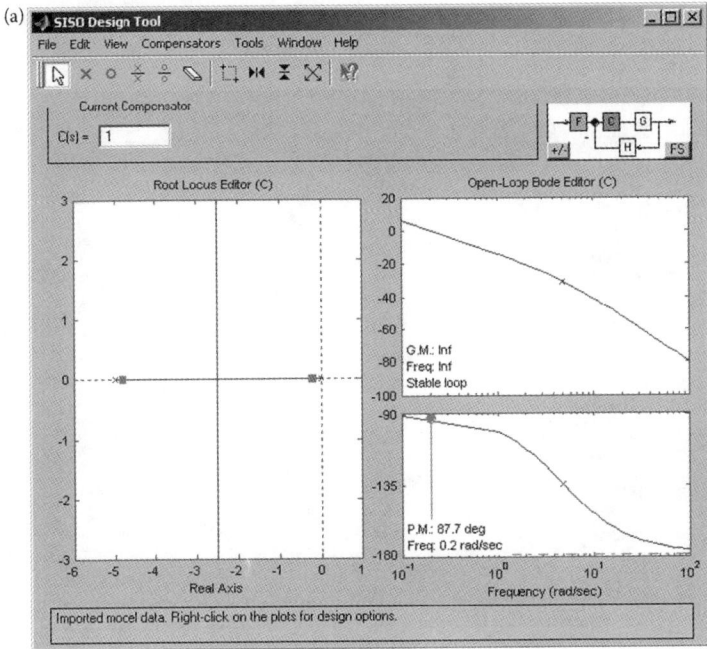

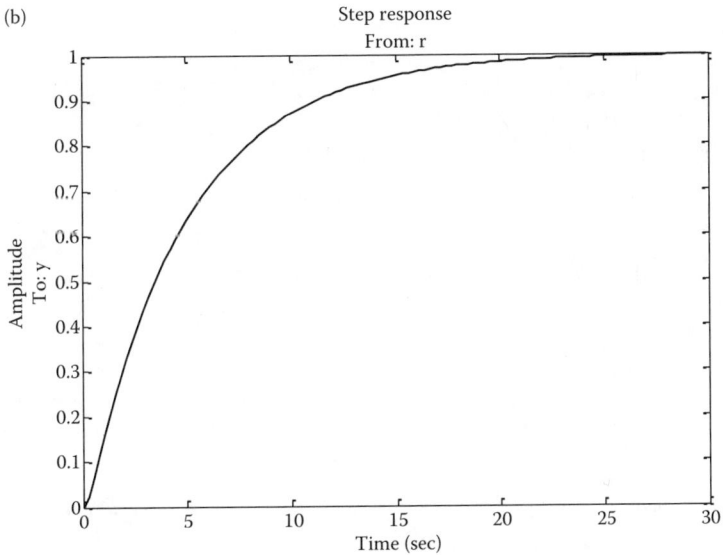

FIGURE B.4
(a) Root locus and Bode plots of the rolling mill system without compensation. (b) Step response of the closed-loop system without compensation. (c) Root locus of the compensated system. (d) Step response of the compensated closed-loop system.

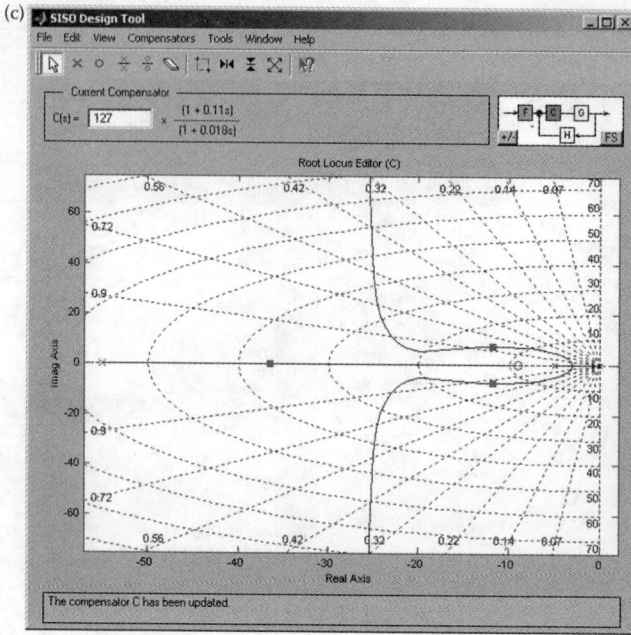

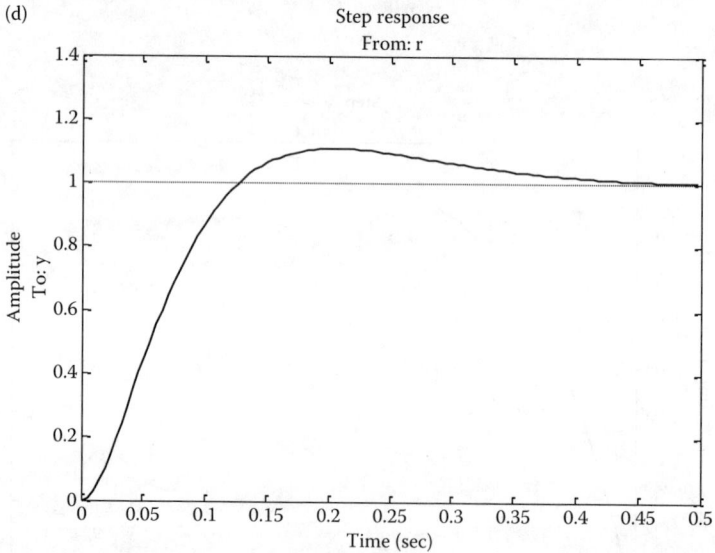

FIGURE B.4   (continued)

The open-loop poles and zeros are obtained using the following MATLAB® commands:
```
>> A = [0 1 0; 0 0 1; 0 0 −1];
>> B = [0; 0; 1];
>> C = [1 0 0];
```

```
>> D=[0];
>> sys_open=ss(A,B,C,D);
>> [nat_freq_open,damping_open,poles_open]=damp(sys_open)
>> pzmap(sys_open)
```

The open-loop poles are: $[0\ 0\ -1]^T$.
The step response of the open-loop system is obtained using the command:
```
>> step(sys_open)
```
The result is shown in Figure B.5a. Clearly, the system is unstable.

With the desired damped natural frequency $\omega_d=1$ and damping ratio $\zeta=1/\sqrt{2}$, we get the undamped natural frequency $\omega_n=\sqrt{2}$ and hence, $\zeta\omega_n=1$. It follows that we need to place two poles at $-1 \pm j$. Also the third pole has to be far from these two on the left half plane (LHP); say, at -10. The corresponding control gain K can be computed using the "place" command in MATLAB:
```
>> p=[-1+j -1-j -10];
>> K=place(A,B,p)
place: ndigits= 15
K =
    20.0000 22.0000 11.0000
```

The corresponding step response of the closed-loop system is shown in Figure B.5b.

B.3.4.2 Linear Quadratic Regulator (LQR) for a Third Order Plant

For the third order plant in the previous example, we design a LQR, which has a state feed-back controller (see Chapter 11), using MATLAB® Control Systems Toolbox. The MATLAB command $K=lqr(A,B,Q,R)$ computes the optimal gain matrix K such that the state-feedback law $u=-Kx$ minimizes the quadratic cost function

$$J = \int_0^\infty (x^T Q x + u^T R u)dt$$

The weighting matrices Q and R are chosen to apply the desired weights to the various states and inputs. The MATLAB commands for designing the controller are:
```
>> A=[0 1 0; 0 0 1; 0 0 -1];
>> B=[0; 0; 1];
>> C=[1 0 0];
>> D=[0];
>> Q=[2 0 0 ;0 2 0 ; 0 0 2];
>> R=2;
>> Klqr=lqr(A,B,Q,R)
>> lqr_closed=ss(A-B*Klqr,B,C,D);
>> step(lqr_closed)
```

The step response of the system with the designed LQR controller is shown in Figure B.6.

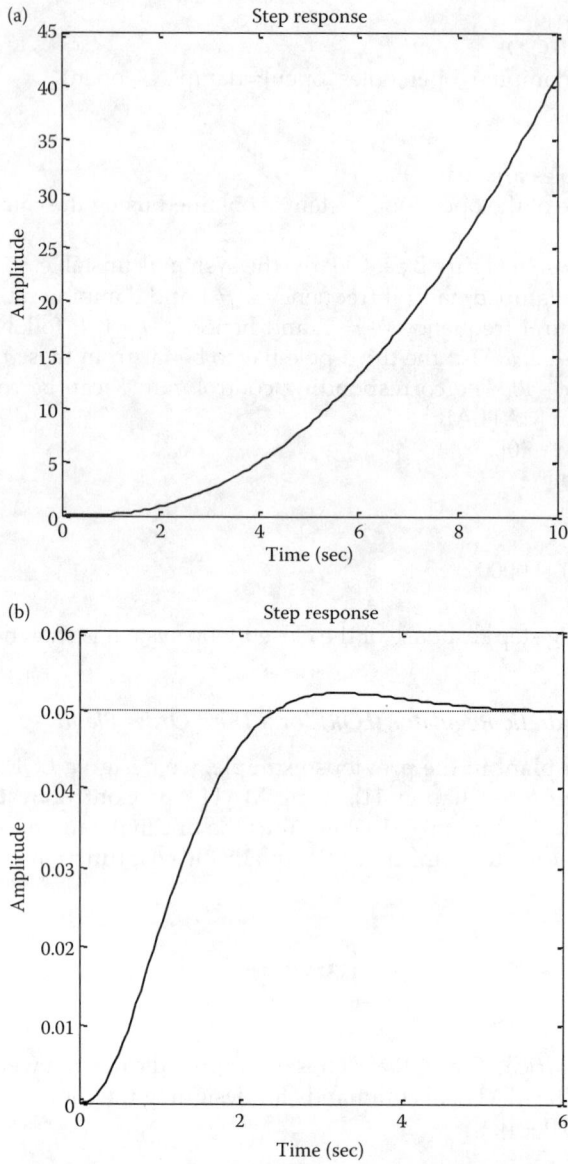

FIGURE B.5
(a) Step response of the open-loop system. (b) Step response of the third order system with pole-placement control.

B.3.4.3 Pole Placement of an Inverted Pendulum on Mobile Carriage

The system is described in Example 11.27, Chapter 11, Figure 11.20. The linearized state-space model is

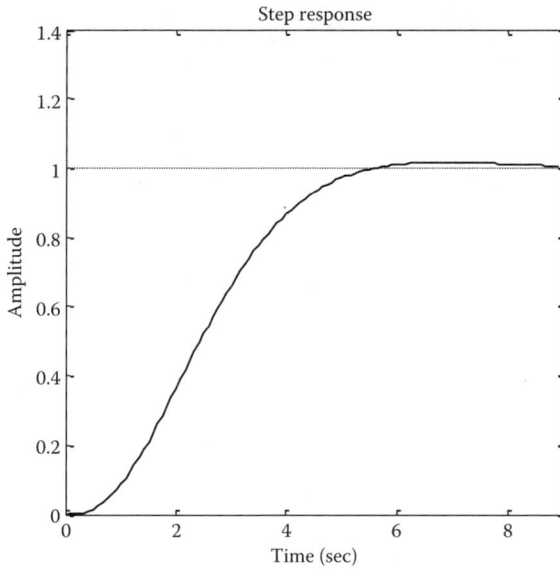

FIGURE B.6
Step response of the third order system with LQR control.

$$\dot{x} = \begin{bmatrix} 0 & 1 & 0 & 0 \\ 0 & 0 & -1 & 0 \\ 0 & 0 & 0 & 1 \\ 0 & 0 & 11 & 0 \end{bmatrix} x + \begin{bmatrix} 0 \\ 1 \\ 0 \\ -1 \end{bmatrix} u$$

As before, the open-loop poles and zeros are obtained using the MATLAB® commands:

```
>> A=[0 1 0 0; 0 0 -1 0; 0 0 0 1; 0 0 11 0];
>> B-[0; 1; 0; -1];
>> C-[1 0 0 0; 0 0 1 0];
>> D=0;
>> sys_open=ss(A,B,C,D);
>> [nat_freq_open,damping_open,poles_open]=damp(sys_open)
>> pzmap(sys_open)
```

Open-loop poles are: $[0 \quad 0 \quad 3.3166 \quad -3.3166]^T$. Note that the system is unstable. The impulse response of the open-loop system is obtained using the command:

```
>> impulse(sys_open)
```

The response is shown in Figure B.7a.

Let the desired closed-loop poles be $\Gamma = [-1 \quad -2 \quad -1 \quad -j \quad -1+j]$. Use the feedback controller

$$u = u_{ref} - Kx$$

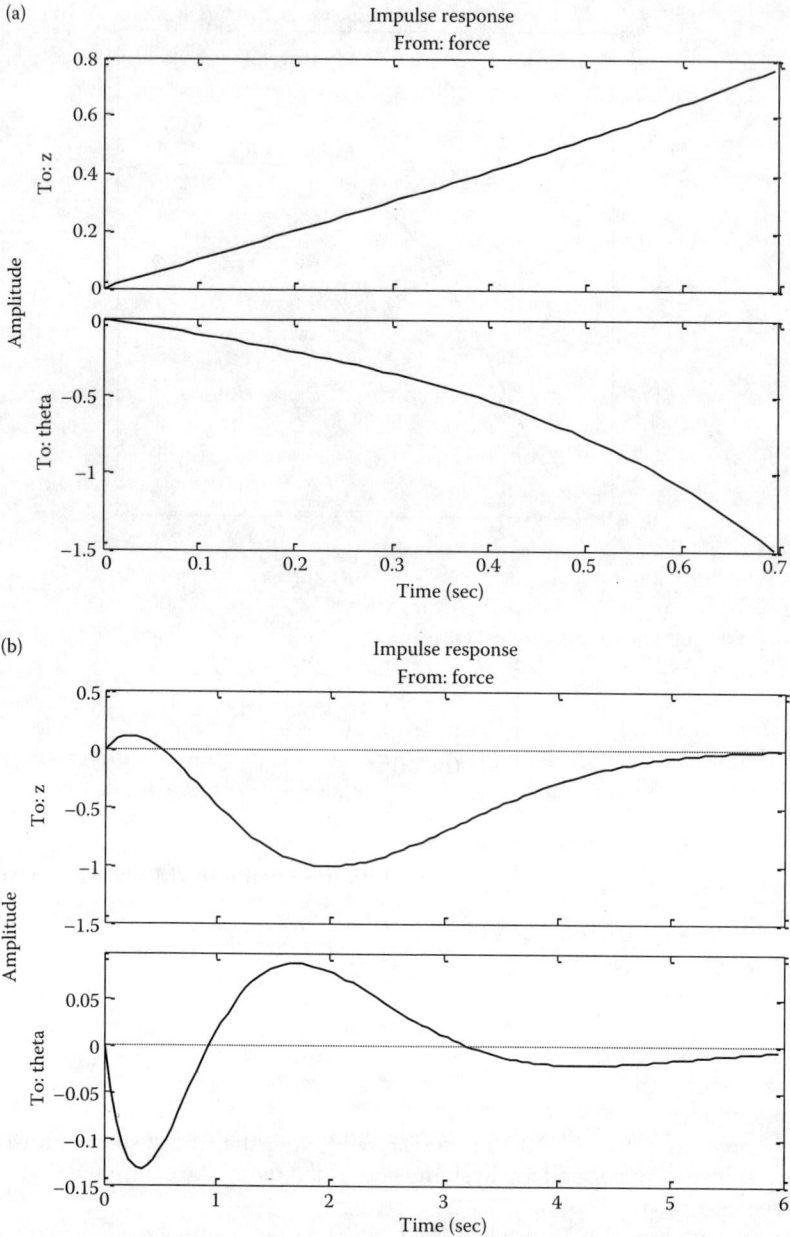

FIGURE B.7
(a) Impulse response of the inverted pendulum (theta) on a moving carriage (z). (b) Impulse response of the pole-placement controlled inverted pendulum on carriage.

where the feedback gain matrix is $K = [k_1 \quad k_2 \quad k_3 \quad k_4]$. As before, K is computed using the "place" command:

```
>> p=[-1 -2 -1+j -1-j];
>> K=place(A,B,p)
```

place: ndigits = 15
K =
    −0.4000 −1.0000 −21.4000 −6.0000

The corresponding impulse response of the closed-loop system is shown in Figure B.7b.

Note that, with the assigned poles, the inverted pendulum balances and the car returns to the initial position.

B.3.4.4 LQG Controller for an Inverted Pendulum Mounted with Mobile Carriage

The LQR is designed using the MATLAB® Control Systems Toolbox, as before (for the third order system). The commands are:

```
>> A=[0 1 0 0; 0 0 −1 0; 0 0 0 1; 0 0 11 0];
>> B=[0; 1; 0; −1];
>> C=[1 0 0 0; 0 0 1 0];
>> D=0;
>> Q=[2 0 0 0;0 2 0 0; 0 0 2 0; 0 0 0 2];
>> R=2;
>> Klqr=lqr(A,B,Q,R)
>> lqr_closed=ss(A−B*Klqr,B,C,D);
>> impulse(lqr_closed)
```

The impulse response of the controlled system is shown in Figure B.8.

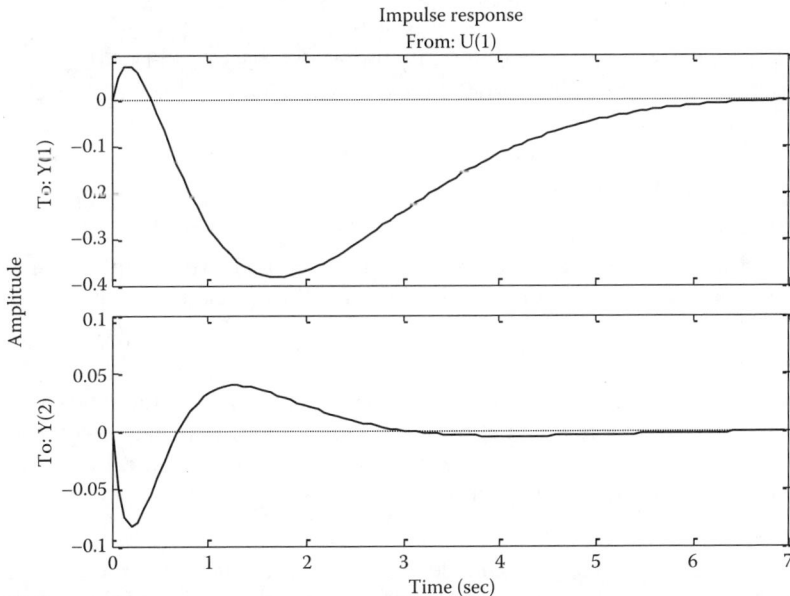

FIGURE B.8
Impulse response of the LQR controlled inverted pendulum on carriage.

B.4 Fuzzy Logic Toolbox

The use of fuzzy logic in intelligent control has been discussed in Chapter 11. The Fuzzy Logic Toolbox of MATLAB® is quite useful in this regard. Using it we create and edit fuzzy decision-making systems (for control and other applications) by means of interactive graphical tools or command-line functions. Simulink® can be used to simulate the developed fuzzy system. The-time workshop can create portable C code from a Simulink environment for use in real-time and nonreal-time applications. The toolbox also provides source codes in C for implementing a stand-alone fuzzy inference engine. The stand-alone C-code fuzzy inference engine can read an FIS file (the file format for saving the fuzzy engine in MATLAB). In other words, it is able to parse the stored information, to perform fuzzy inference directly, or it can be embedded in other external applications. The design process of a fuzzy decision-making system involves the following general steps, as discussed in Chapter 11: Input Data, Fuzzification, Implication (or, Fuzzy Rules), Aggregation (or, Composition), and Inference Defuzzification.

B.4.1 Graphical Editors

There are five primary graphic user interface (GUI) tools for building, editing, and observing fuzzy inference systems in the MATLAB® Fuzzy Logic Toolbox: the FIS Editor, the Membership Function Editor, the Rule Editor, the Rule Viewer and the Surface Viewer. The FIS Editor handles the high-level issues for the system; e.g., number of inputs, outputs and names. The Membership Function Editor is used to define the shapes of the membership functions associated with each variable. The Rule Editor is used for editing the rules in the fuzzy knowledge base, which describes defines the knowledge of the application (control knowledge in the case of fuzzy control). The Rule Viewer and the Surface Viewer are used for observing (not editing) the designed FIS. Try the example on tipping in a restaurant, by clicking on the file menu and loading FIS from disk. The tipper. fis is located at:

/Matlabr12/toolbox/fuzzy/fuzzydemos/tipper.fis

1. **FIS Editor**

   The FIS Editor displays general information about a fuzzy inference system. Double click on an icon to open and, carry out editing related that particular item, and save the results.

2. **Membership Function Editor**

   The Membership Function Editor shares some features with the FIS Editor. It is a menu driven interface, which allows the user to open/display and edit the membership functions for the entire fuzzy inference system; specifically the membership functions of inputs and outputs.

3. **Rule Editor**

   The Rule Editor contains an editable text field for displaying and editing rules. It also has some landmarks similar to those in the FIS Editor and the Membership Function Editor, including the menu bar and the status line. The pop-up menu **Format** is available from the pull-down menu **Options** in the top menu bar. This is used to set the format for the display.

4. Rule Viewer

The Rule Viewer displays a roadmap of the entire fuzzy inference process. It is based on the fuzzy inference diagram. The user will see a single figure window with seven small plots nested in it. The two small plots across the top of the figure represent the antecedent and the consequent of the first rule. Each rule is a row of plots, and each column is a variable.

5. Surface Viewer

This allows the user to view the overall decision-making surface (the control surface, as discussed in Chapter 11). This is a nonfuzzy representation of the fuzzy application, and is analogous to a look-up table albeit continuous.

B.4.2 Command Line Driven FIS Design

A predesigned FIS may be loaded into the MATLAB® workspace by typing:

>> **myfis**=read**fis('name_of_file.fis')**

Typing the **showfis(myfis)** command will enable us to see the details of the FIS. Use the **getfis** command to access information of the loaded FIS. For example,

>> **getfis(myfis)**
>> **getfis(myfis, 'Inlabels')**
>> **getfis(myfis, 'input', 1)**
>> **getfis(myfis, 'output', 1)**

The command **setfis** may be used to modify any property of an FIS. For example,

>> **setfis(myfis, 'name', 'new_name');**

The following three functions are used to display the high-level view of a fuzzy inference system from the command line:

>> **plotfis(myfis)**
>> **plotmf(myfis, 'input', input_number)**
      or **plotmf(myfis, 'output', output_number)**
>> **gensuf(myfis)**

To evaluate the output of a fuzzy system for a given input, we use the function:

>> **evalfis([input matrix], myfis)**

For example, **evalfis([1 1], myfis)** is used for single input evaluation, and **evalfis([1 1; 2 3] myfis)** for multiple input evaluation.

Note that we may directly edit a previously saved .fis file, besides manipulating a fuzzy inference system from the toolbox GUI or from the MATLAB workspace through the command line.

B.4.3 Practical Stand-Alone Implementation in C

The MATLAB® Fuzzy Logic Toolbox allows you to run your own stand-alone C programs directly, without the need for Simulink®. This is made possible by a stand-alone Fuzzy Inference Engine that reads the fuzzy systems saved from a MATLAB session. Since the C source code is provided, you can customize the stand-alone engine to build fuzzy inference into your own code. This procedure is outlined in Figure B.9.

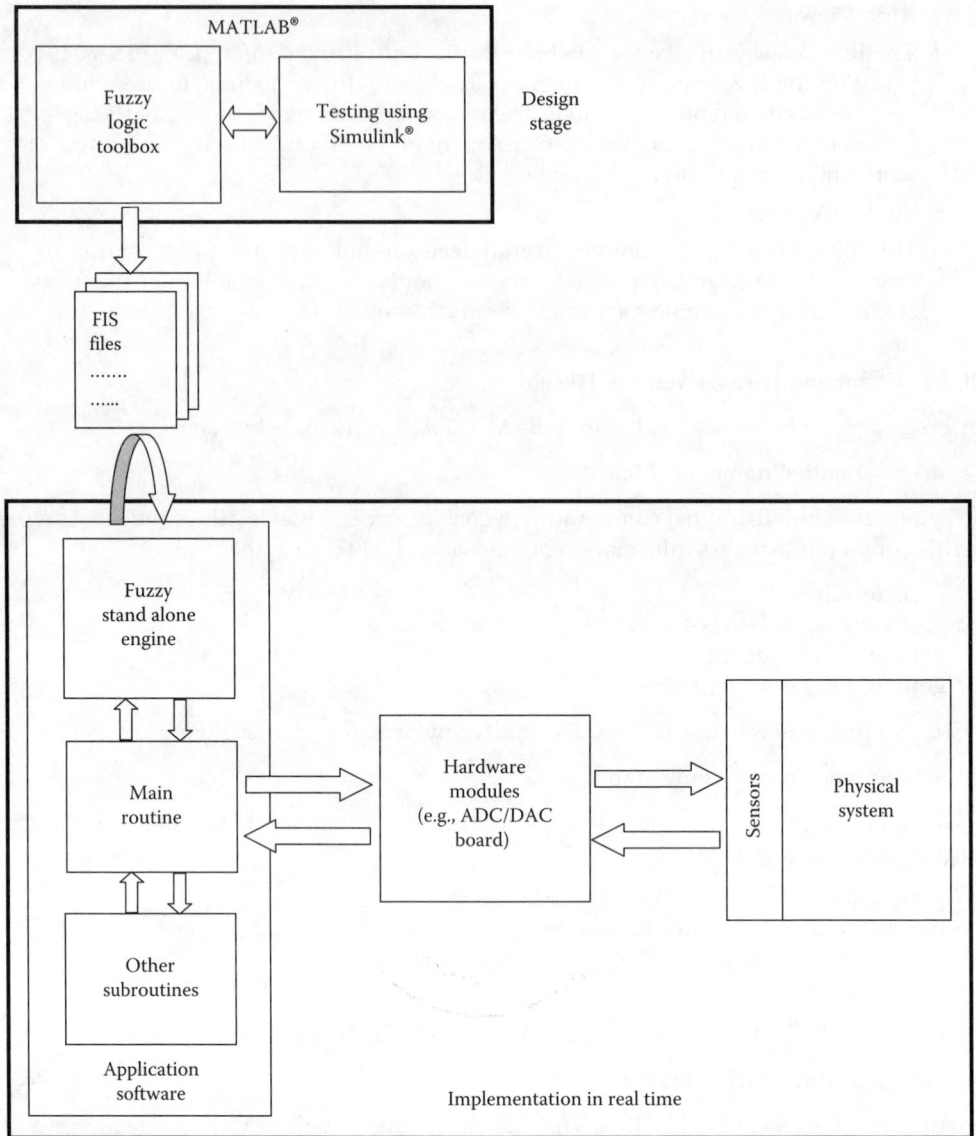

FIGURE B.9
Target implementation of a fuzzy system.

B.5 LabVIEW®†

LabVIEW or Laboratory Virtual Engineering Workbench is a product of National Instruments. It is a software development environment for data acquisition, instrument control, image acquisition, motion control, and presentation. LabVIEW is a complied graphical environment, which allows the user to create programs graphically through wired icons similar to creating a flowchart.

B.5.1 Introduction

LabVIEW® is a general programming language like high-level programming languages such as C or Basic, but LabVIEW is a higher-level. LabVIEW programs are called virtual instruments (VIs) which use icons to represent subroutines. It is similar to flow charting codes as you write them. The LabVIEW development environment uses the graphical programming language G.

B.5.2 Some Key Concepts

Block Diagram	Pictorial description or representation of a program or algorithm. In a G program, the block diagram consists of executable icons called nodes and wires that carry data between the nodes.
G programming	G is a convenient graphical data flow programming language on which LabVIEW® is based. G simplifies scientific computation, process monitoring and control, and applications of testing and measurement.
Control	Front panel object such as a knob, push button, or dial for entering data to a VI interactively or by programming.
Control terminal	Terminal linked to a control on the front panel, through which input data from the front panel passes to the block diagram.
Front panel	Interactive user interface of a VI. The front panel appearance imitates physical instruments, such as oscilloscopes and multimeters.
Indicator	Front panel object that displays output, such as a graph or turning on an LED.
Waveform chart	Indicator that plots data points at a certain rate.
While loop	Loop structure that repeats a code section until a given condition is met. It is comparable to a Do loop or a Repeat-Until in conventional programming languages.
Wire	Data path between nodes.

B.5.3 Working with LabVIEW®

As a software centered system, LabVIEW® resides in a desktop computer, laptop or PXI as an application where it acts as a set VIs, providing the functionality of traditional hardware instruments such as oscilloscopes. Comparing to physical instruments with fixed functions, LabVIEW VIs are flexible and can easily be reconfigured to different applications.

---

† For details see LabVIEW User Manual Glossary and G programming Reference Manual Glossary, which are available online at http://www.ni.com/pdf/manuals/320999b.pdf and http://www.ni.com/pdf/manuals/321296b.pdf, respectively.

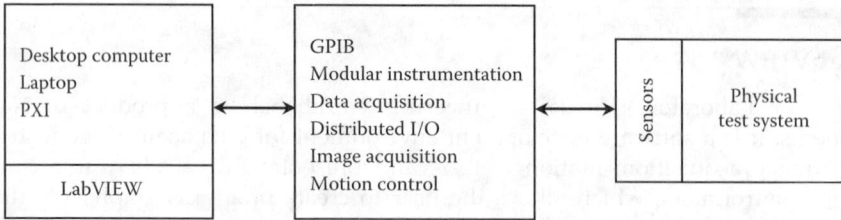

FIGURE B.10
Modular solution of LabVIEW.

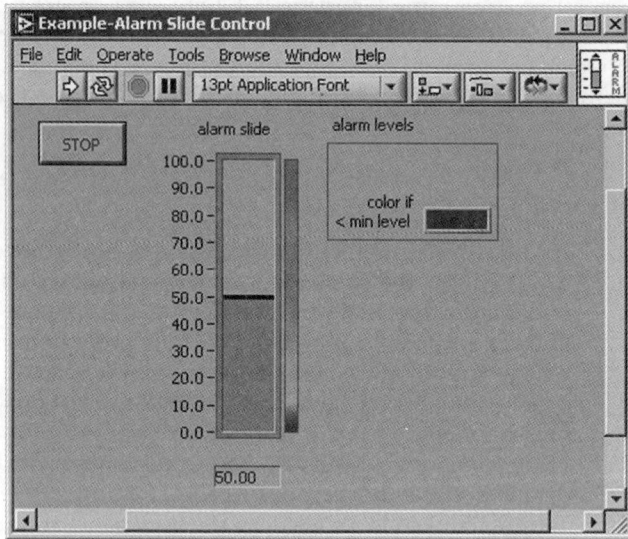

FIGURE B.11
Front panel of the alarm slide control example.

It is able to interface with various hardware devices such as GPIB, data acquisition modules, distributed I/O, image acquisition, and motion control, making it a modular solution. This utility is shown in Figure B.10.

B.5.3.1 Front Panel

Upon launching LabVIEW®, you will be able to create or open an existing VI where the layout of the GUI can be designed. Figure B.11 shows the front panel of the simple alarm slide control (alarmsld.lib) example included with LabVIEW suite of examples. This is the first phase in developing a VI. Buttons, indicators, I/O, and dialogs are placed appropriately. These control components are selected from the "Controls Palette," which contains a list of prebuilt library or user-customized components.

A component is selected from the controls palette by left-clicking the mouse on the particular control icon, and can be placed on the front panel by left-clicking again. Then the component can be resized, reshaped or moved to any desired position. A component property such as visibility, format, precision, labels, data range, or action can be changed by right-clicking, with the cursor placed anywhere on the selected component, to bring up the pop-up menu.

B.5.3.2 Block Diagrams

After designing the GUI in the front panel, the VI has to be programmed graphically through the block diagram window in order to implement the intended functionality of the VI. The block diagram window can be brought forward by clicking on the "Window" pull menu and selecting "Show Diagram." For every control component created on the front panel, there is a corresponding terminal automatically created in the block diagram window. Figure B.12 shows the block diagram for the alarm slide control example provided with LabVIEW®.

The terminal is labeled automatically according to the data type of each control. For example, the stop button has a terminal labeled TF, which is a Boolean type. The vertical level indicator has a DBL type terminal, indicating double-precision number. Other common controls with a DBL terminal include various numeric indicators, sliders, and graphs.

LabVIEW uses the G-programming language to implement the functionality of a VI. It provides an extensive library of basic conditional and looping structures, mathematical operators, Boolean operators, comparison operators, and more advanced analysis and conditioning tools, provided through the Functions Palette. A function may be placed on the block diagram window similar to how a control component is placed on the front panel. Depending on the required flow of execution, they are then wired together using the connect wire tool in the tools palette. In order to wire two terminals together, first click on the connect wire icon in the tools palette, then move the cursor to the input-output hotspot of one terminal, left-click to make the connection, and then move the cursor to the output-input hotspot of the other terminal and left-click again to complete the connection. The corresponding control component on the front panel can be selected by double clicking on the terminal block.

The general flow of execution is to first acquire the data, then analyze, followed by the presentation of results. The terminals and functional components are wired in such a way that data flows from the sources (e.g., data acquisition) to the sinks (e.g., presentation).

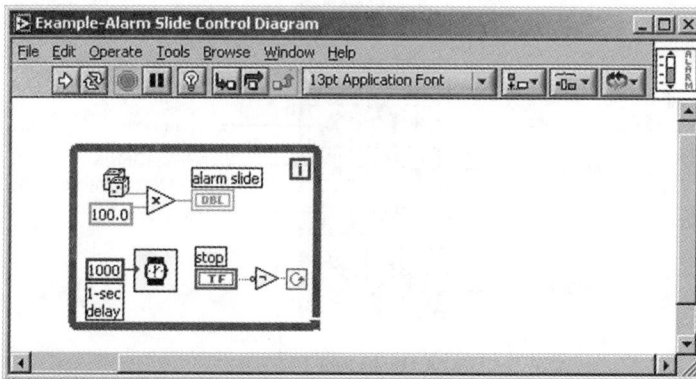

FIGURE B.12
Block diagram of the alarm slide control example.

LabVIEW executes its G-programming code in data flow manner, executing an icon as data becomes available to it through connecting wires.

The dice terminal is a random number generator and its output is multiplied by a constant using the multiplier operator (see Figure B.12). The multiplication result is connected to the input of the alarm slide, which will show up as the level in the vertical indicator on the front panel during VI execution. The gray box surrounding the terminals is the while loop in which all the flow within the gray box will run continuously until the loop is terminated by the stop button with the corresponding Boolean terminal. When the stop terminal is true, the while loop terminates upon reading a false through the not operator. The wait terminal (watch icon) controls the speed of the while loop. The wait terminal input is given in milliseconds. In the figure, the loop runs at an interval of one second since a constant of 1000 is wired to the wait terminal. In order to run the VI, left-click on the arrow icon on the top rows of icons or click on "Operate" and then select "Run." No compilation is required.

Note the remove broken wire command found in the edit pull-down menu. This command cleans up the block diagram of any unwanted or incomplete wiring. The debugging pop-up window that appears when an erroneous VI is executed is very helpful in troubleshooting of the VI. Double-clicking on the items in the errors list will automatically highlight the problematic areas or wires or terminals in the diagram.

B.5.3.3 Tools Palette

LabVIEW® has three main floating palettes for creating VIs. They are the tools palette, controls palette and functions palette. The tools palette, shown in Figure B.13, is the general editing palette with tools for editing components in the front panel and block diagram panel, modifying the position, shape and size of components, labeling, wiring of terminals in the block diagram panel, debugging, and coloring. When manipulating the front panel and the block diagram panel, note which tool icon is selected. For example, the values of a control or terminal cannot be selected or edited when the positioning icon is selected.

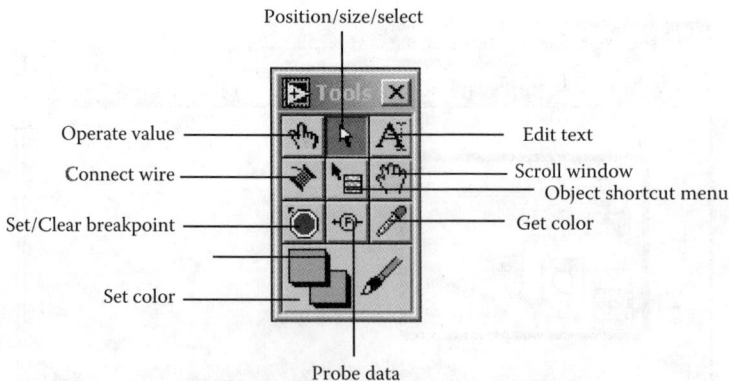

FIGURE B.13
LabVIEW tools palette.

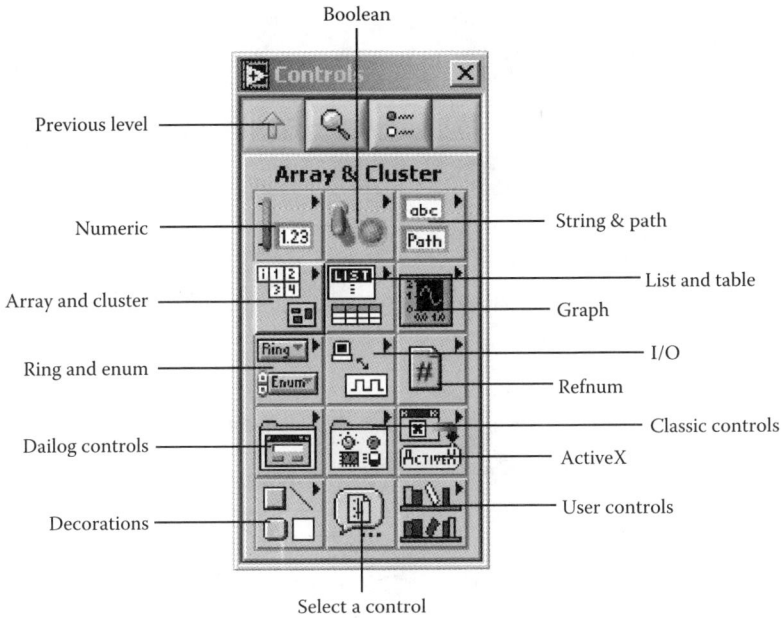

FIGURE B.14
LabVIEW controls palette.

B.5.3.4 Controls Palette

Figure B.14 shows the controls palette, which contains the pre-built and user-defined controls to create a GUI. This palette will be available when the front panel is selected. If it is not showing, click on the "Window" pull-down menu and select the "Show Controls Palette" option. The figure shows the main group of top-level components available in its pre-built library. Clicking on the appropriate top-level icons will bring up the subpalettes of the available controls and indicators. To go back to the top-level icons, click on the up arrow icon on the top-left of the controls palette.

B.5.3.5 Functions Palette

When the block diagram panel is selected, the functions palette is shown as in Figure B.15, enabling you to program the VI. The functions palette contains a complete library of necessary operations for developing the functionality of the VI. Similar to the controls palette, the top-level icons show the grouping of different sub-functions available for the programmer. Several commonly used groups are indicated below:

- Structures: The structures icon consists of the usual programming language sequences, conditional statements and conditional loops. These structures are in the form of boxes where the terminals within the boxes are executed when the statements or loops are invoked. In addition, there is a formula node where custom text-based formulas can be included if you prefer the traditional text-based equations. There are also variable declaration nodes where local and global variables can be declared.

Numeric

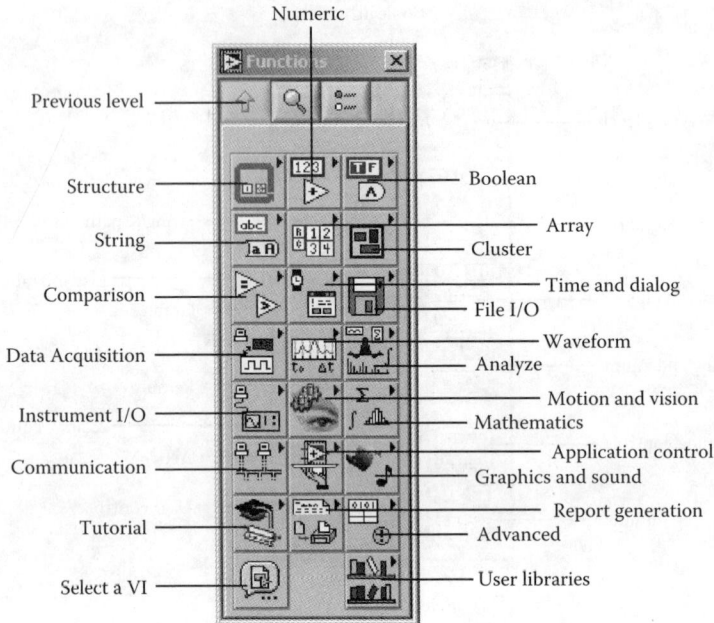

FIGURE B.15
LabVIEW functions palette.

- Numeric: The elementary operators such as summation, subtraction, multiplication, division, and power, are grouped under this icon.
- Boolean: This icon contains the Boolean operators required for logic manipulation.
- Array: The array grouping consists of tools for array manipulation.
- Comparison: Operators for numerical comparison, which provide Boolean outputs, are found under this icon.
- Analyze: This icons contains the more advanced analysis tools such as FFT spectrum, power spectrum, filters, triggering, and waveform generation.
- Mathematics: Under this icon, the tools for mathematical manipulation such as calculus, statistics and probability, linear algebra, optimization, and numeric functions are found.

B.6 LabVIEW® Sound and Vibration Tools

B.6.1 Sound and Vibration Toolkit

This section introduces the capabilities of the LabVIEW Sound and Vibration Toolkit (SVT). In particular, data can be simulated using the VIs located on the Generation palette as well as with other VIs, and can be analyzed using various function tools.

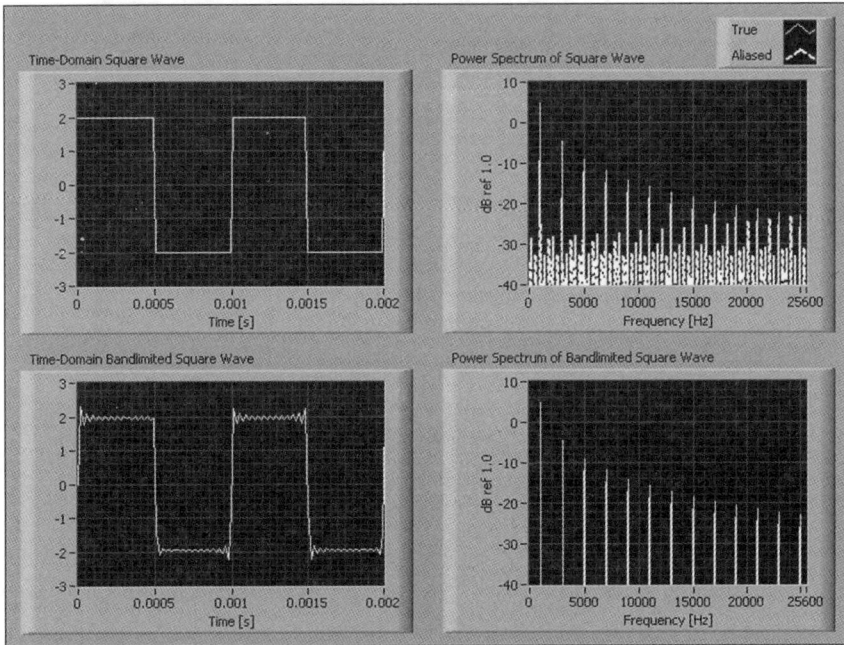

FIGURE B.16
Removal of aliasing through digital filtering.

B.6.2 Signal Acquisition and Simulation

Data may be obtained through a data acquisition (DAQ) device such as the National Instruments (NI) PXI-4461 or may be simulated by generating a VI of LabVIEW®. The NI PXI-4461 employs both digital and analog low-pass filters to provide antialiasing. Digital filtering of a square wave signal to remove aliasing is illustrated in Figure B.16.

B.6.2.1 Integration

The SVT contains the following integration VIs:

- SVT Integration VI located on the Integration palette for time-domain integration
- SVT Integration (frequency) VI located on the Frequency Analysis»Extended Measurements palette for frequency-domain integration

B.6.2.2 Vibration-Level Measurements

Vibration-level measurements can be made by using the Vibration Level VIs located on the Vibration Level palette. In particular, the following measurements can be made:

1. Root Mean Square (RMS) Level.
2. Peak Level.
3. Crest Factor (the ratio: peak value/RMS value).

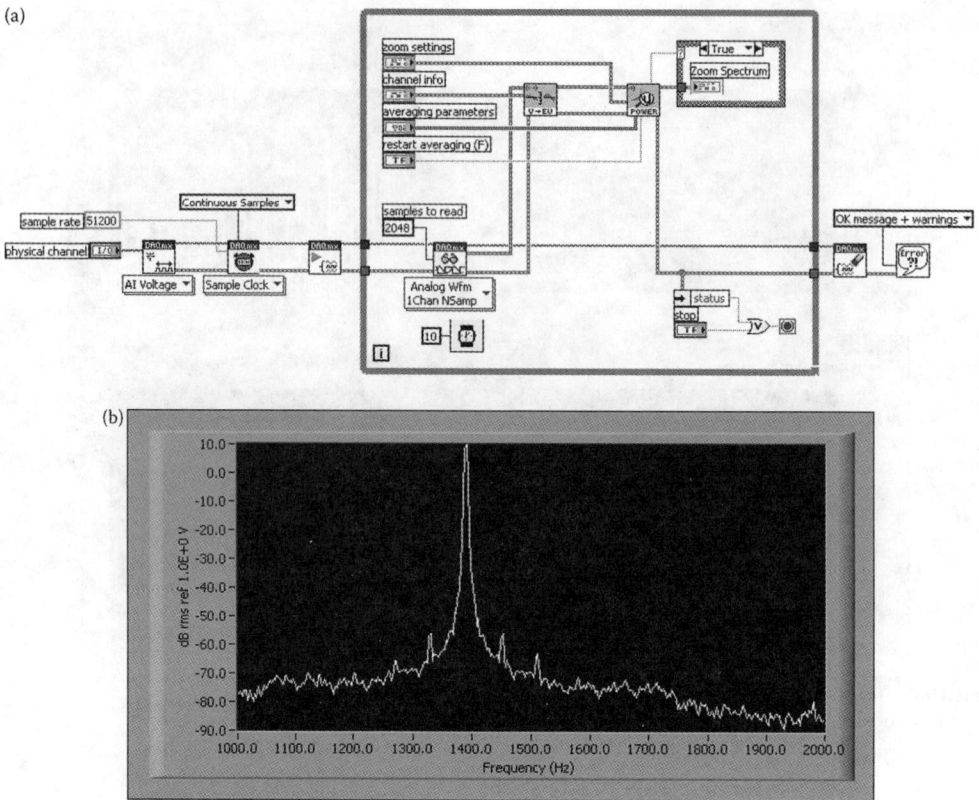

(a)

(b)

FIGURE B.17
(a) The use of the zoom FFT VI. (b) The power spectrum measured using zoom FFT.

B.6.2.3 Frequency Analysis

The Frequency Analysis VIs located on the Frequency Analysis palette may be used for windowing, averaging, and performing frequency-domain analysis. This is based on the discrete Fourier transform (DFT), specifically using FFT. The frequency resolution of the results can be improved by zooming into a required spectral region to observe the details of that spectral region. The use of the Zoom FFT VI for this purpose is illustrated in Figure B.17a. The resulted power spectrum is shown in Figure B.17b.

B.6.2.4 Transient Analysis

Transient analysis of a signal may be performed using the Transient Analysis VIs located on the Transient Analysis palette. 1. Use the STFT for signals in which the frequency content changes relatively slowly with time; 2. Use the shock response spectrum (SRS) for shock waves. Short-time Fourier transform (STFT) is carried out on sliding short intervals (sliding window) of a transient signal, and presented as the spectrum evolves with time. In rotating machine where the rotating speed is acquired simultaneously with the signal of interest, the STFT VIs can provide the frequency information as a function of

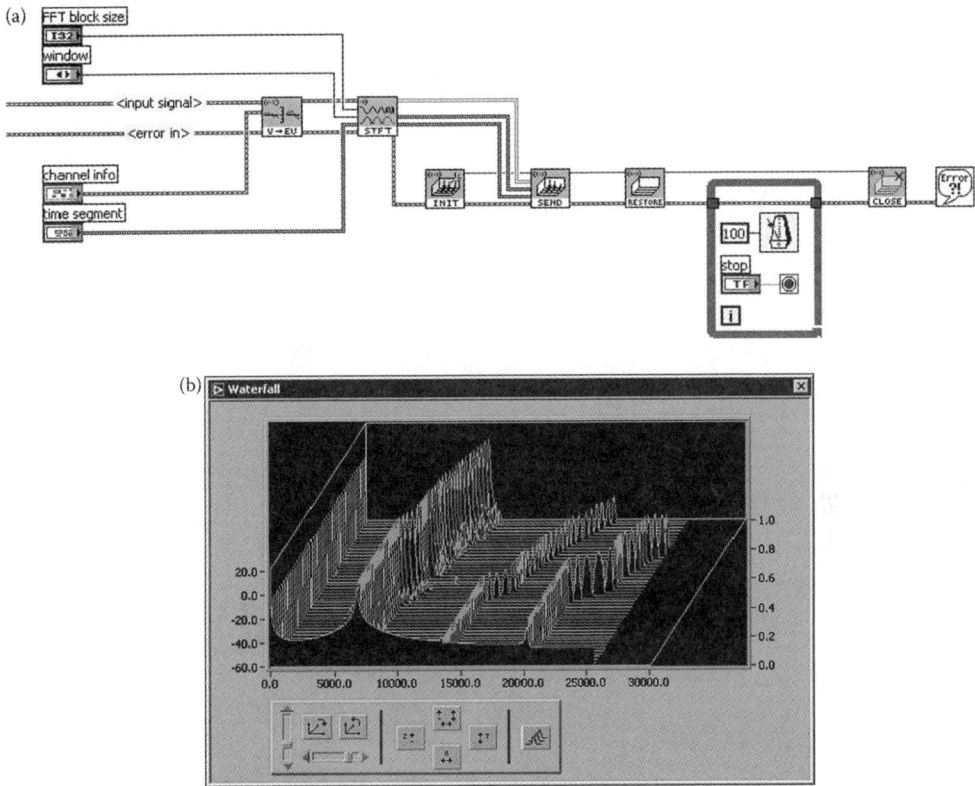

FIGURE B.18
(a) Generation of an STFT waterfall display. (b) A waterfall display.

the rotational speed. These results are typically displayed on a waterfall display or on a color map. Generation of a waterfall display is shown in Figure B.18a. A generated result is shown in Figure B.18b.

Appendix C: Review of Linear Algebra

Linear algebra, the algebra of *sets*, *vectors*, and *matrices*, is useful in the study of control systems in general and the state-space approach in particular. In practical engineering systems, interactions among various components are inevitable. There are many response variables associated with many excitations. Then, it is convenient to consider all excitations (inputs) simultaneously as a single variable and also all responses (outputs) as a single variable. Use of *linear algebra* makes the analysis of such a system convenient. The subject of linear algebra is complex and is based on a rigorous mathematical foundation. In this appendix we will review the basics of vectors and matrices.

C.1 Vectors and Matrices

In the analysis of control systems, vectors and matrices will be useful in both time and frequency domains. First, consider the time domain formulation of a mechanical system. For a single-degree-of-freedom (single-DoF) system with a single forcing excitation $f(t)$ and a corresponding single displacement response y, the dynamic equation would be

$$m\ddot{y} + c\dot{y} + ky = f(t) \tag{C.1}$$

In this single-DoF case, the quantities f, y, m, c, and k are *scalars*. If the system has n DoF, with excitation forces $f_1(t), f_2(t), \ldots, f_n(t)$ and associated displacement responses $y_1, y_2, \ldots, y_n$, the equations of motion may be expressed as

$$M\ddot{y} + C\dot{y} + Ky = f(t) \tag{C.2}$$

in which:

$$y = \begin{bmatrix} y_1 \\ y_2 \\ \vdots \\ y_n \end{bmatrix} = \text{displacement vector (} n\text{th order column vector)}$$

$$f = \begin{bmatrix} f_1 \\ f_2 \\ \vdots \\ f_n \end{bmatrix} = \text{forcing excitation vector (} n\text{th order column vector)}$$

$$M = \begin{bmatrix} m_{11} & m_{12} & \cdots & m_{1n} \\ m_{21} & m_{22} & \cdots & m_{2n} \\ \vdots & & & \\ m_{n1} & m_{n2} & \cdots & m_{nn} \end{bmatrix} = \text{mass matrix } (n \times n \text{ square matrix})$$

$$C = \begin{bmatrix} c_{11} & c_{12} & \cdots & c_{1n} \\ c_{21} & c_{22} & \cdots & c_{2n} \\ \vdots & & & \\ c_{n1} & c_{n2} & \cdots & c_{nn} \end{bmatrix} = \text{damping matrix } (n \times n \text{ square matrix})$$

$$K = \begin{bmatrix} k_{11} & k_{12} & \cdots & k_{1n} \\ k_{21} & k_{22} & \cdots & k_{2n} \\ \vdots & & & \\ k_{n1} & k_{n2} & \cdots & k_{nn} \end{bmatrix} = \text{stiffness matrix } (n \times n \text{ square matrix})$$

In this manner, vectors and matrices are introduced into the formulation of a multiDoF mechanical system. Further vector–matrix concepts will enter into the picture in subsequent analysis of the system; for example, in modal analysis.

Next consider the frequency-domain formulation. In the single-input–single-output (SISO) case, the system equation may be given as

$$y = Gu \tag{C.3}$$

in which:

$u = $ frequency spectrum (Fourier spectrum) of the forcing excitation (input)
$y = $ frequency spectrum (Fourier spectrum) of the response (output)
$G = $ frequency transfer function (frequency response function) of the system.

The quantities u, y and G are *scalars* because each one is a single quantity, and not a collection of several quantities.

Next, consider a multiinput–multioutput (MIMO) system having two excitations u_1 and u_2, and two responses y_1 and y_2; each y_i now depends on both u_1 and u_2. It follows that we need four transfer functions to represent all the excitation-response relationships that exist in this system. We use the four transfer functions (G_{11}, G_{12} G_{21} and G_{22}). For example, the transfer function G_{12} relates the excitation u_2 to the response y_1. The associated two equations that govern the system are:

$$y_1 = G_{11}u_1 + G_{12}u_2$$
$$y_2 = G_{21}u_1 + G_{22}u_2 \tag{C.4}$$

Instead of considering the two excitations (two inputs) as two separate quantities, we can consider them as a single "vector" u having the two components u_1 and u_2. As before, we can write this as a column consisting of the two elements:

$$u = \begin{bmatrix} u_1 \\ u_2 \end{bmatrix}$$

In this case we have a "column vector." Alternately, we can write a "row vector" as:

$$u = [u_1, u_2]$$

But, the column vector representation is more common.

Similarly, we can express the two outputs y_1 and y_2 as a vector y. Consequently, we have the column vector:

$$y = \begin{bmatrix} y_1 \\ y_2 \end{bmatrix}$$

or the row vector: $y = [y_1, y_2]$

It should be kept in mind that the order in which the components (or elements) are given is important since the vector $[u_1, u_2]$ is not equal to the vector $[u_2, u_1]$. In other words, a vector is an "ordered" collection of quantities.

Summarizing, we can express a collection of quantities, in an orderly manner, as a single vector. Each quantity in the vector is known as a *component* or an *element* of the vector. What each component means will depend on the particular situation. For example, in a dynamic system it may represent a quantity such as voltage, current, force, velocity, pressure, flow rate, temperature, or heat transfer rate. The number of components (elements) in a vector is called the *order*, or *dimension* of the vector.

Next let us introduce the concept of a matrix using the frequency domain example given above. Note that we needed four transfer functions to relate the two excitations to the two responses. Instead of considering these four quantities separately we can express them as a single matrix G having four elements. Specifically, the *transfer function matrix* for the present example is:

$$G = \begin{bmatrix} G_{11} & G_{12} \\ G_{21} & G_{22} \end{bmatrix}$$

This matrix has two rows and two columns. Hence the size or order of the matrix is 2×2. Since the number of rows is equal to the number of columns in this example, we have a *square matrix*. If the number of rows is not equal to the number of columns, we have a *rectangular matrix*. Actually, we can interpret a matrix as a collection of vectors. Hence, in the previous example, the matrix G is an assembly of the two column vectors

$$\begin{bmatrix} G_{11} \\ G_{21} \end{bmatrix} \quad \text{and} \quad \begin{bmatrix} G_{12} \\ G_{22} \end{bmatrix}$$

or, alternatively, an assembly of the two row vectors: $[G_{11}, G_{12}]$ and $[G_{21}, G_{22}]$.

C.2 Vector–Matrix Algebra

The advantage of representing the excitations and the responses of a control system as the vectors u and y, and the transfer functions as the matrix G is clear from the fact that the excitation-response (input–output) equations can be expressed as the single equation

$$y = Gu \tag{C.5}$$

instead of the collection of scalar equations (Equation C.4).

Hence the response vector y is obtained by "premultiplying" the excitation vector u by the transfer function matrix G. Of course, certain rules of vector–matrix multiplication have to be agreed upon in order that this single equation is consistent with the two scalar equations given by Equation C.4. Also, we have to agree upon rules for the addition of vectors or matrices.

A vector is a special case of a matrix. Specifically, a third-order column vector is a matrix having three rows and one column. Hence it is a 3×1 matrix. Similarly, a third-order row vector is a matrix having one row and three columns. Accordingly, it is a 1×3 matrix. It follows that we only need to know matrix algebra, and the vector algebra will follow from the results for matrices.

C.2.1 Matrix Addition and Subtraction

Only matrices of the same size can be added. The result (sum) will also be a matrix of the same size. In matrix addition, we add the corresponding elements (i.e., the elements at the same position) in the two matrices, and write the results at the corresponding places in the resulting matrix.

As an example, consider the 2×3 matrix: $A = \begin{bmatrix} -1 & 0 & 3 \\ 2 & 6 & -2 \end{bmatrix}$

and a second matrix: $B = \begin{bmatrix} 2 & 1 & -5 \\ 0 & -3 & 2 \end{bmatrix}$

The sum of these two matrices is given by: $A + B = \begin{bmatrix} 1 & 1 & -2 \\ 2 & 3 & 0 \end{bmatrix}$

The order in which the addition is done is immaterial. Hence:

$$A + B = B + A \tag{C.6}$$

In other words, matrix addition is *commutative*.

Matrix subtraction is defined just like matrix addition, except the corresponding elements are subtracted. An example is given below:

$$\begin{bmatrix} -1 & 2 \\ 3 & 0 \\ -4 & 1 \end{bmatrix} - \begin{bmatrix} 4 & 2 \\ 2 & -1 \\ -3 & 0 \end{bmatrix} = \begin{bmatrix} -5 & 0 \\ 1 & 1 \\ -1 & 1 \end{bmatrix}$$

C.2.2 Null Matrix

The null matrix is a matrix whose elements are all zeros. Hence when we add a null matrix to an arbitrary matrix the result is equal to the original matrix. We can define a *null vector* in a similar manner. We can write

$$A+0=A \tag{C.7}$$

As an example, the 2×2 null matrix is:

$$\begin{bmatrix} 0 & 0 \\ 0 & 0 \end{bmatrix}$$

C.2.3 Matrix Multiplication

Consider the product AB of the two matrices A and B. Let us write this as:

$$C=AB \tag{C.8}$$

We say that B is *premultiplied* by A or, equivalently, A is *postmultiplied* by B. For this multiplication to be possible, the number of columns in A has to be equal to the number of rows in B. Then, the number of rows of the product matrix C is equal to the number of rows in A, and the number of columns in C is equal to the number of columns in B.

The actual multiplication is done by multiplying the elements in a given row (say the ith row) of A by the corresponding elements in a given column (say the, jth column) of B and summing these products. The result is the element c_{ij} of the product matrix C. Note that c_{ij} denotes the element that is common to the ith row and the jth column of matrix C. So, we have:

$$c_{ij} = \sum_k a_{ik} b_{kj} \tag{C.9}$$

As an example, suppose:

$$A = \begin{bmatrix} 1 & 2 & -1 \\ 3 & -3 & 4 \end{bmatrix}; B = \begin{bmatrix} 1 & -1 & 2 & 4 \\ 2 & 3 & -4 & 2 \\ 5 & -3 & 1 & 0 \end{bmatrix}$$

Note that the number of columns in A is equal to three and the number of rows in B is also equal to three. Hence we can perform the premultiplication of B by A. For example:

$$c_{11}=1\times1+2\times2+(-1)\times5=0$$

$$c_{12}=1\times(-1)+2\times3+(-1)\times(-3)=8$$

$$c_{13}=1\times2+2\times(-4)+(-1)\times1=-7$$

$$c_{14}=1\times4+2\times2+(-1)\times0=8$$

$$c_{21}=3\times1+(-3)\times2+4\times5=17$$

$$c_{22}=3\times(-1)+(-3)\times3+4\times(-3)=-24$$

and so on. The product matrix is:

$$C = \begin{bmatrix} 0 & 8 & -7 & 8 \\ 17 & -24 & 22 & 6 \end{bmatrix}$$

It should be noted that both products AB and BA are not always defined, and even when they are defined, the two results are not equal in general. Unless both A and B are square matrices of the same order, the two product matrices will not be of the same order.

Summarizing, matrix multiplication is not commutative:

$$AB \neq BA \tag{C.10}$$

C.2.4 Identity Matrix

An identity matrix (or unity matrix) is a square matrix whose diagonal elements are all equal to 1 and all the remaining elements are zeros. This matrix is denoted by I. For example, the third-order identity matrix is:

$$I = \begin{bmatrix} 1 & 0 & 0 \\ 0 & 1 & 0 \\ 0 & 0 & 1 \end{bmatrix}$$

It is easy to see that when any matrix is multiplied by an identity matrix (provided, of course, that the multiplication is possible) the product is equal to the original matrix. Thus

$$A I = I A = A \tag{C.11}$$

C.3 Matrix Inverse

An operation similar to scalar division can be defined in terms of the inverse of a matrix. A proper inverse is defined only for a square matrix, and even for a square matrix, an inverse might not exist. The inverse of a matrix is defined as follows:

Suppose that a square matrix A has the inverse B. Then these must satisfy the equation:

$$AB = I \tag{C.12}$$

or, equivalently

$$BA=I \tag{C.13}$$

where I is the identity matrix, as defined before.

The inverse of A is denoted by A^{-1}. The inverse exists for a matrix if and only if (iff) the *determinant* of the matrix is nonzero. Such matrices are termed *nonsingular*. We shall discuss the determinant in a later subsection of this Appendix. But, before explaining a method for determining the inverse of a matrix let us verify that:

$$\begin{bmatrix} 2 & 1 \\ 1 & 1 \end{bmatrix} \text{ is the inverse of } \begin{bmatrix} 1 & -1 \\ -1 & 2 \end{bmatrix}$$

To show this we simply multiply the two matrices and show that the product is the second order unity matrix. Specifically,

$$\begin{bmatrix} 1 & -1 \\ -1 & 2 \end{bmatrix} \begin{bmatrix} 2 & 1 \\ 1 & 1 \end{bmatrix} = \begin{bmatrix} 1 & 0 \\ 0 & 1 \end{bmatrix}$$

or

$$\begin{bmatrix} 2 & 1 \\ 1 & 1 \end{bmatrix} \begin{bmatrix} 1 & -1 \\ -1 & 2 \end{bmatrix} = \begin{bmatrix} 1 & 0 \\ 0 & 1 \end{bmatrix}$$

C.3.1 Matrix Transpose

The transpose of a matrix is obtained by simply interchanging the rows and the columns of the matrix. The transpose of A is denoted by A^T.

For example, the transpose of the 2×3 matrix: $A = \begin{bmatrix} 1 & -2 & 3 \\ -2 & 2 & 0 \end{bmatrix}$

is the 3×2 matrix: $A^T = \begin{bmatrix} 1 & -2 \\ -2 & 2 \\ 3 & 0 \end{bmatrix}$

Note: The first row of the original matrix has become the first column of the transposed matrix, and the second row of the original matrix has become the second column of the transposed matrix.

If $A^T=A$ then we say that the matrix A is *symmetric*. Another useful result on the matrix transpose is expressed by

$$(AB)^T=B^T A^T \tag{C.14}$$

It follows that the transpose of a matrix product is equal to the product of the transposed matrices, taken in the reverse order.

C.3.2 Trace of a Matrix

The trace of a square matrix is given by the sum of the diagonal elements. The trace of matrix A is denoted by $\mathrm{tr}(A)$.

$$\mathrm{tr}(A) = \sum_i a_{ii} \tag{C.15}$$

For example, the trace of the matrix: $A = \begin{bmatrix} -2 & 3 & 0 \\ 4 & -4 & 1 \\ -1 & 0 & 3 \end{bmatrix}$

is given by: $\mathrm{tr}(A) = (-2) + (-4) + 3 = -3$.

C.3.3 Determinant of a Matrix

The determinant is defined only for a square matrix. It is a scalar value computed from the elements of the matrix. The determinant of a matrix A is denoted by $\det(A)$ or $|A|$.

Instead of giving a complex mathematical formula for the determinant of a general matrix in terms of the elements of the matrix, we now explain a way to compute the determinant.

First consider the 2×2 matrix: $A = \begin{bmatrix} a_{11} & a_{12} \\ a_{21} & a_{22} \end{bmatrix}$

Its determinant is given by: $\det(A) = a_{11}a_{22} - a_{12}a_{21}$

Next consider the 3×3 matrix: $A = \begin{bmatrix} a_{11} & a_{12} & a_{13} \\ a_{21} & a_{22} & a_{23} \\ a_{31} & a_{32} & a_{33} \end{bmatrix}$

Its determinant can be expressed as: $\det(A) = a_{11}M_{11} - a_{12}M_{12} + a_{13}M_{13}$
where, the *minors* of the associated matrix elements are defined as:

$$M_{11} = \det \begin{bmatrix} a_{22} & a_{23} \\ a_{32} & a_{33} \end{bmatrix}; M_{12} = \det \begin{bmatrix} a_{21} & a_{22} \\ a_{31} & a_{32} \end{bmatrix}; M_{13} = \det \begin{bmatrix} a_{21} & a_{22} \\ a_{31} & a_{32} \end{bmatrix}$$

Note that M_{ij} is the determinant of the matrix obtained by deleting the ith row and the jth column of the original matrix. The quantity M_{ij} is known as the *minor* of the element a_{ij} of the matrix A. If we attach a proper sign to the minor depending on the position of the corresponding matrix element, we have a quantity known as the *cofactor*. Specifically, the cofactor C_{ij} corresponding to the minor M_{ij} is given by

$$C_{ij} = (-1)^{i+j} M_{ij} \tag{C.16}$$

Hence the determinant of the 3×3 matrix may be given by:

$$\det(A) = a_{11}C_{11} + a_{12} + C_{12} + a_{13} C_{13}$$

In the two formulas given above for computing the determinant of a 3×3 matrix, we have expanded along the first row of the matrix. We get the same answer, however, if we expand along any row or any column. Specifically, when expanded along the ith row we have:

$$\det(A) = a_{i1}C_{i1} + a_{i2}C_{i2} + a_{i3} C_{i3}$$

Similarly, if we expand along the jth column we have:

$$\det(A) = a_{1j}C_{1j} + a_{2j} C_{2j} + a_{3j} C_{3j}$$

These ideas of computing a determinant can be easily extended to 4×4 and higher-order matrices in a straightforward manner. Hence, we can write

$$\det(A) = \sum_j a_{ij}C_{ij} = \sum_i a_{ij}C_{ij} \tag{C.17}$$

C.3.4 Adjoint of a Matrix

The adjoint of a matrix is the transpose of the matrix whose elements are the cofactors of the corresponding elements of the original matrix. The adjoint of matrix A is denoted by adj(A).

As an example, in the 3×3 case we have:

$$\text{adj}(A) = \begin{bmatrix} C_{11} & C_{12} & C_{13} \\ C_{21} & C_{22} & C_{23} \\ C_{31} & C_{32} & C_{33} \end{bmatrix}^T = \begin{bmatrix} C_{11} & C_{21} & C_{31} \\ C_{12} & C_{22} & C_{32} \\ C_{13} & C_{23} & C_{33} \end{bmatrix}$$

In particular, it is easily seen that the adjoint of the matrix: $A = \begin{bmatrix} 1 & 2 & -1 \\ 0 & 3 & 2 \\ 1 & 1 & 1 \end{bmatrix}$

is given by: $\text{adj}(A) = \begin{bmatrix} 1 & 2 & -3 \\ -3 & 2 & 1 \\ 7 & -2 & 3 \end{bmatrix}^T$

Accordingly we have: $\text{adj}(A) = \begin{bmatrix} 1 & -3 & 7 \\ 2 & 2 & -2 \\ -3 & 1 & 3 \end{bmatrix}$

Hence, in general:

$$\text{adj}(A) = [C_{ij}]^T \tag{C.18}$$

C.3.5 Inverse of a Matrix

At this juncture it is appropriate to give a formula for the inverse of a square matrix. Specifically :

$$A^{-1} = \frac{\text{adj}(A)}{\det(A)} \tag{C.19}$$

Hence in the 3×3 matrix example given before, since we have already determined the adjoint, it remains only to compute the determinant in order to obtain the inverse. Now expanding along the first row of the matrix, the determinant is given by

$$\det(A) = 1 \times 1 + 2 \times 2 + (-1) \times (-3) = 8$$

Accordingly, the inverse is given by:

$$A^{-1} = \frac{1}{8} \begin{bmatrix} 1 & -3 & 7 \\ 2 & 2 & -2 \\ -3 & 1 & 3 \end{bmatrix}$$

For two square matrices A and B we have:

$$(AB)^{-1} = B^{-1} A^{-1} \tag{C.20}$$

As a final note, if the determinant of a matrix is zero, the matrix does not have an inverse. Then we say that the matrix is *singular*. Some important matrix properties are summarized in Box C.1.

BOX C.1 SUMMARY OF MATRIX PROPERTIES

Addition:	$A_{m \times n} + B_{m \times n} = C_{m \times n}$
Multiplication:	$A_{m \times n} + B_{n \times r} = C_{m \times r}$
Identity:	$AI = IA = A \Rightarrow I$ is the identity matrix
Note:	$AB = 0 \nRightarrow A = 0$ or $B = 0$ in general
Transposition:	$C^T = (AB)^T = B^T A^T$
Inverse:	$AP = I = PA \Rightarrow A = P^{-1}$ and $P = A^{-1}$
	$(AB)^{-1} = B^{-1} A^{-1}$
Commutativity:	$AB \neq BA$ in general
Associativity:	$(AB)C = A(BC)$
Distributivity:	$C(A+B) = CA + CB$
Distributivity:	$(A+B)D = AD + BD$

C.4 Vector Spaces

C.4.1 Field (*F*)

Consider a set of scalars.

If for any α and β from the set, $\alpha + \beta$ and $\alpha\beta$ are also elements in the set and if:

1. $\alpha + \beta = \beta + \alpha$ and $\alpha\beta = \beta\alpha$ (Commutativity)
2. $(\alpha + \beta) + \gamma = \alpha + (\beta + \gamma)$ and $(\alpha\beta)\gamma = \alpha(\beta\gamma)$ (Associativity)
3. $\alpha(\beta + \gamma) = \alpha\beta + \alpha\gamma$ (Distributivity)

are satisfied,
and if:

1. Identity elements 0 and 1 exist in the set such that $\alpha + 0 = \alpha$ and $1\alpha = \alpha$
2. Inverse elements exist in the set such that $\alpha + (-\alpha) = 0$

and $\alpha \cdot \alpha^{-1} = 1$
then, the set is a field.

Example: The set of real numbers.

C.4.2 Vector Space (*L*)

Properties:

1. Vector addition $(x + y)$ and scalar multiplication (αx) are defined.
2. Commutativity: $x + y = y + x$
   Associativity: $(x + y) + z = x + (y + z)$
   are satisfied.
3. Unique null vector $\mathbf{0}$ and negation $(-x)$ exist such that $x + \mathbf{0} = x$

$$x + (-x) = \mathbf{0}.$$

4. Scalar multiplication satisfies

$\alpha(\beta x) = (\alpha\beta)x$ (Associativity)

$$\left.\begin{array}{l} \alpha(x + y) = \alpha x + \beta y \\ (\alpha + \beta)x = \alpha x + \beta x \end{array}\right\} \text{(Distribuitivity)}$$

$1x = x, \ 0x = \mathbf{0}.$

Special Case: Vector space L^n has vectors with n elements from the field F.

$$\text{Consider } x = \begin{bmatrix} x_1 \\ x_2 \\ \cdot \\ \cdot \\ \cdot \\ x_n \end{bmatrix}, \; y = \begin{bmatrix} y_1 \\ y_2 \\ \cdot \\ \cdot \\ \cdot \\ y_n \end{bmatrix}$$

Then

$$x + y = \begin{bmatrix} x_1 + y_1 \\ \cdot \\ \cdot \\ \cdot \\ x_n + y_n \end{bmatrix} = y + x \quad \text{and} \quad \alpha x = \begin{bmatrix} \alpha x_1 \\ \cdot \\ \cdot \\ \cdot \\ \alpha x_n \end{bmatrix}$$

C.4.3 Subspace S of L

1. If x and y are in S then $x + y$ is also in S.
2. If x is in S and α is in F then αx is also in S.

C.4.4 Linear Dependence

Consider the set of vectors: $x_1, x_2, \ldots, x_n$

They are linearly independent if any one of these vectors cannot be expressed as a linear combination of one or more remaining vectors.

Necessary and sufficient condition for linear independence:

$$\alpha_1 x_1 + \alpha_2 x_2 + \ldots \alpha_n x_n = 0 \tag{C.21}$$

gives $\alpha = 0$ (trivial solution) as the only solution.

Example: $x_1 = \begin{bmatrix} 1 \\ 2 \\ 3 \end{bmatrix}; x_2 = \begin{bmatrix} 2 \\ -1 \\ 1 \end{bmatrix}; x_3 = \begin{bmatrix} 5 \\ 0 \\ 5 \end{bmatrix}$

These vectors are not linearly independent because, $x_1 + 2x_2 = x_3$.

C.4.5 Bases and Dimension of a Vector Space

1. If a set of vectors can be combined to form any vector in L then that set of vectors is said to *span* the vector space L (i.e., a generating system of vectors).
2. If the spanning vectors are all linearly independent, then this set of vectors is a *basis* for that vector space.
3. The number of vectors in the basis = dimension of the vector space.

Note: Dimension of a vector space is not necessarily the order of the vectors.

Example: Consider two intersecting third-order vectors. The will form a basis for the plane (two dimensional) that contains the two vectors. Hence, the dimension of the vector space=2, but the order of each vector in the basis=3.

Note: L^n is spanned by n linearly independent vectors $\Rightarrow \dim(L^n)=n$

$$\text{Example:} \begin{bmatrix} 1 \\ 0 \\ 0 \\ \cdot \\ \cdot \\ 0 \end{bmatrix}, \begin{bmatrix} 0 \\ 1 \\ 0 \\ \cdot \\ \cdot \\ 0 \end{bmatrix}, \cdots, \begin{bmatrix} 0 \\ 0 \\ \cdot \\ \cdot \\ 0 \\ 1 \end{bmatrix}$$

C.4.6 Inner Product

$$(x,y)=y^H x \tag{C.22}$$

where "H" denotes the *hermitian transpose* (i.e., complex conjugate and transpose). Hence $y^H=(y^*)^T$ where $(\ )^*$ denotes complex conjugation.

Note:

1. $(x,x) \geq 0$ and $(x,x)=0$ iff $x=0$
2. $(x,y)=(y,x)^*$
3. $(\lambda x,y)=\lambda(x,y)$
   $(x,\lambda y)=\lambda^*(x,y)$
4. $(x,y+z)=(x,y)+(x,z)$

C.4.7 Norm

Properties:

$||x|| \geq 0$ and $||x||=0$ iff $x=0$

$||\lambda x||=|\lambda|\,||x||$ for any scalar λ

$||x+y|| \leq ||x||+||y||$

Example: Euclidean norm: $||x||=x^H x = \left(\sum_{i=1}^{n} x_i^2 \right)^{1/2}$ (C.23)

Unit Vector: $||x||=1$

Normalization: $\dfrac{x}{||x||}=\hat{x}$

Angle Between Vectors: We have $\cos\theta = \dfrac{(x,y)}{||x||\,||y||}=(\hat{x},\hat{y})$ (C.24) where θ is the angle between x and y.

Orthogonal Vectors: iff $(x,y)=0$ (C.25)

Note: n orthogonal vectors in L^n are linearly independent and span L^n, and form a basis for L^n

C.4.8 Gram–Schmidt Orthogonalization

Given a set of vectors $x_1,x_2,...,x_n$ that are linearly independent in L^n, we construct a set of orthonormal (orthogonal and normalized) vectors $\hat{y}_1,\hat{y}_2,...\hat{y}_n$ which are linear combinations of $\hat{x}_i$

Start $\hat{y}_1 = \hat{x}_1 = \dfrac{x_1}{\|x_1\|}$

Then $y_i = x_i - \displaystyle\sum_{j=1}^{i-1}(x_i,\hat{y}_j)\hat{y}_j$ for $i=1,2,3,...,n$

Normalize y_i to produce $\hat{y}_i$.

C.4.9 Modified Gram–Schmidt Procedure

In each step compute new vectors that are orthogonal to the just-computed vector.

Step 1: $\hat{y}_1 = \dfrac{x_1}{\|x_1\|}$ as before.

Then $x_i^{(1)} = x_i - (\hat{y}_1,x_i)\hat{y}_1$ for $i=1,2,3,...,n$

$$\hat{y}_i = \frac{x_i^{(1)}}{\|x_i^{(1)}\|} \text{ for } i=2,3,...n$$

and $x_i^{(2)} = x_i^{(1)} - (\hat{y}_2,x_i^{(1)})\hat{y}_2$, $i=3,4,..., n$ and so on.

C.5 Determinants

Now, let us address several analytical issues of the determinant of a square matrix. Consider the matrix:

$$A = \begin{bmatrix} a_{11} & \cdot & \cdot & a_{1n} \\ & \cdot & & \\ & & \cdot & \\ a_{n1} & \cdot & \cdot & a_{nn} \end{bmatrix}$$

Minor of $a_{ij}=M_{ij}=$ determinant of matrix formed by deleting the ith row and the jth column of the original matrix.

Cofactor of $a_{ij}=C_{ij}=(-1)^{i+j}M_{ij}$

$\text{cof}(A)=$ cofactor matrix of A

$\text{adj}(A)=$ adjoint $A=(\text{cof }A)^T$

C.5.1 Properties of Determinant of a Matrix

1. Interchange two rows (columns) $\Rightarrow$ Determinant sign changes.
2. Multiply one row (column) by $\alpha \Rightarrow \alpha \det ()$.
3. Add a $[\alpha \times$ row (column)] to a second row (column) $\Rightarrow$ determinant unchanged.
4. Identical rows (columns) $\Rightarrow$ zero determinant.
5. For two square matrices A and B, $\det(AB)=\det(A)\det(B)$.

C.5.2 Rank of a Matrix

Rank $A=$ number of linearly independent columns $=$ number of linearly independent rows $=$ dim (column space) $=$ dim (row space)

Here "dim" denotes the "dimension of."

C.6 System of Linear Equations

Consider the set of linear algebraic equations:

$$a_{11}x_1 + a_{12}x_2 + \cdots + a_{1n}x_n = c_1$$
$$a_{21}x_1 + a_{22}x_2 + \cdots + a_{2n}x_n = c_2$$
$$\vdots$$
$$a_{m1}x_1 + a_{m2}x_2 + \cdots + a_{mn}x_n = c_m$$

We need to solve for $x_1, x_2, \ldots, x_n$.

This problem can be expressed in the vector–matrix form:

$$A_{m \times n}x_n = c_m \quad B=[A,c]$$

Solution exists iff rank $[A, c]=$ rank $[A]$

Two cases can be considered:

Case 1: If $m \geq n$ and rank $[A]=n \Rightarrow$ unique solution for x.

Case 2: If $m \geq n$ and rank $[A]=m \Rightarrow$ infinite number of solutions for x.

$$x=A^H(AA^H)^{-1}C \Leftarrow \text{minimum norm form}$$

Specifically, out of the infinite possibilities, this is the solution that minimizes the norm $x^H x$.

Note: The superscript "H" denotes the "hermitian transpose," which is the transpose of the complex conjugate of the matrix:

Example: $A = \begin{bmatrix} 1+j & 2+3j & 6 \\ 3-j & 5 & -1-2j \end{bmatrix}$

$$\text{Then } A^H = \begin{bmatrix} 1-j & 3+j \\ 2-3j & 5 \\ 6 & -1+2j \end{bmatrix}$$

If the matrix is real, its hermitian transpose is simply the ordinary transpose.
In general if rank $[A] \leq n \Rightarrow$ infinite number of solutions.
The space formed by solutions $Ax=0 \Rightarrow$ is called the *null space*
dim (null space)$=n-k$ where rank $[A]=k$.

C.7 Quadratic Forms

Consider a vecor x and a square matrix A. Then the function $Q(x)=(x, Ax)$ is called a quadratic form. For a real vector x and a real and symmetric matrix A:

$$Q(x)=x^T Ax$$

Positive Definite Matrix: If $(x, Ax)>0$ for all $x \neq 0$, then A is said to be a positive definite matrix. Also, the corresponding quadratic form is also said to be positive definite.

Positive Semidefinite Matrix: If $(x, Ax) \geq 0$ for all $x \neq 0$, then A is said to be a positive semidefinite matrix. Note that in this case the quadratic form can assume a zero value for a nonzero x. Also, the corresponding quadratic form is also said to be positive semidefinite.

Negative Definite Matrix: If $(x, Ax)<0$ for all $x \neq 0$, then A is said to be a negative definite matrix. Also, the corresponding quadratic form is also said to be negative definite.

Negative Semidefinite Matrix: If $(x, Ax) \leq 0$ for all $x \neq 0$, then A is said to be a negative semidefinite matrix. Note that in this case the quadratic form can assume a zero value for a nonzero x. Also, the corresponding quadratic form is also said to be negative semidefinite.

Note: If A is positive definite, then $-A$ is negative definite. If A is positive semidefinite, then $-A$ is negative semidefinite.

Principal Minors: Consider the matrix:

$$A = \begin{bmatrix} a_{11} & a_{12} & \cdots & a_{1n} \\ a_{21} & a_{22} & \cdots & a_{2n} \\ \vdots & & & \\ a_{n1} & a_{n2} & \cdots & a_{nn} \end{bmatrix}$$

Its principal minors are the determinants of the various matrices along the principal diagonal, as given by:

$$\Delta_1=a_{11}, \Delta_2 = \det \begin{bmatrix} a_{11} & a_{12} \\ a_{21} & a_{22} \end{bmatrix}, \Delta_3 = \det \begin{bmatrix} a_{11} & a_{12} & a_{13} \\ a_{21} & a_{22} & a_{23} \\ a_{31} & a_{32} & a_{33} \end{bmatrix}, \text{ and so on.}$$

Sylvester's Theorem: A matrix is positive if definite if all its principal minors are positive.

C.8 Matrix Eigenvalue Problem

C.8.1 Characteristic Polynomial

Consider a square matrix A. The polynomial: $\Delta(s) = \det[sI - A]$ is called the characteristic polynomial of A.

C.8.2 Characteristic Equation

The polynomial equation: $\Delta(s) = \det[sI - A] = 0$ is called the characteristic equation of the square matrix A.

C.8.3 Eigenvalues

The roots of the characteristic equation of a square matrix A are the eigenvalues of A. For an $n \times n$ matrix, there will be n eigenvalues.

C.8.4 Eigenvectors

The eigenvalue problem of a square matrix A is given by: $Av = \lambda v$
where, the objective is to solve for λ and the corresponding nontrivial (i.e., nonzero) solutions for v. The problem can be expressed as:

$$(\lambda I - A)v = 0$$

Note: If v is a solution of this equation, then any multiple av of it is also a solution. Hence, an eigenvector is arbitrary up to a multiplication factor.
For a nontrivial (i.e., nonzero) solution to be possible for v, one must have

$$\det[\lambda I - A] - 0$$

Since this is the characteristic equation of A, as defined above, it is clear that the roots of λ are the eigenvalues of A. The corresponding solutions for v are the eigenvectors of A. For an $n \times n$ matrix, there will be n eigenvalues and n corresponding eigenvectors.

C.9 Matrix Transformations

C.9.1 Similarity Transformation

Consider a square matrix A and a nonsingular (and square) matrix T. Then, the matrix obtained according to:

$$B = T^{-1}AT$$

is the similarity transformation of A by T. The transformed matrix B has the same eigenvalues as the original matrix A. Also, A and B are said to be similar.

C.9.2 Orthogonal Transformation

Consider a square matrix A and another square matrix T. Then, the matrix obtained according to:

$$B = T^T A T$$

is the orthogonal transformation of A by T.

If $T^{-1} = T^T$ then the matrix T is said to be an orthogonal matrix. In this case, the similarity transformation and the orthogonal transformation become identical.

C.10 Matrix Exponential

The matrix exponential is given by the infinite series:

$$\exp(At) = I + At + \frac{1}{2!} A^2 t^2 + \dots \tag{C.26}$$

exactly like the scalar exponential:

$$\exp(\lambda t) = 1 + \lambda t + \frac{1}{2!} \lambda^2 t^2 + \dots \tag{C.27}$$

The matrix exponential maybe determined by reducing the infinite series given in Equation C.26 into a finite matrix polynomial of order $n - 1$ (where, A is $n \times n$) by using the Cayley–Hamilton theorem.

Cayley–Hamilton Theorem: This theorem states that a matrix satisfies its own characteristic equation. The characteristic polynomial of A can be expressed as:

$$\Delta(\lambda) = \det(A - \lambda I) = a_n \lambda^n + a_{n-1} \lambda^{n-1} + \dots + a_0 \tag{C.28}$$

in which det( ) denotes determinant. The notation:

$$\Delta(A) = a_n A^n + a_{n-1} A^{n-1} + \dots + a_0 I \tag{C.29}$$

is used. Then, by the Cayley–Hamilton theorem, we have:

$$0 = a_n A^n + a_{n-1} A^{n-1} + \dots + a_0 I \tag{C.30}$$

C.10.1 Computation of Matrix Exponential

Using Cayley–Hamilton theorem, we can obtain a finite polynomial expansion for $\exp(At)$. First we express Equations C.26 and C.27 as:

$$\exp(At) = S(A) \cdot \Delta(A) + \alpha_{n-1} A^{n-1} + \alpha_{n-2} A^{n-2} + \ldots + \alpha_0 I \tag{C.31}$$

$$\exp(\lambda t) = S(\lambda) \cdot \Delta(\lambda) + \alpha_{n-1}\lambda^{n-1} + \alpha_{n-2}\lambda^{n-2} + \ldots + \alpha_0 \tag{C.32}$$

in which $S(.)$ is an appropriate infinite series, which is the result of dividing the exponential (infinite) series by the characteristic polynomial $\Delta(.)$.

Next, since $\Delta(A) = 0$ by the Cayley–Hamilton theorem, Equation C.31 becomes:

$$\exp(At) = \alpha_{n-1}A^{n-1} + \alpha_{n-2}A^{n-2} + \ldots + \alpha_0 I \tag{C.33}$$

Now it is just a matter of determining the coefficients $\alpha_0, \alpha_1, \ldots, \alpha_{n-1}$, which are functions of time. This is done as follows. If $\lambda_1, \lambda_2, \ldots, \lambda_n$ are the eigenvalues of A, however, then, by definition:

$$\Delta(\lambda_i) = \det(A - \lambda_i I) = 0 \quad \text{for} \quad i = 1, 2, \ldots, n \tag{C.34}$$

Thus, from Equation C.32, we obtain:

$$\exp(\lambda_i t) = \alpha_{n-1}\lambda_i^{n-1} + \alpha_{n-2}\lambda_i^{n-2} + \cdots + \alpha_0 \quad \text{for} \quad i = 1, 2, \ldots, n \tag{C.35}$$

If the eigenvalues are all distinct, Equation C.35 represents a set of n independent algebraic equations from which the n unknowns $\alpha_0, \alpha_1, \ldots, \alpha_{n-1}$ could be determined. If some eigenvalues are repeated, the derivatives of the corresponding equations (Equation C.35) have to be used as well.

Index